Medicinal Plants in Asia for Metabolic Syndrome

Medicinal Plants in Asia for Metabolic Syndrome

Natural Products and Molecular Basis

Christophe Wiart

CRC Press is an imprint of the
Taylor & Francis Group, an **informa** business

CRC Press
Taylor & Francis Group
6000 Broken Sound Parkway NW, Suite 300
Boca Raton, FL 33487-2742

First issued in paperback 2021

ISBN-13: 978-1-138-03759-5 (hbk)
ISBN-13: 978-0-367-24693-8 (pbk)

Library of Congress Cataloging-in-Publication Data

Names: Wiart, Christophe, author.
Title: Medicinal plants in Asia for metabolic syndrome : natural products and molecular basis / Christophe Wiart.
Description: Boca Raton : CRC Press, [2017] | Includes bibliographical references and index.
Identifiers: LCCN 2017013444| ISBN 9781138037595 (hardback : alk. paper) | ISBN 9781315177755 (ebook)
Subjects: LCSH: Metabolic syndrome--Treatment. | Medicinal plants--Therapeutic use--Asia. | Materia medica, Vegetable--Asia.
Classification: LCC RC662.4 .W53 2017 | DDC 616.3/99--dc23
LC record available at https://lccn.loc.gov/2017013444

Visit the Taylor & Francis Web site at
http://www.taylorandfrancis.com

and the CRC Press Web site at
http://www.crcpress.com

Dedication

If we want real peace in this world, we should start educating children.

Mahatma Gandhi

Contents

Foreword I

Metabolic syndrome is manifested as elevated fasting blood glucose level, increase in triglyceride level in blood, abdominal obesity, low high-density lipoprotein cholesterol, and high blood pressure. It poses high health risk to humans and is a major predisposing factor for life-threatening disorders such as type 2 diabetes, cardiovascular diseases, and cancer. Concerning its prevalence, the International Diabetes Federation estimates that one-quarter of the world's population has developed metabolic syndrome, and 20% of adults in the Western world have developed metabolic syndrome. Natural products obtained from terrestrial plants, marine organisms, and microorganisms have been successfully harnessed in providing therapeutic agents as well as drug leads for an array of illnesses. Furthermore, natural products have long been a source of prophylactic medicines/preventive remedies, particularly against metabolic disorders. A large number of traditional herbs have proven effective as antimetabolic syndrome medicines in animal models and humans. A myriad of phytochemicals (from different classes such as flavonoids, phenylpropanoids, phenylheptanoids, xanthones, and other polyphenols), steroids, organosulfur compounds, and alkaloids have reportedly exhibited the ability to reduce hyperglycemia, attenuate hypertension, lower hyperlipidemia, and help weight control. Molecular mechanisms involved in the prevention of the metabolic syndrome include antioxidant and anti-inflammatory actions, modulation of key signal transduction cascades, glucose transport, inhibition or stimulation of enzymatic activity, regulation of mitochondrial function, modulation of protein expression, and regulation of transcription factors, in addition to other mechanisms of action. Exploitation of natural products against metabolic syndrome and the associated diseases has been the subject of extensive investigations over the past few decades and currently witnesses growing interest.

Ikhlas A. Khan
University of Mississippi

Foreword

Foreword II

Metabolic syndrome is known for a cluster of conditions—hypertension, elevated blood sugar, excess body fat around the waist, and abnormal cholesterol or triglyceride levels—that occur together, thus increasing oxidative stress and increasing the probability of heart disease, stroke, and diabetes. It is closely linked to overweight or obesity and inactivity. In recent times, the number of people affected with metabolic syndrome is rising at an alarming rate. According to the World Health Organization estimates for 2014, there are 600 million clinically obese and 1.9 billion overweight adults worldwide, and there are more than 415 million people with diabetes. Allopathic medicines cannot cure but merely offer symptomatic relief. Moreover, such medicines are costly and not available or affordable to the poorer sections of the population of a country or people residing in remote regions. Some conditions such as arterial blockages (which can result from high cholesterol and lead to stroke) may need surgery, which is expensive and substantially decreases the quality of lifestyle of the patient. Despite an array of medications to decrease blood sugar levels, there are no medications through which diabetes can be cured. This disease, with the progress of time, can lead to further complications such as cardiovascular disorders, diabetic retinopathy, diabetic nephropathy, and diabetic neuropathy. As such, effective medicines to treat metabolic syndrome is a necessity and cries for attention from scientists.

Plants have always been a source for new and effective drugs. Apart from Brazil, the various countries of Asia in between them contain a huge number of diverse floristic species. These species, most of which remain unexplored from the pharmacological point of view, contain thousands of phytochemicals, which need to be researched as potential sources of new drugs. From that viewpoint, this book is exceptional. Dr. Christophe Wiart has done a magnificent job in exploring the vast medicinal plant wealth of Asia toward identifying possible plants and their secondary metabolites along with their mechanism of action, which can be of immense benefit to scientists and researchers, and help find possible drugs against metabolic syndrome.

Using plants for curing or alleviating metabolic syndrome is a concept that dates back possibly thousands of years ago. The ancient Indian system of medicine, Ayurveda, describes a set of complex clinical disorders, collectively called Prameha, that are characterized by frequent abnormal micturition. The clinical conditions as described in ancient Ayurvedic texts for Prameha correlate in many ways with obesity, metabolic syndrome, and diabetes mellitus. A number of plant-based monoherbal and polyherbal formulations are used in Ayurveda to treat Prameha. However, it cannot be denied that more effective medicines may be essential to treat metabolic syndrome than those that are available in Ayurveda for treatment of Prameha, and the active ingredients in these plant-based medicines are identified. Modern-day scientists are recognizing the importance of plants in treating obesity, hypertension, and diabetes either alone or in combination as in metabolic syndrome.

A simple search of recent scientific literature demonstrates a variety of plants, which are reportedly active against metabolic syndrome or at least some of its symptoms. *Cissus quadrangularis*, a common plant in the Indian subcontinent but also found in other Asian countries, also known as veldt grape in English, is more known for its bone fracture healing abilities. However, recent research has shown that the plant can reduce weight as well as improve blood parameters associated with metabolic syndrome. Red orange juice has proved effective in reducing insulin resistance and systolic blood pressure. Tea, prepared from the leaves of *Camellia sinensis*, has been shown to reduce body weight, alleviate metabolic syndrome, and prevent diabetes and cardiovascular diseases in animal models and humans. Grapes and particularly grape seeds have proved effective in inhibiting hyperlipidemia, hyperglycemia, and hypertension. Plants such as red ginseng or *Hibiscus sabdariffa*, *Rosmarinus officinalis*, and *Hylocereus polyrhizus*, to name only a few, have also shown efficacy against metabolic syndrome. The evidences already present in the scientific literature suggest two things: (1) plants may prove to be the effective remedy against metabolic syndrome and

(2) possibly polyherbal formulations will be necessary to treat the multiple disorders present in metabolic syndrome more effectively.

It is in this context that this book gains importance. The large number of plants discussed in the book can make scientific studies more relevant and also enable potential scientists to combine plants in a manner to treat metabolic syndrome more effectively without any adverse effects from interaction between the various plants that may be used. Thus, the book is not only useful to scientists and researchers, but also to the average persons in knowing more about this metabolic disorder affecting human beings.

Mohammed Rahmatullah
University of Development Alternative

Foreword III

In this rapidly changing world there are always new challenges for production of food, drug, and also for health care at large. Natural products play a major role in health care throughout the globe. Several governments of individual countries and international organizations have advocated the use of traditional medicines in primary health care. This should be more effective when the evidences for use of those traditional medicines are being documented scientifically, which is an urgent need for evidence-based validation of traditional medicines to make them available for the treatment of a large community.

In Asian region, the traditional knowledge has been recorded in books or old scriptures that are thousand years old but still plays an important role in health care. This has led to the development of several new approaches supported with the new economic realities. Ayurveda is one of the holistic health care systems, which has been recognized as an ancient science of life.

बहुता तत्रयोग्यत्वमनेकविधकल्पना|
सम्पच्चेति चतुष्कोऽयं द्रव्याणां गुण उच्यते||

(च .सू – ९/७)

[Devanagari Script]

Bahuta tatrayogyatwamanekvidh kalpana |
Sampaaccheti chatushkoayam dravyanam guna uchhyate ||
(***Charka Samhita Sutrasthana*** - 9/7)

[Diacritical Script]

Available in abundance, affectivity, various pharmaceuticals forms, and having appropriate properties are the four qualities of drugs

Ayurveda is getting global acceptance primarily due to its age-old therapeutic practice and profound conceptual basis. The philosophy of treating a system or body as a whole is gaining relevance during transition from reductionist approach to "systems" approach in the post-genomic era. Ayurveda describes obesity as a disease of "medadhatu" (adipose tissue), which leads to hugeness (sthoulyam) and referred as "medoroga." Chikitsa (therapy) for obesity comprised of elimination of nature's waste (purification), dietary composition, energy expenditure, and reduction of hormonal stress with yoga. In Ayurvedic pharmacology, many plants and formulations have been reported for drug interventions in obesity management. Efficacy and potential of Ayurvedic medicines is also evident from many recent scientific publications, for example, study on Ashwagandha (*Withania somnifera*) that lead to the discovery of a novel therapeutic strategy for Alzheimer's disease reversal. Ayurveda, apart from the therapeutic potential also has a predictive, preventive, and personalized approach to health and management of disease, which has been extensively documented in original texts of Charaka and Sushruta Samhita. This potential has not been harnessed effectively in drug-discovery programs.

The metabolic syndrome is a collective term that refers to obesity-associated metabolic abnormalities. There is several health risks associated with obesity. Utilization of plant components and its derived products has a prospective future for controlling the prevalence of metabolic syndrome. Several evidences are exploring to support the use of herbs as an alternative way of obesity control and weight management. Diet-based therapies and herbal supplements are among the most common complementary and alternative medicine modalities for weight loss. A large number of populations in Asia depend on traditional practitioners and their prescription of medicinal plants to assemble health care needs. Hence, it is really obvious that plants may offer an efficient option for the treatment of metabolic syndrome.

For commercialization of botanical products, the assurance of safety, quality, and efficacy of medicinal herbs and botanical products has become an important issue. The regulations of several countries including the National Centre for Complementary and Alternative Medicine, Bethesda, Maryland, and WHO stress the importance of qualitative and quantitative methods for characterizing botanical samples, quantification of the biomarkers and/or chemical markers, and the fingerprint profiles. Different approaches can be used for chemical standardization such as pretreatment that involves drying and grinding; selection of a suitable method of extraction; analysis of compounds using suitable chromatographic or spectroscopic methods; the analysis of data based on bioactive or marker compounds; quality control; elucidation of the properties of absorption distribution metabolism excretion (ADME) and metabolomics evaluation of medicinal plants. In addition, there is a need to develop scientific proof and clinical validation with chemical standardization, biological assays, animal models, and clinical trials for botanicals.

There has been a global increase in the prevalence of chronic and complex diseases with many lifestyle disorders. Majority of chronic diseases require lifetime medications and in many cases, resistance to drugs is a common problem. Most of the diseases are multifactorial involving complex interplay of a network of genes and nongenetic environmental factors. It is being realized that we need to evolve a systems'-based approach for comprehensive understanding of biology and move toward a network approach in medicine. With the advent of genomics, drug discovery and development program are targeting on the understanding of disease biology in target identification and also aspires to identify responder populations.

This book *Medicinal Plants in Asia for metabolic Syndrome: Natural Products and Molecular Basis* highlights several aspects on the use of the medicinal plants of Asia that are useful in metabolic syndrome particularly against obesity, type 2 diabetes, hypertension, vascular dysfunction, and hyperlipidemia. I appreciate the efforts of Dr. Christophe Wiart for compiling this document, which I am sure will be useful for the researchers and the users of natural medicines to go further with their therapeutic potentials.

Pulok K. Mukherjee
Jadavpur University

Foreword IV

Before the 1950s, natural products derived from medicinal plants played a pivotal role in the development of new human medicines. For example, opium poppy alkaloids such as morphine continued to be important drugs for pain relief, quinine from the bark of *Cinchona cordifolia* was used for many years to treat malaria, reserpine from the roots of *Rauwolfia serpentina* has been used in the treatment of hypertension, and digitoxin from the leaves of *Digitalis purpurea* and related cardiac glycosides are used to treat congestive heart failure and cardiac arrhythmia. Plant-derived natural products have also been identified as having activity against various metabolic diseases. For example, the dihydrochalcone glucoside phlorizin has been a prototype for a new series of type-2 diabetes drugs called the *gliflozins*. Phlorizin was first isolated from the root bark of the apple tree in 1835 and later was found to decrease glucose plasma levels and improve insulin resistance levels through inhibition of sodium glucose cotransporters (SGLTs). As phlorizin was not able to be developed as a drug *per se* due to poor intestinal absorption and rapid enzymatic inactivation, medicinal chemists began to synthesize phlorizin analogs to overcome these issues. This research has culminated in the recent clinical approval of six gliflozin drugs: (1) dapagliflozin (launched in 2012), (2) canagliflozin (launched in 2013), (3) ipragliflozin (launched in 2014), (4) luseogliflozin (launched in 2014), (5) empagliflozin (launched in 2014), and (6) tofogliflozin (launched in 2017) with other analogs that are also currently being clinically investigated. Another interesting example occurred when a herbal medicine from *Galega officinalis*, which had been widely used to treat diabetes mellitus, was found to contain galegine. Both galegine and guanidine were evaluated as antidiabetic agents in the 1920s but their use was discontinued due to toxicity issues. In the 1950s, the biguanidine metformin was introduced into the clinic and is still used today. Although the discovery of metformin was not based on galegine, the structural similarity between these two compounds is striking. Interestingly, metformin is also approved for use in combination with the previously discussed dapagliflozin and empagliflozin. Although natural plant products were important in early drug discovery, the development of the fungal-derived penicillin antibiotic in the early 1940s followed by the plethora of novel compounds derived from bacteria in the 1950s and 1960s changed the focus of new lead discovery. Since then, methods such as high throughput screening, fragment and computational screening, and drug design have become mainstay of lead discovery in industry. However, novel lead compounds have become very difficult to identify, especially in the area of metabolic diseases, which continue to be a major health burden all over the world. As a consequence, this book is very important as it catalogs medicinal plants with activity in this important disease area. May be the next important drug lead will come from one of the plants described in this book?

Mark Butler
Institute for Molecular Bioscience
The University of Queensland

Foreword IV

Preface

Humans lived on earth by hunting animals and gathering plants for food for about 90,000 years. Only recently, have they been exposed to industrial food, urbanization, pollution, and lack of physical activity explaining the recrudescence of obesity, type 2 diabetes, cardiovascular diseases, and other noncommunicable pathologies. In parallel, human knowledge on medicinal plants, or what is called materia medica or pharmacognosy, is disappearing. It is in fact looked down as "an obscure subject" by some accreditation boards lobbied by the pharmaceutical industry. The last traditional healers are aging, and there is a dangerous trend to remove the teaching of pharmacognosy from our comptemporary "Schools of Pharmacy." In fact, graduating pharmacy students (soon to be replaced by dispensing machines) in 2017 are not often getting trained on medicinal plants to the point of ignoring what is opium, cumin if not pepper. The current "late capitalist era," as termed by some, favors profitability and aims at financial benefits of huge corporations. In fact, universities are being themselves often transformed into businesses, resulting in a collapse of academic freedom and a dearth of academic elites. Corporations and financial benefits are also responsible for the destruction of our natural environment, exemplified by the eco-genocide caused by palm oil in Southeast Asia. It can be said that, by the end of this century, many of the medicinal plants provided by Mother Nature and their pharmacological potentials would have been vanished by smoke. The pharmaceutical industry being apparently concerned about its financial benefits does not have much interest in medicinal plants. In fact, it can be said with confidence that a biological feedback will soon occur to force the corporations to change their policies. This is sadly exemplified by the emergence of bacterial resistance and the end of the golden age of antibiotics. It seems that we should be able to live longer and healthier, but it is not the case. In Asia, a wealth of medicinal plants, known since the beginning of time, remains practically unused for the well-being of humans. The purpose of this book is to shed light on the pharmacological properties of carefully selected medicinal plants used in Asia in regard to what has been termed "metabolic syndrome." This book is the result of almost 20 years of medicinal plant research conducted in Southeast Asia. It is principally intended to students, researchers, and academics who have interest in the subject of discovering drugs from Asian medicinal plants for the treatment or prevention of the metabolic syndrome. Medicinal plants, natural products, and their mode of activities are being organized into five chapters corresponding to the major sites of the activity in the body. The plants are listed according to the Takhtajan system of plant classification published in 2008, which allows making chemotaxonomic considerations that are useful to understand the pharmacological activity of medicinal plants. Hundreds of carefully selected bibliographical references are provided and the potentials of the most interesting plants are discussed. It is my hope that this book will create some interest in medicinal plant research and contribute to the discovery of new drugs to fight metabolic syndrome. This book was written in very difficult working conditions, and it would have been impossible to complete it without the support, love, and sacrifices of my family, and particularly my mother, Madam Flora Monllor.

Christophe Wiart
University of Nottingham

Preface

About the Author

Christophe Wiart was born in Saint Malo, France. He received his Pharm D from the Faculty of Pharmacy, University of Rennes, Rennes, France, in 1997 and his PhD from University Pertanian Malaysia, Malaysia in 2001. He served as lecturer and later as associate professor at the University of Malaya, Kuala Lumpur, Malaysia, from 2001 to 2007 and is currently associate professor at the University of Nottingham Malaysia Campus, Selangor, Malaysia, where he teaches pharmacy undergraduates and supervises master's and PhD students. Dr. Wiart appeared on HBO's Vice (television series) in season 3, episode 6 (episode 28 of the series) titled "The Post-Antibiotic World & Indonesia's Palm Bomb." This episode aired on April 17, 2015. It highlighted the need to find new treatments for infections that were previously treatable with antibiotics, but are now resistant to multiple drugs. "The last hope for the human race's survival, I believe, is in the rainforests of tropical Asia," said ethnopharmacologist Dr. Christophe Wiart. "The pharmaceutical wealth of this land is immense." He was invited at TedEx on June 4, 2016. He was the guest at "Inside Story" Aljazeera on September 21, 2016, and interviewed by Adrian Finighan about the rise of superbugs and the chemotherapeutic potentials of medicinal plants in Asia. Dr. Wiart has authored more than 80 publications and 11 academic books on the pharmacological potentials of medicinal plants in Asia. He is the general secretary of the Asian Society of Pharmacognosy and the editor in chief of the *Asian Journal of Pharmacognosy.*

About the Author

Introduction

Processed industrial "foods" and "drinks" are associated with noncommunicable diseases of which obesity, the prevalence of which has more than doubled since 1980.[1,2] Simply put, obesity is an accumulation of triglycerides in adipose tissues to the point that the ratio of the body weight (kg) to the height (m^2) is equal to or more than 30 kg/m^2.[3] Besides ponderal surcharge and aesthetic consideration, visceral adiposity favors the development of insulin resistance, atherogenic dyslipidemia, and hypertension, which are interrelated cardiovascular risk factors collectively referred to as the "metabolic syndrome."[4,5] As for yet, the anti-obesity arsenal is ridiculously limited, and in fact no drug exists yet to efficiently and quickly remove visceral adipose tissues in obese patients who are left with bariatric surgery, strict control of diet, and regular physical exercise. There is therefore a need to develop drugs to prevent or delay the progression of metabolic syndrome in obese patients. In Asia, medicinal plants have been used to treat conditions linked to hyperlipidemia, insulin resistance, type 2 diabetes, hypertension, and cardiovascular diseases since the beginning of mankind, and the systematic pharmacological study of these plants should lead to the discovery of natural products to prevent or manage metabolic syndrome. Today, no single book dedicated to natural products from medicinal plants in Asia for metabolic syndrome exists, and the purpose of this volume is precisely to fill this gap.

REFERENCES

1. Moodie, R., Stuckler, D., Monteiro, C., Sheron, N., Neal, B., Thamarangsi, T., Lincoln, P., Casswell, S. and Lancet NCD Action Group, 2013. Profits and pandemics: Prevention of harmful effects of tobacco, alcohol, and ultra-processed food and drink industries. *The Lancet*, *381*(9867), 670–679.
2. WHO. Obesity and overweight. 2006. http://www.who.int/mediacentre/factsheets/fs311/en/ (accessed September 2016).
3. Kopelman, P.G., Caterson, I.D. and Dietz, W.H. Eds., 2009. *Clinical Obesity in Adults and Children*. Chichester: John Wiley & Sons.
4. Grundy, S.M., 2008. Metabolic syndrome pandemic. *Arteriosclerosis, Thrombosis, and Vascular Biology*, *28*(4), 629–636.
5. Alberti, K., Eckel, R.H., Grundy, S.M., Zimmet, P.Z., Cleeman, J. and Donato, K., 2009. Harmonizing the metabolic syndrome. A joint interim statement of the IDF Task Force on Epidemiology and Prevention; NHL and Blood Institute; AHA; WHF; IAS; and IA for the Study of Obesity. *Circulation*, *120*(16), 1640–1645.

Introduction

1 Inhibiting the Absorption of Dietary Carbohydrates and Fats with Natural Products

Insulin resistance in metabolic syndrome results, at least, from the overconsumption of dietary carbohydrates, cholesterol, and triglycerides leading to the formation of visceral adiposity, increased plasma-free fatty acids, and secretion of pro-inflammatory cytokines, which at cellular level decrease insulin receptor functionality also known as insulin resistance.[1,2] Once insulin resistance is established, increased postprandial glycemia, according to genetic susceptibility, introduces the development of type 2 diabetes and cardiovascular insults.[3–6] Thus, inhibiting the absorption of dietary carbohydrates and fats (cholesterol and triglycerides) with natural products or extracts of medicinal plants constitutes one therapeutic strategy to prevent or manage insulin resistance in metabolic syndrome.

1.1 *Saururus chinensis* (Lour.) Baill.

Synonyms: *Saururopsis chinensis* (Lour.) Turcz.; *Saururopsis cumingii* C. DC.; *Saururus cernuus* Thunb.; *Saururus cumingii* C. DC.; *Saururus loureiri* Decne.; *Spathium chinense* Lour.
Common name: san bai cao (Chinese)
Subclass Magnoliidae, Superorder Piperanae, Order Piperales, Family Saururaceae
Medicinal use: wounds (Cambodia)

The hydrolysis of dietary triglycerides into glycerol and fatty acids is catalyzed by lingual, gastric, and pancreatic lipases.[7] There is a massive bulk of experimental evidence to demonstrate that extracts of medicinal plants in Asia, and most often polar extracts including aqueous, ethanol, and methanol, extracts have the ability to inhibit *in vitro* the enzymatic activity of lipase. For instance, ethanol extracts of *Saururus chinensis* (Lour.) Baill. (Figure 1.1) inhibited the enzymatic activity of pancreatic lipase with IC_{50} equal to 81 μg/mL.[8] Oral administration of aqueous extract of this plant to rats on high-fat diet evoked a decrease in plasma triglycerides and an increase in fecal triglycerides, which suggests inhibition of triglyceride intestinal absorption.[9]

FIGURE 1.1 *Saururus chinensis* (Lour.) Baill.

1.2 *Piper longum* L.

Synonym: *Chavica roxburghii* Miq.
Common names: bi ba (Chinese); long pepper
Subclass Magnoliidae, Superorder Piperanae, Order Piperales, Family Piperaceae
Medicinal use: facilitates digestion (China)
History: The plant was known to Hippocrates, Greek physician (circa 460–370 BC)

The main dietary carbohydrate is starch from plants that consists of amylose and amylopectin composed of linear chains glucose joined by α-1,4-glycosidic linkages, which are, especially for the later, branched by α-1,6-linkages.[10] Decrease in plasma glucose may be produced by decreased intestinal absorption of starch, and flavones in medicinal plants have the ability to hamper the enzymatic decomposition of starch and starch-derived products and sucrose. Apigenin-7,4′-dimethyl ether (Figure 1.2) isolated from the fruits of *Piper longum* L. Figure 1.2 inhibited α-amylase

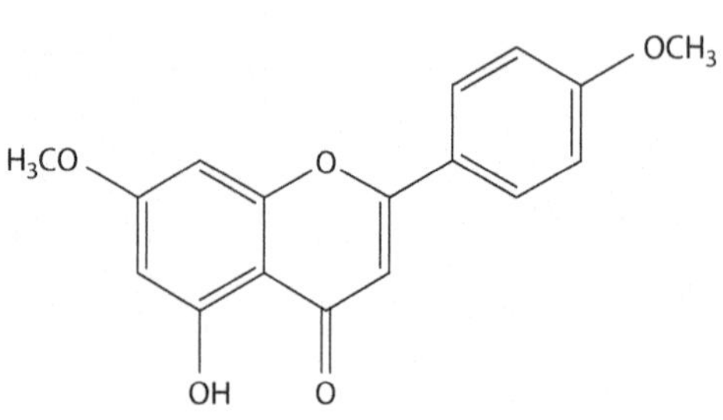

FIGURE 1.2 Apigenin-7,4'-dimethyl ether.

in vitro with IC_{50} value 98.1 μg/mL.[11] Acarbose used in therapeutic strategies to decrease postprandial hyperglycemia inhibited α-amylase with IC_{50} 45.2 μg/mL.[11]

1.3 *Nelumbo nucifera* Gaertn.

Synonyms: *Nelumbium nuciferum* Gaertn.; *Nelumbo speciosa* Willd.; *Nymphaea nelumbo* L.
Common names: lian (Chinese); sacred lotus (English)
Subclass Ranunculidae, Superorder Proteanae, Order Nelumbonales, Family Nelumbonaceae
Medicinal use: anxiety (China)

Pancreatic α-amylase hydrolyzes starch α-1,4-linkages to yield maltose, maltotriose, and α-limit dextrin and vast body of pharmacological evidence suggest that flavonoids in medicinal plants account for the inhibition of pancreatic α-amylase.[12] For instance, ethanolic extract from leaves of *Nelumbo nucifera* Gaertn. inhibited the enzymatic activity of α-amylase and lipase with IC_{50} values equal to 0.8 and 0.4 mg/mL, respectively *in vitro*.[13] The flavonoids quercetin 3-*O*-alpha-arabinopyranosyl-(1→2)-β-galactopyranoside, rutin, catechin, hyperoside, isoquercitrin, quercetin, astragalin, hyperin, kaempferol, and myricetin present in this plant may account for these effects.[14] Myricetin (Figure 1.3) inhibited α-amylase activity with an IC_{50} value of 30.2 μM.[15] Liu et al. (2013) reported the ability of a total flavonoid fraction of leaves of *Nelumbo nucifera* Gaertn. to inhibit yeast α-amylase, yeast α-glucosidase, and porcine lipase with IC_{50} values of 2.2, 1.8, and 0.3 mg/mL, respectively.[16] In this experiment, acarbose used as positive standard inhibited yeast α-amylase and α-glucosidase with IC_{50} values of 0.4 and 0.6 mg/mL, respectively.[16] Quercetin-3-*O*-β-D-arabinopyranosyl-(1→2)-β-D-galactopyranoside and quercetin-3-*O*-β-D-glucuronide (Figure 1.4) isolated from this plant inhibited porcine pancreatic lipase with IC_{50} values of 52.9 and 17.1 μg/mL, respectively.[17] Total flavonoid fraction of leaves of this aquatic plant given orally at a dose of 240 mg/kg/day to Wistar rats, which are often used for metabolic studies, on high-fat diet for 2 weeks decreased plasma triglycerides from 2.5 to 1.2 mmol/L.[16] The leaves

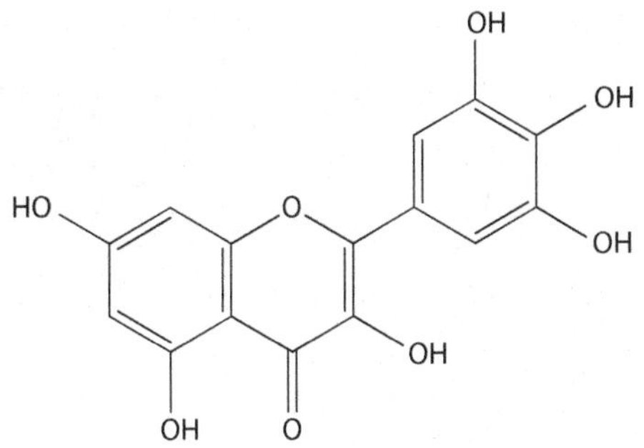

FIGURE 1.3 Myricetin.

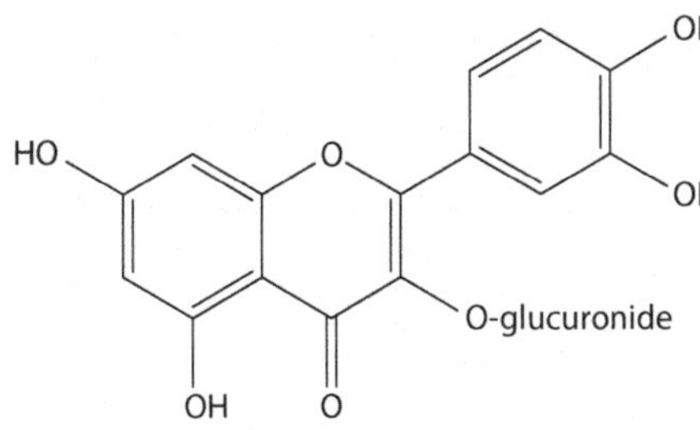

FIGURE 1.4 Quercetin-3-*O*-β-D-glucuronide.

of *Nelumbo nucifera* Gaertn. contain the isoquinoline alkaloids (6*R*,6aR)-roemoerine-N_{β}-oxide, liriodenine, pronuciferine, oleracein E as well as the phenolics trans-*N*-coumaroyltyramine, cis-*N*-coumaroyltyramine, trans-*N*-feruloyltyramine, cis-*N*-feruloyltyramine, which inhibited the enzymatic activity of pancreatic lipase *in vitro* (ii).[18] Being able to inhibit the absorption of carbohydrates and triglycerides, the leaves of this astringent medicinal plant, if not toxic, could be conceptually seen as a dietetic material of interest for metabolic syndrome. Clinical studies in this direction are warranted.

1.4 *Coptis chinensis* Franch.

Common name: huang lian (Chinese)
Subclass Ranunculidae, Superorder Ranunculanae, Order Ranunculales, Family Ranunculaceae
Medicinal use: fever (China)

As a consequence of insulin resistance, postprandial glycaemia in metabolic syndrome is elevated and high concentration of circulating glucose that could be referred to a state of "*glucotoxicity*" contribute to the development of type 2 diabetes, cardiovascular diseases, and all that cause mortality.[19] *Coptis chinensis* Franch. (Figure 1.5) elaborates the alkaloid berberine which when given orally at a

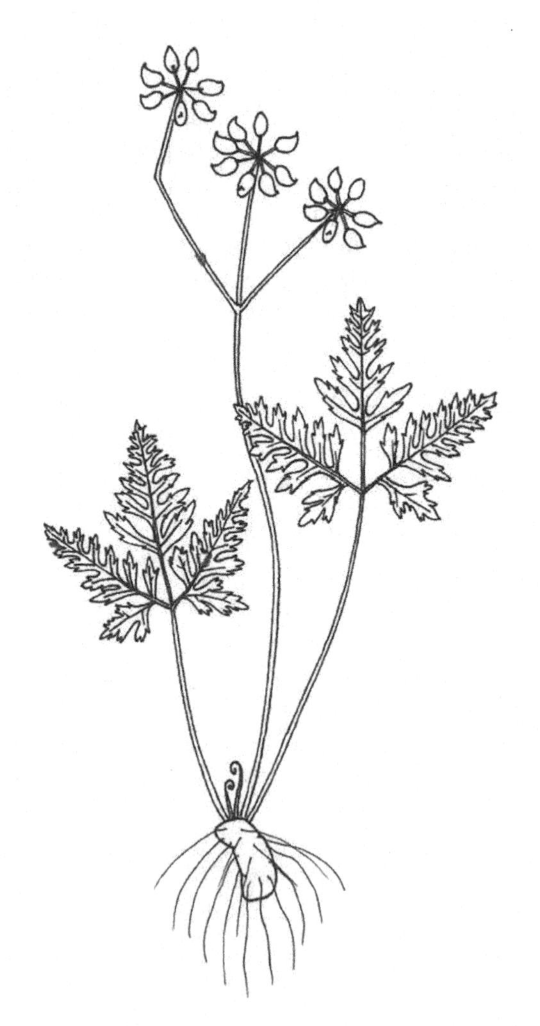

FIGURE 1.5 *Coptis chinensis* Franch.

dose of 200 mg/kg once daily reduced the glycemia of diabetic rodents and inhibited the enzymatic activity of sucrase and maltase.[20] In a clinical study involving type 2 diabetes outpatients, the intake of 500 mg of berberine 3 times daily evoked a reduction in blood glucose.[21] Thus, being relatively nontoxic, and poorly absorbed, berberine could conceptually be seen as a potential agent to mitigate glucose absorption in metabolic syndrome.

1.5 *Tinospora crispa* (L.) Hook. f. & Thomson

Synonyms: *Menispermum crispum* L.; *Tinospora gibbericaulis* Hand.-Mazz.; *Tinospora mastersii* Diels; *Tinospora rumphii* Boerl.; *Tinospora thorelii* Gagnep.

Common names: bo ye qing niu dan (Chinese); akar putarwali (Malay); makabuhay (Philippines); boraphet (Thai)

Subclass Ranunculidae, Superorder Ranunculanae, Order Menispermales, Family Menispermaceae

Medicinal use: jaundice (Vietnam)

Degradation products of starch are hydrolyzed in the jejunum into free absorbable glucose by 4 brush border α-glucosidases arranged into 2 enzymatic complexes termed as sucrase–isomaltase and maltase–glucoamylase.[22] Members of the family Menispermaceae often accumulate isoquinoline alkaloids that hamper glucose absorption by inhibiting enterocyte membrane bound α-glucosidases. As an example, *Tinospora crispa* (L.) Hook. f. & Thomson synthetize palmatine, jatrorrhizine, and magnoflorine that inhibited the enzymatic activity of sucrase with IC_{50} of 36.2, 23.4, and 9.8 μg/mL, respectively.[23] In the same experiment, palmatine, jatrorrhizine, and magnoflorine inhibited the enzymatic activity of maltase with IC_{50} values equal to 22, 38.4, and 7.6 μg/mL.[23] Magnoflorine at a dose of 20 mg/kg mitigated the raise in glycaemia induced by oral administration of 2 g/kg of glucose to rodents.[23] Magnoflorine is known to induce hypotension when parenterally administered and to be nontoxic in animals when given orally.[24]

1.6 *Nigella sativa* L.

Common names: Krishna jiraka (India); habbatus sauda (Malay); fennel flower seeds

Subclass Ranunculidae, Superorder Ranunculanae, Order Menispermales, Family Ranunculaceae

Medicinal use: in Malaysia, the seeds are ingested to invigorate

History: Known of Hippocrates Greek physician (circa 460–370 BC) as tonic spice

Glucose released from maltose, maltotriose, dextrin, and sucrose, it is actively engulfed in jejunal brush border by integral sodium-dependent glucose transporter-1 (SGLT-1) located in the apical cytoplasmic membrane of enterocytes.[25] *Nigella sativa* L. contains natural product(s), yet to be identified, with the ability to attenuate intestinal glucose absorption by inhibiting enterocytes integral membrane Na^{+}-glucose transporter 1 (SGLT1).[26] Aqueous extract from seeds of *Nigella sativa* L. given orally for 6 weeks to Sprague–Dawley rats at a dose of 0.2 g/kg/day improved glycaemia as well as body weight as efficiently as metformin at a dose of 300 mg/kg/day.[26] *In vitro*, this extract at a dose of 1 ng/mL prophylactically inhibited glucose intake by sodium-dependent glucose transporter-1 (SGLT-1) of isolated jejunal mucosa by 81.8%.[26] Besides, methanol extract from seeds of *Nigella sativa* L. at a concentration of 2.5 mg/mL completely inhibited porcine pancreatic lipase *in vitro*.[27] Being relatively nontoxic, consumption of seeds of *Nigella sativa* L. may limit glucose and fatty acids absorption in metabolic syndrome. Like most medicinal plants used since time immemorial, and with the disappearance of pharmacognosy and herbalism from what we call today *Schools of Pharmacy* (?), the exact dose of these seeds to be taken seems unknown.

1.7 *Celosia argentea* L.

Synonym: *Celosia cristata* L.
Common names: qing xiang (Chinese); barhichuda (India); bayam (Malay); palonpalongan (Philippines); wild cockscomb
Subclass Caryophyllidae, Superorder Caryophyllanae, Order Caryophyllales, Family Amaranthaceae
Medicinal use: dysentery (Malaysia)

Evidence supports the view that medicinal plants in the family Amaranthaceae Juss. inhibit the enzymes of carbohydrate intestinal absorption on account of their triterpenoid saponins. On such medicinal plant is *Celosia argentea* L., an ethanolic extract of which inhibited porcine pancreatic amylase and yeast α-glucosidase *in vitro* with IC_{50} values of 1.6 and 1 mg/mL, respectively (acarbose: 0.1 and 0.9 mg/mL, respectively).[28,29]

1.8 *Kochia scoparia* (L.) Schrad.

Synonym: *Chenopodium scoparium* L.
Common names: ti fu (Chinese); fire weed
Subclass Caryophyllidae, Superorder Caryophyllanae, Order Caryophyllales, Family Chenopodiaceae
Medicinal use: promote urinations (China)

In the stomach, dietary triglycerides and cholesteryl esters are dispersed in coarse oil globules, which are emulsified by bile acids in the duodenum into small droplets.[30] Pancreatic cholesteryl ester esterase and lipase catalyse the hydrolysis of cholesteryl ester and triglycerides to form mixed micelles which are then absorbed by the apical cytoplasmic membrane of brush border enterocytes.[30] Saponins, which are abundant in members of the family Amaranthaceae Juss., are amphiphilic and disrupt the formation of mixed micelles and subsequent absorption of fatty acids and cholesterol by enterocytes.[31] Ethanol extract from fruits of *Kochia scoparia* (L.) Schrad. (Figure 1.6) at concentration of 2 mg/mL inhibited the enzymatic activity of lipase *in vitro* by 50%. The extract given orally once at a dose of 250 mg/kg to rats abrogated plasma triglycerides 2 hours peak after oral administration of a lipid emulsion.[32] Mice fed with high-fat diet with 3% of this extract emitted high triglycerides in their feces providing evidence of nonabsorption of dietary triglyceides.[32] The same regimen prolonged for 9 weeks brought body weights close to those observed for rodents fed with normal diet.[2] Further, treated mice, compared to untreated group, had a reduction in parametrial adipose tissue mass from 1.8 to 1.2 g and a reduction of hepatic total cholesterol from 12.9 to 7.4 μmol/L.[32] Vinarova et al. (2015) studied the effects of various saponins on the cholesterol bioaccessibility from emulsions stabilized by Tween 80 and found that saponins decrease cholesterol bioaccessibility by displacing cholesterol from mixed micelles.[30] These findings raise the question whether the intake of saponin containing Asian medicinal plants could effectively decrease the absorption of triglycerides and cholesterol and prevent or manage metabolic syndrome.

FIGURE 1.6 *Kochia scoparia* (L.) Schrad.

1.9 *Rheum ribes* L.

Common name: warted-leaved rhubarb
Subclass Caryophyllidae, Superorder Polygonanae, Order Polygonales, Family Polygonaceae
Nutritional use: food (Turkey)
History: The plant was known to Serapion (twelfth century), Arabic physician as astringent and cold and prescribed for the treatment of cholera and hemorrhoids

Aqueous liquid extract from roots of *Rheum ribes* L. (10g/100 mL) a concentration of 50 mg/mL halved glucose liberation from starch by α-amylase by 50%.[33] At a dose of 125 mg/kg, this extract decreased peak glycaemia at 45 minutes in oral starch tolerance test similar to 3 mg/kg of acarbose in Sprague–Dawley rats at 5 mM.[33] The petioles of members of the genus *Rheum* are particularly rich in fibers, and it must be recalled that dietary fiber adsorb bile acids and cholesteryl ester and promote bile formation and cholesterol fecal excression. Such fibers are particularly present in the leaf stalks of *Rheum officinale* Baill., which given at a dose of 27 g/day to hypercholesterolaemic volunteers for 4 weeks following habitual diets had no effect on the body mass index but induced a mild decrease of total cholesterol and triglycerides from 2.1 to 1.8 mmol/L.[34]

1.10 *Camellia sinensis* (L.) Kuntze

Synonym: *Thea chinesis* L.
Common names: cha (Chinese); tea
Subclass Dillenidae, Superorder Ericanae, Order Theales, Family Theaeae
Medicinal use: tonic (China)
History: Used since time immemorial in China and listed in the penst'sao kang mu

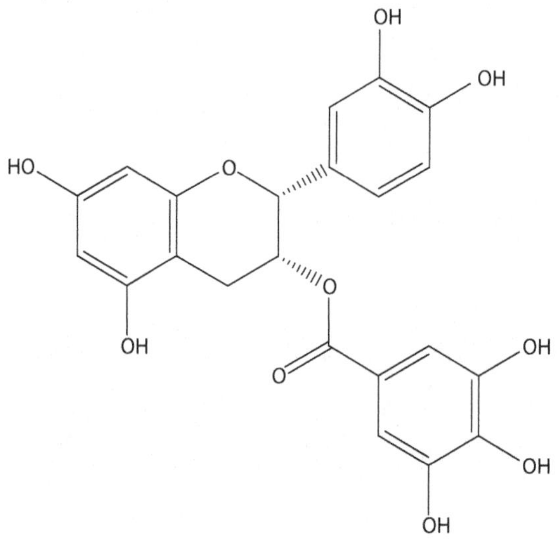

FIGURE 1.7 (–)-Epigallocatechin-3-gallate.

Evidence suggests beneficial effect of green tea catechins on metabolic syndrome (partly on account of lipase, α-amylase, and α-glucosidase inhibition).[35] Wild mice and rats, contrary to human, do not spontaneously develop metabolic syndrome. C57BL/6J mice are used to assess the metabolic effects of medicinal plants because these genetically engineered rodents on high-fat diet develop obesity, hyperlipidemia, hyperinsulinemia, hyperglycemia, insulin resistance, and glucose intolerance.[36] C57BL/6J mice on high-fat diet had a 5.4% weight reduction following the inclusion of 0.3% of green tea (−)-epigallocatechin-3-gallate in diet for 7 weeks.[37] This regimen increased by 29.4% fecal triglycerides as a result of pancreatic lipase inhibition.[37] (−)-Epigallocatechin-3-gallate (Figure 1.7) inhibited *in vitro* the enzymatic activity of pancreatic lipase with an IC_{50} value equal to 7.5 μmol/L.[37] Fei et al. (2014) provided evidence that phenolic fraction of *Camellia sinensis* (L.) Kuntze could inhibit the enzymatic activity of pancreatic α-amylase *in vitro* with an IC_{50} of 0.3 μg/mL.[38] From this fraction, (−)-epigallocatechin gallate and (−)-epigallocatechin 3-*O*-(3-*O*-methyl) gallate inhibited the enzymatic activity of pancreatic α-amylase with an IC_{50} of 0.3 and 0.5 μg/mL, respectively.[38] Green tea or the unfermented leaves of *Camellia sinensis* (L.) Kuntze is consumed daily in Asia and should be recommended, at normal dose, in metabolic syndrome.

1.11 *Garcinia mangostana* L.

Synonym: *Mangostana garcinia* Gaertn.
Common names: mangustan (Malay); mangosteen
Subclass Dillenidae, Superorder Ericanae, Order Hypericales, Family Clusiaceae
Medicinal use: diarrhoea (Malaysia)
History: By the year 1880, the husk of fruits of *Garcinia mangostana* L. was exported from Malaysia as a reputed astringent remedy to treat diarrhea

In the jejunum, sucrase–isomaltase hydrolyses α-1,4-linkages of maltose and sucrose and maltase–glucoamylase hydrolyses α-1,4-linkages of maltose, maltotriose, and limited dextrins.[22] Evidence suggests that prenylated xanthones elaborated by members of the family Clusiaceae have the ability to inhibit α-glucosidases. Ethanol extract from fruit rinds of *Garcinia mangostana* L. inhibited of α-glucosidase with an IC_{50} value of 3.2 μg/mL.[39] This extract given orally and prophylactically at a single dose of 100 mg/kg to streptozotocin-induced diabetic Sprague–Dawley rats reduced glycaemia by 40% during maltose oral challenge.[39] From this extract, the prenylated xanthones

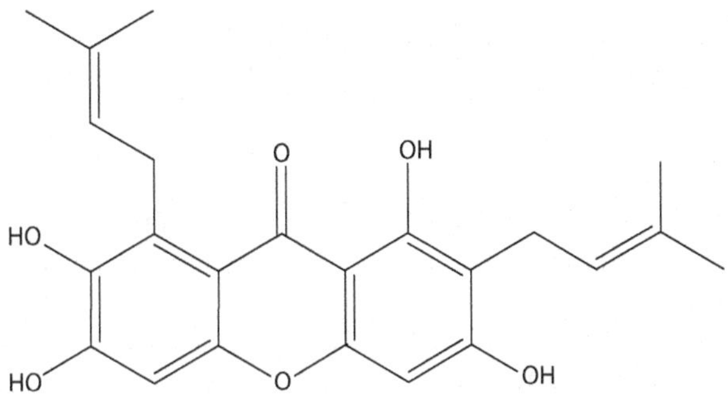

FIGURE 1.8 γ-Mangostin.

β-mangostin, allanxanthone E, α-mangostin, mangostingone, γ-mangostin (Figure 1.8), gartanin, and smeaxanthone A inhibited the enzymatic activity of α-glucosidase with IC_{50} values below 40 μM, respectively.[39]

1.12 *Barringtonia racemosa* (L.) Spreng.

Subclass Dillenidae, Superorder Ericanae, Order Lecythidales, Family Lecythidaceae
Medicinal use: rheumatism (Philippines)

α-Glucosidases are inhibited by oleanane and lupane triterpenes that occur in members of the family Lecythidaceae such as *Barringtonia racemosa* (L.) Spreng.[40] Defatted methanol extract from fruits of *Barringtonia racemosa* (L.) Spreng inhibited α-glucosidase activity with an IC_{50} equal to 26.9 μg/mL. From this extract, the polyhydroxy oleanane triterpenes racemosol C and D isolated inhibited α-glucosidase with IC_{50} values of 5.6 and 45.3 μM, respectively.[41] In the same experiment, the triterpene betulinic acid (Figure 1.9) inhibited α-glucosidase with an IC_{50} value equal to 7.8 μM. 3β-acetoxy-16β-hydroxybetulinic acid, isolated from another plant, inhibited α-glucosidase with an IC_{50}

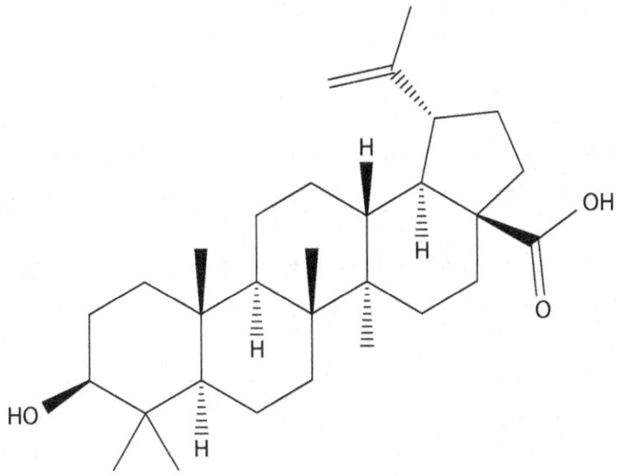

FIGURE 1.9 Betulinic acid.

value equal to 7.6 μM implying that the triterpene backbone is sufficient for activity.[42] Comparatively, acarbose inhibited the enzymatic activity of yeast α-glucosidase with IC_{50} value of 780 μM.[40]

1.13 *Embelia ribes* Burm.f.

Common names: bai hua suan teng guo (Chinese); vidanga (India)
Subclass Dilleniidae, Superorder Primulanae, Order Primulales, Family Myrsinaceae
Medicinal use: jaundice (India)
History: *Embelia ribes* Burm.f. was known of Sushruta (circa 600 BC) an Ayurvedic physician, notably to expel intestinal worms

α-Glucosidases can be inhibited *in vitro* by various types of phenolic compounds explaining the observation that ethanol, aqueous, or methanol extracts of medicinal plants inhibit this group of enzymes as a result of a synergistic effect. *Embelia ribes* Burm.f. elaborates in its leaves the flavonoids kaempferol and quercitrin, and the lignans (+)-syringaresinol-β-D-glucoside, and (+)-syringaresinol that inhibited yeast α-glucosidase with IC_{50} below 90 μM, respectively (acarbose: IC_{50} of 214.5 μM).[43] From the stems, embeliphenol A, 5-(8′Z-heptadecenyl)-resorcinol, 1-(3,5-dihydrophenyl)nonan-1-one, 3-methoxyl-5-pentylphenol (Figure 1.10), and eupomatenoid-8 inhibited yeast α-glucosidase with IC_{50} of 47.4, 41.2, 10.4, 66.9, and 65.7 μM, respectively (acarbose: IC_{50} of 214.5 μM).[44] The antidiabetic use of the plant could be, at least partly, due to α-glucosidase inhibition.

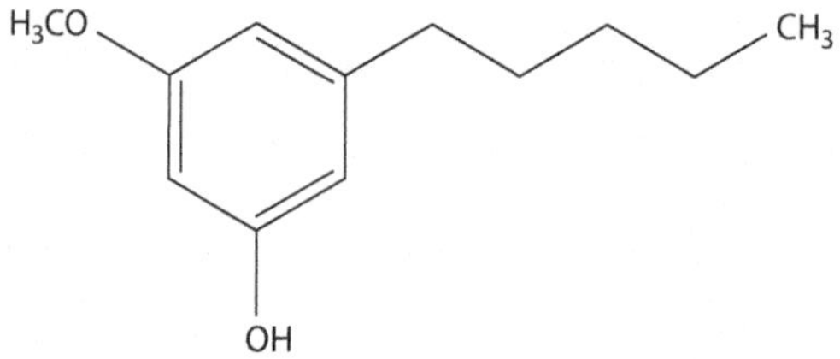

FIGURE 1.10 3-Methoxyl-5-pentylphenol.

1.14 *Gynostemma pentaphyllum* Makino

Common name: jiao gu lan (Chinese)
Subclass Dilleniidae, Superorder Violanae, Order Cucurbitales, Family Cucurbitaceae
Medicinal use: tonic (China)
Pharmacological target: atherogenic hyperlipidemia

Orlistat is a specific and potent pancreatic lipase inhibitor derived from lipstatin, a β-lactone isolated from *Streptomyces toxytricini* used in therapeutic to reduce triglyceride absorption in obese patients.[45,46] However, orlistat offers about 30% efficacy and is responsible for gastro-intestinal, nervous, endocrine, and renal system side effects, justifying the development of safer, natural pancreatic lipase inhibitor.[46] Extract of *Gynostemma pentaphyllum* Makino (containing more than 90% of saponins termed gypenosides) given orally to obese Zucker fatty rats at a dose of 250 mg/kg/day for 3 weeks and administered 1 hour before oral olive oil administration decreased postprandial triglyceridemia by 18% after 5 hours, suggesting pancreatic lipase inhibition.[47] This extract given to Sprague–Dawley at a concentration of 125 mg/kg concomitantly with oral loading of sucrose had no effect of postprandial glycemia but inhibited yeast α-glucosidase activity *in vitro* with an IC_{50} value of 42.8 μg/mL (acarbose: 53.9 μg/mL).[47] That result suggests that inhibition of yeast α-glucosidase *in vitro* by saponins is not correlated with *in vivo* with intestinal α-glucosidase because triterpene glycosides are metabolized by bacteria in the guts. In a subsequent study, gypenosides from

Gynostemma pentaphyllum Makino given orally for 5 weeks at a dose of 200 mg/kg/day to Wistar rats fed with high-fat diet reduced plasma cholesterol and triglycerides to about 40% and 60%.[48] These dammarane saponins normalized hepatic cholesterol and hepatic triglycerides as efficiently as simvastatin at a dose of 10 mg/kg/day and halved the enzymatic activity of 3-hydroxy-3-methylglutaryl-coenzyme A reductase.[48] This protein is a rate limiting enzyme in the synthesis of cholesterol.[49] Su et al. (2016) made the demonstration that gypenosides from *Gynostemma pentaphyllum* Makino at a concentration of 0.2 mg/mL inhibited porcine pancreatic lipase activity to about 40%, whereas orlistat evoked the same concentration approximately 95% inhibition.[50] These saponins did not bind to the catalytic pocket of lipases but instead inhibited cholesterol in mixed micelles via increase in size of mixed micelles.[50] It must be recalled that obese Zucker fatty rats are genetic model of metabolic syndrome due to mutated leptin receptor developing hypercholesterolaemia, hypertriglyceridemia, adipocyte hyperplasia, obesity, hyperglycemia, hyperinsulinemia, and glucose intolerance.[51] *Gynostemma pentaphyllum* Makino's ability to prevent triglyceride and cholesterol absorption in obese Zucker fatty rats by compromising mixed micelle formation and lipase inhibition could conceptually be of value to prevent hypercholesterolemia and hypertriglycemia in metabolic syndrome. Clinical trials are warranted.

1.15 *Lagenaria siceraria* (Mol.) Standl.

Synonyms: *Cucumis mairei* H. Lév.; *Cucurbita lagenaria* L.; *Cucurbita leucantha* Duchesne; *Cucurbita siceraria* Molina; *Lagenaria vulgaris* Ser.
Common names: hu lu (Chinese); kalubay (Philippines); bottle gourd
Subclass Dilleniidae, Superorder Violanae, Order Cucurbitales, Family Cucurbitaceae
Medicinal use: cough (Philippines)

Plant sterols also known as phytosterols, because of their hydrophobicity and high affinity for mixed micelles in the small intestine, displace cholesterol at intestinal micelles levels and inhibit cholesterol absorption resulting in it fecal excression.[52,53] Medicinal plants in the family Cucurbitaceae produce phytosterol and for instance, a mixture of fucosterol, stigmasterol, and stigmasta 7,22-dien-3β,4β-diol isolated from *Lagenaria siceraria* (Molina) Standl. given orally to hyperlipidemic Wistar rats at a dose of 30 mg/kg/day for 30 days decreased cholesterol and triglycerides from 269 to 146.6 mg/dL and from 175.5 to 136.6 mg/dL, respectively.[54]

1.16 *Siraitia grosvenorii* (Swingle) C. Jeffrey ex A.M. Lu & Z.Y. Zhang

Synonyms: *Momordica grosvenorii* Swingle; *Thladiantha grosvenorii* (Swingle) C. Jeffrey
Common name: luo han guo (Chinese)
Subclass Dilleniidae, Superorder Violanae, Order Cucurbitales, Family Cucurbitaceae
Medicinal use: bronchitis (China)

Delaying intestinal glucose absorption is an important therapeutic strategy to fight metabolic syndrome because it decreases postprandial glycemia, decreases insulin resistance, evokes mild loss of body weight, and improves serum lipid profiles.[55] Aqueous extract from fruits of *Siraitia grosvenorii* (Swingle) C. Jeffrey ex A.M. Lu & Z.Y. Zhang given intragastrically and prophylactically to Wistar rats at a single dose of 0.1 g/kg decreased postprandial glycaemia when administered with oral load of maltose.[56] From this extract, a mixture of triterpene glycosides at a dose of 0.1 g/kg decreased maltose-induced, postprandial glycaemia to 70% after 30 minutes.[56] This fraction inhibited *in vitro* the enzymatic activity of rat intestinal maltase with an IC_{50} value equal to 5 mg/mL.[56] From this fraction, the triterpene saponins mogroside V. (Figure 1.11) inhibited rat-intestinal maltase *in vitro* with an IC_{50} of 18 mg/mL.[56] It should be noted that peak blood glucose values in rats are obtained much earlier (15–45 minutes) than in human subjects (around 60 minutes).[57] Being nontoxic, the fruits of

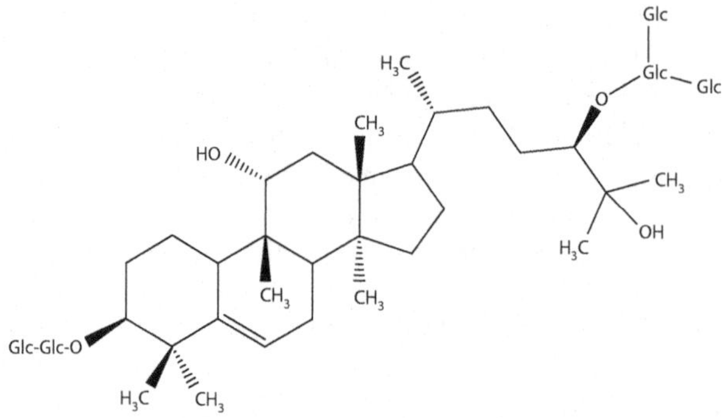

FIGURE 1.11 Mogroside V.

Siraitia grosvenorii (Swingle) C. Jeffrey ex A.M. Lu & Z.Y. Zhang could, be incorporated in the diet of subjects with metabolic syndrome. Clinical studies in this direction are needed.

1.17 *Brassica oleracea* L.

Synonyms: *Crucifera brassica* E.H.L. Krause; *Napus oleracea* (L.) K.F. Schimp. & Spenn.
Common names: ye gan lan (Chinese); cabbage
Subclass Dilleniidae, Superorder Capparanae, Order Capparales, Family Brassicaceae
Medicinal use: carminative (China)

Anthocyanin-rich extract of *Brassica oleracea* L. (Figure 1.12) given orally to Charles Foster rats for 8 weeks at a dose of 100 mg/kg/day reduced plasma cholesterol from 216.7 to 92.1 mg/dL and

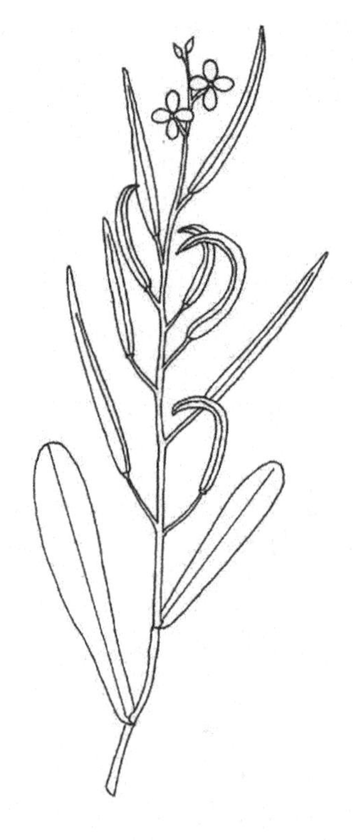

FIGURE 1.12 *Brassica oleracea* L.

triglycerides from 90.5 to 69.6 mg/dL, low density lipoproteins from 230.8 to 67.3 mg/dL, and very low-density lipoproteins from 18.1 to 13.2 mg/dL.[58] This treatment increased triglyceride faeces from 5 to 12.3 mg/g, increased faeces cholesterol from 5.4 to 9 mg/g, and boosted the fecal excression of cholic acid and deoxycholic acid implying the inhibition of cholesterol and triglycerides intestinal absorption.[58]

1.18 *Cotylelobium melanoxylon* (Hook. f.) Pierre

Synonym: *Anisoptera melanoxylon* Hook. f.
Common names: resak bukit (Indonesia); khiam (Thailand)
Subclass Dillenidae, Superorder Malvanae, Order Malvales, Family Dipterocarpaceae
Medicinal use: diabetes (Indonesia)

In type 2 diabetic patients the activity of sucrase–isomaltase is abnormally high.[59] Oligostilbenes have the ability to inhibit α-glucosidase *in vitro* and *in vivo*. These phenolic natural products are accumulated in members of the family Dipterocarpaceae. One such oligostilbene is vaticanol G which at a single oral dose of 50 mg/kg administered 30 minutes before oral loading of sucrose decreased postprandial glycemia of rats from 188.3 to 127.2 mg/dL at 30 minutes, suggesting α-glucosidase inhibition (acarbose 10 mg/kg: 114.8 mg/mL).[60] From the same plant, vaticanol A and E at a single dose of 50 mg/kg administered 30 minutes before oral loading of sucrose decreased the postprandial glycemia from 188.3 to 147.3 mg/dL and from 188.3 to 156.8 mg/dL at 30 minutes, respectively (acarbose 10 mg/kg: 114.8 mg/mL).[60] Further, vaticanol A at a single oral dose of 200 mg/kg lowered plasma triglycerides from 535.6 to 368.6 mg/dL, 2 hours after olive-oil loading in ddY mice indicating pancreatic lipase inhibition (orlistat at 20 mg/kg: 198.6 mg/dL).[60] Vaticanol E at a single oral dose of 200 mg/kg decreased plasma triglycerides from 535.6 to 326.3 mg/dL, 2 hours after olive-oil loading in male ddY mice (orlistat at 20 mg/kg: 198.6 mg/dL).[60] Vaticanol G at a single oral dose of 200 mg/kg decreased plasma triglycerides from 535.6 to 245.7 mg/dL 2 hours after olive-oil loading in male ddY mice.[60] *In vitro*, vaticanol A inhibited the enzymatic activity of maltase, sucrase, and lipase with IC_{50} values of 218, 148, and 52 μM, respectively (acarbose: 2 and 1.7 μM orlistat: 0.05 μM).[60] *In vitro*, vaticanol E inhibited the enzymatic activity of maltase, sucrase, and lipase with IC_{50} values of 342, 89, and 86 μM.[60] *In vitro*, vaticanol E inhibited the enzymatic activity of maltase, sucrase with IC_{50} superior to 400 μM and inhibited lipase with IC_{50} values of and 59 μM.[60]

1.19 *Shorea roxburghii* G. Don

Common names: jalari (India); Meranti temak nipis (Malay); phayom (Thai); Talooralac tree
Subclass Dilleniidae, Superorder Malvanae, Order Malvales, Family Diperocarpaceae
Medicinal use: diarrhoea (Thailand)

Methanol extract from barks of *Shorea roxburghii* G. Don given orally at a single dose of 250 mg/kg to rodents decreased postprandial glycaemia 30 minutes after sucrose loading from 229.9 to 207.6 mg/dL.[61] From this extract, hemsleyanol D, (+)-α-viniferin and (−)-balanocarpol at a dose of 200 mg/kg reduced, in the same experiment, glycaemia from 232.9 to 142.4, 153.5, and 169.2 mg/dL, respectively.[61] *In vitro*, hemsleyanol D and (+)-α-viniferin inhibited the enzymatic activity of maltase with IC_{50} values equal to 266 and 172 μM and sucrase with IC_{50} values equal to 218 and 234 μM, respectively.[61] In regards to lipase, methanol extract from bark of *Shorea roxburghii* G. Don inhibited the enzymatic activity of lipase with an IC_{50} value equal to 31.6 μg/mL. From this extract, phayomphenol A2, (−)-hopeaphenol, (+)-isohopeaphenol, hemsleyanol D, (+)-α-viniferin, and (−)-balanocarpol administered orally to fasted ddY mice 30 minutes prior to olive-oil intake decreased plasma triglycerides from 546.7 to 217.5, 269.5, 237.2, 274.6, 266.9, and 240.5 mg/dL, respectively, whereas orlistat reduced triglyceridaemia to 203.8 mg/dL.[62] *In vitro*, (−)-hopeaphenol, (+)-isohopeaphenol, hemsleyanol D, (+)-α-viniferin inhibited the enzymatic activity with IC_{50} values below 50 μM, whereby (−)-balanocarpol was inactive and orlistat had an IC_{50} value equal 0.05 μM.[62]

1.20 *Broussonetia kazinoki* Siebold & Zucc.

Synonym: *Broussonetia monoica* Hance
Common name: chu (Chinese)
Subclass Dilleniade, Superorder Malvanae, Order Urticales, Family Moraceae
Medicinal use: aphrodisiac (Korea)

The stem bark of *Broussonetia kazinoki* Siebold & Zucc. shelter the diphenylpropanes broussonone A (Figure 1.13) broussonin A and B, the flavans 7,4′-dihydroxyflavan and 3′,7-dihydroxy-4′-methoxyflavan, which at a concentration of 100 μM inhibited the enzymatic activity of pancreatic lipase by 71.9%, 50.7%, 40.4%, 55.6%, and 24.8%, respectively, whereas orlistat at 1 μM evoked 60.4% inhibition.[63]

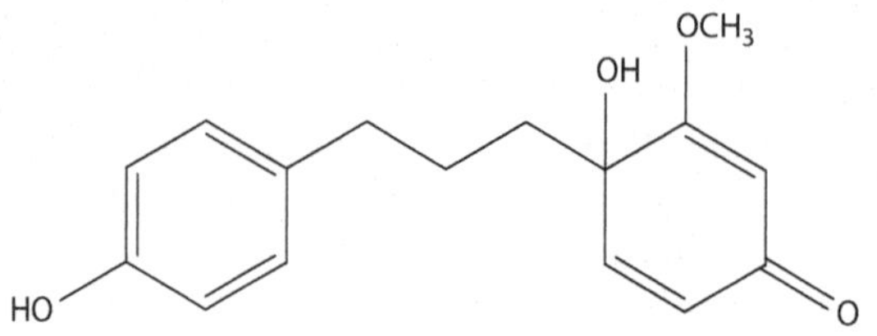

FIGURE 1.13 Broussonone A.

1.21 *Ficus deltoidea* Jack

Synonyms: *Ficus diversifolia* Blume; *Ficus ovoidea* Jack
Common names: ara burong (Malay); Mistletoe Fig
Subclass Dillenidae, Superorder Malvanae, Order Urticales, Family Moraceae
Medicinal use: diabetes (Malaysia)

In man, throughout a 24-hour period, arterial plasma glucose average approximatively 90 mg/dL.[64] Physiologically, the consumption of 75 g of glucose evokes, after 2 hours, a rise in plasmatic glucose concentration that remains below 140 mg/dL (7.8 mmol/L).[64,65] In obese patients, a 2-hour post oral 75 g glucose intake glycemia ranging from 7.8 mmol/L (140 mg/dL) to 11.1 mmol/L (200 mg/dL) evidences a state of impaired glucose tolerance as defined by the World Health Organization (2006). In diabetic patients, 2-hour plasma glucose is equal to or superior to 200 mg/dL (11.1 mmol/L).[65] In order to suppress postprandial hyperglycemia, a number of α-glucosidase inhibitors delaying the absorption of glucose from dietary carbohydrates have been developed including acarbose, miglitol, and voglibose, which have unpleasant side effects.[66] The flavones C-glycosides vitexin and isovitexin (Figure 1.14) isolated from *Ficus deltoidea* Jack given orally at a dose of 100 mg/kg decreased 30 minutes postprandial glycemia in normal mice or diabetic rats loaded with sucrose via α-glucosidase inhibition.[67] In a subsequent study, Yang et al. (2014) provided evidence that vitexin and isovitexin inhibited α-amylase activity with Ki values of 569.6 and 75.8 μg/mL, respectively.[68] The antidiabetic activity of *Ficus deltoidea* Jack could at least be imparted via inhibition of glucose absorption from dietary carbohydrates. The toxicity of the plant appears to be unknown and preclinical studies are needed.

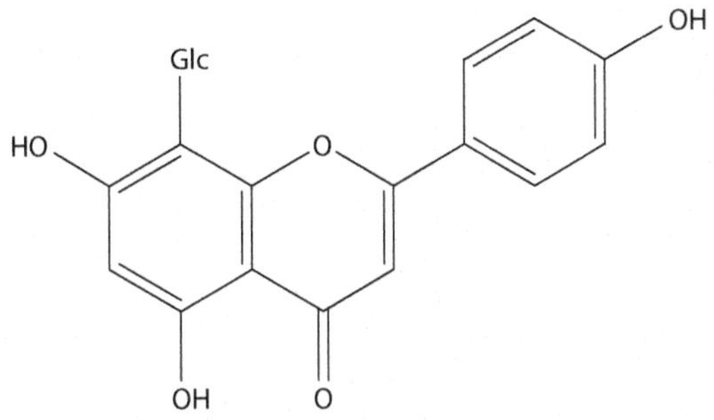

Glc
HO
O
OH
OH O
Vitexin

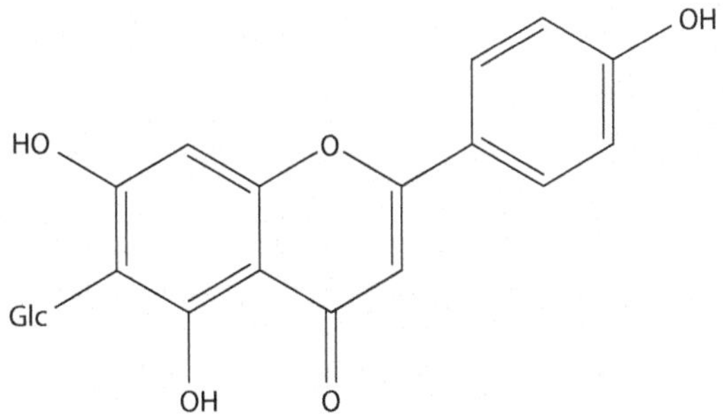

OH
HO
O
Glc
OH O
Isovitexin

FIGURE 1.14 Flavones C-glycosides from *Ficus deltoidea* Jack.

1.22 *Phyllanthus reticulatus* Poir.

Synonyms: *Glochidion microphyllum* Ridl.; *Phyllanthus dalbergioides* Wall. ex J.J. Sm.; *Phyllanthus erythrocarpus* Ridl.
Common names: xiao guo ye xia zhu (Chinese); kayu darah belut (Malay); matang bulud (Philippines)
Subclass Dillenidae, Superorder Euphorbianae, Order Euphorbiales, Family Phyllanthaceae
Medicinal use: sore throat (Malaysia)

Ethanol extract of leaves of *Phyllanthus reticulatus* Poir. given orally to alloxan-induced diabetic Swiss mice at dose of 1 g/kg decreased after 24 hours plasma glucose from 291.3 to 206.3 mg%.[69] Given daily for 21 days, this extract at dose of 1 g/kg decreased plasma glucose from 291.8 to 186 mg%, whereas untreated animals had a variation of glycaemia from 387 to 325 mg%.[69] The astringency of this plant is most probably owed to ellagitannins and gallic acid that may inhibit α-amylase and/or α-glucosidase.[70] Methyl gallate (Figure 1.15) inhibits α-glucosidase *in vitro*.[71] It is tempting to speculate that ellagitannins and their derivatives in *Phyllanthus reticulatus* Poir. could inhibit pancreatic lipase. Maruthappan and Shree reported that the intake of 500 mg/kg/day of aqueous extract from the plant to rats for 45 days on high-fat diet decreased plasma triglycerides.[72]

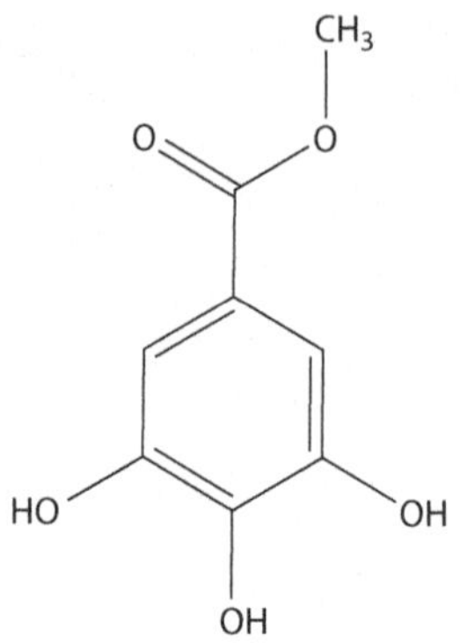

FIGURE 1.15 Methyl gallate.

1.23 *Euphorbia thymifolia* L.

Synonyms: *Anisophyllum thymifolium* (L.) Haw.; *Chamaesyce thymifolia* (L.) Millsp.; *Euphorbia philippina* J. Gay ex Boiss.
Common names: qian gen cao (Chinese); Laghu dugdhi (India); Thyme-leaved Spurge
Subclass Dillenidae, Superorder Euphorbianae, Order Euphorbiales, Family Euphorbiaceae
Medicinal use: diabetes (Bangladesh)
History: The plant was known to Sushruta (600 BC) Ayurvedic physician

Methanol extract of *Euphorbia thymifolia* L. given orally to Swiss albino mice at a single dose of 400 mg/kg 1 hour before oral administration of glucose decreased glycaemia to 60.5%, whereas glibenclamide at 10 mg/kg evoked a 48.6% fall in glycemia.[73] Note that the plant accumulates ellagitannins and quercetin glycosides.[74,75,78,79] Ellagitannins in this plants may inhibit carbohydrate and triglycerides intestinal absorption through inhibition of α-amylase, α-glucosidase, and pancreatic lipase. However, decrease in postprandial glycemia during oral loading of glucose is by itself independent of α-amylase or α-glucosidase but may result from inhibition of glucose by enterocytes increased insulin secretion or increased uptake of glucose in skeletal muscles. Further pharmacological and toxicological studies on the benefits of *Euphorbia thymifolia* L. for metabolic syndrome are needed.

1.24 *Sinocrassula indica* (Decne.) A. Berger

Synonyms: *Crassula indica* Decne.; *Sedum indicum* (Decne.) Raym.-Hamet
Common name: shi lian (Chinese)
Subclass Rosidae, Superoder Rosanae, Order Saxifragales, Family Crassulaceae
Medicinal use: cough (India)

Methanol extract of *Sinocrassula indica* (Decne.) A. Berger (containing flavonoids including quercetin, luteolin, kaempferol) given orally to rats at a single dose of 500 mg/kg decreased postprandial glycaemia from 166.3 to 121.9 mg/dL at 30 minutes during oral sucrose challenge (tolbutamide: 25 mg/kg: 138.1 mg/dL).[76] The same regimen applied to rats challenged with oral glucose decreased glycaemia but had no activity against intraperitoneal glucose loading, indicating an activity elicited at intestinal level.[76] It must be noted that glucose released from maltose, maltotriose, dextrin, and sucrose is actively absorbed in brush border enterocytes by integral sodium-dependent glucose transporter-1 (SGLT-1) located in the apical cytoplasmic membrane.[77] From the cytoplasm of enterocytes, glucose is released in the general circulation via, at least, glucose transporter 2 located in the basolateral cytoplasmic membrane of enterocytes. The sodium gradient necessary for SGLT1 activity is maintained by a basolateral Na^+/K^+ ATPase.[78] In spontaneous type 2 diabetic obese KK-Ay mice, the extract given orally at a dose of 500 mg/kg/day decreased nonfasting glycaemia by 28% and triglycerides by 14% ,whereas cholesterolaemia and serum-free fatty acids were not affected.[79] Quercetin inhibited yeast α-glucosidase with IC_{50} value of 58.9 μM (acarbose: 130.7 μM).[80]

1.25 *Terminalia bellirica* (Gaertn.) Roxb.

Synonyms: *Myrobalanus bellirica* Gaertn.; *Terminalia attenuata* Edgew.; *Terminalia eglandulosa* Roxb. ex C.B. Clarke; *Terminalia gella* Dalzell; *Terminalia laurinoides* Teijsm. & Binn.; *Terminalia punctata* Roth

Common names: pi li le (Chinese); vibhitaka (India); belliric myrobalan

Subclass Rosidae, Superorder Myrtanae, Order Myrtales, Family Combretaceae

Medicinal use: fever (India)

History: The plant was known to Sushruta (circa 600 BC) Ayurvedic physician

Sabu et al. (2009) provided evidence that methanol extract from fruits of *Terminalia bellirica* (Gaertn.) Roxb. (Figure 1.16) given orally to alloxan-induced diabetic Wistar rats at a dose of 100 mg/kg/day for 12 days reduced glycaemia by 37.5%.[81] This regimen brought to normal serum

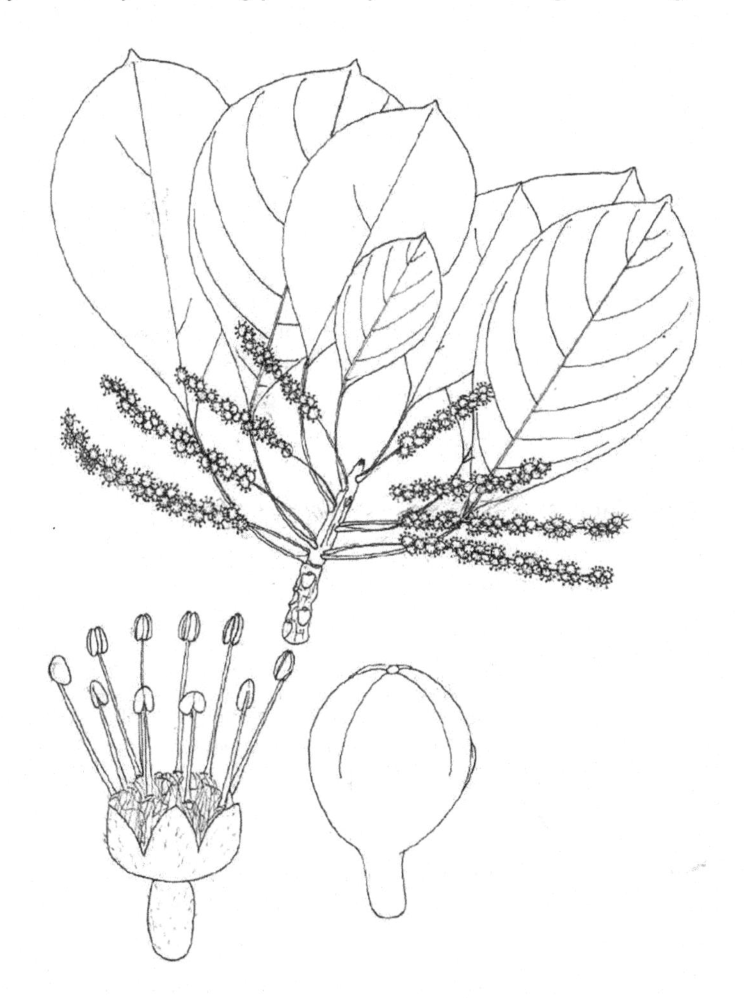

FIGURE 1.16 *Terminalia bellirica* (Gaertn.) Roxb.

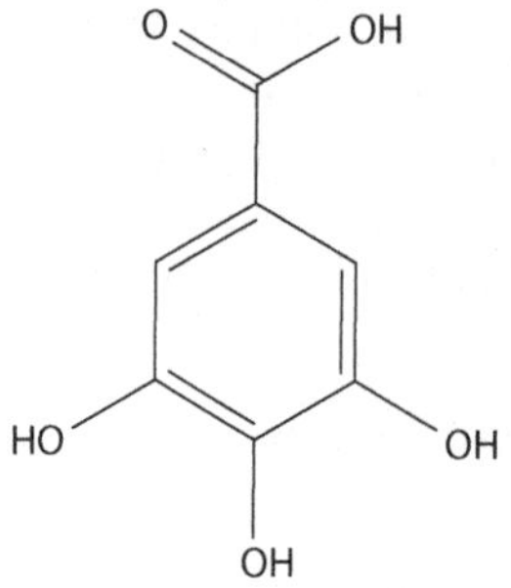

FIGURE 1.17 Gallic acid.

and hepatic lipid peroxidation and glutathione, whereas catalase, glutathione peroxidase, and superoxide dismutase enzymatic activities were increased.[81] Aqueous extract from fruits of *Terminalia bellirica* (Gaertn.) Roxb. given at 3% of diet to spontaneous type 2 diabetic Tsumara Suzuki Obese Diabetes (TSOD) mice for 8 weeks evoked a mild reduction of body weight and weight of visceral, mesenteric, and subcutaneous fat without reduction of food intake.[82] This supplementation improved glucose tolerance as evidenced by a decrease of peak glycaemia from about 450 to 325 mg/dL after 30 minutes in oral glucose tolerance test.[82] The extract decreased fasting insulinaemia as well as insulin resistance and decreased hepatic triglycerides.[82] In ddY mice, the extract at a dose of 1 g/kg halved peak plasma triglycerides at 4 hours in olive-oil loading test suggesting pancreatic lipase inhibition.[82] The extract inhibited the enzymatic activity pancreatic lipase *in vitro* with an IC_{50} of 65.7 μg/mL and gallic acid (Figure 1.17) isolated from it inhibited the enzymatic activity of pancreatic lipase with an IC_{50} of 3.9 μg/mL.[82] In a subsequent study, gallic acid and methyl gallate, which are produced by members of the genus *Terminalia* L,. inhibited *in vitro* the enzymatic activity of α-glucosidase with IC_{50} values of 5.2 and 11.5 μM, respectively.[83] Gallic acid and methyl gallate are derived from ellagitannins suggesting that α-glucosidase and/or lipase inhibition upon oral loading of ellagitannins can be elicited by gastro-intestinal metabolites.[71] In fact, Espin et al. (2007) fed Iberian pigs with ellagitannins and observed the release of ellagic acid in the jejunum, which was directly absorbed in the first portions of the gastrointestinal tract.[71] The intestinal bacterial commensal flora metabolizes nonabsorbed ellagic acid into benzopyranone derivatives such as urolithin A, which are absorbed.[71]

1.26 *Vaccinium myrtillus* L.

Synonyms: *Vaccinium oreophilum* Rydb.
Common names: hei guo yue ju (Chinese); bilberry
Subclass Dillenidae, Superorder Ericanae, Order Ericales, Family Ericaceae
Nutritional use: food (China)

The fruits of *Vaccinium myrtillus* L. accumulate series of anthocyanosides of which delphinidin 3-*O*-β-D-glucopyranoside, cyanidin 3-*O*-β-D-glucopyranoside, peonidin 3-*O*-β-D-glycopyranoside, and malvidin 3-*O*-β-D-glucopyranoside.[84] Methanol extract from fruits of *Vaccinium myrtillus* L. inhibited α-amylase and α-glucosidase with IC_{50} values of 61.3 and 138.4 μg/mL, respectively.[86] Tadeka et al. (2006) provided evidence that cyanidin (Figure 1.18) at a concentration of 500 μM inhibited rat intestinal α-glucosidase and porcine pancreatic α-amylase activity by 6% and 37%, respectively.[86] In this experiment, yeast α-glucosidase was inhibited by 99% at a concentration of 200 μM.[86] In enterocytes, cholesterol is re-esterified in cholesteryl ester by acyl-coenzyme

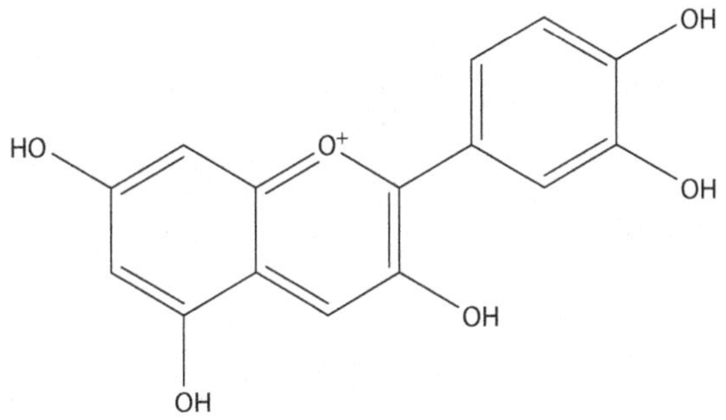

FIGURE 1.18 Cyanidin.

A:cholesterol *O*-acyltransferase-2 (ACAT-2).[87] Anthocyanin fraction of blueberry containing mainly cyanidin-3-*O*-glucoside (Figure 1.18) and petunidin-3-*O*-glucoside added at 1% of diet to Golden Syrian hamsters for 6 weeks decreased plasma cholesterol from 6.6 to 5.8 mmol/L and increased fecal cholesterol implying inhibition of dietary cholesterol absorption as a result of decreased intestinal expression of ACAT-2.[88] In this experiment, plasma triglyceride intestinal absorption was not affected by anthocyanin. Hamsters are good animal models for the study of lipid metabolism because cholesterol metabolism in hamster closely resembles that in human in contrast to rats and mice.[89] The fruits of *Vaccinium myrtillus* L. could be conceptually seen as beneficial ingredient for the diet of subjects with metabolic syndrome.[90,91]

1.27 *Lagerstroemia speciosa* (L.) Pers.

Synonyms: *Lagerstroemia flos-reginae* Retz.; *Lagerstroemia reginae* Roxb.; *Munchausia speciosa* L.
Common names: banaba (Philippines); Queen crape-myrtle
Subclass Rosidae, Superorder Myrtanae, Order Myrtales, Family Lythraceae
Medicinal use: diabetes (Philippines)

Faustino Garcia reported in 1941 that dried leaves or ripe fruits of *Lagerstroemia speciosa* (L.) Pers. (Figure 1.19) known in the Philippines as *banaba* at a dose of 20 g in the form decoction had the same activity as 7 units of insulin in decreasing blood glucose. The flowers at the same dose had activity equivalent to 5 units of insulin.[92] Aqueous extract from leaves given orally at a dose of 150 mg/kg/day to streptozotocin-induced diabetic mice for 2 months had no effect up to 10 days treatment but decreased, after 60 days, glycaemia from 119.7 to 63 mg/dL, a value close to 58.1 mg/dL in normoglycaemic rodents.[93] This treatment brought to normal values hepatic lipid peroxidation, glutathione-S-transferase, superoxide dismutase, and glutathione contents.[93] From the leaves of *Lagerstroemia speciosa* (L.) Pers. the triterpenes oleanolic acid, arjunolic acid, asiatic acid, maslinic acid, corosolic acid, and 23-hydroxyursolic acid inhibited α-glucosidase with IC_{50} values below 35 μg/mL.[94,95] Out of these triterpene, corosolic acid inhibited α-amylase with an IC_{50} value of 100 μg/mL.[94,95] Corosolic acid (Figure 1.20) given to spontaneous type 2 diabetic KK-Ay as 0.023% part of a high cholesterol diet for 10 weeks maintained plasma cholesterol to the level of control whereby it had no effect on weight gain.[94] This treatment halved hepatic cholesterol content and decreased cholesterolaemia in oral cholesterol test to about 10% at 4 hours on probable account of ACAT-2 inhibition.[94] Clinical trials are warranted.

FIGURE 1.19 *Lagerstroemia speciosa* (L.) Pers.

FIGURE 1.20 Corosolic acid.

1.28 *Punica granatum* L.

Common names: shi liu (Chinese); dhalim (India); pomegranate
Subclass Rosidae, Superorder Myrtanae, Order Myrtales, Family Lythraceae
Medicinal use: diabetes (India)
History: The plant was known of twelfth century Arabic physician Serapion as astringent

Ellagitannins in the fruits of *Punica granatum* L. inhibit *in vitro* the intestinal enzymes in charge of carbohydrate and triglyceride absorption. Methanol extract from seeds at a concentration of 2.5 mg/mL inhibited α-amylase activity by 94.5% *in vitro* (IC_{50}: 1.1 mg/mL; acarbose: 1.3 μg/mL).[96] Methanol extract from husk of *Punica granatum* L. seeds at a concentration of 2.5 mg/mL inhibited porcine pancreatic lipase by 100% *in vitro* (IC_{50}: 0.1 mg/mL; orlistat: 0.1 ng/mL).[96] Punicalagin, punicalin, and ellagic acid isolated from *Punica granatum* L. inhibited *in vitro* rat α-glucosidase with IC_{50} values of 140.2, 191.4, and 380 μmol/L, respectively.[97] Methanol fractions of flowers of *Punica granatum* L. inhibited the enzymatic activity of recombinant human maltase–glucoamylase, rat maltase, and rat sucrase with IC_{50} values equal to 567, 87, and 324 μg/mL.[98] In the same experiment, a methanol fraction of arils inhibited the enzymatic activity of recombinant human maltase–glucoamylase, rat maltase, and rat sucrase with IC_{50} values equal to 393.3, 527, and 486 μg/mL, respectively.[98] From the methanol fraction of aril, oenothein B and punicalagin inhibited human maltase–glucoamylase, rat maltase, and rat sucrase with IC_{50} values equal to 174, 290, 213, 305, 535, and 369 μM, respectively.[98] Consumption fruits' juice of *Punica granatum* L. could be of value for metabolic syndrome.

1.29 *Trapa japonica* Flerow

Synonym: *Trapa litwinowii* V.N. Vassil.
Subclass Rosidae, Superorder Myrtanae, Order Myrtales, Family Trapaceae
Medicinal use: diabetes (India)

A single oral 40 mg/kg administration of a polyphenolic extract isolated from the husk of *Trapa japonica* Flerow to ICR mice receiving a load of starch halved postprandial glycaemia after 30 minutes and reduced plasma insulin.[99] From this extract, eugeniin, 1,2,3,6-tetra-*O*-galloyl-β-D-glucopyranose and (−)-epigallocatechin gallate inhibited the enzymatic activity of human salivary α-amylase with IC_{50} of 42, 58, and 53 μM, respectively.[99] Eugeniin, 1,2,3,6-tetra-*O*-galloyl-β-D-glucopyranose and (−)-epigallocatechin gallate inhibited also maltase with IC_{50} of 69, 83, and 107 μM, respectively, and sucrase with IC_{50} of 333, 260, and 268 μM, respectively.[99]

1.30 *Cassia auriculata* L.

Synonym: *Senna auriculata* (L.) Roxb.
Common names: er ye jue ming (Chinese); avartaki (India); tanner's cassia
Subclass Rosidae, Superorder Fabanae, Subclass Rosiidae, Family Fabaceae
Medicinal use: diabetes (India)

Aqueous extract from leaves of *Cassia auriculata* L. given to streptozotocin-induced diabetic Wistar rats at a dose of 400 mg/kg for 21 days reduced fasting glycaemia from 214.2 to 113.8 mg/dL, a value close to a normoglycaemia (82 mg/dL).[100] Further, this treatment normalized serum lipid peroxides, erythrocytes superoxide dismutase, catalase, and glutathion.[100] Ethanol extract of aerial parts *Cassia auriculata* L. inhibited *in vitro* the enzymatic activity of lipase with an IC_{50} value equal to 6 μg/mL.[101] From this plant, kaempferol-3-*O*-rutinoside (Figure 1.21), quercetin, luteolin, and rutin inhibited *in vitro* the enzymatic activity of porcine pancreatic lipase with IC_{50} values equal to 1.7, 49.3, 76.5, and 91 μg/mL, respectively.[101]

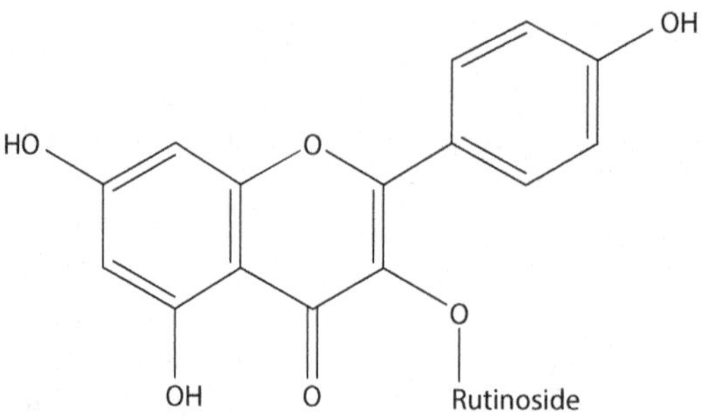

FIGURE 1.21 Kaempferol-3-*O*-rutinoside.

1.31 *Mucuna pruriens* (L.) DC.

Synonym: *Dolichos pruriens* L.
Common names: atmagupta (India); common cowitch
Subclass Rosidae, Superorder Fabanae, Subclass Rosiidae, Family Fabaceae
Medicinal use: tonic (India)

Members of the family Fabaceae synthetize isoflavonoids and pterocarpans that inhibit α-glucosidase *in vitro*. The isoflavanones mucunone A and B, the pterocarpan (6a*R*,11a*R*)-medicarpin, the isoflavanone parvisoflavone B (Figure 1.22), the isoflavans (3R)-vestitol, and 8-methoxyvestitol isolated from the roots of *Mucuna pruriens* (L.) DC. inhibited α-glucosidase with IC_{50} values below 120 μM (acarbose: 7.9 μM).[102]

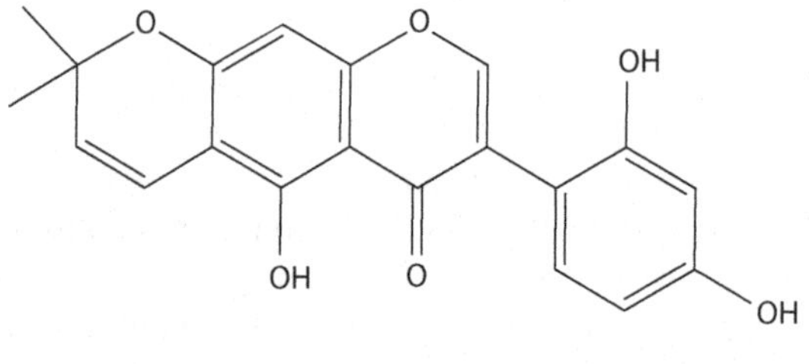

FIGURE 1.22 Parvisoflavone B.

1.32 *Pterocarpus marsupium* Roxb.

Synonyms: *Lingoum marsupium* (Roxb.) Kuntze; *Pterocarpus bilobus* Roxb. ex G. Don
Common names: ma la ba zi tan (Chinese); kum kusrala (India); kino
Subclass Rosidae, Superorder Fabanae, Subclass Rosiidae, Fabaceae
Medicinal use: diabetes (India)
History: The plant was known to Sushruta

The plant yields an exudate called East Indian kino that has been used for the treatment of diarrhea. Abesunadara et al. (2004) made the demonstration that the exudate of *Pterocarpus marsupium*

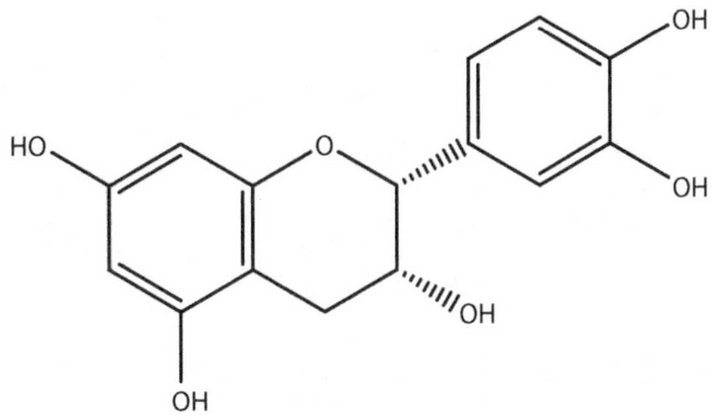

FIGURE 1.23 (–)-Epicatechin.

Roxb. was able to inhibit α-glucosidase *in vitro* on probable account of (–)-epicatechin (Figure 1.23) and catechin[104]. (–)-Epicatechin inhibited *in vitro* the enzymatic activity of α-glucosidase with an IC_{50} of 5.8 µg/mLs.[103,104]

1.33 *Polygala aureocauda* Dunn

Synonym: *Polygala fallax* Hemsl.
Common name: huang hua dao shui lian (Chinese)
Subclass Rosidae, Superorder Fabanae, Order Polygalales, Family Polygalaceae
Nutritional use: food (China)

Reinioside C from the roots of *Polygala aureocauda* Dunn given at a dose of 16 mg/kg/day for 30 days orally to Kunming mice on hyperlipidemic diet attenuated plasma cholesterol from 5.6 to 4 mmol/L, normalized plasma triglycerides from 1.1 to 0.8 mmol/L.[105,107] Besides, this pentacyclic triterpene saponins lowered hepatic cholesterol and brought hepatic triglycerides to normal values and these effects were comparable to simvastatin (4 mg/kg/days).[105] Decrease in serum and hepatic cholesterol is, at least, an indication of decreased absorption of cholesterol in small intestine or increased fecal excression of bile acids in the feces. Triterpene saponins and steroidal glycosides found in medicinal plants in Asia have the tendency to form insoluble stoichiometric complexes with cholesterol *in vitro* and interact with bile acid micelles expelling cholesterol from them, thereby inhibiting cholesterol absorption and decreasing serum cholesterol.[106]

1.34 *Citrus limon* (L.) Osbeck

Synonym: *Citrus limonum* Risso
Common names: limau (Malay); lemon
Subclass Rosidae, Superorder Rutanae, Order Rutales, Family Rutaceae
Medicinal use: high cholesterol (Malaysia)

Kawaguchi et al. (1997) provided evidence that hesperidin and neohesperidine that occur in the peels of fruits of *Citrus limon* (L.) Osbeck (Figure 1.24) inhibited porcine pancreatic lipase with IC_{50} values of 32 and 46 µM, respectively, whereas narirutin and narigin were inactive.[108] Hesperidin given as part of 10% diet to rats had no effect on body weight, increased fecal lipids from 0.09 to 1 g/3days and decreased plasma triglycerides from 89.5 to 64.1 mg/dL.[108]

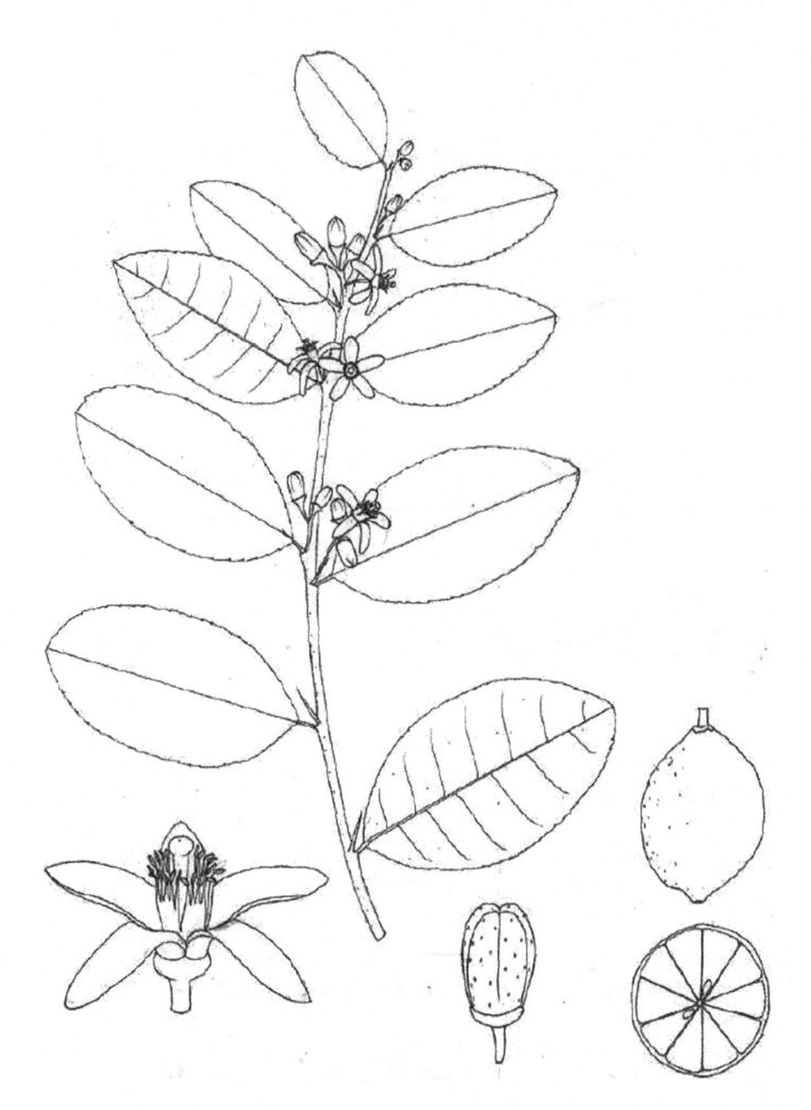

FIGURE 1.24 *Citrus limon* (L.) Osbeck.

1.35 *Murraya koenigii* (L.) Spreng.

Synonyms: *Bergera koenigii* L.; *Chalcas koenigii* (L.) Kurz
Common names: tiao liao jiu li xiang (Chinese); karivepu (India); daun kari (Malay); curry leaf tree
Subclass Rosidae, Superorder Rutanae, Order Rutales, Family Rutaceae
Medicinal use: indigestion (India)

Carbazole alkaloids elaborated by members of the family Rutaceae have the ability to inhibit α-glucosidase and/or intestinal lipase. The dimeric carbazole alkaloids bisgerayafolines A, B, and C isolated from the fruits of *Murraya koenigii* (L.) Spreng. inhibited the enzymatic activity of α-glucosidase with IC_{50} values equal to 45.4, 41.2, and 69 μM, respectively.[109] Dichloromethane

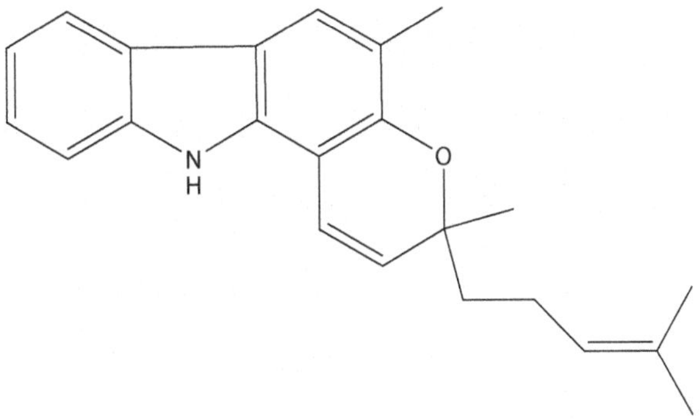

FIGURE 1.25 Mahanimbine.

extract from leaves of *Murraya koenigii* (L.) Spreng. given orally at a dose of 300 mg/kg for 2 weeks to rodents fed with high-fat diet reduced weight gain from 64.2 to 14.6 g compared to control.[110] In the same experiment serum total cholesterol and triglycerides from 117.8 and 178.3 mg/dL to 79.7 mg/dL and 121.9 mg/dL, respectively, whereby glycaemia was unchanged.[110] Mahanimbine from this extract at a dose of 30 mg/kg/day inhibited weight gain in high-fat fed rodents and reduced plasma cholesterol and triglycerides to 98.2 and 130.2 mg/dL.[110] Mahanimbine (Figure 1.25) and koenimbin from *Murraya koenigii* (L.) Spreng. inhibited the enzymatic activity of lipase with IC_{50} values equal to 17.9 and 168.6 µM, respectively.[111]

1.36 *Zanthoxylum piperitum* DC.

Subclass Rosidae, Superorder Rutanae, Order Rutales, Family Rutaceae
Common name: Japanese pepper
Medicinal use: indigestion (China)

In enterocytes, dietary cholesterol is re-esterified into cholesteryl ester by acyl-CoA:cholesterol *O*-acyltransferase-2.[87] The aliphatic amides β-Sanshool and γ-sanshool isolated from the stems of *Zanthoxylum piperitum* DC. (Figure 1.26) inhibited the enzymatic activity of human acyl-CoA:cholesterol *O*-acyltransferase-2 with IC_{50} values of 79.7 and 82.6 µM, respectively.[112]

FIGURE 1.26 *Zanthoxylum piperitum* DC.

1.37 *Cedrela odorata* L.

Subclass Rosidae, Superorder Rutanae, Order Rutales, Family Meliaceae

Ethanolic extract of inner stembark of *Cedrela odorata* L. (containing gallic acid, (−)-gallocatechin and (+)-catechin) inhibited α-glucosidase with an IC_{50} of 84.7 μg/mL (acarbose 5.1 μg/mL).[113] Given to streptozotocin-induced diabetic Wistar rats at a single oral dose of 500 mg/kg, 30 minutes prior to oral lead of glucose decreased postprandial glycemia from about 500 to 255 mg/dL and delayed peak glycaemia from 45 to 90 minutes.[113] Given to streptozotocin-induced diabetic Wistar rats at a single oral dose of 500 mg/kg 30 minutes prior to oral lead of sucrose or starch decreased postprandial glycemia at 30 minutes to a lesser extent, suggesting that the extract may also be inhibiting glucose transporters in the intestine by blocking inhibited Na^+-glucose cotransporter-1 (SGLT1).[113] However, 500 mg/kg given daily for 30 days had no beneficial effects on glycaemia.[113]

1.38 *Mangifera indica* L.

Common names: am (India); mango
Subclass Rosidae, Superoder Rutanae, Order Rutales, Family Anacardiaceae
Medicinal use: diabetes (India)

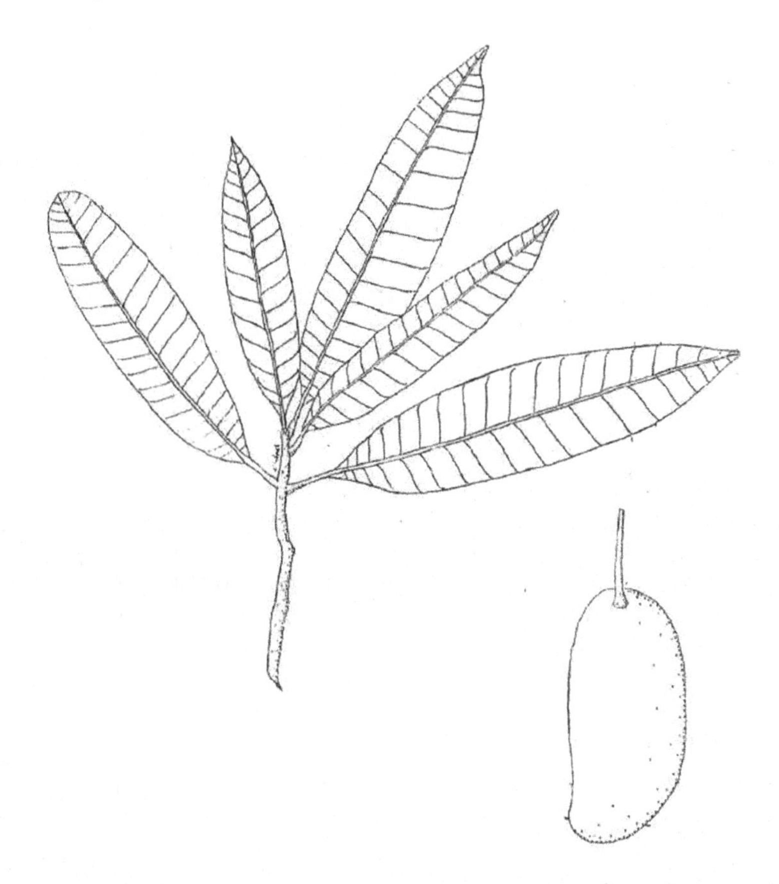

FIGURE 1.27 *Mangifera indica* L.

Ethanol extract from bark of *Mangifera indica* L. (Figure 1.27) at a concentration of 500 μg/mL inhibited α-glucosidase activity by 64.9% (IC_{50}: 314 μg/mL), whereby acarbose at 0.8 μg/mL evoked a 62.4% inhibition.[114] The natural product involved here is to date apparently unknown but one could suggest the involvement of phenolics of which possibly the xanthone glycoside mangiferin or gallotannin, which abounds in the plant.[114]

1.39 *Pistacia chinensis* Bunge

Synonyms: *Pistacia formosana* Matsumura; *Pistacia. Philippinensis* Merrill & Rolfe; *Rhus argyi* H. Léveillé; *Rhus. Gummifera* H. Léveillé.
Common names: huang lian mu (Chinese); karkata (India)
Subclass Rosidae, Superoder Rutanae, Order Rutales, Family Anacardiaceae
Medicinal use: dysentery (India)

Galls of *Pistacia chinensis* Bunge contains the triterpene pistagremic acid (Figure 1.28) that inhibited yeast α-glucosidase and rat intestinal α-glucosidase activities with IC_{50} values equal to 89.1 and 62.4 μM, respectively. This triterpene was more potent than acarbose (IC_{50}: 780.2 and 38.9 μM) against yeast α-glucosidase and rat-intestinal α-glucosidase.[115] Aqueous extract from aerial parts of a member of the genus *Pistacia* at 50 mg/mL inhibited glucose liberation from starch by α-amylase and α-amyloglucosidase by 60%.[116] This extract intragastrically given to Sprague–Dawley rats at a dose of 500 mg/kg decreased peak glycaemia at 45 minutes in oral

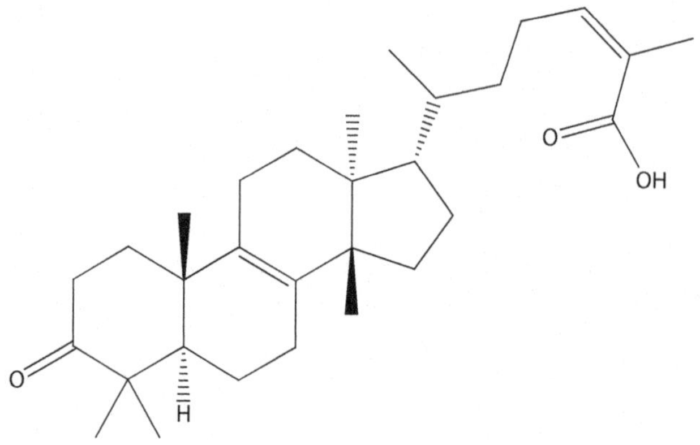

FIGURE 1.28 Pistagremic acid.

starch tolerance test from about 6.5 to 5.5 mM, and this effect was close to 3 mg/kg of acarbose (5 mM).[116] At the same dosage, this extract decreased 90 minutes glycaemia peak from about 6.3 to 5.5 mmol/L.[116]

1.40 *Salacia oblonga* Wall.

Subclass Rosiidae, Superorder Celastranae, Order Celastrales, Family Celastraceae
Medicinal use: diabetes (India)

William et al. (2007) observed that extract of *Salacia oblonga* Wall. given at a dose of 480 mg/kg during meal tolerance test decreased postprandial glycaemia peak at 120 minutes by 27% in patients with type 2 diabetes and decreased peak serum insulin by 12%.[117] In a subsequent study, aqueous extract of roots of *Salacia oblonga* Wall. (containing mangiferin) given orally to mice at a single dose of 100 mg/kg 1 hour before sucrose loading decreased postprandial plasma glucose at 30 minutes more efficiently than acarbose at 200 mg/kg and had no effect of postprandial glycemia following glucose loading.[118] The extract inhibited α-glucosidase activity *in vitro* with an IC_{50} of 5.2 μg/mL, whereas mangiferin and acarbose show much weaker effects with IC_{50} of 22.7 and 53.9 μg/mL, respectively.[118] From this extract kotalagenin 16 acetate, maytenfolic acid, 3β, 22α-dihydroxyoleanane-12-en-29-oic acid, 19-hydroxyferruginol, and lambertic acid inhibited α-glucosidase.[119]

1.41 *Salacia reticulata* Wight

Subclass Rosiidae, Superorder Celastranae, Order Celastrales, Family Celastraceae
Medicinal use: diabetes (Sri Lanka)

Karunanayake et al. (1984) administered aqueous extract from root bark of *Salacia reticulata* Wight at a single dose of 1 mL/100 g to Sprague–Dawley rats and observed a fasting blood glucose decrease by 30%, 1 hour after administration implying at least, an increase of insulin secretion, inhibition of liver secretion of glucose, or increase uptake of glucose by peripheral tissues.[120]

Aqueous extract from roots of *Salacia reticulata* Wight. given orally to Zucker fatty rats at a dose of 125 mg/kg/day for 27 days evoked a decrease in body weight of 14%.[121] From this extract, (−)-epigallocatechin, (−)-epicatechin-(4β→8)-(−)-4-*O*-methylepigallocatechin, and lambertic acid inhibited porcine pancreatic lipase with IC_{50} values of 88, 68, and 225 mg/mL, respectively, *in vitro*.[121] Aqueous extract from leaves of *Salacia reticulata* Wight. at a concentration of 400 μg/m inhibited *in vitro* intestinal rat α-glucosidase by 78.5%.[122] In ddY mice, the extract given orally at a single dose of 1 mg with 160 mg of maltose or sucrose decreased postprandial glycaemia.[122] In the same experiment performed with 160 mg of glucose, the extract had no effect on postprandial glycaemia.[122] In streptozotocin-induced diabetic mice, the extract mixed with drinking water to 0.01% for 4 days lowered glycaemia and the enzymatic activity of intestinal maltase and sucrase.[122] *Salacia reticulata* Wight. proven nontoxic could be of value for the treatment of metabolic syndrome.

1.42 *Viscum album* L.

Subclass Rosiidae, Superorder Santalanae, Order Santalales, Family Viscaceae
Medicinal use: atherosclerosis (Turkey)

Ethanol extract from *Viscum album* L. inhibited *in vitro* the enzymatic activity of pancreatic lipase with an IC_{50} value equal to 33.3 μg/mL.[8] Aqueous extract from the plant given orally at a dose of 100 mg/kg to Swiss albino mice on high-cholesterol diet decreased plasma cholesterol from 218.4 to 139.4 mg/dL and decreased plasma triglycerides from 194.2 to 63.6 mg/dL.[124] This extract also decreased glycemia from 79.8 to 54.6 mg/dL.[123] This parasitic plant well-known of Celts elaborates β-Amyrin acetate, oleanolic acid, betulinic acid, phytosterol, as well as quercetin methyl ethers.[124,125]

1.43 *Viburnum dilatatum* Thunb.

Synonyms: *Viburnum brevipes* Rehder; *Viburnum fulvotomentosum* P.S. Hsu
Common name: jia mi (Chinese)
Subclass Asteridae, Superorder Cornanae, Order Dipsacales, Family Viburnaceae
Medicinal use: sores (China)

Lyophilized fruits' juice of *Viburnum dilatatum* Thunb. (Figure 1.29) given to streptozotocin-induced diabetic Sprague–Dawley rats in drinking water at a concentration of 16.8 mg/mL for 10 weeks had no effects on food consumption but attenuated body weight loss.[126] This supplementation decreased plasma glucose from 2 to 1.5 mmol/L (normal: 1.3 mmol/L), normalized plasma cholesterol from 1.1 to 0.5 mg/mL (normal: 0.6 mg/mL), and triglycerides from 1.6 to 1.2 mg/mL (normal: 1.3 mg/mL).[126] The regimen had no effect on insulin.[126] Lyophilized fruits' juice of *Viburnum dilatatum* Thunb. given to streptozotocin-induced diabetic Sprague–Dawley rats orally at a dose of 500 mg/kg/day for 4 weeks had no effect on body weight loss, decreased postprandial glycemia in oral glucose tolerance test, and had no effect on plasma insulin.[127] From this juice, cyanidin 3-sambubioside inhibited rat sucrase, maltase, isomaltase, glucoamylase, and porcine pancreatic a-amylase with IC_{50} values below 15 mM.[127] Also from this juice, cyanidin 3-*O*-glucoside inhibited rat sucrase, maltase, isomaltase, glucoamylase, and porcine pancreatic α-amylase with IC_{50} values below 110 mM.[127] 5-Caffeoyl quinic acid from the juice inhibited rat sucrase, maltase, isomaltase, glucoamylase, and porcine pancreatic a-amylase with IC_{50} values of 1.4, 24.8, 23.4, 5, and 37.1 mM, respectively.[127]

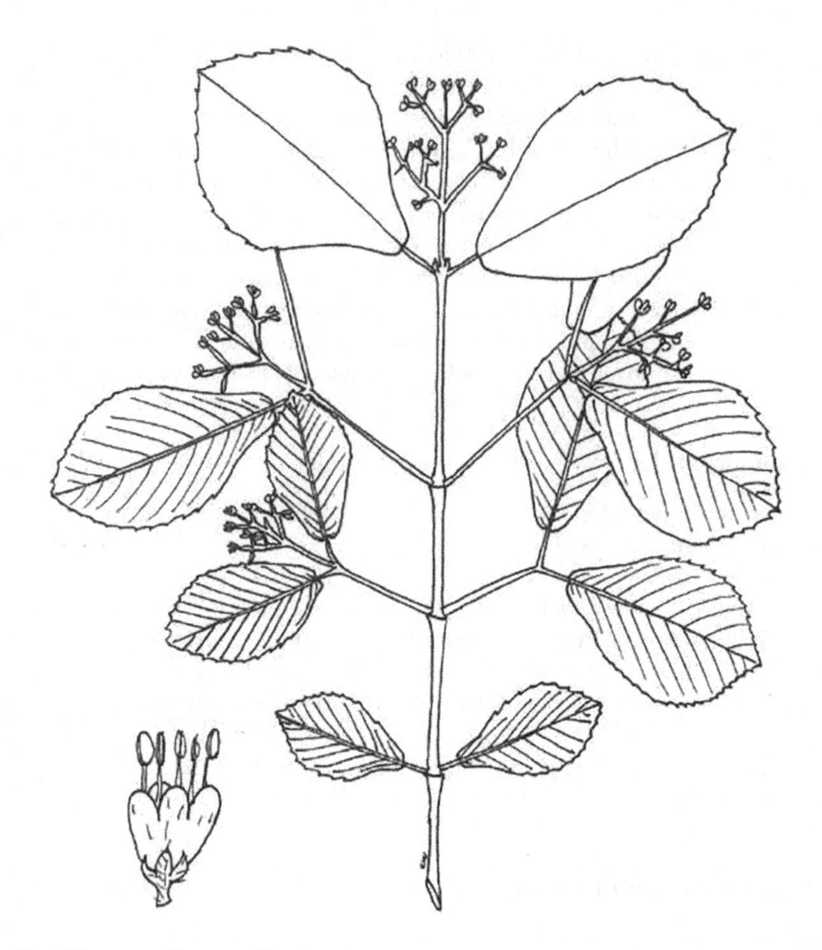

FIGURE 1.29 *Viburnum dilatatum* Thunb.

1.44 *Lonicera coerulea* L.

Subclass Asteridae, Superorder Cornanae, Order Dipsacales, Family Caprifoliaceae
Medicinal use: inflammation (China)

Anthocyanin fraction of fruits of *Lonicera coerulea* L. (containing 87.5 mg/100 mg of cyanidin 3-glucoside) given to C57BL/6 mice at a dose of 200 mg/kg of high-fat diet for 16 weeks had no effect on food intake, evoked a reduction of body weight gain by 24.1% compared to untreated animals (orlistat 100 mg/kg: 16.9%), and evoked a mild reduction of epididymal fat mass.[128] This regimen decreased plasma glucose and triglycerides, had no effect of total plasma cholesterol and decreased parameter of liver injury.[128] This fraction decreased hepatic triglycerides.[128] The supplementation decreased plasma insulin as efficiently as orlistat, halved plasma leptin and decreased insulin resistance to normal levels.[128] In general, a decrease in plasma insulin implies an increase in insulin sensitivity. This set of data suggests that the consumption of fruits of *Lonicera coerulea* L. could assist in treating metabolic syndrome. Clinical studies are needed.

1.45 *Ilex cornuta* Lindl. & Paxton

Subclass Asteridae, Superorder Cornananae, Order Aquifoliales, Family Aquifoliaceae
Medicinal use: fatigue (China)

Triterpenes have the tendency to inhibit acyl-CoA:cholesterol transferase-2 which regulates cholesterol absorption in enterocytes.[87] In fact triterpenes are structurally close to cholesterol. For instance, *Ilex cornuta* Lindl. & Paxton contains the lupane triterpene lupeol (Figure 1.30), which at a concentration of 100 μM inhibited the enzymatic activity of acyl-CoA:cholesterol transferase-2 (hACAT-2) by 48.2%.[129,130] In a subsequent study, Baek et al. (2010) tested lupeol against acyl-CoA:cholesterol transferase-2 and found an IC_{50} of 13.8×10^{-2} mM, whereas lupan-type triterpene betulinic acid had IC_{50} of 13.8×10^{-2} mM. In this experiment, the oleanane-type triterpene oleanolic acid was mildly active with 22% inhibition at a concentration of 50 μg/mL compared to untreated group.[131] Lupeol inhibited α-glucosidase with an IC_{50} value equal to 6.2 μg/mL.[131]

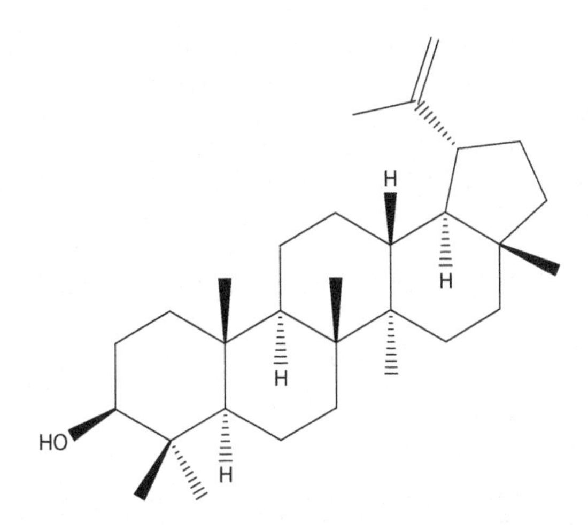

FIGURE 1.30 Lupeol.

1.46 *Acanthopanax senticosus* (Rupr. ex Maxim.) Harms

Synonyms: *Eleutherococcus senticosus* (Rupr. ex Maxim.) Maxim.; *Hedera senticosa* Rupr. ex Maxim.
Common names: ci wu jia (Chinese); Siberian ginseng
Subclass Asteridae, Superoder Cornanae, Order Apiales, Family Araliaceae
Medicinal use: fatigue (China)

Acanthopanax senticosus (Rupr. ex Maxim.) Harms is an example of medicinal plant producing a broad array of natural products with inhibitory activity on intestinal enzymes of carbohydrate and triglycerides absorption. The lupane-type saponin 22α-hydroxychiisanoside and the flavanol (+)-afzelechin (Figure 1.31) isolated from the leaves of *Acanthopanax senticosus* (Rupr. ex Maxim.) Harms inhibited α-glucosidase *in vitro* with IC_{50} values equal to 819, and 186 μM, respectively (acarbose IC_{50} 788.6 μM).[132] Silphioside F, copteroside B, hederagenin

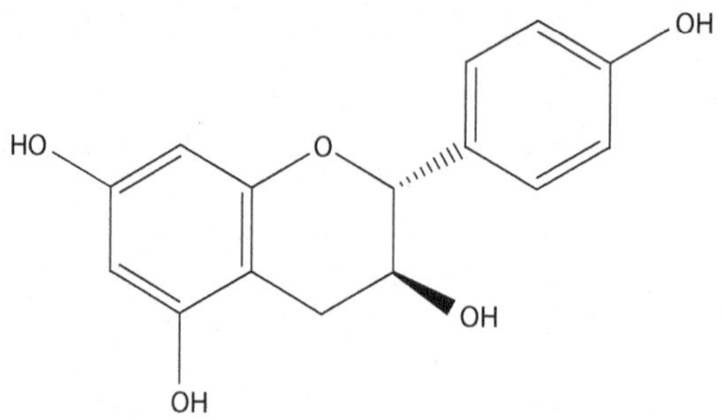

FIGURE 1.31 (+)-Afzelechin.

3-*O*-β-D-glucuronopyranoside 6′-*O*-methyl ester, and gypsogenin 3-*O*-β-D-glucuronide isolated from the fruits of this plant inhibited *in vitro* porcine pancreatic lipase more efficiently than orlistat.[133] From the same plant, erythro-7 *E*-4′,9′-dihydroxy-4,5′-dimethoxy-5,8′-oxyneolign-7-en-9-al isolated inhibited the enzymatic activity of diacylglycerol acyltransferase-1 with an IC_{50} value of 66.5 mM and was inactive against diacylglycerol acyltransferase-2.[134] In brush border enterocytes, short-chain fatty acids penetrate freely, whereby long-chain fatty acids are transported via fatty-acid translocase and fatty-acid transporter protein.[135] In enterocytes, monoacylglycerol transferase catalyzes the formation of diacylglycerol from monoacylglycerol and fatty acids and diacylglycerol acyltransferase-1 catalyze the formation of triglycerides from diacylglycerol. Triglycerides, cholesteryl ester, and apolipoprotein B48 are then packed into chylomicrons via the microsomal transfer protein, which are secreted into the lymphatic system.[135]

1.47 *Panax japonicus* (Nees) C.A Meyer

Synonym: *Aralia japonica* (Nees) Makino
Common name: zhu jie shen (Chinese)
Subclass Asteridae, Superorder Cornanae, Order Apiales, Family Araliaceae
Medicinal use: cough (Japan)

The polyacetylene (3S,10S)-panaxydiol (Figure 1.32) isolated from the roots of *Panax japonicas* (Nees) C.A Meyer inhibited yeast α-glucosidase with an IC_{50} of 22.2 μM (acarbose: IC_{50}: 677.9 μM).[136] Such compounds are common in members of the Family Apiaceae, Araliaceae and Asteraceae.

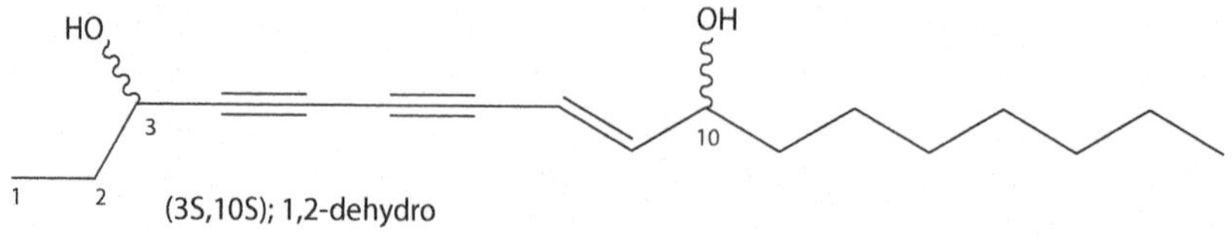

(3S,10S); 1,2-dehydro

FIGURE 1.32 (3S,10S)-Panaxydiol.

1.48 *Centella asiatica* (L.) Urb.

Synonyms: *Centella biflora* (P. Vell.) Nannf.; *Hydrocotyle asiatica* L.; *Hydrocotyle biflora* P. Vell.
Common names: pegaga (Malay/Indonesian); Asiatic pennywort
Subclass Cornanae, Superoder Cornanae, Order Apiales, Family Apiaceae
Nutritional use: Vegetable (Malaysia)

Ethanol extract of *Centella asiatica* (L.) Urb. inhibited *in vitro* porcine pancreatic lipase, porcine pancreatic α-amylase and yeast α-glucosidase with IC_{50} of 759.1 μg/mL (orlistat: 0.6 μg/mL) 536.5 μg/mL (acarbose: 113.2 μg/mL) and 42.2 μg/mL (acarbose: 34 μg/mL), respectively.[137] Rutin isolated from this extract inhibited *in vitro* porcine pancreatic lipase, porcine pancreatic α-amylase, and yeast α-glucosidase with IC_{50} value of 1412.2 μg/mL (orlistat: 0.6 μg/mL), 513 μg/mL (acarbose: 113.2 μg/mL), and 47 μg/mL (acarbose: 34 μg/mL), respectivey.[137] Following oral load of a lipid emulsion to Wistar rats, the extract at a single oral dose of 1000 mg/kg or rutin (Figure 1.33) lowered postprandial increase in serum triglycerides and total cholesterol.[137]

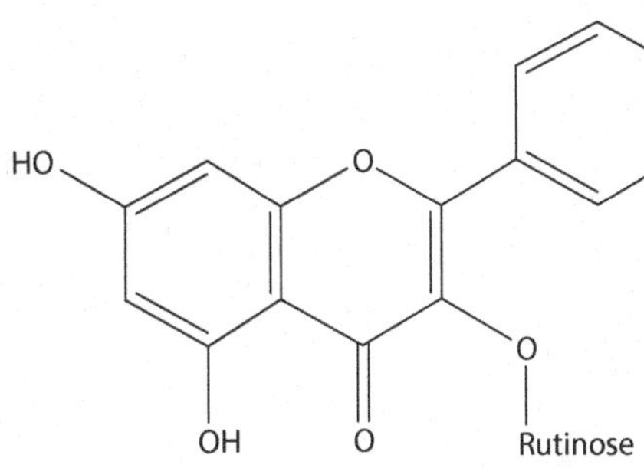

FIGURE 1.33 Rutin.

1.49 *Cnidium officinale* Makino

Subclass Asteranae, Superorder Cornanae, Order Apiales, Family Apiaceae
Medicinal use: blood stasis (Korea)

The phthalide derivative senkyunolide B from the rhizome of *Cnidium officinale* Makino inhibited the enzymatic activity of porcine pancreatic lipase with IC_{50} value equal to 86.4 μM.[138] Another example of phtalide derivative of Apiaceae acting on carbohydrate absorption is 3-(Z)-butylidenephthalide that given to rodent orally at a dose of 56 mg/kg inhibited sucrose absorption by about 55% at 30 minutes peak.[139] 3-(Z)-butylidene phthalide inhibited the enzymatic activity of yeast α-glucosidase with a Ki of 4.8 mM (acarbose 0.4 mM).[139]

1.50 *Ducrosia anethifolia* DC.

Subclass Asteridae, Superorder Cornanae, Order Apiales, Family Apiaceae
Medicinal use: fatigue (Pakistan)

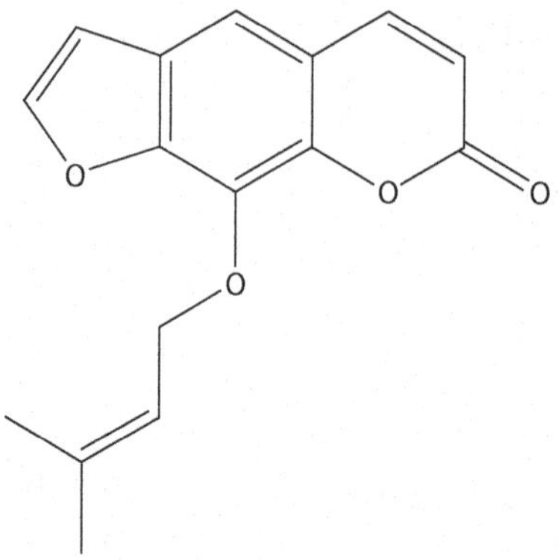

FIGURE 1.34 Imperatonin.

Defatted ethanol extract of aerial parts of *Ducrosia anethifolia* DC. at a concentration of 10 μg/mL inhibited *in vitro* α-amylase and α-glucosidase by 31.2% and 28.8%, respectively (acarbose 10 μg/mL: 32.2% and 29.9%, respectively).[140] From this extract, imperatorin (Figure 1.34) at a concentration of 10 μg/mL inhibited *in vitro* α-amylase and α-glucosidase by 28.2% and 28.8%, respectively.[140] The extract given orally to streptozotocin-induced diabetic rats (fasting blood glucose >300 mg/dL) at a daily dose of 500 mg/kg for 45 days decreased glycaemia from 365 to 165.6 mg/dL (normal: 111.5 mg/dL) and ameliorated serum cholesterol and triglycerides.[140] Rats with fasting blood glucose between 120 and 250 mg/dL are considered as mildly diabetic, whereas rats with a fasting blood glucose value of 300 mg/dL or more are severely diabetic.[141] Severe diabetes in rats suggests massive pancreatic insults by alloxan and streptozotocin.

1.51 *Peucedanum japonicum* Thunb.

Synonym: *Anethum japonicum* (Thunb.) Koso-Pol.
Common name: bin hai qian hu (Chinese)
Subclass Asteridae, Superorder Cornanae, Order Apiales, Family Apiaceae
Medicinal use: cough (Japan)

Ethanol extract of leaves and stems of *Peucedanum japonicum* given to C57BL/6 mice as part of 0.8% of diet for 4 weeks had no effect on food intake, decreased white adipose tissue from 8.3 to 5 g and plasma triglyceride from 60.2 to 39.3 g.[142] Liver triglycerides were reduced from 34.9 to 21.4 mg/dL and fecal triglycerides were increased from 0.3 to 0.5 mg/day.[142] This extract inhibited the activity of pancreatic lipase by 70% at a concentration of 3 mg/mL.[142]

1.52 *Platycodon grandiflorus* (Jacq.) A. DC.

Synonyms: *Platycodon glaucum* (Thunb.) Nak.
Common name: jie geng (Chinese)
Medicinal use: cough (Korea)

Saponin fraction of roots of *Platycodon grandiflorus* (Jacq.) A. DC. (containing Platycodin D 25.1 mg/g) given as part of diet (0.5 g/100 g diet) for 6 weeks reduced food intake, prevented weight loss, decreased fasting plasma glucose by 37%, and improved glucose tolerance in diabetic rodents (db/db mice).[143] This regimen reduced the activity of maltase and sucrase by 41%.[143] *In vitro*, the fraction inhibited yeast α-glucosidase activity by 79% at concentrations of 10 mg/mL. In addition, the fraction was a more effective α-glucosidase inhibitor than acarbose at the same concentration and this effect was superior to acarbose at 5 mg/mL.[143] Db/db mice have a mutated leptin-receptor gene resulting in the increase of food intake and used as a model of obesity and diabetes. These mice are obese, hyperglycemic, hyperlipidemic, have increased plasma insulin and insulin resisitance.[144] *Platycodon grandiflorus* (Jacq.) A. DC. could be of value in the treatment of metabolic syndrome and clinical trials are warranted.

1.53 *Artemisia herba-alba* Asso

Common names: sheeh (Pakistan); worm wood
Subclass Asteridae, Superorder Asteranae, Order Asterales, Family Asteraceae
Medicinal use: fatigue (Pakistan)

Extract of *Artemisia herba-alba* Asso given orally to C57BL/6J on high-fat diet mice for 18 weeks at a dose of 2 g/kg decreased glycaemia from about 230 to 139.5 mg/mL (normal: 120 mg/mL), reduced weight gain, reduced plasma insulin from 3.3 to 1.7 ng/mL, and reduced plasma triglycerides and cholesterol to normal values.[145] The plant shelters chlorogenic acid, 4,5-di-caffeoylquinic acid, 3,5-di-caffeoylquinic acid, 4-caffeoylquinic acid, as well as vicenin-2 and isovitexin.[146]

1.54 *Carthamus tinctorius* L.

Common names: hong hua (Chinese); kusum (India); safflower
Subclass Asteridae, Superorder Asteranae, Order Asterales, Family Asteraceae
Medicinal use: blood stasis (China).

The seeds of *Carthamus tinctorius* L. contains *N*-p-coumaroyl serotonin and *N*-feruloyl serotonin that inhibited yeast α-glucosidase with IC_{50} values equal to 47.2 and 100 μM, respectively.[147] In the same experiment, serotonin inhibited the enzymatic activity of α-glucosidase by 25.6% at 300 mM suggesting that the aforementioned property is owed to the phenolic moiety.[147]

1.55 *Chromolaena odorata* (L.) R.M. King & H. Rob.

Synonym: *Eupatorium odoratum* L.
Common names: fei ji cao (Chinese); Siam weed
Subclass Asteridae, Superorder Asteranae, Order Asterales, Family Asteraceae
Medicinal use: diabetes (India)

16-Kauren-19-oic acid (Figure 1.35) isolated from the roots of *Chromolaena odorata* (L.) R.M. King & H. Rob. inhibited yeast α-glucosidase with an IC_{50} value of 23.7 μM (In fact most inhibitors of α-glucosidase isolated so far from medicinal plants are phenolics and triterpenes acarbose IC_{50}: 780 μM).[148]

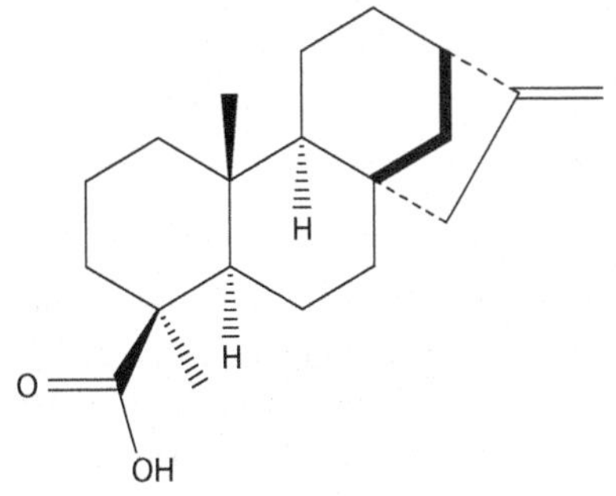

H
H
O
OH

FIGURE 1.35 16-Kauren-19-oic acid.

1.56 *Cichorium intybus* L.

Common names: ju ju (Chinese); kaasani (India); chicory
Subclass Asteridae, Superorder Asteranae, Order Asterales, Family Asteraceae
Medicinal use: jaundice (India)

18α,19β-20(30taraxasten-)-3β,21α-diol and vanillic acid (Figure 1.36) isolated from the seeds of *Cichorium intybus* L. inhibited yeast α-glucosidase with IC_{50} values of 51.9 and 69 μM, respectively.[149] Roots of *Cichorium intybus* L. contain inulin-type fructans, and in rats, a decrease in plasma triglycerides and cholesterol have been reported after oral administration of fructans.[150,151]

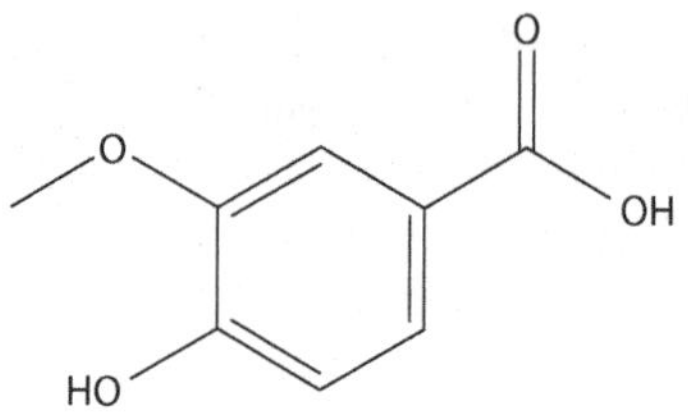

O
O
OH
HO

FIGURE 1.36 Vanillic acid.

1.57 *Chrysanthemum morifolium* Ramat

Synonyms: *Dendranthema grandiflorum* (Ramat.) Kitam.; *Tanacetum morifolium* Kitam.
Common names: ju hua (Chinese); chrysanthemum
Subclass Asteridae, Superorder Asteranae, Order Asterales, Family Asteraceae
Medicinal use: fever (China)

10α-Hydroxy-1α,4α-peroxide-2-guaien-12,6α-olide, acacetin-7-*O*-β-D-glucopyranoside, acacetin-7-*O*-α-L-rhamnopyranoside flowers of *Chrysanthemum morifolium* inhibited a-glucosidase with IC_{50} values of 229.3, 451.8, and 362.5 μM (acarbose: IC_{50} value of 1907 μM).[152] Eriodictyol, acacetin-7-*O*-β-D-glucopyranoside, acacetin-7-*O*-α-L-rhamnopyranoside inhibited α-amylase with IC_{50} values of 318.2, 337.1, and 112.5 μM (acarbose: IC_{50} value 732.4 μM).[152] 10α-Hydroxy-1α,4α-peroxide-2-guaien-12,6α-olide inhibited porcine pancreatic lipase with an IC_{50} value of 161 μM (orlistat: 108.3 μM).[152]

1.58 *Cynara scolymus* L.

Synonym: *Cynara cardunculus* L.
Common name: artichoke
Subclass Asteridae, Superorder Asteranae, Order Asterales, Family Asteraceae
Nutritional use: Vegetable (Turkey)
History: The plant was known Dioscorides

Methanol extract of leaves of *Cynara scolymus* L. given orally to mice at a single dose of 500 mg/kg 30 minutes before olive oil loading reduced plasma triglycerides after 2 hours from about 300 to 100 mg/dL (normal: about 110 mg/dL; orlistat 250 mg/kg: 100 mg/dL).[153] From this extract, the sesquiterpenes aguerine B, grosheimin, cynaropicrin, and the flavone glycoside luteolin

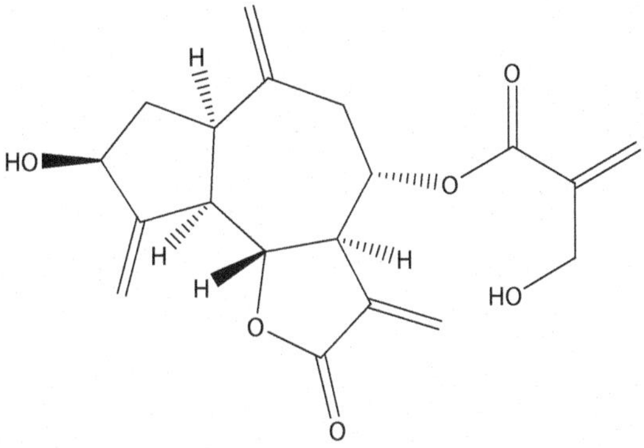

FIGURE 1.37 Cynaropicrin.

7-*O*-β-D-glucopyranoside given orally to mice at a single dose of 100 mg/kg 30 minutes before olive-oil loading reduced plasma triglycerides after 2 hours from about 450 to 150 mg/dL, 500 to 200 mg/dL, 500 to 150 mg/dL, and 500 to 300 mg/dL.[153] Aguerine B, grosheimin, cynaropicrin (Figure 1.37) and luteolin 7-*O*-β-D-glucopyranoside were, in this study, not active against pancreatic lipase but delayed gastric emptying in oral olive-oil load.[153] In contradiction to this, a subsequent study reported that ethanol extract from leaves of *Cynara scolymus* L. at a concentration of 100 μg/mL inhibited porcine pancreatic lipase activity by approximately 20% at a concentration of 100 μg/mL (Orlistat IC_{50} of 0.8 μM) (vi).[154] *Cynara scolymus* L. appears as beneficial for metabolic syndrome.[155]

1.59 *Elephantopus mollis* Kunth

Synonym: *Elephantopus scaber* L.
Common names: di dan cao (Chinese); tutup bumi (Malay); malatabako (Philippines)
Subclass Asteridae, Superorder Asteranae, Order Asterales, Family Asteraceae
Medicinal use: liver intoxication (Malaysia)

3,4-di-*O*-caffeoyl quinic acid isolated from the whole *Elephantopus mollis* Kunth inhibited α-glucosidase with an IC_{50} value of 241.8 μg/mL (acarbose IC_{50}: 7.3 μg/mL).[156]

1.60 *Puchea indica* (L.) Less

Synonyms: *Baccharis indica* L.; *Erigeron denticulatum* Burm. f.
Common names: beluntas (Malay); luntas (Indonesia); tulo-lalaki (Philippines); Indian fleabane (India)
Subclass Asteridae, Superorder Asteranae, Order Asterales, Family Asteraceae
Medicinal use: dysentery (Indonesia)

3,4,5-tri-*O*-caffeoylquinic acid methyl ester, 3,4,5-tri-*O*-caffeoylquinic acid, and 1,3,4,5-tetra-*O*-caffeoylquinic acid from the leaves of *Pluchea indica* (L.) Less. (Figure 1.38) inhibited rat-intestinal maltase with IC_{50} values of 2, 13, and 11 μM, respectively.[157]

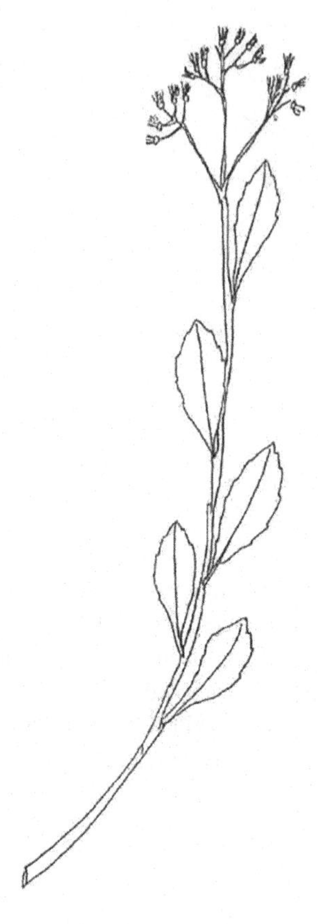

FIGURE 1.38 *Pluchea indica* (L.) Less.

1.61 *Silybum marianum* (L.) Gaertn.

Synonyms Carduus marianus L.; *Carthamus maculatum* (Scop.) Lam. *Cirsium maculatum* Scop.
Common name: milk thistle
Subclass Asteridae, Superorder Asteranae, Order Asterales, Family Asteraceae
Medicinal use: jaundice (India)

Ethanol extract from fruits of *Sylibum marianum* (L.) Gaertn. had no effect of bacterial α-glucosidase but inhibited porcine pancreatic lipase activity by approximately 30% at a concentration of 100 μg/mL (orlistat IC_{50} of 0.8 μM).[158] Sylimarin (a fraction composed of flavonolignans, sylibin, silychristin, and silydianin) from this plant given at a dose of 200 mg to patients with type 2 diabetes 3 times per day before meals for 4 months reduced fasting glycemia from 188 to 133 mg/dL, and had no effect on plasma insulin.[159]

1.62 *Spilanthes acmella* (L.) L.

Synonyms: *Bidens acmella* (L.) Lam.; *Bidens ocymifolia* Lam.; *Pyrethrum acmella* (L.) Medik.; *Spilanthes ocymifolia* (Lam.) A.H. Moore; *Verbesina acmella* L.
Common names: hin ka la (Burmese); krishnarjaka (Sri Lanka); pokok getang kerbau (Malay); biri (Philippines); tooth ache plant

Subclass Asteridae, Superorder Asteranae, Order Asterales, Family Asteraceae
Medicinal use: diuretic (Sri Lanka)

Ethanol extract from flower buds of *Spilanthes acmella* (L.) L. at a concentration of 2 mg/mL inhibited human pancreatic lipase by about 44%.[160] This plant contains series of isobutylamides which comprise spilanthol as well as *trans*-ferulic acid, *trans*-isoferulic acid, and scopolelin.[161] It must be noted that the flowers contain *N*-isobutyl amides: spilanthol, undeca-2E,7Z,9E-trienoic acid isobutylamide, and undeca-2E-en-8,10-diynoic acid isobutylamide, which may account for pancreatic lipase inhibition.[162]

1.63 *Taraxacum officinale* F.H. Wigg.

Synonyms: *Leontodon taraxacum* L.
Common names: kanphool (Pakistan); dandelion
Subclass Asteridae, Superorder Asteranae, Order Asterales, Family Asteraceae
Medicinal use: jaundice (Pakistan)

from leaves of *Taraxacum officinale* F.H. Wigg. inhibited porcine pancreatic lipase with an IC_{50} value of 78.2 μg/mL, whereas orlistat had IC_{50} value of 0.22 μg/mL.[163] The extract given orally at a single dose of 400 mg/kg to ICR mice challenged with oral administration of corn oil reduced postprandial plasma triglycerides from 76.9 mg/dL to about 60 mg/dL at 180 minutes.[163] Pancreatic lipase inhibition was confirmed by Villiger et al. (2015) who reported that ethanol extract of roots of *Taraxacum officinale* F.H. Wigg. inhibited porcine pancreatic lipase with an IC_{50} value of 78.2 μg/mL (orlistat IC_{50}: 0.8 μM).[158] The plant shelters flavonoids which comprise of quercetin, luteolin, and luteolin-7-*O*-glucoside (chlorogenic acid, chicoric acid, and cichorin).[164,165] This plant may be of value for metabolic syndrome.

1.64 *Tussilago farfara* L.

Common names: kuan dong (Chinese); colt's-foot
Subclass Asteridae, Superorder Asteranae, Order Asterales, Family Asteraceae
Medicinal use: difficult breathing (China)

It must be recalled that such phenolics are common in members of the Family Asteraceae. *Tussilago farfara* L. elaborates series of caffeoylquinic derivatives that are able to inhibit carbohydrates and triglycerides absorption. 3,4-Dicaffeoylquinic acid, 3,5-dicaffeoylquinic acid, and 4,5-dicaffeoyllquinic acid isolated from the flower buds of *Tussilago farfara* L. inhibited rat-intestinal α-glucosidase *in vitro* with IC_{50} values of 0.9, 0.9, and 0.8 mM, respectively, whereas cholorogenic acid was inactive.[166] At a concentration of 1 mM, 3,4-dicaffeoylquinic acid, 3,5-dicaffeoylquinic acid, and 4,5-dicaffeoylquinic acid inhibited rat-intestinal maltase by 65, 64, and 62%, respectively, *in vitro,* but these caffeic acid derivatives showed weak inhibitory activity against sucrase, isomaltase, and porcine pancreatic α-amylase.[166] From the same plant, rutin inhibited α-glucosidase by 41% at a concentration of 1 mM.[166] Positive standard 1,2,3,4,6-penta-*O*-galloyl-β-D-glucopyranose gave IC_{50} value of 0.1 mM.[166] Park et al. (2008) provided evidence that tussilagone, 7β-(3-ethyl-cis-crotonoyloxy)-1α-(2-methylbutyryloxy)-3,14-dehydro-Z-notonipetranone and the bisabolane sesquiterpene 8-angeloyloxy-3,4-epoxy-bisabola-7(14),10-dien-2-one inhibited human diacylglycerol acyltransferase-1, which catalyze the formation of triglycerides from diacylglycerol in enterocytes, with IC_{50} values of 49.1, 160.7, and 294.4 μM.[167] This plant should not be used in therapeutic strategies as it contains hepatotoxic pyrrolizidine alkaloids.[168]

1.65 *Gardenia jasminoides* J. Ellis

Synonyms: *Gardenia augusta* Merr.; *Gardenia florida* L.; *Varneria augusta* L.
Common names: zhi zi (Chinese); karinga (India); cape jasmine
Subclass Lamiidae, Superorder Lamianae, Order Rubiales, Family Rubiaceae
Medicinal use: jaundice (China)

Based on its ability to reduce the absorption of dietary triglycerides, orlistat (Xenical®, Roche) is used as an adjunct treatment of obesity in conjunction with mild diet restriction.[169] It is taken in adults at a dose of 120 mg before, during, or up to each main meal but has many side effects, hence the need to develop new leads.[169] The carotenoid glycoside crocin and its aglycone crocetin isolated from the fruits of *Gardenia jasminoides* J. Ellis (Figure 1.39) inhibited pancreatic lipase with IC_{50} values of 2.7 and 2.1 mg/mL (orlistat: IC_{50}: 0.8 mg/mL).[170] Crocin and crocetin at doses of 50 mg/kg/day given orally to high-fat diet ICR mice for 5 weeks reduced triglycerides from 160.4 to 114.6 and 111.9 mg/dL (normal: 74.9 mg/dL; orlistat 10 mg/kg/day: 81.1 mg/dL); total cholesterol from 248 mg/dL to 170.5 mg/dL to 159.1 mg/dL (normal: 93.1 mg/dL; orlistat 10 mg/kg/day: 159.6 mg/dL).[170] This finding was confirmed by Sheng et al. (2006) who reported that crocin evoked 50% inhibition of pancreatic lipase at a concentration of 28.6 μmol/L *in vitro*. Crocin given orally to rats at single dose 100 mg/kg with a lipid emulsion reduced postprandial plasma

FIGURE 1.39 *Gardenia jasminoides* J. Ellis.

triglycerides peak at 6 hours from about 250 to 75 mg/dL and plasma cholesterol peak at 9 hours from about 290 to 140 mg/dL, respectively.[171] In line, rats on high-fat diet given crocin orally at a dose of 100 mg/kg/day for 2 days had increased fecal secretion of cholesterol.[171] These experimental evidences lend support to the suggestion that the fruits of *Gardenia jasminoides* J. Ellis, if not toxic, could be of value for the treatment of metabolic syndrome.

1.66 *Uncaria laevigata* Wall. ex G. Don

Synonym: *Nauclea laevigata* (Wall. ex G. Don) Walp.
Common name: ping hua gou teng (Chinese)
Subclass Lamiidae, Superorder Lamianae, Order Rubiales, Family Rubiaceae
Medicinal use: hypertension (China)

Ursolic acid and 3β-hydroxy-30-methoxy-6-oxo-urs-12,19(20)-dien-28-oic acid isolated from the stem bark of *Uncaria laevigata* Wall. ex G. Don inhibited yeast α-glucosidase with IC_{50} values of 16 and 49 μM, respectively.[172]

1.67 *Swertia kouitchensis* Franch.

Synonym: *Swertia elongata* T.N. Ho & S.W. Liu
Common name: gui zhou zhang ya cai (Chinese)
Subclass Lamiidae, Superorder Lamianeae, Order Rubiales, Family Gentianaceae
Medicinal use: diabetes (China)

The xanthones kouitchenside B, kouitchenside D, kouitchenside, and kouitchenside F isolated from *Swertia kouitchensis* Franch. inhibited yeast α-glucosidase with IC_{50} values of 383, 360, 371, and 184 μM, respectively (acarbose: 627 μM).[173] Ethanol extract of *Swertia kouitchensis* inhibited porcine pancreas α-amylase and yeast α-glucosidase with IC_{50} values of 0.1 and 0.9 mg/mL, respectively (acarbose: 0.04 and 0.7 mg/mL, respectively).[173] In oral starch tolerance test, a single oral dose of 500 mg/kg of extract decreased glucose area under the curve by 16.7% (acarbose: 23.4%). Ethanol extract given to high-fat and fructose diet-streptozotocin-induced diabetic Balb/c mice (fasting blood glucose superior or equal to 11.1 mmol/L) at a dose of 500 mg/kg/day reduced glycaemia from about 16 to 12 mmol/L (normal: 5 mmol/L; glicazide at 15 mg/kg/day: about 7 mmol/L).[174] This plant contains the iridoid swertiamarin (Figure 1.40) and its intestinal metabolite erythrocentaurin, inhibited the activity of α-amylase and α-glucosidase *in vitro* with IC_{50} values of 0.1 and 10 mg/mL, respectively.[175]

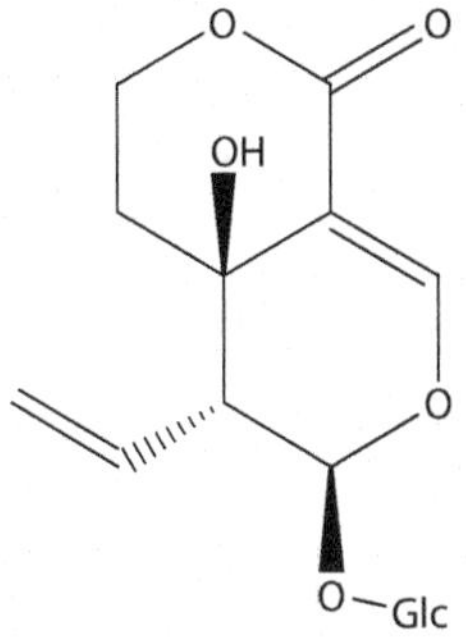

FIGURE 1.40 Swertiamarin.

1.68 *Alstonia macrophylla* Wall. ex G. Don

Common names: da ye tang jiao shu (Chinese); batino (Philippines)
Subclass Lamiidae, Superorder Lamianeae, Order Rubiales, Family Apocynaceae
Medicinal use: dysmenorrhea (Philippines)

The indole alkaloids alstiphyllanines E and F inhibited Na^+ glucose cotransporter -1 (SGLT1) by 60.3% and 65.2%, respectively. From the same plant, 10-methoxy-N(1)-methylburnamine-17-*O*-veratrate alstiphyllanine D inhibited Na^+ glucose cotransporter -1 (SGLT1) by 95.8% and 89.9%, respectively.[176]

1.69 *Carissa carandas* L.

Synonyms: *Arduina carandas* (L.) K. Schum.; *Carissa congesta* Wight.
Common names: ci huang guo (Chinese); karonda (India); Bengal current
Subclass Lamiidae, Superorder Lamianeae, Order Rubiales, Family Apocynaceae
Medicinal use: thirst (India)

Ethanol extracts from leaves of *Carissa carandas* L. inhibited yeast α-glucosidase with an IC_{50} value of 21.1 μg/mL, whereby an IC_{50} value of 117.2 μg/mL for acarbose was recorded.[177] The plant contains series of triterpenes of which of betulinic acid, oleanolic acid, and ursolic acid, which are known inhibitors of α-glucosidase.[178]

1.70 *Gymnema sylvestre* (Retz.) R.Br. ex Schult.

Synonyms: *Gymnema affine* Decne. *Gymnema alterniflorum* (Lour.) Merr.; *Gymnema formosanum* Warb. *Periploca sylvestris* Retz.
Common name: gurmar (India)
Subclass Lamiidae, Superorder Lamianeae, Order Rubiales, Family Asclepiadaceae
Medicinal use: diabetes (India)
History: The plant was known of Sushruta or its antidiabetic properties

Saponin fraction of leaves of *Gymnema sylvestre* given to Wistar rats on high-fat diet for 8 weeks at a dose of 100 mg/kg/day reduced body weight from about 300 to 250 g (rats fed on normal chow: about 245 g).[179] This fraction decreased food intake from 23.7 g/day to 18.4 g (normal diet: 19.4 g) reduced plasma triglycerides, cholesterol, and decreased glycaemia.[179] *In vitro*, the fraction at inhibited dose dependently the release of oleic acid from triolein catalyzed by pancreatic lipase with a maximum activity at 400 mg/dL.[179] Methanol extracts from *Gymnema sylvestre* inhibited by 48% glucose uptake by sodium-dependent glucose transporter 1 (SGLT1) in *Xenopus laevis* oocytes.[179] From *Gymnema sylvestre*, gymnemic acid V and gymnemic acid XV inhibited SGLT1 activity with IC_{50} values of 5.9 and 0.1 μM, respectively (phlorizin: 0.2 μM).[180] Baskaran et al. (1990) provided evidence of the usefulness of this plant for metabolic syndrome.[181]

1.71 *Holarrhena antidysenterica* (L.) Wall. ex A. DC.

Common names: kurchi (India); kurchi tree
Subclass Lamiidae, Superorder Lamianeae, Order Rubiales, Family Apocynaceae
Medicinal use: dysentery (India).
History: The plant was known of Sushruta

Methanol extract from seeds of *Holarrhena antidysenterica* (L.) Wall. ex A. DC. inhibited α-glucosidase of rats with an IC_{50} value of 0.5 mg/mL.[182] The extract given orally at a single dose of 400 mg/kg to rats 30 minutes before oral load of starch reduced peak postprandial glycaemia at 1 hour from approximately 225 to 125 mg/dL (acarbose 3 mg/kg: approximately 125 mg/dL).[182]

1.72 *Ipomoea batatas* (L.) Lam.

Synonyms: *Batatas edulis* (Thunb.) Choisy; *Convolvulus batatas* L.; *Convolvulus edulis* Thunb.
Common names: mitha alu (India); ubi keledek (Malay); kamote (Philippines); man thet (Thai); sweet potato
Subclass Lamiidae, Superoder Lamianae, Order Solanales, Family Convolvulaceae
Medicinal use: diabetes (India)

Aqueous extract from peel of roots of *Ipomoea batatas* (L.) Lam. (Figure 1.41) given to obese Zucker rats orally at a dose of 100 mg/kg/day for 8 weeks decreased plasma insulin from about 753 to 384 μU/mL and attenuated of plasma glucose.[183] This regimen reduced plasma triglycerides, free fatty acids, and had no effect on cholesterol.[183] This extract improved glucose tolerance in oral glucose tolerance test at the end of the regimen.[183] In a subsequent study, *Ipomoea batatas* (L.) Lam.

FIGURE 1.41 *Ipomoea batatas* (L.) Lam.

given to type 2 diabetic patients at a dose of 4 g/day for 3 months decreased, fasting blood glucose from 138.2 to 128.5 mg/dL, reduced total cholesterol from 248.7 to 214.6 mg/dL, and decreased plasma triglycerides.[184] This treatment decreased glucose 2 hours after oral glucose tolerance test from 181 to 162.8 mg/dL.[184] 48.3% of treated diabetic patients achieved a mean fasting blood glucose below the upper normal limit (126 mg/dL) after 3 months versus 7.7% in placebo group.[184] A single oral administration of anthocyanin fraction extracted from the tubers of *Ipomoea batatas* (L.) Lam. at a dose of 400 mg/kg to Sprague–Dawley rats 5 minutes before oral leading of maltose decreased 30 minutes peak glycaemia from 170.3 to 143.8 mg/dL and decreased serum insulin from 2.8 to 1.1 ng/mL.[185] The extract had no effect on sucrose postprandial glycaemia.[185] From this fraction the diacetylated anthocyanin YGM-6 given at a single oral administration of 100 mg/kg to Sprague–Dawley rats 5 minutes before oral loading of maltose reduced 30 minutes plasma glucose by 25%, whereby acarbose at 3 mg/dL evoked about 45% reduction of 30 minutes glycaemia peak.[185] In parallel this anthocyanin reduced serum insulin from 2.8 to 1.6 ng/mL.[185] YGM-6 had no effect on sucrose or glucose postprandial glycaemia.[185] The leaves of *Ipomoea batatas* (L.) Lam. contains 3,4,5-tricaffeoylquinic acid inhibited rat-intestinal maltase, rat-intestinal sucrase, and human saliva α-amylase with an IC_{50} value of 24 μM (acarbose: 0.4 μM), 574 μM (acarbose: 1.2 μM), and 634 μM, respectively.[185]

1.73 *Ipomoea aquatica* Forssk.

Common names: weng cai (Chinese); kalambi (India); kangkong
Subclass Lamiidae, Superoder Lamianae, Order Solanales, Family Convolvulaceae
Nutritional use: food (Sri Lanka)

Evidence is accumulating in favor of a beneficial effect of *Ipomoea aquatica* Forssk. on postprandial glycemia. Aqueous decoction of the plant given orally to Wistar rats at a single oral dose of 3 g/kg, 30 minutes before oral load of glucose, lowered postprandial glycaemia by 33.6% after 120 minutes.[186] Stems and leaves of *Ipomoea aquatica* Forssk. given at a dose of 3.4 g/kg/day to streptozotocin-induced diabetic Wistar rats (blood glucose > 250 mg/dL) for 7 days reduced glycaemia by 48.6%.[187] Aqueous juice made with 100 g of stems and leaves given to type 2 diabetic patients 30 minutes before oral load of glucose decreased 2 hours peak glycaemia by 29.4%.[187] Sokeng et al. (2007) provided evidence that aqueous extract of leafy stem of *Ipomoea aquatica* Forssk. perfused at a single dose of 160 mg/kg to *ex vivo* preparation of rat intestines inhibited intestinal glucose by about 30%.[188] Clinical trials are warranted.

1.74 *Echium vulgare* L.

Common names: lan ji (Chinese); viper bugloss
Subclass Lamiidae, Superorder Lamianae, Order Boraginales, Family Boraginaceae
Medicinal use: fissures of hands (Turkey)
History: The plant was known to Paulus Aegineta (625–690 AD), Greek physician

Ethanol extract of *Echium vulgare* L. inhibited porcine α-amylase with an IC_{50} value of 69.1 μg/mL (acarbose: IC_{50} 50 μg/mL) and inhibited porcine pancreatic lipase by 41% at a concentration of 2.5 mg/mL (orlistat: IC_{50} 18 μg/mL).[189] Methanol extract from leaves of *Echium vulgare* L. at a concentration of 2.5 mg/mL inhibited α-amylase activity by 71.7% and inhibited porcine pancreatic lipase activity by 92.4% *in vitro*.[96] The plant contains hepatotoxic pyrrolizidine alkaloids and is of no use in therapeutic strategies. This is often the case with members of the family

Boraginaceae. However, the seeds are enriched with polyunsaturated fatty acids, stearidonic acid, and γ-linolenic acid that given at 2g/100g diet to African green monkeys for 6 weeks improved glucose tolerance.[190] Plant oils rich in γ-linolenic acid have some beneficial effects on diabetic complications because in diabetic patients there is a defect in the desaturation steps in the metabolism of linoleic acid.[186] Oil-enriched γ-linolenic acid (4 g) plus 2.4 g of sardine oil given to hospitalized obese noninsulin-dependent diabetes for 4 weeks induced a reduction of urinary excression of 11-dehydro-thromboxane B2 by 32%.[189] Added at 6% of diet for 32 weeks to In F344/DuCrj rats oil-enriched γ-linolenic acid reduced the occurrence of ventricular tachycardia and inhibited the duration of ventricular tachycardia induced by experimental and acute coronary artery occlusion compared to rodents receiving sheep fat at 6% of diet.[189] This regimen afforded a complete protection against ventricular fibrillation in ischaemic state.[189] It must be recalled that this plant is poisonous owing to pyrrolizidine alkaloids and cannot be used itself in therapeutic strategies. Its seed oil, however, may have some beneficial effects on metabolic syndrome.

1.75 *Heliotropium zeylanicum* Lam.

Synonym: *Heliotropium linifolium* Lehm.
Common name: Hasthishundi (India)
Subclass Lamiidae, Superoder Lamianae, Order Boraginales, Family Boraginaceae
Medicinal use: inflammation (India)

Methanol fraction of *Heliotropium zeylanicum* given orally to streptozotocin-induced diabetic Wistar rats (plasma glucose > 225 mg/dL) at a daily dose of 300 mg/kg for 14 days prevented weight loss, decreased food and water intake closely to tolbutamide at a dose of 10 mg/kg/day.[192] This extract lowered glycaemia from 312 to 118.2 mg/dL (normal: 85.2 mg/dL; tolbutamide 10 mg/kg/day: 112.5 mg/dL), cholesterol from 148.1 to 116.7 mg/dL (normal: 100 mg/dL; tolbutamide 10 mg/kg/day: 105.5 mg/dL), and triglycerides from 185.6 to 146.9 mg (normal: 97.5 mg/dL; tolbutamide 10 mg/kg/day: 108.3 mg/dL).[192] This regimen decreased hepatic lipid peroxidation and increased hepatic glutathione to values close to normal group.[192] This extract also increased superoxide dismutase activities in the liver of treated diabetic rodents.[192] Schoental and Frayn (1976) administered the pyrrolizidine alkaloid heliotrine, which occurs in members of the genus *Heliotropium* L. to white weanling rats at a dose of 300 mg/kg orally observed increased plasma insulin levels owed to severe pancreatic insults.[193] Again, these alkaloids are toxic and of no use in therapeutic.

1.76 *Lithospermum erythrorhizon* Siebold & Zucc.

Synonym: *Lithospermum officinale* var. *erythrorhizon* (Siebold & Zucc.) Maxim.
Common names: zi cao (Chinese); murasaki (Japanese); Chinese groomwell
Medicinal use: wounds (Japan)

Members of the family Boraginaceae synthetize naphthoquinones that inhibit acyl-CoA:cholesterol acyltransferase an enzyme that reduces plasma lipid levels by inhibiting intestinal cholesterol absorption.[191] Such naphthoquinones are acetylshikonin (Figure 1.42), isobutyrylshikonin, and β-hydroxyisovalerylshikonin isolated from the roots of *Lithospermum erythorhizon* Siebold & Zucc. which inhibited human acyl-CoA:cholesterol acyltransferase with IC_{50} values of 112.2, 57.5, and 169.8 μM, respectively.[194] Inhibition of acyl-CoA:cholesterol acyltransferase also prevents the progression of atherosclerotic lesions by inhibiting the accumulation of cholesteryl ester in macrophages.[195] Acetylshikonin and β-hydroxyisovalerylshikonin inhibited human acyl-CoA:cholesterol acyltransferase-1 with IC_{50} values of 128.9 and 186.9 μM.[194]

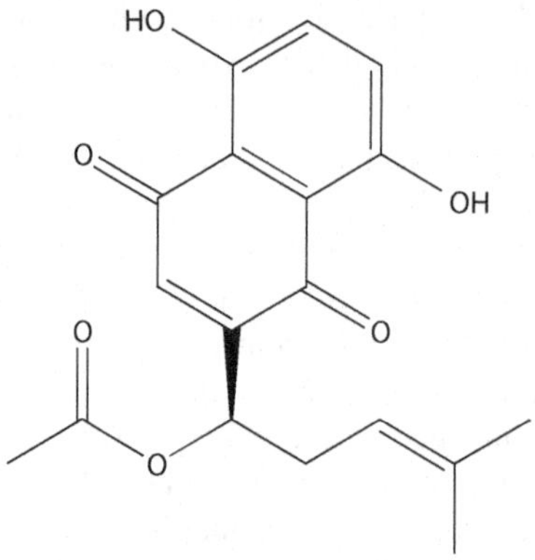

FIGURE 1.42 Acetylshikonin.

1.77 *Olea europaea* L.

Common name: olive
Subclass Lamiidae, Superorder Lamianae, Order Oleales, Family Oleaceae
Medicinal use: nodules (Turkey)

Komaki et al. (2003) made the demonstration that aqueous extract from leaves of *Olea europea* L. could inhibit human pancreatic α-amylase activity *in vitro* with an IC_{50} value of 70.2 mg/mL. From this extract, luteolin-7-*O*-β-glucoside and luteolin-4′-*O*-β-glucoside and oleanolic acid (Figure 1.43) inhibited human pancreatic α-amylase activity with IC_{50} values of 0.5, 0.3, and 0.1 mg/mL.[196] In this experiment, luteolin inhibited human pancreatic α-amylase activity with IC_{50} value of 0.01 mg/mL.[196] Oleanolic acid at 1 mg/kg or luteolin at 0.1 mg/kg given orally to type 2 diabetes Goto–Kakizaki/Jcl rats with starch decreased postprandial blood glucose levels from about 140 to 60 mg/dL after 120 minutes.[196] In healthy subject, the consumption of 1 g of leave powder with 300 g of cooked rice had no effect on postprandial glycemia, but given to borderline subject, leaf powder consumption reduced 1 hour peak glycemia evoked a decrease of postprandial glycemia from about 225 to 180 mg/dL at 1 hour.[196] In subsequent experiments, methanol

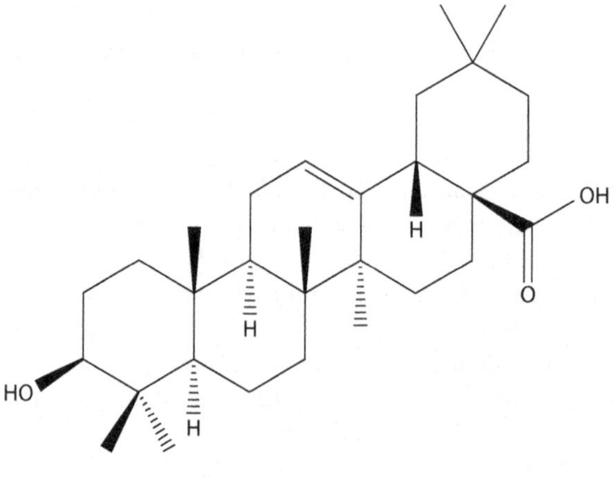

FIGURE 1.43 Oleanolic acid.

extract of leaves at a concentration of 2.5 mg/mL inhibited α-amylase activity by 64.3% *in vitro* (IC_{50}: 0.8 mg/mL; acarbose: 1.3 μg/mL).[96] Methanol extract of leaves of *Olea europea* L. at a concentration of 2.5 mg/mL inhibited porcine pancreatic lipase activity by 100% *in vitro* (IC_{50}: 0.1 mg/mL; orlistat: 0.1 ng/mL).[96] In another study luteolin inhibited α-amylase with an IC_{50} value of 18.4 μM.[15] If proven safe for consumption, the leaves of *Olea euroapea* L. could be used for metabolic syndrome.

1.78 *Dolichandrone falcata* Seem.

Subclass Lamiidae, Superorder Lamianae, Order Lamiales, Family Bignoniaceae
Medicinal use: body pain (India)

The phenylpropanoid glycoside dolichandroside A isolated from *Dolichandrone falcata* Seem. inhibited yeast and rat α-glucosidase with IC_{50} values of 39.7 and 18.7 μg/mL, respectively.[197] From the same plant, 3,8-dihydroxy-1-methyl-9,10-anthraquinone inhibited yeast and rat α-glucosidase. Acarbose inhibited rat-intestinal α-glucosidase with an IC_{50} value of 8.7 μg/mL.[197] Verbascoside was inactive in this study although being found active in other studies, exemplifying the variability of activities between rats and yeast α-glucosidase. Tadera et al. tested a series of flavonoids against yeast and rats α-glucosidase.[198] In yeast model, a concentration of 200 μM of kaempferol, naringenin, epigallocatechin, and cyanidin inhibited α-glucosidase activity by more than 70%.[198] The same flavonoids at a concentration of 0.5 mM were inactive and had no activity against rat-intestinal α-glucosidase suggesting that rat α-glucosidase should be prioritized for the testing of natural products.[198]

1.79 *Stereospermum colais* (Buch.-Ham ex Dillwyn) Mabb.

Subclass Lamiidae, Superorder Lamianae, Order Lamiales, Family Bignoniaceae
Medicinal use: In India, the roots are used to promote urination

Ursolic acid (Figure 1.44), lapachol, and pinoresinol isolated from the roots of *Stereospermum colais* (Buch.-Ham ex Dillwyn) Mabb. inhibited yeast α-glucosidase with IC_{50} values of 12.4, 11, and 45.6 nM, respectively (acarbose IC_{50}: 55.6 nM).[199]

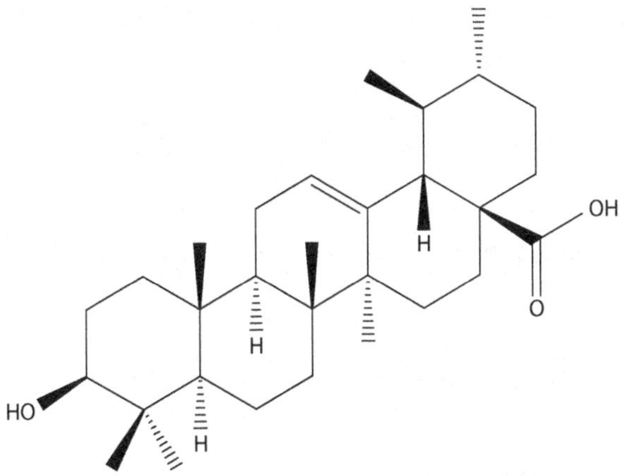

FIGURE 1.44 Ursolic acid.

1.80 *Sesamum indicum* L.

Synonym: *Sesamum orientale* L.
Common names: zhi ma (Chinese); taila (India); sesame
Subclass Lamiidae, Superorder Lamianeae, Order Lamiales, Family Pedaliaceae
Medicinal use: ulcers (India)

The leaves of *Sesamum indicum* L. contains epigallocatechin inhibited α-amylase with an IC_{50} value of 303.9 μM (acarbose: 124 μM).[200]

1.81 *Adhatoda vasica* Nees

Synonym: *Justicia adhatoda* L.
Common names: Sinha muki (India); Malabar nut tree
Subclass Lamiidae, Superoder Lamianae, Order Lamiales, Family Acanthaceae
Medicinal use: bronchitis (India)
History: The plant was known to Sushruta

Quinazoline alkaloids vasicine and vasicinone isolated from the leaves *Adhatoda vasica* Nees (Figure 1.45) inhibited rat sucrase by 93% and 81%, respectively, at a concentration of 1 mM.[201] Vasicine and vasicinone inhibited rat maltase by about 32% and 29%, and inhibited pancreatic α-amylase by about 12% and 19%.[201]

FIGURE 1.45 *Adhatoda vasica* Nees.

1.82 *Clerodendrum bungei* Steud.

Synonym: *Clerodendrum foetidum* Bunge
Common name: xiu mu dan (Chinese)
Subclass Lamiidae, Superorder Lamiidae, Order Lamiales, Family Lamiaceae
Medicinal use: hypertension (China)

Verbascoside, leucosceptoside, and isoacteoside isolated from the roots of *Clerodendrum bungei* Steud. inhibited yeast α-glucosidase with IC_{50} values of 0.5, 0.7, and 0.1 mM, respectively (acarbose IC_{50}: 14.4 mM).[202]

1.83 *Duranta repens* L.

Synonym: *Duranta erecta* L.
Common name: golden dewdrop
Subclass Lamiidae, Superorder Lamiidae, Order Lamiales, Family Verbenaceae
Medicinal use: dysmenorrhea (Indonesia)

7-*O*-α-D-glucopyranosyl-3,5-dihydroxy-3′-(4″-acetoxyl-3″-methylbutyl)-6,4′-dimethoxyflavone, 3,7,4′-trihydroxy-3′-(8″acetoxy-7″-methyloctyl)-5,6-dimethoxyflavone and (–)-6β-hydroxy-5β, 8β, 9β, 10α-cleroda-3,13-dien-16,15-olid-18-oic acid isolated from *Duranta repens* inhibited yeast α-glucosidase with IC_{50} values of 65.5, 757.8, and 577.7 μg/mL (deoxynojirimycin IC_{50}: 425.6 μg/mL).[203] The plant is poisonous.

1.84 *Premna tomentosa* Kurz

Common name: bastard teak
Subclass Lamiidae, Superorder Lamiidae, Order Lamiales, Family Verbenaceae
Medicinal use: liver disorder (India)

8,11,13-Icetexatriene-10-hydroxy-11,12,16-tri acetoxyl, 8,11,13-icetexatriene-7,10,11-dihydroxy-12,13-dihydrofuran and acetoxy syringaldehyde isolated from the roots of *Premna tomentosa* inhibited rat-intestinal α-glucosidase with IC_{50} values of 22.5, 9.5, and 18.4 μg/mL, respectively.[204]

1.85 *Tectona grandis* L.f.

Synonyms: *Tectona theka* Lour.; *Theka grandis* (L.f.) Lam.
Common names: you mu (Chinese); malapangit (Philippines); teak
Subclass Lamiidae, Superorder Lamiidae, Order Lamiales, Family Verbenaceae
Medicinal use: sore throat (Philippines)

Methanol extract from flowers of *Tectona grandis* L.f. given to nicotinamide-streptozotocin-induced type 2 diabetic Wistar albino rats (glycaemia > 200 mg/dL) at a dose of 200 mg/kg/day orally for 4 weeks prevented weight loss and decreased glycaemia to about 100 mg/dL.[205] This extract inhibited α-amylase with an IC_{50} value of 2.2 μg/mL (acarbose: 219.5 μg/mL) and α-glucosidase with an IC_{50} value of 229.2 μg/mL (acarbose: 0.3 μg/mL).[205] The flowers shelter phenolic constituents including ellagic acid, quercetin, and rutin that are known inhibitors of α-glucosidase *in vitro*.[205] Quercetin inhibited α-amylase activity with an IC_{50} value of 21.4 μM.[15]

1.86 *Calamintha officinalis* Moench

Synonyms: *Calamintha nepeta* (L.) Savi; *Melissa calamintha* L.
Common name: calamint
Subclass Lamiidae, Superorder Lamiidae, Order Lamiales, Family Lamiaceae
Nutritional use: seasoning (Turkey)

In healthy individual, plasma glucose levels reach a peak not exceeding 160 mg/dL from 30 to 60 minutes after oral ingestion of 75 g of glucose and gradually return to postabsorptive values by 3–4 hours and insulin resistance in metabolic syndrome elevates that peak, posing the threat of toxic plasma levels of glucose also called "glucotoxicity".[64] *Calamintha officinalis* Moench given orally to alloxan-induced diabetic Wistar rats orally at a dose of 400 mg/kg/day for 14 days decreased plasma glucose from 248.6 to 117.7 mg/dL.[206] Rosmarinic acid (Figure 1.46) or caffeic acid isolated from this extract given orally at a dose of 10 mg/kg/day to alloxan-induced diabetic Wistar rats decreased glycaemia from 248.6 to 97.3 mg/dL and 105.4 mg/dL, respectively, whereby glibenclamide at 10 mg/kg brought glycaemia down to 115.1 mg/dL (normal: 87.5 mg/dL).[207] From the extract, rosmarinic acid and caffeic acid decreased triglycerides and normalized plasma cholesterol.[207] Caffeic acid which is common in members of the family Lamiaceae inhibited yeast α-glucosidase with IC_{50} values of 27.4 μM, respectively (acarbose: and 38.3 μM).[208] Rosmarinic acid common in this family inhibits α-glucosidase *in vitro*.[209]

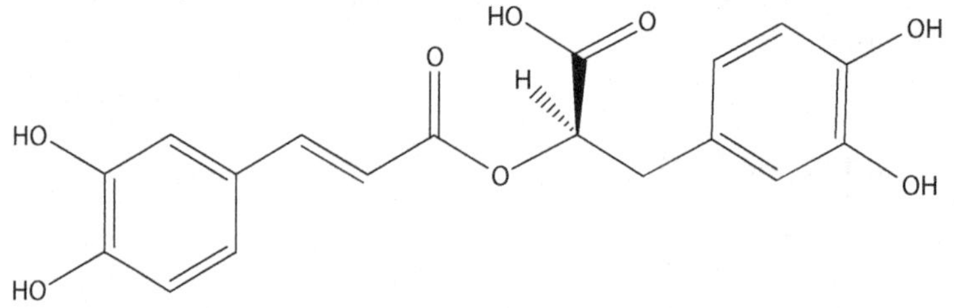

FIGURE 1.46 Rosmarinic acid.

1.87 *Hyssopus officinalis* L.

Synonym: *Thymus hyssopus* (L.) E.H.L. Krause
Common names: shen xiang cao (Chinese); jupha (India); hyssop
Subclass Lamiidae, Superorder Lamiidae, Order Lamiales, Family Lamiaceae
Medicinal use: indigestion (India)
History: The plant was known of Pliny the Elder (23–79 AD). Roman scholar

The phenolic glycosides (7S,8S)-syringoylglycerol-9-*O*-(6′-*O*-cinnamoyl)-β-D-glucopyranoside and (7S,8S)-syringoylglycerol-9-*O*-β-D-glucopyranoside from *Hyssopus officinalis* L. inhibited the enzymatic activity of rat-intestinal α-glucosidase by 54% and 53% at a concentration of 3×10^{-3} M.[209]

1.88 *Melissa officinalis* L.

Synonym: *Melissa bicornis* Klokov
Common names: xiang feng hua (Chinese); lemon balm
Subclass Lamiidae, Superorder Lamiidae, Order Lamiales, Family Lamiaceae

Medicinal use: arteriosclerosis (Turkey)
History: The plant was known of Dioscarides as an antidote for snake bites.

Methanol extract from leaves of *Melissa officinalis* L. at a concentration of 2.5 mg/mL inhibited porcine pancreatic lipase activity by 90% *in vitro*.[96] The active constituents involved in pancreatic lipase inhibition are apparently unknown and would be worth being isolated. A pharmacologic inhibition of the absorption of triglycerides has been used as a clinical strategy in the treatment of obesity.[46] Such inhibitor is orlistat that inhibited porcine lipase with an IC_{50} value of 0.2 mg/mL (Liu et al. 2013).[16] *Melissa officinalis* L. could be beneficial for metabolic syndrome management and further clinical studies in this direction are needed.

1.89 *Ocimum basilicum* L.

Synonym: *Ocimum thyrsiflorum* L.
Common name: luo le (China); kali tulasi (India); basil
Subclass Lamiidae, Superorder Lamiidae, Order Lamiales, Family Lamiaceae
Medicinal use: diuretic (India)

After a meal, postprandial glycaemia usually do not exceed 165 mg/dL but that value is increased in case of insulin resistance.[210] Inhibitors of amylase decrease postprandial glycemia.[211] Aqueous extract from leaves of *Ocimum basilicum* L. inhibited rat-intestinal amylase, rat-intestinal maltase, and porcine pancreatic amylase with IC_{50} values of 36.7, 21.3, and 42.5 mg/mL (acarbose IC_{50} of 0.03 μg/mL).[212]

1.90 *Origanum majorana* L.

Synonyms: *Majorana vulgaris* (L.) Gray; *Thymus majorana* (L.) Kuntze
Common name: sweet marjoram
Subclass Lamiidae, Superorder Lamiidae, Order Lamiales, Family Lamiaceae
Nutritional use: seasoning (Turkey)

Origanum majorana L. elaborates the monoterpene carvacrol (Figure 1.47) which inhibited lipase of ddY mice isolated from mice plasma with an IC_{50} value of 4 mM (orlistat 0.09 mM).[213,214] Carvacrol given orally at a single dose of 300 mg/kg to ddY mice oral olive-oil load decreased, after 180 mon, blood triglycerides from about 900 to 300 mg/dL and this effect was comparable with orlistat at 10 mg/kg.[214] Scutellarein (or 6-hydroxyluteolin) and 6-hydroxyluteolin-7-*O*-β-D-glucopyranoside isolated from *Origanum marjorana* inhibited rat-intestinal α-glucosidase with an IC_{50} values of 12 and 300 μM, respectively.[215] In the same experiment, the isoflavones, flavanols, and flavanones tested had no activity. Scutellarein inhibited α-amylase *in vitro* with an IC_{50} value of 9.6 μM.[15]

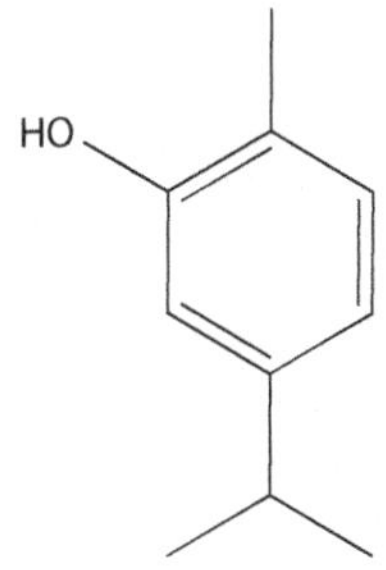

HO

FIGURE 1.47 Carvacrol.

1.91 *Orthosiphon stamineus* Benth.

Synonyms: *Clerodendranthus spicatus* (Thunb.) C.Y. Wu ex H.W. Li; *Orthosiphon aristatus* (Blume) Miq.
Common names: misai kunching (Malay); Java tea
Subclass Lamiidae, Superorder Lamiidae, Order Lamiales, Family Lamiaceae
Medicinal use: diuretic (Malaysia)

Tetramethylscutellarein and 3,7,4′-tri-O-methylkaempferol isolated from this plant inhibited yeast α-glucosidase with IC_{50} values of 6.3 and 0.7 μM, respectively.[216] From this plant, orthosiphol A selectively inhibited intestinal maltase with an IC_{50} value of 6.5 μM.[216]

1.92 *Rosmarinus officinalis* L.

Common names: mi die xiang (Chinese); romero (Philippines); rosemary
Subclass Lamiidae, Superorder Lamiidae, Order Lamiales, Family Lamiaceae
Medicinal use: tonic (Philippines)

Ninomiya et al. (2004) provided evidence that carnosic acid and carnosol from *Rosmarinus officinalis* L. inhibited porcine pancreatic lipase *in vitro* with IC_{50} values of 12 and 4.4 μg/mL, respectively.[217] Methanol extract from the leaves at a concentration of 2.5 mg/mL inhibited porcine pancreatic lipase activity by 100% *in vitro* (IC_{50}: 0.1 mg/mL; orlistat: 0.1 ng/mL).[96] A fraction of *Rosmarinus officinalis* containing 38.9% carnosic acid given to obese Zucker rats orally as part of diet at 0.5% for 64 days did not reduce food consumption and evoked a mild reduction of body weight and increased fecal weight.[218] This supplementation reduced plasma triglycerides and had no effect on plasma cholesterol and glycemia or glycaemia.[218] The fraction inhibited gastric lipase activity by about 80% and had a mild effect on intestinal pancreatic lipase activity.[218] In a subsequent study, a fraction extracted from *Rosmarinus officinalis* L. containing 80% carnosic acid added to high-fat diet at 0.2% given to C57BL/6L for 16 weeks reduced body weight from about 50 to 35 g (normal diet: about 32.5 g), normalized liver mass, increased epididymal fat, reduced mesenteric fats, reduced retroperitoneal fat, and decreased total fat from 4.8 to 3.4 g (normal diet: 2.1 g).[219] This supplementation prevented rise in fasting glycaemia, decreased plasma insulin by 90% and insulin resistance by 96.4%, bringing both parameters to normal diet group.[219] This fraction reduced liver triglycerides by 109.4%, and free fatty acids by 106.7%.[219] The regimen attenuated enlargement and vacuolization of hepatocytes (steatosis).[219] The fraction decreased lipoperoxidation in the plasma and liver.[219] This fraction increased lipid fecal secretion implying, at least, inhibition of triglyceride absorption confirming pancreatic inhibition.[219] The plant contains rosmarinic acid and inhibits α-glucosidase *in vitro*.[208]

1.93 *Salvia miltiorrhiza* Bunge

Common name: dan shen (Chinese)
Subclass Lamiidae, Superorder Lamiidae, Order Lamiales, Family Lamiaceae
Medicinal use heart diseases (China)

Isosalvianolic acid C methyl ester, tanshinone IIA, rosmarinate acid, rosmarinic acid methyl ester, salvianolic acid A methyl ester, salvianolic acid C methyl ester isolated from *Salvia miltiorrhiza*

Bunge inhibited α-glucosidase activity with IC_{50} value of 111.9 × 10^{-3} μM, 230.2 × 10^{-3} μM, 224.1 × 10^{-3} μM, 142.6 × 10^{-3} μM, 180.6 × 10^{-3} μM, and 42.1 × 10^{-3} μM, respectively (acarbose: IC_{50} of 5832.4 × 10^{-3} μM).[220]

1.94 *Salvia officinalis* L.

Common names: sa er wei ya (Chinese); Salbia sefakuss (India); sage
Subclass Lamiidae, Superorder Lamiidae, Order Lamiales, Family Lamiaceae
Medicinal use: indigestion (India)

Ethanol extract from leaves of *Salvia officinalis* L. given orally to type 2 diabetic patients at a dose of 500 mg 3 times per day for 3 months decreased fasting plasma glucose by 25.8%, total cholesterol by 17.7%, triglycerides by 32.2%, low density lipoprotein-cholesterol by 19.2%, and increased high-density lipoprotein-cholesterol by 34.8%.[221] Carnosic acid, carnosol, royleanonic acid, 7-methoxyrosmanol, and oleanolic acid isolated from *Salvia officinalis* L. inhibited porcine pancreatic lipase with IC_{50} values below 85 μg/mL.[222] In ddY mice, carnosic acid at a single oral dose of 20 mg/kg 30 minutes before oral loading of olive oil, reduced 2 hours serum triglycerides from 571 to 220 mg/100 mL (orlistat: 177 mg/100 mL) whereby carnosol, royleanonic acid, 7-methoxyrosmanol, and oleanolic acid up to 200 mg/kg had no effect.[222] Carnosic acid given orally at a dose of 10 mg/kg/day for 14 days to ddY mice on high-fat diet, had no effect on body weight, decreased serum triglycerides from 126 to 78 mg/100 mL (normal: 118 mg/100 mL) and decreased epididymal fat pad from 1472 mg/mouse to 1018 mg/mouse (normal: 839 mg/mouse).[222] The consumption of sage tea could be beneficial in metabolic syndrome.

1.95 *Scutellaria baicalensis* Georgi

Synonyms: *Scutellaria lanceolaria* Miq.; *Scutellaria macrantha* Fisch.
Common name: huang qin (Chinese)
Subclass Lamiidae, Superorder Lamiidae, Order Lamiales, Family Lamiaceae
Medicinal use: fever (China)

Members of the genus *Scutellaria* L. synthetizes a broad array of flavones glycosides, of which luteolin 7-*O*-glucoside, luteolin 7-*O*-glucuronide, and diosmetin 7-*O*-glucuronide which inhibited *in vitro* porcine α-amylase with IC_{50} values of 81.7, 61.5, and 76.3 μM, respectively (acarbose IC_{50}: 43.4 μM).[223] Luteolin 7-*O*-glucoside, luteolin 7-*O*-glucuronide, and diosmetin 7-*O*-glucuronide inhibited *in vitro* yeast α-glucosidase with IC_{50} values of 18.3, 14.7, and 17.1 μM, respectively (acarbose IC_{50}: 16.1 μM).[223] *Scutellaria baicalensis* Georgi shelters baicalein that inhibited rat-intestinal sucrase with an IC_{50} value of 3.5 × 10^{-5} M.[224] Aqueous extract from roots of *Scutellaria baicalensis* Georgi at 18 mg/tube inhibited pancreatic lipase by 66%.[225]

1.96 *Chlorophytum borivilianum* Santapau & R.R. Fern.

Common name: shweta Musali (India)
Subclass Lillidae, Superorder Lilianae, Order Asparagales, Family Asparagaceae
Medicinal use: sex impotance (India)

Root powder of *Chlorophytum borivilianum* Santapau & R.R. Fern. given to Wistar rats on high-fat diet at a dose of 1.5 g/rat/day for 4 weeks decreased plasma cholesterol from 363.1 to 265 mg/dL (normal: 119.9 mg/dL) and triglycerides from 55.9 to 44.7 mg/dL (normal: 43.5 mg/dL).[226] At the hepatic level, the supplementation decreased cholesterol, triglycerides, 3-hydroxy-3-methylglutaryl-coenzyme A reductase activity, and increased bile acids from 6.5 to 8.7 mg/g.[226] The root powder increased fecal cholesterol, neutral fecal sterol, and fecal bile acids suggesting a decrease of intestinal cholesterol absorption. Root powder of *Chlorophytum borivilianum* Santapau & R.R. Fern. contains inulin-type fructans and saponins.[227] In rats, a decrease in plasma triglycerides and cholesterol have been reported after oral administration of fructans.[150,151]

1.97 *Dendrobium loddigesii* Rolfe

Synonyms: *Callista loddigesii* (Rolfe) Kuntze
Common name: mei hua shi hu (China)
Subclass Liliidae, Superorder Lilianae, Order Orchidales, Family Orchidaceae
Medicinal use: indigestion (China)

Loddigesiinol G, H, I, J, and crepidatuol B isolated from the stems of *Dendrobium loddigesii* inhibited α-glucosidase with IC_{50} values below 20 μM (trans-resveratrol: 27.9 μM).[228]

1.98 *Dioscorea bulbifera* L.

Synonyms: *Dioscorea sativa* Thunb.; *Helmia bulbifera* (L.) Kunth
Common names: huang du (Chinese); eeloom poom paw (Thai); potato yam
Subclass Liliidae, Superorder Dioscoreanae, Order Dioscoreales, Family Dioscoreaceae
Medicinal use: boils (China)

Diosgenin (Figure 1.48) isolated from the bulbs of *Dioscorea bulbifera* L. at a concentration of 100 μg/mL inhibited porcine pancreatic α-amylase by 70.9%, crude murine pancreatic α–amylase by 39.5%, and yeast α-glucosidase by 81.7%, and crude murine intestinal α-glucosidase by 70.7%.[229] It would be of interest to assess the activity of this plant on pancreatic lipase as 3,3′,5-trihydroxy-2′-methoxybibenzyl isolated from another member of the genus *Dioscorea* inhibited lipase with an IC_{50} value of 8.8 μM.[230]

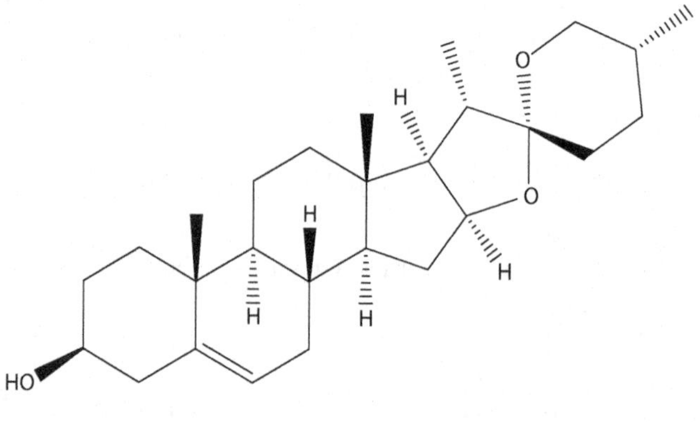

FIGURE 1.48 Diosgenin.

1.99 *Alpinia officinarum* Hance

Synonym: *Languas officinarum* (Hance) Farw.
Common names: gao liang jiang (Chinese); galangal
Subclass Commelinidae, Superoder Zingiberanae, Order Zingiberales, Family Zingiberaceae
Medicinal use: indigestion (China)
History: The plant was known of Avicenna

Shin et al. (2003) provided evidence that 3-methylethergalangin (Figure 1.49) isolated from the rhizome of *Alpinia officinarum* Hance (Figure 1.50) inhibited pancreatic lipase activity with an IC_{50}

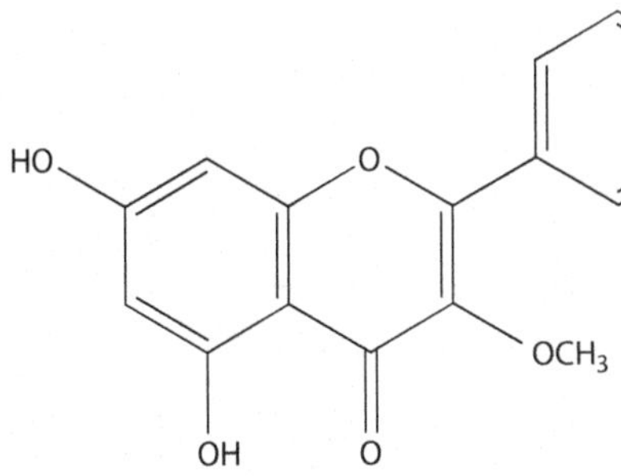

FIGURE 1.49 3-Methylethergalangin.

FIGURE 1.50 *Alpinia officinarum* Hance.

value of 1.3 mg/mL (orlistat, IC_{50}: 0.8 mg/mL).[231] This flavone given orally at a dose of 20 mg/kg/day to corn oil induced hyperlipidemic ICR mice for 5 days decreased triglycerides from 88 to 39.3 mg/dL (normal: 48.1 mg/dL; orlistat 20 mg/kg/day: 31.5 mg/dL) and increased cholesterol from 167.5 to 182.9 mg/dL (normal: 175 mg/dL; orlistat 20 mg/kg/day: 172.2 mg/dL).[231] From the same plant, 5-hydroxy-7-(4′-hydroxy-3′-methoxyphenyl)-1-phenyl-3-heptanone inhibited pancreatic lipase activity with an IC_{50} value of 1.5 mg/mL (orlistat, IC_{50}: 0.8 mg/mL).[232] This curcumoid given orally at a dose of 100 mg/kg/day to corn oil induced hyperlipidemic ICR mice for 5 days decreased triglycerides from 188 to 116.8 mg/dL (normal: 97.3 mg/dL; orlistat 50 mg/kg/day: 63.9 mg/dL), cholesterol from 137.5 to 112.6 mg/dL (normal: 123.3 mg/dL; orlistat 50 mg/kg/day: 126.4 mg/dL), and had no effect on high-density lipoprotein.[232] Rhizomes of *Alpinia officinarum* Hance given to Syrian hamsters at 10% of high-fat diet for 9 weeks attenuated food intake, reduced weight gain from 44 to 34.6 g (normal diet: 35 g), prevented liver weight gain, decreased serum cholesterol from 319 to 116 mg/dL (normal: 138 mg/dL), triglycerides from 223 to 94 mg/dL (normal: 98 mg/dL), low-density lipoprotein cholesterol from 108 to 40 mg/dL (normal: 40 mg/dL), and lowered high-density lipoprotein cholesterol from 194 to 167 mg/dL (normal: 168 mg/dL).[233] This supplementation increased serum superoxide dismutase, decreased catalase, decreased serum lipid peroxides, and increased glutathione.[233] Increase of high-density lipoprotein-cholesterol by 10 mg/dL in human corresponds to a 19% decrease in coronary artery disease death.[234] Clinical trials are warranted.

1.100 *Curcuma longa* L.

Synonym: *Curcuma domestica* Valeton
Common names: jiang huang (Chinese); dilau (Philippines); turmeric
Subclass Commelinidae, Superoder Zingiberanae, Order Zingiberales, Family Zingiberaceae
Medicinal use: diabetes (Philippines)

Curcumin (Figure 1.51), demethoxycurcumin, and bisdemethoxycurcumin, which are not absorbed in the small intestine, isolated from the rhizomes of *Curcuma longa* L. inhibited yeast α-glucosidase with IC_{50} values of 37.2, 42.7, and 23 mM, respectively.[235]

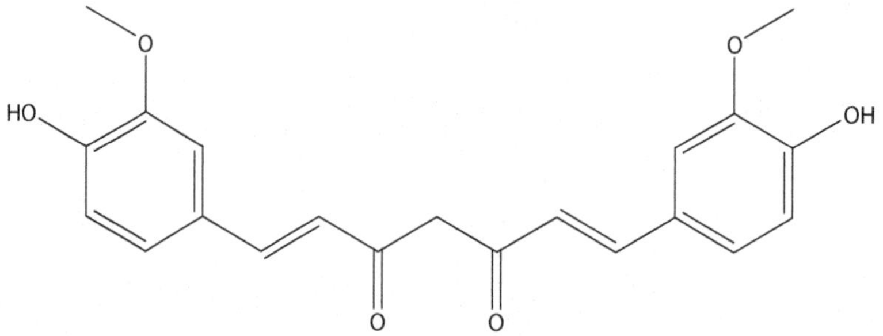

HO OH O O O O

FIGURE 1.51 Curcumin.

1.101 *Hedychium spicatum* Buch.-Ham. ex Sm.

Synonyms: *Hedychium Coronarium* J. Koenig
Common name: cao guo yao (Chinese)
Subclass Commelinidae, Superoder Zingiberanae, Order Zingiberales, Family Zingiberaceae
Medicinal use: indigestion (Taiwan)

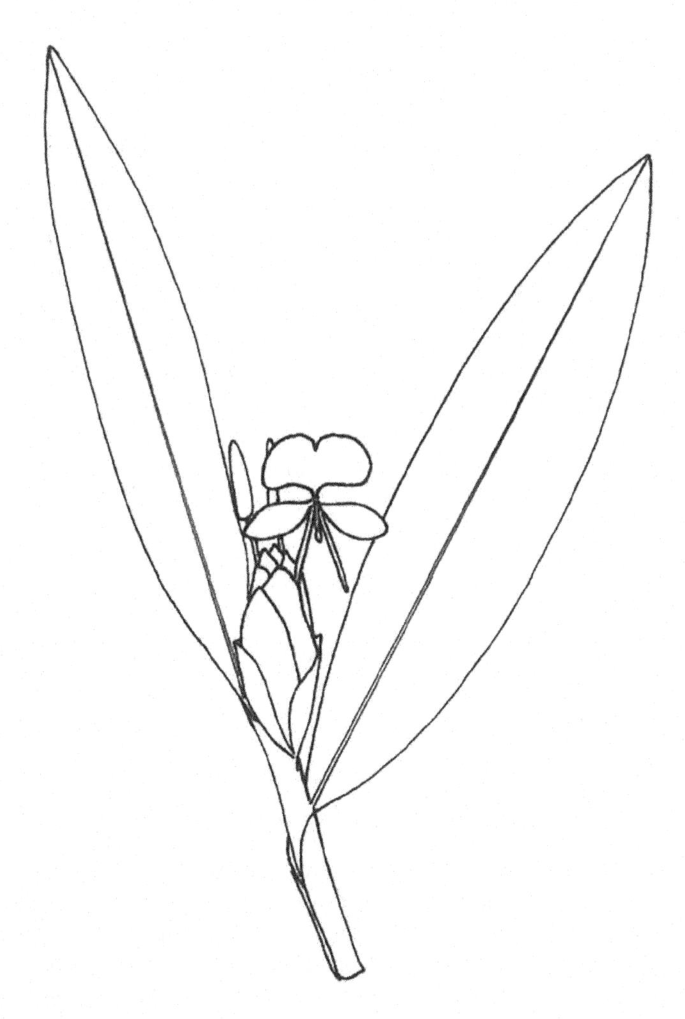

FIGURE 1.52 *Hedychium spicatum* Buch.-Ham. ex Sm.

Spicatanol methyl ether and hedychenone isolated from the rhizome of *Hedychium spicatum* Buch.-Ham. ex Sm. (Figure 1.52) at a concentration of 100 μg/mL inhibited rat-intestinal α-glucosidase by more than 50%.[236] Spicatanol inhibited rat-intestinal α-glucosidase with an IC_{50} value of 34.1 μM (acarbose: IC_{50} of 23.8 μM).[236]

1.102 *Kaempferia parviflora* Wall. ex Baker

Synonyms: *Kaempferia rubromarginata* (S.Q. Tong) R. J. Searle; *Stahlianthus rubromarginatus* S.Q. Tong
Common name: kalahalood (Bangladesh)
Subclass Commelinidae, Superoder Zingiberanae, Order Zingiberales, Family Zingiberaceae
Medicinal use: diarrhea (India)

Methoxyflavones fraction of rhizome of *Kaempferia parviflora* Wall. ex Baker given to Tsumara Suzuki Obese Diabetes (TSOD) mice at 1% of diet for 8 weeks had no effect on food intake, evoked a decrease in body weight gain and visceral fat mass.[237] The fraction lowered glycemia from 216 to 152 mg/dL (normal mice: 152 mg/dL), it had no effect on cholesterol and triglycerides and evoked a decrease in plasma insulin from 15.6 to 5.9 ng/mL (normal mice: 1.1 ng/mL).[237] In oral glucose tolerance test performed at the end of the treatment, the extract reduced 30 minutes peak postprandial glycemia from about 500 to 325 mg/dL. The extract reduced hepatic cholesterol and triglycerides from 14.1 to 8.1 mg and 1.8 to 1.3 mg.[237] The systolic blood pressure was reduced from 112.9 to 102.3 mmHg (i). In a parallel study, powder of rhizome of *Kaempferia parvifolia* given to Tsumara Suzuki Obese Diabetes (TSOD) mice for 8 weeks at 3% of diet had no effect on food intake, decreased body weight gain, and reduced visceral fat accumulation (not subcutaneous).[238] This regimen reduced plasma glucose from 184 to 159 mg/dL, plasma cholesterol from 228 to 172 mg/dL, triglycerides from 234 to 166 mg/dL, it had no effect on low-density lipoprotein, reduced high-density lipoprotein from 123 to 104 mg/dL, and decreased insulin from 10.3 to 2.9 ng/mL.[238] In line, the treatment reduced 30 minutes postprandial peak glycemia in oral glucose tolerance test from about 650 to 500 mg/dL.[238] From the extract, 5-hydroxy-3,7-dimethoxyflavone, 5-hydroxy-3,7,4′-trimethoxyflavone, 5-hydroxy-7.4′-dimethoxyflavone, and 5-hydroxy-7-methoxyflavone inhibited porcine pancreatic lipase with IC_{50} values below 550 μg/mL.[238]

1.103 *Commelina communis* L.

Synonym: *Commelina coreana* H. Lév.; *Commelina ludens* Miq.
Common name: ya zhi cao (Chinese)
Subclass Commelinidae Superorder Commelinanae, Order Commelinales, Family Commelinaceae
Medicinal use: fever (China)

Kim et al. (1999) isolated from *Commelina communis* L. the polyhydroxylated piperidine alkaloids, 1-deoxymannojirimycin, 1-deoxynojirimycin, and α-homonojirimycin.[239] Aqueous extracts from leaves of *Commelina communis* L. and whole plant at a dose of 10 mg/mL inhibited the enzymatic activity of α-glucosidase by 77% and 62.1%, respectively.[239] Administration of extract at a dose of 100 mg/kg for 10 days evoked a mild reduction of fasting blood glucose. In healthy mice, the leaf extract at a dose of 100 mg/kg halved the postprandial hyperglycemia caused by starch loading at 2 g/kg.[240] 1-Deoxynojirimycin (Figure 1.53), which is produced by both *Streptomyces* and flowering plants from various taxons (symbionts?), is a glucose analogue with an amine group substituting for the oxygen atom in the pyranose ring, has been shown to inhibit intestinal α-glucosidases and pancreatic α-amylase both *in vitro* and *in vivo*. This molecule has been used for the development of miglitol, a drug given to lower postprandial glycemia in type 2 diabetes.[241] 1-Deoxynojirimycin inhibits rat-intestinal maltase and isomaltase with IC_{50} of 0.3 mM whereas, 1-deoxymannojirimycin inhibits rat intestine maltase with IC_{50} of 150 mM.[241] As for α-homonojirimycin, it is known to inhibit α-glucosidase.[242] Insulin resistance is associated with hyperglycaemia, a risk factor for cardiovascular disease and this plant, if not toxic, may be of value in metabolic syndrome.[243]

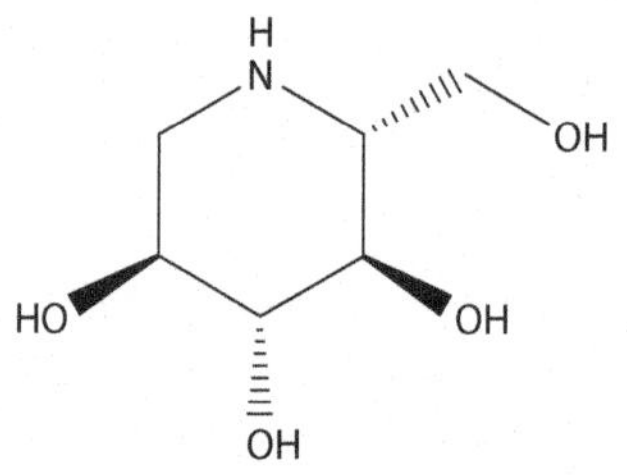

H
N
OH
HO
OH
OH

FIGURE 1.53 1-Deoxynojirimycin.

APPENDIX

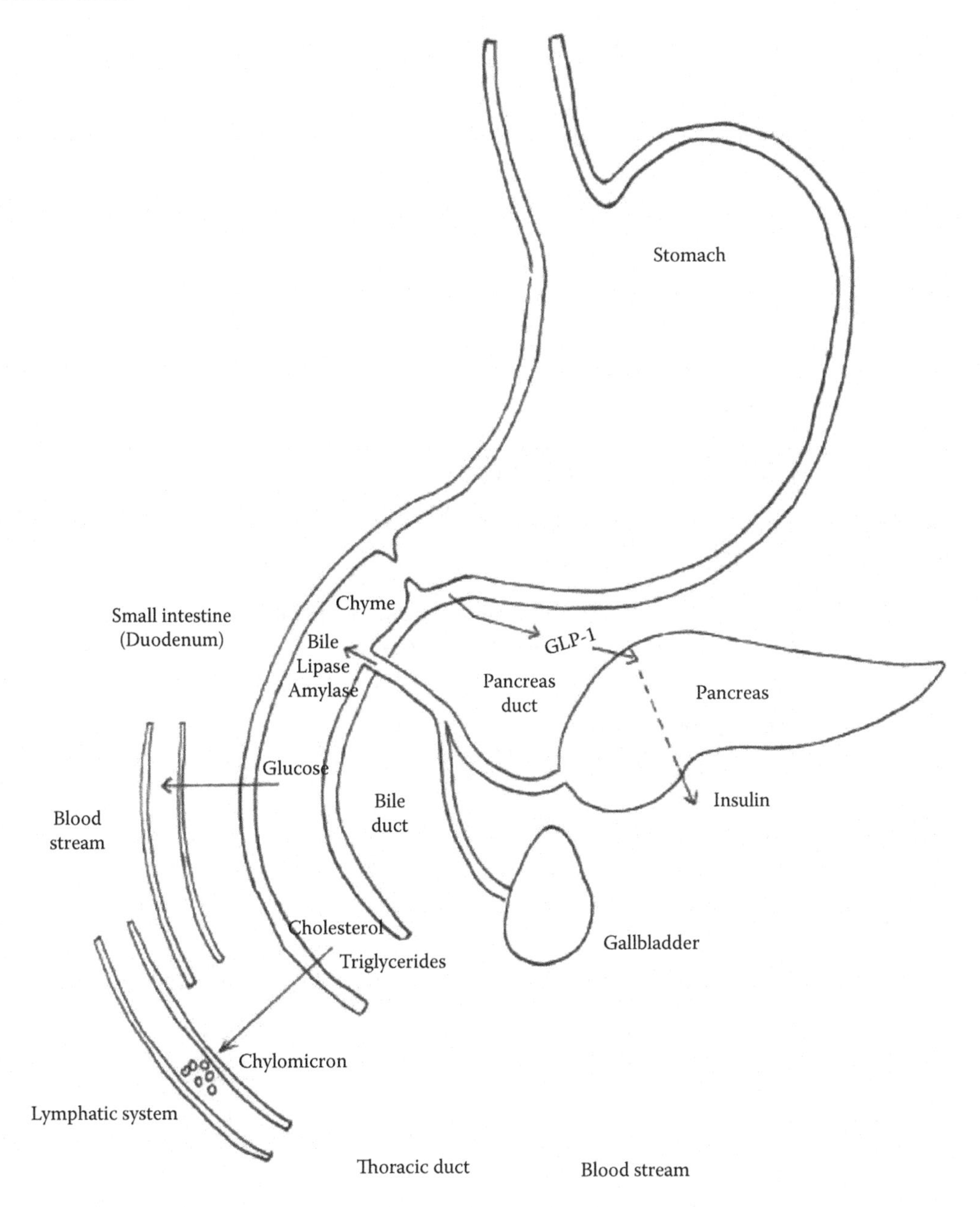

APPENDIX 1.1 Absorption of glucose, cholesterol and triglycerides.

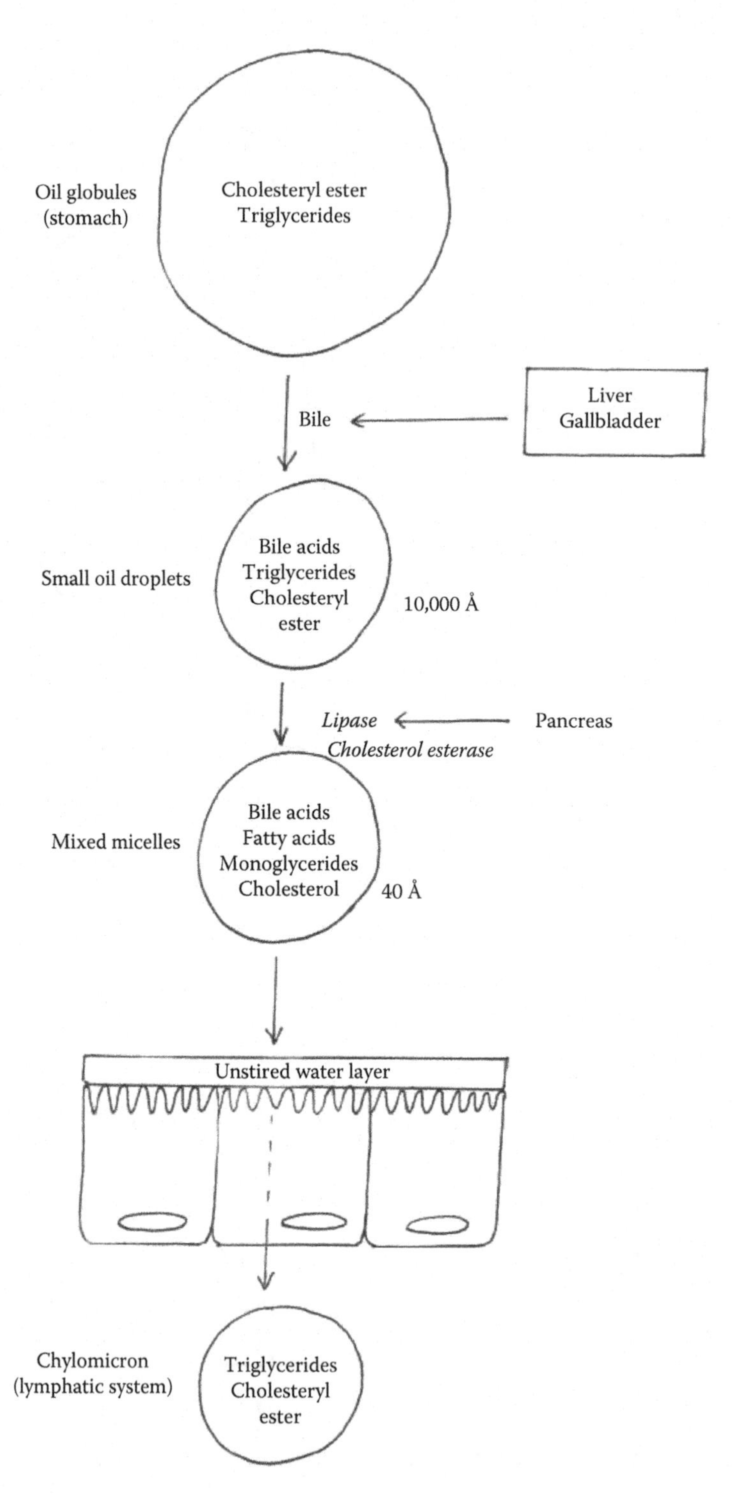

APPENDIX 1.2 Emulsification of dietary fats.

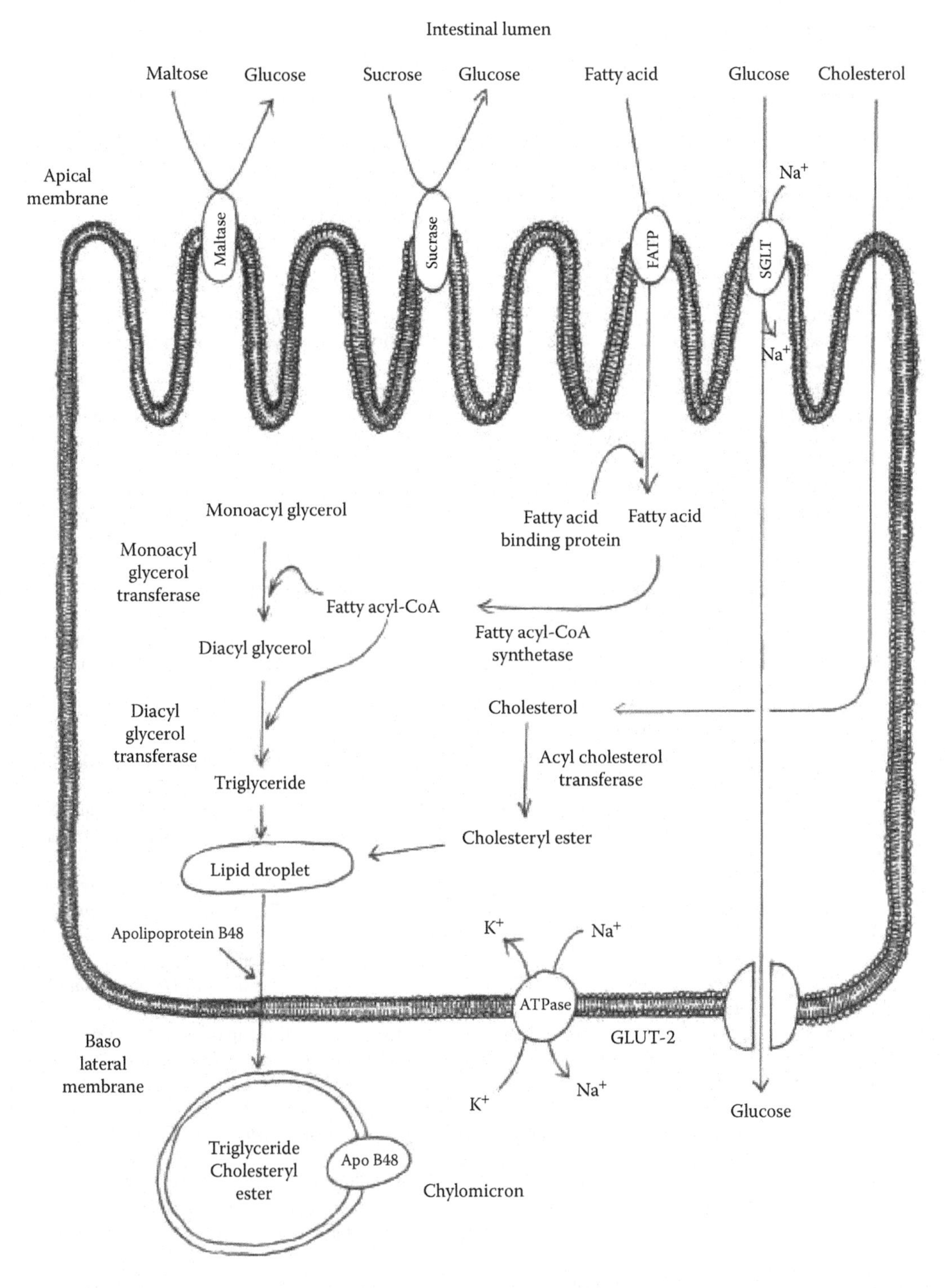

APPENDIX 1.3 Absorption of glucose, fatty acids and cholesterol by enterocytes.

REFERENCES

1. Ceriello, A., Cavarape, A., Martinelli, L., Da Ros, R., Marra, G., Quagliaro, L., Piconi, L., Assaloni, R. and Motz, E., 2004. The post-prandial state in type 2 diabetes and endothelial dysfunction: Effects of insulin aspart. *Diabetic Medicine*, 21(2), 171–175.
2. Delarue, J. and Magnan, C., 2007. Free fatty acids and insulin resistance. *Current Opinion in Clinical Nutrition & Metabolic Care*, 10(2), 142–148.
3. Fonseca, V., 2003. Clinical significance of targeting postprandial and fasting hyperglycemia in managing type 2 diabetes mellitus. *Current Medical Research and Opinion*, 19(7), 635–631.
4. Kahn, S.E., Hull, R.L. and Utzschneider, K.M., 2006. Mechanisms linking obesity to insulin resistance and type 2 diabetes. *Nature*, 444(7121), 840–846.
5. Liu, S. and Manson, J.E., 2001. Dietary carbohydrates, physical inactivity, obesity, and the 'metabolic syndrome' as predictors of corlonary heart disease. *Current Opinion in Lipidology*, 12(4), 395–404.
6. Shoelson, S.E., Herrero, L. and Naaz, A., 2007. Obesity, inflammation, and insulin resistance. *Gastroenterology*, 132(6), 2169–2180.
7. Hamosh, M., 1989. Lingual and gastric lipases. *Nutrition*, 6(6), 421–428.
8. Lee, Y.M., Kim, Y.S., Lee, Y., Kim, J., Sun, H., Kim, J.H. and Kim, J.S., 2012. Inhibitory activities of pancreatic lipase and phosphodiesterase from Korean medicinal plant extracts. *Phytotherapy Research*, 26(5), 778–782.
9. Yun, Y.R., Kim, M.J., Kwon, M.J., Kim, H.J., Song, Y.B., Song, K.B. and Song, Y.O., 2007. Lipid-lowering effect of hot water-soluble extracts of *Saururus chinensis* bail on rats fed high-fat diets. *Journal of Medicinal Food*, 10(2), 316–322.
10. Lin, A.H.M., Lee, B.H., Nichols, B.L., Quezada-Calvillo, R., Rose, D.R., Naim, H.Y. and Hamaker, B.R., 2012. Starch source influences dietary glucose generation at the mucosal α-glucosidase level. *Journal of Biological Chemistry*, 287(44), 36917–36921.
11. Krishna, M.S., Joy, B. and Sundaresan, A., 2015. Effect on oxidative stress, glucose uptake level and lipid droplet content by Apigenin 7, 4′-dimethyl ether isolated from *Piper longum* L. *Journal of Food Science and Technology*, 52(6), 3561–3570.
12. Robyt, J. and French, D., 1963. Action pattern and specificity of an amylase from *Bacillus subtilis*. *Archives of Biochemistry and Biophysics*, 100(3), 451–467.
13. Ono, Y., Hattori, E., Fukaya, Y., Imai, S. and Ohizumi, Y., 2006. Anti-obesity effect of *Nelumbo nucifera* leaves extract in mice and rats. *Journal of Ethnopharmacology*, 106(2), 238–244.
14. Ohkoshi, E., Miyazaki, H., Shindo, K., Watanabe, H., Yoshida, A., Yajima, H., 2007. Constituents from the leaves of *Nelumbo nucifera* stimulate lipolysis in the white adipose tissue of mice. *Planta Medica*, 73(12):1255–1259.
15. Lo Piparo, E., Scheib, H., Frei, N., Williamson, G., Grigorov, M. and Chou, C.J., 2008. Flavonoids for controlling starch digestion: Structural requirements for inhibiting human α-amylase. *Journal of Medicinal Chemistry*, 51(12), 3555–3561.
16. Liu, S., Li, D., Huang, B., Chen, Y., Lu, X. and Wang, Y., 2013. Inhibition of pancreatic lipase, α-glucosidase, α-amylase, and hypolipidemic effects of the total flavonoids from *Nelumbo nucifera* leaves. *Journal of Ethnopharmacology*, 149(1), 263–269.
17. Zhu, Y.T., Jia, Y.W., Liu, Y.M., Liang, J., Ding, L.S. and Liao, X., 2014. Lipase ligands in *Nelumbo nucifera* leaves and study of their binding mechanism. *Journal of Agricultural and Food Chemistry*, 62(44), 10679–10686.
18. Ahn, J.H., Kim, E.S., Lee, C., Kim, S., Cho, S.H., Hwang, B.Y. and Lee, M.K., 2013. Chemical constituents from *Nelumbo nucifera* leaves and their anti-obesity effects. *Bioorganic & Medicinal Chemistry Letters*, 23(12), 3604–3608.
19. Tuomilehto, J., Lindström, J., Eriksson, J.G., Valle, T.T., Hämäläinen, H., Ilanne-Parikka, P., Keinänen-Kiukaanniemi, S., Laakso, M., Louheranta, A., Rastas, M. and Salminen, V., 2001. Prevention of type 2 diabetes mellitus by changes in lifestyle among subjects with impaired glucose tolerance. *New England Journal of Medicine*, 344(18), 1343–1350.
20. Liu, L., Yu, Y.L., Yang, J.S., Li, Y., Liu, Y.W., Liang, Y., Liu, X.D., Xie, L. and Wang, G.J., 2010. Berberine suppresses intestinal disaccharidases with beneficial metabolic effects in diabetic states, evidences from in vivo and in vitro study. *Naunyn-Schmiedeberg's Archives of Pharmacology*, 381(4), 371–381.
21. Yin, J., Xing, H. and Ye, J., 2008. Efficacy of berberine in patients with type 2 diabetes mellitus. *Metabolism*, 57(5), 712–717.
22. Lieberman, M., Marks, A.D. and Peet, A., 2013. *Marks' Basic Medical Biochemistry*. Philadelphia, PA: Wolters Kluwer Health/Lippincott Williams & Wilkins.

23. Patel, M.B. and Mishra, S.M., 2012. Magnoflorine from *Tinospora cordifolia* stem inhibits α-glucosidase and is antiglycemic in rats. *Journal of Functional Foods*, 4(1), 79–86.
24. Guang, J., Yi, D., Ming, Y. and Duan, Z., 1964. Pharmacological studies on magnoflorine, a hypotensive principle from Tu Qing Mu Xiang. *Acta Pharmaceutica Sinica*, 1, 5.
25. Rhoades, R.A. and Bell, D.R. (Eds.), 2012. *Medical Phisiology: Principles for Clinical Medicine*. Philadelphia, PA: Lippincott Williams & Wilkins.
26. Meddah, B., Ducroc, R., Faouzi, M.E.A., Eto, B., Mahraoui, L., Benhaddou-Andaloussi, A., Martineau, L.C., Cherrah, Y. and Haddad, P.S., 2009. Nigella sativa inhibits intestinal glucose absorption and improves glucose tolerance in rats. *Journal of Ethnopharmacology*, 121(3), 419–424.
27. Buchholz, T. and Melzig, M.F., 2016. Medicinal plants traditionally used for treatment of obesity and diabetes mellitus—screening for pancreatic lipase and α-amylase inhibition. *Phytotherapy Research*, 30(2), 260–266.
28. Wu, Q., Wang, Y. and Guo, M., 2011. Triterpenoid saponins from the seeds of celosia argentea and their anti-inflammatory and antitumor activities. *Chemical and Pharmaceutical Bulletin*, 59(5), 666–671.
29. Telagari, M. and Hullatti, K., 2015. In-vitro α-amylase and α-glucosidase inhibitory activity of *Adiantum caudatum Linn.* and *Celosia argentea Linn.* extracts and fractions. *Indian Journal of Pharmacology*, 47(4), p. 425.
30. Bauer, E., Jakob, S. and Mosenthin, R., 2005. Principles of physiology of lipid digestion. *Asian-Australasian Journal of Animal Sciences*, 18(2), 282–295.
31. Vinarova, L., Vinarov, Z., Atanasov, V., Pantcheva, I., Tcholakova, S., Denkov, N. and Stoyanov, S., 2015. Lowering of cholesterol bioaccessibility and serum concentrations by saponins: In vitro and in vivo studies. *Food & Function*, 6(2), 501–512.
32. Han, L.K., Nose, R., Li, W., Gong, X.J., Zheng, Y.N., Yoshikawa, M., Koike, K., Nikaido, T., Okuda, H. and Kimura, Y., 2006. Reduction of fat storage in mice fed a high-fat diet long term by treatment with saponins prepared from *Kochia scoparia* fruit. *Phytotherapy Research*, 20(10), 877–882.
33. Kasabri, V., Afifi, F.U. and Hamdan, I., 2011. In vitro and in vivo acute antihyperglycemic effects of five selected indigenous plants from Jordan used in traditional medicine. *Journal of Ethnopharmacology*, 133(2), 888–896.
34. Goel, V., Ooraikul, B. and Basu, TK., 1997. Cholesterol lowering effects of rhubarb stalk fiber in hypercholesterolemic men. *The American College of Nutrition*, 16(6), 600–604.
35. Thielecke, F. and Boschmann, M., 2009. The potential role of green tea catechins in the prevention of the metabolic syndrome–a review. *Phytochemistry*, 70(1), 11–24.
36. Surwit, R.S., Kuhn, C.M., Cochrane, C., McCubbin, J.A. and Feinglos, M.N., 1988. Diet-induced type II diabetes in C57BL/6J mice. *Diabetes*, 37(9), 1163–1167.
37. Grove, K.A., Sae-Tan, S., Kennett, M.J. and Lambert, J.D., 2012. (–)– Epigallocatechin-3-gallate inhibits pancreatic lipase and reduces body weight gain in high fat-fed obese mice. *Obesity*, 20(11), 2311–2313.
38. Fei, Q., Gao, Y., Zhang, X., Sun, Y., Hu, B., Zhou, L., Jabbar, S. and Zeng, X., 2014. Effects of oolong tea polyphenols, EGCG, and EGCG3 Me on pancreatic α-amylase activity in vitro. *Journal of Agricultural and Food Chemistry*, 62(39), 9507–9514.
39. Ryu, H.W., Cho, J.K., Curtis-Long, M.J., Yuk, H.J., Kim, Y.S., Jung, S., Kim, Y.S., Lee, B.W. and Park, K.H., 2011. α-Glucosidase inhibition and antihyperglycemic activity of prenylated xanthones from *Garcinia mangostana*. *Phytochemistry*, 72(17), 2148–2154.
40. Gowri, P.M., Tiwari, A.K., Ali, A.Z. and Rao, J.M., 2007. Inhibition of α-glucosidase and amylase by bartogenic acid isolated from *Barringtonia racemosa* Roxb. seeds. *Phytotherapy Research*, 21(8), 796–799.
41. Ponnapalli, M.G., Sukki, S., Annam, S.C.V.A.R., Ankireddy, M., Tirunagari, H., Tuniki, V.R. and Bobbili, V.P., 2015. α-Glucosidase inhibitory monoacylated polyhydroxytriterpenoids from the fruits of *Barringtonia racemosa*. *Tetrahedron Letters*, 56(12), 1570–1574.
42. Mbaze, L.M.A., Poumale, H.M.P., Wansi, J.D., Lado, J.A., Khan, S.N., Iqbal, M.C., Ngadjui, B.T. and Laatsch, H., 2007. α-Glucosidase inhibitory pentacyclic triterpenes from the stem bark of *Fagara tessmannii* (Rutaceae). *Phytochemistry*, 68(5), 591–595.
43. Dang, P.H., Nguyen, N.T., Nguyen, H.X., Nguyen, L.B., Le, T.H., Van Do, T.N., Van Can, M. and Nguyen, M.T.T., 2015. α-Glucosidase inhibitors from the leaves of *Embelia ribes*. *Fitoterapia*, 100, 201–207.
44. Dang, P.H., Nguyen, H.X., Nguyen, N.T., Le, H.N.T. and Nguyen, M.T.T., 2014. α-Glucosidase inhibitors from the stems of *Embelia ribes*. *Phytotherapy Research*, 28(11), 1632–1636.
45. Weibel, E.K., Hadvary, P., Hochuli, E., Kupfer, E. and Lengsfeld, H., 1987. Lipstatin, an inhibitor of pancreatic lipase, produced by *Streptomyces toxytricini*. *The Journal of Antibiotics*, 40(8), 1081–1085.

46. Ballinger, A. and Peikin, S.R., 2002. Orlistat: Its current status as an anti-obesity drug. *European Journal of Pharmacology*, 440(2), 109–117.
47. Megalli, S., Davies, N.M. and Roufogalis, B.D., 2006. Anti-hyperlipidemic and hypoglycemic effects of *Gynostemma pentaphyllum* in the zucker fatty rat. *Journal of Pharmacy and Pharmaceutical Sciences*, 9(3), 281–291.
48. Yang, Y.H., Yang, J. and Jiang, Q.H., 2013. Hypolipidemic effect of gypenosides in experimentally induced hypercholesterolemic rats. *Lipids in Health and Disease*, 12(1), 154.
49. Kwiterovich, P.O., 2012. *The John Hopkins Textbook of Dyslipidemia*. Philadelphia, PA: Lippincott Williams & Wilkins.
50. Su, J., Wang, H., Ma, C., Liu, C., Rahman, M.T., Gao, C. and Nie, R., 2016. Hypolipidemic mechanism of gypenosides via inhibition of pancreatic lipase and reduction in cholesterol micellar solubility. *European Food Research and Technology*, 242(3), 305–312.
51. Hau, J. and Schapiro, S.J. (Eds.), 2010. *Handbook of Laboratory Animal Science, Volume I: Essential Principles and Practices*. Boca Raton, FL: CRC Press.
52. Ikeda, I. and Sugano, M., 1998. Inhibition of cholesterol absorption by plant sterols for mass intervention. *Current Opinion in Lipidology*, 9(6), 527–531.
53. Plat, J. and Mensink, R.P., 2001. Effects of plant sterols and stanols on lipid metabolism and cardiovascular risk. *Nutrition, Metabolism, and Cardiovascular Diseases: NMCD*, 11(1), 31–40.
54. Kalsait, R.P., Khedekar, P.B., Saoji, A.N. and Bhusari, K.P., 2011. Isolation of phytosterols and antihyperlipidemic activity of *Lagenaria siceraria*. *Archives of Pharmacal Research*, 34(10), 1599–1604.
55. Chiasson, J.L., Josse, R.G., Gomis, R., Hanefeld, M., Karasik, A., Laakso, M. and STOP-NIDDM Trial Research Group, 2002. Acarbose for prevention of type 2 diabetes mellitus: The STOP-NIDDM randomised trial. *The Lancet*, 359(9323), 2072–2077.
56. Suzuki, Y.A., Murata, Y., Inui, H., Sugiura, M. and Nakano, Y., 2005. Triterpene glycosides of *Siraitia grosvenori* inhibit rat intestinal maltase and suppress the rise in blood glucose level after a single oral administration of maltose in rats. *Journal of Agricultural and Food Chemistry*, 53(8), 2941–2946.
57. Ali, L., Khan, A.K.A., Mamun, M.I.R., Mosihuzzaman, M., Nahar, N., Nur-e-Alam, M. and Rokeya, B., 1993. Studies on hypoglycemic effects of fruit pulp, seed, and whole plant of *Momordica charantia* on normal and diabetic model rats. *Planta Medica*, 59(05), 408–412.
58. Sankhari, J.M., Thounaojam, M.C., Jadeja, R.N., Devkar, R.V. and Ramachandran, A.V., 2012. Anthocyanin-rich red cabbage (*Brassica oleracea* L.) extract attenuates cardiac and hepatic oxidative stress in rats fed an atherogenic diet. *Journal of the Science of Food and Agriculture*, 92(8), 1688–1693.
59. Dyer, J., Wood, I.S., Palejwala, A., Ellis, A. and Shirazi-Beechey, S.P., 2002. Expression of monosaccharide transporters in intestine of diabetic humans. *American Journal of Physiology-Gastrointestinal and Liver Physiology*, 282(2), G241–G248.
60. Matsuda, H., Asao, Y., Nakamura, S., Hamao, M., Sugimoto, S., Hongo, M., Pongpiriyadacha, Y. and Yoshikawa, M., 2009. Antidiabetogenic constituents from the Thai traditional medicine *Cotylelobium melanoxylon*. *Chemical and Pharmaceutical Bulletin*, 57(5), 487–494.
61. Morikawa, T., Chaipech, S., Matsuda, H., Hamao, M., Umeda, Y., Sato, H., Tamura, H., Ninomiya, K., Yoshikawa, M., Pongpiriyadacha, Y. and Hayakawa, T., 2012. Anti-hyperlipidemic constituents from the bark of *Shorea roxburghii*. *Journal of Natural Medicines*, 66(3), 516–524.
62. Morikawa, T., Chaipech, S., Matsuda, H., Hamao, M., Umeda, Y., Sato, H., Tamura, H., Kon'i, H., Ninomiya, K., Yoshikawa, M. and Pongpiriyadacha, Y., 2012. Antidiabetogenic oligostilbenoids and 3-ethyl-4-phenyl-3, 4-dihydroisocoumarins from the bark of *Shorea roxburghii*. *Bioorganic & Medicinal Chemistry*, 20(2), 832–840.
63. Ahn, J.H., Liu, Q., Lee, C., Ahn, M.J., Yoo, H.S., Hwang, B.Y. and Lee, M.K., 2012. A new pancreatic lipase inhibitor from *Broussonetia kanzinoki*. *Bioorganic & Medicinal Chemistry Letters*, 22(8), 2760–2763.
64. Shrayyef, M.Z. and Gerich, J.E., 2010. Normal glucose homeostasis. In *Principles of Diabetes Mellitus*, Poretsky, L., (Ed.). New York: Springer, pp. 19–35.
65. Leu, J.P. and Zonszein, J., 2010. Diagnostic criteria and classification of diabetes. In *Principles of Diabetes Mellitus,* Poretsky, L., (Ed.). New York: Springer, pp. 107–115.
66. Herzlinger, S. and Martin, J.A., 2010. Treating type 2 diabetes mellitus. In *Principles of Diabetes Mellitus*, Poretsky, L., (Ed.). New York: Springer, pp. 731–747..
67. Choo, C.Y., Sulong, N.Y., Man, F. and Wong, T.W., 2012. Vitexin and isovitexin from the leaves of *Ficus deltoidea* with in-vivo α-glucosidase inhibition. *Journal of Ethnopharmacology*, 142(3), 776–781.
68. Yang, J.P., He, H. and Lu, Y.H., 2014. Four flavonoid compounds from *Phyllostachys edulis* leaf extract retard the digestion of starch and its working mechanisms. *Journal of Agricultural and Food Chemistry*, 62(31), 7760–7770.

69. Kumar, S., Kumar, D., Deshmukh, R.R., Lokhande, P.D., More, S.N. and Rangari, V.D., 2008. Antidiabetic potential of *Phyllanthus reticulatus* in alloxan-induced diabetic mice. *Fitoterapia*, 79(1), 21–23.
70. Pojchaijongdee, N., Sotanaphun, U., Limsirichaikul, S. and Poobrasert, O., 2010. Geraniinic acid derivative from the leaves of *Phyllanthus reticulatus. Pharmaceutical Biology*, 48(7), 740–744.
71. Espín, J.C., González-Barrio, R., Cerdá, B., López-Bote, C., Rey, A.I. and Tomás-Barberán, F.A., 2007. Iberian pig as a model to clarify obscure points in the bioavailability and metabolism of ellagitannins in humans. *Journal of Agricultural and Food Chemistry*, 55(25), 10476–10485.
72. Maruthappan, V. and Shree, K.S., 2010. Effects of Phyllanthus reticulatus on lipid profile and oxidative stress in hypercholesterolemic albino rats. *Indian Journal of Pharmacology*, 42(6), 388.
73. Rahmatullah, M., Hasan, S.K., Ali, Z., Rahman, S. and Jahan, R., 2012. Antihyperglycemic and antinociceptive activities of methanolic extract of *Euphorbia thymifolia* L. whole plants. *Journal of Chinese Integrative Medicine*, 10(2): 228–232.
74. Amaral, A.C.F., Kuster, R.M., Gonçalves, J.L.S. and Wigg, M.D., 1999. Antiviral investigation on the flavonoids of *Chamaesyce thymifolia. Fitoterapia*, 70(3), 293–295.
75. Lee, S.H., Tanaka, T., Nonaka, G.I. and Nishioka, I., 1990. Hydrolysable tannins from *Euphorbia thymifolia. Phytochemistry*, 29(11), 3621–3625.
76. Yoshikawa, M., Wang, T., Morikawa, T., Xie, H. and Matsuda, H., 2007. Bioactive constituents from Chinese natural medicines. XXIV. Hypoglycemic effects of *Sinocrassula indica* in sugar-loaded rats and genetically diabetic KK-Ay mice and structures of new acylated flavonol glycosides, sinocrassosides A1, A2, B1, and B2. *Chemical and Pharmaceutical Bulletin*, 55(9), 1308–1315.
77. Wright, E.M., Martín, M.G. and Turk, E., 2003. Intestinal absorption in health and disease—sugars. *Best Practice & Research Clinical Gastroenterology*, 17(6), 943–956.
78. Hediger, M.A., Turk, E. and Wright, E.M., 1989. Homology of the human intestinal Na+/glucose and *Escherichia coli* Na+/proline cotransporters. *Proceedings of the National Academy of Sciences*, 86(15), 5748–5752.
79. Wright, E.M., Martín, M.G. and Turk, E., 2003. Intestinal absorption in health and disease—sugars. *Best Practice & Research Clinical Gastroenterology*, 17(6), 943–956.
80. Islam, M.N., Jung, H.A., Sohn, H.S., Kim, H.M. and Choi, J.S., 2013. Potent α-glucosidase and protein tyrosine phosphatase 1B inhibitors from *Artemisia capillaris. Archives of Pharmacal Research*, 36(5), 542–552.
81. Sabu, M.C. and Kuttan, R., 2009. Antidiabetic and antioxidant activity of *Terminalia belerica.* Roxb. *Indian Journal of Experimental Biology*, 47, 270–275.
82. Makihara, H., Shimada, T., Machida, E., Oota, M., Nagamine, R., Tsubata, M., Kinoshita, K., Takahashi, K. and Aburada, M., 2012. Preventive effect of *Terminalia bellirica* on obesity and metabolic disorders in spontaneously obese type 2 diabetic model mice. *Journal of Natural Medicines*, 66(3), 459–467.
83. Wansi, J.D., Lallemand, M.C., Chiozem, D.D., Toze, F.A.A., Mbaze, L.M.A., Naharkhan, S., Iqbal, M.C., Tillequin, F., Wandji, J. and Fomum, Z.T., 2007. α-Glucosidase inhibitory constituents from stem bark of *Terminalia superba* (Combretaceae). *Phytochemistry*, 68(15), 2096–2100.
84. Bruneton, J., 1987. *Elements de phytochimie et de pharmacognosy.* Paris, France: Lavoisier.
85. Madhavi, D.L., Bomser, J., Smith, M.A.L. and Singletary, K., 1998. Isolation of bioactive constituents from *Vaccinium myrtillus* (bilberry) fruits and cell cultures. *Plant Science*, 131(1), 95–103.
86. Tadera, K., Minami, Y., Takamatsu, K. and Matsuoka, T., 2006. Inhibition of α-glucosidase and α-amylase by flavonoids. *Journal of Nutritional Science and Vitaminology*, 52(2), 149–153.
87. Rudel, L.L., Lee, R.G. and Cockman, T.L., 2001. Acyl coenzyme A: Cholesterol acyltransferase types 1 and 2: Structure and function in atherosclerosis. *Current Opinion in Lipidology*, 12(2), 121–127.
88. Liang, Y., Chen, J., Zuo, Y., Ma, K.Y., Jiang, Y., Huang, Y. and Chen, Z.Y., 2013. Blueberry anthocyanins at doses of 0.5 and 1% lowered plasma cholesterol by increasing fecal excretion of acidic and neutral sterols in hamsters fed a cholesterol-enriched diet. *European Journal of Nutrition*, 52(3), 869–875.
89. Martínez, I., Wallace, G., Zhang, C., Legge, R., Benson, A.K., Carr, T.P., Moriyama, E.N. and Walter, J., 2009. Diet-induced metabolic improvements in a hamster model of hypercholesterolemia are strongly linked to alterations of the gut microbiota. *Applied and Environmental Microbiology*, 75(12), 4175–4184.
90. Ichiyanagi, T., Shida, Y., Rahman, M.M., Hatano, Y. and Konishi, T., 2006. Bioavailability and tissue distribution of anthocyanins in bilberry (*Vaccinium myrtillus* L.) extract in rats. *Journal of Agricultural and Food Chemistry*, 54(18), 6578–6587.
91. Güder, A., Gür, M. and Engin, M.S., 2015. Antidiabetic and antioxidant properties of bilberry (*Vaccinium myrtillus Linn.*) fruit and their chemical composition. *Journal of Agricultural Science and Technology*, 17(2), 387–400.

92. Garcia, F., 1941. Distribution and deterioration of insulin-like principle in *Lagerstroemia speciosa* (banaba). *Acta Med Philippina*, 3, 99–104.
93. Saumya, S.M. and Basha, P.M., 2011. Antioxidant effect of *Lagerstroemia speciosa Pers* (Banaba) leaf extract in streptozotocin-induced diabetic mice. *Indian Journal of Experimental Biology*, 49(2), 125–131.
94. Takagi, S., Miura, T., Ishihara, E., Ishida, T., and Chinzei, Y., 2010. Effect of corosolic acid on dietary hypercholesterolemia and hepatic steatosis in KK-Ay diabetic mice. *Biomedical Research*, 31(4): 213–218.
95. Hou, W., Li, Y., Zhang, Q., Wei, X., Peng, A., Chen, L. and Wei, Y., 2009. Triterpene acids isolated from *Lagerstroemia speciosa* leaves as α-glucosidase inhibitors. *Phytotherapy Research*, 23(5), 614–618.
96. Buchholz, T. and Melzig, M.F., 2016. Medicinal plants traditionally used for treatment of obesity and diabetes mellitus–screening for pancreatic lipase and α-amylase inhibition. *Phytotherapy Research*, 30(2), 260–266.
97. Bellesia, A., Verzelloni, E. and Tagliazucchi, D., 2015. Pomegranate ellagitannins inhibit α-glucosidase activity in vitro and reduce starch digestibility under simulated gastro-intestinal conditions. *International Journal of Food Sciences and Nutrition*, 66(1), 85–92.
98. Kawakami, K., Li, P., Uraji, M., Hatanaka, T. and Ito, H., 2014. Inhibitory effects of pomegranate extracts on recombinant human maltase–glucoamylase. *Journal of Food Science*, 79(9), H1848–H1853.
99. Yasuda, M., Yasutake, K., Hino, M., Ohwatari, H., Ohmagari, N., Takedomi, K., Tanaka, T. and Nonaka, G.I., 2014. Inhibitory effects of polyphenols from water chestnut (Trapa japonica) husk on glycolytic enzymes and postprandial blood glucose elevation in mice. *Food Chemistry*, 165, 42–49.
100. Gupta, S., Sharma, S.B., Prabhu, K.M. and Bansal, S.K., 2009. Protective role of *Cassia auriculata* leaf extract on hyperglycemia-induced oxidative stress and its safety evaluation. *Indian Journal of Biochemistry & Biophysics,* 46(5), 371–377.
101. Habtemariam, S., 2013. Antihyperlipidemic components of *Cassia auriculata* aerial parts: Identification through in vitro studies. *Phytotherapy Research*, 27(1), 152–155.
102. Dendup, T., Prachyawarakorn, V., Pansanit, A., Mahidol, C., Ruchirawat, S. and Kittakoop, P., 2014. α-Glucosidase inhibitory activities of isoflavanones, isoflavones, and pterocarpans from *Mucuna pruriens*. *Planta Medica*, 80(7), 604–608.
103. Abesundara, K.J., Matsui, T. and Matsumoto, K., 2004. α-Glucosidase inhibitory activity of some Sri Lanka plant extracts, one of which, *Cassia auriculata*, exerts a strong antihyperglycemic effect in rats comparable to the therapeutic drug acarbose. *Journal of Agricultural and Food Chemistry*, 52(9), 2541–2545.
104. Maurya, R., Ray, A.B., Duah, F.K., Slatkin, D.J. and Schiff Jr, P.L., 1984. Constituents of *Pterocarpus marsupium*. *Journal of Natural Products*, 47(1), 179–181.
105. Bai, Y.P., Zhang, G.G., Shi, R.Z., Li, Y.J., Tan, G.S. and Chen, J., 2006. Inhibitory effect of reinioside C on LOX-1 expression induced by OX-LDL. *Journal of Central South University Medical Sciences*, 31(5), 659–662.
106. Ros, E., 2000. Intestinal absorption of triglyceride and cholesterol. Dietary and pharmacological inhibition to reduce cardiovascular risk. *Atherosclerosis*, 151(2), 357–379.
107. Li, H., Wang, Q.J., Zhu, D.N. and Yang, Y., 2008. Reinioside C, a triterpene saponin of *Polygala aureocauda* Dunn, exerts hypolipidemic effect on hyperlipidemic mice. *Phytotherapy Research*, 22(2), 159–164.
108. Kawaguchi, K., Mizuno, T., Aida, K. and Uchino, K., 1997. Hesperidin as an inhibitor of lipases from porcine pancreas and Pseudomonas. *Bioscience, Biotechnology, and Biochemistry*, 61(1), 102–104.
109. Uvarani, C., Sankaran, M., Jaivel, N., Chandraprakash, K., Ata, A. and Mohan, P.S., 2013. Bioactive dimeric carbazole alkaloids from *Murraya koenigii*. *Journal of Natural Products*, 76(6), 993–1000.
110. Birari, R., Javia, V. and Bhutani, K.K., 2010. Antiobesity and lipid lowering effects of *Murraya koenigii* (L.) Spreng leaves extracts and mahanimbine on high-fat diet induced obese rats. *Fitoterapia*, 81(8), 1129–1133.
111. Birari, R., Roy, S.K., Singh, A. and Bhutani, K.K., 2009. Pancreatic lipase inhibitory alkaloids of *Murraya koenigii* leaves. *Natural Product Communications*, 4(8), 1089–1092.
112. Park, Y.D., Lee, W.S., An, S. and Jeong, T.S., 2007. Human acyl-CoA: Cholesterol acyltransferase inhibitory activities of aliphatic acid amides from *Zanthoxylum piperitum* DC. *Biological and Pharmaceutical Bulletin*, 30(1), 205–207.
113. Giordani, M.A., Collicchio, T.C.M., Ascêncio, S.D., de Oliveira Martins, D.T., Balogun, S.O., Bieski, I.G.C., da Silva, L.A., Colodel, E.M., de Souza, R.L., de Souza, D.L.P. and de França, S.A., 2015. Hydroethanolic extract of the inner stem bark of *Cedrela odorata* has low toxicity and reduces hyperglycemia induced by an overload of sucrose and glucose. *Journal of Ethnopharmacology*, 162, 352–361.

114. Prashanth, D., Amit, A., Samiulla, D.S., Asha, M.K. and Padmaja, R., 2001. α-Glucosidase inhibitory activity of *Mangifera indica* bark. *Fitoterapia*, 72(6), 686–688.
115. Uddin, G., Rauf, A., Al-Othman, A.M., Collina, S., Arfan, M., Ali, G. and Khan, I., 2012. Pistagremic acid, a glucosidase inhibitor from *Pistacia integerrima*. *Fitoterapia*, 83(8), 1648–1652.
116. Kasabri, V., Afifi, F.U. and Hamdan, I., 2011. In vitro and in vivo acute antihyperglycemic effects of five selected indigenous plants from Jordan used in traditional medicine. *Journal of Ethnopharmacology*, 133(2), 888–896.
117. Williams, J.A., Choe, Y.S., Noss, M.J., Baumgartner, C.J. and Mustad, V.A., 2007. Extract of Salacia oblonga lowers acute glycemia in patients with type 2 diabetes. *The American Journal of Clinical Nutrition*, 86(1), 124–130.
118. Li, Y., Peng, G., Li, Q., Wen, S., Huang, T.H.W., Roufogalis, B.D. and Yamahara, J., 2004. *Salacia oblonga* improves cardiac fibrosis and inhibits postprandial hyperglycemia in obese zucker rats. *Life Sciences*, 75(14), 1735–1746.
119. Matsuda, H., Murakami, T., Yashiro, K., Yamahara, J. and Yoshikawa, M., 1999. Antidiabetic principles of natural medicines. IV. Aldose reductase and α-glucosidase inhibitors from the roots of *Salacia oblonga* wall. (Celastraceae): Structure of a new friedelane-type triterpene, kotalagenin 16-acetate. *Chemical and Pharmaceutical Bulletin*, 47(12), 1725–1729.
120. Karunanayake, E.H., Welihinda, J., Sirimanne, S.R. and Adorai, G.S., 1984. Oral hypoglycaemic activity of some medicinal plants of Sri Lanka. *Journal of Ethnopharmacology*, 11(2), 223–231.
121. Yoshikawa, M., Shimoda, H., Nishida, N., Takada, M. and Matsuda, H., 2002. *Salacia reticulata* and its polyphenolic constituents with lipase inhibitory and lipolytic activities have mild antiobesity effects in rats. *The Journal of Nutrition*, 132(7), 1819–1824.
122. Yoshino, K., Miyauchi, Y., Kanetaka, T., Takagi, Y. and Koga, K., 2009. Anti-diabetic activity of a leaf extract prepared from *Salacia reticulata* in mice. *Bioscience, Biotechnology, and Biochemistry*, 73(5), 1096–1104.
123. Avcı, G., Kupeli, E., Eryavuz, A., Yesilada, E. and Kucukkurt, I., 2006. Antihypercholesterolaemic and antioxidant activity assessment of some plants used as remedy in Turkish folk medicine. *Journal of Ethnopharmacology*, 107(3), 418–423.
124. Fukunaga, T., Kajikawa, I., Nishiya, K., Watanabe, Y., Takeya, K. and Itokawa, H., 1987. Studies on the constituents of the European mistletoe, *Viscum album* L. *Chemical and Pharmaceutical Bulletin*, 35(8), 3292–3297.
125. Wollenweber, E., Wieland, A. and Haas, K., 2000. Epicuticular waxes and flavonol aglycones of the European mistletoe, *Viscum album* L. *Zeitschrift Für Naturforschung C*, 55(5–6), 314–317.
126. Iwai, K., Onodera, A. and Matsue, H., 2004. Inhibitory effects of *Viburnum dilatatum* Thunb.(gamazumi) on oxidation and hyperglycemia in rats with streptozotocin-induced diabetes. *Journal of Agricultural and Food Chemistry*, 52(4), 1002–1007.
127. Iwai, K., Kim, M.Y., Onodera, A. and Matsue, H., 2006. α-Glucosidase Inhibitory and antihyperglycemic effects of polyphenols in the fruit of *Viburnum dilatatum* Thunb. *Journal of Agricultural and Food Chemistry*, 54(13), 4588–4592.
128. Wu, T., Yu, Z., Tang, Q., Song, H., Gao, Z., Chen, W. and Zheng, X., 2013. Honeysuckle anthocyanin supplementation prevents diet-induced obesity in C57BL/6 mice. *Food & Function*, 4(11), 1654–1661.
129. Gau, S.W., Chen, C.C., Chen, Y.P. and Hsu, H.Y., 1983. Constituents of Ilex species I. On the leaves of *Ilex cornuta* Lindl. *Journal of the Chinese Chemical Society*, 30(3), 185–188.
130. Im, K.R., Jeong, T.S., Kwon, B.M., Baek, N.I., Kim, S.H. and Kim, D.K., 2006. Acyl-CoA: Cholesterol acyltransferase inhibitors from *Ilex macropoda*. *Archives of Pharmacal Research*, 29(3), 191–194.
131. Baek, M.Y., Cho, J.G., Lee, D.Y., Ann, E.M., Jeong, T.S. and Baek, N.I., 2010. Isolation of triterpenoids from the stem bark of *Albizia julibrissin* and their inhibition activity on ACAT-1 and ACAT-2. *Journal of the Korean Society for Applied Biological Chemistry*, 53(3), 310–315.
132. Wang, Z.B., Jiang, H., Xia, Y.G., Yang, B.Y. and Kuang, H.X., 2012. α-Glucosidase inhibitory constituents from *Acanthopanax senticosus* harm leaves. *Molecules*, 17(6), 6269–6276.
133. Li, F., Li, W., Fu, H., Zhang, Q. and Koike, K., 2007. Pancreatic lipase-inhibiting triterpenoid saponins from fruits of *Acanthopanax senticosus*. *Chemical and Pharmaceutical Bulletin*, 55(7), 1087–1089.
134. Li, J.L., Li, N., Lee, H.S., Xing, S.S., Qi, S.Z., Tuo, Z.D., Zhang, L., Wang, X.B. and Cui, L., 2016. Inhibition effect of neo-lignans from *Eleutherococcus senticosus* (Rupt. & maxim.) Maxim on diacylglycerol acyltransferase (DGAT). *Phytochemistry Letters*, 15, 147–151.
135. Durrington, P., 2007. *Hyperlipidaemia: Diagnosis and Management*, 3rd ed. Boca Raton, FL: CRC Press.

136. Chan, H.H., Sun, H.D., Reddy, M.V.B. and Wu, T.S., 2010. Potent α-glucosidase inhibitors from the roots of *Panax japonicus* CA Meyer var. major. *Phytochemistry*, 71(11), 1360–1364.
137. Supkamonseni, N., Thinkratok, A., Meksuriyen, D. and Srisawat, R., 2014. Hypolipidemic and hypoglycemic effects of *Centella asiatica* (L.) extract in vitro and in vivo. *Indian Journal of Experimental Biology*, 52(10), 965–971.
138. Mo, E.J., Yang, H.J., Jeong, J.Y., Kim, S.B., Liu, Q., Hwang, B.Y. and Lee, M.K., 2016. Pancreatic lipase inhibitory phthalide derivatives from the rhizome of *Cnidium officinale*. *Records of Natural Products*, 10(2), 148.
139. Brindis, F., Rodríguez, R., Bye, R., González-Andrade, M. and Mata, R., 2010. (Z)-3-Butylidenephthalide from *Ligusticum porteri*, an α-glucosidase inhibitor. *Journal of Natural Products*, 74(3), 314–320.
140. Shalaby, N.M., Abd-Alla, H.I., Aly, H.F., Albalawy, M.A., Shaker, K.H. and Bouajila, J., 2014. Preliminary in vitro and in vivo evaluation of antidiabetic activity of *Ducrosia anethifolia* Boiss. and its linear furanocoumarins. *BioMed Research International*, 2014, 13.
141. Gupta, R.K., Kesari, A.N., Murthy, P.S., Chandra, R., Tandon, V. and Watal, G., 2005. Hypoglycemic and antidiabetic effect of ethanolic extract of leaves of *Annona squamosa* L. in experimental animals. *Journal of Ethnopharmacology*, 99(1), 75–81.
142. Nukitrangsan, N., Okabe, T., Toda, T., Inafuku, M., Iwasaki, H. and Oku, H., 2012. Effect of *Peucedanum japonicum* Thunb extract on high-fat diet-induced obesity and gene expression in mice. *Journal of Oleo Science*, 61(2), 89–101.
143. Lee, J.S., Choi, M.S., Seo, K.I., Lee, J., Lee, H.I., Lee, J.H., Kim, M.J. and Lee, M.K., 2014. Platycodi radix saponin inhibits α-glucosidase in vitro and modulates hepatic glucose-regulating enzyme activities in C57BL/KsJ-db/db mice. *Archives of Pharmacal Research*, 37(6), 773–782.
144. Owen, J.B., Treasure, J.L. and Collier, D.A. (Eds.), 2001. *Animal Models-Disorders of Eating Behaviour and Body Composition*. Dordrecht, the Netherlands: Kluwer Academic Publishers.
145. Hamza, N., Berke, B., Cheze, C., Le Garrec, R., Lassalle, R., Agli, AN., Robinson, P., Gin, H. and Moore, N., 2011. Treatment of high-fat diet induced type 2 diabetes in C57BL/6J mice by two medicinal plants used in traditional treatment of diabetes in the east of Algeria. *Journal of Ethnopharmacology*, 133(2), 931–933.
146. Boudjelal, A., Siracusa, L., Henchiri, C., Sarri, M., Abderrahim, B., Baali, F. and Ruberto, G., 2015. Antidiabetic effects of aqueous infusions of artemisia herba-alba and ajuga iva in alloxan-induced diabetic rats. *Planta Medica*, 81(9), 696–704.
147. Takahashi, T. and Miyazawa, M., 2012. Potent α-glucosidase inhibitors from safflower (*Carthamus tinctorius* L.) seed. *Phytotherapy Research*, 26(5), 722–726.
148. Wafo, P., Kamdem, R.S., Ali, Z., Anjum, S., Begum, A., Oluyemisi, O.O., Khan, S.N., Ngadjui, B.T., Etoa, X.F. and Choudhary, M.I., 2011. Kaurane-type diterpenoids from *Chromoleana odorata*, their X-ray diffraction studies and potent α-glucosidase inhibition of 16-kauren-19-oic acid. *Fitoterapia*, 82(4), 642–646.
149. Zareen, S., Choudhary, M.I., Akhtar, M.N. and Khan, S.N., 2008. α-Glucosidase inhibitory activity of triterpenoids from *Cichorium intybus*. *Journal of Natural Products*, 71(5), 910–913.
150. Trautwein, E.A., Rieckhoff, D. and Erbersdobler, H.F., 1998. Dietary inulin lowers plasma cholesterol and triacylglycerol and alters biliary bile acid profile in hamsters. *The Journal of Nutrition*, 128(11), 1937–1943.
151. Roberfroid, M.B. and Delzenne, N.M., 1998. Dietary fructans. *Annual Review of Nutrition*, 18(1), 117–143.
152. Nguyen T.D., Le Hoang, T., Nguyen T.L., Pham T.B., Tran T.H.H., Chau V.M. and Nguyen H.D., 2013. Inhibitors of a-glucosidase, a-amylase and lipase from *Chrysanthemum morifolium*, 54(1), 74–77.
153. Shimoda, H., Ninomiya, K., Nishida, N., Yoshino, T., Morikawa, T., Matsuda, H. and Yoshikawa, M., 2003. Anti-hyperlipidemic sesquiterpenes and new sesquiterpene glycosides from the leaves of artichoke (*Cynara scolymus* L.): Structure requirement and mode of action. *Bioorganic & Medicinal Chemistry Letters*, 13(2), 223–228.
154. Villiger, A., Sala, F., Suter, A. and Butterweck, V., 2015. In vitro inhibitory potential of *Cynara scolymus*, *Silybum marianum*, *Taraxacum officinale*, and *Peumus boldus* on key enzymes relevant to metabolic syndrome. *Phytomedicine*, 22(1), 138–144.
155. Rondanelli, M., Opizzi, A., Faliva, M., Sala, P., Perna, S., Riva, A., Morazzoni, P., Bombardelli, E. and Giacosa, A., 2014. Metabolic management in overweight subjects with naive impaired fasting glycaemia by means of a highly standardized extract from *Cynara scolymus*: A double-blind, placebo-controlled, randomized clinical trial. *Phytotherapy Research*, 28(1), 33–41.

156. Ooi, K.L., Muhammad, T.S.T., Tan, M.L. and Sulaiman, S.F., 2011. Cytotoxic, apoptotic and anti-α-glucosidase activities of 3, 4-di-O-caffeoyl quinic acid, an antioxidant isolated from the polyphenolic-rich extract of *Elephantopus mollis* Kunth. *Journal of Ethnopharmacology*, 135(3), 685–695.
157. Arsiningtyas, I.S., Gunawan-Puteri, M.D., Kato, E. and Kawabata, J., 2014. Identification of α-glucosidase inhibitors from the leaves of *Pluchea indica* (L.) Less., a traditional Indonesian herb: Promotion of natural product use. *Natural Product Research*, 28(17), 1350–1353.
158. Villiger, A., Sala, F., Suter, A. and Butterweck, V., 2015. In vitro inhibitory potential of *Cynara scolymus*, *Silybum marianum*, *Taraxacum officinale*, and *Peumus boldus* on key enzymes relevant to metabolic syndrome. *Phytomedicine*, 22(1), 138–144.
159. Huseini, H.F., Larijani, B., Heshmat, R.A., Fakhrzadeh, H., Radjabipour, B., Toliat, T. and Raza, M., 2006. The efficacy of *Silybum marianum* (L.) Gaertn.(silymarin) in the treatment of type II diabetes: A randomized, double-blind, placebo-controlled, clinical trial. *Phytotherapy Research*, 20(12), 1036–1039.
160. Ekanem, A.P., Wang, M., Simon, J.E. and Moreno, D.A., 2007. Antiobesity properties of two African plants (*Afromomum meleguetta* and *Spilanthes acmella*) by pancreatic lipase inhibition. *Phytotherapy Research*, 21(12), 1253–1255.
161. Prachayasittikul, V., Prachayasittikul, S., Ruchirawat, S. and Prachayasittikul, V., 2013. High therapeutic potential of *Spilanthes acmella*: A review. *EXCLI Journal*, 12, 291.
162. Ramsewak, R.S., Erickson, A.J. and Nair, M.G., 1999. Bioactive N-isobutylamides from the flower buds of *Spilanthes acmella*. *Phytochemistry*, 51(6), 729–732.
163. Zhang, J., Kang, M.J., Kim, M.J., Kim, M.E., Song, J.H., Lee, Y.M. and Kim, J.I., 2008. Pancreatic lipase inhibitory activity of *Taraxacum officinale* in vitro and in vivo. *Nutrition Research and Practice*, 2(4), 200–203.
164. Williams, C.A., Goldstone, F. and Greenham, J., 1996. Flavonoids, cinnamic acids and coumarins from the different tissues and medicinal preparations of *Taraxacum officinale*. *Phytochemistry*, 42(1), 121–127.
165. Schütz, K., Carle, R. and Schieber, A., 2006. Taraxacum—a review on its phytochemical and pharmacological profile. *Journal of Ethnopharmacology*, 107(3), 313–323.
166. Gao, H., Huang, Y.N., Gao, B., Xu, P.Y., Inagaki, C. and Kawabata, J., 2008. α-Glucosidase inhibitory effect by the flower buds of *Tussilago farfara* L. *Food Chemistry*, 106(3), 1195–1201.
167. Park, H.R., Yoo, M.Y., Seo, J.H., Kim, I.S., Kim, N.Y., Kang, J.Y., Cui, L., Lee, C.S., Lee, C.H. and Lee, H.S., 2008. Sesquiterpenoids isolated from the flower buds of *Tussilago farfara* L. inhibit diacylglycerol acyltransferase. *Journal of Agricultural and Food Chemistry*, 56(22), 10493–10497.
168. Culvenor, C.C.J., Edgar, J.A., Smith, L.W. and Hirono, I., 1976. The occurrence of senkirkine in *Tussilago farfara*. *Australian Journal of Chemistry*, 29(1), 229–230.
169. Leung, W.Y., Thomas, G.N., Chan, J.C. and Tomlinson, B., 2003. Weight management and current options in pharmacotherapy: Orlistat and sibutramine. *Clinical Therapeutics*, 25(1), 58–80.
170. Lee, I.A., Lee, J.H., Baek, N.I. and Kim, D.H., 2005. Antihyperlipidemic effect of crocin isolated from the fructus of *Gardenia jasminoides* and its metabolite crocetin. *Biological and Pharmaceutical Bulletin*, 28(11), 2106–2110.
171. Sheng, L., Qian, Z., Zheng, S. and Xi, L., 2006. Mechanism of hypolipidemic effect of crocin in rats: Crocin inhibits pancreatic lipase. *European Journal of Pharmacology*, 543(1), 116–122.
172. Wang, Z.W., Wang, J.S., Luo, J. and Kong, L.Y., 2013. α-glucosidase inhibitory triterpenoids from the stem barks of *Uncaria laevigata*. *Fitoterapia*, 90, 30–37.
173. Wan, L.S., Min, Q.X., Wang, Y.L., Yue, Y.D. and Chen, J.C., 2013. Xanthone glycoside constituents of *Swertia kouitchensis* with α-glucosidase inhibitory activity. *Journal of Natural Products*, 76(7), 1248–1253.
174. Wan, L.S., Chen, C.P., Xiao, Z.Q., Wang, Y.L., Min, Q.X., Yue, Y. and Chen, J., 2013. In vitro and in vivo anti-diabetic activity of *Swertia kouitchensis* extract. *Journal of Ethnopharmacology*, 147(3), 622–630.
175. Mir, S.R., Ahamad, J., Hassan, N. and Amin, S., 2014. Management of post prandial hyperglycemia with swertiamarin isolated from *Enicostemma littorale* Blume. *Planta Medica*, 80(10), PG13.
176. Arai, H., Hirasawa, Y., Rahman, A., Kusumawati, I., Zaini, N.C., Sato, S., Aoyama, C., Takeo, J. and Morita, H., 2010. Alstiphyllanines E–H, picraline and ajmaline-type alkaloids from *Alstonia macrophylla* inhibiting sodium glucose cotransporter. *Bioorganic & Medicinal Chemistry*, 18(6), 2152–2158.
177. Elya, B., Basah, K., Mun'im, A., Yuliastuti, W., Bangun, A. and Septiana, E.K., 2011. Screening of α-glucosidase inhibitory activity from some plants of *Apocynaceae*, *Clusiaceae*, *Euphorbiaceae*, and *Rubiaceae*. *BioMed Research International*, 2012, 6.
178. Begum, S., Syed, S.A., Siddiqui, B.S., Sattar, S.A. and Choudhary, M.I., 2013. Carandinol: First isohopane triterpene from the leaves of *Carissa carandas* L. and its cytotoxicity against cancer cell lines. *Phytochemistry Letters*, 6(1), 91–95.

179. Reddy, R.M.I., Latha, P.B., Vijaya, T. and Rao, D.S., 2012. The saponin-rich fraction of a *Gymnema sylvestre* R. Br. aqueous leaf extract reduces cafeteria and high-fat diet-induced obesity. *Zeitschrift Für Naturforschung C*, 67(1–2), 39–46.
180. Wang, Y., Dawid, C., Kottra, G., Daniel, H. and Hofmann, T., 2014. Gymnemic acids inhibit sodium-dependent glucose transporter 1. *Journal of Agricultural and Food Chemistry*, 62(25), 5925–5931.
181. Baskaran, K., Ahamath, B.K., Shanmugasundaram, K.R. and Shanmugasundaram, E.R.B., 1990. Antidiabetic effect of a leaf extract from *Gymnema sylvestre* in non-insulin-dependent diabetes mellitus patients. *Journal of Ethnopharmacology*, 30(3), 295–305.
182. Ali, K.M., Chatterjee, K., De, D., Jana, K., Bera, T.K. and Ghosh, D., 2011. Inhibitory effect of hydro-methanolic extract of seed of *Holarrhena antidysenterica* on alpha-glucosidase activity and postprandial blood glucose level in normoglycemic rat. *Journal of Ethnopharmacology*, 135(1), 194–196.
183. Kusano, S., 2000. Antidiabetic activity of white skinned sweet potato (*Ipomoea batatas* L.) in obese zucker fatty rats. *Biological and Pharmaceutical Bulletin*, 23(1), 23–26.
184. Ludvik, B., Neuffer, B. and Pacini, G., 2004. Efficacy of *Ipomoea batatas* (Caiapo) on diabetes control in type 2 diabetic subjects treated with diet. *Diabetes Care*, 27(2), 436–440.
185. Matsui, T., Ebuchi, S., Kobayashi, M., Fukui, K., Sugita, K., Terahara, N. and Matsumoto, K., 2002. Anti-hyperglycemic effect of diacylated anthocyanin derived from *Ipomoea batatas* cultivar Ayamurasaki can be achieved through the α-glucosidase inhibitory action. *Journal of Agricultural and Food Chemistry*, 50(25), 7244–7248.
186. Malalavidhane, T.S., Wickramasinghe, S.N. and Jansz, E.R., 2000. Oral hypoglycaemic activity of *Ipomoea aquatica*. *Journal of Ethnopharmacology*, 72(1), 293–298.
187. Malalavidhane, T.S., Wickramasinghe, S.M.D.N., Perera, M.S.A. and Jansz, E.R., 2003. Oral hypoglycaemic activity of *Ipomoea aquatica* in streptozotocin-induced, diabetic wistar rats and type II diabetics. *Phytotherapy Research*, 17(9), 1098–1100.
188. Sokeng, S.D., Rokeya, B., Hannan, J.M.A., Junaida, K., Zitech, P., Ali, L., Ngounou, G., Lontsi, D. and Kamtchouing, P., 2007. Inhibitory effect of *Ipomoea aquatica* extracts on glucose absorption using a perfused rat intestinal preparation. *Fitoterapia*, 78(7), 526–529.
189. Charnock, J.S., 2000. Gamma-linolenic acid provides additional protection against ventricular fibrillation in aged rats fed linoleic acid rich diets. *Prostaglandins, Leukotrienes and Essential Fatty Acids (PLEFA)*, 62(2), 129–134.
190. Kavanagh, K., Flynn, D.M., Jenkins, K.A., Wilson, M.D. and Chilton, F.H., 2013. Stearidonic and γ-linolenic acids in Echium oil improves glucose disposal in insulin resistant monkeys. *Prostaglandins, Leukotrienes and Essential Fatty Acids (PLEFA)*, 89(1), 39–45.
191. Horrobin, D.F., 1993. Fatty acid metabolism in health and disease: The role of delta-6-desaturase. *The American Journal of Clinical Nutrition*, 57(5), 732S–736S.
192. Murugesh, K., Yeligar, V., Dash, D.K., Sengupta, P., Maiti, B.C. and Maity, T.K., 2006. Antidiabetic, antioxidant and antihyperlipidemic status of *Heliotropium zeylanicum* extract on streptozotocin-induced diabetes in rats. *Biological and Pharmaceutical Bulletin*, 29(11), 2202–2205.
193. Schoental, R. and FRAYN, K.N., 1976. The effects of the pyrrolizidine alkaloid, heliotrine, and nicotinamide on the concentration of plasma insulin and blood glucose in the rat.
194. An, S., Park, Y.D., Paik, Y.K., Jeong, T.S. and Lee, W.S., 2007. Human ACAT inhibitory effects of shikonin derivatives from *Lithospermum erythrorhizon*. *Bioorganic & Medicinal Chemistry Letters*, 17(4), 1112–1116.
195. Accad, M., Smith, S.J., Newland, D.L., Sanan, D.A., King, L.E., Linton, M.F., Fazio, S. and Farese, R.V., 2000. Massive xanthomatosis and altered composition of atherosclerotic lesions in hyperlipidemic mice lacking acyl CoA: Cholesterol acyltransferase 1. *The Journal of Clinical Investigation*, 105(6), 711–719.
196. Komaki, E., Yamaguchi, S., Isafumi, M.A.R.U., Kinoshita, M., Kakehi, K. and Tsukada, Y., 2003. Identification of anti-α-amylase components from olive leaf extracts. *Food Science and Technology Research*, 9(1), 35–39.
197. Aparna, P., Tiwari, A.K., Srinivas, P., Ali, A.Z., Anuradha, V. and Rao, J.M., 2009. Dolichandroside A, a new α-glucosidase inhibitor and DPPH free-radical Scavenger from *Dolichandrone falcata* seem. *Phytotherapy Research*, 23(4), 591–596.
198. Tadera, K., Minami, Y., Takamatsu, K. and Matsuoka, T., 2006. Inhibition of α-glucosidase and α-amylase by flavonoids. *Journal of Nutritional Science and Vitaminology*, 52(2), 149–153.
199. Rani, M.P., Raghu, K.G., Nair, M.S. and Padmakumari, K.P., 2014. Isolation and identification of α-glucosidase and protein glycation inhibitors from *Stereospermum colais*. *Applied Biochemistry and Biotechnology*, 173(4), 946–956.

200. Dat, N.T., Dang, N.H. and Thanh, L.N., 2016. New flavonoid and pentacyclic triterpene from *Sesamum indicum* leaves. *Natural Product Research*, 30(3), 311–315.
201. Gao, H., Huang, Y.N., Gao, B., Li, P., Inagaki, C. and Kawabata, J., 2008. Inhibitory effect on α-glucosidase by *Adhatoda vasica* Nees. *Food Chemistry*, 108(3), 965–972.
202. Liu, Q., Hu, H.J., Li, P.F., Yang, Y.B., Wu, L.H., Chou, G.X. and Wang, Z.T., 2014. Diterpenoids and phenylethanoid glycosides from the roots of *Clerodendrum bungei* and their inhibitory effects against angiotensin converting enzyme and α-glucosidase. *Phytochemistry*, 103, 196–202.
203. Iqbal, K., Malik, A., Mukhtar, N., Anis, I., Khan, S.N. and Choudhary, M.I., 2004. α-Glucosidase inhibitory constituents from *Duranta repens*. *Chemical and Pharmaceutical Bulletin*, 52(7), 785–789.
204. Ayinampudi, S.R., Domala, R., Merugu, R., Bathula, S. and Janaswamy, M.R., 2012. New icetexane diterpenes with intestinal α-glucosidase inhibitory and free-radical scavenging activity isolated from *Premna tomentosa* roots. *Fitoterapia*, 83(1), 88–92.
205. Ramachandran, S. and Rajasekaran, A., 2014. Blood glucose-lowering effect of *Tectona grandis* flowers in type 2 diabetic rats: A study on identification of active constituents and mechanisms for antidiabetic action. *Journal of Diabetes*, 6(5), 427–437.
206. Singh, P.P., Jha, S. and Irchhaiya, R., 2012. Antidiabetic and antioxidant activity of hydroxycinnamic acids from *Calamintha officinalis* Moench. *Medicinal Chemistry Research*, 21(8), 1717–1721.
207. Jabeen, B., Riaz, N., Saleem, M., Naveed, M.A., Ashraf, M., Alam, U., Rafiq, H.M., Tareen, R.B. and Jabbar, A., 2013. Isolation of natural compounds from *Phlomis stewartii* showing α-glucosidase inhibitory activity. *Phytochemistry*, 96, 443–448.
208. Ma, H.Y., Gao, H.Y., Sun, L., Huang, J., Xu, X.M. and Wu, L.J., 2011. Constituents with α-glucosidase and advanced glycation end-product formation inhibitory activities from *Salvia miltiorrhiza* Bge. *Journal of Natural Medicines*, 65(1), 37–42.
209. Matsuura, H., Miyazaki, H., Asakawa, C., Amano, M., Yoshihara, T. and Mizutani, J., 2004. Isolation of α-glusosidase inhibitors from hyssop (*Hyssopus officinalis*). *Phytochemistry*, 65(1), 91–97.
210. Rizza, R.A., Gerich, J.E., Haymond, M.W., Westland, R.E., Hall, L.D., Clemens, A.H. and Service, F.J., 1980. Control of blood sugar in insulin-dependent diabetes: Comparison of an artificial endocrine pancreas, continuous subcutaneous insulin infusion, and intensified conventional insulin therapy. *New England Journal of Medicine*, 303(23), 1313–1318.
211. Heacock, P.M., Hertzler, S.R., Williams, J.A. and Wolf, B.W., 2005. Effects of a medical food containing an herbal α-glucosidase inhibitor on postprandial glycemia and insulinemia in healthy adults. *Journal of the American Dietetic Association*, 105(1), 65–71.
212. El-Beshbishy, H.A. and Bahashwan, S.A., 2012. Hypoglycemic effect of basil (*Ocimum basilicum*) aqueous extract is mediated through inhibition of α-glucosidase and α-amylase activities: An in vitro study. *Toxicology and Industrial Health*, 28(1), 42–50.
213. Baser, K.H.C., Kirimer, N. and Tümen, G., 1993. Composition of the essential oil of *Origanum majorana* L. from Turkey. *Journal of Essential Oil Research*, 5(5), 577–579.
214. Yamada, K., Murata, T., Kobayashi, K., Miyase, T. and Yoshizaki, F., 2010. A lipase inhibitor monoterpene and monoterpene glycosides from *Monarda punctata*. *Phytochemistry*, 71(16), 1884–1891.
215. Kawabata, J., Mizuhata, K., Nishioka, T., Aoyama, Y. and Kasai, T., 2003. 6-Hydroxyflavonoids as α-glucosidase inhibitors from marjoram (*Origanum majorana*) leaves. *Bioscience, Biotechnology, and Biochemistry*, 67(2), 445–447.
216. Damsud, T., Grace, M.H., Adisakwattana, S. and Phuwapraisirisan, P., 2014. Orthosiphol A from the aerial parts of *Orthosiphon aristatus* is putatively responsible for hypoglycemic effect via alpha-glucosidase inhibition. *Natural Product Communications*, 9(5), 639–641.
217. Ninomiya, K., Matsuda, H., Shimoda, H., Nishida, N., Kasajima, N., Yoshino, T., Morikawa, T. and Yoshikawa, M., 2004. Carnosic acid, a new class of lipid absorption inhibitor from sage. *Bioorganic & Medicinal Chemistry Letters*, 14(8), 1943–1946.
218. Vaquero, M.R., Yáñez-Gascón, M.J., Villalba, R.G., Larrosa, M., Fromentin, E., Ibarra, A., Roller, M., Tomás-Barberán, F., de Gea, J.C.E. and García-Conesa, M.T., 2012. Inhibition of gastric lipase as a mechanism for body weight and plasma lipids reduction in Zucker rats fed a rosemary extract rich in carnosic acid. *PLoS One*, 7(6), e39773.
219. Zhao, Y., Sedighi, R., Wang, P., Chen, H., Zhu, Y. and Sang, S., 2015. Carnosic acid as a major bioactive component in rosemary extract ameliorates high-fat-diet-induced obesity and metabolic syndrome in mice. *Journal of Agricultural and Food Chemistry*, 63(19), 4843–4852.
220. Ma, H.Y., Gao, H.Y., Sun, L., Huang, J., Xu, X.M. and Wu, L.J., 2011. Constituents with α-glucosidase and advanced glycation end-product formation inhibitory activities from *Salvia miltiorrhiza* Bge. *Journal of Natural Medicines*, 65(1), 37–42.

221. Kianbakht, S. and Dabaghian, F.H., 2013. Improved glycemic control and lipid profile in hyperlipidemic type 2 diabetic patients consuming *Salvia officinalis* L. leaf extract: A randomized placebo. Controlled clinical trial. *Complementary Therapies in Medicine*, 21(5), 441–446.
222. Ninomiya, K., Matsuda, H., Shimoda, H., Nishida, N., Kasajima, N., Yoshino, T., Morikawa, T. and Yoshikawa, M., 2004. Carnosic acid, a new class of lipid absorption inhibitor from sage. *Bioorganic & Medicinal Chemistry Letters*, 14(8), 1943–1946.
223. Asghari, B., Salehi, P., Sonboli, A. and Ebrahimi, S.N., 2015. Flavonoids from *Salvia chloroleuca* with α-amylsae and α-glucosidase inhibitory effect. *Iranian Journal of Pharmaceutical Research: IJPR*, 14(2), 609.
224. Nishioka, T., Kawabata, J. and Aoyama, Y., 1998. Baicalein, an α-glucosidase inhibitor from *Scutellaria baicalensis*. *Journal of Natural Products*, 61(11), 1413–1415.
225. Shimura, S., Tsuzuki, W., Kobayashi, S. and Suzuki, T., 1992. Inhibitory effect on lipase activity of extracts from medicinal herbs. *Bioscience, Biotechnology, and Biochemistry*, 56(9), 1478–1479.
226. Visavadiya, N.P. and Narasimhacharya, A.V.R.L., 2007. Ameliorative effect of Chlorophytum borivilianum root on lipid metabolism in hyperlipaemic rats. *Clinical and Experimental Pharmacology and Physiology*, 34(3), 244–249.
227. Thakur, G.S., Bag, M., Sanodiya, B.S., Debnath, M., Zacharia, A., Bhadauriya, P., Prasad, G.B.K.S. and Bisen, P.S., 2009. *Chlorophytum borivilianum*: A white gold for biopharmaceuticals and neutraceuticals. *Current Pharmaceutical Biotechnology*, 10(7), 650–666.
228. Lu, Y., Kuang, M., Hu, G.P., Wu, R.B., Wang, J., Liu, L. and Lin, Y.C., 2014. Loddigesiinols G–J: α-glucosidase inhibitors from *Dendrobium loddigesii*. *Molecules*, 19(6), 8544–8555.
229. Ghosh, S., More, P., Derle, A., Patil, A.B., Markad, P., Asok, A., Kumbhar, N., Shaikh, M.L., Ramanamurthy, B., Shinde, V.S. and Dhavale, D.D., 2014. Diosgenin from *Dioscorea bulbifera*: Novel hit for treatment of type II diabetes mellitus with inhibitory activity against α-amylase and α-glucosidase. *PLoS One*, 9(9), e106039.
230. Yang, M.H., Chin, Y.W., Yoon, K.D. and Kim, J., 2014. Phenolic compounds with pancreatic lipase inhibitory activity from Korean yam (*Dioscorea opposita*). *Journal of Enzyme Inhibition and Medicinal Chemistry*, 29(1), 1–6.
231. Shin, J.E., Han, M.J. and Kim, D.H., 2003. 3-Methylethergalangin isolated from *Alpinia officinarum* inhibits pancreatic lipase. *Biological and Pharmaceutical Bulletin*, 26(6), 854–857.
232. Shin, J.E., Han, M.J., Song, M.C., Baek, N.I. and Kim, D.H., 2004. 5-Hydroxy-7-(4′-hydroxy-3′-methoxyphenyl)-1-phenyl-3-heptanone: A pancreatic lipase inhibitor isolated from *Alpinia officinarum*. *Biological and Pharmaceutical Bulletin*, 27(1), 138–140.
233. Lin, L.Y., Peng, C.C., Yeh, X.Y., Huang, B.Y., Wang, H.E., Chen, K.C. and Peng, R.Y., 2015. Antihyperlipidemic bioactivity of *Alpinia officinarum* (Hance) farw zingiberaceae can be attributed to the coexistance of curcumin, polyphenolics, dietary fibers and phytosterols. *Food & Function*, 6(5), 1600–1610.
234. Kenchaiah, S., Evans, J.C., Levy, D., Wilson, P.W., Benjamin, E.J., Larson, M.G., Kannel, W.B. and Vasan, R.S., 2002. Obesity and the risk of heart failure. *New England Journal of Medicine*, 347(5), 305–313.
235. Du, Z.Y., Liu, R.R., Shao, W.Y., Mao, X.P., Ma, L., Gu, L.Q., Huang, Z.S. and Chan, A.S., 2006. α-Glucosidase inhibition of natural curcuminoids and curcumin analogs. *European Journal of Medicinal Chemistry*, 41(2), 213–218.
236. Reddy, P.P., Tiwari, A.K., Rao, R.R., Madhusudhana, K., Rao, V.R.S., Ali, A.Z., Babu, K.S. and Rao, J.M., 2009. New Labdane diterpenes as intestinal α-glucosidase inhibitor from antihyperglycemic extract of *Hedychium spicatum* (Ham. Ex Smith) rhizomes. *Bioorganic & Medicinal Chemistry Letters*, 19(9), 2562–2565.
237. Shimada, T., Horikawa, T., Ikeya, Y., Matsuo, H., Kinoshita, K., Taguchi, T., Ichinose, K., Takahashi, K. and Aburada, M., 2011. Preventive effect of *Kaempferia parviflora* ethyl acetate extract and its major components polymethoxyflavonoid on metabolic diseases. *Fitoterapia*, 82(8), 1272–1278.
238. Akase, T., Shimada, T., Terabayashi, S., Ikeya, Y., Sanada, H. and Aburada, M., 2011. Antiobesity effects of *Kaempferia parviflora* in spontaneously obese type II diabetic mice. *Journal of Natural Medicines*, 65(1), 73–80.
239. Kim, H.S., Kim, Y.H., Hong, Y.S., Paek, N.S., Lee, H.S., Kim, T.M., Kim, K.W. and Lee, J.J., 1999. α-Glucosidase inhibitors from *Commelina communis*. *Planta Medica*, 65(05), 437–439.
240. Youn, J.Y., Park, H.Y. and Cho, K.H., 2004. Anti-hyperglycemic activity of *Commelina communis* L.: Inhibition of α-glucosidase. *Diabetes Research and Clinical Practice*, 66, S149–S155.
241. Asano, N., Oseki, K., Kizu, H. and Matsui, K., 1994. Nitrogen-in-the-ring pyranoses and furanoses: Structural basis of inhibition of mammalian glycosidases. *Journal of Medicinal Chemistry*, 37(22), 3701–3706.

242. Kite, G.C., Fellows, L.E., Fleet, G.W., Liu, P.S., Scofield, A.M. and Smith, N.G., 1988. α-Homonojirimycin [2, 6-dideoxy-2, 6-imino-D-glycero-L-gulo-heptitol] from *Omphalea diandra* L.: Isolation and glucosidase inhibtion. *Tetrahedron Letters*, 29(49), 6483–6485.
243. Levitan, E.B., Song, Y., Ford, E.S. and Liu, S., 2004. Is nondiabetic hyperglycemia a risk factor for cardiovascular disease?: A meta-analysis of prospective studies. *Archives of Internal Medicine*, 164(19), 2147–2155.

2 Protecting Pancreatic β-cells from Metabolic Insults

Metabolic syndrome encompasses hyperglycemia and insulin resistance, which can lead to type 2 diabetes in genetically predisposed individuals. Insulin resistance is evidenced by impaired fasting glucose, defined as a fasting glucose level above 100 mg/dL or impaired glucose tolerance defined as a glucose level above 140 mg/dL, for 120 minutes after ingestion of 75 g of glucose load during an oral glucose tolerance test.[1,2] Once insulin resistance is in full blown, β-cells in the pancreas need to secrete more insulin in an attempt to decrease glycemia, and increased postprandial glycemia constitutes a status of "*glucotoxicity*" for pancreatic β-cell leading to the generation of reactive oxygen species and ultimately type 2 diabetes.[3] In parallel, insulin resistance favors the secretion of unesterified fatty acids from visceral adipose tissues, constituting a state of "lipotoxicity," which contribute further into β-cell dysfunctionement.[4,5] Therefore, protecting pancreatic β-cells from metabolic insults, preserving the ability of β-cells to secrete insulin and preventing β-cell exhaustion constitute a therapeutic strategy to prevent metabolic syndrome.

2.1 *Annona squamosa* L.

Common names: fan li zhi (Chinese); custard apple
Subclass Magnoliidae, Superorder Magnolianae, Order Annonales, Family Annonaceae
Medicinal use: fever (Philippines)

The endocrine pancreas is composed of the islets of Langerhans in which β-cells produce and secrete insulin. Alloxan is a cytotoxic cyclic urea derivative used to induce diabetes mellitus in rats and mice by causing the selective formation of oxygen reactive species targeting DNA and increasing in cytosolic calcium in pancreatic β-cells.[6–8] Natural products scavenging reactive oxygen species decrease alloxan toxicity.[9] Alloxan-induced diabetic albino rabbits treated with ethanol extract of leaves of *Annona squamosa* L. at dose of 200 or 300 mg/kg reduced fasting blood glucose by 9.8% and 18.2%, respectively, within 90 minutes.[10] During glucose tolerance test, a single oral administration of extract at a dose of 350 mg/kg reduced glycemia from circa 350 mg/dL to circa 150 mg/kg after 1 hour.[10] In severely diabetic rabbits, the extract corrected postprandial glucose from 436.7 to 257 mg/dL after 1 hour.[10] Increased plasma glucose after meal ingestion in man that evokes an increase in plasma insulin within 30 to 60 minutes plays a critical role in maintaining normal postprandial glucose homeostasis.[11] The ability of ethanol extract of leaves of *Annona squamosa* L. to reduce glycemia within 90 minutes in rabbits suggests the involvement of insulin secretion from residual β-cells. In general, natural products given orally in rodents evoking a decrease of glycemia at less than 6 hours and typically within 30 minutes to 2 hours are indicative of, at least, insulinotropic effect. Isoquinolines are probably involved in this effect.

2.2 *Nelumbo nucifera* Gaertn.

Synonyms: *Nelumbium nuciferum* Gaertn.; *Nelumbo speciosa* Willd.; *Nymphaea nelumbo* L.
Common names: lian (Chinese); sacred lotus
Subclass Ranunculidae, Superorder Proteanae, Order Nelumbonales, Family Nelumbonaceae
Medicinal use: anxiety (China)

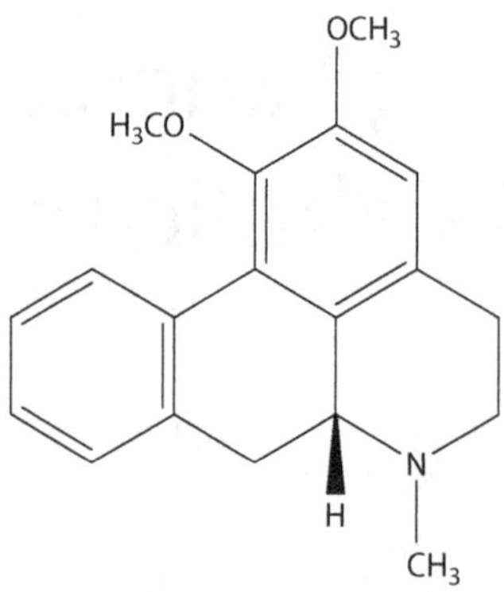

FIGURE 2.1 Nuciferine

Nuciferine (Figure 2.1) from the leaves of *Nelumbo nucifera* at a concentration of 40 μM increased the secretion of insulin by INS1-E cells exposed to gaentn 3.3 mM of glucose from about 50 to 300 ng/mg protein.[12] This isoquinolines alkaloid increased the secretion of insulin by INS1-E cells exposed to glucose 16.7 mM of glucose from about 180 to 560 ng/mg protein.[12] During stress, norepinephrine released by sympathetic nerves innervating the islets bind to α2 adrenoreceptors hence membrane hyperpolarization, via opening of ATP-dependent potassium channels preventing the opening of voltage-gated calcium channels and inhibiting therefore glucose stimulated release of insulin.[13]

2.3 *Stephania tetrandra* S. Moore

Synonyms: *Stephania disciflora* Hand.-Mazz.; *Stephania tetrandra* var. *glabra* Maxim.
Common name: fen fang ji (Chinese)
Subclass Ranunculidae, Superorder Ranunculanae, Order Menispermales, Family Menispermaceae
Medicinal use: tuberculosis (Taiwan)

In metabolic syndrome, elevated plasma glucose stimulates to the production of reactive oxygen species and downstream activation of nuclear factor-κB in of β-cells which ultimately leads, in predisposed individual into type 2 diabetes.[14] An interesting characteristic of *Stephania tetrandra* S. Moore (Figure 2.2) is that contains bisbenzyl-isoquinoline alkaloids with the ability to protect β-cells in rodents. One such alkaloid is tetrandrine which given at a dose of 20 mg/kg/day for 85 days decreased the development of insulin dependent diabetes in BB rats by 95% with improved pancreatic cytoarchitecture especially in term of reduction on insulinitis.[15,16] The findings of Sun et al. (1994) provided further evidence for the protective effect of tetrandrine on β-cells. These authors reported that tetrandrine given at a dose of 100 mg/kg intraperitoneally to alloxan-induced diabetic rats decreased plasma glucose from 25.4 to 7.6 mmol/L while it increased serum insulin up to 11.3 μU/mL and prevented alloxan-induced β-cells damages.[17] Additional support for the insulinotropic activity of *Stephania tetrandra* S. Moore is the evidence that tetrandrine 2′-*N*-β-oxide, tetrandrine 2′-*N*-α-oxide, tetrandrine 2-*N*-β-oxide, fangchinoline, fanchinoline 2′-*N*-α-oxide, 2′-*N*-norfanchinoline or cycleanine at a single intraperitoneal dose of 1 mg/kg lowered the fasting glycemia of streptozotocin-induced diabetic ddY mice after 6 hours by more than 40%, respectively.[17] Of note, tetrandrine in this experiment evoked a mild decreased in blood glucose advocating, at least, in favor of a more cytoprotective than strictly insulinotropic effect. Tetrandrine blocks voltage-gated L-type calcium channels large-conductance calcium-activated potassium channels and intracellular calcium pumps and inhibits nuclear factor-κB activation in macrophages

FIGURE 2.2 *Stephania tetrandra* S. Moore.

challenged with lipopolysaccharides.[17,18] It is reasonable therefore to consider that the protective effect of tetrandrine on β-cell protection might be, at least, by way of interfering with calcium signaling and nuclear factor-κB activation.

2.4 *Tinospora cordifolia* (Willd.) Miers ex Hook. f. & Thomson

Synonym: *Menispermum cordifolium* Willd.
Common name: goorcha (India)
Subclass Ranunculidae, Superorder Ranunculanae, Order Menispermales, Family Menispermaceae
Medicinal use: jaundice (India)

Tinospora cordifolia Miers ex Hook. f. & Thomson accumulates series of alkaloids of which jatrorrhizine, which given at a single oral dose of 100 mg/kg to Kunming mice reduced glycemia from 5.9 to 4.6 mmol/L and increased hepatic glycogen contents.[19] Jatrorrhizine at a single oral dose of 100 mg/kg/day for 4 days to alloxan-induced diabetic Kunming mice reduced glycemia from 21.7 to 19.4 mmol/L.[19] Reduction of plasma glucose in both normal and alloxan intoxicated mice is an indication that jatrorrhizine increases insulin secretion. In a subsequent study, that effect could not be reproduced as alkaloid fraction of stems of *Tinospora cordifolia* Miers ex Hook. f. & Thomson given orally to Wistar rats at a single oral dose of 200 mg/kg had no effect on glycemia.[20]

From this extract, palmatine, jatrorrhizine, and magnoflorine at a concentration of given orally 40 mg/kg did not cause any reduction in serum glucose in Wistar rats.[20] In oral glucose tolerance test, palmatine, jatrorrhizine and magnoflorine at 40 mg/kg lowered 1 hour postprandial glycemia from 7.4 to 5.7, 5.6, and 5.4 mM, respectively, and increased serum insulin in Wistar rats.[20] *In vitro*, palmatine, jatrorrhizine, and magnoflorine from the stems of *Tinospora cordifolia* Miers ex Hook. f. & Thomson increased the secretion of insulin secretion in RINm5F cells in absence or presence (16 mM) of glucose at a dose of 80 μg/ml.[20] The difference of activity between the first and the second study could be owed to the dose difference as jatrorrhizine is not well absorbed orally.[21] These observations support the speculation that palmatine, jatrorrhizine, and magnoflorine at high oral dose increase insulin secretion in rodents by direct action on β-cells. The mode of action of these protoberberine alkaloids is as for yet unknown. Pancreatic β-cell potassium channels include Kir6.2/SUR1 subunits whereas vascular smooth muscle potassium channels are formed by Kir6.1/SUR2B subunits whereby Kir 6.1 and Kir 6.2 constitute the channel pores and SUR subunits are regulatory sulfonylurea receptors.[22] It should be noted that the isoquinoline alkaloid protopine at a dose of 100 μM inhibited K_{ATP} channel in Kir6.1/SUR1 coexpressed HEK-293 cells from 17.4 to 6.2 pA/pF at −60 mV.[23] Clinical trials are warranted.

2.5 *Trianthema decandra* L.

Synonym: *Zaleya decandra* Burm.
Common name: punarnavi (India)
Subclass Caryophyllidae, Superorder Caryophyllanae, Order Caryophyllales, Family Aizoaceae
Medicinal use: hepatitis
History: The plant was mentioned in Tamil Sastrums

β-cells are particularly vulnerable to alloxan because of low levels of intracellular antioxidant enzymes such as glutathione peroxidase and catalase.[8] Ethanol extract of roots of *Trianthema decandra* L. given orally at a dose of 200 mg/kg/day for 15 days to alloxan-induced diabetic Wistar lowered glycemia from 11.5 to 6.2 mmol/L (normal: 5.8 mmol/L; glibenclamide at 1.2 mg/kg/day: 6 mmol/L).[24] In normal Wistar rats, the extract did not lower plasma glucose. Furthermore, in alloxan-induced diabetic Wistar this extract decreased to normal values plasma cholesterol, triglycerides, urea, and creatinine. The extract in alloxan-induced Wistar rats lowered hepatic lipid peroxidation, increased catalase, superoxide dismutase, and glutathione peroxidase confirming the traditional hepatoprotective effect of this herb.[24] The extract prevented alloxan induced hepatic and pancreatic necrosis.[24] The active principles and mechanism of action accounting for the hypoglycemic effect of *Trianthema decandra* L. are unknown. The plant being used for hepatitis may hold some yet unexplored anti-inflammatory properties that may account for the observed pancreato-protective effect.

2.6 *Opuntia ficus-indica* (L.) Mill.

Synonym: *Cactus ficus-indica* L.
Common names: guda (India); Indian fig
Subclass Caryophyllidae, Superorder Caryophyllanae, Order Caryophyllales, Family Cactaceae
Medicinal use: indigestion (India)

Total extract of cladodes of a member of the genus *Opuntia* given orally at a single dose of 100 mg/kg to Wistar rats poisoned with streptozotocin subjected to maltose tolerance test reduced glycemia from 290 to 141 mg/dL after 90 minutes being more efficient that acarbose given orally at a dose of

3 mg/kg via a mechanism independent of α-glucosidase activity.[25] *Opuntia ficus-indica* (L.) Mill. may have potentials in metabolic syndrome.

2.7 *Alternanthera paronychioides* A. St.-Hil.

Synonyms: *Alternanthera ficoidea* (L.) R. Br.; *Gomphrena ficoidea* L.
Common name: hua lian zi cao (Chinese)
Subclass Caryophyllidae, Superorder Caryophyllanae, Order Caryophyllales, Family Amaranthaceae
Nutritional use: vegetable (India)

Evidence suggests that reactive oxygen species activate in β-cells nuclear factor-κB, NH2-terminal Jun kinase/stress activated protein kinases, p38 mitogen-activated protein kinase that lead to decreased insulin secretion.[26] HIT-T15 pancreatic β-cells exposed to high glucose concentration *in vitro* for 72 hours underwent apoptosis via reactive oxygen species increase, Bax expression, mitochondrial insults, caspase 3 and 9 activation, and this effect was nearly abrogated by an ethanol extract of *Althernanthera paronychioides* A. St.-Hil. at 50 μg/mL.[27] In RIN-m5F cells this extract at 50 μg/mL inhibited pancreatic and duodenal homeobox-1 nuclear translocation and subsequently sustained insulin.[27] In response to oxidative stress, pancreatic and duodenal homeobox-1, is translocated from the nucleus to the cytoplasm of β-cell-derived HIT cells through activation of c-Jun NH2-terminal kinase.[28] Extract of *Alternanthera paronychioides* A. St.-Hil. may protect β-cells simply via antioxidant effect and/or inhibition of c-Jun NH2-terminal kinase. Note that aqueous extract of roots of *Achyranthes japonica* (Miq.) Nakai (family Amaranthaceae) inhibited lipopolysaccharide-induced c-Jun NH2-terminal kinase activation in macrophages.[29]

2.8 *Amaranthus esculentus* Besser ex Moq.

Subclass Caryophyllidae, Superorder Caryophyllanae, Order Caryophyllales, Family Amaranthaceae
Nutritional use: food (Korea)

Streptozotocin is produced by *Streptomyces achromogenes* and causes in experimental animals irreversible destruction of β-cells.[30] This antibiotic is preferentially absorbed by β-cells via glucose transporter-2 because of its glucose moiety.[31] In β-cells, streptozotocin is an alkylating agent which causes DNA strand breaks and activates poly(ADP-ribose)synthetase which induces severe depletion of NAD^+.[30,31] This aminoglycoside at doses of 50–65 mg/kg lead to hyperglycemia (20–30 mM) but severe ketosis does not develop even if insulin is not administered. Higher doses (75 mg/kg and above) result in spontaneous ketosis and death within days if insulin is not given.[30,31] The expression and activity of major antioxidant enzymes including superoxide dismutase, catalase and glutathione peroxidase are lower in islets of Langerhans than other tissues and β-cells undergo apoptosis via oxidative insults.[31] Thus, natural products which are antioxidant or increasing the activity of antioxidant enzymes including dismutase and catalase or increasing nonenzymatic antioxidant glutathione protect β-cells against streptozotocin-induced injuries in rats and may conceptually prevent the apparition of type 2 diabetes in human.[28] *Amaranthus esculantus* Besser ex Moq. seeds fixed oil given in a diet at a dose 100 mg/kg for 3 weeks to streptozotocin-induced diabtetic Sprague–Dawley rats lowered plasma glucose from 496.5 to 145.1 mg/dL.[32] That oil increased insulinaemia.[32] This treatment normalized the enzymatic activity of hepatic superoxide, catalase, and glutathione peroxidase.[32] The seeds of *Amaranthus esculentus* Besser ex Moq. accumulate substantial amounts of squalene.[33] This linear triterpene scavenges free radicals in the skin, protecting human skin surface from lipid peroxidation due to exposure to UV.[34] One could speculate that squalene intake in rats may assuage streptozotocin-induced oxidative insults at β-cells levels to favor insulin secretion and protect the liver against high glucose supply. Natural products which are antioxidant or increasing the activity of antioxidant enzymes such superoxide dismutase and catalase or increasing nonenzymatic antioxidant glutathione protect β-cells against streptozotocin-induced injuries in rats and may conceptually prevent the apparition of type 2 diabetes in human.[28]

2.9 *Amaranthus viridis* L.

Synonyms: *Amaranthus gracilis* Desf. ex Poir.; *Amaranthus gracilis* Desf.

Subclass Caryophyllidae, Superorder Caryophyllanae, Order Caryophyllales, Family Amaranthaceae

Medicinal use: dysentery (China)

Normal venous fasting plasma glucose is inferior to 100 mg/dL (5.6 mmol/L) and a value equal to or superior to 126 mg/dL (7.0 mmol/L) denotes type 2 diabetes.[1] A single oral dose of a methanolic extract of *Amaranthus viridis* L. to alloxan-induced diabetic Wistar rats (plasma glucose > 200 mg/dL) lowered the glycemia by circa 40%, 30%, and 20% after 1–3 hours, respectively, whereby gibenclamide 10 mg/kg evoked about 30%, 40%, and 55% reduction after 1–3 hours, respectively.[35] The administration of 400 mg/kg per day of this extract to alloxan-induced diabetic Wistar rats for 15 days evoked a lowering of blood glucose from 342.8 to 95.7 mg/dL (normal: 84.9 mg/dL; gibenclamide 10 mg/kg/day: 86.6 mg/dL) and improved glucose tolerance in oral glucose tolerance test.[35] Plasma cholesterol and triglycerides were decreased to normal values by the 15 days treatment.[35] Furthermore, the treatment restored hepatic glutathione, lowered malondialdehyde (a marker of lipid peroxidation), and increased catalase activity in the liver compared with untreated diabetic animals.[35] The extract inhibited the enzymatic activity of α-amylase *in vitro* with an IC_{50} value equal to 10.1 μg/mL (acarbose: 0.3 μg/mL).[35] The position of *Amaranthus viridis* L. in the Amaranthaceae and its Medicinal use suggest the involvement of pentacyclic triterpene saponins with anti-inflammatory activity. It is possible to speculate that such saponins/and or metabolites my favors pancreatic recovery after alloxan-induced oxidative insults by decreasing macrophage activity and proinflammatory cytokine secretion at the site of pancreatic necrosis. In fact, Krishnamurthy et al. (2011) reported an improvement of pancreatic histoarchitecture in streptozotocin-induced diabetic rats following a 400 mg/kg/day intake of methanol extract for 21 days.[36] After a meal postprandial glycemia usually do not exceed 165 mg/dL and insulin resistance elevates this value contributing to pancreatic and vascular insults. Triterpene saponins by virtue of α-amylase inhibition may lower postprandial glycemia attenuating glucotoxicity at β-cell levels.[37]

2.10 *Celosia argentea* L.

Synonym: *Celosia cristata* L.

Common names: qing xiang (Chinese); barhichuda (India); bayam (Malay); palonpalongan (Philippines); wild cockscomb

Subclass Caryophyllidae, Superorder Caryophyllanae, Order Caryophyllales, Family Amaranthaceae

Medicinal use: dysentery (Malaysia)

Ethanol extract of defatted seeds of *Celosia argentea* L. given orally once at a dose 500 mg/kg orally to alloxan-induced diabetic Swiss albino mice evoked a decrease of fasting glycemia by 38.8% after 6 hours and this effect was comparable to tolbutamide at 100 mg/kg.[38] Daily oral administration for 15 days at 500 mg/kg prevented body weight loss, decreased glycemia by 54.4% and prevented weight loss.[38] To date, the active principle and mechanism behind the hypoglycemic activity of *Celosia argentea* seed are unknown. Tolbutamide that has been used for the treatment of type 2 diabetes with a peak of action at 4–6 hours. Sulphonylurea lower blood glucose in normal subjects and type 2 diabetic patients capable on insulin secretion. These hypoglycemic agents increase insulin secretion hence lowering plasma glucose.[39] The seeds of this plant accumulate substantial amounts of anti-inflammatory oleanane triterpene saponins, of natural for which evidence is available to demonstrate insulinotropic activity.[40] The relatively late onset of hypoglycemic activity of the extract also suggests the possibility of increased uptake of glucose by skeletal muscles and adipose mass. In fact, the oleanane saponin isolated from the root bark of *Aralia* a member of the genius Aralia (family Araliaceae) given orally to streptozotocin/nicotinamide-induced type 2 diabetes mellitus evoked a decrease in plasma glucose via increased glucose consumption in skeletal muscles.[41]

2.11 *Camellia sinensis* (L.) Kuntze

Synonym: *Thea chinesis* L.
Common names: cha (Chinese); tea
Subclass Dillenidae, Superorder Ericanae, Order Theales, Family Theaeae
Medicinal use: Tonic (China)
History: Used since time immemorial in China and listed in the Benst'sao Kang Mu (1590 AD)

Metabolic syndrome is associated with chronic exposure of β-cells to high concentration of glucose inducing to the production of reactive oxygen species, the activation of nuclear factor-κB, pancreatic and duodenal homeobox-1 inhibition and subsequent reduction of insulin secretion.[42] Liquid extracts of tea made by soaking 20 g in 100 ml of boiling water given intragastrically at a dose of 2 ml/100 g body weight/day for 3 weeks to Sprague–Dawley rats poisoned with streptozotocin reduced glycemia by about 40% after 1 week of treatment.[43] The extracts given prophylactically for 2 weeks prevented hyperglycemia induced by streptozotocin.[43] The plant contains theophylline which has the ability to inhibit cyclic nucleotide phosphodiesterase increasing thus the cytosolic contents of β-cells in cyclic adenosine monophosphate and subsequent activation of protein kinase A and cyclic adenosine monophosphate-regulated guanine nucleotide exchange factor II leading to altered ion channel activity, elevation of cytoplasmic calcium contents and exocytosis of insulin-containing vesicles.[44,45] Protein kinase A-mediated phosphorylation of the SUR1 subunit induces the closing of K_{ATP} channel closure via ADP-dependent mechanism.[46] Caffeine present in this plant has been shown at a single oral dose of 5 mg/kg to increase insulin secretion in healthy volunteers during oral glucose tolerance test as evidenced by a 60% in insulin area under the curve.[47] Besides, the oral administration of (−)-epicatechin, (−)-epigallocatechin gallate and (−)-epicatechin gallate from *Camellia sinensis* (L.) Kuntze are known to inhibit the absorption of carbohydrates and improve the antioxidant capacity of rodents.[48] This set of experimental evidence reinforce the notion that the intake of *Camellia sinensis* (L.) Kuntze infusions could be beneficial in the management of metabolic syndrome. It must be recalled here that green tea infusions are better than concentrated extracts sold for weight loss which have incurred acute liver failure.[49]

2.12 *Embelia ribes* Burm.f.

Common names: bai hua suan teng guo (Chinese); vidanga (India)
Subclass Dilleniidae, Superorder Primulanae, Order Primulales, Family Myrsinaceae
Medicinal use: jaundice (India)
History: *Embelia ribes* Burm.f. was known of Sushruta, notably to expel intestinal worms

Natural products protecting β-cells against streptozotocin-induced oxidative injuries in rodent may potentially, but not systematically, prevent the apparition of type 2 diabetes in human.[28] Aqueous extract of fruits of *Embelia ribes* Burm.f. given orally to streptozotocin-induced diabetic albino rats (fasting glycemia > 200 mg/dL) given at a daily dose of 100 mg/kg for 40 days lowered heart rate and systolic blood pressure, lowered glycemia from 344 to 83.2 mg/dL (normal: 69.3 mg/dL; gliclazide 25 mg/kg/day: 79 mg/dL) and normalized plasma glutathione.[50] In the pancreas of diabetic rodents, the extract reduced lipid peroxidation, increased catalase activity and glutathione contents and normalized β-cells cytoarchitecture.[50] In β-cells, streptozotocin induces the generation of nitric oxide and reactive oxygen species which contribute to DNA fragmentation and other oxidative injuries.[51,52] Embelin that abounds in the seeds is antioxidant.[53] This benzoquinone given orally to streptozotocin-high fat-induced type 2 diabetic Wistar rats (fasting blood glucose > 250 mg/dL) at a dose of 50 mg/kg/day 28 days brought activities of superoxide dismutase, catalase and glutathione peroxidase up to normal values in pancreas of diabetic animals and normalized pancreatic cytoarchitecture.[54]

2.13 *Casearia esculenta* Roxb.

Synonym: Guidonia esculenta (Roxb.) Baill.
Common name: kunda jungura (India)
Subclass Dilleniidae, Superorder Violanae, Order Violales, Family Salicaceae
Medicinal use: constipation (India)

In some obese subjects, insulin compensation fails owing to β-cell functional decline on account of genetic abnormalities, chronic hyperglycemia or excessive exposure to fatty acids also termed "*lipotoxicity*."[55] Aqueous extract of roots of *Casearia esculenta* Roxb. given orally to male Wistar albino rats at a dose of 300 mg/kg reduced fasting blood glucose from 69 to 45.6 mg/dL after 3 hours.[56] In an oral glucose challenge involving oral administration of glucose at a dose of 2 g/kg the same extract at dose of 300 mg/kg lowered glycemia from 122 to 83.3 mg/dL after 90 minutes.[56] In streptozotocin-induced diabetic rats the extract at a dose of 300 mg/kg lowered glycemia from 253 to 224 mg/dL after 3 hours in a way similar to insulinotropic agent at a dose of 0.6 mg/kg.[56] When given daily for 45 days this extract sustained body weight, reduced glycemia from 311 to 135.7 mg/dL, and increased plasma glutathione.[56] 3-Hydroxymethyl xylitol from *Casearia esculenta* Roxb. given intragastrically at a dose of 40 mg/kg/day for 45 days to diabetic Wistar rats lowered total cholesterol from 196 to 116.8 mg/dL, triglycerides from 128.8 to 81.1 mg/dL, and fatty acids from 142.3 to 91.1 mg/dL.[57] It is tempting to speculate that 3-hydroxymethyl xylitol account for the aforementioned effect based on cyto-protective effects at the pancreatic level.

2.14 *Cucurbita ficifolia* Bouché

Synonym: *Cucurbita melanosperma* Gasp. *Pepo ficifolius* (Bouché) Britton
Common name: fig leaf gourd
Subclass Dilleniidae, Superorder Violanae, Order Cucurbitales, Family Cucurbitaceae
Nutritional use: vegetable (Philippines)

Evidence is accumulating that the fruits of *Cucurbita ficifolia* Bouché contains natural product(s) the oral administration of which decrease glycemia in diabetic rodents via increased insulin secretion by remaining functional β-cells. Oral administration of *Cucurbita ficifolia* Bouché diabetic rabbits with moderate hyperglycemia (8.33–16.67 mM; 150–300 mg/dl) evoked a decrease of glycemia and this hypoglycemic action was not found in diabetic rabbits with severe hyperglycemia (above16.67 mM/300 mg/dl) implying the involvement of remaining functional β-cells.[58] The juice of fruits given to type 2 diabetic patients with moderate hyperglycemia (fasting glucose levels from 8.3 to 16.6 mM/150–300 mg/dL) at a single dose of 4 mL/kg decreased plasma glucose by 31% after 5 hours (Acosta-Patiño et al. 2005).[59] Methanol extract of fruits of *Cucurbita ficifolia* Bouché given by gavage to streptozotocin-induced diabetic Sprague–Dawley rats at a dose of 300 mg/kg/day for 30 days lowered glycemia from about 400 to 120 mg/dL (normal: about 100 mg/dL) and increased insulinemia by 36%.[60] This regimen lowered pancreatic lipid peroxidation by 28% and augmented insulin positive cells per islets by 50%.[60] Aqueous extract of fruits of *Cucurbita ficifolia* Bouché given to streptozotocin-induced diabetic male CD-1 mice at a dose of 200 mg/kg/day for 30 days lowered polyphagia, halved water intake and reduced glycemia by 78%.[61] This treatment lowered plasmatic oxidative status, increased hepatic and pancreatic glutathione and lowered the activity of hepatic and pancreatic glutathione peroxidase and glutathione reductase.[61] Aqueous extract of fruits of *Cucurbita ficifolia* Bouché at a concentration of 72 μg/mL stimulated the secretion of insulin by RINm5F by 46.9% (glibenclamide at 500 μg/mL: 37%) via activation of inositol 1,4,5-trisphosphate (IP3) receptor in endoplasmic reticulum and increase in cytosolic concentration of calcium.[62] The insulinotropic constituents are unknown but members of the genus *Cucurbita* have the tendency to elaborate trigonelline and nicotinic acid. Trigonelline (Figure 2.3) given orally to high-carbohydrate and high-fat streptozotocin-induced type 2 diabetes Wistar rats at a dose of for 4 weeks at a dose of 50 mg/kg/day increased pancreas islets close to normal, lowered glycemia from 423.8 to 125 mg/dL (normal: 116.9 mg/dL; glibenclamide at 0.3 mg/kg/day: 122.7 mg/dL), increased pancreatic insulin contents and increased pancreatic glutathione catalase and superoxide dismutase activity.[63] As for nicotinic acid, Kahn et al. (1989) provided evidence that oral intake for this to volunteers increased the secretory capacity of β-cells.[64]

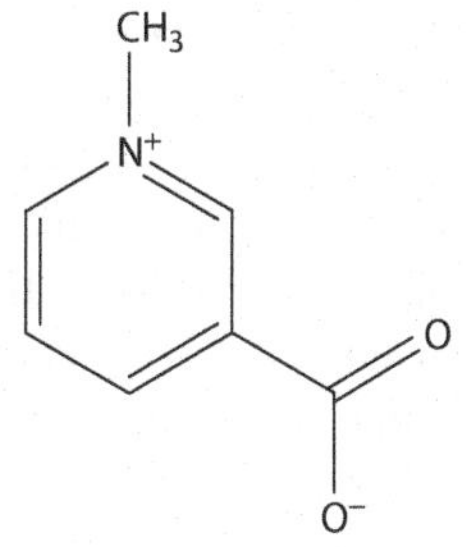

CH_3

N^+

O

O^-

FIGURE 2.3 Trigonelline

2.15 *Citrullus colocynthis* (L.) Schrad.

Synonyms: *Colocynthis vulgaris* Schrad.; *Cucumis colocynthis* L.
Common names: visala (India); bitter apple
Subclass Dilleniidae, Superorder Violanae, Order Cucurbitales, Family Cucurbitaceae
Medicinal use: jaundice (India)

Evidence is accumulating to demonstrate that the seeds of *Citrullus colocynthis* (L.) Schrad. contain polar natural products, yet to be identified, that enhance insulin secretion and protect β-cells functionality *in vitro*. The seeds methanol extract of defatted seeds *Citrullus colocynthis* (L.) Schrad. at a concentration of 500 μg/mL protected Wistar rats isolated islets against streptozotocin-induced reduction of glucose (16.7 mM) stimulated insulin secretion as evidenced by about a 20% increase in insulin secretion.[65] Isopropanol extract of defatted seeds of *Citrullus colocynthis* (L.) Schrad. perfused at 0.1 mg/mL to isolated pancreas of Wistar rats exposed to 8.3 mM glucose concentration increased insulin secretion by 40% at 5 minutes.[66] Methanol extract of defatted seeds *Citrullus colocynthis* (L.) Schrad. at a concentration of 500 μg/mL protected Wistar rats isolated islets against streptozotocin-induced reduction of glucose (16.7 mM) stimulated insulin secretion as evidenced by about a 20% increase in insulin secretion.[65] The seeds contain tryptophan (Akobundu et al. 1982) which is known to induce insulin secretion *in vitro*.[67] It must be recalled that the fruits and aerial part of this plant contain flavone C-glucoside of which isovitexin, isooriantin and isooriantin 3′-methylether.[68] Isovitexin given at a single oral dose of 15 mg/kg to Wistar rats in oral glucose tolerance tests lowered postprandial glycemia from about 200 to 170 mg/dL at 30 minutes with concurrent increase in insulin secretion.[69] In alloxan-induced diabetic Wistar rats with high hyperglycemia (>400 mg/dL) this C-glucosyl flavone had no effect implying the need of remaining viable β-cell function for activity.[69]

2.16 *Cucumis trigonus* Roxb.

Synonym: *Cucumis melo* L.
Common names: indravaruni (India); bitter gourd
Subclass Dilleniidae, Superorder Violanae, Order Cucurbitales, Family Cucurbitaceae
Medicinal use: diabetes (India)

Aqueous extract of fruits of *Cucumis trigonus* Roxb. given orally at a dose of 500 mg/kg/day for 21 days to streptozotocin-induced diabetic Wistar rats reduced glycemia from 390 to 188.3 mg/dL and this effect was more effective than treating animals with insulinotropic.[70] The extract increased body weight by 15%, as well as insulinemia and liver glycogen by 50% and 130%, respectively.[70] The extract lowered cholesterol by 37.9%, low-density lipoprotein by 57.2%, very low-density lipoprotein by 34.4% and triglycerides by 28.1% whereby high-density lipoprotein was not affected.[70] The liver secretes cholesterol and triglycerides in the form of very low-density lipoprotein which delivers triglycerides to the peripheral tissues.[71] In obese patients with insulin resistance or in type 2 diabetic patients inhibition of endothelial lipoprotein lipase and increased activity of hormone sensitive lipase in adipose tissues account for an excess of free fatty acids delivery to the liver and therefore increased secretion of very low-density lipoproteins which are precursor of atherogenic low-density lipoprotein.[71] Reduced plasma triglycerides and very low-density lipoprotein can be explained by increased plasmatic lipoprotein lipase activity and/or reduction of hepatic very low-density lipoprotein synthesis.[71]

2.17 *Gynostemma pentaphyllum* (Thunb.) Makino

Common name: jiao gu lan (Chinese)
Subclass Dilleniidae, Superorder Violanae, Order Cucurbitales, Family Cucurbitaceae
Medicinal use: Tonic (China)

The secretion of insulin by β-cells is proportionate to the rate at which glucose is metabolized in β-cells.[72] Glucose at stimulatory concentration, enters β-cells via cytoplasmic membrane facilitated glucose transporter-2 and is phosphorylated in the cytosol is by glucokinase into glucose 6 phosphate which represents the first step of glycolysis and mitochondrial pyruvate oxidation yielding ATP and subsequent insulin degranulation.[73] Phanoside from *Gynostemma pentaphyllum* Makino at a concentration of 500 μM enhanced the secretion of insulin from pancreatic islets cultured *in vitro* and exposed to nonstimulatory or potentiatory glucose concentration.[74] This dammarane-type saponin given orally and prophylactically at a single dose of 80 mg/kg to normoglycemic Wistar rats challenged with intraperitoneal injection of glucose lowered glycemia by 45% after 30 minutes and this effect was accompanied by an increase in insulinemia from about 16 to 26 μU/mL.[74] Phanoside at a concentration of 150 μM stimulated the release of insulin from

Wistar rat islets from 4.3 to 32 μU/islets at 3.3 mM of glucose and from 15.2 to 57.7 μU/islets at 16.7 mM by a mechanism independent of K-ATP channel closing.[75]

2.18 *Melothria heterophylla* (Lour.) Cogn.

Synonym: Solena amplexicaulis (Lam.) Gandhi
Common name: tarali (India)
Subclass Dilleniidae, Superorder Violanae, Order Cucurbitales, Family Cucurbitaceae
Medicinal use: indigestion (India)

Ethanol extract of aerial parts of *Melothria heterophylla* (Lour.) Cogn. given orally at a dose of 400 mg/kg/day to streptozotocin-diabetic Male Swiss albino rat normalized fasting glycemia from 480 to 97.4 mg/dL.[76] In the same experiment, rutin and gallic acid given at a dose of 4 mg/kg reduced fasting glycemia from 399 and 358.9 to 108.7 mg/dL and 100.7 mg/dL.[76] From the extract, rutin and gallic acid decreased weight loss and augmented insulinemia of diabetic rodents.[76]

2.19 *Momordica charantia* L.

Common names: ku gua (Chinese); periah (Malay); karela (Pakistan); balsam-apple
Subclass Dilleniidae, Superorder Violanae, Order Cucurbitales, Family Cucurbitaceae
Medicinal use: diabetes (Sri Lanka)

In a healthy individual, plasma glucose levels reach a peak not exceeding 160 mg/dL 30 minutes to 60 minutes after oral ingestion of 75 g of glucose and gradually return to post-absorptive values by 3 to 4 hours.[77] In a state of insulin resistance, postprandial glycemia exceeds this value and pose a threat of "*glucotoxicity*," which can be prevented by the consumption of medicinal plants. Juice squeezed from the fruits of *Momordica charantia* L. given orally to Sprague–Dawley rats at a single dose of 1 mL/100 g lowered fasting glycemia by 45%, 4 hours after administration and this hypoglycemia perdured 5 hours.[78] This juice given at a single dose of 1 ml/100g 30 minutes before an oral loading of glucose (1 ml/100 g body weight, 50% w/v) lowered by 80% peak glycemia at 2 hours compared with untreated animals.[78] Aqueous extract of pulp of fruits of *Momordica charantia* L. 0.2% (w/v) promoted the proliferation of hamster pancreatic β-cells HIT-T15 cells by 35.5% as well as insulin secretion by 29.4%.[79] In HIT-T15 cells damaged by alloxan, the extract at 0.2% (w/v) achieved a proliferation rate of 22.1% and augmented insulin secretion by 12%.[79] Momordicoside X isolated from the fruits at a concentration increased insulin secretion by MIN6 β-cells.[80] Saponin-rich fraction of the fruits of *Momordica charantia* L. at a concentration of 125 μg/mL induced the secretion of insulin by MIN6 β-cells *in vitro*.[81] From this fraction, momordicine II and kuguaglycoside G at a concentration of 10 μg/mL.[81] The exact molecular pathways involved the insulinotropic activity of these cucurbitanes saponins is as for yet unknown. Clinical trials are warranted.

2.20 *Momordica cymbalaria* Fenzl ex Naudin

Common name: kadavanchi (India)
Subclass Dilleniidae, Superorder Violanae, Order Cucurbitales, Family Cucurbitaceae
Medicinal use: abortion (India)

Fruit powder of *Momordica cymbalaria* Fenzl ex Naudin given orally to alloxan-diabetic Wistar rats at a dose of 0.25 g/kg/day for 15 days reduced fasting glycemia from 295 to 225 mg/dL, increased body weight by 7.8 g and reduced cholesterolemia.[82] Aqueous extract of fruits of *Momordica cymbalaria* Fenzl ex Naudin given to Wistar rats at a dose of 500 mg/kg/day for 6 weeks had no effect on plasma glucose and insulinemia.[83] The extract given to alloxan-induced diabetic rats (glycemia > 250 mg/mL) at a dose of 500 mg/kg/day for 6 weeks lowered glycemia from 273.8 to 110.3 mg/dL and increased plasma insulin from 8.3 to 18 µU/mL.[83] Saponin fraction of roots *Momordica cymbalaria* Hook. at a concentration of 1 mg/mL increased the secretion of insulin from RIN-5F cells challenged with streptozotocin from 1 to 6.9 µIU/mL, and this effect was reduced by coadministration with adrenaline or nifedipine.[84] Further studies on the principles involved in the insulinotropic principles of *Momordica cymbalaria* Fenzl ex Naudin are needed.

2.21 *Siraitia grosvenorii* (Swingle) C. Jeffrey ex A.M. Lu & Z.Y. Zhang

Synonyms: *Momordica grosvenorii* Swingle; *Thladiantha grosvenorii* (Swingle) C. Jeffrey
Common name: luo han guo (Chinese)
Subclass Dilleniidae, Superorder Violanae, Order Cucurbitales, Family Cucurbitaceae
Medicinal use: bronchitis (China)

The Goto–Kakizaki rat is a spontaneous-onset, non-obese type 2 diabetic animal model with mild fasting hyperglycemia, elevated postprandial hyperglycemia, impaired insulin secretion, progressive reduction of β-cell mass and development of long-term diabetic complications.[85] Aqueous extract of fruits of *Siraitia grosvenori* Swingle given to spontaneous type 2 diabetic Goto–Kakizaki rats at 0.4% of diet for 13 weeks evoked from the 8th week a mild decrease in nonfasting blood glucose.[86] An oral glucose tolerance test performed on the 7th week revealed a sharp and transient increase in insulinemia after 15 minutes and a decrease in glycemia after 2 hours.[86] The treatment increased insulin contents of fasting pancreas, lowered urine secretion and albuminuria and decreased hepatic lipid peroxidation.[86]

2.22 *Carica papaya* L.

Common name: papaya
Subclass Dilleniidae, Superorder Capparanae, Order Caricales, Family Caricaceae
Medicinal use: wounds (Philippines)

In obese, chronically elevated levels of circulating fatty acids may have an etiologic role in β-cells dysfunction and the pathophysiology of type 2 diabetes whereby increase uptake of fatty acids by β-cells leads to islet lipid deposition contributing to a decline of insulin secretion.[4,5] Therefore limiting plasmatic concentration of free fatty acids may prevent the development of type 2 diabetes in obese. Aqueous extract of leaves of *Carica papaya* L. administered for 4 weeks at a dose of 3 g/100 mL in the drinking water of Wistar rats poisoned with streptozotocin decreased water consumption by 30%, mildly decreased body weight, reduced glycemia from 434.7 to 250 mg/dL, cholesterolemia from 75.6 to 57.3 mg/dL, triacylglycerols from 232.1 to 93.7 mg/dL whereby insulinemia was unchanged.[87] This regimen mitigated liver weight, decreased hepatic triglycerides and protected pancreas against islets damages.[4] Canini et al. (2007) reported the presence of caffeic acid, kaempferol, and quercetin in the leaves of this plant, which are able to inhibit the enzymes of carbohydrate decomposition *in vitro* as well as p-coumatic acid.[88] Besides, the leaves contain non negligible amount of protocatechuic acid, which according to Hanini et al. (2010) decreased glycemia when administered orally to streptozotocin-induced diabetic Wistar rats at a dose of 100 mg/kg/day for 45 days and prevented decrease in insulin as efficiently as glibenclamide.[87–89] The mode of action of this antioxidant benzoic acid derivative, like most simple phenolics with insulinitropic effect, is unknown. It must be recalled here that simple phenolics are often beneficial for metabolic syndrome in rodents.

2.23 *Moringa oleifera* Lam.

Synonym: *Moringa pterygosperma* Gaertn.
Common names: la mu (Chinese); murugai (India); moringa (Philippines); horse raddish
Subclass Dillenidae, Superorder Capparanae, Order Moringales, Family Moringaceae
Nutritional use: vegetable (Philippines)

Methanol extract of leaves of *Moringa oleifera* Lam. given orally to alloxan-induced diabetic rats (blood glucose ≥ 250 mg/dL) at a dose of 600 mg/kg for 6 weeks (metformin: 100 mg/kg) had no effect on food intake, and prevented weight loss.[90] This supplementation improved glucose tolerance in diabetic rodents as efficiently as metformin given orally at a daily dose of 100 mg/kg/day.[90] The extract lowered glycemia from about 275 to 125 mg/dL (normal about 100 mg/dL; metformin at 100 mg/kg/day: about 125 mg/dL) and increased insulinemia close to normal.[90] The extract normal cholesterol and low-density lipoprotein, lowered triglycerides, and brought high-density lipoprotein above normal values.[90] In liver and skeletal muscles of treated rodents, the extract increased glycogen synthetase activity and glycogen contents close to normal.[90] Gastrocnemius muscle preparations from rats treated this the extract had increased glucose uptake capacity compared with muscle preparations from untreated diabetic group.[90] All of these plants with activity on sterptozotocin or alloxan may be simply anti-inflammatory to suspect antioxidant. One could draw the inference that principles and/or metabolites of this tree protect β-cells against oxidative insults in human but the appreciation of pancreatic recovery in streptozotocin or alloxan induced diabetic in rodents requires careful consideration. Wild-type rats or mice do not develop type 2 diabetes and have a β-cells replication rate of about 2.5%, whereas in human it is only around 0.2%.[91] Thus, regeneration of damaged β-cell and islets neoformation seems to take place via mitosis of endocrine islet cells and cell proliferation from tubular structures.[92,93] *Moringa oleifera* Lam. is consumed almost daily in the Island of Cebu where type 2 diabetes incidence seems comparatively low to other Southeast Asian countries. Clinical trials to assess the beneficial effect of *Moringa oleifera* Lam. consumption in metabolic syndrome is warranted.

2.24 *Brassica juncea* (L.) Coss.

Synonyms: *Brassica integrifolia* (H. West) O.E. Schulz; *Sinapis integrifolia* H. West; *Sinapis juncea* L.
Subclass Dillenidae, Superorder Capparanae, Order Capparales, Family Brassicaeae
Common name: jie cai (Chinese); kadugu (India); brown mustard
Medicinal use: chest affection (India)

The seeds of *Brassica juncea* (L.) Coss. (Figure 2.4) given prophylactically to albino rats as part of 15% of diet for 7 days mitigated hyperglycemia induced by alloxan.[94] This treatment given 5 weeks evoked a fall of glucose from 200.6 to 160.3 mg/dL.[94] This treatment given to streptozotocin-diabetic albino rats for 5 weeks evoked a fall of glucose from 275 to 264.3 mg/dL.[94]

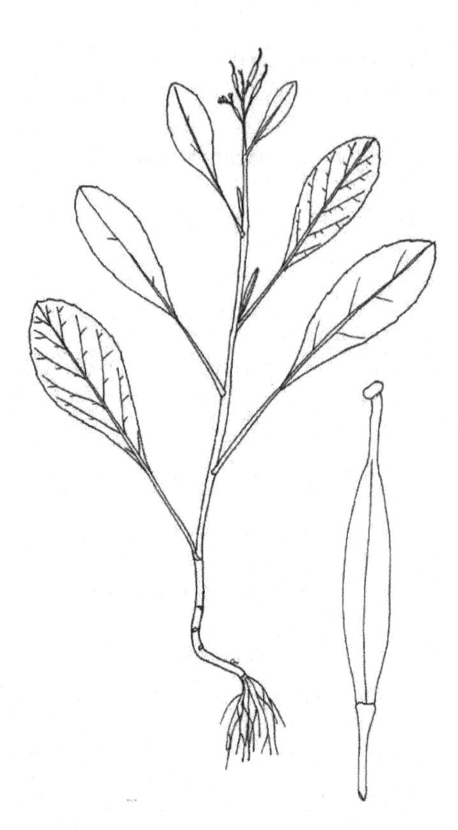

FIGURE 2.4 *Brassica juncea* (L.) Coss.

2.25 *Eruca sativa* Mill.

Synonym: *Brassica eruca* L.
Common name: taramira (India)
Subclass Dillenidae, Superorder Capparanae, Order Capparales, Family Brassicaceae
Medicical use: indigestion (India)

Oil of seeds of *Eruca sativa* Mill. given by gavage to Wistar rats 2 weeks before and 2 weeks after alloxan reduced glycemia by 45%, increased insulinemia, lowered serum triglycerides by 45%, and cholesterol by 20%.[95] At the hepatic level, the oil mitigated the depletion glycogen, glutathione, and superoxide dismutase and reduced glucose-6-phosphatase and lipid peroxidation.[95] The seed oil of *Eruca sativa* Mill. accumulates linolenic acid which might protect β-cells against alloxan-induced oxidative injuries in rats.[96] Further study on the insulino-protective effect of unsaturated fatty acid is required.

2.26 *Ceiba pentandra* (L.) Gaertn.

Synonym: *Bombax pentandrum* L.
Common names: kutasalmali (India); white silk cotton tree
Subclass Dillenidae, Superorder Malvanae, Order Malvalles, Family Bombacaceae
Medicinal use: fever (India)

Ethanol extract of bark of *Ceiba pentandra* (L.) Gaertn. given at a dose of 200 mg/kg orally to diabetic Wistar rats subjected to an oral glucose tolerance test reduced glycemia from 472.8 to

269.3 mg/dL after 30 minutes.[97] When given for 21 days at a dose of 200 mg/kg/day this extract lowered glycemia from 502.8 to 318.5 mg/dL, increased insulin from 158.6 to 248.8 pmol/L, increased liver glycogen from 2.5 to 3.9 mg/g liver, reduced total cholesterol from 241.3 to 200 mg/dL, and reduced triglycerides from 212.5 to 189.3 mg/dL.[97] The active principle(s) involved here are unknown.

2.27 *Ficus benghalensis* L.

Synonym: *Ficus indica* L.
Common names: vatah (India); banyan
Subclass Dillenidae, Superorder Malvanae, Order Urticales, Family Moraceae
Medicinal use: diabetes (India)

The bark of *Ficus benghalensis* L. contains 3′,5,7-trimethyl ether of leucocyanidin, 3′,5-dimethyl ether of leucocyanidin 3-*O*-β-D-galactosylcellobioside, 3′,5,7-trimethyl ether of delphinidin-3-O-a-L rhamnoside, 5,7-dimethoxy of leucopelargonidin-3-*O*-α-L-rhamnoside.[98] Dimethoxy ether of leucopelargonidin-3-*O*-α-L rhamnoside isolated from the bark of the *Ficus benghalensis* L. given orally to alloxan-induced diabetic dogs at a dose of 100 mg/kg evoked a decrease of glycemia.[99] This finding was reinforced by the study of Singh et al. (2009) in which it was found that aqueous extract of aerial roots of *Ficus benghalensis* L. given to Wistar rats orally at a dose of 300 mg/kg reduced fasting blood glucose after 6 hours from 79 to 44.4 mg/dL and glucose tolerance test after 3 hours from 78.5 to 48.2 mg/dL.[100] In sub-diabetic rats challenged with glucose tolerance test this extract reduced glycemia from about 275 to 200 mg/dL and as potently as glipizide given at a dose of 2.5 mg/kg.[100] In diabetic rodent subjected to glucose tolerance test, the extract reduced glycemia from about 400 to 250 mg/dL.[100] The insulinotropic mechanism of action of dimethoxy ether of leucopelargonidin-3-*O*-α-L rhamnoside and/or first pass metabolites is unknown. Jayaprakasam et al. (2005) tested the ability of anthocyanins to stimulate secretion of insulin by INS-1 cells *in vitro* and found that cyanidin-3-glucoside and delphinine-3 glucoside at a concentration of 50 μg/mL enhanced insulin secretion in the presence of 4 mM or 10 mM glucose.[101] Cyanidin-3-galactoside was only able to increase insulin secretion at stimulatory concentration of glucose (10 mM) whereas pelargonidin was only able to increase insulin secretion at basal concentration of glucose (4 mM).[101]

2.28 *Ficus religiosa* L.

Synonym: *Urostigma religiosum* (L.) Gasp.
Common names: pu ti shu (Chinese); pipala (India); sacred fig
Subclass Dillenidae, Superorder Malvanae, Order Urticales, Family Moraceae
Medicinal use: dysentery (India)

Aqueous extract of bark of *Ficus religiosa* L. given orally to streptozotocin-induced neonatal type 2 diabetes (glycemia> 160 mg/dL) at a dose of 200 mg/kg/day for 4 weeks reduced plasma gucose from approximately 190 to 120 mg/dL (gliclazide at 10 mg/kg/day: about 100 mg/dL).[102] This regimen increased body weight gain, decreased superoxide dismutase activity, increased catalase, glutathione peroxidase and glutathione and decreased malondialdehyde in the plasma.[102] Aqueous extract of stem bark given orally to Wistar at a dose of 50 mg/kg reduced glycemia by 24.7% 3 hours after administration.[103] During glucose oral tolerance test, the extract evoked a decrease in glycemia 1 20.1%.[103] In rats poisoned with streptozotocin, the extract at a dose of 50 mg/kg given daily preserved body weight of diabetic rodents and induced the secretion of insulin.[103] Furthermore, this extract reduced serum triglycerides and cholesterol in diabetic rodents and mitigated glycogen loss in liver and skeletal muscles.[103] The bark contains lupen-3-one (Swami et al., 1989) as well as β-sitosteryl-D-glucoside.[104,105] β-sitosteryl-D-glucoside lowered the plasma glucose of rodents.[104] β-sitosteryl-D-glucoside given orally to administration streptozotocin-induced diabetic rats with severe diabetes had no effect on plasma insulin. This suggest that the mode of action of this glucoxide is based on surviving β-cells. In isolated islets of rats, this sterol increased insulin secretion in nonstimulatory glucose concentration and was inactive in the presence of 16 mmol/L.[106] Farnesoid X receptor and other steroid nuclear receptors are interesting targets for medicinal plants phytosterols. Bile acids such as taurochenodeoxycholate acid at a concentration of 10 μmol/L increased insulin secretion by mouse islets, in the presence of 15 mmol/L of glucose via farnesoid X receptor activation, increased calcium concentration in the cytoplasm, closure of K_{ATP} channels implying that natural products with agonistic property on farnesoid X receptor could increase insulin secretion.[107] What is the mechanism of β-sitosteryl-D-glucoside and/or plasma metabolites on β-cells farnesoid X receptor. It should be remembered that bile acids activate TGR5 expressed by enteroendocrine cells to promote the secretion of glucagon like peptide-1 release and subsequent secretion of insulin by pancreatic β-cells.[108] Oleanolic acid and lithocholic acid at a concentration of 50 μM increased the secretion of insulin by MIN6 cells the presence of 3 mM and 25 mM glucose via TGR5 stimulation and downstream increase of cytosolic cyclic adenosine monophosphate, phospholipase C epsilon (PLCε), inositol-3 phosphate (IP3) and increased cytoplasmic calcium concentration.[108] The effect of aqueous extract of bark of *Ficus religiosa* L. and constituents on TGR5 stimulation should be assessed further.

2.29 *Aporosa lindleyana* (Wight) Baill.

Synonym: *Scepa lindleyana* Wight
Common name: kebella (India)
Subclass Dillenidae, Superorder Euphorbianae, Order Euphorbiales, Family Phyllanthaceae
Medicinal use: jaundice (India)

Aqueous extract of roots of *Aporosa lindleyana* (Wight) Baill. given at a single dose of 100 mg/kg to Wistar rats lowered fasting glycemia from 80.4 to 69.8 mg% after 3 hours.[109] In alloxan-induced Wistar rats, this extract lowered fasting glycemia from 306 to 160 mg%.[109] A feature of type 2 diabetes is a chronic inflammatory state during which inflammatory cytokines such as tumor necrosis factor-α generate DNA strand break in β-cells.[110] Furthermore, islets of many patients with type 2 diabetes have the feature of an inflammatory process reflected by the presence of cytokines and macrophages suggesting that inhibition of intra-islets inflammatory mediators would be able to prevent insulitis in type 2 diabetes.[111] Macrophages secrete interleukin-1β, tumor necrosis factor-α, and interferon-γ which lead to islet dysfunction and apoptotic death of β-cells by a mechanism involving nitric oxide.[112] The antidiabetic mechanism of *Aporosa lindleyana* (Wight) Baill. is unknown but evidence is available to demonstrate that methanol extract of root given orally to rodents afforded antioxidant effects in carbon tetrachloride-induced hepatic insults.[113] Furthermore, Ali et al (2014) made the demonstration that methanol fraction of bark given orally at a dose of 300mg/kg protected rodents against carrageenan induced hind paw edema with decrease of tumor necrosis factor-α and nitric oxide, and decreased expression of nuclear factor-κB and cyclooxygenase-2, lowering of lipid peroxides, and increased activity of antioxidant enzymes.[114] It is therefore reasonable to suggest that ellagitannins (which abound in the family Phyllanthaceae) and or plasma metabolites may afford antioxidant and anti-inflammatory effects at the pancreatic level of streptozotocin-induced diabetic rodents decreasing thus interleukin 1β and tumor necrosis factor α which cause the production of nitric oxide and prostaglandin E_2 through inducible nitric oxide synthase (iNOS) and cyclo-oxygenase-2.[115] interleukin-1β, interleukin-6 and tumor necrosis factor-α generate DNA strand break in β-cells of pancreatic islets.[110]

2.30 *Phyllanthus emblica* L.

Synonym: *Emblica officinalis* Gaertn.
Common names: yu gan zi (Chinese); amla (India); Indian gooseberry
Subclass Dillenidae, Superorder Euphorbianae, Order Euphorbiales, Family Phyllanthaceae
Medicinal use: diabetes (Bangladesh)
History: The plant was known of Sushruta

Glibenclamide is a nonselective K_{ATP} channel inhibitor which is used in type 2 diabetes.[116] Methanolic extract of fruits of *Emblica officinalis* given orally to streptozotocin-induced type 2 diabetic rats (fasting blood glucose between 7.6 mmol/L and 11 mmol/L) at a dose of 500 mg/kg/day for 29 days lowered fasting glucose from 9.2 to 5.4 mmol/L (normal 4.3 mmol/L; glibenclamide at 5 mg/kg/day: 5.5 mmol/L) and increased serum insulin from 41.2 to 89.4 pmol/L (normal 132 pmol/L; glibenclamide at 5 mg/kg/day: 91.1 pmol/L).[117] The extract increased β-cell area from 6522 to 10,525 μm^2 (normal 19,506 μm^2) in diabetic rodents.[117] In oral glucose tolerance test, ellagic acid at a single oral dose of 100 mg/kg lowered peak glycemia at 45 minutes by 23.2%.[117] From this extract, ellagic acid (Figure 2.5) at a concentration of 100 μmol/L increased the secretion of insulin by β-cells, by unknown mechanism, in the presence of high concentration of glucose and not a low glucose concentration suggesting that in hyperglycemic state, ellagic acid might promote insulin secretion.[117] Clinical studies to ascertain the beneficial effect of *Emblica officinalis* Gaertn. in metabolic syndrome are warranted.

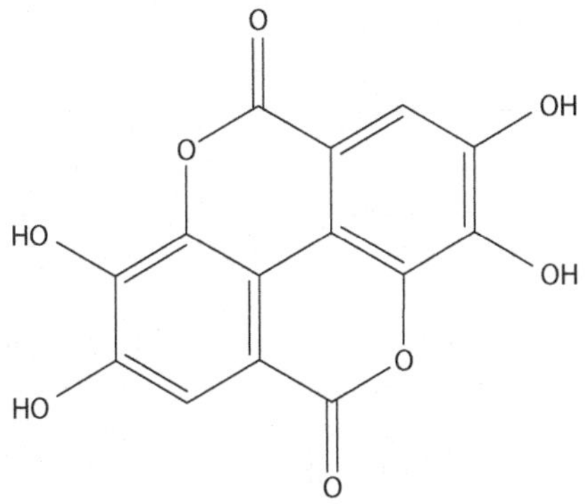

FIGURE 2.5 Ellagic acid.

2.31 *Phyllanthus simplex* Retz.

Synonym: *Phyllanthus virgatus* G. Forst.
Subclass Dillenidae, Superorder Euphorbianae, Order Euphorbiales, Family Phyllanthaceae
Medicinal use: gonorrhea (India)

Aqueous extract of *Phyllanthus simplex* Retz. given orally and daily at a dose of 150 mg/kg to normoglycemic Charles Foster albino rats for 21 days lowered fasting blood glucose from 83.5 to 63.5 mg/mL.[118] The same experiment conducted with alloxan-induced diabetic Charles Foster albino rats lowered fasting blood glucose from 252.3 to 146.5 mg/mL, prevented weight loss and normalized and enzymatic activity of catalase and lipid peroxidation in both liver and kidneys.[118] Furthermore, this regimen brought total cholesterol, triglycerides, and liver glycogen to nondiabetic values.[118] The mode of action of this plant is unknown but an hypoglycemic activity in both normal and diabetic animals suggests an effect mediated by insulin secretion. The plant contains brevifolin and 8,9-epoxy brevifolin which protected rodents against carbon tetrachloride-induced hepatic oxidative insults.[119]

2.32 *Aleurites moluccana* (L.) Willd.

Synonyms: *Aleurites triloba* J.R. Forst. & G. Forst.; *Jatropha moluccana* L.
Common names: shi li (Chinese); buah keras (Malay); lumbang (Philippines); candle berry tree
Subclass Dillenidae, Superorder Euphorbianae, Order Euphorbiales, Family Euphorbiaceae
Medicinal use: constipation (Philippines)

Methanol extract of leaves of *Aleurites moluccana* (L.) Willd. given at a dose of 300 mg/kg/day orally to Wistar rats on high-fat diet for 30 days lowered total cholesterol from 167.8 to 74.6 mg/dL, low-density lipoprotein from 113.7 to 61.9 mg/dL, triglycerides from 231.8 to 109.5 mg/dL and increased high-density lipoprotein–cholesterol from 81.5 to 101.4 mg/dL.[120] This regimen lowered body weight without reduction in food consumption and reduced glycemia by about 30%.[120] The exact mechanism involved here is yet unknown. Methanol extract of leaves of *Aleurite moluccana* inhibited the activity of porcine pancreatic lipase *in vitro*.[121] The plant contains swertisin, which at a single oral dose of 15 mg/kg lowered glycemia by 20% after 30 minutes in oral glucose tolerance due to increased insulin secretion in rodents.[69] Cesca et al. (2012) reported anti-inflammatory effects and wound healing properties from this plant in rodents.[122] The plant is too toxic for being used in therapeutic.

2.33 *Croton klotzschianus* (Wight) Twaites

Synonym: *Oxydectes klotzschianus* (Thwaites) Kuntze
Subclass Dillenidae, Superorder Euphorbianae, Order Euphorbiales, Family Euphorbiaceae
Medicinal use: diabetes (India)

Ethanol extract of aerial part of *Croton klozchianus* (Wight) Twaites given orally to streptozotocin-induced diabetic Sprague–Dawley rats at a dose of 300 mg/kg/day for 3 weeks lowered glycemia by 44.3% and normalized plasma triglycerides and cholesterol.[123] *In vitro*, this extract at 2 mg/mL boosted insulin secretion from MIN6 cells challenged with glucose.[123] The active constituent(s) involved here are to date unknown. Members of the genus *Croton* L. are known to elaborate diterpenes such as trans-dehydrocrotonin, which given orally at a single dose of 50 mg/kg to Wistar rats 1 hour before streptozotocin administration reduced glycemia by 61% after 72 hours.[124] Likewise, a single oral administration of trans-dehydrocrotonin 48 hours after streptozotocin administration evoked a transient fall of glycemia from about 450 to 250 mg/dL.[124] Trans-dehydrocrotonin from *Croton cajucara* Benth., evoked gastroprotective activity of on account of histamine H2 and muscarinic receptors antagonism.[125] In rodents, muscarinic M3 and M1 activation induces the release of insulin whereas in human isolates, M3 muscarinic receptors are the most abundant.[126–128] Gautam et al. (2006) made the interesting demonstration that mutant mice lacking the muscarinic receptor M3 in pancreatic β-cells had impaired glucose tolerance and reduced insulin release. In contrast, mutant mice overexpressing M3 receptors in pancreatic β-cells had increased glucose tolerance and insulinotropic potencies conferring resistance to diet-induced glucose intolerance and hyperglycemia.[129] In a subsequent study, Kong et al. (2010), provided additional evidence that M3 muscarinic receptor stimulation not only enhances insulin release via protein kinase C activation and increase in intracellular calcium but induces 3-Phosphoinositide-dependent protein kinase 1 activation which augments insulin release by promoting trafficking of secretory vesicles to the plasma membrane.[130] Thus, natural products with the ability to stimulate M3 muscarinic receptors in β-cells stimulate insulin secretion and the hypothesis of a cholinergic insulinotropic effect by a trans-dehydrocrotonin-like diterpene from *Croton klotzschianus* (Wight) Twaites need to be explored. Human islets are known to express histamine receptors (Amisten et al., 2013) and the parenteral administration of histamine H1 and H2 antagonists in healthy volunteers result in a decrease of insulin secretion compared with placebo.[126,131] Therefore it would be of interest to explore further the insulinotropic mechanisms of *Croton klotzschianus* (Wight) Twaites and clerodane diterpene constituents with special emphasis on the type of receptors involved.

2.34 *Graptopetalum paraguayense* (N.E. Br.) E. Walther

Synonym: *Cotyledon paraguayensis* N.E. Br.
Common name: ghost plant
Subclass Rosidae, Superorder Rosanae, Order Saxifragales, Family Crassulaceae
Nutritional use: vegetable (Taiwan)

Ethanol extract of leaves of *Graptopetalum paraguayense* (N.E. Br.) E. Walther given intraperitoneally to C57BL/6 mice at a dose of 300 mg/kg/day for 12 weeks lowered plasma glucose and insulinemia during oral glucose tolerance test.[132] In addition, this regimen applied to C57BL/6 mice with pancreas dysfunction induced by carboxymethyllysine, halved peak plasma glucose and boosted insulinemia at 60 minutes during oral glucose tolerance test.[132] At pancreatic level, this extract downregulated CCAAT/enhancer binding protein-β and elevated both peroxisome proliferator-activated receptor-γ and pancreatic and duodenal homeobox-1 expression.[132] The expression and function of mature β-cell is regulated by the pancreatic transcription factor pancreatic and duodenal homeobox-1.[133] Furthermore, the extract increased nuclear factor-erythroid 2-related factor 2 (Nrf2) and glutamate–cysteine ligase, in pancreas as well as the non enzymatic antioxidant glutathione.[132] This extract promoted the phosphorylation of liver and skeletal muscle Akt and therefore glucose uptake.[132] Clinical trails are warranted.

2.35 *Combretum micranthum* G. Don

Subclass Rosidae, Superorder Myrtanae, Order Myrtales, Family Combretaceae
Medicinal use: fever (Vietnam)

Aqueous extract of leaves of *Combretum micranthum* G. Don given to Wistar rats at a single dose of 100 mg/kg lowered glycemia from 110 to 83 mg/dL 1 hour after oral glucose challenge.[134] This extract given to alloxan-induced diabetic Wistar rats at a dose of 100 mg/kg lowered glycemia from 432 to 229 mg/dL 1 hour after glucose challenge and this effect was superior to as gliblenclamide at a dose of 0.6 mg/kg.[134] The leaves contain gallic acid and series of C-glycosylflavones of which vitexin, isovitexin, and orientin.[135,136] Vitexin and gallic acid are known to induce insulin secretion and may participate in the oral hypoglycemic activity of *Combretum micranthum* G. Don.

2.36 *Terminalia catappa* L.

Synonyms: *Myrobalanus catappa* (L.) Kuntze; *Terminalia myrobalana* Roth
Common names: lan ren shu (Chinese); kalisay (Philippines); Indian almond
Subclass Rosidae, Superorder Myrtanae, Order Myrtales, Family Combretaceae
Medicinal use: diuretic (Philippines)

Sulfonylureas and meglitinides bind to and close ATP sensitive potassium channel (potassium-ATP channel) thereby enhancing glucose induced secretion of insulin by β-cells.[137] These secretagogues are used for the treatment of type 2 diabetes.[137] A bulk of experimental evidence exists to demonstrate that in rats, that oral intake of aqueous extracts of medicinal plants lower plasma glucose as potently as glibenclamide. For instance, aqueous extract of fruits of *Terminalia catappa* L. given orally to alloxan-induced diabetic Wistar rats at a dose of 42 mg/kg/day for 21 days lowered fasting glycemia from about 270 to about 120 mg/dL and as potently as glibenclamide 10 mg/kg/day.[138] This regimen and evoked a mild increase of body weight.[138] This treatment normalized cholesterolemia, low-density lipoprotein, high-density lipoprotein to nondiabetic values, triglycerides from 200.8 to 115.5 mg/dL, and reduced serum alkaline phosphatase and improved pancreatic histoarchitecture with notably partial restoration of p cellular population and enlarged size of β-cells with hyperplasia (ii).[138] The fruits of *Terminalia catappa* L. contains cyanidin-3-glucoside, corilagin, ellagic acid, and brevifolin-carboxylic acid.[138] Cyanidin-3-glucoside and ellagic acid could account for the hypoglycemic effect of *Terminalia catappa* L.

2.37 *Lagerstroemia speciosa* (L.) Pers.

Synonyms: *Lagerstroemia flos-reginae* Retz.; *Lagerstroemia reginae* Roxb.; *Munchausia speciosa* L.
Common names: banaba (Philippines); Queen crape-myrtle
Subclass Rosidae, Superorder Myrtanae, Order Myrtales, Family Lythraceae
Medicinal use: diabetes (Philippines).

Aqueous extract of leaves of *Lagerstroemia speciosa* at 50 μg/mL protected Syrian hamster insulin-secreting HIT-T15 cells against oxidative insults with concomitant diminution of intracellular reactive oxygen species, lipid peroxidation and increase of superoxide dismutase activity, glutathione peroxidase and catalase enzymatic activities.[139] Upon treatment with this extract, the cells were able to keep the ability to secrete insulin despite 3-morpholinosydnonimine induced injuries.[139] The leaves contain ellagitannins including largerstroemin, flosin B, and reginin A which could conceptually contribute the protective effect on β-cells against 3-morpholinosydnonimine-induced oxidative insults by simple antioxidant property.[140] Ellagitannins are decomposed in the guts and largerstroemin, flosin B or reginin A would not reach intact and at pharmacological concentration β-cells, skeletal muscles or adipocytes via the plasma. It would be of interest to assess the effect of leave water extract in rodent poisoned with alloxan or streptozotocin and to identify the main circulating phenolic plasma metabolite. The use of *Lagerstroemia speciosa* (L.) Pers. to treat type 2 diabetes in Philippines since a remote period of time and preliminary evidence of pancreato-protective activity *in vitro* warrant further studies to establish with certainty the potential of this plant for metabolic syndrome.

2.38 *Woodfordia fruticosa* (L.) Kurz

Synonyms: *Lythrum fruticosum* L.; *Woodfordia floribunda* Salisb.
Common names: xia zi hua (Chinese); dhavani (India); fire-flamme bush
Subclass Rosidae, Superorder Myrtanae, Order Myrtales, Family Lythraceae
Medicinal use: dysentery (India)
History: The plant was known to Sushruta

Ethanol extract of flowers of *Woodfordia fruticosa* (L.) Kurz given at a single dose of 250 mg/kg to Wistar rats lowered glycemia by 53.2% at 120 minutes.[141] This extract given at a dose of 500 mg/kg/day for 21 days orally to streptozotocin-induced type 1 diabetic Wistar rats lowered fasting glycemia from about 17.5 mM/L to normal value and brought insulinemia up to normal value.[141] This treatment mitigated weight loss, and increased hepatic catalase activity, superoxide dismutase activity, glutathione peroxidase activity, elevated glutathione contents and reduced lipid peroxidation.[141] Furthermore, this extract increased hepatic glucose-6-phosphate dehydrogenase, the first regulatory enzyme of pentose phosphate pathway, hexokinase, lowered glucose-6-phosphatase and increased glycogen contents.[141] Histological study of pancreas of treated rats evidenced prevention islet necrosis and fibrosis.[141] Note that *Woodfordia fruticosa* (L.) Kurz elaborates series of ellagitannins such as woodfordin C, which are responsible for the astringent property of the plant. Note that ellagic acid is released upon intestinal hydrolysis of ellagitannins which abound in the plant. Ellagic acid and gallic acid s probably account for the hypoglycemic effect of this plant. Excessive amounts of tannins in diet are toxic.

2.39 *Jussiaea suffruticosa* L.

Synonym: *Ludwigia suffruticosa* (L.) M. Gómez
Subclass Rosidae, Superorder Myrtanae, Order Myrtales, Family Onagraceae
Medicinal use: astringent (Cambodia)

Methanol extract of *Jussiaea suffruticosa* L. given orally to Wistar rats at a single dose of 400 mg/kg lowered after 24 hours glycemia from 112.1 to 75.6 mg/dL.[142] In alloxan-induced diabetic Wistar rats, this extract at the same dose evoked after 24 hours a fall of glycemia from 296.6 to 173.6 mg/dL.[142] Acute hypoglycemic activity in both normal and alloxan-induced diabetic rats suggest at least, an insulinotropic effect. This aquatic herb is elaborates series of triterpenes including β-amyrin acetate and β-amyrin palmitate, oleanolic acid and ursolic acid, tormentic acid, as well as gallic acid and ellagic acid.[143–145] These may account for the acute hypoglycemic effect of *Jussiaea suffruticosa.* L. via insulinoptropic effects.

2.40 *Syzygium cumini* (L.) Skeels

Synonyms: *Eugenia cumini* (L.) Druce; *Eugenia jambolana* Lam.; *Myrtus cumini* L.; *Syzygium jambolanum* (Lam.) DC.
Common names: wu mo (Chinese); jamu (India); duhat (Philippines); java plum
Subclass Rosidae, Superorder Myrtanae, Order Myrtales, Family Myrtaceae
Medicinal use: diabetes (India)

Methanol extract of bark of *Syzygium cumini* (L.) Skeels given orally at a dose of 5 mg/20 g to Swiss Webster mice concurrently with oral glucose loading, lowered postprandial 30 minutes plasma glucose from 13.5 to 9.2 mmol/L whereby glipizide at 0.023 mg/20 g lowered 30 minutes to 9.7 mmol/L.[146] In a previous study, Schossler et al. (2004) gave extract of bark of *Syzygium cumini* (L.) Skeels to alloxan-induced diabetic Wistar rats (fasting glucose > 180 mg/dL) at a dose of 1 g/kg for 30 days resulting in improvement of pancreatic cytoarchitecture with regeneration of positive insulin staining cells in the epithelia of the pancreatic duct.[147] Ethanol extract containing ellagitannins which at a concentration of 100 mg/mL increased the secretion of insulin by INS-1E cells exposed to 3.3 mM from about 10 to 27 ng/mL/4 h and double the secretion of insulin by INS-1E cells exposed to16.7 mM of glucose.[148] It must be noted that ellagitannins, which are common in members of the order Myrtales, are decomposed in the intestines into ellagic acid which promote insulin secretion by β-cells *in vitro.* The insulinotropic activity and cytoprotective effect of *Syzygium cumini* (L.) Skeels could also be owed to gallic acid which is released upon intestinal hydrolysis of ellagitannins and which is known to elevate plasma insulin in streptozotocin-diabetic Male Swiss albino rat. In rats, not in human, mature β-cells have a life span of a few weeks and are replaced by the replication of pre-existing β-cells or by the differentiation and proliferation of precursor cells present in the pancreatic ducts.[149]

2.41 *Boswellia serrata* Roxb. ex Colebr.

Common name: salai (India); Indian olibanum
Subclass Rosidae, Superorder Rutanae, Order Rutales, Family Burseraceae
Medicinal use: diuretic (India)

Defatted alcoholic extract of gum resin of *Boswellia serrata* Roxb. ex Colebr. (Figure 2.6) given intraperitoneally to streptozotocin-induced diabetic wild-type (WT) BK +/+ mice at a dose of 150 mg/kg/day for 10 days lowered plasma glucose, prevented lymphocyte infiltration into pancreatic

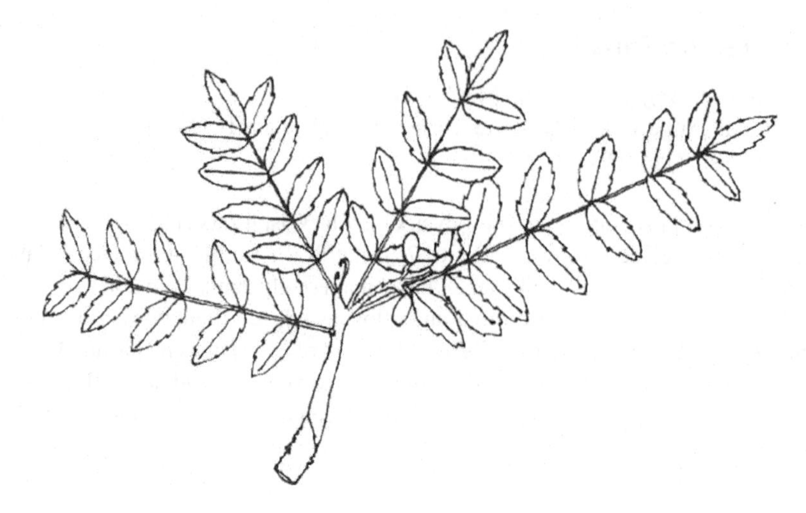

FIGURE 2.6 *Boswellia serrata* Roxb. ex Colebr.

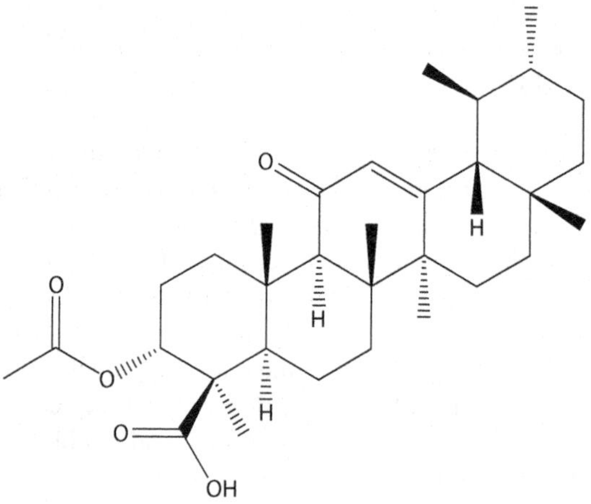

FIGURE 2.7 3-*O*-Acetyl-11-keto-β-boswellic acid.

islets, islet shrinkage and apoptosis in peri-insular tissue. This extract decreased serum levels of proinflammatory cytokines inteleukin-6, interleukin-β and tumor necrosis factor-α induced by streptozotocin.[150] From the gum resin of *Boswellia serrata* Roxb. ex Colebr, 11-keto-β-boswellic acid and 3-*O*-acetyl-11-keto-β-boswellic acid (Figure 2.7) given intraperitoneally at a dose of 15 mg/kg/day for 10 days prevented streptozotocin-induced pancreatic injuries as evidenced by attenuated proinflammatory cytokines in the blood, reduced infiltration of lymphocytes into pancreatic islets, and decrease of glycemia.[151] 11-Keto-β-boswellic acid and acetyl-11-keto-β-boswellic acid have the ability to inhibit 5-Lipoxygenase *in vitro* with an IC_{50} value of 2.8 µM and 1.5 µM, respectively.[152] This enzyme catalyzes the synthesis of leukotriene B4 from arachidonic acid which is one of the first leukocyte chemoattractants generated to promotes leukocytes migration in response to tissue injuries. This eicosanoid contributes to the development of type 2 diabetes in obese.[153,154] 3-*O*-acetyl-11-Keto-β-boswellic acid administered parenterally to rodents at a dose of 100 μmol/kg/day for 7 days inhibited nuclear factor-κB activation induced by lipopolysaccharide in mice.[155] Gerbeth et al. (2011) provided evidence that in healthy volunteers ingesting 300 mg of gum resin of *Boswellia serrata* Roxb. ex Colebr. 3-*O*-acetyl-11-keto-β-boswellic acid could ne be detected in the plasma whereas plasmatic concentration of 11-keto-β-boswellic acid ranged between 6.4 and 247 ng/mL.[156] Further pre-clinical studies are needed to appraise the potentiality of gum resin of *Boswellia serrata* Roxb. ex Colebr to prevent type 2 diabetes in metabolic syndrome.

2.42 *Commiphora mukul* (Hook. ex Stocks) Engl.

Synonym: *Commiphora wightii* (Arn.) Bhandari
Common names: guggul (India); Indian Bdellium
Subclass Rosidae, Superorder Rutanae, Order Rutales, Family Burseraceae
Medicinal use: ulcers (India)

Guggulsterone from the gum resin of *Commiphora mukul* (Hook. ex Stocks) Engl. at a concentration of 25 μM increased the survival of rat pancreatic β RIN cells challenged with inteleukin-1β and interferon-γ by 80% and prevented cytokine-induced reduction of cell proliferation.[157] This treatment suppressed cytokine-induced activation of nuclear factor-κB and subsequent expression of inducible nitric oxide synthetase, nitric oxide generation, cyclo-oxygenase-2 expression and prostaglandin-E_2 production.[157] Besides, guggulsterone attenuated cytokine-induced phosphorylation and nuclear translocation of STAT-1 and STAT-3 and prevented the downregulation of SOCS-3.[157] At a concentration of 25 μM, this sterol protected *in vitro* rat pancreatic islets against interleukin-1β and interferon-γ induced decrease in glucose-induced insulin secretion.[157] In β-cells, interferon-γ induce the activation of STAT-1 and downstream activation of inducible nitric oxide synthetase. Besides, SOCS-3 is a known inhibitor of nuclear factor-κB and downstream expression of inducible nitric oxide synthetase.[157] Guggulsterone at a concentration of 50 mM has been shown to decreases growth factor induced STAT 1/3 phosphorylation and therefore their translocation to the nucleus and to inhibit phosphoinositide 3-kinase in cancer cell.[158] Human exposure to the gum resin of *Commiphora mukul* (Hook. ex Stocks) Engl. include skin rashes (Kölönte et al. 2006) as well as acute liver failure.[159,160]

2.43 *Garuga pinnata* Roxb.

Common names: yu ye bai tou shu (Chinese); golika (India); garuga
Subclass Rosidae, Superorder Rutanae, Order Rutales, Family Burseraceae
Medicinal use: asthma (India)

Aqueous extract of bark of *Garuga pinnata* Roxb. Wistar rats given at a single oral dose of 500 mg/kg to Wistar rats lowered 30 min plasma glucose concentration from 110 to 70 mg/dl from and shifted maximal glycemia from 30 minutes to 1 hour suggesting delayed absorption of glucose.[161] This extract given at a dose of 500 mg/kg per day for 15 days to streptozotocin-nicotinamide induced type 2 diabetic Wistar rats lowered fasting glycemia from 242.5 to 110.1 mg/dL (p value 75.3 mg/dL).[161] This regimen increased serum insulin from 55 to 114.1 μU/mL (normal 130.2 μU/mL), and increased liver glycogen from 5.2 to 14.1 mg/g (above normal values 12.5 mg/g).[161] This extract decreased triglycerides and total cholesterol in diabetic rodents and increased high-density lipoprotein.[161] The extract reduced weight loss in diabetic animals.[161] Histological evaluation of pancreas of treated animals revealed that the extract prevented streptozotocin-nicotinamide pancreatic insults The *medicinal use* of the plant and its position in the family Burseraceae suggest the involvement of anti-inflammatory triterpenes.[161] The bark of *Garuga pinnata* Roxb. contains 13α,14β,17α-lanosta-7,24-diene-1β,3β-diol and 13α,14β,17α-lanosta-7,24-diene-3β,1β,16α–triol that would be worth being assessed with regard to their insulinotropic and or anti-inflammatory effects.[162]

2.44 *Rhus verniciflua* Stokes

Synonym: *Toxicodendron vernicifluum* (Stokes) F.A. Barkley
Common name: qi shu (Chinese)
Subclass Rosidae, Superorder Rutanae, Order Rutales, Family Anacardiaceae
Medicinal use: bleeding (China)

The pathophysiology of type 2 diabetes entails a chronic inflammatory state during which proinflammatory cytokines such as tumor necrosis factor-α, interleukin-6 and interleukin-1β generate DNA strand break in β-cells.[110] Furthermore, islets of many patients with type 2 diabetes have the feature of an inflammatory process reflected by the presence of cytokines, immune cells and apoptosis.[111] This suggests that inhibition of intra-islets inflammatory mediators would be able to prevent insulitis in type 2 diabetes. In pancreatic islets, interleukin-1β, tumor necrosis factor α cause the production of nitric oxide and prostaglandin E_2 through inducible nitric oxide synthase (iNOS) and cyclo-oxygenase-2.[115] Butein (Figure 2.8) from *Rhus verniciflua* Stokes at a concentration of 30 μM prophylactically protected pancreatic β-cell INS-1 *in vitro* against interleukin-1β with an increase in viability by 80.1% with simultaneous inhibition of nuclear factor-κB, decrease in inducible nitric oxide synthase expression and nitric oxide production.[163] This chalcone protected *in vitro* pancreatic islets isolated from Sprague–Dawley rats against interleukin-1β and interferon-γ against induction of inducible nitric oxide synthase and subsequent overproduction of nitric oxide.[163] In addition, butein prevented inhibition of insulin secretion induced by interleukin-1β and interferon-γ *in vitro* γ.[163] Sung and Lee provided evidence that butein inhibited the nuclear translocation of nuclear factor-κB, nitric oxide synthase expression and nitric oxide production in RAW264.7 macrophages challenged lipopolysaccharide. This phenolic compound enhanced the expression of heme oxygenase-1 in RAW264.7 macrophages through probable involvement of the Nrf2/ARE (see p.) and phosphoinositide 3-kinase/Akt pathways.[164,165]

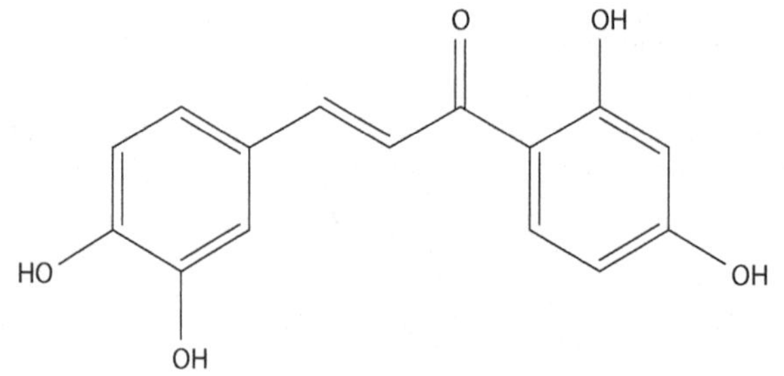

FIGURE 2.8 Butein.

2.45 *Spondias pinnata* (L.f.) Kurz.

Synonym: *Mangifera pinnata* L. f.; *Spondias acuminata* Roxb.
Common name: bing lang qing (Chinese)
Subclass Rosidae, Superorder Rutanae, Order Rutales, Family Anacardiaceae
Medicinal use: diabetes (Philippines)

Aqueous decoction of stem bark of *Spondias pinnata* (L.f.) Kurz. given orally at a single dose of 2 g/kg to alloxan-induced diabetic Wistar rats 30 minutes before an oral 3 g/kg dose of glucose lowered total area under glucose tolerance curve from 70.3 to 50 (normal 25.3).[166] The natural products and mechanism involved in this acute anti-hyperglycemic effect are unknown but one could suggest the probable involvement of at least gallotannins which abound in since members of the family Anacardiaceae. Gallotannins given to streptozotocin to CD-1 mice favored β-cells regeneration.[167] Regeneration of β-cells upon streptozotocin insults is improved by exogenous insulin-induced euglycemia with faster removal of necrotic β-cells and accelerated recovery of islets.[168] Therefore it could be hypothesized that metabolites of gallotannins produced *Spondias pinnata* (L.f.) Kurz. could enhance β-cell recovery and insulin secretion in rats by lowering plasma glucose, reactive oxygen species.[169]

2.46 *Desmodium gangeticum* (L.) DC.

Synonyms: *Desmodium cavaleriei* H. Lév.; *Hedysarum gangeticum* L.
Common names: da ye shan ma huang (Chinese); saalpernie (India)
Subclass Rosidae, Superorder Fabanae, Subclass Rosiidae, Family Fabaceae
Medicinal use: difficulties of breathing (India)
History: The plant was known of Sharf Khan, 18th physician of Mughal emperor Shah Alain

An ethanol extract of aerial parts of *Desmodium gangeticum* (L.) DC. given orally to Sprague–Dawley rats at a dose of 200 mg/kg during oral glucose tolerance inhibited raise of glycemia compared with untreated animals.[170] This extract given orally to streptozotocin-induced diabetic Sprague–Dawley rats at a dose of 200 mg/kg for 2 weeks lowered glycemia from 246.4 to 127.1 mg/dL, triglycerides from 181.2 to 101.2 mg/dL, total cholesterol from 119 to 82.1 mg/dL, and increased high-density lipoprotein from 29.8 to 80.7 mg/dL.[170] The extract at a dose of 2 mg/mL induced the secretion of insulin by MIN6 cells in the presence of calcium chelator.[170] The extract induced a transient and significant secretion of insulin by mouse MIN6 β-cells pseudoislets exposed to the extract (at a dose of 2 mg/mL).[170] An interesting feature of *Desmodium gangeticum* (L.) DC. is the presence of 5-methoxy-*N*, *N*-dimethyltryptamine, *N*, *N*-dimethyltryptamine, *N*-methyl-tetrahydroharman and 6-methoxy-β-carbolinium as well as pterocarpanoids such as gangetin, gangetinin, and desmodin.[171,172] The precise molecular basis of insulinotropic activity is unknown. Note that tryptophan at a single intraperitoneal dose of 750 mg/kg induced a profound hypoglycemia in both fed or starved rodents with maximum effect at 6 hours.[173] This effect was abrogated with the tryptophane hydroxylase inhibitor p-chlorophenylalanine, the aromatic L-amino acid decarboxylase inhibitor MK-486, or methysergide and boosted by the monoamine oxidase inhibitor pargyline suggesting 5-hydroxytryptamine (serotonine) to be responsible for the aformentionned activity.[173] In human, β-cells in Langerhans islets of the pancrease express serotonin receptors 5-hydroxytryptamin-1A, 5-hydroxytryptamin-1B, 5-hydroxytryptamin-1F, 5-hydroxytryptamin-2A, 5-hydroxytryptamin-6, and 5-hydroxytryptamin-7.[126] 5-Hydroxytryptamin-1A, 5-hydroxytryptamin-1B, and 5-hydroxytryptamin-1F receptor agonist sumatripan inhibit insulin, glucagon and somatostatin secretion in human.[174] It would be therefore of interest to appraise the insulinotropic effect of 5-methoxy-*N*, *N*-dimethyltryptamine, *N*, *N*-dimethyltryptamine, *N*-methyl-tetrahydroharman, and 6-methoxy-β-carbolinium in rodent models of type 2 diabetes. As for gangetin, gangetinin, and desmodin, their pharmacological activities are unexplored yet. Rathi et al. (2004) provided evidence that administration of aqueous extract of aerial part of this plant at a dose of 20 mk/kg prevented carageenin-induced inflammation in rodents.[175] One could speculate that the plant contains anti-inflammatory and/or antioxidant water soluble principles that would, at least partly, improve pancreatic function in streptozotocin-induced diabetic rodents. Such principles need to be identified and their mechanisms clarified.

2.47 *Pterocarpus marsupium* Roxb.

Synonyms: *Lingoum marsupium* (Roxb.) Kuntze; *Pterocarpus bilobus* Roxb. ex G. Don
Common names: ma la ba zi tan (Chinese); kum kusrala (India); kino
Subclass Rosidae, Superorder Fabanae, Subclass Rosiidae, Family Fabaceae
Medicinal use: diabetes (India)
History: The plant was known of Sushruta

This tree accumulates series of polyphenolic substances of which (−)-epicatechin is bioavailable in human and probably responsible, to some extent, for the antidiabetic activity of the plant.[175,176] *In vitro*, this flavanol increased glucose-induced insulin secretion by islets of Langerhans of rats at a concentration of 1 mM.[177,178] Mohankumar et al. (2012) provided further evidence that aqueous extract of wood of this tree enhanced *in vitro* insulin secretion from mouse pancreatic tissues.[179] Aqueous extract of the plant given orally at a daily dose of 200 mg/kg for 4 weeks lowered plasma glucose in streptozotocin-induced diabetic rats with concurrent reduction of circulating tumor necrosis factor-α (Halagappa et al. 2010).[180] The anti-inflammatory principle involved here is yet unknown. Note that pterostilbene isolated from the wood given intraperitoneally lowered plasma glucose in hyperglycemic rats via unknown mechanism.[181] Pterostilbene is anti-inflammatory and antioxidant and absorbed orally in human (Hougee et al., 2005; Kosuru et al., 2016) making it an interesting candidate for clinical trials for the prevention or treatment of metabolic syndrome.[182,183]

2.48 *Lupinus albus* L.

Synonym: *Lupinus termis* Forssk.
Common names: bai yu shan dou (China); white lupine
Subclass Rosidae, Superorder Fabanae, Order Rosiidae, Family Fabaceae
Nutritional use: food (China)

Evidence is available to demonstrate that quinolizidine alkaloids from members of the genus *Lupinus* L. decrease glycemia by promoting insulin secretion. Sparteine administered intravenously at a single bolus of 15 mg/kg and 90 mg over 60 minutes in type 2 diabetic patients evoked a decreased of glycemia and increase of plasma insulin.[184] Multiflorine (Figure 2.9), isolated from *Lupinus albus* L. given intraperitoneally at a dose of 30 mg/kg halved the postprandial 60 minutes peak glycemia of ICR mice subjected to oral glucose tolerance test.[185] Sparteine and thionosparteine at a dose of 8.6 mg/kg and lupanine at a dose of 22.1 mg/kg given intraperitoneally to rodents poisoned with

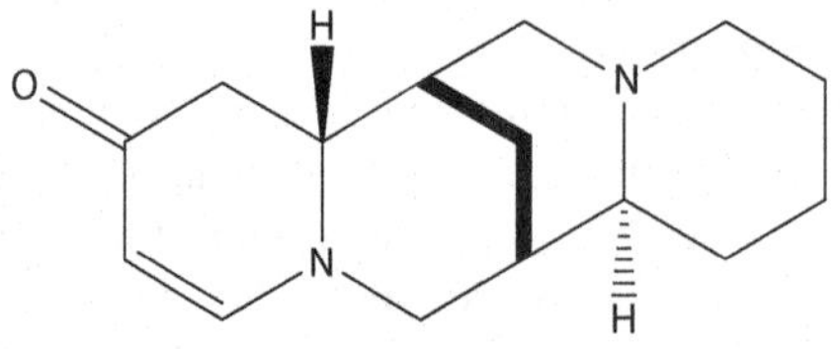

FIGURE 2.9 Multiflorine

streptozotocin lowered glycemia at 90 minutes as potently as glibenclamide given orally at a dose of 3 mg/kg.[186] Furthermore, 2-thionospareine, lupanine, 13-α-hydrohylupanine or 17-oxo-lupanine at a dose of 0.5 mM enhanced the secretion of insulin by isolated pancreatic islets challenged with glucose *in vitro* and this effect was inhibited by the opener of K_{ATP}-sensitive channels diazoxide.[187] Total alkaloids extract of seeds of from *Lupinus mutabilis* Sweet given at a dose or at a dose of 2.5 mg/kg to type 2 diabetic subjects on low dose of metformin evoked after 90 minutes a 9.9% decrease in glycemia.[188] Lupanin (Figure 2.10) which given orally at a single dose of 20 mg/kg to streptozotocin-induced diabetic Wistar rats 30 minutes before oral glucose lowered postprandial glycemia at 60 minustes.[189] In healthy rats, oral administration of lupanin had no effect.[189] Lupanin at a concentration of 0.5 mmol/L enhanced the secretion of insulin by rat islets exposed to high (15 mmol/L) but not low (8 mmol/L) glucose by about 140%.[189] This quinolizidine alkaloid inhibited K_{ATP} current, increased the frequency of calcium action potential. Insulin secretion by lupanine was increased by coadministration of arginine.[189] In human pancreatic α-cells secrete acetylcholine as paracrine transmitter when plasma glucose is low to sensitize β-cells to forthcoming increase in glycemia.[190] In rodents, acetylcholine is released from parasympathetic nerves and β-cell express muscarinic receptor M1, the activation of which induces insulin secretion.[128] While glucose is the primary stimulus for insulin secretion, neurotransmitters and hormones bind to specific cell surface receptors.[73] Vagal activation releases acetylcholine which binds to M3 muscarinic receptors, activate phospholipase C via G protein-coupled mechanism, stimulating phosphoinositide hydrolysis to release diacylglycerol and inositol 3-sphosphate resulting in insulin secretion.[191] Quinolizidines have affinities for muscarinic receptors.[192] Deciphering the precise insulinotropic and, if any, cytoprotective pathways involved in *Lupinus* alkaloids may participate in the development of new leads for metabolic syndrome and associated type 2 diabetes. Furthermore, studies on the beneficial effect of *Lupinus albus* L. seeds consumptions in metabolic syndrome are needed.

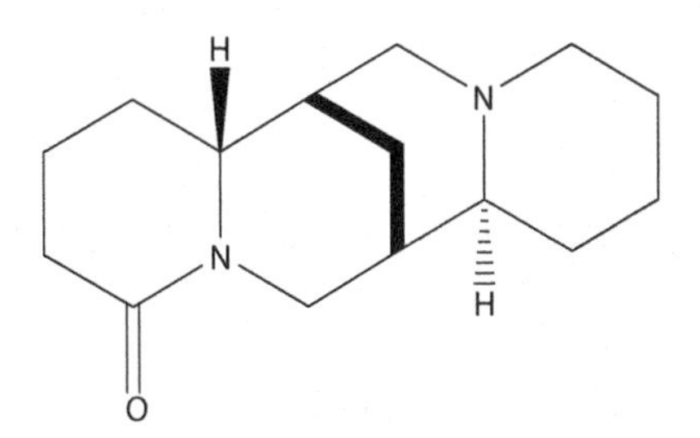

FIGURE 2.10 Lupanine.

2.49 *Aegle marmelos* (L.) Corrêa

Synonym: *Crataeva marmelos* L.
Common names: mu ju (Chinese); bael (India); Bengal quince
Subclass Rosidae, Superorder Rutanae, Order Rutales, Family Rutaceae
Medicinal use: constipation (India)
History: Sacred tree of the Hindus, known to Sushruta

The phenylethyl cinnamides anhydromarmeline, anhydroaegeline, and anhydromarmeline from the leaves of *Aegle marmelos* (L.) Corrêa (Figure 2.11) inhibited at a concentration of 10 μg/mL the enzymatic activity of α-glucosidase by 30.1%, 56.5%, and 36%, respectively.[193] Aqueous extract bark of *Aegle marmelos* (L.) Corrêa given to Sprague–Dawley rats intragastrically at a dose of 1 mL/100 g lowered fasting blood glucose by 44%, 3 hours after administration and this hypoglycemia perdured

FIGURE 2.11 *Aegle marmelos* (L.) Corrêa.

5 hours (42%).[78] This extract given at a single doses of 1 ml/100 g 30 minutes before an oral loading of glucose (1 ml/100 g body weight, 50% w/v) lowered by 30% peak glycemia at 1 hour compared with untreated animals.[78] Methanol extract of bark given to streptozotocin-induced diabetic Wistar rats at a daily oral dose of 400 mg/kg for 15 days lowered peak glycemia from 398 to 283.2 mg/dL (normal 128.3 mg/dL; glibenclamide at 2.5 mg/kg: 287.3 mg/dL) at 30 minutes.[194] After 28 days of treatment, this regimen lowered fasting glycemia from about 400 mg/dL to 125 mg/kg (normal about 75 mg/dL; glibenclamide at 2.5 mg/kg: 125 mg/kg).[194] This regimen decreased weight loss induced by diabetes, increased insulinemia from 6.6 to 15.6 μU/mL (normal 16.7 μU/mL; glibenclamide: 15.6 μU/mL).[194] Furthermore, this extract and improved pancreatic histoachitecture.[194] The bark contains series of lignan glucosides including (−)-lyoniresinol 2α-*O*-β-D-glucopyranoside with antioxidant activity which may participate in β-cell protection against streptozotocin-induced oxidative insults in rats.[195,196] Chronic exposure of β-cells to elevated plasmatic glucose may favor to the production of reactive oxygen species, the activation of nuclear factor-κB and subsequent reduction of insulin secretion and intake of *Aegle marmelos* (L.) Corrêa may prevent the development of pancreatic insulinic exhaustion leading to type 2 diabetes. Besides, the bark contains skimmianine, fagarine, and marmin.[197] It should be noted that skimmianine inhibits 5-phosphodiesterase and could as such promote insulin secretion. Phosphodiesterase inhibitors such as theophylline

increase β-cell contents in cyclic adenosine monophosphate and promotes insulin release induced by glucose.[198] Norepinephrine inhibits the formation of cyclic adenosine monophosphate and distal exocytotic machinery.[199] Physiologically, the insulinotropic gut hormone glucagon-like peptide-1 produced by L-type enterocytes of the small intestine during digestion of mixed meals bind to its G-protein-coupled receptors on β-cells and promotes glucose-induced secretion of insulin by activating adenylate cyclase, increasing cyclic adenosine monophosphate hence activation of protein kinase A and cyclic adenosine monophosphate-regulated guanine nucleotide exchange factor II leading to altered ion channel activity, elevation of cytoplasmic calcium contents and exocytosis of insulin-containing vesicles.[44] Clinical trials are warranted.

2.50 *Citrus limon* (L.) Osbeck

Synonym: *Citrus limonum* Risso
Common names: limau (Malay); lemon
Subclass Rosidae, Superorder Rutanae, Order Rutales, Family Rutaceae
Medicinal use: high cholesterol (Malaysia)

Nomilin from *Citrus* seeds given at 0.2% of a high-fat diet for 77 days to C57BL/6J lowered serum glucose from 224 to 156 mg/dL, lowered insulin from 1.3 to 0.1 ng/mL, reduced nonesterified fatty acids from 581 to 385 μEq/L and lowered epididymal white adipose tissue from 2 to 1.2 g/mouse.[200] This regimen had no effect on triglycerides and food intake.[200] At 100 μM nomilin (Figure 2.12) and aubacunone induced the activation of TGR-5 activity in HEK293 cells suggesting that this limonoid may stimulate the expression of glucagon-like peptide-1 from enteroendrocrine L-cells.[200] TGR5 is a receptor expressed by enteroendocrine L cells, and its activation induces the secretion of intestinal glucagon-like peptide-1 in mice.[201] Glucagon-like peptide-1 is an hormone that act on β-cells to promotes glucose-induced secretion of insulin by increasing cyclic adenosine monophosphate and known to improve glucose utilization by peripheral tissues.[44,202,203]

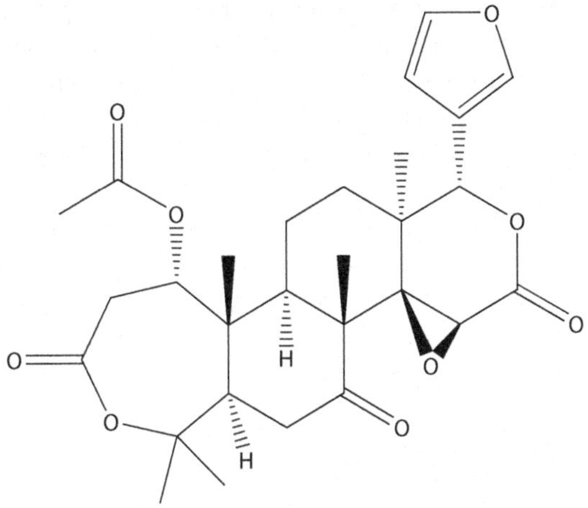

FIGURE 2.12 Nomilin.

2.51 *Clausena anisata* (Willd.) Hook. f. ex Benth.

Synonyms: *Amyris anisata* Willd.; *Clausena inaequalis* (DC.) Benth.
Subclass Rosidae, Superorder Rutanae, Order Rutales, Family Rutaceae

Methanol extract of roots of *Clausena anisata* (Willd.) Hook. f. ex Benth. given orally at a single dose of 800 mg/kg to fasting streptozotocin-induced diabetic Wistar rats evoked after 4 hours a fall of glycemia from about 570 to 270 mg/dL.[204] In the same experiment a subcutaneous injection of insulin (5 mU/kg) lowered glycemia to about 170 mg/dL.[204] In normoglycemics Wistar rats this extract lowered fasting glycemia from about 110 to 45 mg/dL after 8 hours, glibenclamide 0.2 mg/kg lowered fasting glycemia to about 45 mg/dL after 4 hours, and insulin lowered fasting glycemia to about 35 mg/dL after 2 hours.[204] The stem bark contains carbazole alkaloids of which clausenol, clausenin, girinimbine, murrayamine-A, and ekeberginine.[205,206] Carazolol is a β-receptor antagonist with a carbazole moiety used in veterinary medicine.[207] Note that islets of Langerhans express β1 and β2 receptors the blockade of which abrogated insulin secretion.[126,208] Although highly speculative, one could suggest that carbazole alkaloids in this plant acutely stimulate insulin secretion by interacting with β-cells adrenergic receptors. In addition, *O*-methylmurrayamine A from *Murraya koenigii* (L.) Spreng inhibited the release of tumor necrosis factor-α and interleukin-6 by human peripheral mononuclear blood cells challenged *in vitro* with lipopolysaccharides.[209] Methoxylation of murrayamine-A into *O*-methylmurrayamine A during first pass hepatic metabolism would not be surprising and it would be of interest to study the effect of chronic intake of *Clausena anisata* (Willd.) Hook. f. ex Benth. if not toxic, on type 2 diabetes and metabolic syndrome.

2.52 *Poncirus trifoliata* (L.) Raf.

Synonym: *Citrus trifoliata* L.
Subclass Rosidae, Superorder Rutanae, Order Rutales, Family Rutaceae
Medicinal use: indigestion (China)

Tangeretin (Figure 2.13) isolated from *Poncirus trifoliata* (L.) Raf. given orally to streptozotocin-induced diabetic Wistar rats for 30 days at a dose of 100 mg/kg/day prevented weight loss, lowered food and fluid intake.[210] By the end of the treatment, this flavonoid lowered glycemia from 323.3 to 133.1 mg/dL (glibenclamide 5 mg/kg/day: 128.3 mg/L; normal 90.3 mg/dL).[210] This polymethoxylated flavone increased insulinemia from 6.2 to 15.9 µU/L (glibenclamide 5 mg/kg/day: 16.3 µU/L; normal 17.7 µU/L).[210] Oral glucose tolerance test performed at the end of the treatment revealed that this Tangeretin at 100 mg/kg/day was almost as efficient as insulinotropic agent (5 mg/kg/day) in improved glucose tolerance in diabetic rodents.[210] This regimen, increased the enzymatic activity of hepatic glycogen synthetase and decreased the enzymatic activity of glycogen phosphorylase.[210]

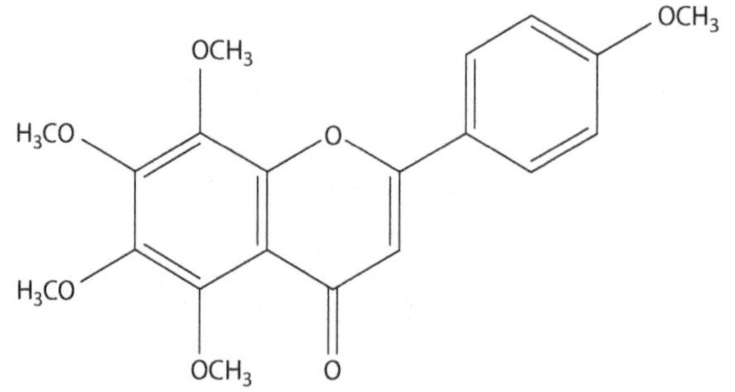

FIGURE 2.13 Tangeretin.

It also evoked the regeneration of islets in streptozotocin-induced diabetic Wistar rats.[210] This flavone given orally to streptozotocin-induced diabetic Wistar rats at a dose of 100 mg/kg/day for 30 days reduced plasma tumor necrosis factor-α, the translocation of nuclear factor-κB, and interleukin-6 close to normal levels implying ant-inflammatory effects.[211] The precise insulinotropic mechanism of tangeretin remains unknown.

2.53 *Toddalia asiatica* (L.) Lam

Synonyms: *Paullinia asiatica* L.; *Toddalia aculeata* Pers.; *Toddalia angustifolia* Lam.; *Toddalia floribunda* Wall.; *Toddalia nitida* Lam.; *Toddalia rubricaulis* Roem. & Schult.; *Toddalia tonkinensis* Guillaumin
Common names: fei long zhang xue (Chinese); dauag (Philippines); wild orange tree
Subclass Rosidae, Superorder Rutanae, Order Rutales, Family Rutaceae
Medicinal use: Tonic (Burma)

Ethylacetate extract of leaves of *Toddalia asiatica* (L.) Lam given orally to Wistar rats at a single dose of 500 mg/kg, 30 minutes before oral glucose loading, lowered 30 minutes postprandial peak glycemia from about 125 to 110 mg/dL (glibenclamide 0.6 mg/kg: 100 mg/dL).[212] This extract given daily at a dose of 500 mg/kg for 28 days to streptozotocin-induced diabetic Wistar rats decreased glycemia from 290.3 to 108.2 mg/dL (glibenclamide at 0.6 mg/kg: 101.3 mg/dL).[212] This regimen prevented weight loss, normalized insulinemia from 4.3 to 13.6 μU/mL; replenished hepatic glycogen, and improved pancreatic histoarchitecture.[212] The leaves of *Toddalia asiatica* (L.) Lam contain series of quinoline alkaloids such as flindersine as well as the coumarin ulopterol.[213–215] Interleukin 1β and tumor necrosis factor α cause the production of nitric oxide and prostaglandin E_2 in β-cells through inducible nitric oxide synthase and cyclo-oxygenase-2.[115] Flindersine inhibits the production of prostaglandin E_2 in 3T3 Swiss albino mouse embryonic fibroblast cells by calcium ionophore with an IC_{50} value of 5 μM.[215] The leaves produce an essential oil comprising β-phellandrene, α-phellandrene cis-β-ocimen and α-pinene, which exhibited anti-inflammatory in rodents.[216,217] Studies aimed at elucidating the insulinotropic mode of action of *Toddalia asiatica* (L.) Lam and the possible role of flindersine

2.54 *Ailanthus excelsa* Roxb.

Common name: maha ruka (India)
Subclass Rosidae, Superorder Rutanae, Order Rutales, Family Simaroubaceae
Medicinal use: fever (India)

Methanol extract of leaves of *Ailanthus excelsa* Roxb. given orally to Wistar rats at a single oral administration of 70 mg/kg had no effect on glycemia.[218] At a single oral dose of 70 mg/kg the lowered glycemia in rats in oral glucose tolerance test from about 130 to 100 mg/dL at 90 minutes whereby peak glycemia at 60 minutes was unchanged.[218] Glymepiride at 5 mg/kg evoked a fall of glycemia at 60 minutes.[218] This extract given daily and orally at a dose of 70 mg/kg/day to streptozotocin-induced diabetic rats 60 days evoked a rapid decrease of glycemia from 423.7 to 113 mg/dL and increased plasma insulin from 4.5 to 13.8 μIU/mL (normal 14.7 μIU/mL).[218] The inability for the methanol extract of leaves to lower the glycemia in non-diabetic rats is an indication that the stimulation of insulin secretion is not probable. The authors record an increase of insulinemia in streptozotocin-induced diabetic rats which suggest recovery of β-cell mass and/or increased secretion of insulin in hyperglycemic plasmatic environment. The leaves are known to contain anti-oxidant flavonoids[219] which are probably attenuating streptozotocin-induced oxidative stress.[218] The plant being a Simaroubaceae contains series of quassinoids of which excelsin, glaucarubine, ailanthinone, glaucarubinone, and glaucarubolone.[220] Zarse et al. (2011) provided evidence that nematodes Bristol N2 *Caenorhabditis elegans* exposed to glaucarubinone at a concentration of 10 nM increased mitochondrial activity, a decrease in triglyceride contents by 21%, and evoked an increase of lifespan.[221] Therefore, one could make an inference that the potent hypoglycemic effects observed diabetic rodents could, at least, result from an increased absorption of glucose by skeletal muscles and adipocytes induced by quassinoids. Increase in cytoplasmic ATP commands the fermeture of ATP sensitive potassium channel (K-ATP channel) hence membrane depolarization, opening of voltage sensitive Ca^{2+}, influx of Ca^{2+}, and secretion of insulin.[199] It would be of interest to assess the effect of glaucarubinone and quassinoids, which are often cytotoxic, in general on β-cells since increased mitochondrial activity is synonym of increased ATP formation hence insulin secretion via blockade of K_{ATP} channels.

2.55 *Brucea javanica* (L.) Merr.

Synonyms: *Rhus javanica* L.; *Brucea sumatrana* Roxb.; *Gonus amarissimus* Lour.
Common names: ya dan zi (Chinese); lada pahit (Malay); Java Brucea
Subclass Rosidae, Superorder Rutanae, Order Rutales, Family Simaroubaceae
Medicinal use: dysentery (China)

Bruceine E isolated from the seeds of *Brucea javanica* (L.) Merr. given at a single oral dose of 2 mg/kg to normoglycemic mice evoked after 8 hours a fall of fasting glycemia and this effect was as potent as glibenclamide.[222] In streptozotocin-induced diabetic Sprague–Dawley rats, bruceine E lowered at a single oral dose of 2 mg/kg fasting glycemia from about 16 to 5 mmol/L after 8 hours, and this effect was as potent as glibenclamide.[222] From the same seeds, bruceine D give orally at a dose of 1 mg/kg to mice lowered fasting glycemia of mice from about 3.5 to 2 mmol/L, and this effect was superior to glibenclamide.[222] Bruceine D at a single oral dose of 1 mg/kg lowered the fasting glycemia of streptozotocin-induced diabetic Sprague–Dawley from about 16 to 2.5 mg/kg, whereby glibenclamide brought fasting glycemia to about 5 mmol/L.[222] The acute hypoglycemic of bruceine D and E suggests the stimulation of insulin secretion by β-cells by a mechanism which is yet elusive. Kuriyama et al. (2005) suggested γ-amino butyric acid receptor antagonism in the nematocidal activity of quassinoids, and also highly speculative, one could draw an inference that inhibition of such receptors in β-cell would participate in insulin secretion.[223] In fact, Braun et al. (2004) made the demonstration that in rat pancreatic islets, the $GABA_B$ receptor antagonist CGP 55845 increased glucose-stimulated insulin secretion whereas the $GABA_B$ receptor agonist baclofen inhibited glucose-stimulated insulin secretion.[224]

2.56 *Panax ginseng* C.A Mey.

Synonyms: *Aralia ginseng* (C.A. Mey.) Baill.; *Panax chin-seng* Nees; *Panax quinquefolius* var. *ginseng* (C.A. Mey.) Regel & Maack
Common names: ren shen (Chinese); ginseng
Subclass Cornanae, Superorder Cornanae, Order Apiales, Family Araliaceae
Medicinal use: tonic (China)

Ethanol extract of roots of *Panax ginseng* C.A Mey. at a concentration of 0.2 mg/mL increased by 281.8% the secretion of insulin by β-cells islets challenged with glucose at a concentration of 8.4 mM *in vitro*.[225] This insulinotropic effect was reduced by L-type calcium channel blocker nifedipine and potassium ATP-channel opener diazoxide.[225] Ginsenoside Rg3 (Figure 2.14) from

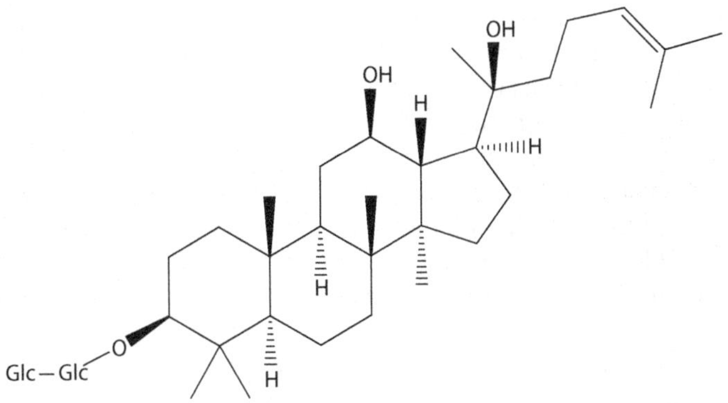

FIGURE 2.14 Ginsenoside Rg3.

ginseng at a concentration of 8 μM augmented the secretion of insulin by hamster pancreatic β-cells HIT-T15 cells induced by glucose and this effect was inhibited by the potassium channel opener diazoxide and L-type calcium channel blocker nifedipine suggesting the involvement of ATP sensitive potassium channel.[226] This saponin given at a single oral dose of 25 mg/kg to ICR mice in oral glucose tolerance test lowered plasma glucose area under the curve by 9% whereas plasma insulin was increased from 21.3 to 28.8 mU/mL at 30 minutes.[226]

2.57 *Azadirachta indica* A. Juss

Synonyms: *Melia azadirachta* L.; *Melia indica* (A. Juss.) Brandis
Common names: numbah (India); neem tree
Subclass Rosidae, Superorder Rutanae, Order Rutales, Family Meliaceae
Medicinal use: enlarged liver (India)

Meliacinolin isolated from the leaves of *Azadirachta indica* A. Juss. given at a single oral dose of 30 mg/kg lowered fasting glycemia from 100.2 to 44.5 mg/dL after 6 hours in p mice whereby glibenclamide at 4 mg/kg lowered glycemia to 74.1 mg/dL at 6 hours.[227] In streptozotocin-induced type 2 diabetic mice, this limonoid lowered fasting blood plasma from 367.4 to 215.2 mg/dL at 6 hours whereby glibenclamide at 4 mg/kg lowered glycemia to 206.3 mg/dL.[227] Being able to lower glycemia on both healthy and diabetic mice, meliconin most probably act at β-cell levels instead of a biguanide-like activity. Biguanides do not produce hypoglycemia in healthy subjects and rodents because the increase in peripheral utilization is compensated by an increase in hepatic glucose output.[228] Given for 28 days to diabetic mice at a daily dose of 20 mg/kg/day, meliacinolin increased body weight gain, lowered food and water intake.[227] Oral glucose tolerance test performed at the end of the treatment reduced of peak glycemia at 60 minutes from 412.8 to 247.3 mg/dL (glibenclamide at 4 mg/kg/day: 221.4 mg/dL).[227] Simultaneously, this limonoid normalized triglycerides, total cholesterol and lowered high-density lipoprotein–cholesterol similarly to glibenclamide.[227] In addition, the terpene normalized lipid peroxidation in the liver and kidneys of diabetic animals.[227] At the hepatic level, diabetic animals had high glucose-6-phosphatasee activity, low glucokinase and hexokinase activity and low glycogen contents and meliacinolin, like glibenclamide, corrected these parameters close to normal.[227] Both meliacinolin and glibenclamide brought close to normal superoxide, catalase, glutathione, and glutathione-peroxidase in liver and kidneys of diabetic animals.[227] This tetranortriterpenoid, increased plasma insulin from 1.4 to 3.3 μU/mL (normal 3.6 μU/mL) and increased pancreatic insulin from 14.8 to 22.4 μU/mL (normal 25.7 μU/mL) and these effects were similar to glibenclamide.[227] Meliacinolin inhibited α-glucosidase and α-amylase *in vitro* with IC_{50} values of 32.1 μg/mL and 46.7 μg/mL, respectively (acarbose: 78.5 μg/mL and 12.2 μg/mL respectively).[227] The precise molecular mechanism accounting for the insulinotropic effect of meliacinolin in rats is unknown. Chattopadhyay (1999) presented evidence that an extract of leaves inhibited serotonin inhibition on insulin secretion mediated by glucose in pancreas of rat suggesting a possible insulinotropic effect via 5-hydroxytryptamine receptors expressed by β-cells.[229] As for the insulinotropic effect on diabetic rats, it is reasonable to suggest at least some antiinflammatory effects, because nimbolide, a limonoid from the same plant, is able to block nuclear factor-κB nuclear translocation in intestinal epithelial cells and macrophages and murine colitis models with downstream inhibition of expression of inflammatory cytokines.[230] Meliacinolin and/or plasmatic metabolites could attenuate streptozotocin-induced nuclear factor-κB activation in β-cells resulting in increased insulin secretion. In metabolic syndrome, chronic exposure of β-cells to elevated plasmatic glucose favors to the production of reactive oxygen species and the activation of nuclear factor-κB, apoptosis and subsequent reduction of insulin secretion leading to type 2 diabetes.

2.58 *Khaya grandifoliola* C. DC.

Subclass Rosidae, Superorder Rutanae, Order Rutales, Family Meliaceae

Aqueous extract of stem bark of *Khaya grandifolia* C. DC. given orally to Wistar rats at a dose of 200 mg/kg/day for 21 days lowered plasma glucose from 128 to 114.9 mg/dL, total cholesterol from 128 to 114 mg/dL, whereby it mildly decreased free fatty acids.[231] At the hepatic level, this regimen had no effect on marker of lipid peroxidation malondialdehyde and decreased glutathione.[231] The stem bark of this timber tree contains series of limonoids of which deacetylkhayanolide E, 6S-hydroxykhayalactone, grandifolide, khayanolide A, anthothecanolide, and 3-*O*-acetylanthothecanolide.[232]

2.59 *Khaya senegalensis* (Desr.) A. Juss.

Synonym: *Swietenia senegalensis* Desr.
Common name: fei zhou lian (Chinese)
Subclass Rosidae, Superorder Rutanae, Order Rutales, Family Meliaceae
Medicinal use: Common street ornamental in Malaysia

Butanol fraction of roots *of Khaya senegalensis* (Desr.) A. Juss. given orally at a dose of 300 mg/kg/day for 4 weeks to fructose-streptozotocin induced type 2 diabetic Sprague–Dawley rats lowered both food and water consumption compared with untreated animals.[233] This regimen did not prevent weight loss, lowered glycemia from about 25 to 10 mmol/L (normal 5 mmol/L).[233] This fraction increased serum insulin from 64.2 to 84.5 pmol/L (normal 153.1 pmol/L), improved the peripheral action of insulin and increased liver glycogen contents.[233] The treatment lowered total cholesterol from 76.2 to 68.7 mg/dL (normal 66.2 mg/dL), increased high-density lipoprotein–cholesterol from 19.8 to 24.2 mg/dL (normal 31), increased low-density lipoprotein–cholesterol from 27.4 to 25.6 (normal 19.8), and decreased plasma triglycerides from 184.7 to 105 mg/dL (normal 102.5 mg/dL). Histopathological examination of pancreas of treated rodents revealed a protective effect on islets.[233] In a parallel study, butanol fraction of root, with potent antioxidant property, inhibited yeast α-glucosidase and porcine α-amylase with IC_{50} values of 2.8 μg/mL and 97.5 μg/mL, respectively, (acarbose: 55.5 μg/mL and 256.6 μg/mL).[234] Inhibition of carbohydrate absorption by the butanol fraction of *Khaya senegalensis* (Desr.) A. Juss. would account for decreased glycemia but does not justify increased insulin secretion. Ethanol extract of stem bark contains polyphenolics, which are antioxidant which may, at least, combat oxidative insults incurred by streptozotocin.[234] The active principle(s) and mode of action remain to be identified.

2.60 *Cornus mas* L.

Synonym: *Macrocarpium mas* (L.) Nakai
Common name: Cornelian cherry wood
Subclass Asteridae, Superorder Cornanae, Order Cornales, Family Cornaceae
Nutritional use: food (Australia)

Anthocyanin extract of fruits of *Cornus mas* L. given at a ratio of 1 g/kg in high-fat diet in high-fat diet to C57BL/6 mice for 6 weeks did not modify food intake, evoked a mild loss of weight gain compared with untreated animals and normalized intraperitoneal glucose tolerance as evidenced by a reduction of 90 minutes glycemia from 363 to 221 mg/dL.[235] This supplementation increased insulinemia, improved pancreatic islets histoarchitecture of high-fat diet to C57BL/6 mice.[235] This treatment reduced hepatic triglycerides from 82.5 to 61 mg/g and lowered cholesterolemia from 156.4 to 123.2 mg/dL.[235] Ursolic acid isolated from the fruits of *Cornus mas* given at a ratio of 0.5 g/kg in high-fat diet to C57BL/6 mice for 6 weeks did not modify food intake nor weight gain compared with untreated animals and normalized intraperitoneal glucose tolerance as evidenced by a reduction of 90 minutes glycemia from 363 to 227 mg/dL.[235] This supplementation increased insulinemia, improved pancreatic islets histoarchitecture and of high-fat diet to C57BL/6 mice.[235] This treatment reduced hepatic triglycerides from 82.5 to 74.3 mg/g.[235] Lyophilizate of fruits of *Cornus mas* L. given for 60 days at a dose of 100 mg/kg to New Zealand rabbits fed with cholesterol enriched diet lowered triglyceridemia by 44% and increased high-density lipoprotein–cholesterol by 13.2% whereby cholesterolemia was unaffected.[236] This supplementation stimulated the expression of hepatic peroxisome proliferator-activated receptor-α, lowered hepatic lipid peroxidation, increased glutathione contents, lowered serum interleukin-6 and tumor necrosis factor α.[236] Further this supplementation prevented the formation of atheromat.[236] In endothelial cell peroxisome proliferator-activated receptor-α agonists inhibit the secretion of inflammatory cytokines interleukin-1, inlerleukin-6, tumor necrosis factor α and adhesion molecule 1 (VCAM-1) and intercellular adhesion molecule 1 (ICAM-1).[237] Besides, peroxisome proliferator-activated receptor-α ligands, via the repression of nuclear factor-κB signaling, lowers the expression of cyclo-oxygenase-2.[238] Clinical trials are warranted.

2.61 *Cornus officinalis* Siebold & Zucc.

Synonyms: *Macrocarpium officinale* (Siebold & Zucc.) Nakai
Common name: shan zhu yu (Chinese)
Subclass Asteridae, Superorder Cornanae, Order Cornales, Family Cornaceae
Medicinal use: tonic (China)

In obese patients with impaired glucose tolerance, the pancreas has to secrete more insulin to normalize postprandial plasma glucose.[239] Chronic stimulation of insulin secretion leads to a progressive deterioration in β-cell function and ultimately type 2 diabetes in genetically predisposed individuals.[55] Methanol fraction of fruits of *Cornus officinalis* Siebold & Zucc. at a concentration of 25 μg/mL improved the viability of pancreatic BRIN-BD11 cells challenged with alloxan, streptozotocin, interleukin-1β or interferon-γ and boosted insulin secretion of pancreatic BRIN-BD11 cells challenged with glucose.[240] 7-*O*-galloyl-D-sedoheptulose from the fruits of *Cornus officinalis* Siebold. & Zucc. given orally at a dose of 100 mg/kg/day for 6 weeks from spontaneous type 2 diabetic obese *db/db* mice lowered glycemia by 10% and insulinemia by 30.1%.[241] This treatment lowered, tumor necrosis factor a and TBARS to p values and lowered interleukin-6 and reactive oxygen species.[241] This treatment increased pancreatic insulin contents.[241] Furthermore, pancreas of treated rodents expressed lower levels of JNK and phosphorylated JNK as well as lower AP-1, nuclear factor-κB, cyclooxygenase-2, and inducible nitric oxide synthetase, transforming growth factor-β1 and fibronectin.[241] The treatment reduced pancreatic fibrosis of db/db mice.[241] Aqueous extract of fruits of *Cornus officinalis* Siebold & Zucc. given orally at a dose of 300 mg/kg/day to streptozotocin-induced diabetic Sprague–Dawley for 4 weeks evoked a reduction of weight gain, fasting glycemia by 35% and triglycerides by 61.7%.[242] This extract increased the number of insulin producing pancreatic β-cells.[242] The iridoid glycoside loganin from the fruits of *Cornus officinalis* Siebold & Zucc. prevented pancreatic damages as evidenced by a decrease in proinflammatory cytokines such as interleukin (IL)-1β and tumor necrosis factor as well as inhibition of nuclear factor-κB.[243] Clinical trials are warranted.

2.62 *Aralia cachemirica* Decne.

Synonym: *Aralia macrophylla* A. Cunn. ex Sweet
Common name: churial
Subclass Asteridae, Superorder Cornanae, Order Apiales, Family Araliaceae

Ethanol extract of roots of *Aralia cachemirica* Decne. given orally at a single dose of 250 mg/kg to fasted Wistar rats lowered glycemia from 89.6 to 69.6 mg/dL after 90 minutes.[244] In the same experiment, gliclazide orally at a dose of 25 mg/kg lowered glycemia to 54.4 mg/dL.[244] Ethanol extract given at a single oral dose of 250 mg/kg, 30 minutes before oral loading of glucose lowered 30 minutes postprandial glycemia from 110.6 to 91.2 mg/dL (gliclazide: 84.2 mg/dL).[244] The plant produces octadec-6-enoic acid, 8-primara-14,15-diene-19-oic acid, aralosides A and B.[245]

2.63 *Aralia taibaiensis* Z.Z. Wang & H.C. Zheng

Synonym: *Aralia elata* (Miq.) Seem var. *elata*
Common name: yuan bian zhong (Chinese)
Subclass Asteridae, Superorder Cornanae, Order Apiales, Family Araliaceae

Total saponins extracted from the root barks of *Aralia taibaiensis* Z.Z. Wang & H.C. Zheng given orally to nicotinamide-streptozotocin-induced type 2 diabetic Wistar rats (fasting plasma glucose > 7 mmol/L) at a daily dose of 320 mg/kg for 28 days prevented weight loss, lowered food intake (polyphagia), and decreased fasting blood glucose from 16.1 to 5.1 mmol/L.[246] This regimen increased serum insulin near normal values (33.9 mlU/L).[246] This saponin extract decreased total cholesterol, triglycerides, low-density lipoprotein–cholesterol, very low-density lipoprotein–cholesterol and atherogenic index.[246] It improved the antioxidant capacity of serum as evidence by an decrease in malondialdehyde by 37%, and an increase in superoxide dismutase by 109% and glutathione by 104%.[246] Histological observation revealed that the total saponin extract enhanced the regeneration of islets of Langerhans in the pancreas of treated rodents.[246] From the plant, chikusetsu saponin IV a given orally to nicotinamide-streptozotocin-induced diabetic Wistar rats at a dose of 180 mg/kg/day for 28 days lowered plasma glucose from about 15 to 6 mmol/L and increased plasma insulin near to normal.[41] The saponin at 40 mM increased insulin secretion by insulin-secreting βTC3 cells exposed to 2 mM or 20 mM glucose via increase stimulation of G protein-coupled receptor GPR40, increase of cytosolic calcium concentration, induction of protein kinase C phosphorylation.[41] The G-protein-coupled receptor (GPR40) is expressed by β-cells as a receptor the stimulation of which by agonists such as fatty acids, couples to the calcium-mobilizing G protein, Gaq hence phospholipase C activation, conversion of plasma membrane phosphatidylinositol-4,5-bisphosphate into inositol trisphosphate and diacylglycerol, the later potentiating insulin secretion by stimulating protein kinase C and exocytosis.[247,248,250] It must be recalled however that triterpene saponins are not well absorbed orally and that the sugar moieties of saponins are being sequentially removed by intestinal commensal bacteria to release triterpenes linked to fewer sugar moieties and/or triterpene aglycones, which are being absorbed in the gut.[248,249] Hence, chikusetsu saponin IV itself may not account for the insulinotropic effects of *Aralia taibaiensis* Z.Z. Wang & H.C. Zheng but instead its aglycone, namely oleanolic acid which is known to be insulinotropic.

2.64 *Dendropanax morbiferum* H. Lév.

Subclass Asteridae, Superorder Cornanae, Order Apiales, Family Araliaceae
Medicinal use: skin disease (Korea)

Aqueous extract of leaves of *Dendropanax morbiferum* H. Lév. given orally to streptozotocin-induced mice at a dose of 200 mg/kg/day for 2 weeks lowered plasma glucose from about 600 mg/dL to 400 mg/dL, increased plasma insulin from approximately 1 to 2.2 μU/mL (normal about 3.8 μU/mL), and improved pancreatic histoarchitecture.[251] This extract elicited anti-oxidant activity *in vitro* and reduced glucose contents in urine of diabetic mices.[251] Dendropanoxide (Figure 2.15) isolated from the leaves of *Dendropanax morbiferum* H. Lév. given orally at a daily dose of 100 mg/kg/day to streptozotocin-induced diabetic Wistar rats (fasting blood glucose > 180 mg/dL) for 14 days decreased glycemia from 543.2 to 311.2 mg/dL (glibenclamide at 600 μg/kg/day: 414.3 mg/dL; normal 93.3 mg/dL) and increased plasma insulin from 1 to 1.9 IU/L (normal 11.3 IU/L).[252] This treatment lowered triglycerides from 114.2 to 71.3 mg/dL (glibenclamide at 600 μg/kg/day: 89.4 mg/dL;

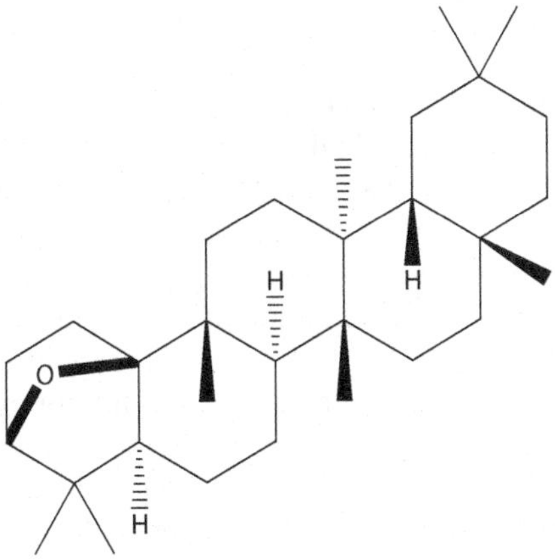

FIGURE 2.15 Dendropanoxide.

normal 73.4 mg/dL) and total cholesterol from 106.7 to 71.6 mg/dL (glibenclamide at 600 μg/kg/day: 84.2 mg/dL; normal 68.8 mg/dL).

2.65 *Cuminum cyminum* L.

Common names: zi ran qin (Chinese); jirakam (India); cumin
Subclass Asteridae, Superorder Cornanae, Order Apiales, Family Apiaceae
Medicinal use: promote digestion (India)
History: The plant was known to Dioscorides (40–90 A.D.), a Greek physician during the time of the Emperor Nero

There is evidence to demonstrate that *Cuminum cymimun* L. could be beneficial for the management of metabolic syndrome. Aqueous extract of seeds of *Cuminum cymimun* L. given orally to Wistar rats at a daily dose of 250 mg/kg for 6 weeks attenuated weight gain and glycemia which decreased from 67.3 to 63.8 mg/dL.[253] Aqueous extract of seeds of *Cuminum cymimun* L. given orally to alloxan-induced diabetic Wistar (glycemia > 200 mg/dL) rats at a daily dose of 250 mg/kg for 6 weeks prevented weight loss, decreased glycemia from 236.5 to 82 mg/dL (normal 67.3 mg/dL; glibenclamide at 0.6 mg/kg/day: 103.3 mg/dL).[253] This regimen, lowered serum cholesterol and normalized triglycerides and free fatty acids.[253] In must be recalled that fatty acids increase basal and glucose-induced insulin secretion. Following glycolysis and Krebs cycle activation, β-cells have increased levels of acetyl-CoA which are converted to malonyl-CoA by acetyl-CoA carboxylase.[254] Malonyl-CoA inhibits carnitine palmitoyltransferase-1 and therefore the entry of fatty acyl-CoA in the mitochondria.[254] Increased cytoplasmic concentrations of fatty acyl-CoA stimulates insulin secretion.[254] In a subsequent study, methanol extract of seeds of *Cuminum cyminum* L. given to streptozotocin-induced diabetic Wistar rats (serum glucose > 250 mg/dL) at a daily dose of 600 mg/kg for 28 days mitigated weight loss, lowered glycemia from 284.2 to 114.8 mg/dL (normal 78.8 mg/dL; glibenclamide at 10 mg/kg: 109.3 mg/dL) and augmented insulinemia from 11.6 to 22.7 μIU/mL, an effect superior to as glibenclamide given at 10 mg/kg/day: 18.8 μIU/mL (normal 30.7 μIU/mL).[255] Of note, cumin treatment lowered kidney advanced glycation end products by 80% achieving thus normalization and this effect was superior to glibenclamide at 10 mg/kg.[255] The extract increased hepatic and skeletal muscle contents in glycogen similarly with glibenclamide and normalized plasma activity of antioxidant enzymes catalase and superoxide dismutase as well as glutathione contents.[255] Patil et al. (2013) presented evidence that in oral glucose tolerance test, the petroleum ether fraction of seeds of *Cuminum cyminum* L. given orally 10 minutes before a load of glucose decreased 60 minutes postprandial glycemia from 30.8 to 23.5 mmol/L (p:5.6 mmol/L; glibenclamide at 2.5 mg/kg: 20.3 mg/dL).[256] This fraction given for 45 days at a dose of 10 mg/kg/day lowered fasting blood glucose from 22 to 7.3 mmol/L, an effect comparable with 8.7 mmol/L obtained with glibenclamide at 2.5 mg/kg/day.[256] This treatment increased insulinemia from 39.8 to 93.4 pmol/L (normal 112.1 pmol/L) and this effect was superior to as glibenclamide.[256] Cuminaldehyde (Figure 2.16) and cuminol isolated from this fraction in oral glucose tolerance test lowered glycemia similarly to glibenclamide at 60 minutes.[256] These monoterpenes at a concentration of 25 μg/mL induced the secretion of insulin by β-cells *in vitro* exposed to 11.8 mmM of glucose and these effects were superior to as glibenclamide at 10 μg/mL.[256] Cuminaldehyde and cuminol did not evoke the secretion of insulin in presence of K^+-ATP channel opener diazoxide nor in the presence of L-type channel blocker nifedipine but increased insulin secretion was obtained in presence of phosphodiesterase inhibitor 3-isobutyl-1-methylxanthine.[256] Furthermore, cuminaldehyde and cuminol at a concentration of 25 μg/mL protected β-cells against streptozotocin-induced nitric production in cells indicating

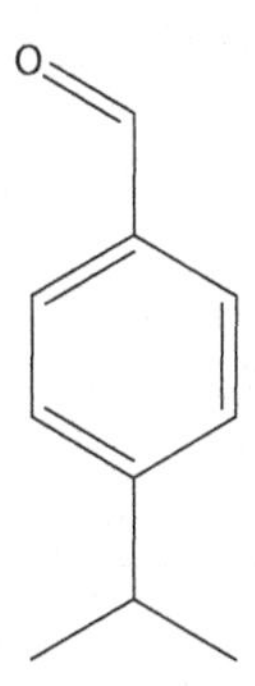

FIGURE 2.16 Cuminaldehyde.

cytoprotective effects.[256] Shimabukuro et al. (1997) provided evidence that free fatty acids induce the generation of nitric oxide in islets of pre-diabetic rats via induction of inducible nitric oxide synthase resulting in both basal and glucose-induced insulin secretion.[257] Streptozotocin induces the generation of nitric oxide and reactive oxygen species which contribute to DNA fragmentation and other oxidative injuries.[51] It should be noted that nitric oxide is a free radical, which inhibits the Krebs-cycle enzyme aconitase and the electron transport chain complexes I and II leading to reduced glucose oxidation rates, ATP generation and insulin production.[258] Triglyceridemia Wistar rats receiving for 8 weeks a high cholesterol diet containing 1.25% of cumin had reduced compared with untreated animals whereby cholesterolemia was unaltered.[259] This treatment lowered hepatic triglycerides by 40%.[258] In obese volunteers, intake of 3 g of a powder of seeds of *Cuminum cymimun* L. per day during meals for 3 months evoked a decrease of body weight, body mass index and fat mass by 7.9%, 7.1%, and 18.1%, respcetively.[260] This supplementation reduced fasting serum triglycerides by 12.8%, cholesterol by 13.9%, low-density lipoprotein–cholesterol by 7.1%, glycemia by 6.1% and increased high-density lipoprotein–cholesterol by 3.4%.[258] In overweight or obese individuals, a fasting plasma glucose between 100 mg/dL (5.6 mmol/L) and 125 mg/dL (6.9 mmol/L) is indicative of impaired fasting glucose.[1] Intake of *Cuminum cymimun* L. by obese patients could be therefore be beneficial.

2.66 *Angelica dahurica* (Fisch.) Benth. & Hook. f.

Common names: bai zhi (Chinese); yoroi-gusa (Japan)
Subclass Asteridae, Subclass Cornanae, Order Apiales, Family Apiaceae
Medicinal use: headache (China)

Hexane extract of *Angelica dahurica* (Fisch.) Benth. & Hook. f. (Figure 2.17) given at a single oral dose of 300 mg/kg given to C56BL6 mice lowered 30 minutes peak glycemia in oral glucose tolerance test and increased plasma insulin.[261] From this extract, imperatorin and phellopterin (Figure 2.18) at a concentration of 100 μM increased the reporter activity in GPR119-CRE-bla CHO-K1 cells, increased glucagon-like peptide-1 secretion in GLUTag cells and increased glucose-stimulated insulin secretion in INS-1 cells.[261] This suggests the ability of these furanocoumarins to stimulates G-protein-coupled receptor 119 (GPR119) expressed in pancreatic β-cells and intestinal endocrine L-cells to promote glucose-stimulated insulin secretion and glucagon-like peptide-1 release.[261] Clinical trials are warranted.

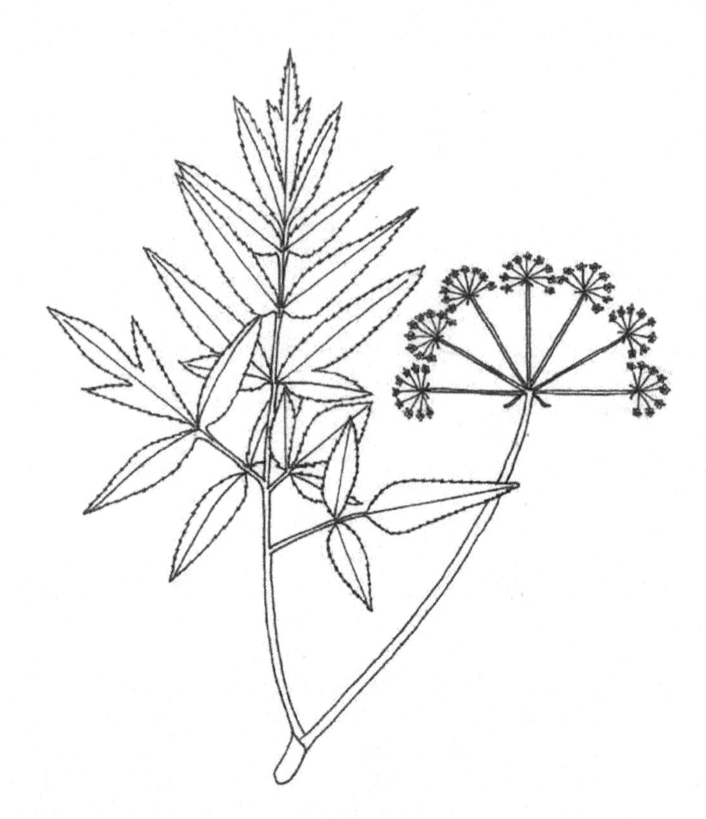

FIGURE 2.17 *Angelica dahurica* (Fisch.) Benth. & Hook. f.

FIGURE 2.18 Dendropanoxide

2.67 *Angelica japonica* A. Gray

Subclass Asteridae, Superorder Cornanae, Order Apiales, Family Apiaceae
Nutritional use: vegetable (China)

Methanol extract of *Angelica japonica* var. *hirsutiflora* (S.L. Liou, C.Y. Chao, & T.I. Chuang) T. Yamaz. at a concentration of 100 μg/mL evoked an increase in glucose-induced secretion of insulin by β-cells-derived-HIT-T15 cells *in vitro* and these insulinotropic effect was more potent in isolated mouse islets of Langerhans.[262] In β-cells-derived-HIT-T15 cells the extract evoked an increase in cytoplasmic calcium, induced the phosphorylation of extracellular signal-regulated kinase-1/2, and this insulinotropic effect was inhibited by PD98059 extracellular signal-regulated kinase-1/2 inhibitor.[262] Intraperitoneal injection of the extract at a single dose of 30 mg/kg prior to oral loading of starch in ICR mice evoked a decrease in postprandial glycemia at 30 minutes and boosted plasma insulin.[262] In high-fat diet-induced diabetic mice, the extract at a single intraperitoneal dose of 10 mg/kg improved glucose tolerance in oral glucose tolerance test.[262] Extracellular signal-regulated kinase-1/2 is activated in β-cells of rats by glucose to stimulate the expression of insulin and to phosphorylate synapsin I to promote insulin exocytosis.[263,264] *Angelica japonica* is known to elaborate series of prenylated coumarins of which osthol, which is known to activate extracellular signal-regulated kinase-1/2.[265,266]

2.68 *Carum carvi* L.

Common name: ge lu zi (Chinese); caraway
Subclass Asteridae, Superorder Cornanae, Order Apiales, Family Apiaceae
Medicinal use: promote digestion (Taiwan)

Eddouks et al. (2004) provided evidence that a single oral dose of 20 mg/kg of an aqueous extract of fruits of *Carum carvi* L. lowered glycemia after 4 hours in streptozotocin-induced diabetic Wistar rats whereby daily administration of this extract for 14 days evoked a 65% fall of glycemia in diabetic animals without increased insulinaemia.[267] In a subsequent study, aqueous extract of seeds of *Carum carvi* given to streptozotocin-induced diabetic Wistar rats (fasting blood glucose > 250 mg/dL) at a dose of 1 g/kg for 21 days reduced weight loss, lowered blood glucose from 330 to 142.8 mg/dL, a value below normal 160.7 mg/dL.[268] This regimen lowered total cholesterol and low-density lipoprotein–cholesterol whereby triglycerides and high-density lipoprotein–cholesterol where not affected.[268] Muruganathan et al. (2013) administered orally carvone isolated from *Carum carvi* L. to streptozotocin-induced diabetic Wistar rats at a dose of 100 mg/kg/day for 30 days and recorded decrease of glycemia from 295.2 to 111.5 mg/dL (normal 78.2 mg/dL; glyclazide at 5 mg/kg/day: 103.2 mg/dL) and increased plasma insulin from 6.9 to 13.9 μUI/mL (normal 16.2 μUI/mL; glyclazide at 5 mg/kg/day: 14.3 μUI/mL).[269] Gamma aminobutyric acid activates INS β-cell and mouse islets growth and survival via calcium-phosphoinositide 3-kinase/Akt and that effect is inhibited by $GABA_A$ antagonist bicuculline.[270] Gamma aminobytric acid, via $GABA_A$ activation, stimulated glucose-induced insulin secretion in the presence of low or physiological concentration of glucose in INS cells but carvone is known to negatively modulate $GABA_A$ receptor in neurones.[271,272] Akiba et al. (2004) provided evidence that capsaicin at concentrations ranging from 10^{-11} M to 10^{-9} M induced the secretion of insulin by RIN β-cells via activation of transient receptor potential vanilloid subfamily 1, also known as vanilloid receptor 1 (VR1) with entry of calcium into the cell.[273] Carvone (Figure 2.19) ability to increase insulin secretion in rats may result from activation of transient receptor potential

O

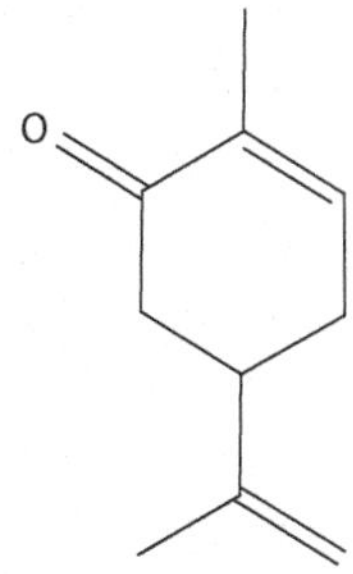

FIGURE 2.19 Carvone.

vanilloid subfamily 1 (TRPV-1) expressed by β-cells.[274] The consumption of caraway could be beneficial in metabolic syndrome.

2.69 *Achillea santolina* L.

Common name: boohemaderan (India); lavender cotton-leaved millefoil
Subclass Asteridae, Superorder Asteranae, Order Asterales, Family Asteraceae
Medicinal use: fatigue (Pakistan)

Ethanol extract of aerial parts of *Achillea santolina* L. given orally and daily to streptozotocin-induced diabetic Wistar rats (glycemia > 15 mmol/L) at a dose equivalent to 0.1 g plant powder/kg for 30 days attenuated body weight loss and lowered blood glucose from 16.3 to 6.7 mmol/L (normal 4.8 mmol/L).[275] At the pancreatic levels, the regimen lowered malondialdehyde, increased glutathione contents, increased the activity of catalase and superoxide dismutase whereas serum nitric oxide was reduced.[275] The extract lowered pancreatic protein carbonyl and advanced oxidation protein products by 66% and 59%, respectively.[275] Aqueous liquid extract of aerial parts of *Achillea santolina* (10g/100 mL) up to 50 mg/mL was unable to inhibit α-amylase and α-amyloglucosidase activity *in vitro*.[276] In Sprague–Dawley rats, this extract at a dose of 125 mg/kg lowered peak glycemia at 45 minutes in oral starch tolerance test from about 6.5 to 5 mM, and this effect was as efficient as acarbose at dose of 3 mg/kg.[276] The plant produces luteolin and quercetin leucodin, 3,9-diacetoxy,13-hydroxy-l(10),4,7(11)-germacratrien-12,6-olide, desacetylmatricarinin, eupatolin, eupatilin 7-methyl ether, and cirsimartin.[277,278]

2.70 *Actium lappa* L.

Synonyms: *Arctium chaorum* Klokov; *Arctium leiospermum* Juz. & Ye. V. Serg.; *Arctium majus* (Gaertn.) Bernh.; *Lappa major* Gaertn.; *Lappa vulgaris* Hill
Common names: niu bang (Chinese); great burdock
Subclass Asteridae, Superorder Asteranae, Order Asterales, Family Asteraceae
Medicinal use: fever (China)
History: The plant was known of Galen (circa 130–200 A.D); Greek physician to the Roman Emperor Marcus Aurelius, to treat ulcers.

A total lignan fraction of fruits of *Arctium lappa* L. given to spontaneously diabetic Goto–Kakizaki rats at a dose of 300 mg/K twice daily before each meal for 12 weeks (nateglinide twice daily before each meal for 12 weeks, 50 mg/kg) lowered fasting blood glucose by 51.4% (nateglinide, 50 mg/kg: 19.6%) close to normal values.[279] This treatment improved glucose tolerance in oral glucose tolerance test done at the end of the treatment with a 35.1% postprandial glycemia at 60 minutes.[279] The fraction boosted postprandial insulinemia in Goto–Kakizaki rats (Xu et al., 2014).[279] This treatment improved pancreatic histoarchitecture. The fraction had no effect on plasma cholesterol or triglycerides.[279] The fraction inhibited the activity of α-glucosidase *in vitro* as potently as acarbose with an IC_{50} of about 150 μg/mL.[279] Arctigenic acid which is the metabolite of arctigenin (Figure 2.20) from fruits of *Arctium lappa* in Goto–Kakizaki rats, given orally at a dose of 50 mg/kg twice daily for 12 weeks lowered fasting glycemia by 37.6% (nateglinide at 50 mg/kg/day: 28.1%) and attenuated body weight gain.[279] The treatment improved glucose tolerance in oral glucose tolerance test.[280] Arctigenin treatment (as well as nateglinide) improved pancreatic histoarchitecture with enhanced regeneration of islets.[279] A fraction of roots of *Arctium lappa* (dicaffeoylquinic acid derivatives 75.4%) at a concentration of 100 μg/mL increased intake of glucose by L6 myotubes by 16% in the presence of insulin and was inactive in absence of insulin.[280] From this extract, 5-*O*-caffeoylquinic acid at 100 μg/mL had similar effect.[280] The fraction and 5-*O*-caffeoylquinic acid at 50 μg/mL inhibited glucagon-induced release of glucose by rat hepatocytes *in vitro* and inhibited glucose 6-phosphatase activity.[280] The fraction and 5-*O*-caffeoylquinic acid had no effect

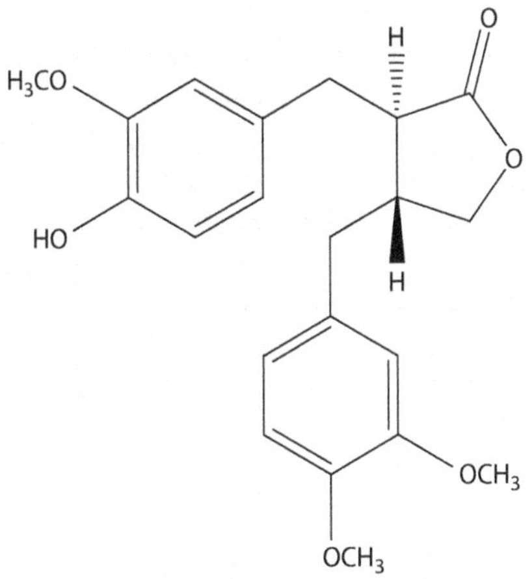

FIGURE 2.20 Arctigenin.

on insulin secretion by INS in the presence of glucose.[280] The fraction given orally at a dose of 15 mg/kg/day orally for 4 days lowered 30 minutes glycemia peak during oral glucose tolerance test from 9.5 to 8.4 mmol/L with a concomitant increase of insulinaemia from 1.7 to 3.6 ng/mL.[281] Clinical trials are warranted.

2.71 *Artemisia dracunculus* L.

Synonyms: *Artemisia aromatica* A. Nelson; *Artemisia glauca* Pall. ex Willd.
Common names: long hao (Chinese); tarragon
Subclass Asteridae, Superorder Asteranae, Order Asterales, Family Asteraceae
Medicinal use: sore throat (Tibet)

Ethanol extract of *Artemisia dracunculus* L. at a concentration of 10 μg/mL increased the secretion of insulin by NIT-1 cells by 2.7 folds compared with cells exposed to 1 mM glucose only.[282] In islets of Langerhans of mice exposed to 1 mM glucose, this extract at 10 μg/mL increased insulin release by 5.3 folds.[282] The extract at a concentration of 10 μg/mL induced the phosphorylation of adenosine monophosphate-activated protein kinase, an energy sensor, and its downstream substrate acetyl-CoA carboxylase as well as Akt, also known as protein kinase B.[282]

2.72 *Bidens pilosa* L.

Synonyms: *Bidens alba* (L.) DC.; *Bidens hirsuta* Nutt.; *Bidens odorata* Cav.
Common names: gui zhen cao (Chinese); pisau-pisau (Philippines); hairy beggarticks
Subclass Asteridae, Superorder Asteranae, Order Asterales, Family Asteraceae
Medicinal use: boils (Philippines)

Ubillas et al. (2000) provided evidence that ethanol fraction of *Bidens pilosa* L. given to genetically obese diabetic *db/db* mice at 0, 8, and 24 hours orally at a dose 1g/kg evoked at 27 hours a 33% fall in glycemia.[283] From this fraction, a mixture containing 2-β-D-glucopyranosyloxy-1-hydroxy-5(*E*)-tridecene-7,9,11-triyne and 3-β-D-glucopyranosyloxy-1-hydroxy-6(*E*)-tetradecene-8,10,12-triyne given to *db/db* mice at 0, 8, and 24 hours orally at a dose of 500 mg/kg evoked at 27 hours, a 41% decrease in glycemia and this effect was equivalent to metformin at 250 mg/kg.[283] Cytopiloyne or 2-β-D-glucopyranosyloxy-1-hydroxytrideca-5,7,9,11-tetrayne isolated from the leaves of *Bidens pilosa* L. administered at single oral dose of 0.5 mg/kg to diabetic *db/db* mice after food intake lowered after 1 hour plasma glucose from 333.1 to 173.4 mg/dL (glimepiride at 1 mg/kg: 270.3 mg/dL), increased plasma insulin from 10.2 to 18.6 ng/mL at 1 hour (glimepiride at 1 mg/kg: 20.3 ng/mL) at 1 hour.[284] This polyacetylenic glucoside given orally at a daily dose of 0.5 mg/kg for 28 days lowered glycemia from about 600 to 350 mg/dL and this effect was superior to as glimepiride 1 mg/kg/day with about 400 mg/dL and increased plasma insulin from about 4 to 16 ng/mL, an effect superior to glimepiride at 1 mg/kg/day.[284] This polyacetylenic glucoside in RIN-m5F cells, increased cytoplasmic calcium, diacylglycerol, and insulin secretion.[285] In must be noted that cytopiloyne attenuated the development of diabetes in non-obese diabetic mice via T cell regulation.[285,286] It implies that this polyacetylene glucoside or *Bidens pilosa* L., if not toxic could be of value for the prevention of type 2 diabetes in metabolic syndrome.

2.73 *Centratherum anthelminticum* (L.) Gamble

Synonym: *Vernonia anthelmintica* (L.) Willd.
Common name: qu chong ban jiu ju (Chinese); somraj (Bengal); purple fleebane
Subclass Asteridae, Superorder Asteranae, Order Asterales, Family Asteraceae
Medicinal use: intestinal worms (Bengal)

Methanol fraction of seeds of *Centratherum anthelminticum* (L.) Gamble at a concentration of 25 μg/mL protected mouse pancreas β-TC6 cells against nuclear factor-κB nuclear translocation induced by hydrogen peroxide.[287] The fraction given orally to streptozotocin-nicotinamide induced type 2 diabetic Sprague–Dawley rats (fasting glycemia between 11 and 14 mmol/L) at a dose of 50 mg/kg/for 12 weeks prevented weight loss and polyphagia, lowered plasma glucose from 23.2 to 5.6 mmol/L (normal 4.6 mmol/L; glibenclamide at 50 mg/kg/day: 7.3 mmol/L).[287] This treatment increased serum insulin.[287] In streptozotocin-nicotinamide induced type 2 diabetic Sprague–Dawley rats total cholesterol, triglycerides, low-density lipoprotein–cholesterol and free fatty acids were increased compared with normal rodents and the fraction lowered these values to normal.[287] In pancreas, liver and kidneys, the treatment increased glutathione and lowered malondialdehyde similarly to glibenclamide.[287] The fraction decreased necrosis factor-α, interleukin-1β, and interleukin-6 in pancreas.[287] In the plasma of treated rodents, the fraction lowered tumor necrosis factor-α, interleukin-1β, and interleukin-6.[287] Methanolic fraction seeds at the concentration of 12.5 μg/mL increased glucose uptake by of mouse β-TC6.[288] In the presence of 12.6 mM of glucose, the fraction at a concentration of 12.5 μg/mL boosted the secretion of insulin by mouse β-TC6 cells.[288] This fraction increased the expression of glucose transporter-2.[288] In streptzotocin-nicotinamide-induced diabetic Sprague–Dawley rats, the fraction given orally at a dose of 100 mg/kg/day for 14 days evoked a decrease in blood glucose by 51.4%, whereas glibenclamide at a dose of 50 mg/kg/day reduced glycemia by 50%.[288] In streptozotocin-induced diabetic Sprague–Dawley rats, the fraction given orally at a dose of 100 mg/kg/day for 14 days evoked a decrease in blood glucose by 23.2% and this effect was superior to insulin (6 U/kg).[288] In oral glucose tolerance test performed at the end of treatment period, the fraction at 100 mg/kg/day for 14 days lowered 90 minutes glycemia by 66.3%, and this effect was superior to glibenclamide-treated rodents with a 60.9% reduction.[288]

2.74 *Chromolaena odorata* (L.) R.M. King & H. Rob.

Synonym: *Eupatorium odoratum L.*
Common name: fei ji cao (Chinese); Siam weed
Subclass Asteridae, Superorder Asteranae, Order Asterales, Family Asteraceae
Medicinal use: diabetes (India)

Ethanol extract of leaves of *Chromolaena odorata* (L.) R.M. King & H. Rob. (containing protocatechuic acid) given to streptozotocin-induced diabetic Wistar rats (plasma glucose > 300 mg/dL) orally at a single oral dose of 400 mg/kg lowered glycemia by 58.8% at 6 hours, and this effect was superior to glibenclamide at 10 mg/kg (about 40%).[289] This extract given 5 weeks orally at a dose of 400 mg/kg/day lowered glycemia by from 530.7 to 249.2 mg/dL (normal 99.8 mg/dL; glibenclamide at 0.5 mg/kg/day: 252. mg/dL) and increased insulinemia from 26.5 to 34.5 μU/mL (normal 34.5 μU/mL; glibenclamide at 0.5 mg/kg/day: 28.5 μU/mL).[289] This regimen reduced insulin resistance from.[289] The plant synthetizes a broad array of flavonoids of which mainly eupatilin, tamarixetin and pentamethoxyflavanone,[290] hesperetin, naringenin, acacetin, salvigenin, kaempferol, aromadendrin 7-methyl ether,[291] quercetin 3-*O*-rutinoside, kaempferol 3-*O*-rutinoside and kaempferol 3-*O*-glucopside,[292] the benzoic acid derivatives 4-hydroxybenzoic acid and protocatechuic acid and p-coumaric acid, as well as triterpenes including 3β-acetyloleanolic acid and ursolic acid.[290] Aqueous extract of the plant given orally to Wistar rats at 200 mg/kg mitigated inflammatory response induced by carrageenan injection,[293] whereas dichloromethane extract of was found to abrogate the activation of nuclear factor-κB in HEK-293/NF-κB-luc cells with an IC_{50} value of 10 μg/mL.[294] Chromomoric acid from this plant is an Nrf2 activator.[295] All of this suggests that combined antiinflammatory and antioxidant effect may participate in the insulinotropic effects of this plant. One could reasonably anticipate, at least, an overall hypoglycemic effect combining increased insulin secretion and increase plasma glucose uptake by skeletal muscles and adipocytes.

2.75 *Elephantopus mollis* Kunth

Synonym: *Elephantopus scaber* L.
Common names: di dan cao (Chinese); tutup bumi (Malay); malatabako (Philippines)
Subclass Asteridae, Superorder Asteranae, Order Asterales, Family Asteraceae
Medicinal use: liver intoxication (Malaysia)

Acetone fraction of *Elephantopus scaber* L. given to streptozotocin-induced diabetic Wistar rats (glycemia ≥ 350 mg/dL) orally at a daily dose of 150 mg/kg for 60 days lowered fasting glycemia from 534.6 to 86.1 mg/dL (normal 85.6 mg/dL; glibenclamide at 0.6 mg/kg/day: 99.2 mg/dL) and increased fasting plasma insulin from 6.1 to 17.8 μU/mL (normal 15.1 μU/mL; glibenclamide at 0.6 mg/kg/day: 12.6 μU/mL).[296] In the same experiment, glibenclamide at 0.6 mg/kg/day lowered glycemia to 99.2 mg/dL and increased plasma insulin to 12.6 mg/dL.[296] From this extract 28Nor-22(R)Witha 2,6,23-trienolide given at a single dose of 2 mg/kg evoked after 2 hour a reduction of glycemia from 532.6 to 156.8 mg/dL (glibenclamide at 0.6 mg/kg: 512.4 mg/dL to 208.5 mg/dL).[296] Withanolides are known to occur mainly in members of the Solanaceae family and the presence of 28Nor-22(R)Witha 2,6,23-trienolide is somewhat odd. Being used to treat ulcers and placed in the Asteraceae, it is tempting to speculate that the active principle involved could be a sesquiterpene lactone with anti-inflammatory activity. Geetha et al. (2012) provided evidence that the sesquiterpene lactones, deoxyelephantopin and isodeoxyelephantopin are anti-inflammatory via inhibition of PHA-stimulated human lymphocytes proliferation.[297]

2.76 *Helichrysum arenarium* (L.) Moench

Synonyms: *Gnaphalium arenarium* L.; *Gnaphalium graveolens* Henning; *Stoechas citrina* Gueldenst.
Common name: sha sheng la ju (Chinese)
Subclass Asteridae, Superorder Asteranae, Order Asterales, Family Asteraceae
Medicinal use: kidney stone (Turkey)

Glucagon-like peptide-1 is released into the circulation by intestinal L-cells upon digestion of mixed meals to stimulate, in a glucose-dependent manner, the synthesis and secretion of insulin by β-cells.[298] The half-life of glucagon-like peptide-1 is short because the enzyme dipeptidyl peptidase IV, which is expressed by endothelial cells, converts glucagon-like peptide-1 into an inactive metabolite.[202] In addition, inhibition of enzyme dipeptidyl peptidase IV improves β cell survival, increase neogenesis and reduce apoptosis.[299] Thus, inhibitors of enzyme dipeptidyl peptidase IV could be of value for the treatment of type 2 diabetes. Methanol extract of flowers of *Helichrysum arenarium* given at a single dose of 2g/kg to ddY mice concurrently with oral sucrose loading, decreased 30 minutes blood glucose from 279.5 to 179.7 mg/dL (normal 126.3 mg/dL; acarbose at 20 mg/kg: 153.8 mg/dL).[300] This extract had no inhibitory effect in rat α-glucosidase and inhibited dipeptidyl peptidase IV with an IC_{50} of 25.4 μg/mL *in vitro*.[300] From this extract, chalconaringenin 2′-*O*-β-D-glucopyrnoside, apigenin 7-*O*-gentiobioside, luteolin 7-*O*-β-D-glucopyranoside, kaempferol 3,7-di-*O*-β-D-glucopyranoside and aureisidin 6-*O*-β-D-glucopyranoside inhibited dipeptidyl peptidase IV with IC_{50} values below 40 μM, respectively (alogliptin: 0.001 μM).[300]

2.77 *Inula japonica* Thunb.

Synonyms: *Inula repanda* Turcz.; *Limbarda japonica* (Thunb.) Raf.
Common name: xuan fu hua (Chinese)
Subclass Asteridae, Superorder Asteranae, Order Asterales, Family Asteraceae
Medicinal use: cough (China)

Aqueous extract of flowers of *Inula japonica* Thunb. given to alloxan-induced diabetic Kunming mice at a dose of 1 g/kg for 18 days decreased water intake and food intake, lowered plasma glucose from 23.1 to 12.8 mmol/L.[301] This regimen lowered kidney weight, lowered triglycerides from 2.4 to 1.6 mmol/L and had no effect on total cholesterol.[301] Plasma insulin was raised by the treatment.[301] In oral glucose tolerance test, the extract at a single oral dose of 1 g/kg lowered postprandial glycemia at 90 minutes.[301]

2.78 *Sphaeranthus indicus* L.

Synonyms: *Sphaeranthus hirtus* Willd.; *Sphaeranthus mollis* Roxb.
Common names: rong mao dai xing cao (Chinese); mundi (India); East Indian globe-thistle
Subclass Asteridae, Superorder Asteranae, Order Asterales, Family Asteraceae
Medicinal use: fatigue (Burma)

Ethanol extract of roots of *Sphaeranthus indicus* L. given orally to streptozotocin-induced diabetic Wistar rats (glycemia between 200 and 300 mg/dL) at 500 mg/kg/day for 15 days attenuated body weight loss, lowered blood glucose from 285 to 104.3 mg/dL (normal 91.8 mg/dL; glibenclamide at 0.2 mg/kg: 89.5 mg/dL).[302] This extract increased serum insulin from 6.6 to 14.2 μU/mL (normal 15.6 μU/mL; glibenclamide at 0.2 mg/kg: 12.1 μU/mL). The extract increased hepatic glycogen.[302] The extract lowered triglycerides and total cholesterol to and increased high-density lipoprotein–cholesterol to normal.[302] In rats, the extract at 500 mg/kg had no effect on postprandial glycemia.[302] The mode of insulinotropic action is unknown. The plant contains series of eudesman including 11α,13-dihydro-3α,7α-dihydroxy-4,5-epoxy-6β,7-eudesmanolide, 11α,13-dihydro-7α-acetoxy-3β-hydroxy-6b,7-eudesm-4-enolide and 3-keto-β-eudesmol.[303] Eudesmane sesquiterpenes have the tendency to inhibit nuclear factor-κB induced by interleukin-1β and tumor necrosis factor-α.[304] In metabolic syndrome, chronic hyperglycemia and free fatty acids lead to an increased production of reactive oxygen species which induce the activation of NFκB, which induced inducible nitric oxide synthetase expression and the production of nitric oxide.[305] This, eudesmane sesquiterpenes from *Sphaeranthus indicus* L. may conceptually the ability of inhibiting streptozotocin-induced reactive oxygen species generation and downstream induction of nuclear factor-κB affording pancreatic cytoprotection and regeneration.

2.79 *Silybum marianum* (L.) Gaertn.

Synonyms: *Carduus marianus* L.; *Carthamus maculatum* (Scop.) Lam. *Cirsium maculatum* Scop.
Common name: milk thistle
Subclass Asteridae, Superorder Asteranae, Order Asterales, Family Asteraceae
Medicinal uses: jaundice (India)

Silymarin given orally at a dose of 200 mg/kg/day to Wistar rats with partial pancreatectomy for 63 days evoke an increased in expression of pancreatic and duodenal homeobox-1 and insulin.[306] This regimen increased β-cell proliferation and increased plasma insulin from 1.1 to 4.5 ng/mL.[306] At day 21, the extract lowered glycemia from 9 to 6.4 mM.[306] Intake of this plant could be of value in metabolic syndrome.

2.80 *Vernonia amygdalina* Delile

Synonyms: *Gymnanthemum amygdalinum* (Delile) Sch. Bip. ex Walp.
Common name: bitter leaf
Subclass Asteridae, Superorder Asteranae, Order Asterales, Family Asteraceae
Medicinal use: diabetes (Malaysia)

Ethanol extract of leaves of *Vernonia amygdalina* Delile (containing chlorogenic acid, 1,5-dicaffeoyl-quinic acid and dicaffeoyl-quinic acid) given orally to streptozotocin-induced diabetic Wistar rats (fasting blood glucose ≥ 250 mg/dL) at a single dose of 400 mg/kg 30 minutes before oral glucose loading lowered 1 hour peak glycemia to almost normal group levels.[307] The extract given orally to streptozotocin-induced diabetic Wistar rats (fasting blood glucose ≥ 250 mg/dL) for 28 days at a daily dose of 400 mg/kg had no effect on food consumption, attenuated body weight gain, lowered fasting blood glucose by 32.1% (metformin at 500 mg/kg/day: 35.3%) and lowered plasma triglycerides by 18.2% and cholesterol by 41%.[307] This regimen increased pancreatic and plasma insulin by 54% and 58%, respectively.[307] The extract increased skeletal muscle glucose uptake normal and increased glycogen to normal.[307] Insulin depletion caused by streptozotocin induced an increase of hepatic glucose-6-phosphatase and the extract decreased glucose-6-phosphatase by 40%, increased glutathione peroxidase activity and restored glutathione level near to normal in the liver.[307] At the pancreatic level, the extract evoked a decrease in vacuolization in islets.[307] Chlorogenic acid at a concentration of 100 μg/mL increased glucose-induced insulin secretion by rat insulinoma-derived INS-1E cells as well as rat ancreatic islets.[308] Chlorogenic acid had no effect on β-cell K_{ATP} channels.[308]

2.81 *Strychnos potatorum* L.f.

Common name: katakah (India); clearing nut tree
Subclass Lamiidae, Superorder Lamianae, Order Rubiales, Family Loganiaceae
Medicinal use: diabetes (Malaysia). The seeds have been used in India since immemorial times to clear muddy drinking water

The seeds are known to contain the indole alkaloids diaboline acetyldiaboline and angustine.[309,310] Diaboline is a weak glycine antagonist.[311] Powder of seeds given orally at a dose of 100 mg/kg/day to streptozotocin-induced diabetic Wistar rats for 12 weeks lowered glycemia from 159.3 to 90.3 mg/dL, an effect superior to insulinotropic agent Glipizide at 40 mg/kg/day with 110.3 mg/dL (normal 71 mg/dL).[312] Ethanol extract of seeds given to nicotinamide-streptozotocin-induced type 2 diabetic Wistar rats (glycemia $>$ 200 mg/dL) at a dose of 400 mg/kg/day orally for 21 days (glibenclamide: 0.6 mg/Lg) lowered glycemia from 212.3 to 118.2 mg/dL (normal 78.8 mg/dL; glibenclamide at 0.6 mg/kg: 94.6 mg/dL), increase of hepatic and kidney activity of glutathione peroxidase, catalase, glutathione-S-transferase and glutathione contents, and improved pancreatic cytoarchitecture.[313] As for the effects of the seeds, it should be remembered that glucagon-like peptide-1 is transiently released by the small intestine in presence of digestion products and accounts for enhancement of insulin secretion, stimulation of insulin gene expression, suppression of glucagon secretion and gastric emptying.[314] The water cleaning property of seeds is owed to galactomannan and a galactan [315,316] which may induce glucagon-like peptide-1 secretion to increase insulin secretion, regenerate pancreatic islets and lower glycemia in streptozotocin-induced diabetic Wistar rats.

2.82 *Gardenia jasminoides* J. Ellis

Synonyms: *Gardenia augusta* Merr.; *Gardenia florida* L.; *Varneria augusta* L.
Common names: zhi zi (Chinese); karinga (India); cape jasmine
Subclass Lamiidae, Superorder Lamianae, Order Rubiales, Family Rubiaceae
Medicinal use: Jaundice (China)

Geniposide from the fruits of *Gardenia jasminoides* J. Ellis at a concentration of 10 µmol/L induced the secretion of insulin *in vitro* by INS-1 rat insulinoma cell line to about 10 ng/mL/h and this effect was dose dependent.[317] This iridoid glucoside at 10 µmol/L potentiated insulin secretion evoked by glucose and this effect was abrogated by glucagon-like peptide-1 receptor antagonist exendin.[318] Geniposide (Figure 2.21) at a concentration of 10 µM protected INS-1 rat insulinoma cells against palmitic acid-induced apoptosis.[317] This effect was abolished by glucagon-like peptide-1 receptor antagonist exendin.[318] Geniposide increased the level of pancreatic and duodenal homeobox-1, and reversed palmitate-induced decreased phosphorylation of Akt, and Foxo1.[318] Note that geniposide is decomposed into genipin by bacterial flora (Aburada et al., 1978) hence a probable lower ability to be as effective orally.[319]

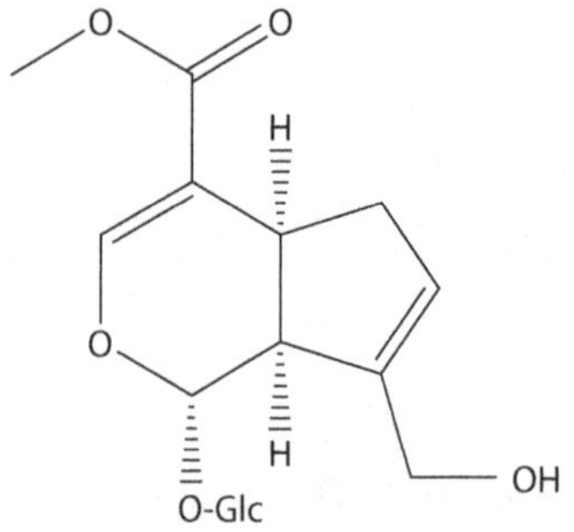

FIGURE 2.21 Geniposide.

2.83 *Centaurium erythraea* Rafn

Common names: kırmızı (Turkey); centaury
Superorder Lamianeae, Order Rubiales, Family Gentianaceae
Medicinal uses: malaria (Turkey)

Aqueous extract of *Centaurium erythraea* Rafn given intraperitoneally to streptozotocin-induced diabetic Wistar rats at a dose of 200 mg/kg/day for 30 days evoked a decrease of glycemia from 288 to 107 mg/dL (normal 81 mg/dL) and increased plasma insulin from 0.4 to 0.8 ng/mL (normal 1.2 ng/mL).[320] At the pancreatic level this treatment lowered malonyldialdehyde, increased gluthathione contents and the enzymatic activity of catalase, glutathione peroxidase and superoxide dismutase.[320] Furthermore, this extract improved pancreatic cytoarchitecture.[320] Measuring the pharmacological activity of plant extract by parenteral administration implies the effect of nonmetabolized natural products in the plasma and as such, a parenteral insulinotropic effect does not always correlates with oral insulinotropic effect and vice versa. In fact, medicinal plants are given orally in traditional medium, most often in aqueous form arguing in favor of oral administration of extract in rodents, and this is particularly important in the context of Gentianaceae. Methanol extract of aerial parts of *Centaurium erythrea* Rafn given to streptozotocin-induced diabetic Wistar rats (fasting blood glucose > 15 mmol/L) orally at a dose of 250 mg/kg/day for 12 days lowered glycemia from about 15 to 5 mmol/L, an effect similar to glibenclamide at 2.5 mg/kg/day.[321] After 20 days, the extract was unable to correct glycemia in diabetic animals. This regimen at the hepatic level lowered glucose concentration, increased to p glycogen contents.[321] This 12 days regimen lowered total cholesterol, triglycerides, high-density lipoprotein–cholesterol and low-density lipoprotein–cholesterol to normal.[321] The plant elaborates series of secoiridoids of which swertiamarin and sweroside,[322] as well as gentiopicroside.[323] Swertiamarin is known to promote insulin secretion from RINm5F cells *in vitro* but is unable when given orally to rodents to induce insulin secretion.[324] In rats the secoiridoid gentiopicroside is metabolized into erythrocentaurin and gentiopicral, the former being metabolized into 3,4-dihydro-5-(hydroxymethyl)-isochroman-1-one.[325] Other constituents in these medicinal plants are xanthones including 1-hydroxy-3-methoxy xanthone and flavonoids of which quercetin 3-O-rhamnoside-7-glycosyl-(1→2)-rhamnoside.[326] The hypoglycemic activity of *Centaurium erythraea* Rafn may account from the activity of secoiridoids as well as xanthones, the later having potentials for protein sulfonylurea receptor 1. *Centaurium erythraea* extract given to C57BL/6J mice on high-fat diet for 18 weeks orally at a dose of 2 g/kg lowered glycemia from about 230 to 143.8 mg/mL (normal 120 mg/mL), lowered weight gain, reduced insulinemia from 3.3 to 0.9 ng/mL (normal: 0.7 ng/mL), decreased triglycerides and total cholesterol, high-density lipoprotein.[327] Ethanol extract of aerial parts *Centaurium erythrea* given orally to C57BL/6J mice on high-fat diet at a dose of 2 g/kg/day for 20 weeks reduced hepatic steatosis by 70% compared with untreated rodents. The extract given from 17th to 35th week at the same dose to the rodents on high-fat diet from the 1st week reduced hepatic steatosis.[321] It must be noted that swertiamarin is metabolized into gentianine by intestinal bacteria.[328]

2.84 *Swertia corymbosa* (Griseb.) Wight ex Clarke

Common name: Kirata tikta (India)
Superorder Lamianeae, Order Rubiales, Family Gentianaceae
Medicinal use: fever (India)

1,2,8-Trihydroxy-6-methoxy xanthone and 1,2-dimethoxy-6-methoxyxanthone-8-*O*-β-D-xylopyranosyl isolated from the aerial parts of *Swertia corymbosa* (Griseb.) Wight ex Clarke given orally to streptozotocin-induced diabetic abino rats (blood glucose > 250 mg/dL) at a dose of 50 mg/kg/day for 28 days prevented body weight loss, lowered glycemia from 545.4 to 256 mg/dL (normal 96.1 mg/dL; glibenclamide at 10 mg/kg/day: 145.2 mg/dL), increased insulinemia from 10.2 to 14.9 μU/mL (normal 17 μU/mL; glibenclamide at 10 mg/kg/day: 16.3 μU/mL), and improved lipid profile and pancreatic cytoarchitecture.[329] From the same plant, 1,2-dimethoxy-6-methoxyxanthone-8-*O*-β-D-xylopyranosyl isolated from the aerial parts of *Swertia corymbosa* (Griseb.) Wight ex Clarke given orally to streptozotocin-induced diabetic albino rats (blood glucose > 250 mg/dL) at a dose of 50 mg/kg/day for 28 days, prevented body weight loss, lowered glycemia from 545.4 to 142.7 mg/dL (p: 96.1 mg/dL; glibenclamide at 10 mg/kg/day: 145.2 mg/dL), increased insulinemia from 10.2 to 16.1 μU/mL (p: 17 μU/mL; glibenclamide at 10 mg/kg/day: 16.3 μU/mL), and improved lipid profile and pancreatic cytoarchitecture.[329] Molecular interaction study of these xanthones indicated binding affinity with sulfonylurea receptor-1 protein target gave favorable binding conformation suggesting the binding of that xanthone with receptor SUR1 closing K channels and increasing insulin production.[329]

2.85 *Swertia kouitchensis* Franch.

Synonym: *Swertia elongata* T.N. Ho & S.W. Liu
Common name: gui zhou zhang ya cai (Chinese)
Superorder Lamianeae, Order Rubiales, Family Gentianaceae
Medicinal use: diabetes (China)

Ethanol extract (containing mangiferin, swertianolin, and bellidifolin) of *Swertia kouitchensis* Franch. dose-dependently increased the secretion of insulin by mouse insulinoma NIT-1 cells induced by glucose at 5.6 mM (basal) and 16.7 mM (stimulatory) with a maximal effect at a concentration of 100 μg/mL similarly to gliclazide at 10 μg/mL.[330] The extract given orally at a single dose of 500 mg/dL to normal and high-fat and fructose diet streptozotocin-induced type 2 diabetic Balb/c mice evoked a mild reduction of glycemia.[330] The extract given to high-fat and fructose diet streptozotocin-induced type 2 diabetic Balb/c mice (fasting blood glucose superior or equal to 11.1 mmol/L) orally at a dose of 500 mg/kg/day for 4 weeks lowered glycemia from about 16 to 12 mmol/L (normal 5 mmol/L; gliclazide at 15 mg/kg/day: about 7 mmol/L).[330] This extract lowered serum cholesterol and triglycerides, increased high-density lipoprotein, and lowered low-density lipoprotein.[330] The extract increased the activity of glucokinase, lowered glucose-6-phosphatase and replenished hepatic glycogen. The treatment increased plasma insulin by 40.4% (gliclazide at 15 mg/kg/day: 83% increase in glycemia.[330] The insuliniotropic mode of action of *Swertia kouitchensis* Franch.and principles involved are not know. What is the effect of containing mangiferin, swertianolin and bellidifolin on β-cells receptor SUR1. Sulfonylureas like glibenclamide or gliclazide close the pancreatic β-cell K_{ATP} channels through an interaction with the SUR-1 receptor, leading to membrane depolarization, opening of voltage-dependent calcium channels, entry of extracellular calcium, increase of cytosolic concentration of calcium in β-cells and exocytosis of insulin.[331] According to this mechanism, sulfonylureas command insulin secretion by β-cells in the presence of basal or stimulatory concentrations of glucose.[332,333]

2.86 *Swertia macrosperma* (C.B. Clarke) C.B. Clarke

Synonym: *Swertia scandens* H. Lév.
Common name: da zi zhang ya cai (Chinese)
Superorder Lamianeae, Order Rubiales, Family Gentianaceae
Medicinal use: hepatitis (Tibet)

Ethanol extract of *Swertia macrosperma* (C.B. Clarke) C.B. Clarke given orally at a dose of 500 mg/kg/day to high-fat and fructose diet streptozotocin induced diabetic Wistar rats (fasting blood glucose > 16.7 mmol/L) for 28 days decreased fasting blood glucose from 25.4 to 9.9 mmol/L, and this effect was superior to metformin (normal 4.7 mmol/L; metformin at 150 mg/kg/day: 10.2 mmol/L).[333] This regimen increased fasting serum insulin levels from 9 to 12.7 μIU/mL (normal 10.7 μIU/mL; metformin at 150 mg/kg/day: 11.6 μIU/mL) and lowered insulin resistance.[333] Oral glucose tolerance test performed after 25 days of treatment evoked a fall glycemia from approximately 30 to 15 mmol/L after 4 hours (normal about 5 mmol/L; metformin: 13 mmol/L).[333] The extract decreased total serum cholesterol, triglycerides, increased high-density lipoprotein–cholesterol and lowered low-density lipoprotein–cholesterol similarly to metformin.[333] At the hepatic level, as a sign of enhanced insulin availability/sensibility, the ethanol extract increased the enzymatic activity of glucokinase and decreased the activity of glucose-6-phosphatase and increased hepatic glycogen similarly to metformin.[333] At the pancreatic level, this treatment attenuated islets shrinkage and vacuolation in pancreatic tissues.[333]

2.87 *Calotropis gigantea* (L.) W.T. Aiton

Synonym: *Asclepias gigantea* L.; *Periploca cochinchinensis* Lour.; *Streptocaulon cochinchinense* (Lour.) G. Don
Common names: niu jiao gua (Chinese); akond mul (Bengal), giant milkweed
Subclass Lamiidae, Superorder Lamianeae, Order Rubiales, Family Asclepiadaceae
Medicinal use: boils (Indonesia)

Rathod et al. (2011) recorded a decrease of plasma glucose from 352 to 249.1 mg/dL (glibenclamide at 10 mg/kg/day: 160 mg/dL) in streptozotocin-induced diabetic Wistar rats receiving orally a chloroform extract of leaves at a dose of 50 mg/kg/day for 27 days.[334] The principle(s) involved here are not known yet. Saratha and Subramanian provided evidence that lupeol from the latex of *Calotropis gigantea* (L.) W.T. Aiton given orally at a dose of 50 mg/kg/day for 4 weeks to rats could reduce Freund's Complete Adjuvant induced arthritis and related increased of proinflammatory cytokines in the plasma.[335] Being anti-inflammatory, lupeol could conceptually promote β-cells regeneration. However, *Calotropis gigantea* (L.) W.T. Aiton contains the cardiac glycoside calotropin which has a median lethal dose of 9.8 mg/kg intraperitoneally in Swiss Webster mice.[336,337] The plant is toxic and of no use in therapeutic.

2.88 *Calotropis procera* (Aiton) W.T. Aiton

Subclass Lamiidae, Superorder Lamianeae, Order Rubiales, Family Asclepiadaceae
Medicinal use: toothache (India)

Ethanol extract of leaves of *Calotropis procera* (Aiton) W.T. Aiton given orally to streptozotocin-induced diabetic Wistar rats (fasting glycemia > 200 mg/dL) at a dose of 300 mg/kg/day for 28 days lowered food intake, reduced water intake, lowered fasting glycemia by 68% (comparable effect with metformin: 500 mg/kg/day), improved glucose tolerance compared with untreated animals.[338] Yadav et al. (2014) reported that methanol extract of roots of this plant given orally at a dose of 100 mg/kg/day for 2 weeks to streptozotocin-induced Wistar rats decreased plasma glucose, increased plasma insulin and improved pancreatic cytoarchitecture compared with untreated rodents.[339] The principle(s) and mechanisms involved here are yet to be determined. It must be noted that chronic oral administration of latex of *Calotropis procera* (Aiton) W.T. Aiton at a dose of 400 mg/kg in alloxan-induced diabetic rats prevented body weight gain and increased water intake, decreased blood glucose and increased liver glycogen and increased the activity of hepatic superoxide dismutase, catalase and glutathione.[340] The plant contains the toxic cardiac glycoside calotropin.[341] One would wonder if the hypoglycemic effect of this cardiotoxic plant would not simply be the consequence of chronic stress in experimental animals inducing the secretion of norepinephrine and glucocorticoids, and subsequent insulin secretion and replication of β-cell in rats.[342]

2.89 *Carissa carandas* L.

Synonyms: *Arduina carandas* (L.) K. Schum.; *Carissa congesta* Wight
Common names: ci huang guo (Chinese); karonda (India); Bengal current
Subclass Lamiidae, Superorder Lamianeae, Order Rubiales, Family Apocynaceae
Medicinal use: thirst (India)

There are several observations that support the notion that the fruits of this plant, which are edible, contain insulinotropic agents. Ethyl acetate fraction of unripe fruits of *Carissa carandas* L. given at a single oral dose of 400 mg/kg to Sprague–Dawley rats in oral glucose tolerance test lowered glycemia from 97 to 82 mg/dL (metformin intraperitoneally at 50 mg/kg: 83 mg/dL) after 120 minutes in oral glucose tolerance tests.[343] This fraction given orally to streptozotocin-induced diabetic Sprague–Dawley rats at a single dose oral dose of 400 mg/kg evoked a fall of glycemia by 64.5%, an effect similar to metformin given intraperitoneally at a dose of 50 mg/kg after 24 hours.[343] The plant contains series of triterpenes of which carandinol, betulinic acid, oleanolic acid, ursolic acid, and 4-hydroxybenzoic acid.[344] 4-Hydroxybenzoic acid administered orally at a dose of 5 mg/kg in oral glucose tolerance test evoked a fall of glycemia by 79.2% and increased liver glycogen content and serum insulin level by 143.1% and 121.9%, respectively (glibenclamide 5 mg/kg: serum glucose decreased by 62.8%, increased the liver glycogen content and serum insulin level to 201.5% and 184.3%).[345] Betulinic acid given orally at a dose of 50 mg/L in drinking water to mice poisoned on high-fat diet for 15 weeks reduced weight gain by 10% compared with untreated animals with a reduction on abdominal fat from 887.6 mg/10 g of body weight to 332.6 mg/10 g of body weight.[346] Furthermore, this triterpene decreased plasma triglycerides and cholesterol, pized glycemia, and increased plasma insulin.[346] Therefore, one could draw an inference that the fruits of *Carissa carandas* L. lower plasma glucose in rats by inducing insulin secretion by at least 4-hydroxybenzoic acid and betulinic and promoting glucose intake by peripheral tissues via its triterpenes. Clinical trials are warranted.

2.90 *Catharanthus roseus* (L.) G. Don

Common name: periwinkle
Subclass Lamiidae, Superorder Lamianeae, Order Rubiales, Family Apocynaceae
Medicinal use: diabetes (Laos)

Dichloromethane-methanol extract of leaves and stems of *Catharanthus roseus* (L.) G. Don given orally to streptozotocin-induced diabetic Sprague–Dawley rats (glycemia > 200 mg/dL) at a dose of 500 mg/kg/day for 15 days lowered glycemia by 57.6%. After 7 days of treatment liver glycogen was increased as well as activity of glycogen synthetase, an enzyme known to be stimulated by insulin.[347] Administration of the extract for 30 days at a dose of 500 mg/kg/day abrogated hyperglycemia induced by a single intraperitoneal injection of streptozotocin.[347] This extract increased glucose-6-phosphate dehydrogenase[347] which indicates insulin stimulation.[347] Vindoline from *Catharanthus roseus* (L.) enhanced the secretion of insulin by MIN6 cells exposed to 16.8 mM of glucose with an IC_{50} equal to 50 μM.[348] This alkaloid at a concentration of 50 μM increased the secretion of *db/db* mice Islets of Langerhans exposed to glucose *in vitro*.[348] The insulinotropic effect of vindoline on by MIN6 insulinoma was inhibited by nifedipine, a specific inhibitor of L-voltage dependent calcium channel and potassium channel opener diazoxide.[348] This alkaloid at prevented apoptosis of β-cells induced by interleukin-1β and interferon-γ and tumor necrosis factor-α.[348] Vindoline given orally for 4 weeks to *db/db* mice at a dose of 40 mg/kg/day lowered fasting glycemia from about 20 to 10 mM, lowered triglycerides and increased insulinemia from about 1 to 2 ng/mL.[348] Vindoline given orally for 4 weeks to high-fat diet-streptozotocin-induced type 2 diabetic Wistar rats at a dose of 20 mg/kg/day lowered fasting glycemia from about 15 to 10 mM (normal about 7 mM), improved glucose tolerance, plasma triglycerides and increased insulinemia from about 0.05 to 0.1 ng/mL (normal 0.3 ng/mL).[348] Vindoline given orally at a dose of 40 mg/kg/day to spontaneous type 2 diabetic obese *db/db* mice for 4 weeks increased insulinaemia by about 50%, halved plasma glucose and reduced plasma triglycerides after 3 weeks of treatment. Sprague–Dawley rats rendered diabetic by high-fat diet for 2 weeks followed by streptozotocin administration and then treated with vindoline at a dose of 20 mg/kg/day for 4 weeks exhibited lower glycemia compared with untreated animals, increased insulinaemia by about 3 folds and reduced serum triglyceride.[348] The precise molecular mechanism behind the insulinotropic effect of vindoline is unknown. Vindoline is an indole alkaloid and as such may be able to interact with β-cell 5-hydroxytryptamine receptors. In healthy rodents or alloxan diabetic rodents, fed or fasting, pretreated with nialamide at a dose 80 mg/kg, 5-hydroxytryptophane given at a dose of 60 mg/kg intravenously commanded hypoglycemia which prevented by methysergide.[348] Vincristine at a dose of 1 mg/kg and *N*-formyl-leurosine at a dose of 3 mg/kg reduced serum lipid by 49.05% and 40.40% in rodents 4 hours after intraperitoneal injection.[349]

2.91 *Caralluma attenuata* Wight

Subclass Lamiidae, Superorder Lamianeae, Order Rubiales, Family Aclepiadaceae
Medicinal use: diabetes (India)

Butanol fraction of the whole plant given to Wistar rats orally at a single dose of 250 mg/kg 30 minutes before oral loading of glucose lowered postprandial glycemia at 30 minutes from 156.7 to 103.6 mg/dL.[351] This extract given to alloxan-induced diabetic Wistar rats at a single dose of 250 mg/kg lowered glycemia from 370.9 to 321.6 mg/dL after 1 hour.[351] This extract given to alloxan-induced diabetic Wistar rats at a single dose of 250 mg/kg/day for 10 days lowered glycemia from 341.2 to 228.6 mg/dL (normal 81.3 mg/dL).[351] Butanol fractions of medicinal plants often concentrate saponins, of which steroidal saponins, which are known to occur in members of the family Asclepiadaceae.

2.92 *Caralluma tuberculata* N.E. Br.

Synonym: *Boucerosia aucheriana* Decne.
Common name: chungan (India)
Subclass Lamiidae, Superorder Lamianeae, Order Rubiales, Family Aclepiadaceae
Medicinal use: diabetes (Pakistan)

Methanol extract of *Caralluma tuberculate* N.E. Br. given orally to streptozotocin-induced diabetic Sprague–Dawley rats (glycemia > 250 mg/dL) for 4 weeks at a dose of 250 mg/kg/days lowered glycemia by 53.3% (gliclazide at 15 mg/kg: 80.6%).[352] This extract increased serum insulin levels from 33.6% to 47.9% (gliclazide at 15 mg/kg: 81.5%) and lowered the enzymatic activity of hepatic glucose-6-phosphatase by about 20%.[352] *In vitro* the methanolic extract at a concentration of 500 g/mL induced glucose absorption by intestinal preparation of rats and enhanced glucose absorption by rat isolated psoas muscle.[352]

2.93 *Cynanchum acutum* L.

Synonyms: *Solenostemma acutum* (L.) Wehmer; *Vincetoxicum acutum* (L.) Kuntze
Common name: sirkenek (Turkey)
Subclass Lamiidae, Superorder Lamianeae, Order Rubiales, Family Apocynaceae

Quercetin 3-*O*-β-galacturonopyranoside isolated from *Cynanchum acutum* L. given orally to alloxan-induced diabetic Swiss albino mice at a daily dose of 50 mg/kg/day for 8 weeks lowered glycemia from 265.8 to 123.2 mg/dL and increased plasma insulin from 8.3 to 26.1 μU/mL.[353] From the same plant, quercetin 7-*O*-β-glucopyranoside at the same dose lowered glycemia from 265.8 to 131.2 mg/dL and increased plasma insulin from 8.3 to 27.2 μU/mL.[353] Likewise, tamarixtin 3-*O*-β-galacturonopyranoside reduced glycemia from 265.8 to 138.2 mg/dL and increased insulinemia from 8.3 to 29.8 μU/mL.[353] In this experiment metformin decreased plasma glucose and increased plasma insulin.[353] Metformin in MIN6 β-cells and human islets of Langerhans and activates adenosine monophosphate-activated protein kinase to inhibit glucose stimulated insulin secretion.[354] Therefore, increased insulin secretion induced by the adenosine monophosphate-activated protein kinase activator metformin could be owed to a regenerative/cytoprotective effect on damaged β-cells.[355] The mechanism underlying insulin secretion by *Cynanchum acutum* L. is unknown and may involve at least antioxidant effects as suggested by increased plasma glutathione close to normal values

2.94 *Gymnema montanum* (Roxb.) Hook. f.

Synonym: *Asclepias montana* Roxb.
Subclass Lamiidae, Superorder Lamianeae, Order Rubiales, Family Asclepiadaceae
Medicinal use: An inflammation (India)

Ethanol extract of leaves of *Gymnema montanum* (Roxb.) Hook. f. given to alloxan-induced diabetic Wistar rats at a dose of 200 mg/kg/day for 3 weeks prevented weight loss and reduced water intake.[356] This regimen brought glycemia from 298 to 86 mg/dL (normal 81 mg/dL; glibenclamide at 600 mg/kg/day: 118 mg/dL) and increased insulinemia from 5.5 to 12 (normal 13.6; glibenclamide at 600 μg/kg/day: 10 μU/mL).[356] *In vitro*, ethanol extract of leaves of *Gymnema montanum* (Roxb.) Hook. f. at a concentration of 5 μg/mL protected RINm5F cell line against alloxan-induced apoptosis.[357] This extract reduced close to normal the cytosolic levels of reactive oxygen species and nitrites evoked by alloxan in RINm5F cells and prevented loss of mitochondrial membrane potential and increased in cytoplasmic calcium and lowered lipid peroxidation.[357] The extract increased the enzymatic activity of superoxide dismutase, catalase, glutathione peroxidase and glutathione S-transferase and increased glutathione, contents in RINm5F cells.[357] The extract inhibited caspase 3 and PARP cleavage in RINm5F cells challenged with alloxan.[357]

2.95 *Gymnema sylvestre* (Retz.) R. Br. ex Schult.

Common name: gurmar (India)
Subclass Lamiidae, Superorder Lamianeae, Order Rubiales, Family Asclepiadaceae
Medicinal use: diabetes (India)
History: The plant was known to Susrhuta for its antidiabetic properties

Murakami et al. (1996) made the demonstration that the leaves of *Gymnema sylvestre* (Retz.) R. Br. ex Schult. produces series of pentacyclic triterpenes saponins including gymnemoside b, gymnemic acid III, gymnemic acid V, and gymnemic acid VII, which given at a single oral dose of 100 mg/kg 30 minutes before oral loading of glucose, lowered 30 minutes postprandial peak glycemia of Wistar rats from 55.5 to 43.9, 40.4, 35.5, and 42.7 mg/dL, respectively.[358] In the same experiment, gymnemic acid I lowered plasma glucose at 2 hour from 11 to 5.7 mg/dL and gymnemic IV was inactive.[358] Gymnemic acid IV at a single intraperitoneal dose of 13.4 mg/kg increased plasma insulin levels in streptozotocin-induced diabetic mice,[359] suggesting its inactivation by first pass metabolism. Acetone extract of *Gymnema sylvestre* (Retz.) R. Br. ex Schult. given orally to streptozotocin-induced diabetic Wistar rats (fasting blood glucose: 280 mg/dL–350 mg/dL) at a dose of 600 mg/kg/day for 45 days lowered plasma glucose from 443 to 114.2 mg/dL (normal 76 mg/dL; insulin 3 IU/kg 87.9 mg/dL.[360] From this extract dihydroxy gymnemic triacetate given orally to streptozotocin-induced diabetic Wistar rats (fasting blood glucose: 280 mg/dL–350 mg/dL) at a dose of 20 mg/kg/day for 45 days lowered plasma glucose from 430 to 121.6 mg/dL (normal 84.5 mg/dL; insulin at 3 IU/kg: 113.5 mg/dL.[360] This saponin increased plasma insulin from 5.6 to 15.2 μU/mL, a value close to normal (17.8 μU/mL).[360] Dihydroxy gymnemic triacetate replenished glycogen stocks in the liver and skeletal muscle of diabetic animals.[360] This triterpene saponin lowered total cholesterol, triglycerides, low-density lipoprotein–cholesterol, normalized high-density lipoprotein–cholesterol.[360] Liu et al (2009) provided evidence that saponin fraction of leaves of *Gymnema sylvestre* at a concentration of 0.2 mg/mL increased the secretion of insulin by MIN6 mouse β-cell line exposed to glucose at 2 mM or 20 mM with concurrent increase in cytosolic concentration of calcium through voltage-operated calcium channels.[361] The fraction at a concentration of 0.1 mg/mL increased glucose stimulated secretion of insulin by human islets *in vitro* exposed to glucose at 2 mM or 20 mM with concurrent increase in cytosolic concentration of calcium through voltage-operated calcium channels.[361] The exact insulinotropic mode of action of saponins of *Gymnema sylvestre* (Retz.) R. Br. ex Schult. remains elusive. Saponins not uncommonly induce insulin secretion *in vitro*.

2.96 *Hemidesmus indicus* (L.) R. Br. ex Schult.

Synonym: *Periploca indica* L.
Common names: sariba (India); India sarsaparilla
Subclass Lamiidae, Superorder Lamianeae, Order Rubiales, Family Apocynaceae
Medicinal use: venereal diseases (India)

β-Amyrin palmitate (Figure 2.22) isolated from the roots of *Hemidesmus indicus* (L.) R. Br. ex Schult. given to Wistar rats at a single oral dose of 100 μg/kg, 30 minutes before oral loading of glucose, lowered 30 minutes peak glycemia from 8.1 to 5.5 mmol/L whereby intraperitoneal administration of this triterpene had no effects suggesting an inhibiting effect of intestinal absorption of glucose or hypoglycemic effect of a β-amyrin after fist pass metabolism.[362] This triterpene given to streptozotocin-induced diabetic Wistar rats (glycemia: 12.2 mmol/L to 13.8 mmol/L) orally for 20 days at a daily dose of 50 μg/kg for 20 days lowered glycemia from about 16 to 7 mmol/L, a value close to normal group value (5.5 mmol/L) and superior to glibenclamide at 500 μg/kg/day (9.5 mmol/L).[362] This treatment, attenuated weight loss, repleted hepatic glycogen, and lowered total cholesterol.[362] β-Amyrin palmitate given to alloxan-induced diabetic Wistar rats (glycemia: 16.6 to 18.8 mmol/L) orally for 15 days at a daily dose of 50 μg/kg lowered glycemia from about 24.5 to 9 mmol/L, a value close to the effect of insulin (at 5 IU/kg: 6 mmol/L; normal 5 mmol/L).[362]

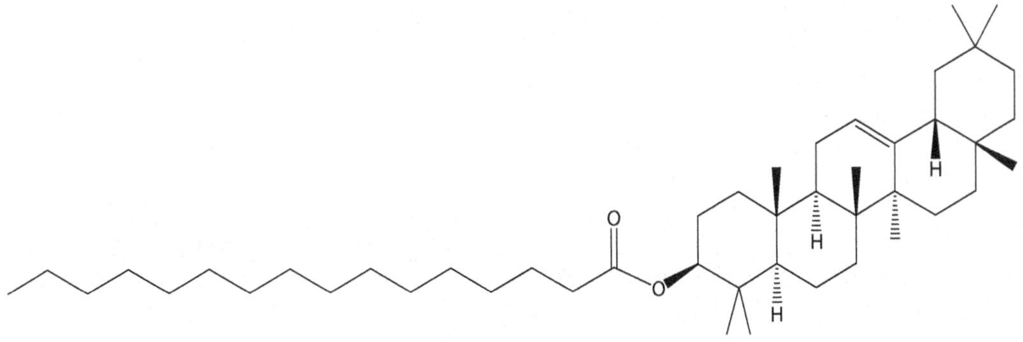

FIGURE 2.22 β-Amyrin palmitate.

This pentacyclic triterpene repleted hepatic glycogen and lowered total cholesterol to normal in alloxan-induced diabetic Wistar rats after 15 days of treatment.[362] The precise mechanism involved here is unknown. Note that α-amyrin palmitate, which not uncommon in members of the family Apocynaceae is antiinflammatory as administered orally at a dose of 56 mg/kg/day daily for 8 days to rodents with complete Freund's adjuvant-induced arthritic rats lowered plasma granulocytes and hyaluronate and blood granulocytes toward normal levels. It is therefore tempting to speculate that the chronic oral administration of β-amyrin palmitate could protect pancreatic β-cells against alloxan-induced inflammation allowing the functionality of β-cells and increased insulin secretion. Further studies on that triterpene of potential value for the treatment of metabolic syndrome are needed. It should be noted that the roots of *Hemidesmus indicus* (L.) R. Br. ex Schult. also contain 2-hydroxy-4-methoxy benzoic acid (Figure 2.23), which given to streptozotocin-induced Wistar rats orally at a dose of 500 µg/kg/day for 7 weeks increased plasma insulin from 6.8 to 16 µU/mL (normal 15.6 µU/mL; tolbutamide: 13.9 µU/mL), lowered plasma glucose from 288 to 75 mg/dL (normal 69.3 mg/dL), replenished hepatic glycogen, pized serum lipid profile as well as alanine aminotransferase, aspartate aminotransferase, and alkaline phosphatase[363] and improved pancreatic histoarchitecture.[364] The root of *Hemidesmus indicus* (L.) R. Br. ex Schult. represents an interesting candidate for the prevention of type 2 diabetes in metabolic syndrome. Further studies are warranted.

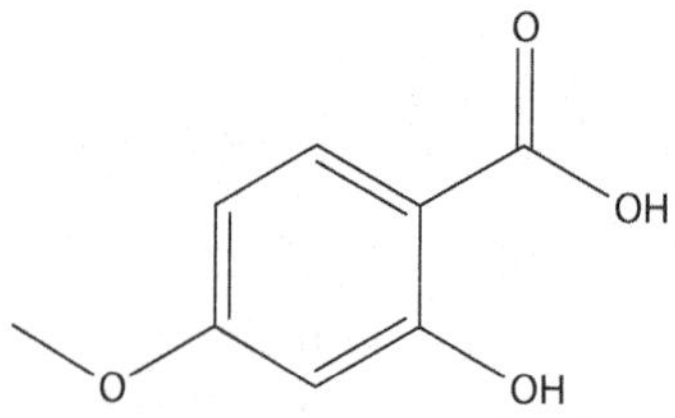

O
OH
O
OH

FIGURE 2.23 2-Hydroxy-4-methoxy benzoic acid

2.97 *Leptadenia reticulata* (Retz.) Wight & Arn.

Subclass Lamiidae, Superorder Lamianeae, Order Rubiales, Family Apocynaceae
Medicinal use: promote lactation (India)

In mild diabetes induced by streptozotocin, some healthy β-cells remain with the capacity to proliferate and secrete insulin in rodents. Natural products induce in rodents the replication of β-cells and differentiation from ductal or intra islet pancreatic precursors cells. Ethanol extract of leaves of *Leptadenia reticulata* (Retz.) Wight & Arn. given to Wistar rats (fasting glucose > 150 mg/dL) at a single oral dose of 200 mg/kg decreased 30 minutes peak glycemia from 175.2 to 133.2 mg/dL (metformin at 50 mg/kg: 127.1 mg/dL).[365] The same extract given to streptozotocin-induced diabetic Wistar rats (fasting glucose > 150 mg/dL) at a single oral dose of 200 mg/kg decreased 30 minutes peak glycemia from 244.1 to 234.2 mg/dL and this effect was superior to as metformin at 50 mg/kg: 236.1 mg/dL.[365] The plant is known to elaborate hentriacontanol, α-amyrin, β-amyrin, and stigmasterol as well as the flavones diosmetin and luteolin.[366] Santos et al. (2012) provided evidence that oral administration of mixture of α-amyrin and β-amyrin at a dose of 100 mg/kg/day for 5 days to albino mice evoked a decrease in plasma glucose and an increase in plasma insulin as efficiently as glibenclamide (at 10 mg/kg/day). Besides, the treatment improved pancreatic cytoarchitecture.[367] The precise molecular basis of insulinotropic activity of α-amyrin and β-amyrin is unknown.

2.98 *Rauvolfia serpentina* Benth.

Common names: sarpagandha (India); Indian snakeroot
Subclass Lamiidae, Superorder Lamianeae, Order Rubiales, Family Apocynaceae
Medicinal use: anxiety (India)
History: This plant was known of Sushruta

Alloxan-induced diabetic mice given standard laboratory diet and receiving orally for 14 days 60 mg/kg/day of *Rauwolfia serpentina* Benth. root methanol extract lost 2.9% of bodyweight whereby untreated diabetic animals lost 11% of body weight.[368] Furthermore, the extract brought close to normal values fasting glycemia, hepatic glycogen and insulinemia after 14 days.[368] With regard to the lipid profile of treated rodents, the extract normalized cholesterolemia, triglyceridemia, and low-density lipoprotein–cholesterol.[368]

2.99 *Tabernaemontana divaricata* (L.) R. Br. ex Roem. & Schult.

Synonyms: *Ervatamia coronaria* (Jacq.) Stapf; *Ervatamia divaricata* (L.) Burkill
Common name: gou ya hua (Chinese)
Subclass Lamiidae, Superorder Lamianeae, Order Rubiales, Family Apocynaceae
Medicinal use: snake bites (China)

Evidence suggests that *Tabernamontana divaricata* (L.) R. Br. ex Roem. & Schult. produces indole alkaloids able to induce β-cells regeneration in rodents. Ethanol extract of leaves containing (2.3 mg/g of conophylline) given orally given to streptozotocin-induced diabetic Sprague–Dawley rats (plasma glucose > 250 mg/dL) at a dose of 200 mg/kg/day for 15 days decreased fasting glycemia from about 440 to 375 mg/dL increased insulinemia from 152 to 311 pg/mL and increased β-cell mass of diabetic rodents.[369] Conophylline is absorbed orally,[369] and when injected subcutaneously at a dose of 5 μg/g for 7 days to neonatal-streptozotocin-induced type 2 diabetic Wistar rats, it prevented rise in plasma glucose, increased pancreatic insulin content and the β-cell mass compared with untreated animals.[370] This indole alkaloid increased the expression of pancreatic and duodenal homeobox-1 in pancreatic ductal *in vitro* suggesting neogenesis of β-cells by acting on the early stage of β-cell differentiation via p38 mitogen-activated protein kinase.[370]

2.100 *Anisodus tanguticus* (Maxim.) Pascher

Synonym: *Scopolia tangutica* Maxim.
Common name: shan lang dang (Chinese)
Subclass Lamiidae, Superorder Lamianae, Order Solanales, Family Solanaceae
Medicinal use: pain (China)

Anisodamine (Figure 2.24) from *Anisodus tanguticus* (Maxim.) Pascher at a concentration 1 mg/mL reduced by 40% the production of malondialdehyde by hamster pancreatic islet beta HIT cells challenged with alloxan.[371] This tropane alkaloid at a concentration of 1 mmol/L enhanced glucose-induced insulin secretion from beta HIT from 4262 μunits/mg protein increased to 6412 μunits/mg.[372] Anisodamine is a nonselective muscarinic antagonist for M1 and M2 receptors.[373] In rodent β-cell lines, M1 activation stimulate insulin secretion.[128]

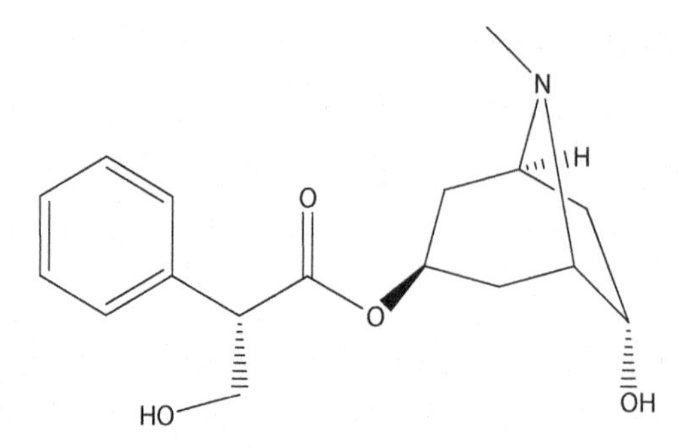

FIGURE 2.24 Anisodamine.

2.101 *Capsicum frutescens* L.

Synonyms: *Capsicum fastigiatum* Bl.; *Capsicum minimum* Roxb.
Common name: ladah padi (Malay)
Subclass Lamiidae, Superorder Lamianae, Order Solanales, Family Solanaceae
Medicinal use: jaundice (Cambodia)

Capsicum frutescens L. (Figure 2.25) produces capsaicin which at a dose of 9 mg intravenously evoked an increase in insulinemia hence a fall of glycemia in dogs subjected to an oral glucose tolerance test.[373] In a subsequent study, powder of fruits of *Capsicum frutescens* L. given to streptozotocin-induced diabetic Sprague–Dawley (≥ 300 mg/dL) rats at 0.5% of diet for 4 weeks evoked a mild reduction of food intake, mitigated body weight loss, lowered fasting blood glucose from 181.7 to 172 mg/dL (normal

FIGURE 2.25 *Capsicum frutescens* L.

113.1 mg/dL), and increased serum insulin from 61.8 to 88 pmol/L (normal 295.6 pmol/L).[374] This regimen improved glucose tolerance in intraperitoneal glucose tolerance with a decrease of glycemia by about 10%.[374] Akiba et al. (2004) made the demonstration that capsaicin at concentrations ranging from 10 to 10^{-9} M induced the secretion of insulin by RIN β-cells *in vitro* via the activation of transient receptor potential vanilloid subfamily 1 (TRVP-1), also known as vanilloid receptor 1 (VR1) with entry of calcium into the cell.[273] Such mechanism would, at least, explain the hypoglycemic effect of *Capsicum frutescens* L. in dogs and streptozotocin-induced diabetic Sprague–Dawley rats. It should be recalled that lycopene which is not uncommon in the fruits in members of the genus *Capsicum* has insulinotropic effects. Streptozotocin-induced diabetic Sprague–Dawley rats (glycemia > 300 mg/dL) treated with lycopene orally at a dose of 90 mg/kg/day for 30 days experienced a reduction of glycemia by 78%, increased plasma insulin by 158% and increased pancreatic insulin by 48%.[375] This regimen lowered plasma hydrogen peroxide, nitric oxide and increased antioxidant capacity.[375] Lycopene treatment increased erythrocyte catalase, superoxide dismutase and glutathione peroxidase activity by 150%, 184%, and 86%, respectively.[375] This regimen lowered serum total cholesterol and triglycerides close to normal and decreased low-density lipoprotein–cholesterol and increased high-density lipoprotein–cholesterol.[375]

2.102 *Datura metel* L.

Synonym: *Datura fastuosa* L.
Common names: yang jin hua (Chinese); dhatura (india); downy thorn-apple
Subclass Lamiidae, Superorder Lamianae, Order Solanales, Family Solanaceae
Medicinal use: diabetes (India)

The seeds of *Datura metel* L. powdered and given to alloxan-induced diabetic Wistar rats at a single dose of 75 mg/kg lowered glycemia from about 20 to 5 mmol/L at 8 hours, and this effect was similar to gliclazide at a single oral dose of 0.5 mg/kg.[376] The powder given to p Wistar rats at a single dose of 75 mg/kg lowered glycemia from about 6 to 4 mmol/L.[376] Induction of hypoglycemia in both healthy and alloxan-induced diabetic rodent indicates an insulinotropic effect. The seeds of that toxic plant contain atropine and scopolamine which are muscarinic receptor antagonists. Natural products that stimulate M3 muscarinic receptors in β-cell increase insulin secretion and conversely muscarinic antagonists reduce insulin secretion. The results presented by Krishna Murthy et al. (2004) provides further understanding of anticholinergic natural products of β-cells.[378] The plant is toxic.

2.103 *Solanum surattense* Burm.f.

Synonyms: *Solanum diffusum* Ruiz & Pav.; *Solanum jacquini* Willd.; *Solanum xanthocarpum* Schrad. & J.C. Wendl.
Common name: kutai (India)
Subclass Lamiidae, Superorder Lamianae, Order Solanales, Family Solanaceae
Medicinal use: cough (India)

β-Sitosterol from the stem bark of *Solanum surattense* Burm.f given to streptozotocin-induced diabetic Wistar rats (glycemia > 250 mg/dL) orally at a dose of 20 mg/kg/day for 21 days lowered glycemia from 271.8 to 134 mg/dL (normal 85.1 mg/dL; glibenclamide at 0.3 mg/kg/day: 123.3 mg/dL) and increased serum insulin from 7.8 to 14.3 μU/mL (normal 17.8 μU/mL; glibenclamide at 0.3 mg/kg/day: 16.8 μU/mL.[378] This regimen prevented glycosuria in diabetic animals.[378] In the pancreas of treated rodents, the extract increased superoxide dismutase, catalase, glutathione, glutathione S-transferase and glutathione peroxidase and decreased lipid peroxidation.[378] β-Sitosterol evoked a rejuvenation of pancreatic β-cells in diabetic rodents.[378]

2.104 *Solanum torvum* Sw.

Common names: shui qie (Chinese); terung pipit (Malay)
Subclass Lamiidae, Superorder Lamianae, Order Solanales, Family Solanaceae
Nutritional use: vegetable (Malaysia)

Methyl caffeate (Figure 2.26) isolated from the fruits of *Solanum torvum* Sw. given orally at a single dose of 40 mg/kg to Wistar rats 30 minutes after oral load of glucose lowered postprandial glycemia from 9.4 to 8.22 mmol/L and this effect was similar to glibenclamide at a dose of 2.5 mg/kg.[379] Methyl caffeate given orally at a dose of 40 mg/kg/day for 28 days to streptozotocin-induced diabetic Wistar rats lowered glycemia by about 70% similarly with glibenclamide at 2.5 mg/kg/day.[379] This regimen increased plasma insulin from 6.5 to 15.6 μU/mL (normal 16.6 μU/mL; glibenclamide at 2.5 mg/kg: 15.5 μU/mL).[379] In diabetic rodents, hexokinase and glycogen were decreased while glucose-6-phosphatase and fructose-1,6-biphosphate were increased.[379] Methyl caffeate treatment increased hepatic glycogen and hexokinase and decreased glucose-6-phophatase and fructose-1,6-biphosphatase.[379] In skeletal muscles, the treatment increased the expression of glucose transporter-4.[379] The treatment increased the number of islets in the pancreas of diabetic rats similarly with glibenclamide.[379] The precise molecular basis of activity of methyl caffeate at β-cells levels in unknown. This phenolic compound is known to inhibit rat intestinal sucrose and maltase *in vitro* with IC_{50} values of 1.5 mM and 2 mM, respectively,[380] but there effects do not explain insulinotropic activity. Methyl caffeate is cytotoxic against various cell types *in vitro* on account of its ability to bind stably to the active sites of poly (ADP-ribose) polymerase.[381] It should be noted that in β-cells, poly (ADP-ribose) polymerase activation causes NAD^+ depletion to form poly (ADP-ribose), resulting in necrosis.[382] Poly (ADP-ribose) polymerase inhibitors induce the regeneration of pancreatic β-cells,[383] and prevent alloxan and streptozotocin-induced diabetes.[382] Although highly speculative, it is here tempting to suggest that PARP inhibition in β-cells may contribute to the insulinotropic effect of methyl caffeate. Clinical studies are warranted.

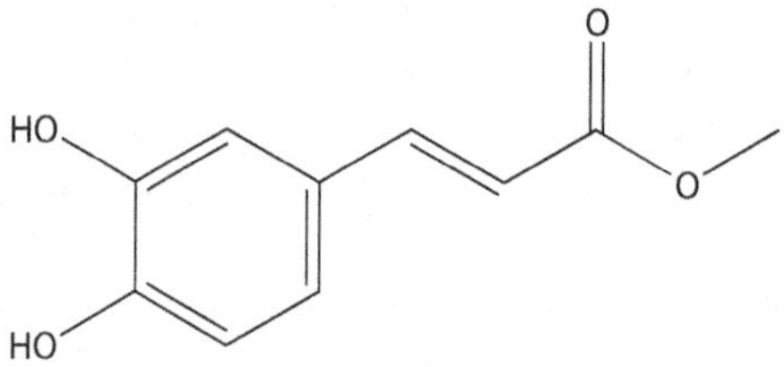

FIGURE 2.26 Methyl caffeate.

2.105 *Withania somnifera* (L.) Dunal

Synonyms: *Physalis somnifera* L. *Withania kansuensis* Kuang & A. M. Lu; *Withania microphysalis* Suess.
Common names: shui qie (Chinese); ashwagandha (India); winter cherry
Subclass Lamiidae, Superorder Lamianae, Order Solanales, Family Solanaceae
Medicinal use: diuretic (India)
History: The plant was known of Sushruta

Powdered root of *Withania somnifera* (L.) Dunal given to patients with type 2 diabetics at a dose of 500 mg 6 times per day for 30 days lowered glycemia by 12% and normalized serum lipid profile and these effect was equivalent to glibenclamide by unknown mechanism.[379] In a subsequent study, aqueous extract of roots given orally to streptozotocin-induced diabetic Wistar rats at a dose of 400 mg/kg/day for 5 weeks lowered blood glucose from 324.6 to 121.2 mg/dL (normal 97.1 mg/dL), lowered pancreatic malondialdehyde contents, and increased pancreatic glutathione contents.[384] This regimen increased the enzymatic activity of pancreatic superoxide dismutase and catalase in the pancreas of diabetic animals and improved pancreatic cytoarchitecture with reduction of fibrosis and regeneration of β-cell.[384] At the pancreatic level, the extract increased the activity of glutathione peroxidase, glutathione reductase and glutathione –S-transferase[384,385] suggesting improvement of pancreatic tissues, β-cells regeneration and recovery of insulin secretion Gorelick et al. (2015) provided further evidence on the pancreatic effect of this plant by reporting that methanol extract of leaves *Withania somnifera* (L.) Dunal at a concentration of 100 μg/mL increased the basal production of insulin in RIN-5 β pancreatic cells *in vitro* by more than 50% in absence of glucose.[386] The precise insulinotropic mechanism evoked here is yet unknown. The plant is known to contain withaferin A, withanone, withanolide A and B, and withanoside IV.[386] Withanolide A inhibits the nuclear translocation of nuclear factor-κB in human mononuclear cells challenged with lipopolysaccharides implying that the protective effect of *Withania somnifera* (L.) Dunal could be owed to nuclear factor-κB inhibition.[387] As for the acute insulinotropic activity observed in RIN-5 cells the natural product involved and mechanisms are unknown. Being relatively safe medicinal plant, *Withania somnifera* (L.) Dunal should attract more interest with regard to its clinical development as drug for the management of metabolic syndrome.

2.106 *Argyreia nervosa* (Burm. f.) Bojer

Synonyms: *Argyreia speciosa* (L. f.) Sweet; *Convolvulus speciosus* L. f.; *Rivea nervosa* (Burm. f.) Hallier f.
Common names: bastantri (India); areuj bohol keboh (Indonesia), sedang-dahon (Philippines) elephant-climber
Subclass Lamiidae, Superorder Lamianae, Order Solanales, Family Convolvulaceae
Medicinal use: diabetes (India)

Methanol extract of stems of *Argyreia nervosa* (Burm. f.) Bojer given orally to Sprague–Dawley rats at a single dose of 750 mg/kg reduced glycemia by 34.2% at 8 hours (Tolbutamide at 40 mg/kg: 29.6%).[388] This extract given orally to alloxan-induced diabetic Sprague–Dawley rats (glycemia ≥250 mg/dL) at a single dose of 750 mg/kg reduced glycemia by 40.4% at 8 hours (Tolbutamide at 40 mg/kg: 35%).[388] In oral glucose tolerance test in rats, the extract at 750 mg/kg lowered 30 minutes peak glycemia from 187.4 to 138.6 mg/dL.[388] In alloxan-induced diabetic Sprague–Dawley rats, the extract at 750 mg/kg lowered 60 minutes peak glycemia from 387.5 to 176.8 mg/dL (Tolbutamide 40 mg/kg: 189.7 mg/dL).[388,389] The plant, and especially the seeds, contain series of ergoline alkaloids such as D-lysergic acid amide and isoergine,[390] which are hallucinogenic rendering the plant unfit for therapeutic purposes. These alkaloids are agonists of dopamine D2 receptors.[391] Garcia-Tornadu et al. (2010) provided evidence that dopamine inhibits insulin secretion by β-cells via dopamine D2 receptors in mice.[392]

2.107 *Ipomoea batatas* (L.) Lam.

Synonyms: *Batatas edulis* (Thunb.) Choisy; *Convolvulus batatas* L.; *Convolvulus edulis* Thunb.
Common names: mitha alu (India); ubi keledek (Malay), kamote (Philippines); man thet (Thai); sweet potato
Subclass Lamiidae, Superorder Lamianae, Order Solanales, Family Convolvulaceae
Medicinal use: diabetes (India)

Ethanol extract of leaves of *Ipomoea batatas* (L.) Lam. given as 3% part of diet given to KK-Ay mice for 5 weeks lowered glycemia from 418.3 to 339.3 mg/dL without changing plasma insulin.[393] Body weight and food intake were not changed compared with untreated diabetic mice.[393] The extract *in vitro* increased the secretion of glucagon-like peptide-1 from a murine enteroendocrine cell line.[393] From this extract, 3,4,5 tricaffeoylquinic acid evoked an increased glucagon-like peptide-1 secretion from about 40 pM to 650 pM *in vitro*.[393] The extract given at a single oral dose of 2 g/kg, 30 minutes before intraperitoneal glucose loading in Sprague–Dawley rats lowered 15 minutes peak glycemia by about 20%, increased 15 minutes insulin secretion and increased plasma glucagon-like peptide-1.[393] In Sprague–Dawley rats, a single oral administration of gliclazide at a dose of 30 mg/kg provoked a quick and prolonged hypoglycemia and increased insulinemia whereby the extract given orally at a dose of 2 g/kg had no hypoglycemic effect nor insulinotropic effects.[393] The insulinotropic gut hormone glucagon-like peptide-1 is released from small intestine endocrine L cells during digestion of mixed meals.[394] This hormone bind to its G protein-coupled receptors on β-cells to promote glucose-induced secretion of insulin by activating adenylate cyclase and increasing cyclic adenosine monophosphate, activating cyclic adenosine monophosphate-regulated guanine nucleotide exchange factor II leading to altered ion channel activity, elevation of cytoplasmic calcium contents and exocytosis of insulin-containing vesicles.[202] The insulinotropic gut hormone glucagon-like peptide-1 also suppresses glucagon secretion[395] inhibits appetite and food intake,[396] and ameliorates pancreatic β-cell dysfunction.[397] glucagon-like peptide-1 secretion is lower in type 2 diabetic patients compared with healthy individuals.[398] The insulinotropic gut hormone glucagon-like peptide-1 activity is decreased in type 2 diabetic patients[398] and could be a strategy to prevent β-cell degeneration in metabolic syndrome. It should be noted that 3,4,5-tricaffeoylquinic acid inhibited rat intestinal maltase, rat intestinal sucrose and human saliva α-amylase with IC_{50} values of 24 μM (acarbose: 0.4 μM), 574 μM (acarbose: 1.2 μM) and 634 μM, respectively,[399] contributing to the hypoglycemic activity of *Ipomoea batatas* (L.) Lam. in KK-Ay mice. Note that chlorogenic acid and 3,5-dicaffeoylquinic acid which are not uncommon in members of the family Solanaceae, Lamiaceae and Asteraceae stimulate glucagon-like peptide-1 secretion in rodents.[400]

2.108 *Tournefortia sarmentosa* Lam.

Subclass Lamiidae, Superorder Lamianae, Order Boraginales, Family Boraginaceae
Medicinal use: ulcers (Philippines)

Evidence is available to suggest that members of the genus *Tournefortia* L. are able to lower glycemia in healthy and streptozotocin-induced diabetic rodents.[401,402] As for *Tournefortia sarmentosa* Lam., it produces allantoin and salicylic acid, and tournefolins A-C.[403] This plant also produces salvianolic acid A, isosalvianolic acid C, lithospermic acid, salvianolic acid F and rosmarinic acid.[404] The hypoglycemic activity of rosmarinic acis is known and lithospermic is antioxidant and cytoprotective.[405] Allantoin at a concentration of 100 μM protected β-cells against streptozotocin-induced apoptosis *in vitro* and this effect was alleviated by the imidazoline 3 receptor antagonist KU14R.[406] Allantoin given daily at a dose of 10 mg/kg intravenously for 8 days to streptozotocin-induced diabetic Wistar rats lowered plasma glucose from about 300 to 225 mg/dL and doubled insulinemia via activation of imidazoline 3 receptors.[406] Imidazoline 3 receptor activation evoke the secretion of insulin by β-cells.[407] *Tournefortia sarmentosa* Lam., like most members of the family Boraginaceae contains the hepatotoxic pyrrolizidine alkaloids such as supinine,[408] and should be avoided in therapeutic.

2.109 *Plantago asiatica* L.

Common names: che qian (Chinese); Asian plantain
Subclass Lamiidae, Superorder Lamianae, Order Lamiales, Family Plantaginaceae
Medicinal use: diuretic (China)

Aucubin from *Plantago asiatica* L. (Figure 2.27) given intraperitoneally to streptozotocin-induced diabetic Wistar rats at a dose of 5 mg/kg twice a day for 5 days followed by 5 mg/kg/day for 10 days

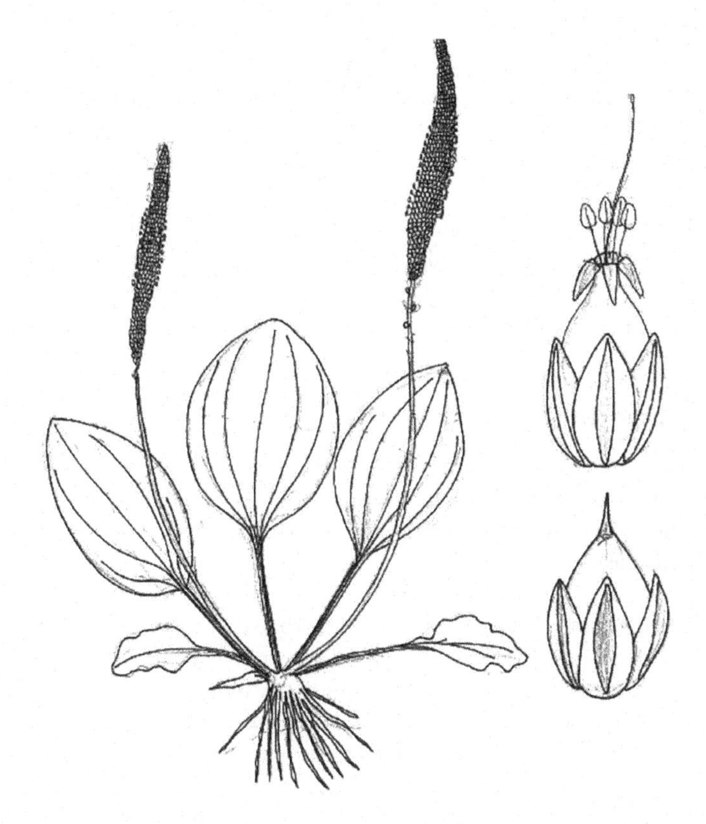

FIGURE 2.27 *Plantago asiatica* L.

reduced body weight loss, lowered glycemia from about 20 to 12.5 mmol/L and improved pancreatic histoarchitecture.[409] It is interesting to speculate that aucubin,[409] may activate glucagon-like peptide-1 receptors in β-cells

2.110 *Pseuderanthemum palatiferum* (Wall.) Radlk.

Synonym: *Justicia palatifera* Wall.
Common name: cay xudn hoa (Vietnamese)
Subclass Lamiidae, Superorder Lamianae, Order Lamiales, Family Acanthaceae
Medicinal use: hypertension (Vietnam)

Ethanol extract of leaves of *Pseuderanthemum palatiferum* (Wall.) Radlk. given to streptozotocin-induced diabetic Wistar rats (fasting glycemia > 126 mg/dL) at a dose of 250 mg/kg/day for 14 days lowered fasting plasma glucose by 36.6% (glibenclamide 0.2 mg/kg/day: 24.3%), increased, like glibenclamide, insulinemia near to normal values.[410] This extract lowered total cholesterol, triglycerides, increased high-density lipoprotein, very low-density lipoprotein, similarly to glibenclamide.[410] The roots of this plant contain series of triterpenes of betulin, lupeol, pomolic acid, epifriedelanol and lupenone.[411] The principle accounting for the insulinotropic property of *Pseuderanthemum palatiferum* (Wall.) Radlk. is unknown. Ethanol extract of leaves (containing gallic acid, rutin, isoquercetin, kaempferol, quercetin, catechin) at a concentration of 250 μg/mL inhibited interleukin-6 and tumor necrosis factor-α production by mouse peritonela macrophages challenged with lipopolysaccharides *in vitro*.[412]

2.111 *Ruellia tuberosa* L.

Common names: lu li cao (Chinese); meadow-weed
Subclass Lamiidae, Superorder Lamianae, Order Lamiales, Family Acanthaceae
Medicinal use: diuretic (India)

Ruellia tuberosa L. produces verbascoside, isoverbascoside, and forsythoside B.[413] Isoverbascoside at a concentration of 0.1 mg/mL increased glucose-induced secretion of insulin by INS-1 cells by 127.4%, respectively (glibenclamide at 0.001 mg/mL: 163.2%).[414] The insulinotropic mode of action of this phenylpropanoid glycoside is unknown.

2.112 *Gmelina arborea* Roxb. ex Sm.

Synonyms: *Gmelina rheedii* Hook.; *Premna arborea* Farw.
Common names: yun nan shi zi (Chinese); katanam (India); candahar tree
Subclass Lamiidae, Superorder Lamiidae, Order Lamiales, Family Verbenaceae
Medicinal use: diabetes (Sri Lanka)

Aqueous extract of stem bark of *Gmelina arborea* Roxb. ex Sm. given orally at a dose of 1 g/kg to streptozotocin-induced diabetic Wistar rats (fasting plasma glucose $\geq$ 12 mmol/L) for 30 days lowered fasting glycemia by 37% (glibenclamide 0.5 mg/kg: 42%).[415] This regime increased serum insulin by 57% and C-peptide by 39%.[415] Furthermore, the extract lowered total cholesterol, low-density lipoprotein–cholesterol, very low-density lipoprotein–cholesterol and triglycerides by 31%, 43%, 25%, and 29%, respectively, whereby high-density lipoprotein–cholesterol was increased by 45%.[415] The extract evoked a regeneration of pancreatic β islet cells.[415] Stem bark contains tyrosol, (+)-balanophonin, 2,6-dimethoxy-p-benzoquinone and 3,4,5-trimethoxyphenol.[416] The iridoid glycosides 6-*O*-(3″-*O*-benzoyl)-α-L-rhamnopyranosylcatalpol, 6-*O*-(3″-*O*-trans-cinnamoyl)-α-L-rhamnopyranosylcatalpol, 6-*O*-(3″-*O*-cis-cinnamoyl)-α-L-rhamnopyranosylcatalpol and 6-*O*-(3″,4″-*O*-dibenzoyl)-α-L-rhamnopyranosylcatalpol.[417] The active constituent(s) responsible for the insulinotropic activity of the plant is unknown but one could reasonably anticipate the involvement of iridoid(s) and simple phenols including 3,4,5-trimethoxyphenol.

2.113 *Gmelina asiatica* L.

Subclass Lamiidae, Superorder Lamiidae, Order Lamiales, Family Verbenaceae
Medicinal use: diuretic (India)

Alcohol extract of bark of *Gmelina asiatica* L. given orally at a single dose of 500 mg/kg to Sprague–Dawley lowered glycemia by 32.4% 6 hours after administration.[418] The experiment repeated with alloxan-induced Sprague–Dawley (glycemia 250 mg/dL-350 mg/dL) lowered 6 hours glycemia by 48.2% whereby tolbutamide at 40 mg/kg evoked a 41.2% reduction of plasma glucose.[418] In oral glucose tolerance test, the extract given at a single dose of 500 mg/kg reduced 30 minutes glycemia from 322.2 to 207.2 mg/dL and this effect was equipotent with tolbutamide (normal 172.3 mg/dL; tolbutamide at 40 mg/kg: 207.1 mg/dL).[418] Jothi et al. (2008) reported that aqueous extract of *Gmelina asiatica* L given orally at a dose of 400 mg/kg/day or a period of 45 days decreased the blood glucose and increased plasma insulin concentration as a result of β-cells rejuvenation and regeneration.[419] The plant contains the quercetagetin,[420] and probably shelters some iridoid glycosides. This 6-hydroxyflavonol had the ability to inhibit *in vitro* c-Jun NH_2-terminal kinase-1 activity with an IC_{50} value of 4.6 μM.[421] Oxidative stress in β-cells induced the activation of c-Jun NH_2-terminal kinase also known as stress-activated protein kinase which suppress pancreatic and duodenal homeobox-1 activity and subsequent suppression of insulin gene transcription.[422] Does 6-hydroxyflavonol increased insulin gene secretion though c-Jun NH_2-terminal kinase inhibition?

2.114 *Lantana camara* L.

Common names: ma ying dan (Chinese); caturangi (India); lantana
Subclass Lamiidae, Superorder Lamiidae, Order Lamiales, Family Verbenaceae
Medicinal use: wounds (India)

Twelve g/kg of leaves given to goats induced lethal liver, kidney, and lung injuries.[423] This plant is extremely poisonous and has no place in therapeutic but it elaborates urs-12-en-3β-ol-28-oic acid 3β-D-glucopyranosyl-4′-octadecanoate which given orally at a dose of 0.3 mg/kg to streptozotocin-induced diabetic Wistar rats (≥ 250 mg/dL) for 21 days decreased glycemia from 375.3 to 118.6 mg/dL (normal 119.5 mg/dL; Glimepiride 0.1 mg/kg: 105.5 mg/dL).[424] It must be recalled that acute insolinotropic effects of toxic plant extract can be owed to pancreatic insults.

2.115 *Tectona grandis* L.f.

Synonyms: *Tectona theka* Lour.; *Theka grandis* (L.f.) Lam.
Common names: you mu (Chinese); malapangit (Philippines); teak
Subclass Lamiidae, Superorder Lamiidae, Order Lamiales, Family Verbenaceae
Medicinal use: sore throat (Philippines)

Methanol extract of flowers of *Tectona grandis* L.f. given to nicotinamide-streptozotocin-induced type 2 diabetic Wistar albino rats (glycemia > 200 mg/dL) at a dose of 200 mg/kg/day orally for 4 weeks prevented weight loss, lowered glycemia to about 100 mg/dL, increased serum insulin, lowered plasma creatinine, lowered serum cholesterol, triglycerides and increased high-density lioprotein-cholesterol.[425] This extract inhibited α-amylase with an IC_{50} of 2.2 μg/mL (acarbose: 219.5 μg/mL) and α-glucosidase with an IC_{50} of 229.2 μg/mL (acarbose: 0.3 μg/mL).[425] The extract contains gallic acid, quercetin, and rutin.[425] Gallic acid and rutin are known to induce insulin secretion after oral administration in rats. Quercetin, rutin and gallic acid have the ability to inhibit the activity of enzymes responsible for carbohydrate absorption. The flowers of this plant may have potencies to manage metabolic syndrome as a form of tea. Studies to examine the safety and efficacy of such preparation are needed.

2.116 *Stachytarpheta indica* (L.) Vahl

Synonyms: *Stachytarpheta angustifolia* (Mill.) Vahl; *Stachytarpheta jamaicensis* (L.) Vahl; *Verbena indica* L.
Common names: rumput ekor kucing (Malay); bastard vervain
Subclass Lamiidae, Superorder Lamiidae, Order Lamiales, Family Verbenaceae
Medicinal use: diarrhea (Indonesia)

Evidence is available to demonstrate that extracts of members of the genus *Stachytarpheta* Vahl lower orally the glycemia of alloxan-induced diabetic rats on account of iridoid glycosides.[426] *Stachytarpheta indica* (L.) Vahl elaborates series of iridoid glycosides such as ipolamiide[426] and verbascoside.[427] Ipolamiide (Figure 2.28), at a concentration of 0.1 mg/mL increased glucose-induced secretion of insulin by INS-1 cells by 124.9% (glibenclamide at 0.001 mg/mL: 163.2%).[427] The precise insulinotropic mode of action of this iridoid glucoside on β-cell is unknown.

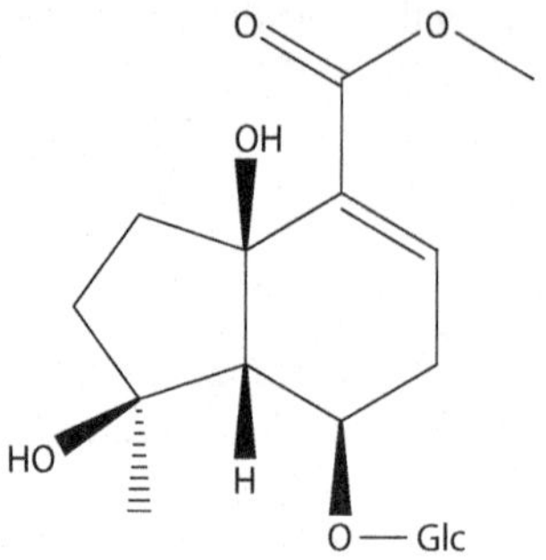

FIGURE 2.28 Ipolamiide.

2.117 *Vitex negundo* L.

Subclass Lamiidae, Superorder Lamiidae, Order Lamiales, Family Verbenaceae
Medicinal use: fever (China)

Iridoid glucoside from *Vitex negundo* L. (Figure 2.29) given at a dose of 50 mg/kg/day orally to streptozotocin-induced diabetic Wistar rats for 30 days lowered glucose from 251.8 to 108.8 mg/dL (normal 93.6 mg/dL; glibenclamide at 5 mg/kg/day: 103.5 mg/dL) and increased plasma insulin from 8.4 to 14.2 μU/mL (normal 16.8 μg/dL; glibenclamide at 5 mg/kg/day: 15.8 μg/dL).[428] Iridoid glucoside is orally active providing evidence that in rats, iridoid glycosides which are not

FIGURE 2.29 *Vitex negundo* L.

uncommon in members of the Superorder Lamianae, or their plasmatic metabolites can reach β-cells to increase plasma insulin.

2.118 *Marrubium vulgare* L.

Synonym: *Marrubium hamatum* Kunth
Common names: ou xia zhi cao (Chinese); horehound
Subclass Lamiidae, Superorder Lamianae, Order Lamiales, Family Lamiaceae
Medicinal use: colds (China)
History: The plant was known to Pliny the elder (27–79 AD), Roman naturalist

Aqueous extract of *Marrubium vulgare* L. (containing principally ballotetroside, verbascoside, and luteolin *O*-glucoside) given to alloxan-induced diabetic Wistar rats at a dose of 300 mg/kg/day orally for 15 days prevented weight loss, lowered glycemia from 484.3 to 142.6 mg/dL (normal 99.1 mg/dL; glibenclamide at 5 mg/kg/day: 129.5 mg/dL).[429] This medicinal plant elaborates marrubiin[430] and marrubenol[431] which inhibit α-glucosidase with IC_{50} values of 16.6 μM and 262.2 μM (acarbose: 64.1 μM).[432] Marrubiin increased the secretion of insulin by INS-1 rat insulinoma cells exposed to 11.1 mM or 33.3 mM glucose *in vitro*.[433] Under hyperglycemic conditions (33.3 mM), this diterpene evoked an increase in cellular oxygen consumption, increase of mitochondrial membrane potential and expression of glucose transporter-2 and increased insulin expression.[433] Marrubiin given to Wistar rats on cafeteria diet 2 weeks at a dose of 50 mg/kg/day for 2 weeks reduced glucose aera under the curve from 1380 to 932 mmol/(min L) (normal 912 mmol/[min L]) in intraperitoneal glucose tolerance test.[433] Intake of this furanic labdane, decreased serum triglycerides are total cholesterol, and index, lowered plasma interleukin-1β and interleukin-6 and elevated plasma insulin from 1.2 to 2.7 ng/mL (normal 0.9 ng/mL).[433] Marrubine and marrubenol relaxed aortic rings contracted by high potassium concentration with IC_{50} values of 24 μM and 7.7 μM, respectively,[434] via, at least, for marrubenol inhibition of voltage-dependent calcium channels.[431,434]

2.119 *Ocimum canum* Sims

Synonym: *Ocimum americanum* L.
Common names: kukka tulasi (India); hoary basil
Subclass Lamiidae, Superorder Lamiidae, Order Lamiales, Family Lamiaceae
Medicinal use: colds (India)

An aqueous extract of leaves of *Ocimum canum* Sims given to obese diabetic C57BL/KsJ *db/db* mice in drinking water (approximately 2.4 g of leaves/group/day) for 13 weeks lowered fasting blood glucose by about 60% and decreased body weight by 2.3%.[435] This regimen had no effect on plasma triglycerides, lowered total cholesterol and low-density lipoprotein–cholesterol and increased high-density lipoprotein–cholesterol.[435] The extract decreased lipoperoxidation in liver homogenates challenged with copper *in vitro* by about 30%.[435] In line, aqueous extract of *Ocimum canum* Sims (120 g/ 1L) given to C57BL/KsJ *db/db* diabetic mice in drinking water for 13 weeks decreased fasting blood glucose from 11.6 to 5.6 mmol/L and evoked weight loss compared with untreated animals.[436] In isolated islets of β-cells exposed to glucose, the extract at a concentration of 0.03 mg/mL evoked a 12 fold increased in insulin release.[436] The plant is known to contain oleanolic acid and ursolic acid, camphor and linalool.[437,438] Ursolic acid is known to protect β-cell against high-fat diet in rodents. Balasubramaniam et al. (2011) administered linalool to streptozotocin-induced diabetic Wistar rats at a dose of 25 mg/kg/day orally for 45 days and observed a decrease in food and water intake, prevented body weight loss, plasma glucose from 15.5 to 5.9 mmol/L (normal 5.1 mmol/L), and an increased plasma insulin from 27 to 76.6 pmol/L (normal 84.5 pmol/L).[439] This monoterpene is a non-NMDA glutamate receptor antagonist,[440] but evidence exist to demonstrate that non-NMDA receptor agonists such as kainate and quisqualate induce insulin secretion.[441] Linalool given to Wistar rats subcutaneously at a single dose of 25 mg/kg/day for days prevented carrageenin-induced paw edema formation,[442] suggesting that this monoterpene could, at least, mitigate streptozotocin-induced pancreatic inflammation. Camphor is known to stimulate TRVP1 which is expressed by MIN6 mouse β-cells.[443] Activation of TRVP1 in mouse β-cells induce insulin secretion.

2.120 *Ocimum sanctum* L.

Synonym: *Ocimum tenuiflorum* L.
Common names: sheng luo le (Chinese); tulasi (India); holy basil
Subclass Lamiidae, Superorder Lamiidae, Order Lamiales, Family Lamiaceae
Medicinal use: heart diseases (India)

Ethanol extract of inflorescence of *Ocimum sanctum* L. given to alloxan-induced diabetic Charles Foster rats at a dose of 750 mg/kg for 2 weeks lowered plasma glucose from 250 to 114 mg/dL.[444] Ethanol extract of leaves given orally at a dose of 200 mg/kg/day for 30 days to albino rats on high fructose diet weight from 180 to 168.1 g (normal 171.7 g), reduced glycemia from 118 to 99.8 mg/dL (normal 84.3 mg/dL), evoked a mild attenuation of insulinemia and had no effect on plasma triglycerides.[445] Ethanol extract of leaves induced insulin secretion by perfused rat pancreas and increased glucose-induced secretion.[446] In isolated islets of exposed to 3 mM (basal) glucose Langerhans, the extract at a concentration of 30 μg/mL increased insulin secretion from 2.9 to 4.8 mg/dL.[446] *In vitro*, the extract at a concentration of 200 μg/mL increased the secretion of insulin by clonal BRIND-BD11 β-cells and this effect was inhibited by K_{ATP}-channel opener diazoxide and voltage-dependent calcium channel blocker verapamil and attenuated by calcium removal from test buffer.[447] The plant is known to produce eugenol (Figure 2.30)[446] which given to streptozotocin-induced diabetic Wistar rats orally at a dose of 10 mg/kg/day for 30 days prevented weight loss, decreased food intake and water intake, lowered plasma glucose from about 300 to 90 mg/dL with a parallel increase of plasma insulin form about 7 to 13 μmol/mL (normal about 16 μmol/mL) and improved glucose tolerance in oral glucose tolerance test.[448] At the hepatic level, eugenol increased the enzymatic activity of hexokinase, pyruvate kinase and glucose-6-phosphate and decreased the activity of gluconeogenic enzymes glucose-6-phosphatase and fructose-1,6-biphosphatase and elevated near to normal hepatic glucogen stocks.[448] This regimen improved pancreatic cytoarchitecture with regeneration of β-cells.[447] It should be noted that eugenol activates vanilloid receptor 1 (VR1) human embryonic kidney (HEK) 293 cells and trigeminal ganglion neurons.[449] It is tempting to speculate that eugenol activates vanilloid receptor 1 expressed by β-cells resulting in increased secretion of insulin, participating in the hypoglycemic property of *Ocimum sanctum* L. Carvone is also present in *Ocimum sanctum* L. contributing further to the insulinotropic activity of the plant as discussed elsewhere. Clinical trials are warranted.

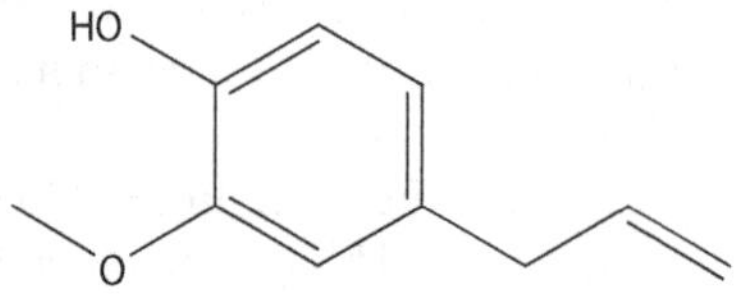

HO

O

FIGURE 2.30 Eugenol.

2.121 *Orthosiphon stamineus* Benth.

Synonyms: *Clerodendranthus spicatus* (Thunb.) C.Y. Wu ex H.W. Li; *Orthosiphon aristatus* (Blume) Miq.
Common names: misai kunching (Malay); Java tea
Subclass Lamiidae, Superorder Lamianae, Order Lamiales, Family Lamiaceae
Medicinal use: diuretic (Malaysia)

Aqueous extract of *Orthosiphon stamineus* Benth. given to Wistar rats and streptozotocin-induced diabetic Wistar rats (glycemia > 300 mg/dL) at a single oral dose of 1 g/kg lowered glycemia by about 25%, 210 minutes after oral glucose loading.[450] The extract given daily at a dose of 0.5 g/kg to streptozotocin-induced diabetic Wistar rats for 7 days lowered fasting plasma glucose from 395.7 to 341.2 mg/dL (normal 81.3 mg/dL; glibenclamide at 5 mg/kg: 326.8 mg/dL), had no effect on serum cholesterol, lowered plasma triglycerides and brought high-density lipoprotein–cholesterol.[450] The extract at a concentration of 100 μg/mL increased glucose-induced secretion of insulin by perfused rat pancreas.[450] Lee et al. provided evidence that a hexane fraction of aqueous extract of leaves of *Orthosiphon stamineus* Benth. at a concentration of 200 μM induced insulin expression and pancreatic and duodenal homeobox-1 expression buy INS-1 cells. Increased insulin expression was accompanied by increased phosphorylated phosphoinositide 3-kinase and Akt phosphorylation.[451] This fraction, at a concentration of 200 μM protected INS-1 cells against high glucose-induced insults. The activation of pancreatic and duodenal homeobox-1 via Akt and phosphoinositide 3-kinase is known to promote the regeneration of β-cells in rodents,[452] and glucose-stimulated preproinsulin gene expression and nuclear trans-location of pancreatic duodenum homeobox-1 require activation of phosphatidylinositol 3-kinase.[453] The active constituent(s) in *Orthosiphon stamineus* Benth. inducing insulin secretion is yet unknown but being soluble in hexane, one can anticipate nonpolar natural product. Yuliana et al. (2009) isolated series of methoxylated flavonoids such as 3′,4′,5,6,7-pentamethoxyflavone which are antagonists for adenosine A1 receptor *in vitro*. In mice, adenosine A1 receptor agonists decrease insulin secretion.[454,455] Other nonpolar constituents in this plant are isopimarane diterpenes such as orthosiphol X which inhibited nitric oxide production by macrophage like J774.1 cells challenged with lipopolysaccharides with an IC_{50} value of 6.4 μM.[456] Nitric oxide interferes with the secretion of insulin and this free radical is known to inhibit Akt.[457] It would be of interest to assess the insulinotropic activity of isopimaranes produced by *Orthosiphon stamineus* Benth. Clinical trials are warranted.

2.122 *Rosmarinus officinalis* L.

Common names: mi die xiang (Chinese); romero (Philippines); rosemary
Subclass Lamiidae, Superorder Lamianae, Order Lamiales, Family Lamiaceae
Medicinal use: tonic (Philippines)

Natural products which are antioxidant or increasing the activity of antioxidant enzymes such superoxide dismutase and catalase or increasing nonenzymatic antioxidant glutathione protect β-cells against streptozotocin-induced injuries in rats and may conceptually prevent the apparition of type 2 diabetes in human.[422] Aqueous extract of leaves of *Rosmarinus officinalis* L. given orally to streptozotocin-induced diabetic albino rats (fasting blood glucose ≥ 250 mg/dL) at a dose of 200 mg/kg/day for 2 weeks lowered glycemia from about 340 to 140 mg/dL (normal about 125 mg/dL, and increased plasma insulin from 4.5 to 12.4 μU/mL (normal 13.4 μU/mL).[458] Govindaraj et al. provided evidence that rosmarinic acid, which is antioxidant, given orally at a dose of 100 mg/kg/day to high-fat diet-streptozotocin-induced type 2 diabetes in Wistar albino rats and increased the activities of pancreatic superoxide dismutase, catalase, glutathione peroxidase and glutathione-S-transferase and contents of glutathione in diabetic rats.[459] In obese patients with glucose intolerance, persistent hyperglycemia ("glucotoxicity") contributes to the pathophysiology of type 2 diabetes. Complications by lowering the activity of antioxidant systems and promoting the generation of reactive oxygen species leading to β-cell dysfunction and necrosis.[460] Furthermore, high levels of plasma free fatty acids increase reactive oxygen species in β-cells inducing necrosis via mitochondrial perturbation.[461,462] Physiologically, superoxide dismutase and catalase are enzymes which protect β-cells against oxygen free radicals by catalyzing the removal of superoxide radical and hydrogen peroxide, respectively, and in type 2 diabetes, the activities of these enzymes is decreased.[463] Nuclear factor-erythroid 2-related factor 2 is a key cellular transcription factor that protects β-cells against oxidative damage-related impairment of insulin secretion by inducing the expression of antioxidant enzymes antioxidant/detoxification enzymes such as catalase and heme oxygenase 1. Rosemary elaborates carnosol (Figure 2.31) which protected HepG2 cells against hydrogen peroxide via activation of nuclear factor-erythroid 2-related factor 2.[465] Hence, incorporation of *Rosmarinus officin*alis L. in the diet of individuals with metabolic syndrome might prevent type 2 diabetes. Clinical trials are warranted.

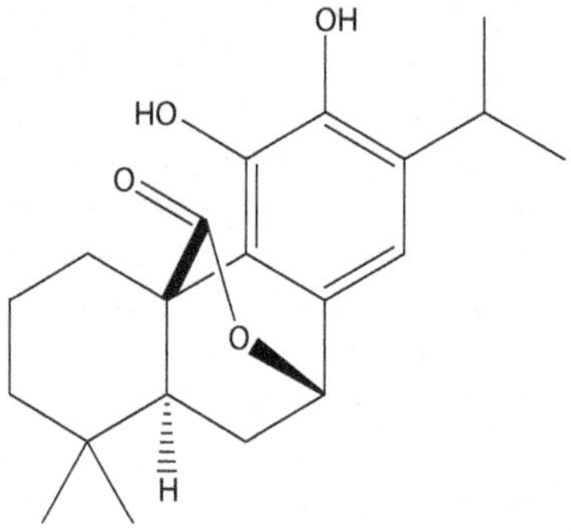

FIGURE 2.31 Carnosol.

2.123 *Teucrium polium* L.

Synonym: *Chamaedrys polium* (L.) Raf.
Common name: golden germander
Subclass Lamiidae, Superorder Lamiidae, Order Lamiales, Family Lamiaceae
Medicinal use: exanthematous disorders (Turkey)
History: The plant was known to Dioscorides.

Aqueous liquid extract of aerial parts of *Teucrium polium* L administered orally at a dose of 125 mg/kg decreased peak glycemia at 45 minutes in oral starch tolerance test from about 6.5 to 5 mM in Sprague–Dawley rats, and this effect identical to of 3 mg/kg of acarbose (5 mM).[276] In oral glucose tolerance test, the extract at 125 mg/kg had no effects on improving glucose tolerance, however, at 500 mg/kg, it lowered glycemia from about 6.5 to 5.5 mM.[276] The extract at 10g/100 mL had no effect on *a*-amylase activity *in vitro*.[276] Ethanol extract of aerial parts of *Teucrium polium* L. given to streptozotocin-induced diabetic Wistar rats at a dose of 1 mL/rat/day for 6 weeks decreased plasma glucose from 22.6 to 13 mmol/L (normal 5.1 mmol/L) and attenuated body weight loss.[466] In addition, blood insulin levels of the treated diabetic rats, after six consecutive weeks of treatments increased from 3.4 to 8.9 μg/L.[466] *In vitro*, the extract at a concentration of 0.1 mg/mL increased insulin secretion by rats islets in the presence of 2 or 16 mmol/L of glucose.[466] Hydroethanolic extract of aerial parts of this plant given to streptozotocin-induced diabetic Wistar rats (plasma glucose > 280 mg/dL), orally at a dose equivalent to 0.5 g plant powder/kg/day for 30 days lowered glycemia from 294 to 98 mg/dL (normal 88 mg/dL).[467] This regimen decreased pancreatic lipid peroxidation by 64%, increased pancreatic activities of catalase, superoxide dismutase and increase contents of glutathione by 52%, 45%, and 105%, respectively.[467] The extract brought plasma levels of nitric oxide down to nondiabetic group values.[467] Golden germander is known to produce the flavones cirsimaritin, eupatorin, apigenin-4′, 7-dimethylether, cirsimaritin and cirsiliol[468] 6 as well as 3,6-dimethoxy-apigenin, 4,7-dimethoxyapigenin, rutin and apigenin which are antioxidant *in vitro*.[469] Rutin and apigenin from *Teucrium polium* L at a concentration of 0.5 mM to 8 mM had no effect of insulin secretion by isolated rat pancreatic islets in the presence of either 5 or 11.1 mM but increased insulin secretion in pancreatic islets challenged with streptozotocin in the presence of either 5 or 11.1 mM of glucose.[470]

2.124 *Acorus calamus* L.

Synonyms: *Acorus americanus* (Raf.) Raf.; *Acorus angustatus* Raf.; *Acorus angustifolius* Schott; *Acorus asiaticus* Nakai; *Acorus cochinchinensis* (Lour.) Schott; *Acorus griffithii* Schott; *Acorus spurius* Schott; *Acorus triqueter* Turcz. ex Schott; *Acorus verus* (L.) Houtt.; *Orontium cochinchinense* Lour

Common names: chang pu (Chinese); sweet-flag

Subclass Alismatidae, Superorder Aranae, Order Arales, Family Acoraceae

Medicinal use: type 2 diabetes (Indonesia)

History: The plant was known to Dioscorides

Ethylacetate fraction of roots of *Acorus calamus* L. at a concentration of 25 μg/mL boosted the secretion of insulin by HIT-T15 cells more potently than gliclazide at 10 μmol/L *in vitro*.[471] This fraction inhibited yeast α-glucosidase with an IC_{50} value of 0.4 mg/mL and this effect was superior to acarbose.[471] In ICR mice, the fraction given orally at a single dose of 800 mg/kg lowered glycemia after 1 hour from about 4.8 to 3.2 mmol/L (gliclazide at 100 mg/kg: about 2.5 mmol/L).[471] In mice receiving intraperitoneal load of glucose, the extract at 100 mg/kg given orally 1 hour before lowered after 1 hour of glucose injection glucose as potently as gliclazide at 100 mg/kg.[471] In mice receiving an oral load of starch, the extract at 100 mg/kg orally lowered 30 minutes postprandial peak.[471] *Acorus calamus* L. accumulates β-asarone.[472]

2.125 *Aloe vera* (L.) Burm.f.

Common names: lu hu (Chinese); ghwar (India); true aloe

Subclass Liliidae, Superorder Lilianae, Order Liliales, Family Liliaceae

Medicinal use: pain in bowels (India)

History: The plant was known of Pliny the elder (circa 27–79 AD), Roman naturalist

Aloe vera (L.) Burm.f. gel given to BKS.Cg-$m^{+/+}Lepr^{db/J}$ *(db/db)* mice orally at a dose of 50 mg/kg/day for 29 days lowered fasting blood glucose from 463 to 277.7 mg/dL and increased insulin contents of pancreatic tissues.[473] From the gel, the phytosterol B12 at a dose of 1 μg/mouse/day for 30 days lowered fasting blood glucose from about 400 to 180 mg/dL.[473] This phytosterol prevented body weight loss.[473] In healthy C57BL/6J mice, phytosterol B12 had no effect on blood glucose.[473] The gel of Aloe vera (L.) Burm.f. could be of value for the treatment or prevention of metabolic syndrome.

2.126 *Anoectochilus roxburghii* (Wall.) Lindl. ex Wall.

Synonym: *Chrysobaphus roxburghii* Wall.
Common names: jin xian lan (Chinese); jewel orchid
Subclass Liliidae, Superorder Lilianae, Order Orchidales, Family Orchidaceae
Medicinal use: diabetes (China)

Kinsenoside (Figure 2.32) from *Anoectochilus roxburghii* (Wall.) Lindl. given to streptozotocin-induced diabetic Wistar rats (glycemia > 11 mmol/L) at a dose of 15 mg/kg/day for 22 days prevented body weight loss, decreased food and water intake, lowered blood glucose from 24.6 to 9 mmol/L (normal 4.9 mmol/L; metformin: 60 mg/kg/day: 11.8 mmol/L).[474] This regimen improved glucose tolerance in oral glucose tolerance test, increased plasma insulin from 11 to 20 μU/mL (normal 22.8 μU/mL; metformin: 60 mg/kg/day: 19.9 μU/mL), increased plasma antioxidant capacity and improved pancreatic histoarchitecture.[474]

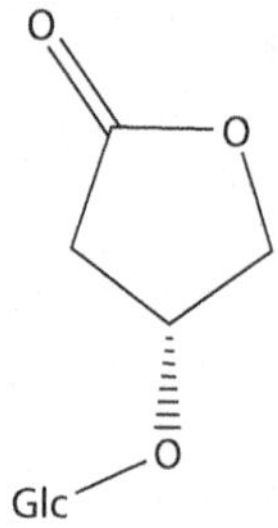

O
O
O
Glc

FIGURE 2.32 Kinsenoside.

2.127 *Crocus sativus* L.

Common names: fan hong hua (Chinese); kesar (India); Saffron
Subclass Liliidae, Superorder Lilianae, Order Iridales, Family Iridaceae
Medicinal use: enlargement of the liver (India)
History: The plant was known to Aulus Cornelius Celsus (25 BC–50 AD), Roman medical writer

Methanol extract of stigma of *Crocus sativus* L. given orally to alloxan-induced diabetic rats (glycemia > 200 mg/dL) orally at a dose of 240 mg/kg/day for 6 weeks lowered blood glucose from 361.5 to 145 mg/dL (normal 68.7 mg/dL; glibenclamide at 5 mg/kg/day: 115.8 mg/dL) and increased plasma insulin from 6.2 to 12.1 μU/mL (normal 14.5 μU/mL; glibenclamide at 5 mg/kg/day: 12.9%.[475] From this extract crocin at 150 mg/kg/day for 6 weeks orally lowered glycemia from 361.5 to 125.6 mg/dL, and increased plasma insulin from insulin from 6.2 to 12.3 μU/mL.[475] From this extract safranal (Figure 2.33) at 0.5 mL/kg/day for 6 weeks orally lowered glycemia from 361.5 to 73.5 mg/dL, and increased plasma insulin from 6.2 to 12.2 μU/mL.[475]

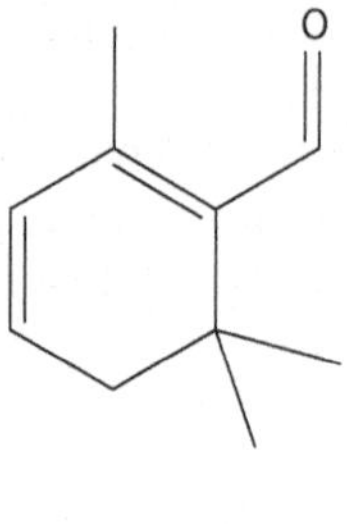

O

FIGURE 2.33 Safranal.

2.128 *Belamcanda chinensis* (L.) Redouté

Synonyms: *Belamcanda punctata* Moench; *Ixia chinensis* L.; *ardanthus chinensis* (L.) Ker Gawl.
Common names: she gan (Chinese); leopard flower
Subclass Liliidae, Superorder Lilianae, Order Iridales, Family Iridaceae
Medicinal use: fever (China)

Aqueous extract (containing swertisin, tectoridin, iristectoriginin A, and iridin) of leaves *Belamcanda chinensis* (L.) Redouté given to Wistar rats orally at a single dose of 1.6 g/kg evoked a fall of blood glucose from about 5.5 to 2.5 mmol/L and the effect perdured about 5 hours with concomitant increase in plasma insulin.[476] The extract given 2 hours before oral starch load lowered 30 min postprandial glycemia.[476] Streptozotocin-induced diabetic Wistar rats given orally at a single dose of 1.6 g/kg evoked a fall of blood glucose from about 30 to 20 mmol/L after 2 hours.[476] The hypoglycemic effect of the extract in p rats was abrogated with concurrent administration of nicorandil, a potassium-ATP channel blocker and nifedipine, a calcium channel blocker suggesting the insulinotropic effect to be mediated by the closure of potassium-ATP channels in β-cells.[476]

2.129 *Asparagus adscendens* Roxb.

Common name: safed musli (India)
Subclass Lillidae, Superorder Lilianae, Order Asparagales, Family Asparagaceae
Medicinal use: fatigue (India)

GPR30, a G-protein coupled oestrogen receptor is expressed in human islets.[477] In mice deletion of GPR30 induces intolerance to glucose and decreased glucose-induced insulin secretion.[478] 17β-estradiol (estrogen) administration in mice stimulates insulin secretion[479] and human islets.[477,480] Steroid hormones have chemical similitude with phytosterol and triterpenes. Aqueous extract of rhizomes of *Asparagus adscendens* Roxb. at a concentration of 5 mg/mL increased insulin secretion by BRIN-BD11 cells exposed to 5.6 mM glucose from 0.6 to 1.4 ng/10^6 cells per 20 minutes.[481] The secretion of insulin was further increased to 4.5 ng/10^6 cells per 20 minutes when glucose concentration was raised to 16.7 mmol/L.[481] These effects were abolished by diazoxide and verapamil.[481] The root is known to accumulate spirostanol steroidal saponins[482] of which sarsasapogenin and diosgenin.[483] A common example of spirostanol is diosgenin or (25R)5-spirosten-3β-ol which has the ability to bind and activate to estrogen receptors,[484] suggesting that *Asparagus adscendens* Roxb. could *in vitro* stimulate insulin secretion via the stimulation of estrogen receptors. Clinical trials are warranted.

2.130 *Asparagus racemosus* Willd.

Common names: chang ci tian men dong (Chinese); indivari (India); India asparagus
Subclass Lillidae, Superorder Lilianae, Order Asparagales, Family Asparagaceae
Medicinal use: Male infertility (India)

Ethanol extract of roots of *Asparagus racemosus* Willd. at a concentration of 30 μg/mL increased insulin secretion from isolated pancreatic islets from 2.9 5 to 5 ng/mg islet protein in the presence of 3 mM glucose and from 5.4 to 6.6 ng/mg islet protein in the presence of 11 mM glucose.[485] This extract booted insulin secretion by BRIN-BD11 cells at basal glucose concentration.[485] The root is known to accumulate spirostanol steroidal saponins.[486] Ethanol extract of roots of *Asparagus racemosus* Willd. given orally to streptozotocin-induced diabetic Wistar rats at a dose of 250 mg/kg/day for 4 weeks lowered plasma glucose from 362.8 to 165.8 mg/dL (normal 79.2 mg/dL) and corrected plasma cholesterol and triglycerides.[487] Ethanol extract of roots of *Asparagus racemosus* Willd. given orally to streptozotocin-induced diabetic Long-Evans rats (glycemia 8-9 mmol/L) at a single oral dose of 1.25 mg/kg lowered 30 minutes glycemia in oral glucose tolerance test.[488]

The extract given orally at dose of 1.25 g/kg/day twice a day for 28 days lowered glycemia from 8.9 to 6.5 mmol/L, increased plasma insulin by 30%.[488] The roots contain spirostane saponins of which diosgenin.[486,489] Diosgenin given to Sprague–Dawley rats on high-cholesterol diet at part of 0.5% of diet for 6 weeks had no effect on food intake but lowered relative liver weight by 24%. This supplementation lowered plasma triglycerides from 50.7 to 42.6 mg/dL, lowered cholesterol by about 32%, evoked a 1.5 fold increase in high-density lipoprotein–cholesterol and a reduction of atherogenic index from 3.4 to 0.9.[490] The regimen lowered hepatic cholesterol and triglycerides.[490] Diosgenin increased plasma and hepatic superoxide dismutase activity.[490]

2.131 *Asparagus officinalis* L.

Synonym: *Asparagus polyphyllus* Steven
Common names: shi diao bai (China); halgun (India); asparagus
Subclass Liliidae, Superorder Lilianae, Order Asparagales, Family Asparagaceae
Medicinal use: diuretic (India)

Methanol extract of seeds of *Asparagus officinalis* L. given to streptozotocin-induced diabetic Wistar rats orally at a dose of 500 mg/kg/day for 28 days for 15 days lowered glycemia from about 8.5 to 5.5 mmol/L and fasting plasma insulin was increased from about 50 to 80 pmol/L.[491] The extract improved pancreatic histoarchitecture.[491] The potentiality for spirostanol saponins to induce insulin secretion is known. It must be noted that these saponins are anti-inflammatory and as such attenuate streptozotocin pancreatic macrophages stimulation, pro-inflammatory cytokines secretion and favor β-cells regeneration.[492] As discussed previously, the stimulation of GPR30 is also possible.

2.132 *Amomum xanthioides* Wall ex. Baker

Common names: suo sha ren (Chinese); elam (India)
Subclass Commelinidae, Superorder Zingiberanae, Order Zingiberales, Family Zingiberaceae
Medicinal use: carminative (India)

In obese patients with impaired glucose tolerance, plasma glucose levels reach a glycemia 2 hours after oral ingestion of 75 g of glucose between 140 mg/dL(7.8 mmol/L) and 199 mg/dL (11 mmol/L).[55] Postprandial hyperglycemia generates a constant inflammatory stress on pancreatic β-cells, leading in genetically predisposed individual to type 2 diabetes.[55] It is a state of plasmatic "glucotoxicity." Aqueous extract of seeds of *Amomum xanthioides* Wall ex. Baker at a concentration of 10 μg/mL increased the viability of RIN cells challenged with pro-inflammatory cytokines interleukin-1β and interferon-γ-by 88.9%.[493] Concomitantly, the extract inhibited the nuclear translocation of nuclear factor-κB in RIN cells hence reduced the expression of inducible nitric oxide synthetase and decreased nitric oxide production.[493] Aqueous extract of seeds of *Amomum xanthioides* Wall ex. Baker at a concentration of 0.4 mg/mL protected Syrian hamster β-cells HIT-T 15 cells against alloxan-induced cytotoxicity with concomitant reduction of cytoplasmic reactive oxygen species and calcium as well as DNA fragmentation and prevented the depletion of ATP.[494] In pancreatic islets of mice challenged with alloxan, the extract prophylactically sustained the secretion of insulin in the presence of 20 mM.[494] Note that interleukin-1β and interferon-γ-mediated destruction of β-cells is caused by activation of inducible nitric oxide synthetase expression and production of nitric oxide.[495] Nitric oxide is a short-lived and highly reactive radical, which inhibits the Krebs-cycle enzyme aconitase and the electron transport chain complexes I and II leading to decreased glucose oxidation rates, ATP generation and insulin production.[496] The cytoprotective natural product involved here is yet unknown. Kitajima et al. (2003) isolated from an aqueous extract of seeds (1R,2S,4R,7S)-vicodiol 9-*O*-β-D-glucopyranoside, (1S,2S,4R,6S)-bornane-2,6-diol 2-*O*-β-D-glucopyranoside, (1R,4S,6S)-6-hydroxycamphor β-D-glucopyranoside, vanillic acid β-D-glucopyranosyl ester.[497]

2.133 *Imperata cylindrica* (L.) P. Beauv.

Synonyms: *Imperata arundinacea* Cirillo; *Lagurus cylindricus* L.; *Saccharum cylindricum* (L.) Lam.
Common name: bai mao (Chinese)
Subclass Commelinidae, Superorder Poanae, Order Poales, Family Poaceae
Medicinal use: diuretic (China)

Methanol extract of roots of *Imperata cylindrical* (L.) P. Beauv., given orally at a dose of 5 mg/20 g to Swiss-Webster mice concurrently with oral glucose loading lowered after 30 minutes blood glucose from 13.5 to 9.9 mmol/L, (glipizide at 0.023 mg/20 g to 9.7 mmol/L).[146] The plant is known to elaborate series of lignans and the sesquiterpene.[498] Cylindrene inhibited the contractions of rabbit aorta preparation induced by norepinephrine.[499] It should recalled that norepinephrine inhibits the formation of cyclic adenosine monophosphate and distal exocytotic machinery in β-cells.[199] In mice, activation of α_{2A}-receptors expressed by β-cells results in decreased secretion on insulin via inhibition of adenylyl cyclase.[500] The precise hypoglycemic mode of action of *Imperata cylindrica* (L.) P. Beauv. remains unknown. More endocrinological studies on that common tropical grass are needed.

APPENDIX

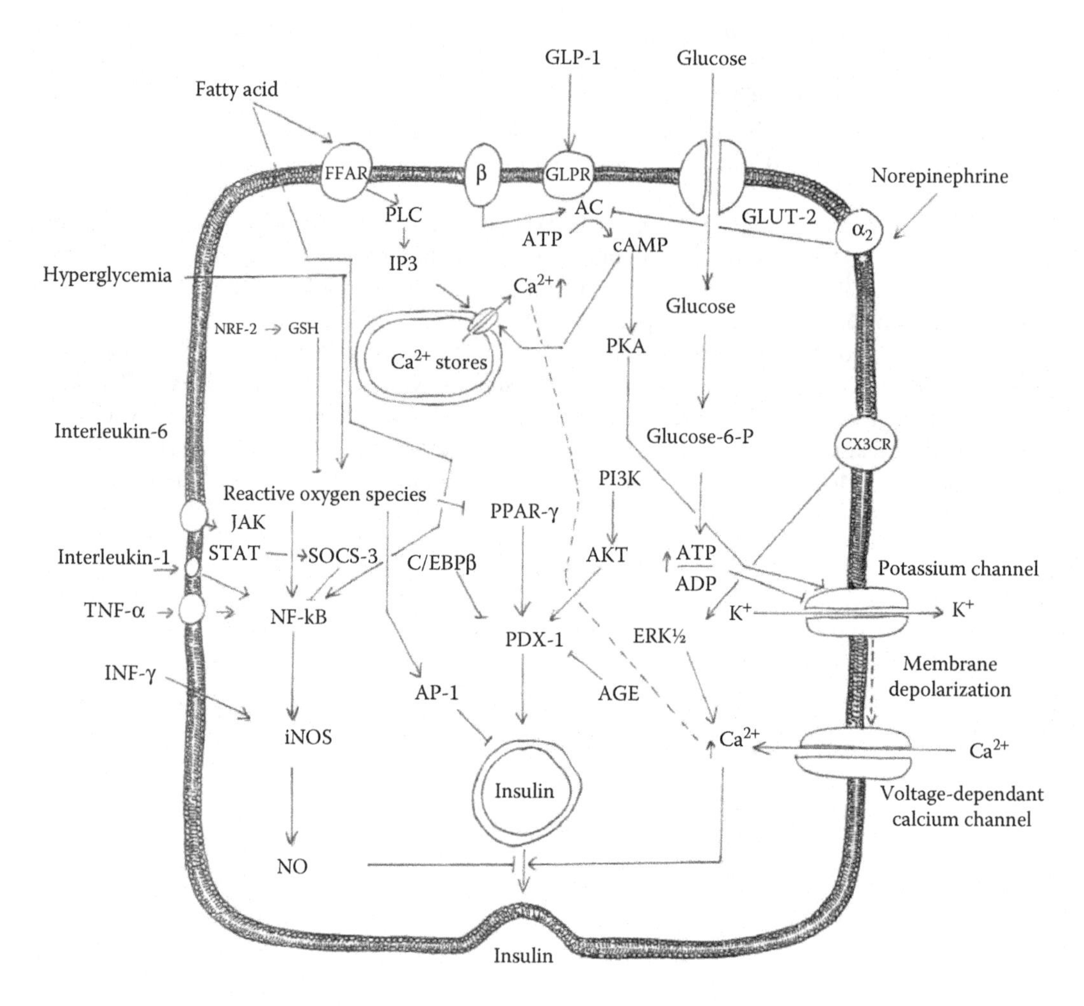

APPENDIX 2.1 Secretion of insulin by β-cells.

α2: alpha receptor AC: adenylate cyclase; Akt: protein kinase B; AGE: advanced glycation end product; AP-1: activating protein-1; β: beta receptor; C/EBPβ: CCAAT/enhancer-binding protein β; CX3CR: CX3C chemokine receptor; ERK1/2: extracellular signal-regulated kinase ½; FFAR: free fatty acids receptor; GLP-1: glucagon-like peptide-1 GSH: glutathion; GLPR: GLP-1: glucagon-like peptide-1 receptor; INF-γ: interferon-γ; iNOS: inducible nitric oxide synthetase; IP3: inositol 1,4,5-trisphosphate; JAK: c-jun NH2 terminal kinase; NO: nitric oxide; NRF-2: nuclear factor-erythroid 2-related factor 2; PI3K: phosphoinositide 3-kinase; PDX-1: pancreatic and duodenal homeobox protein 1;PKA: protein kinase A; PLC: phospholipase C; PPAR-γ: Peroxisome proliferator-activated receptor-γ; STAT: Signal transducer and activator of transcription.

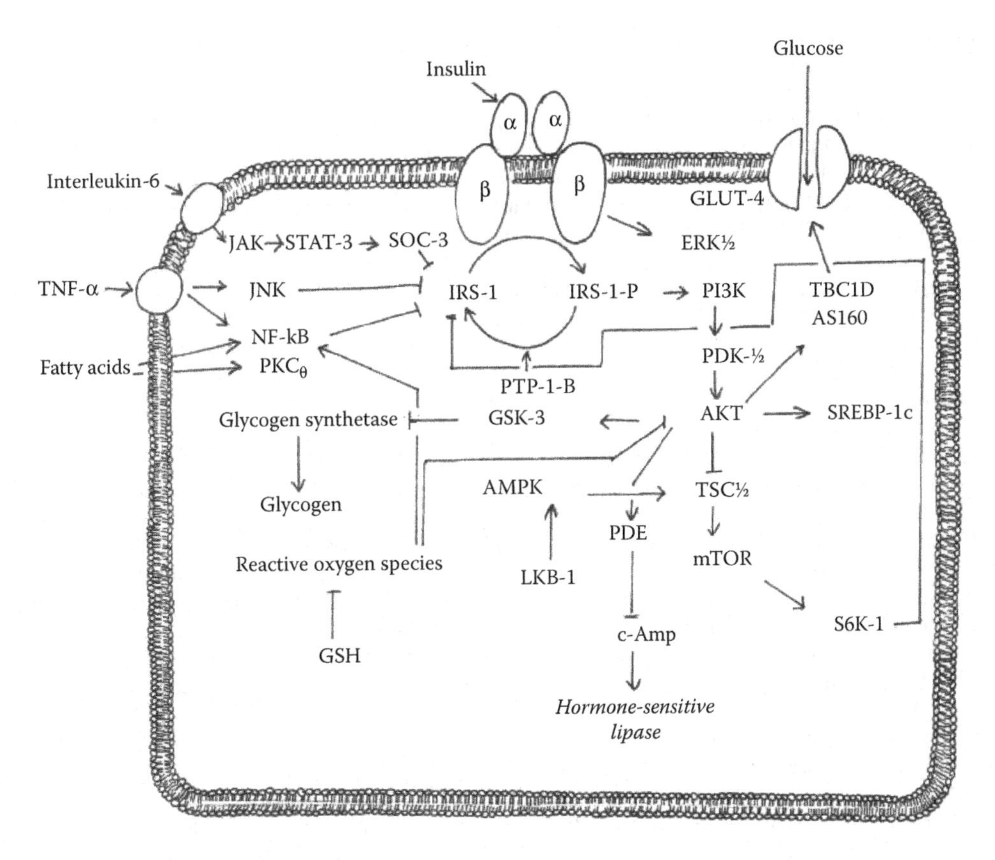

APPENDIX 2.2 Insulin receptor on peripheral cell and entry of glucose.

Akt: protein kinase B; AMPK: adenosine monosphosphate-activated protein kinase; AS160: Akt substrate of 160 kDa (TBC1D); ERK ½: Extracellular-signal regulated kinase (ERK) 1/2 ; GSH: glutathione; GSK-3: Glycogen synthase kinase-3; GLUT-4: glucose transporter-4; GSK-3: IRS: Insulin receptor substrate-1; JNK: c-jun NH2 terminal kinase; LKB1: liver kinase B1; mTOR: mechanistic target of rapamycin; PDK1/2: phosphionsitide-dependent kinase 1/2; PI3K: phosphoinositide 3-kinase; PDE: phosphodiesterase; PTP-1B: Protein-tyrosine phosphatase 1B; PKCθ: protein kinase Cθ; S6K1: ribosomal S6 kinase 1; SOCS-3: suppressor of cytokine signaling 3; STAT: Signal transducer and activator of transcription; SREBP-1c: sterol regulatory element binding protein-1c; SOCS-3: suppressor of cytokine signaling 3; TNF-a: tumor necrosis-a; TSC1/2: tuberous sclerosis 1 protein ½.

REFERENCES

1. Leu, J.P. and Zonszein, J., 2010. Diagnostic criteria and classification of diabetes. In *Principles of Diabetes Mellitus,* Poretsky, L., (Ed.). New York: Springer, pp. 107–115.
2. Fonseca, V., 2003. Clinical significance of targeting postprandial and fasting hyperglycemia in managing type 2 diabetes mellitus. *Current Medical Research and Opinion*, *19*(7), 635–663.
3. Tuomilehto, J., Lindström, J., Eriksson, J.G., Valle, T.T., Hämäläinen, H., Ilanne-Parikka, P., Keinänen-Kiukaanniemi, S. et al., 2001. Prevention of type 2 diabetes mellitus by changes in lifestyle among subjects with impaired glucose tolerance. *New England Journal of Medicine*, *344*(18), 1343–1350.
4. Bergman, R.N. and Ader, M., 2000. Free fatty acids and pathogenesis of type 2 diabetes mellitus. *Trends in Endocrinology & Metabolism*, *11*(9), 351–356.
5. McGarry, J.D. and Dobbins, R.L., 1999. Fatty acids, lipotoxicity and insulin secretion. *Diabetologia*, *42*(2), 128–138.
6. Rerup, C.C., 1970. Drugs producing diabetes through damage of the insulin secreting cells. *Pharmacological Reviews*, *22*(4), 485–518.
7. Kim, H.R., Rho, H.W., Park, B.H., Park, J.W., Kim, J.S., Kim, U.H. and Chung, M.Y., 1994. Role of Ca2+ in alloxan-induced pancreatic β-cell damage. *Biochimica et Biophysica Acta (BBA)-Molecular Basis of Disease*, *1227*(1–2), 87–89.
8. Malaisse, W.J., Malaisse-Lagae, F., Sener, A. and Pipeleers, D.G., 1982. Determinants of the selective toxicity of alloxan to the pancreatic B cell. *Proceedings of the National Academy of Sciences*, *79*(3), 927–930.
9. Gandy, S.E., Buse, M.G. and Crouch, R.K., 1982. Protective role of superoxide dismutase against diabetogenic drugs. *Journal of Clinical Investigation*, *70*(3), 650.
10. Shirwaikar, A., Rajendran, K., Kumar, C.D. and Bodla, R., 2004. Antidiabetic activity of aqueous leaf extract of Annona squamosa in streptozotocin–nicotinamide type 2 diabetic rats. *Journal of Ethnopharmacology*, *91*(1), 171–175.
11. Dinneen, S., Gerich, J. and Rizza, R., 1992. Carbohydrate metabolism in non-insulin-dependent diabetes mellitus. *New England Journal of Medicine*, *327*(10), 707–713.
12. Nguyen, K.H., Ta, T.N., Pham, T.H.M., Nguyen, Q.T., Pham, H.D., Mishra, S. and Nyomba, B.G., 2012. Nuciferine stimulates insulin secretion from beta cells—An in vitro comparison with glibenclamide. *Journal of Ethnopharmacology*, *142*(2), 488–495.
13. Marik, P.E. and Raghavan, M., 2004. Stress-hyperglycemia, insulin and immunomodulation in sepsis. *Intensive Care Medicine*, *30*(5), 748–756.
14. Evans, J.L., Goldfine, I.D., Maddux, B.A. and Grodsky, G.M., 2003. Are oxidative stress– ctivated signaling pathways mediators of insulin resistance and β-cell dysfunction? *Diabetes*, *52*(1), 1–8.
15. Liang, X.C., Hagino, N., Guo, S.S., Tsutsumi, T. and Kobayashi, S., 2002. Therapeutic efficacy of *Stephania tetrandra* S. Moore for treatment of neovascularization of retinal capillary (retinopathy) in diabetes–in vitro study. *Phytomedicine*, *9*(5), 377–384.
16. Lieberman, I., Lentz, D.P., Trucco, G.A., Seow, W.K. and Thong, Y.H., 1992. Prevention by tetrandrine of spontaneous development of diabetes mellitus in BB rats. *Diabetes*, *41*(5), 616–619.
17. Sun, G.R., Zhang, G.F., Wei, Y.J., Yang, D.S., Zhang, J.X. and Tian, Z.B., 1994. Protective effect of tetrandrine on pancreatic islet cells damaged by alloxan in rats. *Sheng Li Xue Bao:* [*Acta Physiologica Sinica*], *46*(2), 161–167.
18. Wang, G., Lemos, J.R. and Iadecola, C., 2004. Herbal alkaloid tetrandrine: From an ion channel blocker to inhibitor of tumor proliferation. *Trends in Pharmacological Sciences*, *25*(3), 120–123.
19. Yan, F., Benrong, H., Qiang, T., Qin, F. and Jizhou, X., 2005. Hypoglycemic activity of jatrorrhizine. *Journal of Huazhong University of Science and Technology [Medical Sciences]*, *25*(5), 491–493.
20. Patel, M.B. and Mishra, S., 201 Hypoglycemic activity of alkaloidal fraction of *Tinospora cordifolia*. *Phytomedicine*, *18*(12), 1045–1052.
21. Cui, H.M., Zhang, Q.Y., Wang, J.L., Chen, J.L., Zhang, Y.L. and Tong, X.L., 2015. Poor permeability and absorption affect the activity of four alkaloids from Coptis. *Molecular Medicine Reports*, *12*(5), 7160–7168.
22. Inagaki, N., Gonoi, T., Clement IV, J.P., Namba, N., Inazawa, J., Gonzalez, G., Aguilar-Bryan, L., Seino, S. and Byran, J., 1995. Reconstitution of IKATP: An inward rectifier subunit plus the sulfonylurea receptor. *Science*, *270*(5239), 1166–117.
23. Jiang, B., Cao, K. and Wang, R., 2004. Inhibitory effect of protopine on K ATP channel subunits expressed in HEK-293 cells. *European Journal of Pharmacology*, *506*(2), 93–100.
24. Meenakshi, P., Bhuvaneshwari, R., Rathi, M.A., Thirumoorthi, L., Guravaiah, D.C., Jiji, M.J. and Gopalakrishnan, V.K., 2010. Antidiabetic activity of ethanolic extract of Zaleya decandra in alloxan-induced diabetic rats. *Applied Biochemistry and Biotechnology*, *162*(4), 1153–1159.

25. Becerra-Jiménez, J. and Andrade-Cetto, A., 2012. Effect of Opuntia streptacantha Lem. on alpha-glucosidase activity. *Journal of Ethnopharmacology*, *139*(2), 493–496.
26. Obata, T., Brown, G.E. and Yaffe, M.B., 2000. MAP kinase pathways activated by stress: The p38 MAPK pathway. *Critical Care Medicine*, *28*(4), N67–N77.
27. Wu, C.H., Hsieh, H.T., Lin, J.A. and Yen, G.C., 2013. Alternanthera paronychioides protects pancreatic β-cells from glucotoxicity by its antioxidant, antiapoptotic and insulin secretagogue actions. *Food Chemistry*, *139*(1), 362–370.
28. Kaneto, H., Nakatani, Y., Kawamori, D., Miyatsuka, T., Matsuoka, T.A., Matsuhisa, M. and Yamasaki, Y., 2006. Role of oxidative stress, endoplasmic reticulum stress, and c-Jun N-terminal kinase in pancreatic β-cell dysfunction and insulin resistance. *The International Journal of Biochemistry & Cell Biology*, *38*(5), 782–793.
29. Bang, S.Y., Kim, J.H., Kim, H.Y., Lee, Y.J., Park, S.Y., Lee, S.J. and Kim, Y., 2012. *Achyranthes japonica* exhibits anti-inflammatory effect via NF-κB suppression and HO-1 induction in macrophages. *Journal of Ethnopharmacology*, *144*(1), 109–117.
30. Matteucci, E. and Giampietro, O., 2008. Proposal open for discussion: Defining agreed diagnostic procedures in experimental diabetes research. *Journal of Ethnopharmacology*, *115*(2), 163–172.
31. Szkudelski, T., 2001. The mechanism of alloxan and streptozotocin action in B cells of the rat pancreas. *Physiological Research*, *50*(6), 537–546.
32. Kim, H.K., Kim, M.J., Cho, H.Y., Kim, E.K. and Shin, D.H., 2006. Antioxidative and anti-diabetic effects of amaranth (*Amaranthus esculantus*) in streptozotocin-induced diabetic rats. *Cell Biochemistry and Function*, *24*(3),195–199.
33. Kelly, G.S., 1999. Squalene and its potential clinical uses. *Alternative Medicine Review: A Journal of Clinical Therapeutic*, *4*(1), 29–36.
34. Becker, R., 1989. Preparation, composition, and nutritional implications of amaranth seed. *Cereal Foods World*, *34*, 950–953.
35. Kumar, B.A., Lakshman, K., Jayaveea, K.N., Shekar, D.S., Khan, S., Thippeswamy, B.S. and Veerapur, V.P., 2012. Antidiabetic, antihyperlipidemic and antioxidant activities of methanolic extract of *Amaranthus viridis* Linn in alloxan induced diabetic rats. *Experimental and Toxicologic Pathology*, *64*(1), 75–79.
36. Krishnamurthy, G., Lakshman, K., Pruthvi, N. and Chandrika, P.U., 2001 Antihyperglycemic and hypolipidemic activity of methanolic extract of *Amaranthus viridis* leaves in experimental diabetes. *Indian Journal of Pharmacology*, *43*(4), 450.
37. Rizza, R.A., Gerich, J.E., Haymond, M.W., Westland, R.E., Hall, L.D., Clemens, A.H. and Service, F.J., 1980. Control of blood sugar in insulin-dependent diabetes: Comparison of an artificial endocrine pancreas, continuous subcutaneous insulin infusion, and intensified conventional insulin therapy. *New England Journal of Medicine*, *303*(23), 1313–1318.
38. Vetrichelvan, T., Jegadeesan, M. and Devi, B.A.U., 2002. Anti-diabetic activity of alcoholic extract of Celosia argentea Linn. seeds in rats. *Biological and Pharmaceutical Bulletin*, *25*(4), 526–528.
39. Lancaster, R., 1980. *Pharmacology in Clinical Practice.* Oxford, UK: Butterworth-Heinemann.
40. Wu, Q., Wang, Y. and Guo, M., 2011. Triterpenoid saponins from the seeds of *Celosia argentea* and their anti-inflammatory and antitumor activities. *Chemical and Pharmaceutical Bulletin*, *59*(5), 666–667.
41. Li, Y., Zhang, T., Cui, J., Jia, N., Wu, Y., Xi, M. and Wen, A., 2015. Chikusetsu saponin IVa regulates glucose uptake and fatty acid oxidation: Implications in antihyperglycemic and hypolipidemic effects. *Journal of Pharmacy and Pharmacology*, *67*(7), 997–1007.
42. Patel, S. and Santani, D., 2009. Role of NF-κB in the pathogenesis of diabetes and its associated complications. *Pharmacological Reports*, *61*(4), 595–603.
43. Gomes, A., Vedasiromoni, J.R., Das, M., Sharma, R.M. and Ganguly, D.K., 1995. Anti-hyperglycemic effect of black tea (Camellia sinensis) in rat. *Journal of Ethnopharmacology*, *45*(3), 223–226.
44. Baggio, L.L. and Drucker, D.J., 2007. Biology of incretins: GLP-1 and GIP. *Gastroenterology*, 132(6), 2131.
45. Turtle, J.R., Littleton, G.K. and Kipnis, D.M., 1967. Stimulation of insulin secretion by theophylline. *Nature*, *213*(5077), 727–728.
46. Light, P.E., Manning, F.J.E., Riedel, M.J. and Wheeler, M.B., 2002. Glucagon-like peptide-1 inhibits pancreatic ATP-sensitive potassium channels via a protein kinase A-and ADP-dependent mechanism. *Molecular Endocrinology*, *16*(9), 2135–2144.
47. Graham, T.E., Sathasivam, P., Rowland, M., Marko, N., Greer, F. and Battram, D., 2001. Caffeine ingestion elevates plasma insulin response in humans during an oral glucose tolerance test. *Canadian Journal of Physiology and Pharmacology*, *79*(7), 559–565.
48. Frei, B. and Higdon, J.V., 2003. Antioxidant activity of tea polyphenols in vivo: Evidence from animal studies. *The Journal of nutrition*, *133*(10), 3275S–3284S.

49. Patel, S.S., Beer, S., Kearney, D.L., Phillips, G. and Carter, B.A., 2013. Green tea extract: A potential cause of acute liver failure. *World Journal of Gastroenterology*, *19*(31), 5174–5177.
50. Bhandari, U. and Ansari, M.N., 2008. Antihyperglycaemic activity of aqueous extract of *Embelia ribes* Burm. in streptozotocin-induced diabetic rats. *Indian Journal of Experimental Biology*, *46*, 607–613.
51. Ohkuwa, T., Sato, Y. and Naoi, M., 1995. Hydroxyl radical formation in diabetic rats induced by streptozotocin. *Life Sciences*, *56*(21), 1789–1798.
52. Kwon, N.S., Lee, S.H., Choi, C.S., Kho, T. and Lee, H.S., 1994. Nitric oxide generation from streptozotocin. *The FASEB Journal*, *8*(8), 529–533.
53. Joshi, R., Kamat, J.P. and Mukherjee, T., 2007. Free radical scavenging reactions and antioxidant activity of embelin: Biochemical and pulse radiolytic studies. *Chemico-Biological Interactions*, *167*(2), 125–134.
54. Gandhi, G.R., Stalin, A., Balakrishna, K., Ignacimuthu, S., Paulraj, M.G. and Vishal, R., 2013. Insulin sensitization via partial agonism of PPARγ and glucose uptake through translocation and activation of GLUT4 in PI3K/p-Akt signaling pathway by embelin in type 2 diabetic rats. *Biochimica et Biophysica Acta (BBA)-General Subjects*, *1830*(1), 2243–2255.
55. Surampudi, P.N., John-Kalarickal, J. and Fonseca, V.A., 2009. Emerging concepts in the pathophysiology of type 2 diabetes mellitus. *Mount Sinai Journal of Medicine: A Journal of Translational and Personalized Medicine*, *76*(3), 216–226.
56. Prakasam, A., Sethupathy, S. and Pugalendi, K.V., 2003. Effect of *Casearia esculenta* root extract on blood glucose and plasma antioxidant status in streptozotocin diabetic rats. *Polish Journal of Pharmacology*, *55*(1), 43–50.
57. Chandramohan, G., Al-Numair, K.S., Sridevi, M. and Pugalendi, K.V., 2010. Antihyperlipidemic activity of 3-hydroxymethyl xylitol, a novel antidiabetic compound isolated from *Casearia esculenta* (Roxb.) root, in streptozotocin-diabetic rats. *Journal of Biochemical and Molecular Toxicology*, *24*(2), 95–100.
58. Román, R.R., Lara, L.A., Alarcón, A.F. and Flores, S.J., 1992. Hypoglycemic activity of some antidiabetic plants. *Archives of Medical Research*, *23*(3), 105–109.
59. Acosta-Patino, J.L., Jimenez-Balderas, E., Juarez-Oropeza, M.A. and Dıaz-Zagoya, J.C., 2001. Hypoglycemic action of *Cucurbita ficifolia* on type 2 diabetic patients with moderately high blood glucose levels. *Journal of Ethnopharmacology*, *77*(1), 99–10.
60. Xia, T. and Wang, Q., 2007. Hypoglycaemic role of *Cucurbita ficifolia* (*Cucurbitaceae*) fruit extract in streptozotocin-induced diabetic rats. *Journal of the Science of Food and Agriculture*, *87*(9), 1753–1757.
61. Díaz-Flores, M., Angeles-Mejia, S., Baiza-Gutman, L.A., Medina-Navarro, R., Hernández-Saavedra, D., Ortega-Camarillo, C., Roman-Ramos, R., Cruz, M. and Alarcon-Aguilar, F.J., 2012. Effect of an aqueous extract of *Cucurbita ficifolia* Bouché on the glutathione redox cycle in mice with STZ-induced diabetes. *Journal of Ethnopharmacology*, *144*(1), 101–108.
62. Miranda-Perez, M.E., Ortega-Camarillo, C., Escobar-Villanueva, M.D.C., Blancas-Flores, G. and Alarcon-Aguilar, F.J., 2016. *Cucurbita ficifolia* Bouché increases insulin secretion in RINm5F cells through an influx of Ca 2+ from the endoplasmic reticulum. *Journal of Ethnopharmacology*, *188*, 159–166.
63. Zhou, J., Zhou, S. and Zeng, S., 2013. Experimental diabetes treated with trigonelline: Effect on β cell and pancreatic oxidative parameters. *Fundamental & Clinical Pharmacology*, *27*(3), 279–287.
64. Kahn, S.E., Beard, J.C., Schwartz, M.W., Ward, W.K., Ding, H.L., Bergman, R.N., Taborsky, G.J. and Porte, D., 1989. Increased β-cell secretory capacity as mechanism for islet adaptation to nicotinic acid-induced insulin resistance. *Diabetes*, *38*(5), 562–568.
65. Benariba, N., Bellakdhar, W., Djaziri, R., Hupkens, E., Louchami, K. and Malaisse, W.J., 2013. Protective action of *Citrullus colocynthis* seed extracts against the deleterious effect of streptozotocin on both in vitro glucose-stimulated insulin release from rat pancreatic islets and in vivo glucose homeostasis. *Biomedical Reports*, *1*(1), 119–112.
66. Nmila, R., Gross, R., Rchid, H., Roye, M., Manteghetti, M., Petit, P., Tijane, M., Ribes, G. and Sauvaire, Y., 2000. Insulinotropic effect of *Citrullus colocynthis* fruit extracts. *Planta Medica*, *66*(05), 418–423.
67. Akobundu, E.N.T., Cherry, J.P. and Simmons, J.G., 1982. Chemical, functional, and nutritional properties of egusi (*Colocynthis citrullus* L.) seed protein products. *Journal of Food Science*, *47*(3), 829–835.
68. Gurudeeban, S., Satyavani, K. and Ramanatan, T., 2010. Bitter apple (*Citrullus colocynthis*): An overview of chemical composition and biochemical potentials. *Asian Journal of Plant Sciences*, *9*(7), 394–340.
69. Folador, P., Cazarolli, L.H., Gazola, A.C., Reginatto, F.H., Schenkel, E.P. and Silva, F.R.M.B., 2010. Potential insulin secretagogue effects of isovitexin and swertisin isolated from *Wilbrandia ebracteata* roots in non-diabetic rats. *Fitoterapia*, *81*(8), 1180–1187.
70. Salahuddin, M.D. and Jalalpure, S.S., 2010. Antidiabetic activity of aqueous fruit extract of *Cucumis trigonus* Roxb. in streptozotocin-induced-diabetic rats. *Journal of Ethnopharmacology*, *127*(2), 565–567.

71. Durrington, P., 2007. *Hyperlipidaemia 3Ed: Diagnosis and Management*. Boca Raton, FL: CRC Press.
72. Newgard, C.B. and McGarry, J.D., 1995. Metabolic coupling factors in pancreatic β-cell signal transduction. *Annual Review of Biochemistry*, *64*(1), 689–719.
73. Brass, B.J., Abelev, Z., Liao, E.P. and Poretsky, L., 2010. Endocrine pancreas. In *Principles of Diabetes Mellitus,* Poretsky, L., (Ed.). New York: Springer, pp. 37–55.
74. Norberg, Å., Hoa, N.K., Liepinsh, E., Van Phan, D., Thuan, N.D., Jörnvall, H., Sillard, R. and Östenson, C.G., 2004. A novel insulin-releasing substance, phanoside, from the plant *Gynostemma pentaphyllum*. *Journal of Biological Chemistry*, *279*(40), 41361–41367.
75. Hoa, N.K., Norberg, A., Sillard, R., Van, P.D., Thuan, N.D., Dzung, D.T., Jörnvall, H. and Ostenson, C.G., 2007. The possible mechanisms by which phanoside stimulates insulin secretion from rat islets. *Journal of Endocrinology, 192*(2), 389–394.
76. Mondal, A., Maity, T.K. and Pal, D., 2012. Hypoglycaemic effect of *Melothria heterophylla* in streptozotocin-induced diabetic rats. *Pharmaceutical Biology*, *50*(9), 1151–1156.
77. Shrayyef, M.Z. and Gerich, J.E., 2010. Normal glucose homeostasis. In *Principles of Diabetes Mellitus,* Poretsky, L., (Ed.). New York: Springer, pp. 19–35.
78. Karunanayake, E.H., Welihinda, J., Sirimanne, S.R. and Adorai, G.S., 1984. Oral hypoglycaemic activity of some medicinal plants of Sri Lanka. *Journal of Ethnopharmacology*, *11*(2), 223–223.
79. Xiang, L.W., Huang, X.N., Chen, L.M., Rao, P.F. and Ke, L.J., 2007. The reparative effects of *Momordica charantia* Linn. extract on HIT-T15 pancreatic β-Cells. *Asia Pacific Journal of Clinical Nutrition*, *16*(S1), 249–252.
80. Ma, J., Whittaker, P., Keller, A.C., Mazzola, E.P., Pawar, R.S., White, K.D., Callahan, J.H., Kennelly, E.J., Krynitsky, A.J. and Rader, J.I., 2010. Cucurbitane-type triterpenoids from *Momordica charantia*. *Planta Medica*, *76*(15), 1758–1176.
81. Keller, A.C., Ma, J., Kavalier, A., He, K., Brillantes, A.M.B. and Kennelly, E.J., 2011. Saponins from the traditional medicinal plant *Momordica charantia* stimulate insulin secretion in vitro. *Phytomedicine*, *19*(1), 32–37.
82. Kameswararao, B., Kesavulu, M.M. and Apparao, C.H., 1999. Antidiabetic and hypolipidemic effects of *Momordica cymbalaria* Hook. fruit powder in alloxan-diabetic rats. *Journal of Ethnopharmacology*, *67*(1), 103–109.
83. Kameswararao, B., Kesavulu, M.M. and Apparao, C.H., 2003. Evaluation of antidiabetic effect of *Momordica cymbalaria* fruit in alloxan-diabetic rats. *Fitoterapia*, *74*(1), 7–13.
84. Koneri, R.B., Samaddar, S. and Ramaiah, C.T., 2014. Antidiabetic activity of a triterpenoid saponin isolated from *Momordica cymbalaria* Fenzl. *Indian Journal of Experimental Biology*, *52*(1), 46–52.
85. Vogel, H. (Ed.), 2007. *Drug Discovery and Evaluation: Pharmacological Assays*. Berlin, Germany: Springer Science & Business Media.
86. Suzuki, Y.A., Tomoda, M., Murata, Y., Inui, H., Sugiura, M. and Nakano, Y., 2007. Antidiabetic effect of long-term supplementation with *Siraitia grosvenori* on the spontaneously diabetic Goto–Kakizaki rat. *British Journal of Nutrition*, *97*(04), 770–775.
87. Juárez-Rojop, I.E., Díaz-Zagoya, J.C., Ble-Castillo, J.L., Miranda-Osorio, P.H., Castell-Rodríguez, A.E., Tovilla-Zárate, C.A., Rodríguez-Hernández, A., Aguilar-Mariscal, H., Ramón-Frías, T. and Bermúdez-Ocaña, D.Y., 2012. Hypoglycemic effect of Carica papaya leaves in streptozotocin-induced diabetic rats. *BMC Complementary and Alternative Medicine*, *12*(1), 236.
88. Canini, A., Alesiani, D., D'Arcangelo, G. and Tagliatesta, P., 2007. Gas chromatography–mass spectrometry analysis of phenolic compounds from *Carica papaya* L. leaf. *Journal of Food Composition and Analysis*, *20*(7), 584–590.
89. Harini, R. and Pugalendi, K.V., 2010. Antihyperglycemic effect of protocatechuic acid on streptozotocin-diabetic rats. *The Journal of Basic and Clinical Physiology and Pharmacology, 21*(1), 79–91.
90. Olayaki, L.A., Irekpita, J.E., Yakubu, M.T. and Ojo, O.O., 2015. Methanolic extract of *Moringa oleifera* leaves improves glucose tolerance, glycogen synthesis and lipid metabolism in alloxan-induced diabetic rats. *Journal of Basic and Clinical Physiology and Pharmacology*, *26*(6), 585–593.
91. Cano, D.A., Rulifson, I.C., Heiser, P.W., Swigart, L.B., Pelengaris, S., German, M., Evan, G.I., Bluestone, J.A. and Hebrok, M., 2008. Regulated β-cell regeneration in the adult mouse pancreas. *Diabetes*, *57*(4), 958–966.
92. Li, W.C., Rukstalis, J.M., Nishimura, W., Tchipashvili, V., Habener, J.F., Sharma, A. and Bonner-Weir, S., 2010. Activation of pancreatic-duct-derived progenitor cells during pancreas regeneration in adult rats. *Journal of Cell Science*, *123*(16), 2792–2802.
93. Srivastava, L.M., Bora, P.S. and Bhatt, S.D., 1982. Diabetogenic action of streptozotocin. *Trends in Pharmacological Sciences*, *3*, 376–378.

94. Grover, J.K., Yadav, S. and Vats, V., 2002. Hypoglycemic and antihyperglycemic effect of *Brassica juncea* diet and their effect on hepatic glycogen content and the key enzymes of carbohydrate metabolism. *Molecular and Cellular Biochemistry*, *241*(1–2), 95–100.
95. El-Missiry, M.A. and El Gindy, A.M., 2000. Amelioration of alloxan induced diabetes mellitus and oxidative stress in rats by oil of *Eruca sativa* seeds. *Annals of Nutrition and Metabolism*, *44*(3), 97–100.
96. Flanders, A. and Abdulkarim, S.M., 1985. The composition of seed and seed oils of Taramira (*Eruca sativa*). *Journal of the American Oil Chemists' Society*, *62*(7), 1134–1135.
97. Satyaprakash, R.J., Rajesh, M.S., Bhanumathy, M., Harish, M.S., Shivananda, T.N., Shivaprasad, H.N. and Sushma, G., 2013. Hypoglycemic and antihyperglycemic effect of *Ceiba pentandra* L. gaertn in normal and streptozotocininduced diabetic rats. *Ghana Medical Journal*, *47*(3), 121–127.
98. Deraniyagala, S.A. and Wijesundera, R.L.C., 2002. *Ficus benghalensis*. Colombo, Sri Lanka: *National Science Foundation*.
99. Augusti, K.T., Daniel, R.S., Cherian, S.H.E.E.J.A., Sheela, C.G. and Nair, C.R., 1994. Effect of leucopelargonin derivative from *Ficus bengalensis* Linn. on diabetic dogs. *The Indian Journal of Medical Research*, *99*, 82–86.
100. Singh, R.K., Mehta, S., Jaiswal, D., Rai, P.K. and Watal, G., 2009. Antidiabetic effect of *Ficus bengalensis* aerial roots in experimental animals. *Journal of Ethnopharmacology*, *123*(1), 110–114.
101. Jayaprakasam, B., Vareed, S.K., Olson, L.K. and Nair, M.G., 2005. Insulin secretion by bioactive anthocyanins and anthocyanidins present in fruits. *Journal of Agricultural and Food Chemistry*, *53*(1), 28–31.
102. Katsuma, S., Hirasawa, A. and Tsujimoto, G., 2005. Bile acids promote glucagon-like peptide-1 secretion through TGR5 in a murine enteroendocrine cell line STC-1. *Biochemical and Biophysical Research Communications*, *329*(1), 386–390.
103. Pandit, R., Phadke, A. and Jagtap, A., 2010. Antidiabetic effect of *Ficus religiosa* extract in streptozotocin-induced diabetic rats. *Journal of Ethnopharmacology*, *128*(2), 462–466.
104. Swami, K.D., Malik, G.S. and Bisht, N.P.S., 1989. Chemical investigation of stem bark of *Ficus-religiosa* and *Prosopis-spicigera*. *Journal of the Indian Chemical Society*, *66*(4), 288–289.
105. Ambike, S.H. and Rao, M.R., 1967. Studies on phytosterolin from the bark of *Ficus religiosa*. *Indian Journal of Pharmacology, 29*, 91–92.
106. Ivorra, M.D., Paya, M. and Villar, A., 1990. Effect of beta-sitosterol-3-beta-D-glucoside on insulin secretion in vivo in diabetic rats and in vitro in isolated rat islets of Langerhans. *Die Pharmazie*, *45*(4), 271–273.
107. Düfer, M., Hörth, K., Wagner, R., Schittenhelm, B., Prowald, S., Wagner, T.F., Oberwinkler, J. et al., 2012. Bile acids acutely stimulate insulin secretion of mouse β-cells via farnesoid X receptor activation and KATP channel inhibition. *Diabetes*, *61*(6), 1479–1489.
108. Kumar, D.P., Rajagopal, S., Mahavadi, S., Mirshahi, F., Grider, J.R., Murthy, K.S. and Sanyal, A.J., 2012. Activation of transmembrane bile acid receptor TGR5 stimulates insulin secretion in pancreatic β cells. *Biochemical and Biophysical Research Communications*, *427*(3), 600–605.
109. Jayakar, B. and Suresh, B., 2003. Antihyperglycemic and hypoglycemic effect of *Aporosa lindleyana* in normal and alloxan induced diabetic rats. *Journal of Ethnopharmacology*, *84*(2–3), 247–249.
110. Rabinovitch, A., 1994. Immunoregulatory and cytokine imbalances in the pathogenesis of IDDM: Therapeutic intervention by immunostimulation? *Diabetes*, *43*(5), 613–621.
111. Boni-Schnetzler, M., Thorne, J., Parnaud, G., Marselli, L., Ehses, J.A., Kerr-Conte, J., Pattou, F., Halban, P.A., Weir, G.C. and Donath, M.Y., 2008. Increased interleukin (IL)-1β messenger ribonucleic acid expression in β-cells of individuals with type 2 diabetes and regulation of IL-1β in human islets by glucose and autostimulation. *The Journal of Clinical Endocrinology & Metabolism*, *93*(10), 4065–4074.
112. Narang, A.S. and Mahato, R.I., 2006. Biological and biomaterial approaches for improved islet transplantation. *Pharmacological Reviews*, *58*(2), 194–243.
113. Badami, S., Rai, S.R. and Suresh, B., 2005. Antioxidant activity of *Aporosa lindleyana* root. *Journal of Ethnopharmacology*, *101*(1–3), 180–184.
114. Ali, Y., Alam, M.S., Hamid, H., Husain, A., Kharbanda, C., Bano, S., Nazreen, S. and Haider, S., 2014. Attenuation of inflammatory mediators, oxidative stress and toxic risk evaluation of *Aporosa lindleyana* Baill bark extract. *Journal of Ethnopharmacology, 155*(3), 1513–1521.
115. Arnush, M., Heitmeier, M.R., Scarim, A.L., Marino, M.H., Manning, P.T. and Corbett, J.A., 1998. IL-1 produced and released endogenously within human islets inhibits beta cell function. *Journal of Clinical Investigation*, *102*(3), 516.

116. Vila-Carriles, W.H., Zhao, G. and Bryan, J., 2007. Defining a binding pocket for sulfonylureas in ATP-sensitive potassium channels. *The FASEB Journal*, *21*(1), 18–25.
117. Fatima, N., Hafizur, R.M., Hameed, A., Ahmed, S., Nisar, M. and Kabir, N., 2015. Ellagic acid in *Emblica officinalis* exerts anti-diabetic activity through the action on β-cells of pancreas. *European Journal of Nutrition*, 1–11.
118. Shabeer, J., Srivastava, R.S. and Singh, S.K., 2009. Antidiabetic and antioxidant effect of various fractions of *Phyllanthus simplex* in alloxan diabetic rats. *Journal of Ethnopharmacology*, *124*(1), 34–38.
119. Niu, X.F., He, L.C., Fan, T. and Li, Y., 2006. Protecting effect of brevifolin and 8, 9-single-epoxy brevifolin of *Phyllanthus simplex* on rat liver injury. *Zhongguo Zhong yao za zhi= Zhongguo zhongyao zazhi= China journal of Chinese materia medica*, *31*(18), 1529–1532.
120. Pedrosa, R.C., Meyre-Silva, C., Cechinel-Filho, V., Benassi, J.C., Oliveira, L.F.S., Zancanaro, V., Dal Magro, J. and Yunes, R.A., 2002. Hypolipidaemic activity of methanol extract of *Aleurites moluccana*. *Phytotherapy Research*, *16*(8), 765–768.
121. Ado, M.A., Abas, F., Mohammed, A.S. and Ghazali, H.M., 2013. Anti-and pro-lipase activity of selected medicinal, herbal and aquatic plants, and structure elucidation of an anti-lipase compound. *Molecules*, *18*(12), 14651–14669.
122. Cesca, T.G., Faqueti, L.G., Rocha, L.W., Meira, N.A., Meyre-Silva, C., de Souza, M.M., Quintão, N.L.M., Silva, R.M.L., Cechinel Filho, V. and Bresolin, T.M.B., 2012. Antinociceptive, anti-inflammatory and wound healing features in animal models treated with a semisolid herbal medicine based on *Aleurites moluccana* L. Willd. Euforbiaceae standardized leaf extract: Semisolid herbal. *Journal of Ethnopharmacology*, *143*(1), 355–362.
123. Govindarajan, R., Vijayakumar, M., Rao, C.V., Pushpangadan, P., Asare-Anane, H., Persaud, S., Jones, P. and Houghton, P.J., 2008. Antidiabetic activity of *Croton klozchianus* in rats and direct stimulation of insulin secretion in-vitro. *Journal of Pharmacy and Pharmacology*, *60*(3), 371–376.
124. Silva, R.M., Santos, F.A., Rao, V.S.N., Maciel, M.A. and Pinto, A.C., 2001. Blood glucose-and triglyceride-lowering effect of trans-dehydrocrotonin, a diterpene from *Croton cajucara* Benth., in rats. *Diabetes, Obesity and Metabolism*, *3*(6), 452–456.
125. Hiruma-Lima, C.A., Spadari-Bratfisch, R.C., Grassi-Kassisse, D.M. and Brito, A.R.S., 1999. Antiulcerogenic mechanisms of dehydrocrotonin, a diterpene lactone obtained from *Croton cajucara*. *Planta Medica*, *65*(4), 325–330.
126. Amisten, S., Salehi, A., Rorsman, P., Jones, P.M. and Persaud, S.J., 2013. An atlas and functional analysis of G-protein coupled receptors in human islets of Langerhans. *Pharmacology & Therapeutics*, *139*(3), 359–391.
127. Gilon, P. and Henquin, J.C., 2001. Mechanisms and physiological significance of the cholinergic control of pancreatic β-cell function. *Endocrine Reviews*, *22*(5), 565–604.
128. Iismaa, T.P., Kerr, E.A., Wilson, J.R., Carpenter, L., Sims, N. and Biden, T.J., 2000. Quantitative and functional characterization of muscarinic receptor subtypes in insulin-secreting cell lines and rat pancreatic islets. *Diabetes*, *49*(3), 392–398.
129. Gautam, D., Han, S.J., Hamdan, F.F., Jeon, J., Li, B., Li, J.H., Cui, Y. et al., 2006. A critical role for β cell M 3 muscarinic acetylcholine receptors in regulating insulin release and blood glucose homeostasis in vivo. *Cell Metabolism*, *3*(6), 449–461.
130. Kong, K.C., Butcher, A.J., McWilliams, P., Jones, D., Wess, J., Hamdan, F.F., Werry, T. et al., 2010. M3-muscarinic receptor promotes insulin release via receptor phosphorylation/arrestin-dependent activation of protein kinase D1. *Proceedings of the National Academy of Sciences*, *107*(49), 21181–21186.
131. Pontiroli, A.E., Petrelli, P.L., Vicari, A., Alberetto, M., Foa, P.P. and Pozza, G., 1982. Different effects of histaminergic H1 and H2 antagonists on basal and stimulated insulin and glucagon release in humans. *Hormone and Metabolic Research*, *14*(09), 496–497.
132. Lee, B.H., Lee, C.C., Cheng, Y.H., Chang, W.C., Hsu, W.H. and Wu, S.C., 2013. Graptopetalum paraguayense and resveratrol ameliorates carboxymethyllysine (CML)-induced pancreas dysfunction and hyperglycemia. *Food and Chemical Toxicology*, *62*, 492–498.
133. Brissova, M., Shiota, M., Nicholson, W.E., Gannon, M., Knobel, S.M., Piston, D.W., Wright, C.V. and Powers, A.C., 2002. Reduction in pancreatic transcription factor PDX-1 impairs glucose-stimulated insulin secretion. *Journal of Biological Chemistry*, *277*(13), 11225–11232.
134. Chika, A. and Bello, S.O., 2010. Antihyperglycaemic activity of aqueous leaf extract of *Combretum micranthum* (Combretaceae) in normal and alloxan-induced diabetic rats. *Journal of Ethnopharmacology*, *129*(1), 34–37.
135. Jentzsch, K., Spiegl, P. and Fuchs, L., 1962. Untersuchungen über die inhaltsstoffe der blätter von *Combretum micranthum g. don*. *Planta Medica*, *10*(1), 1–8.

136. Bassene, E., Olschwang, D. and Pousset, J.-L., 1987. Plantes medicinales africaines XXIII: Flavonoides du *Combretum micranthum, G. Don* (Kinkeliba). *Plantes Medicinales et Phytotherapie, 21*(2), 173–176.
137. Herzlinger, S. and Abrahamson, M.J., 2010. Treating type 2 diabetes mellitus. In *Principles of Diabetes Mellitus,* Poretsky, L., (Ed.). New York: Springer, pp. 731–741.
138. Nagappa, A.N., Thakurdesai, P.A., Rao, N.V. and Singh, J., 2003. Antidiabetic activity of *Terminalia catappa* Linn fruits. *Journal of Ethnopharmacology, 88*(1), 45–50.
139. Song, J.L., Zhao, X., Wang, Q. and Zhang, T., 2013. Protective effects of *Lagerstroemia speciosa* on 3-morpholinosydnonimine (SIN-1)-induced oxidative stress in HIT-T15 pancreatic β cells. *Molecular Medicine Reports, 7*(5), 1607–1612.
140. Hayashi, T., Maruyama, H., Kasai, R., Hattori, K., Takasuga, S., Hazeki, O., Yamasaki, K. and Tanaka, T., 2002. Ellagitannins from *Lagerstroemia speciosa* as activators of glucose transport in fat cells. *Planta Medica, 68*(2), 173–175.
141. Verma, N., Amresh, G., Sahu, P.K., Rao, C.V. and Singh, A.P., 2012. Antihyperglycemic activity of *Woodfordia fruticosa* (Kurz) flowers extracts in glucose metabolism and lipid peroxidation in streptozotocin-induced diabetic rats. *Indian Journal of Experimental Biology, 50*(5), 351–358.
142. Murugesan, T., Rao, B., Sinha, S., Biswas, S., Pal, M. and Saha, B.P., 2000. Anti-diabetic activity of Jussiaea suffruticosa extract in rats. *Pharmacy and Pharmacology Communications, 6*(10), 451–453.
143. Chang, C.I. and Kuo, Y.H., 2007. Oleanane-type triterpenes from *Ludwigia octovalvis. Journal of Asian Natural Products Research, 9*(1), 67–72.
144. Yan, J. and Yang, X.W., 2005. Studies on the chemical constituents in herb of *Ludwigia octovalvis. Zhongguo Zhong yao za zhi= Zhongguo zhongyao zazhi= China Journal of Chinese Materia Medica, 30*(24), 1923–1926.
145. Chang, C.I., Kuo, C.C., Chang, J.Y. and Kuo, Y.H., 2004. Three new oleanane-type triterpenes from *Ludwigia octovalvis* with cytotoxic activity against two human cancer cell lines. *Journal of Natural Products, 67*(1), 91–93.
146. Villasenor, I.M. and Lamadrid, M.R.A., 2006. Comparative anti-hyperglycemic potentials of medicinal plants. *Journal of Ethnopharmacology, 104*(1), 129–131.
147. Deila, R.C.S., Cinthia, M.M., Sônia, C.A.L., Andreane, F., Danívia, P., Aron F.S. and Marcelo, C., 2004. *Syzygium cumini* and the regeneration of insulin positive cells from the pancreatic duct. *Brazilian Journal of Veterinary Research and Animal Science, 41,* 236–239.
148. Sanches, J.R., França, L.M., Chagas, V.T., Gaspar, R.S., dos Santos, K.A., Gonçalves, L.M., Sloboda, D.M. et al., 2016. Polyphenol-rich extract of *Syzygium cumini* leaf dually improves peripheral insulin sensitivity and pancreatic islet function in monosodium l-glutamate-induced obese rats. *Frontiers in Pharmacology,* 7.
149. Wang, R.N., Klöppel, G. and Bouwens, L., 1995. Duct-to islet-cell differentiation and islet growth in the pancreas of duct-ligated adult rats. *Diabetologia, 38*(12), 1405–1411.
150. Shehata, A.M., Quintanilla-Fend, L., Bettio, S., Singh, C.B. and Ammon, H.P.T., 2011. Prevention of multiple low-dose streptozotocin (MLD-STZ) diabetes in mice by an extract from gum resin of *Boswellia serrata* (BE). *Phytomedicine, 18*(12), 1037–1044.
151. Shehata, A.M., Quintanilla-Fend, L., Bettio, S., Jauch, J., Scior, T., Scherbaum, W.A. and Ammon, H.P.T., 2015. 11-Keto-β-boswellic acids prevent development of autoimmune reactions, insulitis and reduce hyperglycemia during induction of multiple low-dose streptozotocin (MLD-STZ) diabetes in mice. *Hormone and Metabolic Research, 47*(6), 463–469.
152. Sailer, E.R., Subramanian, L.R., Rall, B., Hoernlein, R.F., Ammon, H. and Safayhi, H., 1996. Acetyl-11-keto-β-boswellic acid (AKBA): Structure requirements for binding and 5-lipoxygenase inhibitory activity. *British Journal of Pharmacology, 117*(4), 615–618.
153. Harizi, H., Corcuff, J.B. and Gualde, N., 2008. Arachidonic-acid-derived eicosanoids: Roles in biology and immunopathology. *Trends in Molecular Medicine, 14*(10), 461–469.
154. Spite, M., Hellmann, J., Tang, Y., Mathis, S.P., Kosuri, M., Bhatnagar, A., Jala, V.R. and Haribabu, B., 2011. Deficiency of the leukotriene B4 receptor, BLT-1, protects against systemic insulin resistance in diet-induced obesity. *The Journal of Immunology, 187*(4), 1942–1949.
155. Cuaz-Pérolin, C., Billiet, L., Baugé, E., Copin, C., Scott-Algara, D., Genze, F., Büchele, B., Syrovets, T., Simmet, T. and Rouis, M., 2008. Antiinflammatory and antiatherogenic effects of the NF-κB inhibitor acetyl-11-keto-β-boswellic acid in LPS-challenged ApoE−/− mice. *Arteriosclerosis, Thrombosis, and Vascular Biology, 28*(2), 272–277.
156. Gerbeth, K., Meins, J., Kirste, S., Momm, F., Schubert-Zsilavecz, M. and Abdel-Tawab, M., 2011. Determination of major boswellic acids in plasma by high-pressure liquid chromatography/mass spectrometry. *Journal of Pharmaceutical and Biomedical Analysis, 56*(5), 998–1005.

157. Lv, N., Song, M.Y., Kim, E.K., Park, J.W., Kwon, K.B. and Park, B.H., 2008. Guggulsterone, a plant sterol, inhibits NF-kappaB activation and protects pancreatic beta cells from cytokine toxicity. *Molecular and Cellular Endocrinology*, *289*(1–2), 49–59.
158. Macha, M.A., Rachagani, S., Gupta, S., Pai, P., Ponnusamy, M.P., Batra, S.K. and Jain, M., 2013. Guggulsterone decreases proliferation and metastatic behavior of pancreatic cancer cells by modulating JAK/STAT and Src/FAK signaling. *Cancer Letters*, *341*(2), 166–177.
159. Radha Krishna, Y., Mittal, V., Grewal, P., Fiel, M.I. and Schiano, T., 2011. Acute liver failure caused by "fat burners" and dietary supplements: A case report and literature review. *Canadian Journal of Gastroenterology and Hepatology*, *25*(3), 157–160.
160. Kölönte, A., Guillot, B. and Raison-Peyron, N., 2006. Allergic contact dermatitis to guggul extract contained in an anticellulite gel-cream. *Contact Dermatitis*, *54*(4), 226–227.
161. Shirwaikar, A., Rajendran, K. and Barik, R., 2006. Effect of aqueous bark extract of Garuga pinnata Roxb. in streptozotocin-nicotinamide induced type-II diabetes mellitus. *Journal of Ethnopharmacology*, *107*(2), 285–290.
162. Venkatraman, G., Thombare, P.S. and Sabata, B.K., 1994. A tetracyclic triterpenoid from *Garuga pinnata*. *Phytochemistry*, *36*(2), 417–419.
163. Jeong, G.S., Lee, D.S., Song, M.Y., Park, B.H., Kang, D.G., Lee, H.S., Kwon, K.B. and Kim, Y.C., 2011. Butein from *Rhus verniciflua* protects pancreatic β cells against cytokine-induced toxicity mediated by inhibition of nitric oxide formation. *Biological and Pharmaceutical Bulletin*, *34*(1), 97–102.
164. Sung, J. and Lee, J., 2015. Anti-inflammatory activity of butein and luteolin through suppression of NF κ B activation and induction of heme oxygenase-1. *Journal of Medicinal Food*, *18*(5), 557–564.
165. Lee, D.S. and Jeong, G.S., 2016. Butein provides neuroprotective and anti-neuroinflammatory effects through Nrf2/ARE-dependent haem oxygenase 1 expression by activating the PI3K/Akt pathway. *British Journal of Pharmacology*, *173*(19), 2894–2909.
166. Attanayake, A.P., Jayatilaka, K.A., Pathirana, C. and Mudduwa, L.K., 2013. Study of antihyperglycaemic activity of medicinal plant extracts in alloxan induced diabetic rats. *Ancient Science of Life*, *32*(4), 193.
167. Fernandes, A., King, L.C., Guz, Y., Stein, R., Wright, C.V.E. and Teitelman, G., 1997. Differentiation of new insulin-producing cells is induced by injury in adult pancreatic islets 1. *Endocrinology*, *138*(4), 1750–1762.
168. Guz, Y., Nasir, I. and Teitelman, G., 2001. Regeneration of pancreatic β cells from intra-islet precursor cells in an experimental model of diabetes. *Endocrinology*, *142*(11), 4956–4968.
169. Kawamori, D., Kajimoto, Y., Kaneto, H., Umayahara, Y., Fujitani, Y., Miyatsuka, T., Watada, H., Leibiger, I.B., Yamasaki, Y. and Hori, M., 2003. Oxidative stress induces nucleo-cytoplasmic translocation of pancreatic transcription factor PDX-1 through activation of c-Jun NH2-terminal kinase. *Diabetes*, *52*(12), 2896–2904.
170. Govindarajan, R., Asare-Anane, H., Persaud, S., Jones, P. and Houghton, P.J., 2007. Effect of *Desmodium gangeticum* extract on blood glucose in rats and on insulin secretion in vitro. *Planta Medica*, *53*(5), 427–432.
171. Banerjee, P.K. and Ghosal, S., 1969. Simple indole bases of *Desmodium gangeticum* (Leguminosae). *Australian Journal of Chemistry*, *22*(1), 275–277.
172. Mishra, P.K., Singh, N., Ahmad, G., Dube, A. and Maurya, R., 2005. Glycolipids and other constituents from *Desmodium gangeticum* with antileishmanial and immunomodulatory activities. *Bioorganic & Medicinal Chemistry Letters*, *15*(20), 4543–4546.
173. Smith, S.A. and Pogson, C.I., 1977. Tryptophan and the control of plasma glucose concentrations in the rat. *Biochemical Journal*, *168*(3), 495–506.
174. Coulie, B., Tack, J., Bouillon, R., Peeters, T. and Janssens, J., 1998. 5-Hydroxytryptamine-1 receptor activation inhibits endocrine pancreatic secretion in humans. *American Journal of Physiology-Endocrinology and Metabolism*, *274*(2), E317–E320.
175. Maurya, R., Ray, A.B., Duah, F.K., Slatkin, D.J. and SchiffJr, P.L., 1984. Constituents of *Pterocarpus marsupium*. *Journal of Natural Products*, *47*(1), 179–181
176. Scalbert, A., Morand, C., Manach, C. and Rémésy, C., 2002. Absorption and metabolism of polyphenols in the gut and impact on health. *Biomedicine & Pharmacotherapy*, *56*(6), 276–282.
177. Ahmad, F., Khan, M.M., Rastogi, A.K., Chaubey, M. and Kidwai, J.R., 1991. Effect of (-)epicatechin on cAMP content, insulin release and conversion of proinsulin to insulin in immature and mature rat islets in vitro. *Indian Journal of Experimental Biology*, *29*(6):516–520.
178. Rizvi, S.I., Abu, Z.M. and Suhail, M., 1995. Insulin-mimetic effect of (-) epicatechin on osmotic fragility of human erythrocytes. *Indian Journal* of *Experimental Biology*, *33*(10), 791–792.
179. Mohankumar, S.K., O'Shea, T. and McFarlane, J.R., 2012. Insulinotrophic and insulin-like effects of a high molecular weight aqueous extract of *Pterocarpus marsupium* Roxb. hardwood. *Journal of Ethnopharmacology*, *141*(1), 72–79.

180. Halagappa, K., Girish, H.N. and Srinivasan, B.P., 2010. The study of aqueous extract of *Pterocarpus marsupium* Roxb. on cytokine TNF-α in type 2 diabetic rats. *Indian Journal of Pharmacology*, *42*(6), 392.
181. Manickam, M., Ramanathan, M., Farboodniay, J.M.A., Chansouria, J.P.N. and Ray, A.B., 1997. Antihyperglycemic activity of phenolics from *Pterocarpus marsupium*. Journal of *Natural Products*, *60*(6), 609–610.
182. Hougee, S., Faber, J., Sanders, A., de Jong, R.B., van den Berg, W.B., Garssen, J., Hoijer, M.A. and Smit, H.F., 2005. Selective COX-2 inhibition by a *Pterocarpus marsupium* extract characterized by pterostilbene, and its activity in healthy human volunteers. *Planta Medica*, *71*(5), 387–392.
183. Kosuru, R., Rai, U., Prakash, S., Singh, A. and Singh, S., 2016. Promising therapeutic potential of pterostilbene and its mechanistic insight based on preclinical evidence. *European Journal of Pharmacology*, *789*, 229–243.
184. Paolisso, G., Sgambato, S., Passariello, N., Pizza, G., Torella, R., Tesauro, P., Varricchio, M. and d'Onofrio, F., 1988. Plasma glucose lowering effect of spartein sulphate infusion in non-insulin dependent (Type 2) diabetic subjects. *European Journal of Clinical Pharmacology*, *34*(3), 227–232.
185. Kubo, H., Inoue, M., Kamei, J. and Higashiyama, K., 2006. Hypoglycemic effects of multiflorine derivatives in normal mice. *Biological and Pharmaceutical Bulletin*, *29*(10), 2046–2050.
186. Bobkiewicz-Kozłowska, T., Dworacka, M., Kuczyński, S., Abramczyk, M., Kolanoś, R., Wysocka, W., Garcia Lopez, P.M. and Winiarska, H., 2007. Hypoglycaemic effect of quinolizidine alkaloids–lupanine and 2-thionosparteine on non-diabetic and streptozotocin-induced diabetic rats. *European Journal of Pharmacology*, *565*(1–3), 240–244.
187. López, P.M.G., de la Mora, P.G., Wysocka, W., Maiztegui, B., Alzugaray, M.E., Del Zotto, H. and Borelli, M.I., 2004. Quinolizidine alkaloids isolated from *Lupinus* species enhance insulin secretion. *European Journal of Pharmacology*, *504*(1), 139–142.
188. Baldeón, M.E., Castro, J., Villacrés, E., Narváez, L. and Fornasini, M., 2012. Hypoglycemic effect of cooked *Lupinus mutabilis* and its purified alkaloids in subjects with type 2 diabetes. *Nutricion Hospitalaria*, *27*(4):1261–1266.
189. Wiedemann, M., Gurrola-Díaz, C.M., Vargas-Guerrero, B., Wink, M., García-López, P.M. and Düfer, M., 2015. Lupanine improves glucose homeostasis by influencing KATP channels and insulin gene expression. *Molecules*, *20*(10), 19085–19100.
190. Rodriguez-Diaz, R., Dando, R., Jacques-Silva, M.C., Fachado, A., Molina, J., Abdulreda, M.H., Ricordi, C., Roper, S.D., Berggren, P.O. and Caicedo, A., 2011. Alpha cells secrete acetylcholine as a non-neuronal paracrine signal priming beta cell function in humans. *Nature Medicine*, *17*(7), 888–892.
191. Gilon, P. and Henquin, J.C., 2001. Mechanisms and physiological significance of the cholinergic control of pancreatic β-cell function. *Endocrine Reviews*, *22*(5), 565–604.
192. Schmeller, T., Sauerwein, M., Sporer, F., Wink, M. and Müller, W.E., 1994. Binding of quinolizidine alkaloids to nicotinic and muscarinic acetylcholine receptors. *Journal of Natural Products*, *57*(9), 1316–1319.
193. Phuwapraisirisan, P., Puksasook, T., Jong-Aramruang, J. and Kokpol, U., 2008. Phenylethyl cinnamides: A new series of α-glucosidase inhibitors from the leaves of Aegle marmelos. *Bioorganic & Medicinal Chemistry Letters*, *18*(18), 4956–4958.
194. Gandhi, G.R., Ignacimuthu, S. and Paulraj, M.G., 2012. Hypoglycemic and β-cells regenerative effects of *Aegle marmelos* (L.) Corr. bark extract in streptozotocin-induced diabetic rats. *Food and Chemical Toxicology*, *50*(5), 1667–1674.
195. Ohashi, K., Watanatabe, H., Okumura, Y., Uji, T. and Kitagawa, I., 1994. Indonesian medicinal plants. XII. Four isomeric lignan-glucosides from the bark of *Aegle marmelos* (*Rutaceae*). *Chemical and Pharmaceutical Bulletin*, *42*(9), 1924–1926
196. Huang, X.Z., Cheng, C.M., Dai, Y., Fu, G.M., Guo, J.M., Liang, H. and Wang, C., 2012. A novel lignan glycoside with antioxidant activity from *Tinospora sagittata* var. yunnanensis. *Natural Product Research*, *26*(20), 1876–1880.
197. Maity, P., Hansda, D., Bandyopadhyay, U. and Mishra, D.K., 2009. Biological activities of crude extracts and chemical constituents of Bael, *Aegle marmelos* (L.) Corr. *Indian Journal of Experimental Biology*, *47*(11), 849–861.
198. Brisson, G.R., Malaisse-Lagae, F. and Malaisse, W.J., 1972. The stimulus-secretion coupling of glucose-induced insulin release VII. A proposed site of action for adenosine-3′,5′-cyclic monophosphate. *The Journal of clinical Investigation*, *51*(2), 232–241.
199. Howell, S.L., 1999. *The Biology of the Pancreatic Cell*. Amsterdam, the Netherlands: Elsevier.

200. Ono, E., Inoue, J., Hashidume, T., Shimizu, M. and Sato, R., 2011. Anti-obesity and anti-hyperglycemic effects of the dietary citrus limonoid nomilin in mice fed a high-fat diet. *Biochemical and Biophysical Research Communications*, *410*(3), 677–681.
201. Thomas, C., Gioiello, A., Noriega, L., Strehle, A., Oury, J., Rizzo, G., Macchiarulo, A. et al., 2009. TGR5-mediated bile acid sensing controls glucose homeostasis. *Cell Metabolism*, *10*(3), 167–177.
202. Baggio, L.L. and Drucker, D.J., 2007. Biology of incretins: GLP-1 and GIP. *Gastroenterology*, *132*(6), 2131–2157.
203. D'alessio, D.A., Kahn, S.E., Leusner, C.R. and Ensinck, J.W., 1994. Glucagon-like peptide 1 enhances glucose tolerance both by stimulation of insulin release and by increasing insulin-independent glucose disposal. *Journal of Clinical Investigation*, *93*(5), 2263.
204. Ojewole, J.A., 2002. Hypoglycaemic effect of *Clausena anisata* (Willd) Hook methanolic root extract in rats. *Journal of Ethnopharmacology*, *81*(2), 231–237.
205. Chakraborty, A., Chowdhury, B.K. and Bhattacharyya, P., 1995. Clausenol and clausenine—two carbazole alkaloids from *Clausena anisata*. *Phytochemistry*, *40*(1), 295–298.
206. Songue, J.L., Dongo, E., Mpondo, T.N. and White, R.L., 2012. Chemical constituents from stem bark and roots of *Clausena anisata*. *Molecules*, *17*(11), 13673–13686.
207. Bartsch, W., Dietmann, K., Leinert, H. and Sponer, G., 1976. Cardiac action of carazolol and methypranol in comparison with other beta-receptor blockers. *Arzneimittel-Forschung*, *27*(5), 1022–1026.
208. Möricke, R. and Marek, H., 1981. Effect of the beta-receptor blockers propranolol and talinolol on glucose-orciprenaline induced insulin secretion, glucose tolerance and the behavior of nonesterified free fatty acids in metabolically normal probands. *Zeitschrift fur die gesamte innere Medizin und ihre Grenzgebiete*, *36*(5), 263–267.
209. Nalli, Y., Khajuria, V., Gupta, S., Arora, P., Riyaz-Ul-Hassan, S., Ahmed, Z., Ali, A., 2016. Four new carbazole alkaloids from *Murraya koenigii* that display anti-inflammatory and anti-microbial activities. *Organic & Biomolecular Chemistry*, *14*(12), 3322–3332.
210. Sundaram, R., Shanthi, P. and Sachdanandam, P., 2014. Effect of tangeretin, a polymethoxylated flavone on glucose metabolism in streptozotocin-induced diabetic rats. *Phytomedicine*, *21*(6), 793–799.
211. Sundaram, R., Shanthi, P. and Sachdanandam, P., 2015. Tangeretin, a polymethoxylated flavone, modulates lipid homeostasis and decreases oxidative stress by inhibiting NF-κB activation and proinflammatory cytokines in cardiac tissue of streptozotocin-induced diabetic rats. *Journal of Functional Foods*, *16*, 315–333.
212. Irudayaraj, S.S., Sunil, C., Duraipandiyan, V. and Ignacimuthu, S., 2012. Antidiabetic and antioxidant activities of *Toddalia asiatica* (L.) Lam. Leaves in Streptozotocin induced diabetic rats. *Journal of Ethnopharmacology*, *143*(2), 515–523.
213. Duraipandiyan, V. and Ignacimuthu, S., 2009. Antibacterial and antifungal activity of flindersine isolated from the traditional medicinal plant, *Toddalia asiatica* (L.) Lam. *Journal of Ethnopharmacology*, *123*(3), 494–498.
214. Raj, M.K., Balachandran, C., Duraipandiyan, V., Agastian, P. and Ignacimuthu, S., 2012. Antimicrobial activity of Ulopterol isolated from *Toddalia asiatica* (L.) Lam.: A traditional medicinal plant. *Journal of Ethnopharmacology*, *140*(1), 161–165.
215. Banbury, L.K., Shou, Q., Renshaw, D.E., Lambley, E.H., Griesser, H.J., Mon, H. and Wohlmuth, H., 2015. Compounds from *Geijera parviflora* with prostaglandin E 2 inhibitory activity may explain its traditional use for pain relief. *Journal of Ethnopharmacology*, *163*, 251–255.
216. Thirugnanasampandan, R., Jayakumar, R. and Prabhakaran, M., 2012. Analysis of chemical composition and evaluation of antigenotoxic, cytotoxic and antioxidant activities of essential oil of *Toddalia asiatica* (L.) Lam. *Asian Pacific Journal of Tropical Biomedicine*, *2*(3), S1276–S1279.
217. Kavimani, S., Vetrichelvan, T., Ilango, R., and Jaykar, B., 1996. Anti inflammatory activity of the volatile oil of *Toddalia asiatica*. *Indian Journal of Pharmaceutical Sciences*, *58*, 67–70.
218. Cabrera, W., Genta, S., Said, A., Farag, A., Rashed, K. and Sánchez, S., 2008. Hypoglycemic activity of *Ailanthus excelsa* leaves in normal and streptozotocin-induced diabetic rats. *Phytotherapy Research*, *22*(3), 303–307.
219. Said, A., Tundis, R., Hawas, U.W., El-Kousy, S.M., Rashed, K., Menichini, F., Bonesi, M., Huefner, A. and Loizzo, M.R., 2010. In vitro antioxidant and antiproliferative activities of flavonoids from *Ailanthus excelsa* (Roxb.) (*Simaroubaceae*) leaves. *Zeitschrift für Naturforschung C*, *65*(3–4), 180–186.
220. Joshi, B.C., Pandey, A., Sharma, R.P. and Khare, A., 2003. Quassinoids from *Ailanthus excelsa*. *Phytochemistry*, *62*(4), 579–584.

221. Zarse, K., Bossecker, A., Müller-Kuhrt, L., Siems, K., Hernandez, M.A., Berendsohn, W.G., Birringer, M. and Ristow, M., 2011. The phytochemical glaucarubinone promotes mitochondrial metabolism, reduces body fat, and extends lifespan of *Caenorhabditis elegans*. *Hormone and Metabolic Research*, *43*(4), 241–243.
222. NoorShahida, A., Wong, T.W. and Choo, C.Y., 2009. Hypoglycemic effect of quassinoids from *Brucea javanica* (L.) Merr (*Simaroubaceae*) seeds. *Journal of Ethnopharmacology*, *124*(3), 586–591.
223. Kuriyama, T., Ju, X.L., Fusazaki, S., Hishinuma, H., Satou, T., Koike, K., Nikaido, T. and Ozoe, Y., 2005. Nematocidal quassinoids and bicyclophosphorothionates: A possible common mode of action on the GABA receptor. *Pesticide Biochemistry and Physiology*, *81*(3), 176–187.
224. Braun, M., Wendt, A., Buschard, K., Salehi, A., Sewing, S., Gromada, J. and Rorsman, P., 2004. GABAB receptor activation inhibits exocytosis in rat pancreatic β-cells by G-protein-dependent activation of calcineurin. *The Journal of Physiology*, *559*(2), 397–409.
225. Kim, K. and Kim, H.Y., 2008. Korean red ginseng stimulates insulin release from isolated rat pancreatic islets. *Journal of Ethnopharmacology*, *120*(2), 190–195.
226. Park, M.W., Ha, J. and Chung, S.H., 2008. 20 (S)-ginsenoside Rg3 enhances glucose-stimulated insulin secretion and activates AMPK. *Biological and Pharmaceutical Bulletin*, *31*(4), 748–751.
227. Perez-Gutierrez, R.M. and Damian-Guzman, M., 2012. Meliacinolin: A potent α-glucosidase and α-amylase inhibitor isolated from *Azadirachta indica* leaves and in vivo antidiabetic property in streptozotocin-nicotinamide-induced type 2 diabetes in mice. *Biological and Pharmaceutical Bulletin*, *35*(9), 1516–1524.
228. Karam, J.H., 1998. Pancreatic hormones and antidiabetic drugs. In *Basic and Clinical Pharmacology*, Katzung, B.G., (Ed.). New York: Appleton and Lange, p. 684.
229. Chattopadhyay, R.R., 1999. Possible mechanism of antihyperglycemic effect of *Azadirachta indica* leaf extract: Part V. *Journal of Ethnopharmacology*, *67*(3), 373–376.
230. Seo, J.Y., Lee, C., Hwang, S.W., Chun, J., Im, J.P. and Kim, J.S., 2016. Nimbolide inhibits nuclear factor-KB pathway in intestinal epithelial cells and macrophages and alleviates experimental colitis in mice. *Phytotherapy Research*, *30*(10), 1605–1614.
231. Bumah, V.V., Essien, E.U., Agbedahunsi, J.M. and Ekah, O.U., 2005. Effects of *Khaya grandifoliola* (*Meliaceae*) on some biochemical parameters in rats. *Journal of Ethnopharmacology*, *102*(3), 446–449.
232. Zhang, H., Odeku, O.A., Wang, X.N. and Yue, J.M., 2008. Limonoids from the stem bark of *Khaya grandifoliola*. *Phytochemistry*, *69*(1), 271–275.
233. Ibrahim, M.A. and Islam, M.S., 2014. Butanol fraction of *Khaya senegalensis* root modulates β-cell function and ameliorates diabetes-related biochemical parameters in a type 2 diabetes rat model. *Journal of Ethnopharmacology*, *154*(3), 832–838.
234. Ibrahim, M.A., Koorbanally, N.A. and Islam, M.D., 2014. Antioxidative activity and inhibition of key enzymes linked to type 2 diabetes (α-glucosidase and α-amylase) by *Khaya senegalensis*. *Acta Pharmaceutica*, *64*(3), 311–324.
235. Jayaprakasam, B., Olson, L.K., Schutzki, R.E., Tai, M.H. and Nair, M.G., 2006. Amelioration of obesity and glucose intolerance in high-fat-fed C57BL/6 mice by anthocyanins and ursolic acid in Cornelian cherry (*Cornus mas*). *Journal of Agricultural and Food Chemistry*, *54*(1), 243–248.
236. Sozański, T., Kucharska, A.Z., Szumny, A., Magdalan, J., Bielska, K., Merwid-Ląd, A., Woźniak, A., Dzimira, S., Piórecki, N. and Trocha, M., 2014. The protective effect of the *Cornus mas* fruits (cornelian cherry) on hypertriglyceridemia and atherosclerosis through PPARα activation in hypercholesterolemic rabbits. *Phytomedicine*, *21*(13), 1774–1784.
237. Gervois, P. and Mansouri, R.M., 2012. PPARα as a therapeutic target in inflammation-associated diseases. *Expert Opinion on Therapeutic Targets*, *16*(11), 1113–1125.
238. Delerive, P., De Bosscher, K., Besnard, S., Berghe, W.V., Peters, J.M., Gonzalez, F.J., Fruchart, J.C., Tedgui, A., Haegeman, G. and Staels, B., 1999. Peroxisome proliferator-activated receptor α negatively regulates the vascular inflammatory gene response by negative cross-talk with transcription factors NF-κB and AP-1. *Journal of Biological Chemistry*, *274*(45), 32048–32054.
239. Reaven, G.M., Bernstein, R., Davis, B. and Olefsky, J.M., 1976. Nonketotic diabetes mellitus: Insulin deficiency or insulin resistance? *The American Journal of Medicine*, *60*(1), 80–88.
240. Chen, C.C., Hsu, C.Y., Chen, C.Y., Liu, H.K., 2008. Fructus Corni suppresses hepatic gluconeogenesis related gene transcription, enhances glucose responsiveness of pancreatic beta-cells, and prevents toxin induced beta-cell death. *Journal of Ethnopharmacology*, *117*(3), 483–490.
241. Park, C.H., Tanaka, T. and Yokozawa, T., 2013. Anti-diabetic action of 7-O-galloyl-d-sedoheptulose, a polyphenol from *Corni Fructus*, through ameliorating inflammation and inflammation-related oxidative stress in the pancreas of type 2 diabetics. *Biological and Pharmaceutical Bulletin*, *36*(5), 723–732.

242. Han, Y., Jung, H.W. and Park, Y.K., 2014. Selective therapeutic effect of *Cornus officinalis* fruits on the damage of different organs in STZ-induced diabetic rats. *The American Journal of Chinese Medicine*, *42*(5), 1169–1182.
243. Kim, M.J., Bae, G.S., Jo, I.J., Choi, S.B., Kim, D.G., Shin, J.Y., Lee, S.K. et al., 2015. Loganin protects against pancreatitis by inhibiting NF-κB activation. *European Journal of Pharmacology*, *765*, 541–550.
244. Bhat, Z.A., Ansari, S.H., Mukhtar, H.M., Naved, T., Siddiqui, J.I. and Khan, N.A., 2005. Effect of *Aralia cachemirica* Decne root extracts on blood glucose level in normal and glucose loaded rats. *Die Pharmazie-An International Journal of Pharmaceutical Sciences*, *60*(9), 712–713.
245. George, V., Nigam, S.S. and Rishi, A.K., 1984. Isolation and characterization of aralosides and acids from *Aralia cachemirica*. *Fitoterapia*, *55*, 124–126.
246. Weng, Y., Yu, L., Cui, J., Zhu, Y.R., Guo, C., Wei, G., Duan, J.L. et al., 2014. Antihyperglycemic, hypolipidemic and antioxidant activities of total saponins extracted from *Aralia taibaiensis* in experimental type 2 diabetic rats. *Journal of Ethnopharmacology*, *152*(3), 553–560.
247. Gwiazda, K.S., Yang, T.L.B., Lin, Y. and Johnson, J.D., 2009. Effects of palmitate on ER and cytosolic Ca2+ homeostasis in β-cells. *American Journal of Physiology-Endocrinology and Metabolism*, *296*(4), E690–E701.
248. Zhou, Y., Li, W., Chen, L., Ma, S., Ping, L. and Yang, Z., 2010. Enhancement of intestinal absorption of akebia saponin D by borneol and probenecid in situ and in vitro. *Environmental Toxicology and Pharmacology*, *29*(3), 229–234.
249. Ling, Y., Yong, L. and Chang-Xiao, L., 2006. Metabolism and pharmacokinetics of ginsenosides. *Asian Journal of Pharmacodynamics and Pharmacokinetics*, *6*(2), 103–120.
250. Feng, X.T., Leng, J., Xie, Z., Li, S.L., Zhao, W. and Tang, Q.L., 2012. GPR40: A therapeutic target for mediating insulin secretion (review). *International Journal of Molecular Medicine*, *30*(6), 1261–1266.
251. An, N.Y., Kim, J.E., Hwang, D. and Ryu, H.K., 2014. Anti-diabetic effects of aqueous and ethanol extract of *Dendropanax morbifera* Leveille in streptozotocin-induced diabetes model. *Journal of Nutrition and Health*, *47*(6), 394–402.
252. Moon, H.I., 2011. Antidiabetic effects of dendropanoxide from leaves of *Dendropanax morbifera* Leveille in normal and streptozotocin-induced diabetic rats. *Human & Experimental Toxicology*, *30*(8), 870–875.
253. Dhandapani, S., Subramanian, V.R., Rajagopal, S. and Namasivayam, N., 2002. Hypolipidemic effect of *Cuminum cyminum* L. on alloxan-induced diabetic rats. *Pharmacological Research*, *46*(3), 251–255.
254. Prentki, M., 1996. New insights into pancreatic β-cell metabolic signaling in insulin secretion. *European Journal of Endocrinology*, *134*(3), 272–286.
255. Jagtap, A.G. and Patil, P.B., 2010. Antihyperglycemic activity and inhibition of advanced glycation end product formation by *Cuminum cyminum* in streptozotocin induced diabetic rats. *Food and Chemical Toxicology*, *48*(8), 2030–2036.
256. Patil, S.B., Takalikar, S.S., Joglekar, M.M., Haldavnekar, V.S. and Arvindekar, A.U., 2013. Insulinotropic and β-cell protective action of cuminaldehyde, cuminol and an inhibitor isolated from *Cuminum cyminum* in streptozotocin-induced diabetic rats. *British Journal of Nutrition*, *110*(8), 1434–1443.
257. Shimabukuro, M., Ohneda, M., Lee, Y. and Unger, R.H., 1997. Role of nitric oxide in obesity-induced beta cell disease. *Journal of Clinical Investigation*, *100*(2), 290.
258. Corbett, J.A. and McDaniel, M.L., 1994. Reversibility of interleukin-1β-induced islet destruction and dysfunction by the inhibition of nitric oxide synthase. *Biochemical Journal*, *299*(3), 719–724.
259. Ahmad, M., Akhtar, M.S., Malik, T. and Gilani, A.H., 2000. Hypoglycaemic action of the flavonoid fraction of *Cuminum nigrum* seeds. *Phytotherapy Research*, *14*(2), 103–106.
260. Zare, R., Heshmati, F., Fallahzadeh, H. and Nadjarzadeh, A., 2014. Effect of cumin powder on body composition and lipid profile in overweight and obese women. *Complementary Therapies in Clinical Practice*, *20*(4), 297–301.
261. Park, E.Y., Kim, E.H., Kim, C.Y., Kim, M.H., Choung, J.S., Oh, Y.S., Moon, H.S. and Jun, H.S., 2016. *Angelica dahurica* extracts improve glucose tolerance through the activation of GPR119. *PLoS ONE*, *11*(7), e0158796.
262. Leu, Y.L., Chen, Y.W., Yang, C.Y., Huang, C.F., Lin, G.H., Tsai, K.S., Yang, R.S. and Liu, S.H., 2009. Extract isolated from *Angelica hirsutiflora* with insulin secretagogue activity. *Journal of Ethnopharmacology*, *123*(2), 208–212.
263. Longuet, C., Broca, C., Costes, S., Hani, E.H., Bataille, D. and Dalle, S., 2005. Extracellularly regulated kinases 1/2 (p44/42 mitogen-activated protein kinases) phosphorylate synapsin I andregulate insulin secretion in the MIN6 beta-cell line and islets of Langerhans. *Endocrinology*, *146*(2), 643–654.

264. Khoo, S., Griffen, S.C., Xia, Y., Baer, R.J., German, M.S., Cobb, M.H., 2003. Regulation of insulin gene transcription by ERK1 and ERK2 in pancreatic beta cells. *Journal of Biological Chemistry, 278*(35), 32969–32977.
265. Fujioka, T., Furumi, K., Fujii, H., Okabe, H., Mihashi, K., Nakano, Y., Matsunaga, H., Katano, M. and Mori, M., 1999. Antiproliferative constituents from umbelliferae plants. V. A new furanocoumarin and falcarindiol furanocoumarin ethers from the root of *Angelica japonica. Chemical and Pharmaceutical Bulletin, 47*(1), 96–100.
266. Kuo, P.L., Hsu, Y.L., Chang, C.H. and Chang, J.K., 2005. Osthole-mediated cell differentiation through bone morphogenetic protein-2/p38 and extracellular signal-regulated kinase 1/2 pathway in human osteoblast cells. *Journal of Pharmacology and Experimental Therapeutics, 314*(3), 1290–1299.
267. Eddouks, M., Lemhadri, A. and Michel, J.B., 2004. Caraway and caper: Potential anti-hyperglycaemic plants in diabetic rats. *Journal of Ethnopharmacology, 94*(1), 143–148.
268. Haidari, F., Seyed-Sadjadi, N., Taha-Jalali, M. and Mohammed-Shahi, M., 2011. The effect of oral administration of *Carum carvi* on weight, serum glucose, and lipid profile in streptozotocin-induced diabetic rats. *Saudi Medical Journal, 32*(7), 695–700.
269. Muruganathan, U., Srinivasan, S. and Indumathi, D., 2013. Antihyperglycemic effect of carvone: Effect on the levels of glycoprotein components in streptozotocin-induced diabetic rats. *Journal of Acute Disease, 2*(4), 310–315.
270. Soltani, N., Qiu, H., Aleksic, M., Glinka, Y., Zhao, F., Liu, R., Li, Y. et al., 2011. GABA exerts protective and regenerative effects on islet beta cells and reverses diabetes. *Proceedings of the National Academy of Sciences, 108*(28), 11692–11697.
271. Dong, H., Kumar, M., Zhang, Y., Gyulkhandanyan, A., Xiang, Y.Y., Ye, B., Perrella, J. et al., 2006. Gamma-aminobutyric acid up-and downregulates insulin secretion from beta cells in concert with changes in glucose concentration. *Diabetologia, 49*(4), 697–705.
272. Sánchez-Borzone, M., Delgado-Marin, L. and García, D.A., 2014. Inhibitory effects of carvone isomers on the GABAA receptor in primary cultures of rat cortical neurons. *Chirality, 26*(8), 368–372.
273. Akiba, Y., Kato, S., Katsube, K.I., Nakamura, M., Takeuchi, K., Ishii, H. and Hibi, T., 2004. Transient receptor potential vanilloid subfamily 1 expressed in pancreatic islet β cells modulates insulin secretion in rats. *Biochemical and Biophysical Research Communications, 321*(1), 219–225.
274. Gonçalves, J.C.R., Silveira, A.L., de Souza, H.D., Nery, A.A., Prado, V.F., Prado, M.A., Ulrich, H. and Araujo, D.A., 2013. The monoterpene (–)-carvone: A novel agonist of TRPV1 channels. *Cytometry Part A, 83*(2), 212–219.
275. Yazdanparast, R., Ardestani, A. and Jamshidi, S., 2007. Experimental diabetes treated with *Achillea santolina*: Effect on pancreatic oxidative parameters. *Journal of Ethnopharmacology, 112*(1), 13–18.
276. Kasabri, V., Afifi, F.U. and Hamdan, I., 2011. In vitro and in vivo acute antihyperglycemic effects of five selected indigenous plants from Jordan used in traditional medicine. *Journal of Ethnopharmacology, 133*(2), 888–896.
277. Urmanova, F.F. and Komilov, K.M., 1999. Flavonoids of *Achillea santolina. Chemistry of Natural Compounds, 35*(2), 214–214.
278. Balboul, B.A., Ahmed, A.A., Otsuka, H., Bando, M., Kido, M. and Takeda, Y., 1997. A guaianolide and a germacranolide from *Achillea santolina. Phytochemistry, 46*(6), 1045–1049.
279. Xu, Z., Ju, J., Wang, K., Gu, C. and Feng, Y., 2014. Evaluation of hypoglycemic activity of total lignans from Fructus Arctii in the spontaneously diabetic Goto-Kakizaki rats. *Journal of Ethnopharmacology, 151*(1), 548–555.
280. Xu, Z., Gu, C., Wang, K., Ju, J., Wang, H., Ruan, K. and Feng, Y., 2015. Arctigenic acid, the key substance responsible for the hypoglycemic activity of Fructus Arctii. *Phytomedicine, 22*(1), 128–137.
281. Tousch, D., Bidel, L.P., Cazals, G., Ferrare, K., Leroy, J., Faucanié, M., Chevassus, H., Tournier, M., Lajoix, A.D. and Azay-Milhau, J., 2014. Chemical analysis and antihyperglycemic activity of an original extract from burdock root (*Arctium lappa*). *Journal of Agricultural and Food Chemistry, 62*(31), 7738–7745.
282. Aggarwal, S., Shailendra, G., Ribnicky, D.M., Burk, D., Karki, N. and Wang, M.Q., 2015. An extract of *Artemisia dracunculus* L. stimulates insulin secretion from β cells, activates AMPK and suppresses inflammation. *Journal of Ethnopharmacology, 170*, 98–105.
283. Ubillas, R.P., Mendez, C.D., Jolad, S.D., Luo, J., King, S.R., Carlson, T.J., Fort, D.M., 2000. Antihyperglycemic acetylenic glucosides from *Bidens pilosa. Planta Medica, 66*(1), 82–83.
284. Chien, S.C., Young, P.H., Hsu, Y.J., Chen, C.H., Tien, Y.J., Shiu, S.Y., Li, T.H. Et al., 2009. Anti-diabetic properties of three common *Bidens pilosa* variants in Taiwan. *Phytochemistry, 70*(10), 1246–1254.

285. Chang, C.L., Chang, S.L., Lee, Y.M., Chiang, Y.M., Chuang, D.Y., Kuo, H.K. and Yang, W.C., 2007. Cytopiloyne, a polyacetylenic glucoside, prevents type 1 diabetes in nonobese diabetic mice. *The Journal of Immunology*, (11), 6984–6993.
286. Chang, C.L.-T., Liu, H.-Y., Kuo, T.-F., Hsu, Y.-J., Shen, M.-Y., Pan, C.-Y. and Yang, W.-C., 2013. Antidiabetic effect and mode of action of cytopiloyne. *Evidence-Based Complementary and Alternative Medicine*, 2013, 685642.
287. Arya, A., Cheah, S.C., Looi, C.Y., Taha, H., Mustafa, M.R. and Mohd, M.A., 2012. The methanolic fraction of *Centratherum anthelminticum* seed downregulates pro-inflammatory cytokines, oxidative stress, and hyperglycemia in STZ-nicotinamide-induced type 2 diabetic rats. *Food and Chemical Toxicology*, *50*(11), 4209–4220.
288. Arya, A., Looi, C.Y., Cheah, S.C., Mustafa, M.R. and Mohd, M.A., 2012. Anti-diabetic effects of *Centratherum anthelminticum* seeds methanolic fraction on pancreatic cells, β-TC6 and its alleviating role in type 2 diabetic rats. *Journal of Ethnopharmacology*, *144*(1), 22–32.
289. Onkaramurthy, M., Veerapur, V.P., Thippeswamy, B.S., Reddy, T.M., Rayappa, H. and Badami, S., 2013. Anti-diabetic and anti-cataract effects of *Chromolaena odorata* Linn., in streptozotocin-induced diabetic rats. *Journal of Ethnopharmacology*, *145*(1), 363–372.
290. Phan, T.T., Wang, L., See, P., Grayer, R.J., Chan, S.Y. and Lee, S.T., 2001. Phenolic compounds of *Chromolaena odorata* protect cultured skin cells from oxidative damage: Implication for cutaneous wound healing. *Biological and Pharmaceutical Bulletin*, *24*(12), 1373–1379.
291. Hung, T.M., Cuong, T.D., Dang, N.H., Zhu, S., Long, P.Q., Komatsu, K. and Min, B.S., 2011. Flavonoid glycosides from *Chromolaena odorata* leaves and their in vitro cytotoxic activity. *Chemical and Pharmaceutical Bulletin*, *59*(1), 129–131.
292. Zhang, M.L., Irwin, D., Li, X.N., Sauriol, F., Shi, X.W., Wang, Y.F., Huo, C.H., Li, L.G., Gu, Y.C. and Shi, Q.W., 2012. PPARγ agonist from *Chromolaena odorata*. *Journal of Natural Products*, *75*(12), 2076–2081.
293. Owoyele, V.B., Adediji, J.O. and Soladoye, A.O., 2005. Anti-inflammatory activity of aqueous leaf extract of *Chromolaena odorata*. *Inflammopharmacology*, *13*(5–6), 479–484.
294. Tran, T.V.A., Malainer, C., Schwaiger, S., Hung, T., Atanasov, A.G., Heiss, E.H., Dirsch, V.M. and Stuppner, H., 2015. Screening of Vietnamese medicinal plants for NF-κB signaling inhibitors: Assessing the activity of flavonoids from the stem bark of *Oroxylum indicum*. *Journal of Ethnopharmacology*, *159*, 36–42.
295. Heiss, E.H., Tran, T.V.A., Zimmermann, K., Schwaiger, S., Vouk, C., Mayerhofer, B., Malainer, C., Atanasov, A.G., Stuppner, H. and Dirsch, V.M., 2014. Identification of chromomoric acid CI as an Nrf2 activator in *Chromolaena odorata*. *Journal of Natural Products*, *77*(3), 503–508.
296. Daisy, P., Jasmine, R., Ignacimuthu, S. and Murugan, E., 2009. A novel steroid from *Elephantopus scaber* L. an ethnomedicinal plant with antidiabetic activity. *Phytomedicine*, *16*(2–3), 252–257.
297. Geetha, B.S., Nair, M.S., Latha, P.G. and Remani, P., 2012. Sesquiterpene lactones isolated from *Elephantopus scaber* L. inhibits human lymphocyte proliferation and the growth of tumour cell lines and induces apoptosis in vitro. *Journal of Biomedicine & Biotechnology*, 2012, 721285.
298. Mulakayala, N., Ch, U.R., Iqbal, J. and Pal, M., 2010. Synthesis of dipeptidyl peptidase-4 inhibitors: A brief overview. *Tetrahedron*, *66*(27), 4919–4938.
299. Rosenstock, J. and Zinman, B., 2007. Dipeptidyl peptidase-4 inhibitors and the management of type 2 diabetes mellitus. *Current Opinion in Endocrinology, Diabetes and Obesity*, *14*(2), 98–107.
300. Morikawa, T., Ninomiya, K., Akaki, J., Kakihara, N., Kuramoto, H., Matsumoto, Y., Hayakawa, T. et al., 2015. Dipeptidyl peptidase-IV inhibitory activity of dimeric dihydrochalcone glycosides from flowers of *Helichrysum arenarium*. *Journal of Natural Medicines*, *69*(4), 494–506.
301. Shan, J.J., Yang, M. and Ren, J.W., 2006. Anti-diabetic and hypolipidemic effects of aqueous-extract from the flower of *Inula japonica* in alloxan-induced diabetic mice. *Biological and Pharmaceutical Bulletin*, *29*(3), 455–459.
302. Prabhu, K.S., Lobo, R. and Shirwaikar, A., 2008. Antidiabetic properties of the alcoholic extract of *Sphaeranthus indicus* in streptozotocin-nicotinamide diabetic rats. *Journal of Pharmacy and Pharmacology*, *60*(7), 909–916.
303. Pujar, P.P., Sawaikar, D.D., Rojatkar, S.R. and Nagasampagi, B.A., 2000. Eudesmanoids from *Sphaeranthus indicus*. *Fitoterapia*, *71*(3), 264–268.
304. Tamura, R., Chen, Y., Shinozaki, M., Arao, K., Wang, L., Tang, W., Hirano, S. et al., 2012. Eudesmane-type sesquiterpene lactones inhibit multiple steps in the NF-κB signaling pathway induced by inflammatory cytokines. *Bioorganic & Medicinal Chemistry Letters*, *22*(1), 207–211.

305. Newsholme, P., Haber, E.P., Hirabara, S.M., Rebelato, E.L.O., Procopio, J., Morgan, D., Oliveira-Emilio, H.C., Carpinelli, A.R. and Curi, R., 2007. Diabetes associated cell stress and dysfunction: Role of mitochondrial and non-mitochondrial ROS production and activity. *The Journal of Physiology*, *583*(1), 9–24.
306. Soto, C., Raya, L., Juárez, J., Pérez, J. and González, I., 2014. Effect of Silymarin in Pdx-1 expression and the proliferation of pancreatic β-cells in a pancreatectomy model. *Phytomedicine*, *21*(3), 233–239.
307. Ong, K.W., Hsu, A., Song, L., Huang, D. and Tan, B.K.H., 2011. Polyphenols-rich Vernonia amygdalina shows anti-diabetic effects in streptozotocin-induced diabetic rats. *Journal of Ethnopharmacology*, *133*(2), 598–607.
308. Tousch, D., Lajoix, A.D., Hosy, E., Azay-Milhau, J., Ferrare, K., Jahannault, C., Cros, G. and Petit, P., 2008. Chicoric acid, a new compound able to enhance insulin release and glucose uptake. *Biochemical and Biophysical Research Communications*, *377*(1), 131–135.
309. Singh, H., Kapoor, V.K., Phillipson, J.D. and Bisset, N.G., 1975. Diaboline from *Strychnos potatorum*. *Phytochemistry*, *14*(2), 587–588.
310. Quetin-Leclercq, J., Angenot, L. and Bisset, N.G., 1990. South American *Strychnos* species. Ethnobotany (except curare) and alkaloid screening. *Journal of* Ethnopharmacology, *28*(1), 1–52.
311. John, R., 1978. *Receptors in Pharmacology*. New York: Marcel Dekker.
312. Biswas, A., Chatterjee, S., Chowdhury, R., Sen, S., Sarkar, D.I.P.A.K., Chatterjee, M. and Das, J., 2012. Antidiabetic effect of seeds of *Strychnos potatorum* Linn. in a streptozotocin-induced model of diabetes. *Acta Pol Pharm*, *69*(5), 939–943.
313. Mishra, S.B., Verma, A. and Vijayakumar, M., 2013. Preclinical valuation of anti-hyperglycemic and antioxidant action of Nirmali (*Strychnos potatorum*) seeds in streptozotocin-nicotinamide-induced diabetic Wistar rats: A histopathological investigation. *Biomarkers and Genomic Medicine*, *5*(4), 157–163.
314. Drucker, D.J., 1998. The glucagons-like peptides. *Diabetes*, *47*, 159–169.
315. Adinolfi, M., Corsaro, M.M., Lanzetta, R., Parrilli, M., Folkard, G., Grant, W. and Sutherland, J., 1994. Composition of the coagulant polysaccharide fraction from *Strychnos potatorum* seeds. *Carbohydrate Research*, *263*(1), 103–110.
316. Howarth, N.C., Saltzman, E. and Roberts, S.B., 2001. Dietary fiber and weight regulation. *Nutrition Reviews*, *59*(5), 129–139.
317. Liu, J., Yin, F., Xiao, H., Guo, L. and Gao, X., 2012. Glucagon-like peptide 1 receptor plays an essential role in geniposide attenuating lipotoxicity-induced β-cell apoptosis. *Toxicology in Vitro*, *26*(7), 1093–1097.
318. Guo, L.X., Xia, Z.N., Gao, X., Yin, F. and Liu, J.H., 2012. Glucagon-like peptide 1 receptor plays a critical role in geniposide-regulated insulin secretion in INS-1 cells. *Acta Pharmacologica Sinica*, *33*(2), 237–241.
319. Aburada, M., Takeda, S., Shibata, Y. and Harada, M., 1978. Pharmacological studies of gardenia fruit. III. Relationship between in vivo hydrolysis of geniposide and its choleretic effect in rats. *Journal of Pharmacobio-Dynamics*, *1*(2), 81–88.
320. Sefi, M., Fetoui, H., Lachkar, N., Tahraoui, A., Lyoussi, B., Boudawara, T. and Zeghal, N., 2011. *Centaurium erythrea* (Gentianaceae) leaf extract alleviates streptozotocin-induced oxidative stress and β-cell damage in rat pancreas. *Journal of Ethnopharmacology*, *135*(2), 243–250.
321. Hamza, N., Berke, B., Cheze, C., Marais, S., Lorrain, S., Abdouelfath, A., Lassalle, R., Carles, D., Gin, H. and Moore, N., 2015. Effect of *Centaurium erythraea* Rafn, Artemisia herba-alba Asso and *Trigonella foenum-graecum* L. on liver fat accumulation in C57BL/6J mice with high-fat diet-induced type 2 diabetes. *Journal of Ethnopharmacology*, *171*, 4–11.
322. Kumarasamy, Y., Nahar, L., Cox, P.J., Jaspars, M. and Sarker, S.D., 2003. Bioactivity of secoiridoid glycosides from *Centaurium erythraea*. *Phytomedicine*, *10*(4), 344–347.
323. Kumarasamy, Y., Nahar, L. and Sarker, S.D., 2003a. Bioactivity of gentiopicroside from the aerial parts of *Centaurium erythraea.Fitoterapia*, *74*(1–2), 151–154.
324. Patel, M.B. and Mishra, S.H., 2011. Hypoglycemic activity of C-glycosyl flavonoid from *Enicostemma hyssopifolium*. *Pharmaceutical Biology*, *49*(4), 383–391.
325. Wang, Z., Tang, S., Jin, Y., Zhang, Y., Hattori, M., Zhang, H. and Zhang, N., 2015. Two main metabolites of gentiopicroside detected in rat plasma by LC–TOF-MS following 2, 4-dinitrophenylhydrazine derivatization. *Journal of Pharmaceutical and Biomedical Analysis*, *107*, 1–6.
326. Stefkov, G., Miova, B., Dinevska-Kjovkarovska, S., Stanoeva, J.P., Stefova, M., Petrusevska, G. and Kulevanova, S., 2014. Chemical characterization of *Centaurium erythrea* L. and its effects on carbohydrate and lipid metabolism in experimental diabetes. *Journal of Ethnopharmacology*, *152*(1), 71–77.

327. Hamza, N., Berke, B., Cheze, C., Le Garrec, R., Lassalle, R., Agli, A.N., Robinson, P., Gin, H. and Moore, N., 2011. Treatment of high fat diet induced type 2 diabetes in C57BL/6J mice by two medicinal plants used in traditional treatment of diabetes in the east of Algeria. *Journal of Ethnopharmacology*, *133*(2), 931–933.
328. El-Sedawy, A.I., Shu, Y.Z., Hattori, M., Kobashi, K. and Namba, T., 1989. Metabolism of swertiamarin from *Swertia japonica* by human intestinal bacteria. *Planta Medica*, *55*(2), 147–150.
329. Mahendran, G., Manoj, M., Murugesh, E., Kumar, R.S., Shanmughavel, P., Prasad, K.R. and Bai, V.N., 2014. In vivo anti-diabetic, antioxidant and molecular docking studies of 1, 2, 8-trihydroxy-6-methoxy xanthone and 1, 2-dihydroxy-6-methoxyxanthone-8-O-β-d-xylopyranosyl isolated from *Swertia corymbosa*. *Phytomedicine*, *21*(11), 1237–1248.
330. Wan, L.S., Chen, C.P., Xiao, Z.Q., Wang, Y.L., Min, Q.X., Yue, Y. and Chen, J., 2013. In vitro and in vivo anti-diabetic activity of Swertia kouitchensis extract. *Journal of Ethnopharmacology*, *147*(3), 622–630.
331. Aguilar-Bryan, L., Clement, J.P., Gonzalez, G., Kunjilwar, K., Babenko, A. and Bryan, J., 1998. Toward understanding the assembly and structure of KATP channels. *Physiological Reviews*, *78*(1), 227–245.
332. Jennings, A.M., Wilson, R.M. and Ward, J.D., 1989. Symptomatic hypoglycemia in NIDDM patients treated with oral hypoglycemic agents. *Diabetes Care*, *12*(3), 203–208.
333. Wang, Y.L., Xiao, Z.Q., Liu, S., Wan, L.S., Yue, Y.D., Zhang, Y.T., Liu, Z.X. and Chen, J.C., 2013. Antidiabetic effects of *Swertia macrosperma* extracts in diabetic rats. *Journal of Ethnopharmacology*, *150*(2), 536–544.
334. Rathod, N.R., Chitme, H.R., Irchhaiya, R. and Chandra, R., 2011. Hypoglycemic effect of *Calotropis gigantea* Linn. leaves and flowers in streptozotocin-induced diabetic rats. *Oman Medical Journal*, *26*(2), 104.
335. Saratha, V. and Subramanian, S.P., 2012. Lupeol, a triterpenoid isolated from *Calotropis gigantea* latex ameliorates the primary and secondary complications of FCA induced adjuvant disease in experimental rats. *Inflammopharmacology*, *20*(1), 27–37.
336. Kiuchi, F., Fukao, Y., Maruyama, T., Obata, T., Tanaka, M., Sasaki, T., Mikage, M., Haque, M.E. and Tsuda, Y., 1998. Cytotoxic principles of a Bangladeshi crude drug, akond mul (roots of *Calotropis gigantea* L.). *Chemical and Pharmaceutical Bulletin*, *46*(3), 528–530.
337. Benson, J.M., Seiber, J.N., Keeler, R.F. and Johnson, A.E., 1978. Studies on the toxic principle of *Asclepias eriocarpa* and *Asclepias labriformis* [to range animals]. In Keeler, R.F, Van Kamper, K.R., James, L.F. et al. (Eds.) *Effects of Poisonous Plants on Livestock (Joint United States-Australian Symposium on Poisonous Plants). Logan, Utah, June 19–24, 1977*, p. 273 New York: Academia Press.
338. Neto, M.C., de Vasconcelos, C.F., Thijan, V.N., Caldas, G.F., Araújo, A.V., Costa-Silva, J.H., Amorim, E.L., Ferreira, F., de Oliveira, A.F. and Wanderley, A.G., 2013. Evaluation of antihyperglycaemic activity of *Calotropis procera* leaves extract on streptozotocin-induced diabetes in Wistar rats. *Revista Brasileira de Farmacognosia*, *23*(6), 913–919.
339. Yadav, S.K., Nagori, B.P. and Desai, P.K., 2014. Pharmacological characterization of different fractions of *Calotropis procera* (Asclepiadaceae) in streptozotocin induced experimental model of diabetic neuropathy. *Journal of Ethnopharmacology*, *152*(2), 349–357.
340. Roy, S., Sehgal, R., Padhy, B.M. and Kumar, V.L., 2005. Antioxidant and protective effect of latex of *Calotropis procera* against alloxan-induced diabetes in rats. *Journal of Ethnopharmacology*, *102*(3), 470–473.
341. Ibrahim, S.R., Mohamed, G.A., Shaala, L.A., Moreno, L., Banuls, Y., Kiss, R. and Youssef, D.T.A., 2014. Proceraside A, a new cardiac glycoside from the root barks of *Calotropis procera* with in vitro anticancer effects. *Natural product research*, *28*(17), 1322–1327.
342. Assefa, Z., Akbib, S., Lavens, A., Stangé, G., Ling, Z., Hellemans, K.H. and Pipeleers, D., 2016. Direct effect of glucocorticoids on glucose-activated adult rat β-cells increases their cell number and their functional mass for transplantation. *American Journal of Physiology-Endocrinology and Metabolism*, *311*(4), E698–E705.
343. Itankar, P.R., Lokhande, S.J., Verma, P.R., Arora, S.K., Sahu, R.A. and Patil, A.T., 2011. Antidiabetic potential of unripe *Carissa carandas* Linn. fruit extract. *Journal of Ethnopharmacology*, *135*(2), 430–433.
344. Begum, S., Syed, S.A., Siddiqui, B.S., Sattar, S.A. and Choudhary, M.I., 2013. Carandinol: First isohopane triterpene from the leaves of *Carissa carandas* L. and its cytotoxicity against cancer cell lines. *Phytochemistry Letters*, *6*(1), 91–95.
345. Peungvicha, P., Temsiririrkkul, R., Prasain, J.K., Tezuka, Y., Kadota, S., Thirawarapan, S.S. and Watanabe, H., 1998. 4-Hydroxybenzoic acid: A hypoglycemic constituent of aqueous extract of *Pandanus odorus* root. *Journal of Ethnopharmacology*, *62*(1), 79–84.

346. de Melo, C.L., Queiroz, M.G.R., Arruda, F.A.C.V., Rodrigues, A.M., de Sousa, D.F., Almeida, J.G.L., Pessoa, O.D.L. et al., 2009. Betulinic acid, a natural pentacyclic triterpenoid, prevents abdominal fat accumulation in mice fed a high-fat diet. *Journal of Agricultural and Food Chemistry*, *57*(19), 8776–8781.
347. Singh, S.N., Vats, P., Suri, S., Shyam, R., Kumria, M.M.L., Ranganathan, S. and Sridharan, K., 2001. Effect of an antidiabetic extract of *Catharanthus roseus* on enzymic activities in streptozotocin induced diabetic rats. *Journal of Ethnopharmacology*, *76*(3), 269–277.
348. Yao, X.G., Chen, F., Li, P., Quan, L., Chen, J., Yu, L., Ding, H. et al., 2013. Natural product vindoline stimulates insulin secretion and efficiently ameliorates glucose homeostasis in diabetic murine models. *Journal of Ethnopharmacology*, *150*(1), 285–297.
349. Furman, B.L., 1974. The hypoglycaemic effect of 5-hydroxytryptophan. *British Journal of Pharmacology*, *50*(4), 575–580.
350. Kremmer, T., Holczinger, L. and Somfai, R.S., 1979. Serum lipid lowering effect of vinca alkaloids. *Biochemical Pharmacology*, *28*(2), 227–230.
351. Venkatesh, S., Reddy, G.D., Reddy, B.M., Ramesh, M. and Rao, A.A., 2003. Antihyperglycemic activity of *Caralluma attenuata*. *Fitoterapia*, *74*(3), 274–279.
352. Abdel-Sattar, E.A., Abdallah, H.M., Khedr, A., Abdel-Naim, A.B. and Shehata, I.A., 2013. Antihyperglycemic activity of *Caralluma tuberculata* in streptozotocin-induced diabetic rats. *Food and Chemical toxicology*, *59*, 111–117.
353. Fawzy, G.A., Abdallah, H.M., Marzouk, M.S., Soliman, F.M. and Sleem, A.A., 2008. Antidiabetic and antioxidant activities of major flavonoids of *Cynanchum acutum* L. (*Asclepiadaceae*) growing in Egypt. *Zeitschrift für Naturforschung C*, *63*(9–10), 658–662.
354. Leclerc, I., Woltersdorf, W.W., da Silva Xavier, G., Rowe, R.L., Cross, S.E., Korbutt, G.S., Rajotte, R.V., Smith, R. and Rutter, G.A., 2004. Metformin, but not leptin, regulates AMP-activated protein kinase in pancreatic islets: Impact on glucose-stimulated insulin secretion. *American Journal of Physiology-Endocrinology and Metabolism*, *286*(6), E1023–E1031.
355. Lablanche, S., Cottet-Rousselle, C., Lamarche, F., Benhamou, P.Y., Halimi, S., Leverve, X. and Fontaine, E., 2011. Protection of pancreatic INS-1 β-cells from glucose-and fructose-induced cell death by inhibiting mitochondrial permeability transition with cyclosporin A or metformin. *Cell Death & Disease*, *2*(3), e134.
356. Ramkumar, K.M., Ponmanickam, P., Velayuthaprabhu, S., Archunan, G. and Rajaguru, P., 2009. Protective effect of *Gymnema montanum* against renal damage in experimental diabetic rats. *Food and Chemical Toxicology*, *47*(10), 2516–2521.
357. Ramkumar, K.M., Lee, A.S., Krishnamurthi, K., Devi, S.S., Chakrabarti, T., Kang, K.P., Lee, S. et al., 2009. *Gymnema montanum* H. protects against alloxan-induced oxidative stress and apoptosis in pancreatic β-cells. *Cellular Physiology and Biochemistry*, *24*(5–6), 429–440.
358. Murakami, N., Murakami, T., Kadoya, M., Matsuda, H., Yamahara, J. and Yoshikawa, M., 1996. New hypoglycemic constituents in "gymnemic acid" form *Gymnema Sylvestre*. *Chemical and Pharmaceutical Bulletin*, *44*(2), 469–471.
359. Sugihara, Y., Nojima, H., Matsuda, H., Murakami, T., Yoshikawa, M. and Kimura, I., 2000. Antihyperglycemic effects of gymnemic acid IV, a compound derived from *Gymnema sylvestre* leaves in streptozotocin-diabetic mice. *Journal of Asian Natural Products Research*, *2*(4), 321–327.
360. Daisy, P., Eliza, J. and Farook, K.A.M.M., 2009. A novel dihydroxy gymnemic triacetate isolated from *Gymnema sylvestre* possessing normoglycemic and hypolipidemic activity on STZ-induced diabetic rats. *Journal of Ethnopharmacology*, *126*(2), 339–344.
361. Liu, B., Asare-Anane, H., Al-Romaiyan, A., Huang, G., Amiel, S.A., Jones, P.M. and Persaud, S.J., 2009. Characterisation of the insulinotropic activity of an aqueous extract of *Gymnema sylvestre* in mouse β-cells and human islets of Langerhans. *Cellular Physiology and Biochemistry*, *23*(1–3), 125–132.
362. Nair, S.A., Sabulal, B., Radhika, J., Arunkumar, R. and Subramoniam, A., 2014. Promising anti-diabetes mellitus activity in rats of β-amyrin palmitate isolated from *Hemidesmus indicus* roots. *European Journal of Pharmacology*, *734*, 77–82.
363. Gayathri, M. and Kannabiran, K., 2009. Effect of 2-hydroxy-4-methoxy benzoic acid from the roots of *Hemidesmus indicus* on streptozotocin-induced diabetic rats. *Indian Journal of Pharmaceutical Sciences*, *71*(5), 581.
364. Gayathri, M. and Kannabiran, K., 2010. 2-hydroxy 4-methoxy benzoic acid isolated from roots of *Hemidesmus indicus* ameliorates liver, kidney and pancreas injury due to streptozotocin-induced diabetes in rats. *Indian Journal of Experimental Biology*, *48*(2), 159–164.

365. Venkatesan, N. and Smith, A.G.D.A., 2014. Effect of an active fraction isolated from the leaf extract of *Leptadenia reticulata* on plasma glucose concentration and lipid profile in streptozotocin-induced diabetic rats. *Chinese Journal of Natural Medicines*, *12*(6), 455–460.
366. Krishna, P.V.G., Rao, E.V. and Rao, D.V., 1975. Crystalline principles from the leaves and twigs of *Leptadenia reticulata*. *Planta Medica*, *27*(4), 395–400.
367. Santos, F.A., Frota, J.T., Arruda, B.R., de Melo, T.S., de Castro Brito, G.A., Chaves, M.H. and Rao, V.S., 2012. Antihyperglycemic and hypolipidemic effects of α, β-amyrin, a triterpenoid mixture from *Protium heptaphyllum* in mice. *Lipids in Health and Disease*, *11*(1), 98.
368. Azmi, M.B. and Qureshi, S.A., 2012. Methanolic root extract of Rauwolfia serpentina benth improves the glycemic, antiatherogenic, and cardioprotective indices in alloxan-induced diabetic mice. *Advances in Pharmacological Sciences*, *2012*.
369. Fujii, M., Takei, I. and Umezawa, K., 2009. Antidiabetic effect of orally administered conophylline-containing plant extract on streptozotocin-treated and Goto-Kakizaki rats. *Biomedicine & Pharmacotherapy*, *63*(10), 710–716.
370. Ogata, T., Li, L., Yamada, S., Yamamoto, Y., Tanaka, Y., Takei, I., Umezawa, K. and Kojima, I., 2004. Promotion of β-cell differentiation by conophylline in fetal and neonatal rat pancreas. *Diabetes*, *53*(10), 2596–2602.
371. Shu-Lun, Z., Lax, D., Ying, L., Stejskal, E., Lucas, R.V. and Einzig, S., 1990. Anisodamine increases blood flow to the retina-choroid and protects retinal and pancreatic cells against lipid peroxidation. *Journal of Ethnopharmacology*, *30*(2), 121–134.
372. Poupko, J.M., Baskin, S.I. and Moore, E., 2007. The pharmacological properties of anisodamine. *Journal of Applied Toxicology*, *27*(2), 116–121.
373. Tolan, I., Ragoobirsingh, D. and Morrison, E.S.A., 2001. The effect of capsaicin on blood glucose, plasma insulin levels and insulin binding in dog models. *Phytotherapy Research*, *15*(5), 391–394.
374. Islam, M.S. and Choi, H., 2008. Dietary red chilli (*Capsicum frutescens* L.) is insulinotropic rather than hypoglycemic in type 2 diabetes model of rats. *Phytotherapy Research*, *22*(8), 1025–1029.
375. Ali, M.M. and Agha, F.G., 2009. Amelioration of streptozotocin-induced diabetes mellitus, oxidative stress and dyslipidemia in rats by tomato extract lycopene. *Scandinavian Journal of Clinical and Laboratory Investigation*, *69*(3), 371–379.
376. Murthy, B.K., Nammi, S., Kota, M.K., Rao, R.K., Rao, N.K. and Annapurna, A., 2004. Evaluation of hypoglycemic and antihyperglycemic effects of *Datura metel* (Linn.) seeds in normal and alloxan-induced diabetic rats. *Journal of Ethnopharmacology*, *91*(1), 95–98.
377. Temerdashev, A.Z., Kolychev, I.A. and Kiseleva, N.V., 2012. Chromatographic determination of some tropane alkaloids in *Datura metel*. *Journal of Analytical Chemistry*, *67*(12), 960–966.
378. Gupta, R., Sharma, A.K., Dobhal, M.P., Sharma, M.C. and Gupta, R.S., 2011. Antidiabetic and anti-oxidant potential of β-sitosterol in streptozotocin-induced experimental hyperglycemia. *Journal of Diabetes*, *3*(1), 29–37.
379. Andallu, B. and Radhika, B., 2000. Hypoglycemic, diuretic and hypocholesterolemic effect of winter cherry (Withania somnifera, Dunal) root. *Indian Journal of Experimental Biology*, *38*, 607–609.
380. Takahashi, K., Yoshioka, Y., Kato, E., Katsuki, S., Iida, O., Hosokawa, K. and Kawabata, J., 2010. Methyl caffeate as an α-glucosidase inhibitor from *Solanum torvum* fruits and the activity of related compounds. *Bioscience, Biotechnology, and Biochemistry*, *74*(4), 741–745.
381. Balachandran, C., Emi, N., Arun, Y., Yamamoto, Y., Ahilan, B., Sangeetha, B., Duraipandiyan, V. et al., 2015. In vitro anticancer activity of methyl caffeate isolated from *Solanum torvum* Swartz. fruit. *Chemico-biological interactions*, *242*, 81–90.
382. Takasawa, S. and Okamoto, H., 2002. Pancreatic β-cell death, regeneration and insulin secretion: Roles of poly (ADP-ribose) polymerase and cyclic ADP-ribose. *Journal of Diabetes Research*, *3*(2), 79–96.
383. Yonemura, Y., Takashima, T., Miwa, K., Miyazaki, I., Yamamoto, H. and Okamoto, H., 1984. Amelioration of diabetes mellitus in partially depancreatized rats by poly (ADP-ribose) synthetase inhibitors: Evidence of islet B-cell regeneration. *Diabetes*, *33*(4), 401–404.
384. Anwer, T.A.R.I.Q.U.E., Sharma, M., Pillai, K.K. and Khan, G., 2012. Protective effect of *Withania somnifera* against oxidative stress and pancreatic beta-cell damage in type 2 diabetic rats. *Acta Poloniae Pharmaceutica*, *69*(6), 1095–1101.
385. Anwer, T., Sharma, M., Pillai, K.K. and Iqbal, M., 2008. Effect of *Withania somnifera* on insulin sensitivity in non-insulin-dependent diabetes mellitus rats. *Basic & Clinical Pharmacology & Toxicology*, *102*(6), 498–503.
386. Gorelick, J., Rosenberg, R., Smotrich, A., Hanuš, L. and Bernstein, N., 2015. Hypoglycemic activity of withanolides and elicitated *Withania somnifera*. *Phytochemistry*, *116*, 283–289.

387. Singh, D., Aggarwal, A., Maurya, R. and Naik, S., 2007. *Withania somnifera* inhibits NF-κB and AP-1 transcription factors in human peripheral blood and synovial fluid mononuclear cells. *Phytotherapy Research*, *21*(10), 905–913.
388. Latha, E.H., Satyanarayana, T., Ramesh, A., Prasad, Y.D., Routhu, K.V. and Srinivas, R.L.A., 2008. Hypoglycemic and antihyperglycemic effect of *Argyreia speciosa* Sweet. In normal and in alloxan induced diabetic rats. *Journal of Natural Remedies*, *8*(2), 203–208.
389. Paulke, A., Kremer, C., Wunder, C., Wurglics, M., Schubert-Zsilavecz, M. and Toennes, S.W., 2015. Studies on the alkaloid composition of the Hawaiian Baby Woodrose *Argyreia nervosa*, a common legal high. *Forensic Science International*, *249*, 281–293.
390. Chao, J.M. and Der Marderosian, A.H., 1973. Ergoline alkaloidal constituents of Hawaiian baby wood rose, *Argyreia nervosa* (Burm. f.) Bojer. *Journal of Pharmaceutical Sciences*, *62*(4), 588–591.
391. Sibley, D.R. and Creese, I., 1983. Interactions of ergot alkaloids with anterior pituitary D-2 dopamine receptors. *Molecular Pharmacology*, *23*(3), 585–593.
392. García-Tornadu, I., Ornstein, A.M., Chamson-Reig, A., Wheeler, M.B., Hill, D.J., Arany, E., Rubinstein, M. and Becu-Villalobos, D., 2010. Disruption of the dopamine d2 receptor impairs insulin secretion and causes glucose intolerance. *Endocrinology*, *151*(4), 1441–1450.
393. Nagamine, R., Ueno, S., Tsubata, M., Yamaguchi, K., Takagaki, K., Hira, T., Hara, H. and Tsuda, T., 2014. Dietary sweet potato (*Ipomoea batatas* L.) leaf extract attenuates hyperglycaemia by enhancing the secretion of glucagon-like peptide-1 (GLP-1). *Food & Function*, *5*(9), 2309–2316.
394. Elliott, R.M., Morgan, L.M., Tredger, J.A., Deacon, S., Wright, J. and Marks, V., 1993. Glucagon-like peptide-1 (7–36) amide and glucose-dependent insulinotropic polypeptide secretion in response to nutrient ingestion in man: Acute post-prandial and 24-h secretion patterns. *Journal of Endocrinology*, *138*(1), 159–166.
395. Nauck, M.A., Niedereichholz, U., Ettler, R., Holst, J.J., Ørskov, C., Ritzel, R. and Schmiegel, W.H., 1997. Glucagon-like peptide 1 inhibition of gastric emptying outweighs its insulinotropic effects in healthy humans. *American Journal of Physiology-Endocrinology and Metabolism*, *273*(5), E981–E988.
396. Tang-Christensen, M., Larsen, P.J., Goke, R., Fink-Jensen, A., Jessop, D.S., Moller, M. and Sheikh, S.P., 1996. Central administration of GLP-1-(7-36) amide inhibits food and water intake in rats. *American Journal of Physiology-Regulatory, Integrative and Comparative Physiology*, *271*(4), R848–R856.
397. Xu, G., Stoffers, D.A., Habener, J.F. and Bonner-Weir, S., 1999. Exendin-4 stimulates both beta-cell replication and neogenesis, resulting in increased beta-cell mass and improved glucose tolerance in diabetic rats. *Diabetes*, *48*(12), 2270–2276.
398. Vilsbøll, T., Krarup, T., Sonne, J., Madsbad, S., Vølund, A., Juul, A.G. and Holst, J.J., 2003. Incretin secretion in relation to meal size and body weight in healthy subjects and people with type 1 and type 2 diabetes mellitus. *The Journal of Clinical Endocrinology & Metabolism*, *88*(6), 2706–2713.
399. Matsui, T., Ebuchi, S., Fujise, T., Abesundara, K.J., Doi, S., Yamada, H. and Matsumoto, K., 2004. Strong antihyperglycemic effects of water-soluble fraction of Brazilian propolis and its bioactive constituent, 3, 4, 5-tri-O-caffeoylquinic acid. *Biological and Pharmaceutical Bulletin*, *27*(11), 1797–1803.
400. Hussein, G.M.E., Matsuda, H., Nakamura, S., Hamao, M., Akiyama, T., Tamura, K. and Yoshikawa, M., 2011. Mate tea (*Ilex paraguariensis*) promotes satiety and body weight lowering in mice: Involvement of glucagon-like peptide-1. *Biological and Pharmaceutical Bulletin*, *34*(12), 1849–1855.
401. Ortiz-Andrade, R.R., Garcia-Jimenez, S., Castillo-Espana, P., Ramirez-Avila, G., Villalobos-Molina, R. and Estrada-Soto, S., 2007. α-Glucosidase inhibitory activity of the methanolic extract from *Tournefortia hartwegiana*: An anti-hyperglycemic agent. *Journal of Ethnopharmacology*, *109*(1), 48–53.
402. Andrade-Cetto, A., Revilla-Monsalve, C. and Wiedenfeld, H., 2007. Hypoglycemic effect of *Tournefortia hirsutissima* L., on n-streptozotocin diabetic rats. *Journal of Ethnopharmacology*, 112(1), 96–100.
403. Lin, Y.L., Tsai, Y.L., Kuo, Y.H., Liu, Y.H. and Shiao, M.S., 1999. Phenolic compounds from *Tournefortia sarmentosa*. *Journal of Natural Products*, *62*(11), 1500–1503.
404. Lin, Y.L., Chang, Y.Y., Kuo, Y.H. and Shiao, M.S., 2002. Anti-lipid-peroxidative principles from *Tournefortia s armentosa*. *Journal of Natural Products*, *65*(5), 745–747.
405. Chan, K.W.K. and Ho, W.S., 2015. Anti-oxidative and hepatoprotective effects of lithospermic acid against carbon tetrachloride-induced liver oxidative damage in vitro and in vivo. *Oncology Reports*, *34*(2), 673–680.
406. Amitani, M., Cheng, K.C., Asakawa, A., Amitani, H., Kairupan, T.S., Sameshima, N., Shimizu, T., Hashiguchi, T. and Inui, A., 2015. Allantoin ameliorates chemically-induced pancreatic β-cell damage through activation of the imidazoline I3 receptors. *PeerJ*, *3*, e1105.

407. Head, G.A. and Mayorov, D.N., 2006. Imidazoline receptors, novel agents and therapeutic potential. *Cardiovascular & Hematological Agents in Medicinal Chemistry (Formerly Current Medicinal Chemistry-Cardiovascular & Hematological Agents)*, *4*(1), 17–32.
408. Crowley, C. and Culvenor, C.C.J., 1955. Occurrence of supinine in *Tournefortia sarmentosa* Lam. *Australian Journal of Chemistry*, *8*(3), 464–465.
409. Jin, L., Xue, H.Y., Jin, L.J., Li, S.Y. and Xu, Y.P., 2008. Antioxidant and pancreas-protective effect of aucubin on rats with streptozotocin-induced diabetes. *European Journal of Pharmacology*, *582*(1), 162–167.
410. Padee, P., Nualkaew, S., Talubmook, C. and Sakuljaitrong, S., 2010. Hypoglycemic effect of a leaf extract of *Pseuderanthemum palatiferum* (Nees) Radlk. in normal and streptozotocin-induced diabetic rats. *Journal of Ethnopharmacology*, *132*(2), 491–496.
411. Mai, H.D.T., Minh, H.N.T., Pham, V.C., Bui, K.N. and Chau, V.M., 2011. Lignans and other constituents from the roots of the Vietnamese medicinal plant *Pseuderanthemum palatiferum*. *Planta Medica*, *77*(09), 951–954.
412. Sittisart, P., Chitsomboon, B. and Kaminski, N.E., 2016. *Pseuderanthemum palatiferum* leaf extract inhibits the proinflammatory cytokines, TNF-α and IL-6 expression in LPS-activated macrophages. *Food and Chemical Toxicology*, *97*, 11–22.
413. Phakeovilay, C., Disadee, W., Sahakitpichan, P., Sitthimonchai, S., Kittakoop, P., Ruchirawat, S. and Kanchanapoom, T., 2013. Phenylethanoid and flavone glycosides from *Ruellia tuberosa* L. *Journal of Natural Medicines*, *67*(1), 228–233.
414. Adebajo, A.C., Olawode, E.O., Omobuwajo, O.R., Adesanya, S.A., Begrow, F., Elkhawad, A., Akanmu, M.A. et al., 2007. Hypoglycaemic constituents of *Stachytarpheta cayennensis* leaf. *Planta Medica*, *73*(03), 241–250.
415. Attanayake, A.P., Jayatilaka, K.A.P.W., Pathirana, C. and Mudduwa, L.K.B., 2016. *Gmelina arborea* Roxb. (Family: Verbenaceae) extract upregulates the β-cell regeneration in STZ induced diabetic rats. *Journal of Diabetes Research, 2016*.
416. Falah, S., Katayama, T. and Suzuki, T., 2008. Chemical constituents from *Gmelina arborea* bark and their antioxidant activity. *Journal of Wood Science*, *54*(6), 483–489.
417. Tiwari, N., Yadav, A.K., Srivastava, P., Shanker, K., Verma, R.K. and Gupta, M.M., 2008. Iridoid glycosides from *Gmelina arborea*. *Phytochemistry*, *69*(12), 2387–2390.
418. Kasiviswanath, R., Ramesh, A. and Kumar, K.E., 2005. Hypoglycemic and antihyperglycemic effect of *Gmelina asiatica* Linn. in normal and in alloxan induced diabetic rats. *Biological and Pharmaceutical Bulletin*, *28*(4), 729–732.
419. Jothi, G., Ramachandran, D. and Brindha, P., 2008. Antidiabetic potential of *Gmelina asiatica* Linn. on alloxan induced diabetic rats. *Biomedicine*, *28*(4), 278–283.
420. Nair, A.R. and Subramanian, S.S., 1975. Quercetagetin and other flavones from *Gmelina arborea* and G. *asiatica*. *Phytochemistry*, *14*(4), 1135–1136.
421. Baek, S., Kang, N.J., Popowicz, G.M., Arciniega, M., Jung, S.K., Byun, S., Song, N.R. et al., 2013. Structural and functional analysis of the natural JNK1 inhibitor quercetagetin. *Journal of Molecular Biology*, *425*(2), 411–423.
422. Kaneto, H., Xu, G., Fujii, N., Kim, S., Bonner-Weir, S. and Weir, G.C., 2002. Involvement of c-Jun N-terminal kinase in oxidative stress-mediated suppression of insulin gene expression. *Journal of Biological Chemistry*, *277*(33), 30010–30018.
423. Seawright, A.A., 1964. Studies on the pathology of experimental lantana (*Lantana camara* L.) poisoning of sheep. *Pathologia Veterinaria Online*, *1*(6), 504–529.
424. Kazmi, I., Rahman, M., Afzal, M., Gupta, G., Saleem, S., Afzal, O., Shaharyar, M.A., Nautiyal, U., Ahmed, S. and Anwar, F., 2012. Anti-diabetic potential of ursolic acid stearoyl glucoside: A new triterpenic gycosidic ester from *Lantana camara*. *Fitoterapia*, *83*(1), 142–146.
425. Ramachandran, S. and Rajasekaran, A., 2014. Blood glucose-lowering effect of Tectona grandis flowers in type 2 diabetic rats: A study on identification of active constituents and mechanisms for antidiabetic action. *Journal of Diabetes*, *6*(5), 427–437.
426. Roengsumran, S., Sookkongwaree, K., Jaiboon, N., Chaichit, N. and Petsom, A., 2002. Crystal structure of ipolamiide monohydrate from *Stachytarpheta indica*. *Analytical sciences*, *18*(9), 1063–1064.
427. Melita, R.S. and Castro, O., 1996. Pharmacological and chemical evaluation of *Stachytarpheta jamaicensis* (Verbenaceae). *Revista de Biologia Tropical*, *44*(2A), 353–359.
428. Sundaram, R., Naresh, R., Shanthi, P. and Sachdanandam, P., 2012. Antihyperglycemic effect of iridoid glucoside, isolated from the leaves of *Vitex negundo* in streptozotocin-induced diabetic rats with special reference to glycoprotein components. *Phytomedicine*, *19*(3), 211–216.

429. Boudjelal, A., Henchiri, C., Siracusa, L., Sari, M. and Ruberto, G., 2012. Compositional analysis and in vivo anti-diabetic activity of wild Algerian *Marrubium vulgare* L. infusion. *Fitoterapia*, *83*(2), 286–292.
430. Nicholas, H.J., 1964. Isolation of marrubiin, a sterol, and a sesquiterpene from *Marrubium vulgare*. *Journal of Pharmaceutical Sciences*, *53*(8), 895–899.
431. El Bardai, S., Wibo, M., Hamaide, M.C., Lyoussi, B., Quetin-Leclercq, J. and Morel, N., 2003. Characterisation of marrubenol, a diterpene extracted from *Marrubium vulgare*, as an L-type calcium channel blocker. *British Journal of Pharmacology*, *140*(7), 1211–1216.
432. El-Mohsen, M.A., Rabeh, M.A., El-Hefnawi, M., Abou-Setta, L., Elgarf, I., El-Rashedy, A. and Hussein, A.A., 2014. Marrubiin: A potent α-glucosidase inhibitor from *Marrubium alysson*. *International Journal of Applied Research in Natural Products*, *7*(1), 21–27.
433. Mnonopi, N., Levendal, R.A., Mzilikazi, N. and Frost, C.L., 2012. Marrubiin, a constituent of *Leonotis leonurus*, alleviates diabetic symptoms. *Phytomedicine*, *19*(6), 488–493.
434. El Bardai, S., Morel, N., Wibo, M., Fabre, N., Llabres, G., Lyoussi, B. and Quetin-Leclercq, J., 2003. The vasorelaxant activity of marrubenol and marrubiin from *Marrubium vulgare*. *Planta Medica*, *69*(01), 75–77.
435. Nyarko, A.K., Asare-Anane, H., Ofosuhene, M., Addy, M.E., Teye, K. and Addo, P., 2002. Aqueous extract of *Ocimum canum* decreases levels of fasting blood glucose and free radicals and increases anti-atherogenic lipid levels in mice. *Vascular Pharmacology*, *39*(6), 273–279.
436. Nyarko, A.K., Asare-Anane, H., Ofosuhene, M. and Addy, M.E., 2002. Extract of Ocimum canum lowers blood glucose and facilitates insulin release by isolated pancreatic β-islet cells. *Phytomedicine*, *9*(4), 346–351.
437. Weaver, D.K., Dunkel, F.V., Ntezurubanza, L., Jackson, L.L. and Stock, D.T., 1991. The efficacy of linalool, a major component of freshly-milled *Ocimum canum* Sims (Lamiaceae), for protection against postharvest damage by certain stored product *Coleoptera*. *Journal of Stored Products Research*, *27*(4), 213–220.
438. Xaasan, C.C., Cabdulraxmaan, A.D., Passannanti, S., Piozzi, F. and Schmid, J.P., 1981. Constituents of the essential oil of *Ocimum canum*. *Journal of Natural Products*, *44*(6), 752–753.
439. Deepa, B. and Anuradha, C.V., 2011. Linalool, a plant derived monoterpene alcohol, rescues kidney from diabetes-induced nephropathic changes via blood glucose reduction. *Diabetologia Croatica*, *40*, 4.
440. Venâncio, A.M., Marchioro, M., Estavam, C.S., Melo, M.S., Santana, M.T., Onofre, A.S., Guimarães, A.G. et al., 2011. Ocimum basilicum leaf essential oil and (-)-linalool reduce orofacial nociception in rodents: A behavioral and electrophysiological approach. *Revista Brasileira de Farmacognosia*, *21*(6), 1043–1051.
441. Bertrand, G., Gross, R., Puech, R., Loubatières-Mariani, M.M. and Bockaert, J., 1992. Evidence for a glutamate receptor of the AMPA subtype which mediates insulin release from rat perfused pancreas. *British Journal of Pharmacology*, *106*(2), 354–359.
442. Peana, A.T., D'Aquila, P.S., Panin, F., Serra, G., Pippia, P. and Moretti, M.D.L., 2002. Anti-inflammatory activity of linalool and linalyl acetate constituents of essential oils. *Phytomedicine*, *9*(8), 721–726.
443. Uchida, K. and Tominaga, M., 2011. The role of thermosensitive TRP (transient receptor potential) channels in insulin secretion [Review]. *Endocrine Journal*, *58*(12), 1021–1028.
444. Kar, A., Choudhary, B.K. and Bandyopadhyay, N.G., 2003. Comparative evaluation of hypoglycaemic activity of some Indian medicinal plants in alloxan diabetic rats. *Journal of Ethnopharmacology*, *84*(1), 105–108.
445. Grover, J.K., Vats, V. and Yadav, S.S., 2005. Pterocarpus marsupium extract (Vijayasar) prevented the alteration in metabolic patterns induced in the normal rat by feeding an adequate diet containing fructose as sole carbohydrate. *Diabetes, Obesity and Metabolism*, *7*(4), 414–420.
446. Hakkim, F.L., Shankar, C.G. and Girija, S., 2007. Chemical composition and antioxidant property of holy basil (*Ocimum sanctum* L.) leaves, stems, and inflorescence and their in vitro callus cultures. *Journal of Agricultural and Food Chemistry*, *55*(22), 9109–9117.
447. Hannan, J.M.A., Marenah, L., Ali, L., Rokeya, B., Flatt, P.R. and Abdel-Wahab, Y.H.A., 2006. *Ocimum sanctum* leaf extracts stimulate insulin secretion from perfused pancreas, isolated islets and clonal pancreatic β-cells. *Journal of Endocrinology*, *189*(1), 127–136.
448. Srinivasan, S., Sathish, G., Jayanthi, M., Muthukumaran, J., Muruganathan, U. and Ramachandran, V., 2014. Ameliorating effect of eugenol on hyperglycemia by attenuating the key enzymes of glucose metabolism in streptozotocin-induced diabetic rats. *Molecular and Cellular Biochemistry*, *385*(1–2), 159–168.
449. Yang, B.H., Piao, Z.G., Kim, Y.B., Lee, C.H., Lee, J.K., Park, K., Kim, J.S. and Oh, S.B., 2003. Activation of vanilloid receptor 1 (VR1) by eugenol. *Journal of Dental Research*, *82*(10), 781–785.
450. Sriplang, K., Adisakwattana, S., Rungsipipat, A. and Yibchok-Anun, S., 2007. Effects of *Orthosiphon stamineus* aqueous extract on plasma glucose concentration and lipid profile in normal and streptozotocin-induced diabetic rats. *Journal of Ethnopharmacology*, *109*(3), 510–514.

451. Lee, H.J., Choi, Y.J., Park, S.Y., Kim, J.Y., Won, K.C., Son, J.K. and Kim, Y.W., 2015. Hexane extract of *Orthosiphon stamineus* induces insulin expression and prevents glucotoxicity in INS-1 cells. *Diabetes & Metabolism Journal*, *39*(1), 51–58.
452. Watanabe, H., Saito, H., Nishimura, H., Ueda, J. and Evers, B.M., 2008. Activation of phosphatidylinositol-3 kinase regulates pancreatic duodenal homeobox-1 in duct cells during pancreatic regeneration. *Pancreas*, *36*(2), 153.
453. Rafiq, I., da Silva Xavier, G., Hooper, S. and Rutter, G.A., 2000. Glucose-stimulated preproinsulin gene expression and nucleartrans-location of pancreatic duodenum homeobox-1 require activation of phosphatidylinositol 3-kinase but not p38 MAPK/SAPK2. *Journal of Biological Chemistry*, *275*(21), 15977–15984.
454. Yuliana, N.D., Khatib, A., Link-Struensee, A.M.R., Ijzerman, A.P., Rungkat-Zakaria, F., Choi, Y.H. and Verpoorte, R., 2009. Adenosine A1 receptor binding activity of methoxy flavonoids from *Orthosiphon stamineus*. *Planta Medica*, *75*(2), 132–136.
455. Johansson, S.M., Salehi, A., Sandström, M.E., Westerblad, H., Lundquist, I., Carlsson, P.O., Fredholm, B.B. and Katz, A., 2007. A 1 receptor deficiency causes increased insulin and glucagon secretion in mice. *Biochemical Pharmacology*, *74*(11), 1628–1635.
456. Awale, S., Tezuka, Y., Banskota, A.H., Adnyana, I.K. and Kadota, S., 2003. Nitric oxide inhibitory isopimarane-type diterpenes from *Orthosiphon stamineus* of Indonesia. *Journal of Natural Products*, *66*(2), 255–258.
457. Yasukawa, T., Tokunaga, E., Ota, H., Sugita, H., Martyn, J.J. and Kaneki, M., 2005. S-nitrosylation-dependent inactivation of Akt/protein kinase B in insulin resistance. *Journal of Biological Chemistry*, *280*(9), 7511–7518.
458. Ramadan, K.S., Khalil, O.A., Danial, E.N., Alnahdi, H.S. and Ayaz, N.O., 2013. Hypoglycemic and hepatoprotective activity of *Rosmarinus officinalis* extract in diabetic rats. *Journal of Physiology and Biochemistry*, *69*(4), 779–783.
459. Govindaraj, J. and Pillai, S.S., 2015. Rosmarinic acid modulates the antioxidant status and protects pancreatic tissues from glucolipotoxicity mediated oxidative stress in high-fat diet: streptozotocin-induced diabetic rats. *Molecular and Cellular Biochemistry*, *404*(1–2), 143–159.
460. Wolff, S.P., 1993. Diabetes mellitus and free radicals, transition metals and oxidative stress in the aetiology of diabetes mellitus and complications. *British Medical Bulletin*, *49*(3), 642–652.
461. Poitout, V., Amyot, J., Semache, M., Zarrouki, B., Hagman, D. and Fontés, G., 2010. Glucolipotoxicity of the pancreatic beta cell. *Biochimica et Biophysica Acta (BBA)-Molecular and Cell Biology of Lipids*, *1801*(3), 289–298.
462. Shimabukuro, M., Zhou, Y.T., Levi, M. and Unger, R.H., 1998. Fatty acid-induced β cell apoptosis: A link between obesity and diabetes. *Proceedings of the National Academy of Sciences*, *95*(5), 2498–2502.
463. Sözmen, E.Y., Sözmen, B., Delen, Y. and Onat, T., 2001. Catalase/superoxide dismutase (SOD) and catalase/paraoxonase (PON) ratios may implicate poor glycemic control. *Archives of Medical Research*, *32*(4), 283–287.
464. Fu, J., Woods, C.G., Yehuda-Shnaidman, E., Zhang, Q., Wong, V., Collins, S., Sun, G., Andersen, M.E. and Pi, J., 2010. Low-level arsenic impairs glucose-stimulated insulin secretion in pancreatic beta cells: Involvement of cellular adaptive response to oxidative stress. *Environmental Health Perspectives*, *118*(6), 864.
465. Chen, C.C., Chen, H.L., Hsieh, C.W., Yang, Y.L. and Wung, B.S., 2011. Upregulation of NF-E2-related factor-2-dependent glutathione by carnosol provokes a cytoprotective response and enhances cell survival. *Acta Pharmacologica Sinica*, *32*(1), 62–69.
466. Esmaeili, M.A. and Yazdanparast, R., 2004. Hypoglycaemic effect of *Teucrium polium*: Studies with rat pancreatic islets. *Journal of Ethnopharmacology*, *95*(1), 27–30.
467. Ardestani, A., Yazdanparast, R. and Jamshidi, S.H., 2008. Therapeutic effects of *Teucrium polium* extract on oxidative stress in pancreas of streptozotocin-induced diabetic rats. *Journal of Medicinal Food*, *11*(3), 525–532.
468. Verykokidou-Vitsaropoulou, E. and Vajias, C., 1986. Methylated flavones from *Teucrium polium*. *Planta Medica*, *52*(05), 401–402.
469. Sharififar, F., Dehghn-Nudeh, G. and Mirtajaldini, M., 2009. Major flavonoids with antioxidant activity from *Teucrium polium* L. *Food Chemistry*, *112*(4), 885–888.
470. Esmaeili, M.A., Zohari, F. and Sadeghi, H., 2009. Antioxidant and protective effects of major flavonoids from *Teucrium polium* on β-cell destruction in a model of streptozotocin-induced diabetes. *Planta Medica*, *75*(13), 1418–1420.

471. Si, M.M., Lou, J.S., Zhou, C.X., Shen, J.N., Wu, H.H., Yang, B., He, Q.J. and Wu, H.S., 2010. Insulin releasing and alpha-glucosidase inhibitory activity of ethyl acetate fraction of Acorus calamus in vitro and in vivo. *Journal of Ethnopharmacology*, *128*(1), 154–159.
472. Lee, M.H., Chen, Y.Y., Tsai, J.W., Wang, S.C., Watanabe, T. and Tsai, Y.C., 2011. Inhibitory effect of β-asarone, a component of *Acorus calamus* essential oil, on inhibition of adipogenesis in 3T3-L1 cells. *Food Chemistry*, *126*(1), 1–7.
473. Tanaka, M., Misawa, E., Ito, Y., Habara, N., Nomaguchi, K., Yamada, M., Toida, T. et al., 2006. Identification of five phytosterols from Aloe vera gel as anti-diabetic compounds. *Biological and Pharmaceutical Bulletin*, *29*(7), 1418–1422.
474. Zhang, Y., Cai, J., Ruan, H., Pi, H. and Wu, J., 2007. Antihyperglycemic activity of kinsenoside, a high yielding constituent from *Anoectochilus roxburghii* in streptozotocin diabetic rats. *Journal of Ethnopharmacology*, *114*(2), 141–145.
475. Kianbakht, S. and Hajiaghaee, R., 2011. Anti-hyperglycemic effects of saffron and its active constituents, crocin and safranal, in alloxan-induced diabetic rats. *Journal of Medicinal Plants*, *3*(39), 82–89.
476. Wu, C., Li, Y., Chen, Y., Lao, X., Sheng, L., Dai, R., Meng, W. and Deng, Y., 2011. Hypoglycemic effect of *Belamcanda chinensis* leaf extract in normal and STZ-induced diabetic rats and its potential active faction. *Phytomedicine*, *18*(4), 292–297.
477. Kumar, R., Balhuizen, A., Amisten, S., Lundquist, I. and Salehi, A., 2011. Insulinotropic and antidiabetic effects of 17β-estradiol and the GPR30 agonist G-1 on human pancreatic islets. *Endocrinology*, *152*(7), 2568–2579.
478. Martensson, U.E., Salehi, S.A., Windahl, S., Gomez, M.F., Sward, K., Daszkiewicz-Nilsson, J., Wendt, A. et al., 2009. Deletion of the G protein-coupled receptor 30 impairs glucose tolerance, reduces bone growth, increases blood pressure, and eliminates estradiol-stimulated insulin release in female mice. *Endocrinology*, *150*(2), 687–698.
479. Nadal, A., Rovira, J.M., Laribi, O., Leon-quinto, T., Andreu, E., Ripoll, C. and Soria, B., 1998. Rapid insulinotropic effect of 17β-estradiol via a plasma membrane receptor. *The FASEB Journal*, *12*(13), 1341–1348.
480. Al-Majed, H.T., Squires, P.E., Persaud, S.J., Huang, G.C.C., Amiel, S., Whitehouse, B.J. and Jones, P.M., 2005. Effect of 17β-estradiol on insulin secretion and cytosolic calcium in MIN6 mouse insulinoma cells and human islets of Langerhans. *Pancreas*, *30*(4), 307–313.
481. Mathews, J.N., Flatt, P.R. and Abdel-Wahab, Y.H., 2006. *Asparagus adscendens* (*Shweta musali*) stimulates insulin secretion, insulin action and inhibits starch digestion. *British Journal of Nutrition*, *95*(03), 576–581.
482. Jadhav, A.N. and Bhutani, K.K., 2006. Steroidal saponins from the roots of *Asparagus adscendens* Roxb and *Asparagus racemosus* Willd. *Indian Journal of Chemistry*, *45B*, 1515–1524.
483. Sharma, S.C., Chand, R. and Sati, O.P., 1980. Steroidal sapogenins from the fruits of *Asparagus adscendens* roxb. *Pharmazie*, *35*(11), 711–712.
484. Yen, M.L., Su, J.L., Chien, C.L., Tseng, K.W., Yang, C.Y., Chen, W.F., Chang, C.C. and Kuo, M.L., 2005. Diosgenin induces hypoxia-inducible factor-1 activation and angiogenesis through estrogen receptor-related phosphatidylinositol 3-kinase/Akt and p38 mitogen-activated protein kinase pathways in osteoblasts. *Molecular Pharmacology*, *68*(4), 1061–1073.
485. Hannan, J.M.A., Marenah, L., Ali, L., Rokeya, B., Flatt, P.R. and Abdel-Wahab, Y.H., 2007. Insulin secretory actions of extracts of *Asparagus racemosus* root in perfused pancreas, isolated islets and clonal pancreatic β-cells. *Journal of Endocrinology*, *192*(1), 159–168.
486. Hayes, P.Y., Jahidin, A.H., Lehmann, R., Penman, K., Kitching, W. and De Voss, J.J., 2008. Steroidal saponins from the roots of *Asparagus racemosus*. *Phytochemistry*, *69*(3), 796–804.
487. Somani, R., Singhai, A.K., Shivgunde, P. and Jain, D., 2012. *Asparagus racemosus* Willd (Liliaceae) ameliorates early diabetic nephropathy in STZ induced diabetic rats. *Indian Journal of Experimental Biology*, *50*(7), 469–475.
488. Hannan, J.M.A., Ali, L., Khaleque, J., Akhter, M., Flatt, P.R. and Abdel-Wahab, Y.H., 2012. Antihyperglycaemic activity of *Asparagus racemosus* roots is partly mediated by inhibition of carbohydrate digestion and absorption, and enhancement of cellular insulin action. *British Journal of Nutrition*, *107*(9), 1316–1323.
489. Subramanian, S.S. and Nair, A.G.R., 1969. Occurrence of diosgenin in *Asparagus racemosus* leaves. *Current Science*, *17*, 414.
490. Son, I.S., Kim, J.H., Sohn, H.Y., Son, K.H., Kim, J.S. and Kwon, C.S., 2007. Antioxidative and hypolipidemic effects of diosgenin, a steroidal saponin of yam (Dioscorea spp.), on high-cholesterol fed rats. *Bioscience, Biotechnology, and Biochemistry*, *71*(12), 3063–3071.

491. Zhu, X., Zhang, W., Pang, X., Wang, J., Zhao, J. and Qu, W., 2011. Hypolipidemic effect of n-butanol extract from *Asparagus officinalis* L. in mice fed a high-fat diet. *Phytotherapy Research*, *25*(8), 1119–1124.
492. Hafizur, R.M., Kabir, N. and Chishti, S., 2012. *Asparagus officinalis* extract controls blood glucose by improving insulin secretion and β-cell function in streptozotocin-induced type 2 diabetic rats. *British Journal of Nutrition*, *108*(9), 1586–1595.
493. Kwon, K.B., Kim, J.H., Lee, Y.R., Lee, H.Y., Jeong, Y.J., Rho, H.W., Ryu, D.G., Park, J.W. and Park, B.H., 2003. *Amomum xanthoides* extract prevents cytokine-induced cell death of RINm5F cells through the inhibition of nitric oxide formation. *Life Sciences*, *73*(2), 181–191.
494. Lee, J.H., Park, J.W., Kim, J.S., Park, B.H. and Rho, H.W., 2008. Protective effect of Amomi semen extract on alloxan-induced pancreatic β-cell damage. Phytotherapy Research, *22*(1), 86–90.
495. Heitmeier, M.R., Scarim, A.L. and Corbett, J.A., 1997. Interferon-γ increases the sensitivity of islets of Langerhans for inducible nitric-oxide synthase expression induced by interleukin 1. *Journal of Biological Chemistry*, *272*(21), 13697–13704.
496. Welsh, N., Eizirik, D.L., Bendtzen, K. and Sandler, S., 1991. Interleukin-l β-induced nitric oxide production in isolated rat pancreatic islets requires gene transcription and may lead to inhibition of the krebs cycle enzyme aconitase. *Endocrinology*, *129*(6), 3167–3173.
497. Kitajima, J. and Ishikawa, T., 2003. Water-soluble constituents of *Amomum* seed. Chemical and *Pharmaceutical Bulletin*, *51*(7), 890–893.
498. Matsunaga, K., Shibuya, M. and Ohizumi, Y., 1994. Graminone B, a novel lignan with vasodilative activity from *Imperata cylindrica*. *Journal of Natural Products*, *57*(12), 1734–1736.
499. Matsunaga, K., Shibuya, M. and Ohizumi, Y., 1994. Cylindrene, a novel sesquiterpenoid from *Imperata clylindrica* with inhibitory activity on contractions of vascular smooth muscle. Journal of *Natural Products*, *57*(8), 1183–1184.
500. Peterhoff, M., Sieg, A., Brede, M., Chao, C.M., Hein, L. and Ullrich, S., 2003. Inhibition of insulin secretion via distinct signaling pathways in alpha2-adrenoceptor knockout mice. *European Journal of Endocrinology*, *149*(4), 343–350.

3 Inhibiting Insulin Resistance and Accumulation of Triglycerides and Cholesterol in the Liver

Visceral obesity favors the generation of reactive oxygen species, plasmatic nonesterified fatty acids, tumor necrosis factor-α, and interleukin-6 that act synergistically to bring about hepatic insulin resistance.[1] Insulin resistance evokes a subnormal hepatic storage of glucose into glycogen and increases glucose production by the liver from glycogen, as well as de novo synthesis of glucose (gluconeogenesis) accounting for a fasting glycaemia, above 6.1 mmol/L (110 mg/dL) and below 6.9 mmol/L (125 mg/dL).[2] In a state of insulin resistance, insulin is unable to suppress lipolysis in adipocytes and to activate adipose tissues endothelial lipoprotein lipase resulting in increased plasma nonesterified fatty acid supply to the liver that translates into increased hepatic production of very low-density lipoprotein levels and atherogenic hyperlipidemia.[3–5] According to the National Cholesterol Education Program Adult Treatment Panel (ATP) III definition, metabolic syndrome will be present if at least three of the following criteria are met: waist circumference more than 40 inches (men) or 35 inches (women), blood pressure more than 130/85 mmHg, fasting triglyceride level superior to 150 mg/dl, fasting high-density lipoprotein–cholesterol below 40 mg/dL (men) or 50 mg/dL (women), and fasting blood sugar above 100 mg/dL.[6] Excess of plasma glucose and triglycerides in metabolic syndrome result is fatty acid accumulation in the liver disrupting hepatocytes function leading to nonalcoholic steatohepatitis.[7] Thus, inhibiting insulin resistance and accumulation of triglycerides and cholesterol in the liver with natural products constitute one therapeutic strategy to prevent or manage insulin resistance in metabolic syndrome.

3.1 *Myristica fragrans* Hout.

Synonyms: *Myristica aromatica* Lam.; *Myristica moschata* Thunb.; *Myristica officinalis* L.f.
Common names: buah pala (Malay); ru du ku (Chinese); nutmeg
Subclass Magnoliidae, Superorder Magnolianae, Order Myristicales, Family Myristicaceae
Medicinal use: facilitate digestion (Malaysia)
Pharmacological targets: atherogenic hyperlipidemia; insulin resistance

Macelignan from *Myristica fragrans* Hout. at a concentration of 5 μM protected HepG2 cells against tert-butyl hydroperoxide increasing their viability by 91.2% as evidenced by a decrease in reactive oxygen species and malondialdehyde, which is a marker of lipid peroxidation.[8] Insulin binding to its hepatic receptor stimulates the expression of the transcription factor sterol regulatory element-binding protein-1c in the liver mediating most of insulin effects on fatty acid synthesis.[9] *Meso*-dihydroguaiaretic acid (Figure 3.1) from this plant at a concentration of 10 μM repressed the transcription factor sterol regulatory element-binding protein-1c and consequently fatty acid synthetase and acetyl-CoA carboxylase in HepG2 cells.[10] Furthermore, this lignan reduced by more than 50% of triglyceride accumulation in HepG2 cells pretreated with insulin.[10] Simultaneously, this lignan induced the expression of peroxisome proliferator-activated receptor-α and downstream carnitine palmitoyltransferase-1 and uncoupling protein-2,[10] which catalyze fatty acids oxidation in

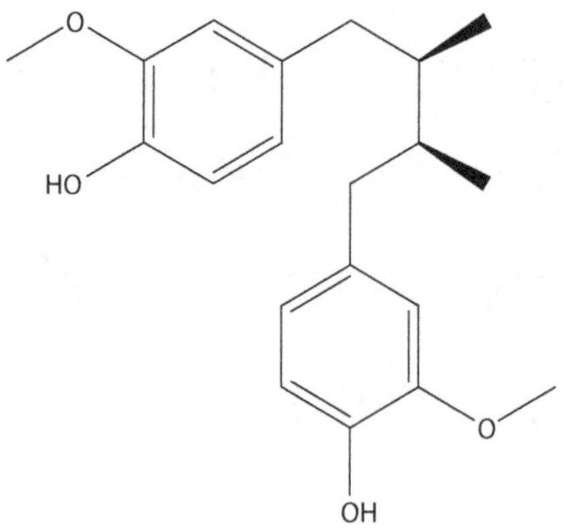

FIGURE 3.1 Meso-dihydroguaiaretic acid.

the liver.[11] In hepatocytes, protein tyrosine phosphatase 1B is a negative regulator in insulin signal transduction by dephosphorylating the activated insulin receptor or insulin receptor substrates.[12] Meso-dihydroguaiaretic acid and otobaphenol inhibited the enzymatic activity of this enzyme with IC_{50} values equal to 19.6 and 48.9 mM, respectively,[13] implying increased insulin sensitivity. In 32D cells, meso-dihydroguaiaretic acid at a concentration of 10 μM enhanced the phosphorylation of insulin receptor tyrosine resulting from insulin binding and increased insulin sensitivity.[13] For every 30 mg/dL, reduction in plasmatic low-density lipoprotein, the relative risk of developing cardiovascular diseases is lowered by approximately 30%.[14] *Myristica fragrans* tetrahydrofuran lignans mixture given orally to C57BL/6 at a dose of 200 mg/kg/day poisoned with a high-fat diet for 6 weeks resulted in a 30% reduction of epididymis fat compared with untreated animals, a mild reduction in food intake, low-density lipoprotein–cholesterol, cholesterol, and glyceamia.[15] Adenosine monophosphate-activated protein kinase is a heterotrimeric protein consisting of a catalytic subunit (α) and 2 noncatalytic subunits (β and γ). In response to elevated AMP/ATP (State of energy deprivation) ratios, adenosine monophosphate-activated protein kinase is phosphorylated in the α-subunit.[16] Activated adenosine monophosphate-activated protein kinase phosphorylates and inhibit acetyl-CoA carboxylase, which is the rate-limiting enzyme in fatty acid synthesis.[17] Fatty acid synthetase is the rate-limiting enzyme in fatty acid synthesis by catalyzing the final step.[18] Adenosine monophosphate-activated protein kinase promotes fatty acid oxidation by upregulating the expression of peroxisome proliferator-activated receptor-α and carnitine palmitoyltransferase-1.[19] Activated adenosine monophosphate-activated protein kinase inhibits the synthesis of cholesterol, via the suppression of 3-hydroxy-3-methylglutaryl-coenzyme A reductase.[20] The lignans tetrahydrofuroguaiacin B, nectandrin B, and nectandrin A isolated from this plant at a concentration of 5 μM induced the activation of adenosine monophosphate-activated protein kinase and downstream inhibition of acetyl-CoA carboxylase.[14] Lipid peroxidation in the liver is linked with insulin resistance.[21]

3.2 *Cinnamomum burmannii* (Nees & T. Nees) Blume

Synonyms: *Cinnamomum chinense* Blume; *Cinnamomum dulce* (Roxb.) Sweet; *Laurus burmannii* Nees & T. Nees; *Laurus dulcis* Roxb.

Common names: yin ziang (Chinese); kayu manis (Malay); Indonesian cassia

Subclass Magnoliidae, Superorder Lauranae, Order Laurales, Family Lauraceae

Medicinal use: hypertension (Indonesia)

Pharmacological target: insulin resistance

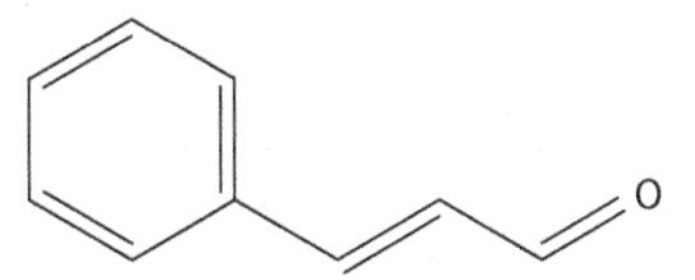

FIGURE 3.2 Cinnamaldehyde.

In the postprandial state, insulin inhibits phosphoenolpyruvate carboxykinase and glucose-6-phosphatase to block the production of glucose also termed gluconeogenesis.[22] *Cinnamomum burmannii* (Nees & T. Nees) Blume contains cinnamaldehyde (Figure 3.2), which given orally at a dose of 20 mg/kg/days for 60 days to streptozotocin-induced diabetic Wistar rats decreased glycaemia from 396 to 152 mg/dL, decreased weight loss and polydipsia and an increased plasma insulin.[23] At the hepatic level, a normalization of phosphoenolpyruvate carboxykinase was observed as a possible consequence of insulin stimulation.[23] Phosphoenolpyruvate carboxykinase is the rate-limiting enzyme of gluconeogenesis leading to the release of glucose from the liver, and the expression of this enzyme is used as an indicator of liver glucose secretion.[24] Ethanol extract of bark given at a single oral dose of 500 mg/kg to hyperglycemic C57Bl/6J mice lowered fasting glycaemia by 18.9% after 6 hours.[25] Decrease of glycaemia in fasting rodents by natural products, if not owed to increased insulin secretion, can be explained by stimulation of glucose uptake by peripheral tissues, correction of insulin resistance, inhibition of liver glucose production, or stimulation of glycogen synthesis by stimulating glycogen synthetase activity. The extract at a dose of 25 μg/mL inhibited glucose production by H4IIE rat hepatoma cells with concomitant repression phosphoenolpyruvate carboxykinase and glucose-6-phosphatase.[25] Aqueous extracts of bark of *Cinnamomum burmanii* inhibited the enzymatic activity of protein tyrosine phosphatase 1B with IC_{50} values 6.2 μg/mL respectively implying increased insulin sensitivity.[26] From this extract, 5′-hydroxy-5-hydroxymethyl-4",5"-methylenedioxy-1,2,3,4-dibenzo-1,2,5-cycloheptatriene and cinnamaldehyde inhibited the enzymatic activity of protein tyrosine phosphatase 1 B with IC_{50} values of 29.7 and 57.6 μM, respectively.[26] Protein tyrosine phosphatase 1 B decreases the sensitivity of insulin to its receptor and contributed to insulin resistance.[27]

3.3 *Cinnamomum zeylanicum* Blume

Synonym: *Cinnamomum verum* J. S. Presl
Common names: cocam (India); true cinnamon
Subclass Magnoliidae, Superorder Lauranae, Order Laurales, Family Lauraceae
Medicinal use: indigestion (India)
History: the plant was known of Hippocrates, Greek physician (circa 460–370 BC)
Pharmacological targets: atherogenic hyperlipidemia; insulin resistance

In hepatocytes, excess of fatty acids brought by the plasmatic circulation from adipose tissues activate peroxisome proliferator-activated receptor-α that binds to the peroxisome proliferators response element of DNA and induce the transcription of genes encoding hepatic fatty acid β-oxidation.[11] Spontaneous type 2 diabetic obese *db/db* mice receiving orally bark powder of *Cinnamomum zeylanicum* Blume at a daily dose of 200 mg/kg for 12 weeks had fasting glucose levels reduced by more than 50% and insulinaemia increased by 74%.[28] This regimen decreased triglyceridaemia, plasma cholesterol, fatty acids, and increased in high-density lipoprotein–cholesterol.[28] Furthermore, the hepatic contents in fatty acids was reduced by 65.6% compared with untreated rodents and histological observation livers evidenced a reduction in lipid droplets.[28] Of note, Cinnamon-treated rodents had increased expression of peroxisome proliferator-activated receptor-α by 11.4%,[28] and this is interesting because activators of peroxisome proliferator-activated receptor-α by induce fatty acid β-oxidation in the liver whence lipid-lowering activity.[29] In a subsequent study, Ranasinghe et al. provided evidence that aqueous extract of *Cinnamomum zeylanicum* Blume given to streptozotocin-induced Sprague–Dawley rats orally at a dose of 600 mg/kg/day for 1 month decreased fasting glycaemia from 320 to 247 mg/dL, lowered plasma cholesterol from 72 to 50 mg/dL, low-density cholesterol from 12.9 to 3.1 mg/dL, elevated high-density lipoprotein–cholesterol from 19.5 to 26.9 mg/dL, and lowered triglycerides from 198.2 to 131 mg/dL.[30] In hepatocytes, fatty acids contribute to insulin resistance by via metabolic competition or through an effect on the insulin-signaling pathway, possibly by activating atypical protein kinase C.[31] Activation of peroxisome proliferator-activated receptor-α increase the oxidation of fatty acids by decreasing liver content of fatty acids and triglycerides to reduce lipotoxicity and increase insulin sensitivity.[32] Spontaneous type 2 diabetic obese *db/db* mice receiving orally cinnamaldehyde at a dose of 20 mg/kg/day for 4 weeks evoked a mild reduction of body weight, fasting blood glucose, insulinaemia, free fatty acids, and increase in high-density lipoprotein–cholesterol, a protective factor against coronary heart disease.[33,34] Furthermore, an increase in phosphorylated Akt and glucose transporter-4 expression in peripheral tissues were observed.[33] The consumption of *Cinnamomum zeylanicum* Blume could be beneficial in metabolic syndrome. Clinical trails are warranted.

3.4 *Lindera strychnifolia* (Siebold & Zucc.) Fern.-Vill.

Common name: wu yao (Chinese)
Subclass Magnoliidae, Superorder Lauranae, Order Laurales, Family Lauraceae
Medicinal use: blood stasis (China)
Pharmacological targets: atherogenic hyperlipidemia; insulin resistance

Bile acids in the liver activate a nuclear receptor termed farnesoid X receptor that controls triglyceride and cholesterol metabolism. With regard to cholesterol metabolism, this nuclear receptor induces the expression of small heterodimer partner (SHP) that inhibits CYP7A1, an enzyme also known as cholesterol-7α-hydroxylase, which catalyze the synthesis of bile acids from cholesterol.[35] Besides, activation of farnesoid X receptor prompts the secretion of bile acids and cholesterol into bile duct via the activation of hepatic ATP-binding cassette (ABC) transporters ABCB11 and ABCG5/8, respectively.[36] Natural products with the ability to activate farnesoid X receptor promote biliary cholesterol secretion and reduces fractional absorption of dietary cholesterol.[37] The roots of *Lindera strychnifolia* (Sieb. et Zucc.) Fern.-Vill. contain isoquinoline alkaloids including boldine (Figure 3.3).[38] Boldine given orally at a dose of 100 mg/kg/day for 8 weeks reduced the glycaemia of rodents poisoned with streptozotocin from 538.4 to 311.4 mg/dL, increased body weight, reduced

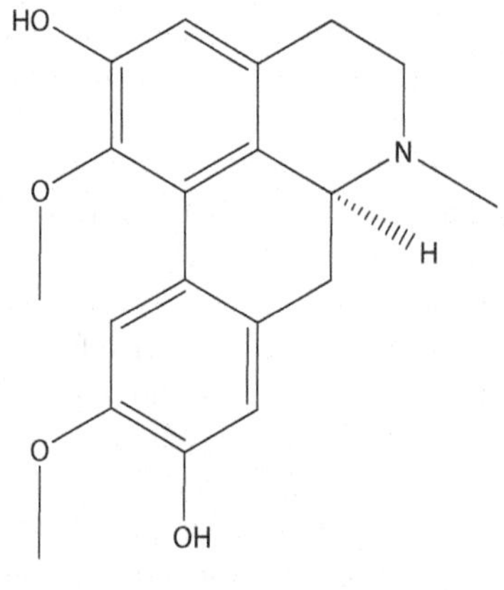

HO
O
N
H
O
OH

FIGURE 3.3 Boldine.

hepatic lipid peroxidation, and increased hepatic glutathione peroxidase activity.[39] Boldine is antioxidant as at a concentration of 100 μM prevented *in vitro* the generation of superoxide and hydrogen peroxide production from hepatic mitochondria challenged with antimycin.[39] In a subsequent study, boldine given to hereditary hypertriglyceridemic rats on as part of 0.2% of high-sucrose diet for 6 weeks induced a decrease of glycaemia from 15 to 14 mmol/L, increased high-density lipoprotein–cholesterol from 0.8 to 0.9 mmol/L, triglycerides from 2.7 to 1.5 mmol/L, and bile acids from 5.5 to 2.7 μmol/L, whereas plasma cholesterol was unchanged.[40] This alkaloid reduced hepatic triglyceride contents in high-sucrose diet rats from 4.9 to 4.2 μmol/L and improved hepatic cytoarchitecture.[40] Boldine increased bile flow and bile acid secretion toward levels seen in control animals, increased hepatic glutathione contents, and increased the expression of transporters for bile acids, ATP-dependent human bile salt export pump (Bsep/ABCB11) and sodium-taurocholate cotransporting polypeptide (Ntcp).[40] Boldine at a concentration of 5 μM evoked the activation of farnesoid X receptor in transfected HepG2 cells.[41] Other hepatoprotective constituents in the roots of *Lindera strychnifolia* (Sieb. et Zucc.) Fern.-Vill. are sesquiterpenes of which bi-linderone and lindestrene.[38] Lindestrene given orally at a dose of 100 mg/kg, twice daily for 3 days, and 1 hour before galactosamine-induced hepatic insults evoked a decrease of plasmatic aspartate aminotransferase and GPT I Wistar rats.[42] Bi-linderone at a concentration of 1 μg/mL protected HepG2 cells against glucosamine-induced inhibition of insulin receptor sensitivity as evidenced by increased expression of phosphorylated insulin receptor and phosphorylated Akt.[43]

3.5 *Persea americana* Mill.

Synonyms: *Laurus persea* L.; *Persea edulis* Raf.; *Persea gratissima* C.F. Gaertn.
Common name: e li (Chinese); avocado
Subclass Magnoliidae, Superorder Lauranae, Order Laurales, Family Lauraceae
Medicinal use: diarrhea (Philippines)
Pharmacological targets: atherogenic hyperlipidemia; insulin resistance

Hashimura et al. provided evidence that persin also known as (2R,12Z,15Z)-12-hydroxy-4-oxo-heneicosa-12,15-dienyl acetate as well as 5E, 12Z, 15Z-2-hydroxy-4-oxo-heneicosa-5,12,15-trienyl actetate or persenone A from the fruits of *Persea americana* Mill. inhibited *in vitro* acetyl-CoA carboxylase with IC_{50} values of 4.9×10^{-6} M and 4×10^{-6} M, respectively.[44] In the liver, the synthesis of fatty acid starts with the building of malonyl-CoA from acetyl-CoA by acetyl-CoA carboxylase.[45] Inhibition of acetyl-CoA carboxylase may account for the fact that aqueous extract of leaves given orally to high-cholesterol diet albino rats at a dose of 10 mg/kg/day for 8 weeks lowered plasma cholesterol by 8%, increased high-density lipoprotein by 85%, and decreased low-density lipoprotein–cholesterol by 19%.[46] In the liver, fatty acids are used for the building of triglycerides, which are packed into very low-density lipoproteins.[47] Besides, the fruits given at 5% of diet to rats for 14 days prevented hepatic damages evoked by D-galactosamine as evidence by a drastic reduction of plasmatic aspartate aminotransferase compared with untreated animals.[48] From the fruits, persenone A, (2E,5E,12Z,15Z)-1-hydroxyheneicosa-2,5,12,15-tetraen-4-one, (2E,12Z,15Z)-1-hydroxyheneicosa-2,12,15-trien-4-one, and (5E,12Z)-2-hydroxy-4-oxoheneicosa-5,12-dien-1-yl acetate given orally at a single dose of 100 mg/kg prevented by D-galactosamine-induced liver injuries with a reduction of markers of liver insults plasma serum aspartate transaminase and alanine transaminase by about 50%.[48] The hepatoprotective mechanism involved here is unknown but it must be recalled that in hepatocytes, insults induce the transcription of lipogenic genes and accumulation of triglycerides in intracellular lipid droplets.[49] In hepatocytes, accumulation of fatty acids forsters the formation of reactive oxygen promoting hepatic insulin resistance,[49] and inhibition of acetyl-CoA carboxylase improve hepatic insulin sentitivity.[50] Hydroalcoholic extract of leaves of *Persea americana* Mill. given at a dose of 0.3 g/kg/day for 28 days to streptozotocin-diabetic rats reduced fasting glucose levels by 71%, improved glucose tolerance, lowered daily water intake, and increased body mass gain compared with untreated diabetic animals.[51] Further, hepatocytes and soleus muscles of treated diabetic rats exhibited an increased expression of phosphorylated Akt by more than 75% suggesting increased glucose intake from skeletal muscles in treated animals.[51] Of note, 2 hours after consumption of a test beef burger, the peripheral arterial tone score of healthy male volunteers was decreased by 27.4% and this was prevented by adding 68 g of avocado into a test beef burger.[52] Peripheral blood mononuclear cells isolated after test burger consumptions evidenced a slight elevation of IkB-α suggesting a decrease in nuclear factor-κB activity and therefore peripheral inflammation.[52] A decrease in serum interleukin-6 was observed 4 hours after burgers consumption.[52] The consumption of avocado could be of benefit in metabolic syndrome.

3.6 *Piper longum* L.

Synonym: *Chavica roxburghii* Miq.
Common names: bi ba (Chinese); long pepper
Subclass Magnoliidae, Superorder Piperanae, Order Piperales, Family Piperaceae
Medicinal use: facilitates digestion (China)
History: the plant was known of the plant was known of Hippocrates
Pharmacological targets: atherogenic hyperlipidemia; insulin resistance

In the liver, the synthesis of triglycerides from fatty acids and glycerol-3-phosphate is catalyzed by glycerol-3-phosphate-acyltransferase and acyl CoA:diacylglycerol acyltransferase and synthetized triglycerides are either stored in cytosolic lipid droplets or packed into very low-density lipoproteins by conjugation to apoB-100 and exported in the blood stream.[47] (2E,4Z,8E)-N-[9-(3,4-Methylenedioxyphenyl)-,4,8-nonatrienoyl]-piperidine and pipernonaline from *Piper longum* L. inhibited acyl CoA:diacylglycerol acyltransferase with IC_{50} values of 29.8 and 37.2 μM.[53] Ethanol extract of fruits of *Piper longum* L. given orally to rats at a dose of 20 mg/kg/day for 15 days lowered plasma cholesterol from 7.5 to 5.8 mmol/L and triglycerides from 0.7 to 0.6 mmol/L.[54] From this extract, piperlonguminine, piperine, and pipernonaline given orally at a dose of 5.6 mg/kg/day for 15 days reduced plasma cholesterol and triglycerides and elevated high-density lipoprotein–cholesterol more potently that simvastatin at 5.6 mg/kg/day for 15 days.[54] In this experiment, piperlonguminine, piperine, and pipernonaline given orally at a dose of 5.6 mg/kg/day for 15 days had no effect of low-density lipoprotein–cholesterol.[54] Piperine given to C57BL/6N mice as part of 0.05% high-fat diet for 10 weeks had no effect on food intake, lowered body weight gain by 67%, significantly reduced plasma triglycerides by 83%, free fatty acids by 81%, cholesterol by 58%, high-density lipoprotein–cholesterol by 38%, and very low-density lipoprotein–cholesterol + low-density lipoprotein–cholesterol by 82%.[55] This regimen decreased lipid deposition in the livers of mice fed a high-fat diet.[55] Piperine administration decreased insulin receptor substrate-1 serine phosphorylation, increased Akt phosphorylation, and elevated glucose transporter-2 membrane translocation.[55] Improved insulin sensitivity was also evidenced by a decreased activity gluconeogenic enzymes namely, glucose 6-phosphatase and phosphoenolpyruvate carboxykinase that are physiologically inhibited by insulin.[55] In hepatocytes, liver X receptor-α is activated by metabolites of cholesterol also known as oxysterols. Activation of this nuclear receptor induce the expression of sterol regulatory element binding protein-1c and subsequent expression of fatty acid synthase and stearoyl coenzyme A desaturase 1 (SCD-1) leading to fatty acid synthesis and increased triglyceridemia.[56] In transfected HEK293 cells, piperine at a concentration of 100 μM inhibited the transcriptional activity of liver X receptor-α.[55] Physiologically, adenosine monophosphate-activated protein kinase is activated by adiponectin via LKB1.[57] Phosphorylated adenosine monophosphate-activated protein kinase inhibits mTORC1, S6 protein kinase 1 and downregulates the expression of liver X receptor-α. At the same time, phosphorylated adenosine monophosphate-activated protein kinase inhibits acetyl-CoA carboxylase lifting malonyl-CoA inhibition on carnitine palmitoyltransferase-1 and promoting thus the entry of fatty acyl-CoA in mitochondria for β oxidation.[58] S6 protein kinase 1 inhibition results in insulin receptor substrate-1 activation and Akt phosphorylation and thereby increased sensitivity to insulin.[59] Piperine given at a dose of 50 mg/kg/day to C57BL/6N mice feeding a high-fat diet for 13 weeks evoked in the liver of treated animals increased expression of adiponectin receptors, phosphorylated adenosine monophosphate-activated protein kinase, and reduction in expression of liver X receptor-α, sterol regulatory element binding protein-1c, and fatty acid translocase as well as decreased phosphorylation of S6K1 implying adenosine monophosphate-activated protein kinase activation.[59] In line, administration of piperine to mice resulted in an increase in glucose transporter-2 found in the membrane, decreased phosphorylation of insulin receptor substrate-1 and increased the phosphorylation of Akt and increased expression of phosphoenolpyruvate carboxykinase and glucose-6-phosphatase, key enzymes involved in hepatic gluconeogenesis implying improved insulin sensitivity.[59] All of these suggest that intake of *Piper longum* L. could be of value in metabolic syndrome to lower hepatic insulin resistance as well as low-density lipoprotein–cholesterol, the elevation of which is a strong indicator of a coronary heart disease in diabetic patients.[60] Clinical trails are warranted.

3.7 *Piper retrofractum* Vahl.

Synonyms: *Chavica officinarum* Miq.; *Piper chaba* Hunter; *Piper officinarum* (Miq.) C. DC.
Common names: jia bi ba (Chinese); litlit (Philippines)
Subclass Magnoliidae, Superorder Piperanae, Order Piperales, Family Piperaceae
Medicinal use: indigestion (Philippines)

Hypercholesterolemia is the critical step in the initiation of atherosclerosis, placing hypercholesterolemic individuals at a greater risk of cardiovascular diseases.[61] Total plasma cholesterol depends on the absorption of cholesterol by the intestine, the synthesis of cholesterol in the liver and the catabolism of cholesterol. In C57BL/6J mice fed with high-fat diet, piperidine alkaloids of which piperidine, dehydropipernonaline, and pipernonaline given at a dose of 300 mg/kg/day for 8 weeks reduced plasma cholesterol by 44.3% and low-density lipoprotein–cholesterol by 57.6%.[62] In the liver of high-fat diet fed C57BL/6J mice piperidine alkaloids evoked the phosphorylation of adenosine monophosphate-activated protein kinase and target acetyl-CoA carboxylase.[62] This regimen also inhibited the expression of sterol regulatory element-binding protein-1c and subsequently fatty acid synthetase and acetyl-CoA carboxylase.[62] In parallel, this regimen increased carnitine palmitoyltransferase-1 and uncoupling protein 2, which are targets of peroxisome proliferator-activated receptor-α hence hepatic lipolysis.[62] Activation of adenosine monophosphate-activated protein kinase in the liver evokes the activation of peroxisome proliferator-activated receptor-γ coactivator 1-α that coactivates peroxisome proliferator-activated receptor-α leading to β-oxidation of fatty acids.[63]

3.8 *Nelumbo nucifera* Gaertn.

Synonyms: *Nelumbium nuciferum* Gaertn.; *Nelumbo speciosa* Willd.; *Nymphaea nelumbo* L.
Common names: lian (Chinese); sacred lotus
Subclass Ranunculidae, Superorder Proteanae, Order Nelumbonales, Family Nelumbonaceae
Medicinal use: anxiety (China)

Visceral adiposity is associated with elevated triglycerides and reduced high-density lipoprotein–cholesterol concentrations in the plasma.[64] Nuciferine given orally at a dose of 15 mg/kg/day to golden Syrian hamsters on high-fat diet for 8 weeks protected rodents against high-fat diet induced obesity for 8 weeks attenuated body weight gain and liver weight gain without decreasing food intake.[65] This regimen decreased toward normal values plasma and liver cholesterol, triglycerides and fatty acids, and decreased plasma low-density lipoprotein–cholesterol without affecting high-density lipoprotein–cholesterol.[65] In the same experiment, nuciferine defended hepatocytes against fat-induced hepatosteatosis and hepatic necroinflammation as evidenced by a decrease in plasma alanine transaminase.[65] In hepatocytes, nuciferine decreased the expression of liver X receptor-α, sterol regulatory element binding protein-1c and subsequently acetyl-CoA carboxylase, fatty acid synthetase, stearoyl coenzyme A desaturase 1.[65] Simultaneously, nuciferine treatment increased the expression of peroxisome proliferator-activated receptor-α and carnitine palmitoyltransferase 1 implying increased fatty acid β-oxidation.[65] This alkaloid repressed the expression of peroxisome proliferator-activated receptor-γ, lipoprotein lipase, and fatty acid translocase suggesting reduced infiltration of plasmatic fatty acids.[65] The isoqunoline enhanced the expression of apolipoprotein-B and microsomal triglyceride transfer protein that account for very low-density lipoprotein secretion.[65] Nuciferine itself may not be responsible for the aforementioned effect as, after oral administration, nuciferine undergoes phase I and phase II metabolism to afford about 10 metabolites, with phase I demethylation being the principal route.[66] It must be recalled that oral LD_{50} of nuciferine in mice and rat is 240 and 280 mg/kg, respectively,[67] and that it evokes decreased motor activity, ptosis, hind-leg spread, and hypotonia in rodents.[68] Having a relatively low LD50 and psychopharmacological effect, nuciferine itself may be of limited use in therapeutic. Instead, intake of leaves as a form of tea, if not toxic, could be possibly envisaged in metabolic syndrome. The current tendency to look for pure compounds and to impose the standardization of extract could be a mistake.

3.9 *Coptis chinensis* Franch.

Common name: huang lian (Chinese)
Superorder Ranunculanae, Order Ranunculales, Family Ranunculaceae
Medicinal use: fever (China)

Evidence has been provided to demonstrate that berberine, coptisine, columbamine, and jatrorrhizine present in the rhizome of *Coptis chinensis* Franch. inhibit the synthesis of triglycerides and cholesterol in hepatocytes *in vitro*. Berberine inhibited the synthesis of cholesterol and triglycerides in HepG2 cells with an IC_{50} value equal to approximately 15 μg/mL *in vitro*,[69] and this effect was later confirmed by Fan et al.[70] and Cao et al.[71] with increased expression of carnitine palmitoyltransferase 1 and medium-chain acyl-CoA dehydrogenase, which are associated with fatty acid oxidation. Berberine induced the expression of low-density lipoprotein receptor in HepG2 cells at dose of 20 μg/mL via with phosphorylation of extracellular signal-regulated kinase-1/2.[72] From this plant columbamine at a concentration of 15 μM reduced triglyceride contents in Hep G2 cells by 35% with concomitant phosphorylation of adenosine monophosphate-activated protein kinase as potently as berberine, and consequently both alkaloids repressed the expression fatty acid synthetase, acetyl-CoA carboxylase, and glycerol-3-phosphate-acyltransferase, which are associated with triglyceride synthesis and of 3-hydroxy-3-methyl-glutaryl-CoA reductase, which catalyzes the synthesis of cholesterol.[71] Columbamine also increased the gene expression of and medium-chain acyl-CoA dehydrogenase, but had no effects on carnitine palmitoyltransferase 1.[71]

Coptisine for this plant at concentration of 0.2 μg/mL reduced the accumulation of triglycerides in HepG2 cells cultured in the presence of fatty acids by 48.9%.[70] Jatrorrhizine at a concentration of 15 μM reduced triglyceride contents in HepG2 cells challenged with fatty acids by 30% with modest effects on adenosine monophosphate-activated protein kinase phosphorylation.[71] As for *in vivo* studies, Brusq et al. administered orally to rodent on high-fat diet berberine given at a dose of 100 mg/kg/day for 10 days and noted a decrease in plasma low-density lipoprotein–cholesterol by 39% and at the hepatic level a reduction of triglycerides, cholesterol, and cholesteryl ester by 23%, 27%, and 41%, respectively.[69] In a subsequent study, Cao et al. provided evidence that an alkaloidal extract of rhizomes of a member of the genus *Coptis* Salisb. given orally to Sprague–Dawley rats on high-fat diet for 14 days at a dose of 100 mg/kg/day reduced plasma cholesterol and low-density lipoprotein–cholesterol and normalized triglycerides and high-density lipoprotein–cholesterol.[73] Furthermore, this regimen doubled the production of bile in the liver and tripled the presence of bile acids in the feces.[73] The extract at a dose of 100 mg/kg/day for 14 days increased the expression of peroxisome proliferator-activated receptor-α and decreased the expression of farnesoid X receptor and therefore increased the expression of cholesterol 7α-hydroxylase also known as CYP7A1, a key enzyme in bile acids synthesis from cholesterol.[73] Jatrorrhizine from *Coptis chinensis* Franch. at a dose of 100 mg/kg to mice induced a reduction of glycaemia from 5.9 to 4.6 mmol/L and a decrease in liver glycogen from 17.4 to 8.4 mg/.[74] In alloxan-induced diabetic mice the glycaemia was reduced by daily administration of jatrorrhizine oral at a dose 100 mg/kg/day for 5 days from 21.6 to 16.4 mmol.[74] The enzymatic activity of succinate dehydrogenase was increased from 6.8 to 11.2 U/mg protein suggesting an increase in aerobic utilization of glucose in hepatocytes.[74] In a subsequent study, this protoberberine given orally to Syrian golden hamsters at a dose of 70 mg/kg/day for 90 days lowered plasma cholesterol, triglycerides, and low-density lipoprotein–cholesterol by 20%, 43%, and 19%, respectively, and increased high-density lipoprotein–cholesterol and bile acids content in feces.[75] Besides, jatrorrhizine upregulated the expression of low-density lipoprotein–cholesterol receptor and cholesterol 7α-hydroxylase but exhibited no effect on the expression of 3-hydroxy-3-methyl-glutaryl-CoA reductase and sodium-dependent bile acid transporter in hamsters.[75] In human, a direct effect of berberine is improbable because oral administration of decoction of rhizomes of a member of the genus *Coptis* Salisb. in healthy volunteers is followed by the presence of jatrorrhizine 3-*O*-β-D-glucuronide, columbamine 2-*O*-β-D-glucuronide, jatrorrhizine 3-*O*-sulfate, and traces of berberine.[76]

3.10 *Agrostemma githago* L.

Synonyms: *Lychnis githago* (L.) Scop.
Common names: mai xian weng (Chinese); katir cicegi (Turkey); corn-rose (English)
Subclass Caryophyllidae, Superorder Caryophyllanae, Order Caryophyllales, Family Caryophyllaceae
Medicinal use: cough (Turkey)

Agrostemma githago given orally at a dose of 100 mg/kg/day for 30 days to Swiss albino mice feeding on diet containing 1% cholesterol, reduced cholesterolaemia from 218.4 to 98.2 mg/dL.[77] Serum high-density lipoprotein–cholesterol was increased by *Agrostemma githago* L. from 25.8 to 36.8 mg/dL.[77] *Agrostemma githago* reduced low-density lipoprotein–cholesterol from 143.0 to 42.0 mg/L[77] triglyceridemia from 194.2 to 117.8 mg/dL and glycaemia from 79.8 to 61 mg/dL.[77] The use of the plant for cough suggests the presence of saponins, which have the tendency to inhibit dietary cholesterol absorption as discussed in chapter 1.

3.11 *Nigella sativa* L.

Common names: Krishna jiraka (India); habbatus sauda (Malay); fennel flower seeds
Subclass Caryophyllidae, Superorder Ranunculanae, Order Menispermales, Family Ranunculaceae
Medicinal use: tonic (Malaysia)
History: known of Hippocrates

Ethanol extract of seeds of *Nigella sativa* L. at a concentration of 200 μg/mL increased *in vitro* the phosphorylation of Akt in H4IIE hepatocytes with increased contents of phosphorylated adenosine monophosphate-activated protein kinase and its downstream substrate acetyl-CoA carboxylase.[78] In addition, the extract evoked uncoupling of oxidative phosphorylation in isolated liver mitochondria.[78] *In vitro*, ethanol extract of seeds of *Nigella sativa* L. reduced the triglycerides levels in high glucose-pretreated HepG2 cells by about 15% at a dose of 10 μg/mL.[79] *In vivo*, this extract given to streptozotocin-induced diabetic Wistar rats at a dose of 300 mg/kg/day for 30 days lowered plasma cholesterol from 283.5 to 171.2 mg/dL, low-density lipoprotein–cholesterol from 186.1 to 85 mg/dL, increased high-density lipoprotein–cholesterol from 59.8 to 63.5 mg/dL, and lowered plasma triglycerides from 185.1 to 113.8 mg/dL.[80] This regimen increased plasma insulin from 4.2 to 11.3 μU/mL.[80] Methanol extract of seeds given orally at a doses of 500 mg/kg/day for 3 days mildly increased food intake, body and liver weights, and reduced liver triglycerides in ddY male mice.[79] One of the major constituent of the seeds is thymoquinone (Figure 3.4), which given to Sprague–Dawley rats on high-cholesterol diet orally at a dose of 100 mg/kg/day for 8 weeks plasma cholesterol from 2.1 to 0.9, low-density lipoprotein–cholesterol from 1.6 to 0.4 and plasma triglycerides from 0.6 to 0.4.[81] At the hepatic level, this regimen increased the expression of low-density lipoprotein–cholesterol receptor and 3-hydroxy-3-methylglutaryl-coenzyme A reductase.[81] Thymoquinone given orally at a dose of 20 mg/kg/day for 6 weeks to Wistar rats on high-fat diet lowered glycaemia and plasma insulin, insulin resistance by 53%, decreased triglyceride from 89 to 61.5 mg/dL, cholesterol from 205.9 to 151.3 mg/dL, and increased high-density lipoprotein–cholesterol from 30.2 to 47.2 mg/dL.[82] High-fat diet elevated the expression of peroxisome proliferator-activated receptor-γ, which was reduced by 167% upon thymoquinone treatment.[82] Male volunteers aged 35–50 years with mild hypertension receiving 200 mg of an aqueous extract of seeds of *Nigella sativa* twice a day for 8 weeks had lower systolic blood pressure and diastolic blood pressure compared to untreated individuals as well as reduced low-density lipoprotein–cholesterol.[82]

3.12 *Corydalis saxicola* Bunting

Synonym: *Corydalis thalictrifolia* Franch.
Common name: yan huang lian (Chinese)
Subclass Ranunculidae, Superorder Ranunculanae, Order Papaverales, Family Fumariaceae
Medicinal use: hepatitis (China)

Koruk et al. observed increased plasmatic malondialdehyde, nitric oxide, and lower activity of superoxide dismutase in patients with nonalcoholic steatohepatitis and suggested an impairment of hepatic antioxidant enzymatic defense system.[83,84] In hepatocytes, superoxide dismutase converts superoxide into hydrogen peroxide, oxygen, and water.[85] Glutathione peroxidase catalyzes reductive destruction of hydrogen peroxide and lipid hydroperoxide using glutathione.[85] Increased intrahepatic levels of fatty acids are a source of oxidative stress, which interferes with insulin sensitivity. *Corydalis saxicola* Bunt elaborates series of hepatoprotective alkaloids of which tetrahydropalmatine,[86] palmatine,[87] protopine,[88] berberine, coptisine, and dehydroapocavidine.[89] Dehydrocavidine given at a dose of 1 mg/kg before or after carbon tetrachloride poisoning, prevented the increase

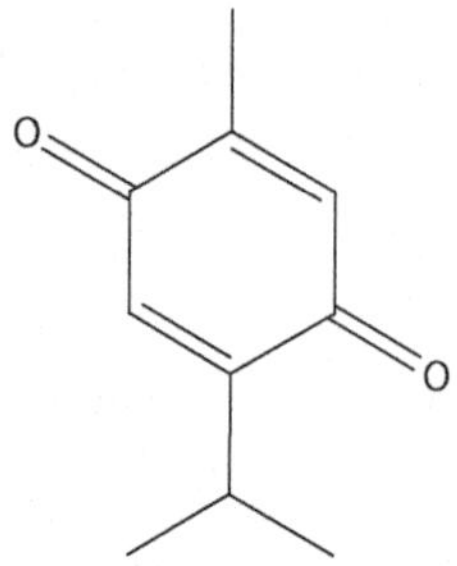

O
O

FIGURE 3.4 Thymoquinone.

of serum alanine aminotransferase, aspartate aminotransferase, alkaline phosphatase and reduced the peroxidation of hepatic lipids as evidenced by a decrease in malondialdehyde, and commanded an increase in hepatic superoxide dismutase and glutathione peroxidase.[90] Histopathological observation of the liver of rodents pretreated with dehydrocavidine evidenced a reduction of liver injuries compared to the untreated group.[90] The metabolism of carbon tetrachloride by cytochrome P450 in the liver leads to the formation of reactive oxygen species resulting in acute hepatic insults.[91] The precise molecular basis for the hepatoprotective effect of protoberberine is not yet determined. Domitrović et al. suggested that free radicals trigger tumor necrosis release factors from Kupffer cells and injured hepatocytes, which further activates nuclear factor-κB, and downstream expression of inducible nitric oxide synthetase and cyclooxygenase-2, which are inhibited by berberine on account of, at least, radical scavenging activity.[92] Clinical trials are warranted.

3.13 *Fumaria parviflora* Lam.

Synonym: *Fumaria indica* Pugsley
Common names: pitpara (India); fumitory
Subclass Ranunculidae, Superorder Ranunculanae, Order Papaverales, Family Fumariaceae
Medicinal use: jaundice (India)

Fumaria parviflora Lam. is hepatoprotective. A single dose of 500 mg/kg of methanol extract of shoots of *Fumaria parviflora* Lam. given orally as a pretreatment reduced the mortality of mice poisoned with paracetamol orally at a dose of 1 g/kg.[93] The same extract given orally twice a day for 2 days before the administration of a bolus of paracetamol at a dose of 640 mg/kg reduced the enzymatic activity of serum alanine aminotransferase and aspartate aminotransferase.[93] Protopine from this plant given orally once a day for a week at a dose of 50 mg/kg protected rodents against galactosamine-induced hepatotoxicity as evidenced by histopathological evidence, decrease in serum alanine aminotransferase and aspartate aminotransferase as well as serum alkaline phosphatase and serum bilirubin as well as reduction of hepatic malondialdehyde and glutathione.[94]

3.14 *Juglans regia* L.

Synonyms: *Juglans duclouxiana* Dode; *Juglans fallax* Dode; *Juglans kamaonia* (C. DC.) Dode; *Juglans orientis* Dode; *Juglans sinensis* (C. DC.) Dode
Common names: hu tao (Chinese); walnut
Subclass Hamamelidae, Superorder Juglandanae, Order Juglandales, Family Juglandaceae
Medicinal use: abscesses (China)

Dietary triglycerides stored in adipose tissues are, in state low plasmatic levels of insulin, hydrolyzed into unesterified fatty acids, which are excreted in the general circulation.[45] Unesterified fatty acids absorbed by hepatocytes are used for the synthesis of triglyceride, which are excreted in the general circulation in the form of very low-density lipoprotein or stored in cytosolic droplets.[45] Evidence suggests that polyunsaturated fatty acids are more difficult to incorporate in very low-density lipoproteins, causing a drop in plasma concentration of very low-density lipoproteins and consequently low-density lipoproteins that are atherogenic.[95] The seeds of *Juglans regia* L. contains predominantly polyunsaturated fatty acids of which linoleic acid, followed by oleic acid and α-linolenic acid.[96] Oil of seeds of *Juglans regia* L. given to volunteers with plasma triglycerides >350 mg/dL or total cholesterol >250 mg/dL at a dose of 3 g/day for 45 days lowered plasma triglycerides from 572 to 461 mg/dL and had no effect on total cholesterol, high-density lipoprotein and low-density lipoprotein.[97] In healthy subjects, the consumption of 43 g of seeds of *Juglans regia* L. per day for 8 weeks evoked a reduction of nonhigh-density lipoprotein–cholesterol and plasma apolipoprotein B, and attenuated total cholesterol, low-density lipoprotein–cholesterol, very low-density lipoprotein–cholesterol, and triglycerides.[98] Ethanol extract of pellicles (containing mainly pedunculagin, tellimagrandin I, tellimagrandin II, and ellagic acid) of seeds of *Juglans regia* L. given to ddY mice on high-fat diet at a dose of 100 mg/kg/day for 13 days evoked a mild reduction of weight gain, lowered liver weight, and hepatic triglycerides. Serum triglycerides was lowered to normal values whereby liver and serum cholesterol were not affected.[99] This regimen increased the expression of peroxisome proliferator-activated receptor-α and its target acyl-CoA oxidase, which is a key enzyme in peroxisomal-β-oxidation and fatty acid β-oxidation in the cytoplasm of hepatocytes.[99] The extract had no effect on plasma triglyceride in olive oil oral loading implying the noninvolvement of pancreatic lipase inhibition.[99] In line, *in vitro*, the extract at a concentration of 100 μg/mL increased the expression of peroxisome proliferator-activated receptor-α, carnitine palmitoyltransferase 1, and acyl-CoA oxidase in hepG2 cells.[99] Ellagitannins in the intestines are decomposed into ellagic acid, which, given to genetically obese type 2 diabetic KK-Ay mice at 0.1% of high-fat diet for 45 diet lowered fasting serum glucose from 3.5 to 2.9 mmol/L, serum triglycerides from 180.5 to 138.1 mg/dL, serum free fatty acid from 1.8 to 1.4 mmol/L, and attenuated cholesterol from 301.6 to 272 mg/dL.[100] This supplementation decreased plasma resistin, improved hepatic cytoarchitecture with decreased steatosis and increased the expression of low-density lipoprotein receptor, apolipoprotein A-I, sterol regulatory element-binding transcription factor 2, fatty acid synthase, carnitine palmitoyltransferase 1A, which is rate-limiting enzyme of fatty acid β-oxidation, and boosted the expression of peroxisome proliferator-activated receptor-α.[100] In hepatocytes, adenosine monophosphate-activated protein kinase upregulating expression of protein involved in fatty acid β-oxidation, including peroxisome proliferator-activated receptor-α.[17] One could draw an inference that adenosine monophosphate-activated protein kinase activation by ellagic acid[101] could account peroxisome proliferator-activated receptor-α activation and subsequent fatty acid oxidation. Activators of peroxisome proliferator-activated receptor-α fibrates command in hepatocytes both mitochondrial and peroxisomal oxidation of fatty acids resulting in decreased plasma triglyceride[102] the elevation of which is closely linked to cardiovascular diseases.[103] Activators of peroxisome proliferator-activated receptor-α induce apolipoprotein A-I and A-II expression and increase high-density lipoprotein-mediated cholesterol efflux from arterial wall macrophages hence decreased atherogenesis.[104] It must be noted that fibrates are low molecular weight phenolic compounds with somewhat structural similarities with plant phenols. Consumption of seeds of *Juglans regia* L. may improve lipid profile in metabolic syndrome and further nutritional studies in that direction are needed.

3.15 *Fagopyrum esculentum* Moench

Synonyms: *Fagopyrum sagittatum* Gilib.; *Polygonum fagopyrum* L.
Common names: qiao mai (Chinese); buckwheat
Subclass Caryophyllidae, Superorder Polygonanae, Order Polygonales, Family Polygonaceae
Medicinal use: diarrhea (China)

Phenolic fraction of the seeds of *Fagopyrum esculentum* Moench (containing catechin, epicatechin, catechin-7-*O*-glucoside, epicatechin 3-*O*-[3,4-di-*O*-methyl]-gallate, and rutin) given orally at a dose of 10 mg/kg/day orally for 3 days to ddY mice prevented the decrease in glycaemia and increase in plasma corticosterone incurred by restrain stress.[105] Furthermore, this phenolic fraction lowered stress-induced cholesterolaemia from 144.3 to 125.4 mg/mL, increased high-density lipoprotein–cholesterol from 72.9 to 82.7 mg/dL, and lowered atherosclerotic index from 0.7 to 0.5.[105] This regimen attenuated plasma oxidative reactive substance and decreased aspartate aminotransferase and alanine aminotransferase activities from 46.4 and 10 IU/L to 37.1 and 7 IU/L, respectively.[105] In the liver of treated animals, this phenolic fraction decreased triglycerides from 26 to 21.2 mg/g, lowered cholesterol from 3.9 to 3.2 mg/g, and reduced lipid peroxidation.[105] In a subsequent study, powder of leaves and flowers of *Fagopyrum esculentum* Moench given to Sprague–Dawley rats on high-fat diet at 5% of diet for 6 weeks had no effect on food intake, lowered plasma cholesterol from 4.2 to 2.7 mmol/L, had no effect on plasma triglycerides, and lowered atherogenic index from 5 to 3.7.[106] The regimen lowered liver triglycerides from 3.2 to 2 mmol/g, decreased liver cholesterol from 3 to 2.3 mmol/g, increased the activity of both hepatic 3-hydroxy-3-methylglutaryl-coenzyme A reductase and acyl-CoA:cholesterol acyltransferase[106] suggesting a depletion of hepatic unesterified cholesterol. This supplementation increased fecal sterols and fecal triglyceride contents[106] on probable account of fibers. In hepatocytes, low-density lipoprotein binding its receptor is followed by lysosomal degradation releasing unesterified cholesterol in the cytoplasm which inhibits 3-hydroxy-3-methylglutaryl-coenzyme A reductase and activates acyl CoA: cholesterol *O*-acyltransferase to stop the synthesis of cholesterol and to convert cholesterol to cholesteryl ester for packing into cytoplasmic droplets.[45] Buckwheat could be beneficial in metabolic syndrome.

3.16 *Fagopyrum tataricum* (L.) Gaertn.

Synonym: *Polygonum tataricum* L.
Common names: ku qiao (Chinese); phaphar (India); Indian wheat
Subclass Caryophyllidae, Superorder Polygonanae, Order Polygonales, Family Polygonaceae
Nutritional use: food (India)

Defatted ethanol extract of envelope of seeds *Fagopyrum tataricum* (L.) Gaertn. given orally for 6 weeks at a dose of 0.5 g/kg/day to Wistar rats on high-fat diet lowered weight gain from 100.8 to 70.7 g.[107] This regimen lowered serum triglycerides from 2.8 to 1.4 mmol/L and a decreased plasma cholesterol from 4 to 2.3 mmol/L.[107] This extract increased high-density lipoprotein–cholesterol, and lowered low-density lipoprotein–cholesterol, apolipoprotein A-I, apolipoprotein B, and atherogenic index.[107] Apolipoprotein B is present in very low-density lipoproteins and low-density lipoproteins and account for the entrapment of low-density lipoprotein in arterial wall and serve as ligand for apoB and apoB, E receptors mediating peripheral uptake of cholesterol.[108] Increased apolipoprotein B is an indicator or ischemic heart disease.[109] Plasma superoxide dismutase, glutathione peroxidase activities were increased by 147.2% and 9.1% and malondialdehyde was decreased by 15.3%.[107] In the liver of treated animals, triglycerides, cholesterol, and malondialdehyde contents was reduced, whereas superoxide dismutase and glutathione peroxidase activities were elevated by 4.9% and 21.6%.[107] Ethanol extract of seeds given orally at a dose of 50 mg/kg/day for 4 weeks to C57BL/6 mice on ethanol-enriched diet lowered serum aspartate transaminase and alanine transaminase, and alkaline phosphatase.[110] This extract improved liver cytoarchitecture, decreased hepatic reactive oxygen species to normal values as well as oxidative reactive substances and boosted glutathione levels.[110] This regimen increased the activity of hepatic catalase, glutathione peroxidase, superoxide dismutase, and glutathione S-transferase and lowered tumor necrosis factor-α, interleukin-1β, and interleukine-6 levels in the liver.[110] Ethanol-enriched diet significantly increased the plasma triglycerides and cholesterol in the plasma and the liver and the extract lowered serum triglycerides and cholesterol levels to 107 and 13.3 mg/dL, and reduced the hepatic triglycerides and cholesterol levels to 142.3 and 2.6 mg/dL, respectively. From this extract, rutin given at a dose of 11.5 mg/kg/day, or quercetin at a dose 3 mg/kg/day evoked similar effects.[110] Indian wheat could be beneficial in metabolic syndrome.

3.17 *Rheum tanguticum* Maxim. ex Balf.

Common name: ji zhua da huang (Chinese)
Subclass Caryophyllidae, Superorder Polygonanae, Order Polygonales, Family Polygonaceae
Medicinal use: constipation (China)
Pharmacological targets: atherogenic hyperlipidemia; insulin resistance

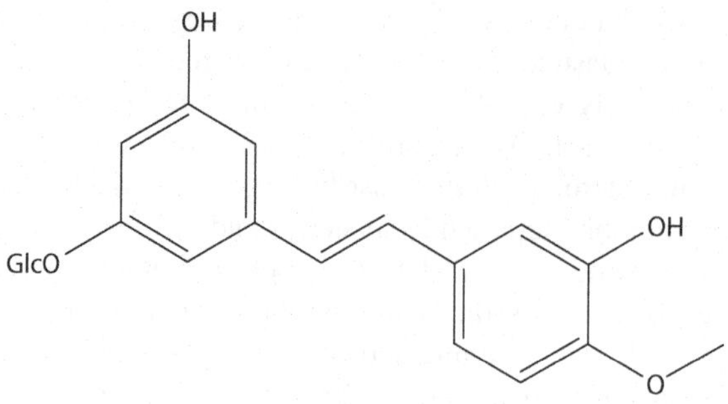

FIGURE 3.5 Rhaponticin.

Rheum tanguticum Maxim. ex Balf. contains rhaponticin (Figure 3.5) that given orally to spontaneous type 2 diabetic obese KK-Ay diabetic mice at a dose of 125 mg/kg/day for 28 days lowered glycaemia and insulinaemia by more than 50%.[111] This stilbene glucoside reduced serum cholesterol and triglycerides and lowered low-density lipoprotein by more than 60% and decreased plasma nonesterified free fatty acids by more than 50%.[111] Furthermore, this treatment increased pancreatic and liver weight and reduced serum aspartate aminotransferase and alanine aminotransferase, whereby hepatic glycogen increased by more than 2 folds.[111] In hepatocytes, glycolysis of glucose yields pyruvic acid which, in mitochondria, is converted into citric acid. In the cytosol, citric acid is converted to acetyl-CoA by ATP citrate lyase.[112] Acetyl-CoA is serves as substrate for the synthesis of malonyl-CoA by acetyl-CoA carboxylase, and fatty acid synthetase use both acetyl-CoA and malonyl-CoA for the construction of fatty acids.[112] Desoxyrhaponticin and rhaponticin from the rhizomes of *Rheum tanguticum* Maxim. ex Balf. contain inhibited fatty acid synthetase with IC_{50} values of 172.6 and 73.2 μM *in vitro*.[113] In MCF-7 cells, these stilbenes decreased fatty acid synthetase expression and enzymatic activities were reduced to 13% and 51%, respectively, at the concentration of 400 μM.[113]

3.18 *Rheum palmatum* L.

Synonyms: *Rheum potaninii* Losinsk.; *Rheum qinlingense* Y.K. Yang, J.K. Wu & D.K. Zhang
Common name: zhang ye da huang (Chinese)
Subclass Caryophyllidae, Superorder Polygonanae, Order Polygonales, Family Polygonaceae
Medicinal use: constipation (China)

Oxysterols such as 22(*R*)-hydroxycholesterol derived from cholesterol in the liver, bind to and activate liver X receptor which heterodimerizes with 9-cis retinoid acid-activated retinoid X receptor.[114] This nuclear receptor and transcription factor controls the expression of cholesterol 7α-hydroxylase (CYP7A1), sterol regulatory-binding element protein-1c in the liver, ABCA1, and G1 and apolipoprotrein E in macrophages.[115] Liver X receptor serves also as a glucose sensor and regulates the expression of phosphoenol carboxykinase, glucose-6-phosphatase, and glucokinase.[116] In adipose tissues, liver X receptor induces the expression of glucose transporter type 4 explaining why liver X receptor agonists improve insulin sensitivity.[117] Rhein from *Rheum palmatum* L. is an antagonist of liver X receptor-α which *K*d values of 46.7 μM.[118] In HepG2, this anthraquinone at a concentration of 25 μM inhibited the expression of adenosine triphosphate-binding cassette protein A1 ABC transporter G1, sterol regulatory element binding protein-1c, fatty acid synthetase, stearoyl coenzyme A desaturase 1, and acetyl-CoA carboxylase induced by liver X receptor agonist GW3965 *in vitro*.[118] Liver X receptor-α antagonism by rhein could at least partially account to the fact that C57BL/6J mice fed a high-fat diet given for 4 weeks, rhein decreased the expression of sterol regulatory element binding protein-1c, fatty acid synthetase, stearoyl coenzyme A desaturase 1 and acetyl-CoA carboxylase and triphosphate-binding cassette protein-A1.[118] In hepatocyte, this anthraquinone increased glucose transporter-2 and decreased 3-hydroxy-3-methylglutaryl-coenzyme A reductase expression.[118]

3.19 *Rheum rhabarbarum* L.

Synonyms: *Rheum franzenbachii* Münter; *Rheum undulatum* L
Common names: bo ye da huang (Chinese); rhubarb
Subclass Caryophyllidae, Superorder Polygonanae, Order Polygonales, Family Polygonaceae
Medicinal use: stop bleeding (China)

Rhapontigenin from the roots of *Rheum rhabarbarum* L. given orally at a dose of 1 mg/kg/day for 4 weeks to Sprague–Dawley rats on high-fat diet prevented weight gain, lowered plasma cholesterol, low-density lipoprotein and triglycerides.[119] At a dose of 5 mg/kg/day, this stilbene prevented the formation of lipid droplets in the cytoplasm of centrilobular hepatocytes.[119] It must be recalled that rhapontigenin is the aglycone of rhaponticin, which is a known inhibitor of fatty acid synthetase as discussed previously.

3.20 *Amaranthus hypochondriacus* L.

Common name: qian sui gu (Chinese)
Subclass Caryophyllidae, Superorder Caryophyllanae, Order Caryophyllales, Family Amaranthaceae
Nutritional use: food (China)
Pharmacological target: atherogenic hyperlipidemia

Bile acids are secreted into the duodenum and almost completly reabsorbed and transported in the liver by active and passive mechanism.[45] This is called the enterohepatic cycle. Bile acid-sequestrating agents like cholestyramine, β-glucans, or pectin from apples bind bile salts in the intestinal lumen and inhibit the reabsorption of bile salts leading to increased fecal bile salts.[45,122] Increased fecal excression of bile salts is compensated by enhanced hepatic synthesis of bile salts from cholesterol via increased activity of cholesterol-7α-hydroxylase.[120] Bile acid-sequestrating agents like cholestyramine, soluble fibers, or pectins lower plasma low density lipoproetin-cholesterol via an increased hepatic expression of low-density lipoprotein expression receptor.[121] A diet containing 200 g/kg of whole seeds of *Amaranthus hypochondriacus* L. given for 4 weeks to White Leghorn pullets lowered total serum cholesterol from 3.8 to 2.9 mmol/L and evoked a mild decrease of high-density lipoprotein–cholesterol, low-density lipoprotein–cholesterol and serum triglycerides.[122,124] This regiment further resulted in a decrease in apolipoprotein AI and apolipoprotein B, a mild decrease in 3-hydroxy-3-methylglutaryl-coenzyme A reductase activity and an increase in enzyme for bile acid biosynthesis cholesterol-7-α-hydroxylase activity.[123] A number of studies have demonstrated that fiber and saponins increase the excretion of endogenous cholesterol and bile acids because they inhibit cholesterol and bile acid intestinal absorbsion.[123] In hepatocytes, bile acids activate farnesoid X receptor which in turn inhibits cholesterol-7-α-hydroxylase.[122] One could reasonably speculate that, fibers and/or saponins in the seeds of *Amaranthus hypochondriacus* L., at least, by inhibiting the reabsorption of biles acids could suppress farnesoid X receptor-induced inhibition of cholesterol 7-α-hydroxylase (CYP7A1).

3.21 *Salicornia herbacea* (L.) L.

Synonym: *Salicornia europaea* L.
Common name: yan jiao cao (China)
Subclass Caryophyllidae, Superorder Caryophyllanae, Order Caryophyllales, Family Chenopodiaceae
Medicinal use: hepatitis (Korea)

ICR mice fed high-fat diet treated with 700 mg/kg of an ethanolic extract of *Salicornia herbacea* (L.) L. for 10 weeks were protected against weight gain by 34%.[125] This regimen lowered fasting glycaemia, had no effect on plasma insulin, and reduced insulin resistance index by 25%.[125] In addition, this extract lowered plasma nonesterified fatty acids, triglycerides, cholesterol, and low-density lipoprotein–cholesterol by 29%, 33%, 27%, and 69%, respectively.[125] In the liver, the extract inhibited the expression of sterol regulatory element binding protein-1a by 64% and therefore attenuated the expression of fatty acid synthetase and glycerol-3-phosphate-acyltransferase and subsequently decreased by 83% hepatic triglycerides compared with untreated group.[125] The extract reduced the expression of liver phosphoenolpyruvate carboxykinase and glucose-6-phosphatase by 75% and 62% compared with untreated group suggesting an amelioration of hepatic insulin sensitivity.[125] 3-Caffeoyl, 4-dihydro-caffeoylquinic acid from this plant at a concentration of 10 μM inhibited the synthesis of triglycerides and cholesterol by HepG2 cells cultured in high glucose concentration via activation of adenosine monophosphate-activated protein kinase via SIRT-1 and LKB1, acetyl-CoA carboxylase phosphorylation and reduced expression of sterol regulatory element-binding protein-1c.[126] In hepatocytes challenged with high glucose, adenosine monophosphate-activated protein kinase phosphorylate and inhibit sterol regulatory element-binding protein-1c[127] and induces the phosphorylation of acetyl-CoA carboxylase.[128]

3.22 *Camellia assamica* (J.W. Mast.) H.T. Chang

Synonym: *Camellia sinensis* var. *assamica* (J.W. Mast.) Kitam.
Common name: pu er cha (China)
Subclass Dillenidae, Superorder Ericanae, Order Theales, Family Theaceae
Medicinal use: tonic (China)

Camellia assamica (J.W. Mast.) H.T. Chang produces methylxanthine alkaloids caffeine, theophylline, theobromine, and theacrine,[129] which are inhibitors of cyclic adenosine monophosphate phosphodiesterase and increase cytosolic cyclic adenosine monophosphate levels in hepatocytes.[130] Increase of cyclic adenosine monophosphate induces phosphoinositide 3-kinase leading to the formation of phosphatidyl inositol phosphate, which activates Ras, Raf, and extracellular signal-regulated kinase, translocation of ATP-dependent human bile salt export pump (Bsep/ABCB11) and consequently the secretion of bile.[131] Theophylline at a dose of 20 mg/kg/h given intravenously decreased bile cholesterol in dogs receiving sodium taurocholate at a dose of 500 mg/h from 282 to 221 μg/mL and increased bile flow from 2.6 to 4.2 mL/15 min.[132] In hepatocytes, the binding of glucagon to its receptor during fasting or the binding of norepinephrine to its receptor during stress activates adenylate cyclase. Increase of cyclic adenosine monophosphate in the cytosol stimulates protein kinase A which activates phosphorylase kinase and subsequently glycogen phosphorylase. This enzyme catalyzes the depolymerization of glycogen into glucose-1-phosphate.[133] Purine alkaloid like theophylline or caffeine inhibit cyclic adenosine monophosphate phosphodiesterase and increase of cyclic adenosine monophosphate resulting in glycogenolysis and increased plasma glucose. Caffeine, theophylline, and theobromine at concentrations of 24, 6.5, and 1 mM commanded *in vitro* a decrease in calcium influx into mitochondria by circa 50% after 2 minutes as well as a diminution of respiration and reduction of

ATP content in rat-liver mitochondria[134] and as such could stimulate adenosine monophosphate-activated protein kinase.[135] In hepatocytes, increase of cyclic adenosine monophosphate induces LKB1 and adenosine monophosphate-activated protein kinase and the activation of adenosine monophosphate-activated protein kinase inhibits 3-hydroxy-3-methylglutaryl-coenzyme A reductase in the liver.[136,137] This is probably why theobromine given at a dose of 700 mg/kg/day given orally to rodents for 4 days with a normal diet lowered plasma cholesterol from 2 to 1.8 mg/dL, increased in high-density lipoprotein–cholesterol from 1 to 1.19 mg/dL, decreased of low-density lipoprotein–cholesterol from 0.9 to 0.5 mg/dL, and decreased plasma triglycerides from 0.6 to 0.3 mg/dL.[138] However, rats fed with a diet including 2.5 g/kg of caffeine for 25 days had limited weight gain, decreased plasma triglycerides from 1.2 to 0.5 mmol/L, and increased plasma cholesterol from 2.3 to 2.5 mmol/L as a result of increased hepatic cholesterogenesis.[139] The same regimen applied 25 days induced reduced body weight gain, decreased plasma triglycerides from 1.1 to 0.7 mmol/L, and unchanged plasma cholesterol.[139] Rodents fed with a cholesterol-enriched diet including 2.5 g/kg of caffeine had a reduction of body weight, an increase in cholesterolaemia from 11.4 to 19.4 mmol/L, increase in hepatic cholesterol from 7.9 to 10.2 g/kg, a decrease of triglyceridaemia from 0.8 to 0.5 mmol/L and an increase in aortic lipogenesis.[139] Surprisingly, 4 month of caffeine treatment at the same dose with a high-cholesterol diet did not evidence any aorthic damages.[139] Caffeine at a dose of 90 mg/kg intravenously in fasting rodents induced an elevation of glycaemia, insulin, and free fatty acids whereby cholesterolaemia was not changed.[140] Coffee consumption in human is correlated with the increased cholesterolaemia as evidenced in an epidemiological study involving 5,858 Japanese men for 6 years.[141] Subjects consuming coffee at a dose of more than 9 cups per day had a cholesterolaemia equal to 220 mg/dL compared with 210 mg/dL for those who do not consume coffee.[141] In the same study, tea consumption was not correlated with hypercholesterolaemia.[141] Theacrine from *Camellia assamica* (J.W. Mast.) H.T. Chang given at a dose of 30 mg/kg/day for 1 week protected the hepatic function of rodents exposed to 18 hours of experimental restrain stress as evidenced by a decrease of plasmatic alanine transaminase from 79.3 to 30.5 U/L and a reduction of serum aspartate transaminase from 84.1 to 48.2 U/L and afforded a reduction of liver inflammation and necrosis.[142] Furthermore, purine alkaloid normalized the enzymatic activities of hepatic superoxide dismutase, glutathione peroxidase, catalase, and glutathione S-transferase in stressed rodents and decreased proinflammatory cytolines interleukin-1, interleukin-6, tumor necrosis factor-α, and IFN-γ. *Camellia assamica* (J.W. Mast.) H.T. Chang could be of value to decrease plasma cholesterol in metabolic syndrome.

3.23 *Garcinia atroviridis* Griff. ex T. Anderson

Common names: asam gelugor (Malay); somkhag (Thai)
Subclass Dillenidae, Superorder Ericanae, Order Hypericales, Family Clusiaceae
Medicinal use: constipation (Thailand)
Pharmacological target: atherogenic hyperlipidemia

In hepatocytes, citrate produced by mitochondria is cleaved in the cytosol by ATP-citrate oxaloacetate lyase into oxaloacetate and acetyl-CoA, the latter being used for the synthesis of fatty acids.[143] The fruits of *Garcinia atroviridis* Griff. ex T. Anderson, which are sour, contain (–)-hydroxycitric acid, which is an ATP-citrate oxaloacetate lyase inhibitor.[144] (–)-Hydroxycitric acid given to Wistar rats at 2% of diet for 15 days decreased food intake, decreased gain in body weight from 55.9 to 28.9 g, and decreased epipdydimal fat mass from 4.3 to 3.1 g.[145] This supplementation decreased plasma triglycerides from 93.5 to 71.9 mg/dL and had no effect on liver fat contents, plasma free fatty acids and plasma cholesterol.[145] Thom evidenced a beneficial of (–)-Hydroxycitric acid given daily at a dose of 1.3 g/day on weight loss in double-blind clinical study.[146]

3.24 *Garcinia dulcis* (Roxb.) Kurz

Common names: bogalot (Philippine); maphut (Thailand)
Subclass Dillenidae, Superorder Ericanae, Order Hypericales, Family Clusiaceae
Nutritional use: jam (Philippines)

The plant contains morelloflavone that inhibited the enzymatic activity of inhibits 3-hydroxy-3-methylglutaryl-coenzyme A reductase with *K*i of 80 μM.[147] It also shelters benzophenones including xanthochymol and guttiferone E,[148] which could have affinity for liver X receptor-α since guttiferone I from a member of the genus *Garcinia* L. is a ligand for liver X receptor and inhibited the binding activity of liver X receptor with an IC_{50} value equal to 3.4 μM *in vitro*.[149] An another constituent of *Garcinia dulcis* (Roxb.) Kurz is α-mangostin,[148] which given orally to C57BL/6 mice on high-fat diet at a dose of 50 mg/kg/day for 6 weeks decreased body weight gain and decreased toward normal plasma alanine aminotransferase and aspartate aminotransferase.[150] At the hepatic level, this regimen decreased hepatic triglycerides from about 100 to 55 μmol/liver (normal: about 25 μmol/liver).[150] This xanthone decreased plasma glucose, triglycerides, free fatty acids, cholesterol, high-density lipoprotein, and low-density lipoprotein.[150] In the liver, α-mangostin (Figure 3.6) increased the expression of silent information regulator T1 (SIRT1) and therefore the phosphorylation of adenosine monophosphate-activated protein kinase and peroxisome proliferator-activated receptor-γ as well as retinoid X receptor-α.[150] In hepatocytes, adenosine monophosphate-activated protein kinase is activated by SIRT1 in response to increased NAD^+.[150] Activated adenosine monophosphate-activated protein kinase induces the activation of peroxisome proliferator-activated receptor-γ that heterodimerizes with retinoid X receptor and binds to peroxisome proliferator-activated receptor response element in the promoters of target genes.[151] Han et al. provided evidence that α-mangostin is not well absorbed orally with preferential distribution in the liver.[152] Clinical trials are warranted.

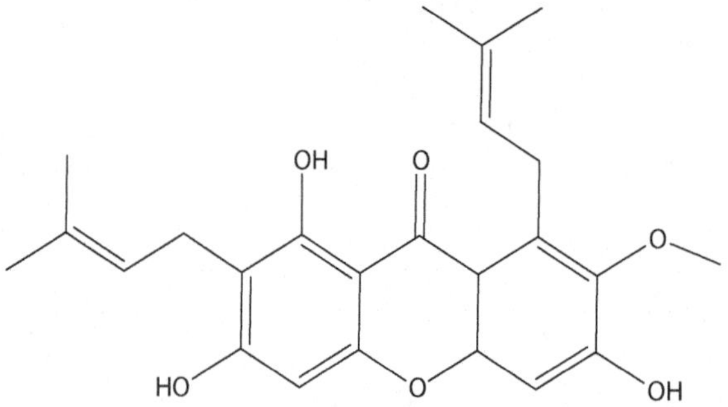

FIGURE 3.6 α-Mangostin.

3.25 *Garcinia mangostana* L.

Synonym: *Mangostana garcinia* Gaertn.
Common names: mangustan (Malay); mangosteen
Subclass Dillenidae, Superorder Ericanae, Order Hypericales, Family Clusiaceae
Medicinal use: diarrhea (Malaysia)
History: By the year 1880, the husk of fruits of *Garcinia mangostana* L. was exported from Malaya is a reputed astringent remedy to treat diarrhea.
Pharmacological target: atherogenic hyperlipidemia

Ethanol extract of pericarps of *Garcinia mangostana* L. inhibited fatty acid synthetase *in vitro* with an IC_{50} value at 1.7 μg/mL.[153] From this extract, α-mangostin, γ-mangostin, garcinone E, and 2,4,6,3′,5′-pentahydroxybenzophenone inhibited fatty acid synthetase with IC_{50} below 10 μM, respectively.[153] In HepG2 cells, γ-mangostin at a concentration of 25 μM induced the expression of acyl-CoA synthetase and carnitine palmitoyltransferase-1 similarly to the peroxisome proliferator-activated receptor-α agonist bezafibrate.[154] In must be recalled that in hepatocytes, fatty acids brought from adipose tissues by the general circulation are acetylated by fatty acyl-CoA synthetase and esterified fatty acid enter mitochondria via carnitine palmitoyltransferase-1 for β-oxidation under peroxisome proliferator-activated receptor-α stimulation.[155] It would be of interest to appraise the peroxisome proliferator-activated receptor-α agonist property of γ-mangostin.

3.26 *Hypericum perforatum* L.

Synonym: *Hypericum nachitschevanicum* Grossh.
Common names: guan ye lian qiao (Chinese); tenturototu (Turkey); St. John's Wort
Subclass Dillenidae, Superorder Ericanae, Order Hypericales, Family Clusiaceae
Medicinal use: stomachache (Turkey)

A flavonoid fraction of *Hypericum perforatum* L. given for 16 weeks orally at a dose of 150 g/kg/day to Wistar rats fed with a diet enriched in cholesterol, lowered plasma cholesterol from about 8 to 4.5 mmol/L, triglycerides from about 3 to 2 mmol/L, halved low-density lipoprotein–cholesterol, and increased high-density lipoprotein–cholesterol.[156] This treatment normalized plasma and liver malondialdehyde and increased the enzymatic activity of superoxide dismutase and catalase.[156] In the liver, pregnane X receptor is a nuclear receptor for bile salts and its activation inhibit the expression of CYP7A1 as well as bile acid transporters resulting in decreased bile salts production and excretion.[157] Hyperforin from *Hypericum perforatum* L. at a concentration of 1 μM activated pregnane X receptor in transfected CV-1 cells.[158]

3.27 *Symplocos racemosa* Roxb.

Common names: lodh (India); lodh tree
Synonyms: *Symplocos intermedia* Brand; *Symplocos macrostachya* Brand; *Symplocos propinqua* Hance
Common name: zhu zi shu (Chinese)
Subclass Dilleniidae, Superorder Primulanae, Order Styracales, Family Symplocaceae
Medicinal use: bleeding gums (India)
Pharmacological target: atherogenic hyperlipidemia
History: the plant was mentioned in the Satapatha Brahmana (700 BC)

Current drugs used to lower plasma cholesterol in obese patients include statins that decrease cholesterol synthesis in hepatocytes, thereby increasing the expression of hepatic low-density lipoprotein receptors and increasing the clearance of atherogenic plasma low-density lipoprotein.[159] Ethanol extract of bark of *Symplocos racemosa* Roxb given orally at a dose of 400 mg/kg/day for 15 days to Sprague–Dawley rats given a high-fat diet evoked a reduction of body weight by 9%, halved total cholesterol and triglycerides, increased high-density lipoprotein–cholesterol, reduced low-density lipoprotein–cholesterol by 60%, and lowered atherogenic index from 9.7 to 3.6.[160] The liver of treated animals lowered hepatic cholesterol as well as thiobarbituric reactive species and increased glutathione, catalase, and superoxide dismutase.[160] Histological observation of hepatic tissues of treated rodents showed a decrease of fat and mononuclear cells infiltrations within lobules.[160] The hypocholesterolemic principles and mechanism of action are unknown. Ursolic acid, corosolic acid, and 2α,3α,19α,23-tetrahydroxyurs-12-en-28-oic acid isolated from this plant inhibited protein-tyrosine phosphatase 1B with IC_{50} values of 3.8, 7.2, and 42.1 μM, respectively.[161] Ursolic acid given to Swiss mice on high-fat diet at 0.05% of drinking water for 15 weeks had no effect on food intake, decreased body weight increase by 10.7% (sibutramine 10.9%).[162] This triterpene lowered plasma glucose from 136 to 78.6 mg/dL (normal: 86.1 mg/dL), had no effect on plasma insulin, and lowered total cholesterol and triglycerides by 25.3% and 17.7%, respectively.[162] This triterpene decreased visceral fat mass by about 50% and attenuated hepatic steatosis.[162]

3.28 *Diospyros kaki* Thunb.

Common names: shi (Chinese); gam (Korean); kaki (Japanese); Japanese persimmon
Subclass Dilleniidae, Superorder Primulanae, Order Styracales, Family Ebenaceae
Medicinal use: astringent (China)
Pharmacological targets: atherogenic hyperlipidemia; insulin resistance

Uripe Fruits of *Diospyros kaki* L. added to the high-fat diet of C57BL16 mice for 14 days had no effect on food intake, weight gain, and decreased triglyceride accumulation in hepatocytes.[163] This supplementation had no effect on glycaemia, decreased plasma cholesterol from about 130 to 100 mg/dL, triglycerides from about 120 to 90 mg/dL, and low-density lipoprotein–cholesterol from about 70 to 50 mg/dL, whereas plasma fatty acids were not affected.[163] In the liver of rodents fed on unripe fruits of *Diospyros kaki* L., the expression of CYP7A1 and 3-hydroxy-3-methylglutaryl-coenzyme A reductase was increased.[163] CYP7A1 and 3-hydroxy-3-methylglutaryl-coenzyme A reductase are in the liver both inhibited by farnesoid X receptor. In the same experiment, feeding on mature fruits had no effect.[163] C57BL/6.Cr mice fed with a diet containing 5% of unripe *Diospyros kaki* L.f. fruits for 10 weeks exhibited reduction of cholesterol, chylomicrons, very low-density lipoprotein, and triglycerides by 23%, 60%, 44%, and 23%, respectively.[163] This supplementation evoked an increased expression of sterol regulatory element-binding protein-2 and low-density lipoprotein receptors and a decrease of liver cholesterol and triglycerides and increased fecal bile acid from 1.3 to 5 μmol/days.[163] Increase of hepatic expression of low-density lipoprotein receptor enhances the plasmatic clearance of atherogenic low-density lipoprotein–cholesterol.[164] In hepatocytes, sterol regulatory element binding proteins-2 is a transcription factor that induces the expression of 3-hydroxy-3-methylglutaryl-coenzyme A reductase and low-density lipoprotein receptor. When levels of cholesterol in hepatocytes are low, sterol regulatory element-binding proteins-2 is activated resulting in cholesterol synthesis and cholesterol uptake from low-density lipoproteins.[165] Conversely, elevated levels of cholesterol in hepatocytes inhibit the transcriptional activity of sterol regulatory element-binding proteins-2 is inhibited resulting in decreased synthesis and uptake of cholesterol.[165] It is therefore reasonable to speculate that intake of fruits of *Diospyros kaki* L.f. in mice may inhibit cholesterol intestinal absorption. In fact, decrease in plasmatic chylomicron is a sign of decreased absorption of dietary cholesterol and triglycerides[45] which is owed to proanthocyanidins present in the fruits.[166] Zou et al. treated orally Sprague–Dawley rats on high-fat diet with condensed tannin fraction from mature fruits of *Diospyros kaki* L.f. at a dose of 100 mg/kg/day for 9 weeks.[167] This regimen had no effect of food intake, body weight gain, decreased plasma triglycerides from 0.9 to 0.7 mmol/L, cholesterol from 7 to 4.2 mmol/L, low-density lipoprotein–cholesterol from 2.7 to mmol/L, fatty acids from 1131.5 to 803.1 μmol/L, increased high-density lipoprotein–cholesterol from 0.3 to 0.4 mmol/L, and lowered atherogenic index from 19 to 8.4. In the liver of treated rats, the fraction lowered triglyceride and cholesterol contents.[167] This regimen increased fecal cholesterol from 18.7 to 20.7 μmol/g feces and cholic acid from 1.6 mg/g feces to 2.3 mg/g feces on probable account of binding with condensed tannins, whereas no increase in fecal triglycerides was observed.[167] This regimen lowered serum aspartate aminotransferase and alanine aminotransferase activities improved hepatic cytoarchitecture.[167] The fraction reduced serum lipoprotein lipase by 40% and increased serum lecithin:cholesterol acyltransferase activity by 24%, whereas this regimen had no effect on hepatic 3-hydroxy-3-methylglutaryl-coenzyme A reductase activities.[167] Rotungenic acid, pomolic acid, 24-hydroxyursolic acid, ursolic acid, 19a,24-dihydroxyurs-12-en-3-on-28-oic acid, oleanolic acid, and spathodic acid isolated from the leaves of *Diospyros kaki* inhibited the enzymatic activity of protein tyrosine phosphatase 1 B *in vitro*.[168] Clinical trials are warranted.

3.29 *Diospyros peregrina* Gürke

Common names: gab (India); gaub persimmon
Subclass Dilleniidae, Superorder Primulanae, Order Styracales, Family Ebenaceae
Medicinal use: diarrhea (Bangladesh)
Pharmacological target: insulin resistance

Aqueous extract of fruits of *Diospyros peregrina* Gürke given orally at a dose of 100 mg/kg/day to nicotinamide-streptozotocin-induced type 2 diabetic Wistar rats (fasting glucose levels between 140 and 200 mg/dL) for 28 days decreased plasma glucose from about 190 to 125 mg/dL (normal: about 75 mg/dL; glibenclamide 1 mg/kg: about 100 mg/dL).[169] This regimen decreased plasma cholesterol and triglycerides toward normal value similarly with glibenclamide at 1 mg/kg.[169] At the hepatic level, the extract decreased oxidative reactive substances lowered hepatic lipid peroxidation, increased glutathione contents, and increased both superoxide dismutase and catalase activity.[169] The active principle involved here as well are precise molecular basis of activity are apparently unknown but one could suggest a reduction of insulin resistance. In hepatocytes, insulin binding to its receptor activates glucokinase and also sterol regulatory element-binding protein-1c, which induces the expression of fatty acid synthetase.[165] Furthermore, the activation of sterol regulatory element-binding protein-1c by insulin is augmented during insulin resistance leading to accumulation of triglycerides on the liver and increased production of very low-density lipoproteins.[170] Glucokinase, also known as hexokinase, in the liver catalyzes the first step of glucose metabolism and determines the rate of glucose use and glycogen synthesis.[171] In addition, the synthesis of fatty acids in the liver and resulting accumulation of fatty acyl-CoA activates nuclear factor-κB as well as protein kinase C_θ from which serine phosphorylation of insulin receptor substrate-1/2, worsening further insulin resistance.[172]

3.30 *Citrullus lanatus* (Thunb.) Matsum. & Nakai

Synonyms: *Citrullus edulis* Spach; *Citrullus vulgaris* Schrad.; *Citrullus colocynthis* var. *lanatus* (Thunb.) Matsum. & Nakai *Colocynthis citrullus* L.) Kuntze; *Cucurbita citrullus* L.; *Momordica lanata* Thunb.
Common names: xi gua (Chinese); tarambuja (India); watermelon
Subclass Dilleniidae, Superorder Violanae, Order Cucurbitales, Family Cucurbitaceae
Medicinal use: jaundice (China)

Juice of fruits of *Citrullus lanatus* (Thunb.) Matsum. & Nakai (Figure 3.7) in 2% drinking water given to low-density lipoprotein receptor deficient mice on high-fat diet for 12 weeks had no effect on food intake, lowered weight gain, and reduced fat mass.[173] This treatment lowered serum cholesterol from approximately 2100 to 1900 mg/dL, lowered intermediate low-density lipoprotein–cholesterol, and lowered aortic atherosclerotis.[173] Furthermore, this fruit juice lowered plasma, interferon-γ, increased anti-inflammatory interleukin-10 and increased the plasmatic concentration of citrulline, which is accumulated in the fruits.[173] Fruits given as 0.3% of high-fat diet to Sprague–Dawley rats decreased plasma triglycerides, cholesterol, and low-density lipoprotein–cholesterol.[174] This regimen lowered oxidative reactive substances and increased plasma superoxide dismutase and catalase.[174] The supplementation decreased plasma aspartate aminotransferase, alanine aminotransferase, and alkaline phosphatase. In the liver of treated rodents, fatty acid synthase, 3-hydroxy-3methyl-glutaryl-CoA reductase, sterol regulatory element-binding protein-1, sterol regulatory element-binding protein 2, and cyclooxygenase-2 gene expression were significantly downregulated.[174] Evidence suggest that

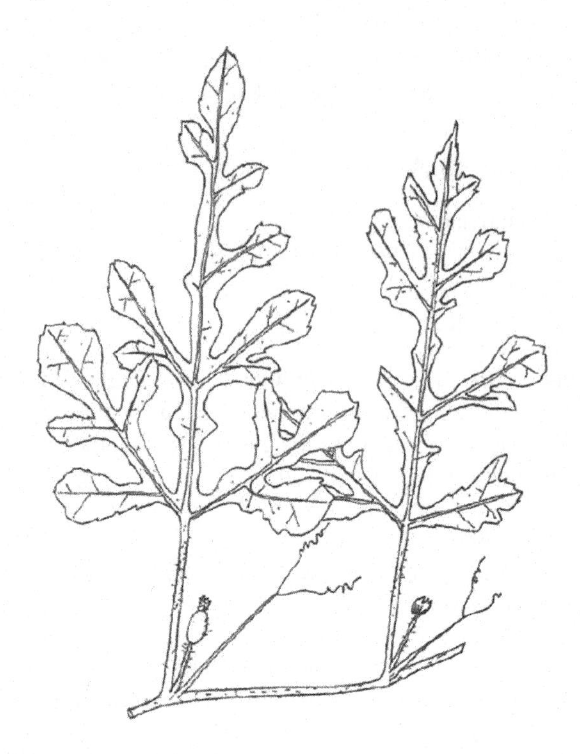

FIGURE 3.7 *Citrullus lanatus* (Thunb.) Matsum. & Nakai.

citrulline could be of value for hepatic steatosis.[175] Citrulline given orally at a dose of 2.5 g/kg/day to C57BL/6J mice on Western diet for 6 weeks had no effect on body weight gain, lowered fasting glycaemia, insulin resistance, and hepatic triglyceride levels.[176] High-fat, high-fructose diet Sprague–Dawley rats receiving orally citrulline at a dose of 1 g/kg/day for 8 weeks developed lower level of hepatic triglycerides, decreased plasma triglycerides, and insulin levels.[177] Watermelon intake could be beneficial in metabolic syndrome.

3.31 *Cucurbita pepo* L.

Common name: xi hu lu (Chinese); pumpkin
Subclass Dilleniidae, Superorder Violanae, Order Cucurbitales, Family Cucurbitaceae
Medicinal use: gastritis (Laos)
Pharmacological target: atherogenic hyperlipidemia

A mixture of seeds of *Cucurbita pepo* L. (Figure 3.8) containing polyunsaturated fatty acids given to Wistar rats fed on high-cholesterol diet at a dose of 333 g/kg of diet reduced weight gain, lowered plasma cholesterol from 1 to 0.8 g/L, normalized plasma triglycerides, halved low-density lipoprotein–cholesterol and reduced atherogenic index from 3.6 to 1.6.[178] The plasma and liver of treated animals registered a decrease in saturated fatty acids and an increase in oleic acid, linoleic acid, and linolenic acid.[178] Besides, this treatment lowered malondialdehyde level in plasma, increased the plasmatic concentration in glutathione, and the elevated the enzymatic activity of activity of superoxide dismutase and glutathione peroxidase.[178] This treatment elicited a reduction of lipid vacuolization in the liver.[178] It must be recalled that polyunsaturated fatty are nonlinear acids hence more difficult to pack in very low-density lipoproteins, causing a decrease in plasma concentration of very low-density lipoproteins and consequently low-density lipoproteins.[95]

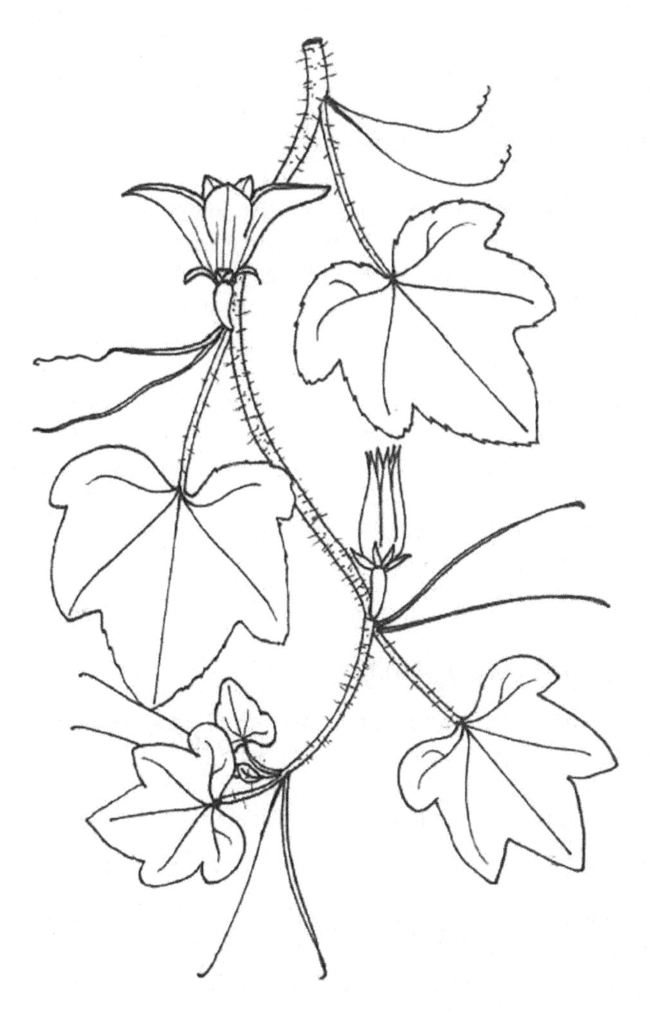

FIGURE 3.8 *Cucurbita pepo* L.

Spontaneous type 2 diabetic Goto–Kakizaki rats fed for 49 days with a diet containing 1% of fruits of *Cucurbita maxima* Duch. exhibited lower postprandial glycaemia in glucose oral tolerance than untreated rodents.[179] This supplementation decreased serum cholesterol from 109 to 93.6 mg/dL, plasma free fatty acids from 0.6 to 0.1 mEq/L, triglycerides from 47.9 to 45.7 mg/dL, and atherogenic index from 0.7 to 0.4.[179] Trigonelline and nicotinic acid isolated from the fruits evoked a mild reduction of glycaemia in Goto–Kakizaki rats given a diet containing 11.2 or 10 mg/20 g/day of these alkaloids respectively for 49 days, improved glucose tolerance, and lowered tumor necrosis factor-α.[179] Trigonelline and nicotinic acid evoked a reduction of serum cholesterol, atherogenic index, serum triglycerides, free fatty acids, and in particular, serum total bile.[179] Zhang et al. provided evidence that trigonelline from given orally to Sprague–Dawley rats at a dose of 40 mg/kg/day for 8 weeks lowered plasma cholesterol from 3.5 to 1.7 mmol/L (normal: 1.2 mmol/L), triglycerides from 1.1 to 1 mmol/L (normal: 0.9 mmol/L), low-density lipoprotein–cholesterol from 0.7 to 0.4 mmol/L (normal: 0.1 mmol/L), and increased high-density cholesterol from 0.6 to 0.7 mmol/L (normal: 1.1 mmol/L).[180] This regimen lowered plasma aspartate aminotransferase and alanine aminotransferase improved hepatic cytoarchitecture with decreased intracellular triglyceride droplets.[180] This regimen decreased liver cholesterol from 1.8 to 1.3 mmol/L (normal: 1.6 mmol/L) and liver triglycerides from 3.8 to 2.5 mmol/L (normal: 1.1 mmol/L), decreased malondialdehyde, and increased the activity of superoxide dismutase.[180] In metabolic syndrome, the lipolysis of triglycerides in adipose tissues increase the delivery of fatty acids to the liver.[181] The binding of nicotinic acid to hydroxycarboxylic acid receptor-2 (HCAR2 or GPR109A) expressed at the surface of adipocytes inhibits adenylate cyclase, decreases cytosolic contents of cyclic adenosine monophosphate hence inactivation of protein kinase A and subsequent inactivation of hormone sensitive lipase, which catalyze the release of unesterified fatty acids from stored triglycerides.[182] As a result, decreased release of free fatty acids from adipose tissues to the general circulation induces a decrease in free fatty acid concentration, hence lower free fatty acid supply to the liver, decreased hepatic synthesis of triglycerides and subsequent decrease in plasma very low-density lipoprotein and low-density lipoprotein.[182] Nicotinic acid (niacin) has been used to lower plasma triglyceride, very low-density lipoprotein–cholesterol, low-density lipoprotein–cholesterol in hyperlipidemic patients and has been associated with unpleasant side effects including flushes as well as hepatotoxicity.[183] Besides, nicotinic acid inhibits the hepatic catabolism of high-density lipoprotein apolipoprotein A-I resulting in plasmatic increases in high-density lipoprotein–cholesterol.[184] N-methylation of nicotinic acid in first-pass metabolism yields trigonelline.[185] Intake of pumpkin could be beneficial in metabolic syndrome.

3.32 *Lagenaria siceraria* (Mol.) Standl.

Synonyms: *Cucumis mairei* H. Lév.; *Cucurbita lagenaria* L.; *Cucurbita leucantha* Duchesne; *Cucurbita siceraria* Molina; *Lagenaria vulgaris* Ser.
Common names: hu lu (Chinese); kalubay (Philippines); bottle gourd
Subclass Dilleniidae, Superorder Violanae, Order Cucurbitales, Family Cucurbitaceae
Medicinal use: cough (Philippines)

Methanol extract of fruits of *Lagenaria siceraria* (Mol.) Standl. given orally at a dose of 300 mg/kg/day orally to Wistar rats on high-fat diet for 30 days reduced cholesterolemia from 290.1 to 228.5 mg/dL, low-density lipoprotein–cholesterol from 195.1 to 120.5 mg/dL, triglycerides from 232.4 to 181.7 mg/dL, very low-density lipoprotein from 46.4 to 36.3 mg/mL, and increased high-density lipoprotein from 48.5 to 71.6 mg/dL.[186] Furthermore, body weight was lowered from 26.8 to 9.6 g, excreted fecal bile acids were increased from 14.6 to 29.6 mg/dL.[186] This hypolipidemic effect was confirmed by Nadeem et al. whereby ethanol extract of fruits of *Lagenaria siceraria* (Mol.) Standl. given orally to high-fat diet Wistar rats at a dose of 300 mg/kg/day for 30 days lowered body weight from 198.8 to 189.8 g.[187] This treatment reduced fasting blood glucose from 157.7 to 92.6 mg/dL, cholesterolemia from 158.3 to 101.9 mg/dL, triglycerides from 244.9 to 197.3 mg/dL, low-density lipoprotein from 69.5 to 47.9 mg/dL, very low-density lipoprotein from 48.9 to 39.4 mg/dL, and tumor necrosis factor-α from 4.3 to 2 pg/mL.[187] The precise molecular mechanism involved here is unknown and accounts probably from cucurbitacins including cucurbitacin B,[188] which protected rodents against carbon tetrachloride hepatotoxicity in rodents.[189,190] In obesity and type 2 diabetes, plasma interleukin-6 levels are maintained in a persistent, chronically elevated state.[191] The binding of interleukin-6 to its receptor at the surface of hepatocytes activates STAT-3 resulting in decreased sensitivity of insulin receptor via SOCS3 and insulin receptor substrate-1/2 phosphorylation.[191,192] Cucurbitacin B is known to inhibit STAT3 signaling,[193] and although too toxic for therapeutic uses,[194] it could serve as a template to design STAT3 inhibitors for metabolic syndrome. Insulin promotes the removal of low-density lipoprotein from the plasma by increasing low-density lipoprotein receptor expression[195] and could be reasonably inferred that hypolipidemic effect of *Lagenaria siceraria* (Mol.) Standl. in Wistar could, at least involve increased hepatic insulin sensitivity via STAT3 inhibition. This point needs to be examined.

3.33 *Momordica charantia* L.

Common names: ku gua (Chinese); periah (Malay); karela (Pakistan); balsam-apple
Subclass Dilleniidae, Superorder Violanae, Order Cucurbitales, Family Cucurbitaceae
Medicinal use: diabetes (Sri Lanka)
Pharmacological targets: atherogenic hyperlipidemia; insulin resistance

Evidence suggests that the fruits of *Momordica charantia* L. accumulate cucurbitane-type triterpenes, the oral administration of which in rodent decrease hepatic insulin resistance. Ethylacetate fraction of fruits of *Momordica charantia* L. given orally to alloxan-induced diabetic male ddY strain mice with oral hypoglycemic activity afforded 5β,19-epoxy-3β,25-dihydroxycucurbita-6,23(E)-diene and 3β,7β,25trihydroxycucurbita-5,23(E)-dien-19-al which given orally at a dose of 200 mg/kg lowered glycaemia from about 90 to 75 mg/dL after 6 hours.[196] 5β,19-Epoxy-25-methoxy-cucurbita-6,23-diene-3β,19-diol isolated from this plant at a dose of 20 μM inhibited the expression of protein tyrosine phosphatase 1B and interleukin-1β by mouse FL83B hepatocytes exposed to tumor necrosis factor-α via inhibition of IKK and IκB phosphorylation.[197] Daily gavage with fruits of *Momordica charantia* L. at a dose of 5 g/kg/day for 16 weeks to C57BL/6 mice fed with high-fat diet reduced liver mass by 30%.[198] This treatment improved glucose tolerance, serum glucose, and lowered insulinemia below value obtained with rodent getting control diet.[198] Besides, the treatment lowered interleukin-6, triglyceridemia, low-density lipoprotein–cholesterol, and halved free fatty acids serum contents.[198] In the liver of treated rodents triglycerides and cholesterol were both reduced as well as the expression of Sterol regulatory element-binding protein-1c and its targets fatty acid synthetase and acetyl-CoA-carboxylase-1.[198] In addition, liver glutathione S-transferase and superoxide dismutase activity were increased.[198] The treatment increased the hepatic expression of mitochondrial fusion-related protein mitofusin-1 and decreased the expression of mitochondrial fission-related protein by more than 50%.[198] In hepatocytes, insulin binding to its receptor as well as activation of liver X receptor activate Sterol regulatory element-binding protein-1c, which induces the expression of acetyl-CoA carboxylase, fatty acid synthase, acetyl-CoA synthetase, glycerol-3-phosphate acyltransferase-1, and stearoyl-CoA desaturase.[165] Besides activation of farnesoid X receptor, by agonists, increases SHP levels, which in turn reduces sterol regulatory element-binding protein-1c expression.[199] It must be recalled that cucurbitane-type triterpenes are somewhat similar in structure with bile acids allowing to suggest that cucurbitacins may act as farnesoid X receptor agonists. In a subsequent study, extract of fruits of *Momordica charantia* L. given orally for 4 weeks at a dose of 0.4 g/kg/day to male C57BL/6J mice on high-fat diet reduced visceral fat diet from about 2.2 to 1.5 g, glycaemia from about 150 to 105 mg/dL, triglycerides from 135 to 105 mg/dL.[200] This regimen lowered total cholesterol from 146 to 113.3 mg/dL and free fatty acids from 1.2 to 0.8 meq/L.[200] Insulin was reduced from 0.9 to 0.8 μg/mL and insulin resistance score was reduced from 7 to 3.3.[200] In the liver, this regimen induced the phosphorylation of adenosine monophosphate-activated protein kinase, decreased glucose-6-phosphate, phosphoenolpyruvate carboxykinase, sterol regulatory element-binding protein-1c, and fatty acid synthetase were lowered as well as apolipoprotein C-III.[200] In the liver, activation of farnesoid X receptor by agonist is known to inhibit sterol regulatory element-binding protein-1c and the expression of apolipoprotein C-III and to activate peroxisome proliferator-activated receptor-α.[201] With regard to glucose metabolism, farnesoid X receptor inhibits glucose-6-phosphatase and phosphoenolpyruvate carboxykinase, key enzymes involved in hepatic gluconeogenesis.[201] Other interesting principles in the fruits of *Momordica charantia* L. is 13-Oxo-9(Z),11(E),15(Z)-octadecatrienoic which induced the expression of genes targeted by peroxisome proliferator-activated receptor-α including carnitine palmitoyltransferase I, acyl-CoA oxidase, acyl-CoA synthetase, and uncoupling protein-2 as well as fatty acid translocase in mouse primary hepatocytes and this effect was abolished by peroxisome proliferator-activated receptor-α-specific antagonist.[202] This unsaturated fatty acid mixed with high-fat diet at 0.05% given to KK-Ay mice had no effect on body mass but lowered serum, liver, and skeletal muscles triglycerides.[202] In the liver of treated rodents, this unsaturated fatty acid increased the hepatic expression of α carnitine-palmitoyl transferase II, acyl-CoA oxidase, fatty acid translocase, acyl-CoA synthetase, and uncoupling protein-2.[202] The regimen did not affect the expression of sterol regulatory element-binding protein-1c nor ABC subfamily A member 1.[202] The treatment lowered glycaemia by 20% and plasma insulin by 30% and increased plasma adiponectin.[202] In physiological conditions, the binding of adiponectin to adiponectin receptor expressed by skeletal muscle cells increases the concentration of calcium in the cytoplasm, activates calcium/calmodulin-dependent protein-kinase kinase (CaMKK), adenosine monophosphate-activated protein kinase and SIRT1 (in response to elevated NAD^+) resulting in enhanced expression of peroxisome proliferator-activated receptor coactivator-1α increasing thus insulin sensitivity.[203] Oral glucose tolerance tests performed on the 3rd week evidenced a mild improvement of glucose tolerance.[202] The fruits of *Momordica charantia* L. should, in moderate amount, be part of diet of patients with metabolic syndrome but the exact dose to take is as for yet unknown. The fruits that are powerfully hypoglycemic and hypoglycemic coma in children have been reported.[204]

3.34 *Sechium edule* (Jacq.) Sw.

Synonyms: *Chayota edulis* (Jacq.) Jacq.; *Sechium americanum* Poir.; *Sicyos edulis* Jacq.
Common names: fo shou gua (Chinese); chayote (Philippines)
Subclass Dilleniidae, Superorder Violanae, Order Cucurbitales, Family Cucurbitaceae
Nutritional use: vegetable (Philippines)
Pharmacological target: atherogenic hyperlipidemia

Extract of shoots of *Sechium edule* (Jacq.) Sw. at a concentration of 1 mg/mL inhibited *in vitro* the accumulation of triglycerides in HepG2 cells cultured with oleic acid.[205] This treatment reduced the expression of sterol regulatory element-binding protein-1 and its targets fatty acid synthetase and glycerol-3-phosphate-acyltransferase.[205] Cholesterol synthesis was decreased as evidenced by a decreased 3-hydroxy-3-methylglutaryl-coenzyme A reductase and low-density lipoprotein receptor.[205] Carnitine palmitoyltransferase I and peroxisome proliferator-activated receptor-α expression were increased as well as phosphorylated adenosine monophosphate-activated protein kinase.[205] In physiological conditions, activation of farnesoid X receptor by agonist in the liver result in decrease in sterol regulatory element-binding protein-1c and downstream fatty acid synthetase and glycerol-3-phosphate-acyltransferase and activates peroxisome proliferator-activated receptor-α.[49] Simultaneously, activation of farnesoid X receptor by agonist lowers low-density lipoprotein receptor expression as well as 3-hydroxy-3-methylglutaryl-coenzyme A reductase lowering thus hepatic cholesterol.[49] In this experiment, sterol regulatory element-binding protein-2 was decreased on probable account of reduced cytosolic concentration of cholesterol.[206] The active principles here are unknown. Consumption of *Sechium edule* (Jacq.) Sw. could be of value in metabolic syndrome.

3.35 *Octomeles sumatrana* Miq.

Common names: binuang (Malay); binonang (Philippines)
Subclass Dilleniida, Superorder Violanae, Order Cucurbitales, Family Datiscaceae
Medicinal use: tonic (Philippines)

Aqueous extract of bark of *Octomeles sumatrana* Miq. (Figure 3.9) given to streptozotocin-induced diabetic Sprague–Dawley rats at a dose of 0.5 g/kg, twice a day for 21 days reduced fasting glycaemia from about 27.5 to 10 mmol/L increased body weight and reduced food consumption.[207] This treatment induced the expression of hepatic glucose transporter-2 and repressed the expression of glucose-6-phosphatase and phosphoenolpyruvate carboxykinase.[207] This medicinal plant, like many others precious species, will disappear soon because

FIGURE 3.9 *Octomeles sumatrana* Miq.

of current burning and logging of the Southeast Asian rainforest for palm oil and the principles involved here may never be known.

3.36 *Brassica oleracea* L.

Synonyms: *Crucifera brassica* E.H.L. Krause; *Napus oleracea* (L.) K.F. Schimp. & Spenn.
Common name: ye gan lan (Chinese); cabbage
Subclass Dilleniidae, Superorder Capparanae, Order Capparales, Family Brassicaceae
Medicinal use: carminative (China)
History: the plant was known to Theophrastus (371–287 BC), Greek philosopher and botanist

Ethanol extract of sprout of *Brassica oleracea* L. given orally to high-fat diet fed Sprague–Dawley rats for 21 days prevented weight gain, reduced cholesterol by 7%, triacylglycerol by 24%.[208] At the liver level, the extract and increased the activity of antioxidant enzymes NAD(P)H:quinone reductase, superoxide reductase, and glutathione S-transferase.[208] *Brassica oleracea* L., like most member of the Brassicaceae, are well known to shelter glucosinolates which upon tissue damage yield isothiocyanates including allyl isothiocyanate.[209] Ernst et al. provided evidence that allyl isothiocyanate at a concentration of 25 μmol/L in NIH3T3 cells induced the phosphorylation of extracellular signal-regulated kinase-1/2 and nuclear translocation of Nrf2 and subsequent expression of glutamyl cysteine synthetase and heme oxygenase-1and NAD(P)H:quinone oxidoreductase 1.[210] It must be recalled that Nrf-2 nuclear translocation is induced by reactive oxygen species, which are generated in cells by isothiocyanates hence hepatic antioxidant activity.[211,212] Daily intake of 150 mL of fresh juice of *Brassica oleraceae acephala* for 3 months by subjects with a cholesterolaemia superior to 200 mg/dL (hypercholesterolaemia) and a triglyceridemia inferior to 150 mg/dL (normal) resulted in an increase of high-density lipoprotein–cholesterol by 27%, a decrease in low-density lipoprotein–cholesterol by 10% and a lowering of atherogenic index but did not reduce cholesterol-aemia nor triglyceridaemia. Kale consumption increases serum glutathione peroxidase by 74.8%.[213] Type 2 diabetic patients with 4 weeks supplementation with broccoli sprouts powder at a dose of 10 g/days for 4 weeks reduced triglycerides from 174 to 135 mg/dL and mildly reduced atherogenic index in type 2 diabetic patients.[214] Oxidized low-density lipoproteins play an important role in the early stage of atherosclerosis.[137] Cabbage intake could be beneficial in metabolic syndrome.

3.37 *Brassica rapa* L.

Synonym: *Brassica campestris* L.
Common names: niuma (India); turnip
Subclass Dilleniidae, Superorder Capparanae, Order Capparales, Family Brassicaceae
Nutritional use: vegetable (India)

Spontaneous type 2 diabetic obese *db/db* mice fed with a diet containing 0.26/100 g of ethanol extract of roots of *Brassica rapa* L. for 5 weeks had lower mass of adipose tissue, liver and increased skeletal muscle weight compared with untreated animals.[215] This diet lowered plasma cholesterol, free fatty acid and triglycerides, as well as plasma insulin.[215] Intraperitoneal glucose tolerance test performed at the end of the diet revealed that the extract improved glucose tolerance after 120 minutes.[215] The extract increased hepatic glucokinase activity and lowered glucose-6-phosphatase activities and increased glycogen contents.[215] The supplementation decreased hepatic triglyceride, decreased the activity of fatty acid synthetase, carnitine palmitoyltransferase, and phosphatidate phosphohydrolase, which is the limiting enzyme for triglyceride synthesis.[215] This extract lowered hepatic cholesterol and inhibited acyl-CoA: cholesterol acyltransferase, which catalyze the esterification of cholesterol into cholesteryl ester, and well as 3-hydroxy-3-methylglutaryl-CoA reductase activities.[215] The active principles and mechanisms are not established with certainty. *Brassica rapa* L. like most members of the Brassicaceae are well known to shelter glucosinolates, which on tissue damage yield isothiocyanates including allyl isothiocyanate.[216] The isothiocyanate sulforaphane that occurs in members of the family Brassicaceae when given orally to C57BL/6N mice at 0.1% of high-fat diet had no effect on food intake, evoked a decrease of body weight, reduced the liver weight by approximately 10.0%, and attenuated the accumulation of triglyceride droplets in the liver.[217] This supplementation decreased serum cholesterol by 10% and liver triglyceride by about 16.2%.[217] Furthermore, sulforaphane was found to induce the phosphorylation of adenosine monophosphate-activated protein kinase.[217] In the liver, phosphorylation of adenosine monophosphate-activated protein kinase command inhibition of 3-hydroxy-3-methylglutaryl-coenzyme reductase.[217] It must be recalled that adenosine monophosphate-activated protein kinase is induced by reactive oxygen species that are generated in cells by isothiocyanates.[128,135]

3.38 *Brassica napus* L.

Synonym: *Brassica rugosa* (Roxb.) L.H. Bailey
Common name: sarshapa (India); rapeseed
Subclass Dilleniidae, Superorder Capparanae, Order Capparales, Family Brassicaceae
Medicinal use: diuretic (India)
Pharmacological target: atherogenic hyperlipidemia

In a double blind clinical trial, 30 g of oil seeds from a member of the family Brassicaceae given daily for 6 weeks to hypercholesterolaemic subjects reduced serums total cholesterol from 6.1 to 5.7 mmol/L and low-density lipoprotein–cholesterol was reduced by 5.4%.[218] The oil expressed from the seeds of *Brassica napus* L. commonly known as rapeseed oil contains notably oleic acid, α-linolenic, and linoleic acid.[219] Polyunsaturated linoleic acid such as α-linolenic are known to lower plasma cholesterol[220] presumably because they are nonlinear and hamper the formation of very low-density lipoproteins in the liver.[95,220] Oleic is the favorite substrate for acyl cholesterol acyl transferase and increase of this monounsaturated fatty acid may decrease unesterified cholesterol hepatocytes contents and as a consequence increase low-density lipoprotein receptor expression and liver X receptor inactivation.[221] Indole-3-carbinol, a common constituent of members in the family Brassicaceae, at a concentration of 50 μg/mL reduced the secretion of apolipoprotein B by HepG2 cells by 40%, halved both synthesis and secretion of triglycerides, reduced the synthesis of cholesterol by 50% and reduced cholesterol secretion by 38%.[222] That alkaloid decreased diacylglycerol acyltransferase-1 and 2 by 56% and 59%, and reduced fatty acid synthetase expression by 25% and acyl-CoA cholesterol acyltransferase expression by 34%.[222] This treatment inhibited the expression of sterol regulatory element-binding protein-1 and sterol regulatory element-binding protein-2 by 44% and 37%, respectively.[222] The enzymatic activity of microsomal triglyceride transfer protein was reduced by 17%.[222]

3.39 *Nasturtium officinale* W.T Aiton

Synonyms: *Sisymbrium nasturtium-aquaticum* L.; *Rorippa nasturtium-aquaticum* (Linnaeus) Hayek.

Common names: dou ban cai (Chinese); watercress

Subclass Dilleniidae, Superorder Capparanae, Order Capparales, Family Brassicaceae

Medicinal use: sore throat (China)

Pharmacological target: atherogenic hyperlipidemia

FIGURE 3.10 *Nasturtium officinale* R. Br.

Ethanol extract of aerial parts of *Nasturtium officinale* R. Br. (Figure 3.10) given orally to high-fat diet rats at a dose of 500 mg/kg/day by gavage for 30 days brought serum total cholesterol, triglycerides and low-density lipoprotein–cholesterol to normal values and lowered alanine aminotransferase, aspartate aminotransferase and alkaline phosphatase.[223] At the hepatic level, this extracts increased glutathione significantly.[223] Furthermore, the enzymatic activities of both hepatic catalase and superoxide dismutase increased in treated animals as well as hepatic malondialdehyde.[223] Isothiocyanates are probably responsible for the observed improvement of hepatic oxidative and lipid profile in high-fat diet fed rats.

3.40 *Raphanus sativus* L.

Common names: luo bo (Chinese); Chinese radish
Subclass Dilleniidae, Superorder Capparanae, Order Capparales, Family Brassicaceae
Medicinal use: diuretic (China)
Pharmacological target: atherogenic hyperlipidemia

Squeezed juice of *Raphanus sativus* L. (Figure 3.11) diluted 10-fold with tap water and given 9 days to high-fat diet fed Wistar rats reduced serum and hepatic lipid peroxidation, increased the enzymatic activity of glutathione peroxidase and normalized their scavenging capacities.[224] Ethanol extract of sprouts given by gavage to Sprague–Dawley rats at a dose of 1.5 g/kg/day for 4 days induced an increase in bile secretion.[225] Juice of tubers of *Raphanus sativus* given at a dose of 0.1 mL/10g for 4 days by gavage to female C57BL/6 mice after 34 days of a diet enriched with cholesterol and cholic acid lowered to normal serum cholesterol and triglycerides and elevated high-density lipoprotein–cholesterol.[226] Isothiocyanates are common in members of the family Brassicaceae.[212] Manley and Ding provided

FIGURE 3.11 *Raphanus sativus* L.

evidence that reactive oxygen species impair farnesoid X receptor inhibition of cholesterol-7-α-hydroxylase which is a key enzyme in bile acids synthesis from cholesterol.[227] Induction of Nrf2 and subsequent expression of antioxidant enzymes in the liver may alleviate this impairment.

3.41 *Morus alba* L.

Synonyms: *Morus atropurpurea* Roxb.; *Morus australis* Poir.; *Morus indica* L.
Common names: sang (Chinese); white mulberry
Subclass Dillenidae, Superorder Malvanae, Order Urticales, Family Moraceae
Medicinal use: diuretic (China)
Pharmacological target: atherogenic hyperlipidemia

Morus alba L. (Figure 3.12) elaborates 1-deoxynojirimycin, a very well known α-glucosidase inhibitor from which the drug miglitol was derived.[228] This polyhydroxy piperidine given to Sprague–Dawley rats at a dose 1 mg/kg of body weight/day orally for 4 weeks experiences a loss of epididymal adipose tissue from 0.8 g/100 g of body weight to 0.6 g/100 g of body weight, a decrease in glycaemia from 104.4 to 93.4 mg/dL, and increased plasma adiponectin.[229] At the hepatic level, this treatment evoked a reduction of hepatic triglycerides from 50.7 to 40.1 μmol/g.[229] This alkaloid evoked an increase in activity and expression of carnitine palmitoyl transferase and acyl-CoA oxidase.[229] This alkaloid elevated the expression of adenosine monophosphate-activated protein kinase and commanded a decrease of hepatic lipid peroxidation.[229] In a subsequent study, 1-deoxynojirimycin given at a dose of 5 mg/kg body weight/day to C57BL/6J mice feeding on a high-fat diet for 12 weeks

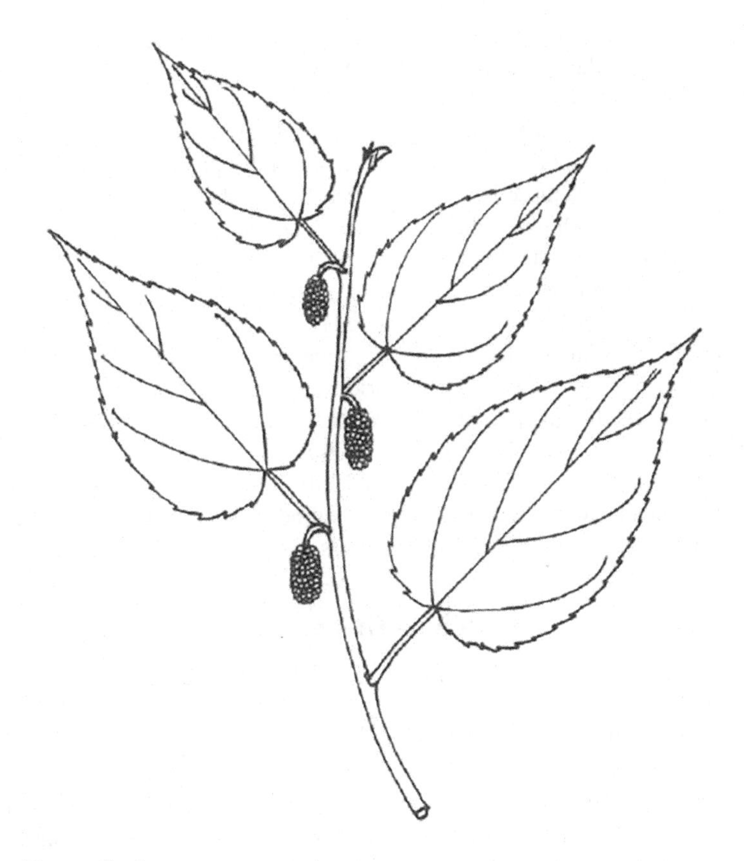

FIGURE 3.12 *Morus alba* L.

evoked an increase in food intake from 1.9 to 2.2 g/day, a decrease in perirenal adipose tissue weight from 1.7 g/100 g of body weight to 0.8 g/100 g of body weight, a decrease in mesenteric adipose tissue from 1.2 g/100 g of body weight to 0.7 g/100 g of body weight, and epididymal adipose tissue from 3.8 g/100 g of body weight to 2.1 g/100 g of body weight.[230] This regimen evoked a decrease in serum triglycerides, cholesterol, glucose, insulin, free fatty acids, but adiponectin increased from 4.9 to 7.1 μg/mL.[230] Liver biochemical parameters evidenced a decrease in triglycerides, total cholesterol and thiobarbituric reactive.[230] Furthermore, in liver of treated rodents, the expression of carnitine palmitoyltransferase, acyl-CoA oxidase, and peroxisome proliferator-activated receptor-α were increased.[230] Physiologically, adenosine monophosphate-activated protein kinase is activated by adiponectin via LKB.[57] In hepatocytes, adenosine monophosphate-activated protein kinase stimulates fatty acid oxidation by upregulating expression of proteins involved in fatty acid β-oxidation, including peroxisome proliferator-activated receptor-α and carnitine palmitoyltransferase-1.[231] Phosphorylated adenosine monophosphate-activated protein kinase directly inhibits of acetyl-CoA carboxylase lifting malonyl-CoA inhibition on carnitine acyltransferase and thus promoting the entry of fatty acyl-CoA in mitochondria for β-oxidation.[58]

3.42 *Urtica dioica* L.

Synonym: *Urtica galeopsifolia* Wierzb. ex Opiz
Common name: yi zhu qian ma (Chinese); isirgan (Turkey); stinging nettle
Subclass Dillenidae, Superorder Malvanae, Order Urticales, Family Urticaceae
Medicinal use: stomachache (Turkey)
Pharmacological target: atherogenic hyperlidemia

Urtica dioica L. given orally at a dose of 100 mg/kg/day for 30 days to Swiss albino mice feeding on diet containing 1% cholesterol, reduced plasma triglycerides, glycaemia from 79.8 to 69.8 mg/dL, increased serum high-density lipoprotein–cholesterol, and reduced low-density lipoprotein–cholesterol from 143.0 to 48.6 mg/L.[77] This regimen reduced aspartate aminotransferase from 75.8 to 21.5 IU/L and alanine aminotransferase from 41.4 to 8.2 IU/L.[77] Aqueous extracts of this plant given for 5 weeks in the drinking water of streptozotocin-diabetic Wistar rats lowered fasting glucose by 13%.[232] In a parallel study, aqueous extract given orally to high-fat diet fed Sprague–Dawley rats at a dose of 150 mg/kg/day for 4 weeks lowered plasma triglycerides from 83 to 63 mg/mL, cholesterol from 97 to 78 mg/dL, had no effect on high-density lipoprotein–cholesterol and lowered low-density lipoprotein–cholesterol from 40 to 27 mg/dL.[233] Simultaneously, this extract lowered plasma apolipoprotein B concentration toward normal values as well as serum enzymes serum aspartate aminotransferase and alanine aminotransferase.[233] This supplement had no effect on triglyceride and cholesterol in feces.[233] This plant contains as caffeic acid, rutin, quercetin, hyperin, and isoquercitrin,[234] kaempferol 3-*O*-glucoside, kaempferol 3-*O*-rutinoside, isorhamnetin 3-*O*-rutinoside,[235] as well as chlorogenic acid and 2-*O*-caffeoyl-malic acid.[236] Quercetin inhibited the enzymatic activity of fatty acid synthetase with IC_{50} of 12 μg/mL *in vitro*.[237] The precise molecular basis and principles behind plasma cholesterol lowering activity of *Urtica dioica* L. are apparently unknown. Clinical trials are warranted.

3.43 *Phyllanthus amarus* Schumach. & Thonn.

Common names: ku wei ye xia zhu (Chinese); San Pedro (Philippines)
Subclass Dillenidae, Superorder Euphorbianae, Order Euphorbiales, Family Phyllanthaceae
Medicinal use: fever (Philippines)

Aqueous extract of *Phyllanthus amarus* Schumach. & Thonn. given to Wistar rats at a daily dose of 600 mg/kg for 30 days evoked a 10.2% reduction in body weight compared with untreated animals, corrected fasting blood glucose from about 65 to 45 mg/dL, lowered triglycerides, total cholesterol, halved low-density lipoprotein–cholesterol, and reduced atherogenic index from 0.5 to 0.3.[238] The same experiment performed with rodents given daily and prophylactically 600 mg/kg of extract before high-fructose diet, reduced weight gain from 15.2% to 9%, fasting blood glucose from about 130 to 55 mg/dL.[238] This regimen lowered plasma insulin and insulin resistance close to normal.[238] In addition, the extract prevented increase serum triglycerides, total cholesterol, low-density lipoprotein–cholesterol and atherogenic index and decreased of high-density lipoprotein–cholesterol and coronary artery risk index.[238] Phyllantin[239] given to Swiss albino mice on high-fat diet at a dose of 4 mg/kg/day orally for 12 weeks lowered plasma triglycerides from 133.9 to 91.6 mg/dL, had no effect on plasma cholesterol, high-density lipoprotein and low-density lipoprotein whereas very low-density lipoprotein was decreased.[240] This regimen improved insulin sensitivity and glucose clearance.[240] This lignan reduced serum tumor necrosis factor-α, interleukin-1β and interleukin-6 levels.[240] At the hepatic level, this regimen decreased triglycerides from 292.55 μg mg^{-1} tissue to 195.5 μg mg^{-1} tissue, lowered thiobarbituric reactive species, increased superoxide dismutase and increased glutathione concentration, repressed nuclear factor-κB and increased the expression of insulin receptor.[240]

3.44 *Phyllanthus urinaria* L.

Synonyms: *Diasperus urinaria* (L.) Kuntze; *Phyllanthus alatus* Blume; *Phyllanthus cantoniensis* Hornem.
Common name: dukong anak (Malay)
Subclass Dillenidae, Superorder Euphorbianae, Order Euphorbiales, Family Phyllanthaceae
Medicinal use: clean the liver (Malaysia)
Pharmacological targets: atherogenic hyperlipidemia; insulin resistance

Accumulation of triglycerides in liver resulting from choline deficiency occurs because choline is required the make phosphatidyl-choline in very low-density lipoproteins.[241] Thus, choline deficient diet is used to include fatty liver in rodents. Extract of *Phyllanthus urinaria* L. given at 2000 ppm in a methionine and choline deficient diet to C57BL/6 mice for 10 days prevented hepatic triglyceride accumulation and lowered lipid peroxidation but was unable to prevent weight loss.[242] Histological observation of liver tissue revealed that the extract mitigated steatohepatitis induced by methionine and choline deficient diet.[242] Furthermore, this extract lowered the expression of tumor necrosis factor-α and interleukin-6 as well as phosphorylated-JNK.[242] Hepatic nuclear translocation of nuclear factor-κB was inhibited by the treatment.[242] *Phyllanthus urinaria* L. given at a dose two tablets of 400 mg three times daily to patients with nonalcoholic steatohepatitis for 24 weeks was not able to improve steatosis, lobular inflammation, ballooning, and fibrosis. *Phyllanthus urinaria* L. produces series of lignans of which phyllantin.[240]

3.45 *Potentilla reptans* L.

Common names: pu fu wei ling cai (Chinese); resatinotu (Turkey); creeping cinquefoil
Subclass Rosiidae, Superorder Rosanae, Order Rosales, Family Rosaceae
Medicinal use: blood circulation (Turkey)
History: the plant was known as astringent remedy by Galen (circa 130–200 AD); Greek physician to the Roman Emperor Marcus Aurelius
Pharmacological target: insulin resistance

Potentilla reptans L. given orally at a dose of 100 mg/kg/day for 30 days to Swiss albino mice feeding on diet containing 1% cholesterol increased serum high-density lipoprotein–cholesterol from 25.8 to 38.6 mg/dL whereas plasma cholesterol, low-density lipoprotein–cholesterol, and triglyceridemia were not affected.[77] Besides the extract reduced serum aspartate aminotransferase from 75.8 to 13 IU/L and serum aspartate aminotransferase from 41.4 to 8 IU/L and increased serum nitric oxide from 21.8 to 49.3 µmol/L.[77] The chemical constituents of *Potentilla reptans* L. appear to be unknown. Members of the genus *Potentilla* L. produce series of ellagitannins of which laevigatin B and F as well as agrimoniin and pedunculagin.[243] Other constituents known to this genus are proanthocyanidins such as procyanidin B3 and pentacyclic triterpenes of which tormentic acid, ursolic, arjunetin, and pomolic acid.[243] Ellagitannins are not absorbed in the gut but decomposed into phenolic metabolites such as gallic acid. Wistar rats on high-fat diet supplemented with 0.2% of gallic acid for 10 weeks had no effect on food intake, lowered body weight gain, decreased liver weight.[244] This regimen lowered plasma triglycerides from 0.9 to 0.8 mmol/L, lowered plasma cholesterol from 3 to 2,4 mmol/L, low-density lipoprotein–cholesterol from 0.6 to 0.4 mmol/L and had no effect on high-density lipoprotein–cholesterol.[244] This regimen lowered serum insulin as well as serum leptin.[244] Leptin from adipocytes acts on hypothalamus to suppress food intake.[245] At the hepatic level, this regimen attenuates hepatic steatosis and lowered hepatic triglycerides whereby ehaptic cholesterol content was unchanged whereas activity of glutathione peroxidase.[244]

3.46 *Rubus alceifolius* Poir.

Synonyms: *Rubus bullatifolius* Merr.; *Rubus hainanensis* Focke; *Rubus roridus* Lindl.
Common name: cu ye xuan gou zi (Chinese)
Subclass Rosiidae, Superorder Rosanae, Order Rosales, Family Rosaceae
Medicinal use: hepatitis (China)
Pharmacological target: insulin resistance

Alkaloid fraction of *Rubus alceifolius* Poir. given at a dose of 1.4 g/kg body weight/day orally to rodents on high-fat diet prevented weight gain, hepatic steatosis and reduced the serum levels of alanine aminotransferase, aspartate transaminase, alkaline phosphatase, cholesterol, triglycerides, and low-density lipoprotein–cholesterol.[246] In addition, this treatment decreased hepatic inflammation as evidenced by a reduction in cyclo-oxygenase-2 and interleukin-6 expression.[246] It must be recalled that in hepatocytes, the expression of cyclo-oxygenase-2 and interleukin-6 is induced by nuclear factor-κB via tumor necrosis factor-α release from Kupffer cells and injured hepatocytes, as well as chronic hyperglycemia and free fatty acids leading to an increased production of reactive oxygen species.[307] Alkaloid fraction of *Rubus alceifolius* Poir. Given orally at a dose of 1.4 g/kg/day for 7 days afforded hepatoprotection against carbon tetrachloride as evidenced by improved cytoarchitecture and decreased serum aspartate aminotransferase and alanine aminotransferase activities toward normal values. This fraction lowered hepatic expression and activity of CYP2E1.[247] Increase of triglycerides in the liver in metabolic syndrome activates CYP2E1 leading to increased reactive oxygen species and subsequent insulin resistance via nuclear factor-κB activation.[248] Activated nuclear factor-κB promote SCOS-3 via interleukin-6 and to inhibit tyrosine phosphorylation of insulin receptor substrate as well as protein kinase C_θ and serine phosphorylation of insulin receptor substrate-1/2 hampering normal insulin receptor functionment.[172] It would be of interest to identify the active alkaloids present in this fraction.

3.47 *Terminalia paniculata* Roth

Common names: asvakarnah (India); flowering mundah
Subclass Rosidae, Superorder Myrtanae, Order Myrtales, Family Combretaceae
Medicinal use: diabetes (India)
Pharmacological target: atherogenic hyperlipidemia

Ethanol extract of bark of *Terminalia paniculata* Roth given orally at a dose of 200 mg/kg/day for 10 weeks to Sprague–Dawley rats on high-fat diet for days lowered body weight and liver weight.[249] This treatment lowered plasma cholesterol by 39%, triglycerides by 42%, and fatty acids by 56%, and these effects were comparable to orlistat at 30 mg/kg/day.[249] This treatment increased hepatic superoxide dismutase and catalase activity by 62% and 53%, respectively and decreased malondialdehyde by 49%.[249] Simultaneously, this extract decreased hepatic expression of fatty acid synthetase, elevated the expression of adenosine monophosphate-activated protein kinase and improved hepatic cytoarchitecture particularly in regards to steatosis.[249] Ramachandran et al. evidenced the presence of gallic acid, ellagic acid, catechin, and epicatechin in the bark of this plant, and it is reasonable to speculate their involvement of at last in the activation of adenosine monophosphate-activated protein kinase as discussed elsewhere.[250]

3.48 *Caesalpinia bonduc* (L.) Roxb.

Synonyms: *Caesalpinia bonducella* (L.) Fleming; *Caesalpinia crista* L.; *Guilandina bonduc* L.; *Guilandina bonducella* L.
Common names: ci guo su mu (Chinese); latakaranja (India); fever nut
Medicinal use: fever (India)
Pharmacological target: atherogenic hyperlipidemia; insulin resistance

Ethanolic extract of defatted seeds of *Caesalpinia bonduc* (L.) Roxb. (Figure 3.13) given to alloxan-induced diabetic Wistar rats at a dose of 300 mg/kg/day orally for 14 days reduced glycaemia from 578 to 324.5 mg/dL, lowered plasma cholesterol from 177 to 143 mg/dL, triglycerides from 107.7 to 88.5 mg/dL, very low-density lipoprotein–cholesterol from 21 to 17.2 mg/dL, and low-density lipoprotein–cholesterol from 127.5 to 67.5 mg/dL.[251] Chakrabarti et al. administered aqueous extract of seeds to streptozotocin-induced diabetic Long Evans rats orally at a dose of 250 mg/kg twice a day for 5 days which resulted in lowered glycaemia from 7.4 to 5.5 mmol/L (metformin 250 mg/kg: 6.1 mmol/L) and decreased plasma cholesterol and triglycerides toward normal values. This regimen increased

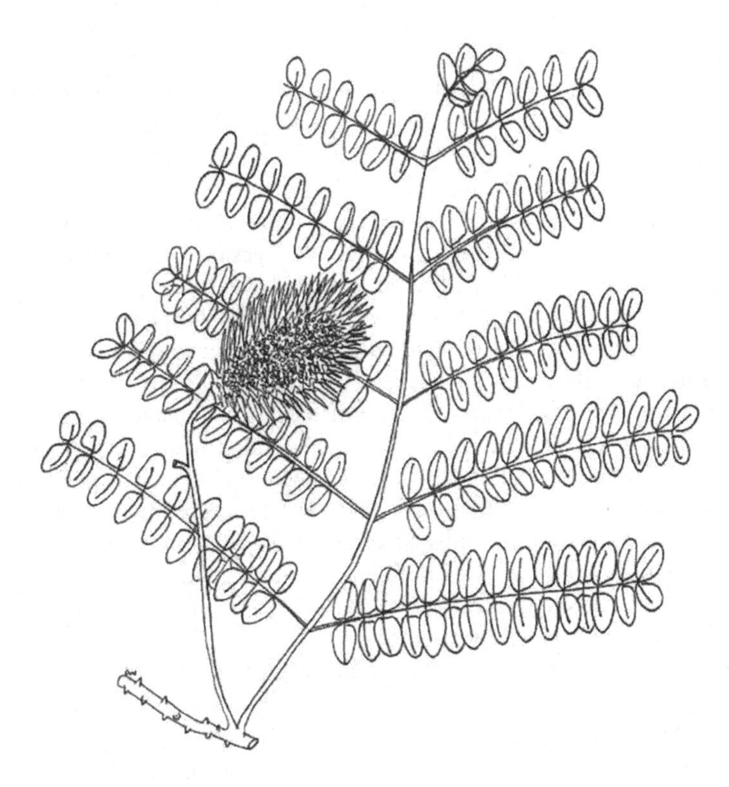

FIGURE 3.13 *Caesalpinia bonduc* (L.) Roxb.

hepatic glycogen from 6.7 to 15.1 mg/g (metformin 250 mg/kg: 11.8 mg/g).[252] In the postprandial state, insulin triggers the storage of plasma glucose in the liver and inhibits the production of glucose by the liver to bring down postprandial hyperglycemia.[253] The binding of this hormone to its hepatic receptor promotes the storage of glucose in the form of glycogen in the liver by successively activating phosphoinositide 3-kinase, Akt, inhibiting glycogen synthetase kinase 3, and activating glycogen synthetase. Insulin also stimulates protein phosphatase 1 and inhibits glucose-6-phosphatase. In a subsequent study, oral administration of aqueous extract of seeds streptozotocin-induced diabetic Long Evans rats orally at a dose of 250 mg/kg twice day decreased glycaemia and brought close to normal plasma lipids whereas increase in plasma insulin was observed.[254] The seeds contain cassane furanoditerpenes of which α-caesalpine and caesalpinia F,[255] which might be involved in insulinotropic properties. Insulin promotes the removal of low-density lipoprotein from the plasma by the liver by increasing hepatic expression of low-density lipoprotein–cholesterol receptor.[195] Insulin also activates lecithin:cholesterol acyltransferase,[256] which is responsible for the conversion of cholesterol from peripheral tissues into cholesteryl ester in reverse cholesterol transport.[257] Insulin reduces the hepatic production of very low-density lipoproteins by inhibiting adipocyte hormone sensitive lipase.[45]

3.49 *Entada phaseoloides* (L.) Merr.

Synonyms: *Acacia scandens* (L.) Willd.; *Entada scandens* (L.) Benth.; *Entada gigas* (L.) Fawc. & Rendle *Lens phaseoloides* L.
Common names: ke teng (China); gannyin (Burma); gogo (Philippines); giant rattle
Subclass Rosidae, Superorder Fabanae, Subclass Rosiidae, Family Fabaceae
Medicinal use: fever (Burma)
Pharmacological target: insulin resistance

Total saponins fraction of seeds of *Entada phaseoloides* (L.) Merr. (Figure 3.14) given orally at a dose of 100 mg/kg/day to streptozotocin-induced type 2 diabetes Sprague–Dawley rats on high-fat diet for 3 weeks induced a diminution of fasting blood glucose and insulinaemia.[258] In oral glucose tolerance test, the extract lowered plasma glucose after 60 minutes by 19.6%.[258] Serum malondialdehyde was reduced by 29.1%, whereby serum glutathione and superoxide dismutase levels were increased by 74% and 19.1%, respectively.[258] This regimen decreased plasma contents in interleukin-6 and tumor necrosis factor-α by 48.4% and 18.1%, respectively.[258] The treatment decreased plasma cholesterol, triglycerides, fatty acids, and low-density lipoprotein–cholesterol, whereas high-density lipoprotein–cholesterol was increased by 24%.[258] Total cholesterol and triglycerides in liver were lowered upon saponins treatment.[258] Zheng et al. provided evidence that saponin fraction of seeds of this climber induced the phosphorylation of adenosine monophosphate-activated protein kinase in HepG2 cells *in vitro* hence decreased glucose production, decreased levels of glucose-6-phosphatase and phosphoenolpyruvate carboxykinase, (key enzymes were involved in hepatic gluconeogenesis).[258] It must be recalled that activation of adenosine monophosphate-activated protein kinase inhibits S6K1-induced inhibition of insulin receptor, improving thus hepatic insulin sensitivity explaining why the saponin fraction activation of adenosine monophosphate-activated protein kinase in HepG2 cells induced the expression of glycogen synthetase via phosphoinositide 3-kinase, Akt and subsequent inhibition of glycogen synthetase kinase-3β.[259] In regards to fatty acids, activation of adenosine monophosphate-activated protein kinase in cultured hepatocytes translates into inhibition of acetyl-CoA carboxylase lifting malonyl-CoA inhibition on carnitine acyltransferase and promoting the entry of fatty acyl-CoA in mitochondria for β-oxidation.[258] The possibility to have intact saponins reaching the liver is thin as the sugar moieties are removed upon intestinal bacterial flora metabolism. Thus, it is possible to suggest that saponin aglycone such as the pentacyclic triterpene entagenic acid[260] or first pass metabolites could decrease insulin resistance in streptozotocin-induced type 2 diabetes Sprague–Dawley rats on high-fat diet. It would be interesting to clarify that point.

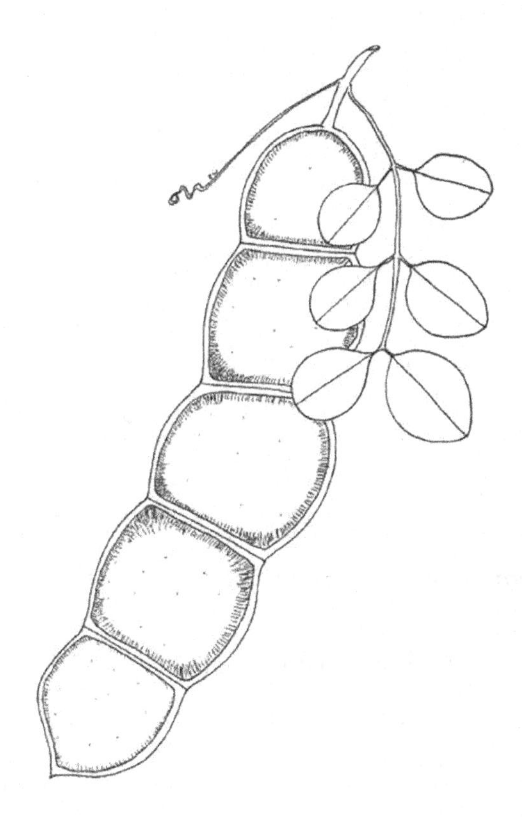

FIGURE 3.14 *Entada phaseoloides* (L.) Merr.

3.50 *Glycyrrhiza glabra* L.

Synonyms: *Glycyrrhiza alalensis* X.Y. Li; *Glycyrrhiza brachycarpa* Boiss.; *Glycyrrhiza glandulifera* Waldst. & Kit.; *Glycyrrhiza hirsuta* Pall.; *Liquiritia officinarum* Medik.
Common names: yang gan cao (Chinese); licorice
Subclass Rosidae, Superorder Fabanae, Subclass Rosiidae, Family Fabaceae
Medicinal use: tonic (China)
History: this plant was known to Theophrastus
Pharmacological target: insulin resistance

Glycyrrhiza glabra L. contains the isoflavonoid glabridin (Figure 3.15) which at a concentration of 100 μM activated peroxisome proliferator-activated receptor-γ in transfected CHO-K1 cells and increased the expression of phosphoenolpyruvate carboxykinase, decreased the expression of fatty acid binding protein-1 and 3-hydroxy-3-methylglutaryl-CoA synthase[261] suggesting peroxisome proliferator-activated receptor-γ activity.[262] An another constituent of this plant, glabrol, inhibited the enzymatic activity of microsomal diacylglycerol acyltransferase prepared from rat liver with an IC_{50} value equal to 8 μM.[263] The triterpene glycyrrhizin from this plant which is a known inhibitor of nuclear factor-κB[264] and as such could ameliorate hepatic sensitivity in Metabolic Syndrome. Sil et al. administered glycyrrhizin intraperitoneally to rats on high-fructose diet at a single dose of 50 mg/kg and noted after 3 weeks a decrease in plasma cholesterol, triglycerides, very low-density lipoprotein, low-density lipoprotein to values close to normal. This treatment decreased serum insulin and improved glucose tolerance.[265]

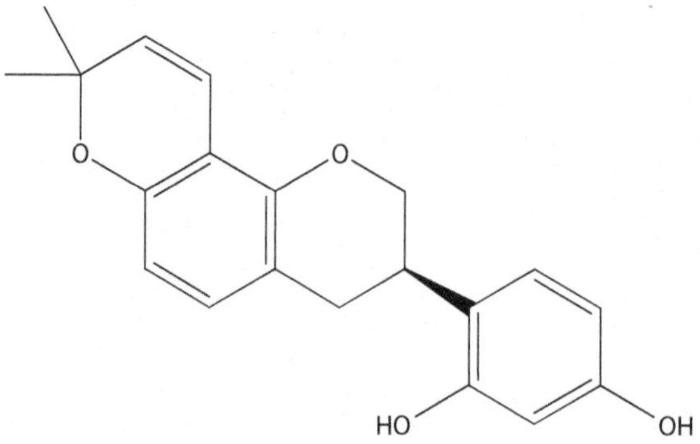

FIGURE 3.15 Glabridin.

3.51 *Pterocarpus santalinus* Buch.-Ham. ex Wall.

Common names: raktasara (India); Red Sandalwood
Subclass Rosidae, Superorder Fabanae, Subclass Rosiidae, Family Fabaceae
Medicinal use: dysentery (India)
Pharmacological target: atherogenic hyperlipidemia; insulin resistance

Elevated plasma triglyceride is a characteristic lipid abnormality in obese patients of type 2 diabetics and is associated with accumulation of triglycerides in the liver.[266] Ethylacetate fraction of barks *Pterocarpus santalinus* Buch.-Ham. ex Wall. given to streptozotocin-induced diabetic Wistar rats at a single oral dose of 150 mg/kg decreased glycaemia from 373.2 to 70.6 mg/dL after 5 hours.[266] This fraction given for 45 days at a dose of 150 mg/kg/day to streptozotocin-induced diabetic Wistar rats lowered glycaemia by 64.3% and increased plasma insulin.[266] On probable account of increased insulin availability, this treatment increased hepatic glycogen, reduced total cholesterol, triglycerides, low-density lipoprotein–cholesterol, and increased high-density cholesterol.[266] This fraction inhibited the enzymatic activity of glucose-6-phosphatase, fructose-1,6-biphosphatase, and induced glucokinase as well as glucose-6-phosphate dehydrogenase.[266] Glucose-6-phosphatase and fructose-1,6-biphosphatase are key enzymes in catalyzing gluconeogenesis and are termed gluconeogenic enzymes.[267] Insulin inhibits hepatic glucose production, also termed as gluconeogenesis, by inhibiting glucose-6-phosphatase and fructose 1,6-biphosphatase activity.[267] In physiological postprandial state, insulin inhibits gluconeogenesis and stimulates glycolysis and glycogen synthesis by controlling several enzymes. One such enzyme is glucose-6-phosphatase, which catalyzes the dephosphorylation of glucose-6-phosphate into glucose, which is then transported across hepatic cytoplasmic membrane to the general blood circulation.[73] As a part of gluconeogenesis, fructose-1,6-biphosphatase catalyzes the dephosphorylation of fructose 1,6-biphosphate to fructose 6-phosphate which is substrate for phosphohexose isomerase to produce glucose-6-phosphate.[73]

3.52 *Sophora flavescens* Aiton

Synonyms: *Sophora angustifolia* Siebold & Zucc.; *Sophora macrosperma* DC.; *Sophora tetragonocarpa* Hayata
Common name: ku shen (Chinese)
Subclass Rosidae, Superorder Fabanae, Subclass Rosiidae, Family Fabaceae
Medicinal use: jaundice (China)
Pharmacological target: insulin resistance

FIGURE 3.16 *Sophora flavescens* Ait.

Evidence suggests that *Sophora flavescens* Ait. (Figure 3.16) produces alkaloids which have the ability to decrease insulin resistance. Matrine given to rodents orally at a dose of 160 mg/kg/day for 28 days to rodents fed with high-fructose lowered insulinemia from 1.3 to 0.7 ng/mL, fasting glucose from 6.2 to 5.5 mM/L and insulin resistance.[269] Furthermore, matrine reduced alanine aminotransferase from 40.8 to 33.1 IU/L and serum aspartate aminotransferase from 130.5 to 118.8 IU/L.[269] The treatment with matrine reduced by more than 50% in hepatic malondialdehyde and liver triglycerides. Matrine (Figure 3.17) increased the enzymatic activity of catalase, superoxide dismutase and glutathione peroxidase by 87.7%, 96.9%, and 51.7%, respectively, as a result of Nrf2 translocation in the nucleus.[269] Furthermore, the expression of heme oxygenase-1 and NQO-1 was increased by 165% and 140%, respectively.[269] Matrine treatment decreased nuclear factor-κB nuclear translocation which account for insulin resistance.[269] Oxymatrine from this plant given at a dose of 150 mg/kg/day orally to streptozotocin-induced diabetic rodent induced a mild weight gain after 11 weeks of treatment together with a reduction of food intake, water intake, and diuresis. After 11 weeks of treatment, oxymatrine evoked a reduction of fasting blood glucose from 21.1 to 12.9 mmol/L, whereby diabetic rodent treated with metformin had a fasting blood glucose value equal to 15.5 mmol/L.[270] Furthermore, 11 weeks of treatment with oxymatrine at a dose of 150 mg/kg/day orally commanded an increase of insulinaemia from 13.7 to 16.4 mU/L and a reduction in

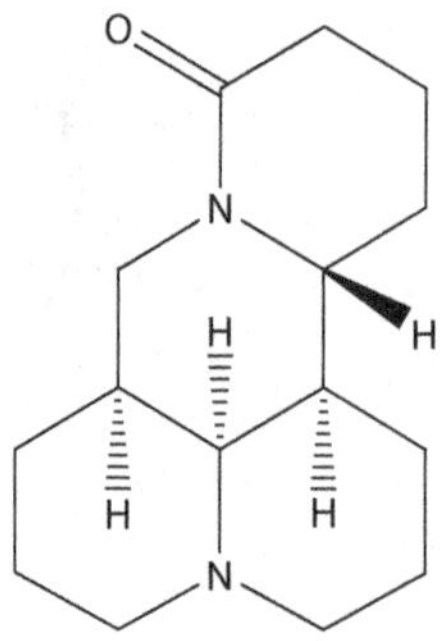

FIGURE 3.17 Matrine.

insulin resistance as well as an increase in glucagon-like peptide-1 and consequently increased liver glycogen from 8.5 to 11.7 mg/g.[270] With regard to serum lipids, oxymatrin evoked a decrease of triglycerides from 2.2 to 1.1 mmol/L, a decrease of total cholesterol from 5.6 to 3.3 mmol/L, a decrease in low-density lipoprotein–cholesterol from 2.5 to 1.2 mmol/L, and an increase in high-density lipoprotein–cholesterol from 0.7 to 1.2 mmol/L, respectively.[270] Histological evaluation of liver tissues from treated animals revealed the disappearance of hepatocellular necrosis and vacuolization observed in diabetic rodents was observed after 11 weeks.[270] In hepatocytes protein tyrosine phosphatase-1B inhibits in insulin signal transduction by dephosphorylating-activated insulin receptors or insulin receptor substrates.[12] Kuraridin, norkurarinone, kurarinone, 2′-methoxykurarinone and kushenol T isolated from the roots of *Sophora flavescens* Ait. inhibited the enzymatic activity of protein tyrosine phosphatase 1B with IC_{50} below 50 μM.[271] In HepG2 cells, kuraridin, norkurarinone, and 2′-methoxykurarinone at a concentration of 10 μM enhanced, in the presence of insulin, the phosphorylation of Akt.[271] Sophocarpine (Figure 3.18) from this plant given daily subcutaneously at a dose of 20 mg/kg injected for 12th week prevented the gain of weight in rodents on high-fat diet together with a decrease in serum alanine aminotransferase, total cholesterol and plasma triglyceride, whereby glycaemia was not corrected.[272] In addition, the aforementioned regimen protected the liver as evidenced histologically with the absence of steatosis, hepatocellular balloonng, Mallory's bodies, inflammation and fibrosis.[272] Sophocarpine inhibited the expression of hepatic interleukin-6, and tumor necrosis factor-α and increased the expression of adiponectin.[272] *In vitro*, sophocarpine at a concentration of 0.4 mmol/L prevented accumulation of triglycerides primary hepatocytes of rodents challenged with oleic acid by more than 50% with increased the phosphorylation of adenosine monophosphate-activated protein kinase and acetyl-CoA carboxylase, and decreased the expression of sterol regulatory element-binding protein-1c (SERBP-1c).[273]

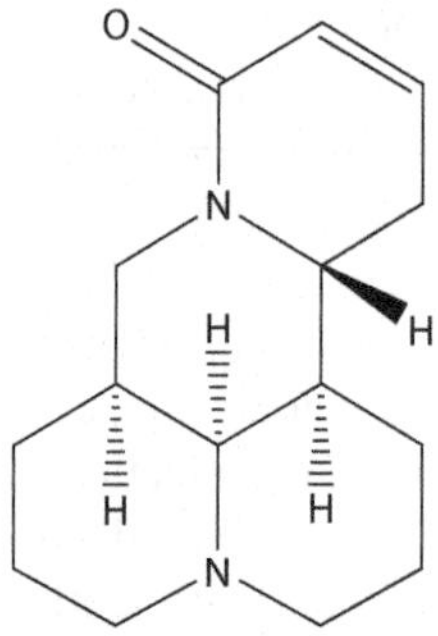

FIGURE 3.18 Sophocarpine.

3.53 *Sophora japonica* L.

Synonyms: *Ormosia esquirolii* H. Lév.; *Styphnolobium japonicum* (L.) Schott
Common names: huai (Chinese); Japanese pagoda tree
Subclass Rosidae, Superorder Fabanae, Subclass Rosiidae, Family Fabaceae
Medicinal use: bleeding (Vietnam)

Sophora japonica L. added 5% of high-fat diet to C57BL/6 male mice reduced weight gain from 0.3 to 0.2 g/day, total cholesterol from 162.7 to 100.9 mg/dL, high-density lipoprotein–cholesterol 81.5 and 60.8 mg/dL and glucose from 344.5 to 235.2 mg/dL.[274] Furthermore, this supplementation reduced hepatic triglycerides from 28.9 to 11 mg/g and attenuated hepatic cholesterol contents.[274] Total fat mass was reduced by 58%.[274]

3.54 *Tephrosia purpurea* (L.) Pers.

Synonyms: *Cracca purpurea* L.; *Galega purpurea* (L.) L.; *Glycyrrhiza mairei* H. Lév.
Common name: hui mao dou (Chinese)
Subclass Rosidae, Superorder Fabanae, Subclass Rosiidae, Family Fabaceae
Medicinal use: facilitates digestion (Burma)

Aqueous extract of leaves of *Tephrosia purpurea* (L.) Pers. given orally at a dose of 600 mg/kg/day to streptozotocin-induced diabetic Wistar rats for 45 days decreased glycaemia from 285.3 to 128.4 mg/dL and increased insulinaemia from 10.6 to 14.1 mU/mL.[275] This treatment lowered plasma cholesterol, triglycerides, fatty acids, low-density lipoprotein–cholesterol, very low-density lipoprotein–cholesterol and increased high-density lipoprotein–cholesterol.[275] Furthermore, this extract increased the enzymatic activity of plasmatic lecithin cholesterol acyltransferase and lipoprotein lipase.[275] Hepatic cholesterol and triglycerides were reduced by the treatment.[275] It must be recalled that in the plasma, cholesteryl ester is formed by lecithin:cholesterol acyltransferase in high-density lipoproteins or lipoproteins containing apolipoprotein B in exchange of triglycerides by cholesteryl ester transfer protein. Apolipoprotein A-I is the main protein in high-density lipoproteins in human and activates lecithin cholesterol acyltransferase and promotes cholesterol efflux from peripheral tissue.[276] Increased high-density lipoprotein–cholesterol and plasmatic apolipoprotein A-I are inversely related to the incidence of coronary artery diseases.[34,277] Conversely, decreased concentration of high-density lipoproteins and apolipoprotein A-I in human indicates atherosclerosis and risk of coronary disease.[278] Besides, inhibition of cholesteryl ester transfer protein leads to an increase in high-density lipoproteins cholesterol thereby decreasing risk of atherogenesis. Mice lack of cholesteryl ester transfer protein and carry majority of plasma lipoprotein–cholesterol in high-density lipoproteins. In $ApoE^{-/-}$, apolipoprotein A-I appears in very low-density lipoproteins and low-density lipoproteins.[279] Members of the family Fabaceae produce series of flavonoids of which notably isoflavonoids and pterocarpans of pharmacological interest. *Tephrosia purpurea* contains flavonoids of which (+)-purpurin, pongamol, lanceolatin B, (–)-maackiain, (–)-3-hydroxy-4-methoxy-8,9-methylenedioxypterocarpan, and (–)-medicarpin which induced the expression of quinone reductase in Hepa 1c1c7 mouse hepatoma cells.[280] Identification of the flavonoid responsible for the hypoglycemic and normolipidemic effects observed in streptozotocin-induced diabetes Wistar rats is warranted.

3.55 *Citrus maxima* (Burm.) Merr.

Synonym: *Citrus* x *paradisi* Macfad.
Common names: you (Chinese); lukban (Philippines); pummelo
Subclass Rosidae, Superorder Rutanae, Order Rutales, Family Rutaceae

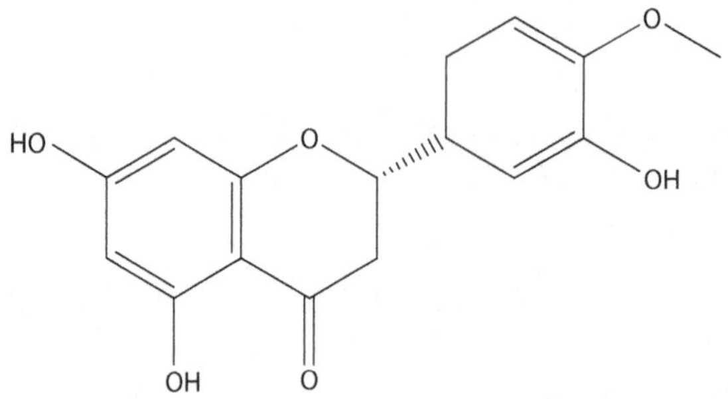

FIGURE 3.19 Hesperetin.

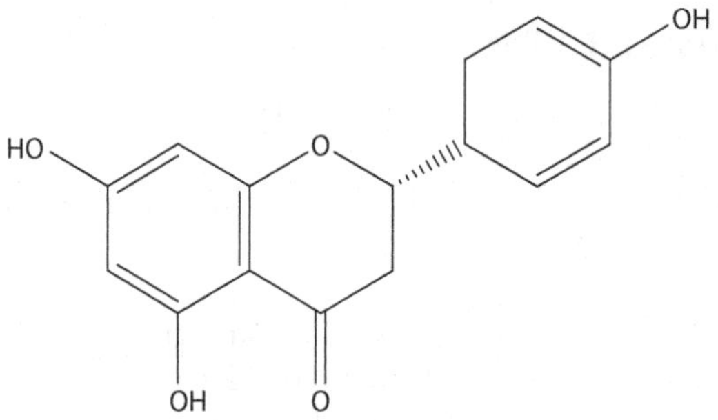

FIGURE 3.20 Naringenin.

Medicinal use: dysentery (Philippines)
Pharmacological targets: atherogenic hyperlipidemia; insulin resistance

The fruits of *Citrus maxima* (Burm.) Merr. contain the flavanones hesperetin (Figure 3.19) and naringenin (Figure 3.20), which are bioavailable in human.[281] Long-Evans rats on high-fat diet containing 0.01% of naringenin for 6 weeks did not reduce their food intake and experienced a lowering of plasma triglycerides from 86.4 to 38.8 mg/dL, reduction of epididymal and perirenal adipose tissues mass by 41.1%.[282] This flavonoid induced the hepatic expression of peroxisome proliferator-activated receptor-α which is a key transcriptional regulator of fatty acid oxidation, and increased downstream expression of carnitine palmitoyltransferase-I.[282] In a subsequent study, extract of *Citrus paradisi* Macfad. given orally at a dose of 0.5 g/kg/day for 6 weeks to diabetic C57BL/6J *db/db* mice decreased glycaemia from 23.7 to 20.1 mmol/L whereby triglycerides were unchanged.[283] This regimen lowered the hepatic expression of tumor necrosis factor-α, cyclo-oxygenase-2 and nuclear factor-κB and increased liver glucokinase expression[283] suggesting an improvement of hepatic insulin sensitivity.[165,284] Juice of fruits of *Citrus paradisi* Macfad. given to C57BL/6J mice on high-fat diet administered at 50% of drinking water for 100 days had no effect on food intake, evoked a net decrease in intra-abdominal fat pad weight and decreased fasting blood glucose by 13% and reduced plasma insulin by serum insulin levels were 72%.[285] This regimen improved glucose tolerance.[285] As evidence of improved insulin sensitivity, this supplementation produced a 1.4-fold increase, respectively, in p-Akt/total Akt ratios in the liver, compared to control.[285] Consumption of this juice reduced hepatic triglycerides by 38% and improved hepatic cytoarchitecture.[285] In addition, this regimen evoked a decrease in a decrease in sterol regulatory element-binding protein-1c, fatty acid synthetase CYP7A1 and peroxisome proliferator-activated receptor coactivator-1α[285] suggesting a possible activation of adenosine monophosphate-activated protein kinase. In the same experiment, naringin at 0.7 mg/day in drinking water had no effect on body weight, decreased blood glucose concentration by about 20%, and improved glucose tolerance.[285] Hesperetin and naringenin at a concentration of 300 μM inhibited glucose production by perfused Wistar rat livers in the presence of lactate and pyruvate by 79.6% and 88.1% whereby hesperidin had no inhibitory activity.[283] In isolated mitochondria, hesperetin and naringenin at a concentration of 300 μM inhibited pyruvate carboxylation by inhibiting the enzymatic activity of pyruvate carboxylase by 82.6% and 86.9%, respectively.[283] Pummelo could be of value for metabolic syndrome.

3.56 *Citrus japonica* Thunb.

Synonyms: *Atalantia hindsii* (Champ. ex Benth.) Oliv. ex Benth.; *Citrus margarita* Lour.; *Fortunella japonica* (Thunb.) Swingle; *Fortunella margarita* (Lour.) Swingle
Common names: jin gan (Chinese); cây quất cảnh. (Vietnam); kumquat
Subclass Rosidae, Superorder Rutanae, Order Rutales, Family Rutaceae
Medicinal use: cough (Vietnam)
Pharmacological target: atherogenic hyperlipidemia

Ethanol extract of fruits of *Citrus japonica* Thunb. mixed with high-fat diet (1%) and given for 8 weeks to C57BL/6 mice mitigated weight gain, lowered glycaemia from about 28 to 25 g (normal diet 22 g), glycaemia from about 6.5 to 5.5 mmol/L (normal 5.2 mmol/L), slightly improved glucose tolerance, and reduced adipocyte size by 44%.[286] This treatment also reduced fasting total cholesterol, triglycerides and low-density lipoprotein–cholesterol, whereby high-density lipoprotein–cholesterol was mildly lowered. In C57BL/6 mice fed on high-fat diet for 3 months, a subsequent treatment with 1% extract in diet lowered serum total cholesterol, triglycerides, low-density lipoprotein–cholesterol and elevated high-density lipoprotein–cholesterol. Further, this treatment reduced hepatic cholesterol and triglycerides and induced hepatic peroxisome proliferator-activated receptor-α and target enzymes such as fatty acid translocase, and CYP4A10 and CYP4A14 (which are both catalyzing ω-oxidation), and acyl-CoA oxidase and stearoyl coenzyme A desaturase 1.[286,287] The fruit of this plant is known to contain flavonoids of which 6,8-di-C-glucosyl apigenin, 3,6-di-C-glucosyl acacetin, 2″-*O*-α-L-rhamnosyl-4′-*O*-methylvitexin, 2″-*O*-α-L-rhamnosyl-4′-*O*-methylisovitexin, 2″-*O*-α-L-rhamnosyl orientin and ponicilin.[288] Flavonoids from members of the genus *Citrus* have the tendency to activate peroxisome proliferator-activated receptor-α.[289]

3.57 *Amoora rohituka* (Roxb.) W. & A.

Synonym: *Aphanamixis polystachya* (Wall.) R. Parker
Subclass Rosidae, Superorder Rutanae, Order Rutales, Family Meliaceae
Medicinal use: jaundice (India)
Pharmacological target: atherogenic hyperlipidemia

Amoora rohituka (Roxb.) W. & A. elaborates rohitukine given orally at a dose of 50 mg/kg/day for 30 days to Charles Foster rats on high-fat diet had no effect on food intake, lowered serum cholesterol from 248.6 to 188.1 mg/dL, triglycerides from 258.2 to 189.6 mg/dL, very low-density lipoprotein from 51.6 to 37.9 mg/dL and low-density lipoprotein from 167.2 to 114 mg/dL and evoked an increase of high-density lipoprotein from 29.8 to 36.1 mg/dL.[290] Rohitukine regimen decreased plasma levels of oxidative reactive species by 23%.[290] Rohitukine (Figure 3.21) treatment lowered hepatic triglyceride and cholesterol, and evoked an increase in fecal excretion of cholic acid and deoxycholic acid.[290] This regimen increased the plasma activity of plasma lecithin cholesterol acyl transferase, lipoprotein lipase, and triglyceride lipase.[290] *In vitro*, this unusual chromone alkaloid inhibited the enzymatic activity of 3-hydroxy-3-methylglutaryl-coenzyme A reductase by 69.8% at a dose of 80 μM.[290] Rohitukine given orally at a dose of 100 mg/kg to golden Syrian hamsters for

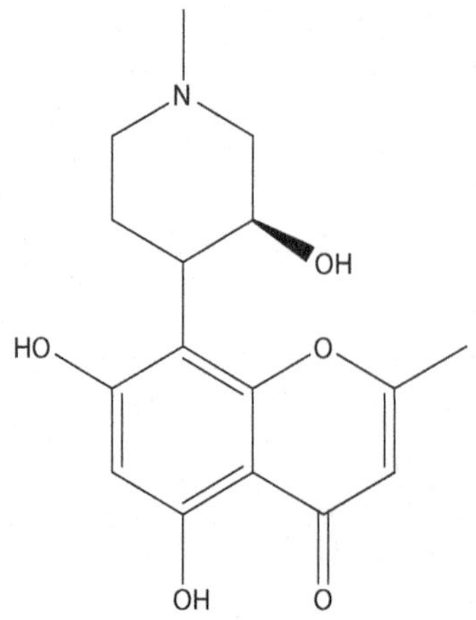

FIGURE 3.21 Rohitukine.

7 days evoked a decrease of plasma cholesterol, triglycerides, low-density lipoprotein, and high-density lipoprotein.[291] At the hepatic level, this regimen increased the expression of liver X receptor and consequently sterol regulatory binding protein-2, the expression of low-density lipoprotein receptor and 3-hydroxy-3-methylglutaryl-coenzyme A reductase. This regimen improved hepatic cytoarchitecture with decreased in triglyceride droplets. When cytosolic cholesterol levels are low, sterol regulatory element-binding protein-2 is activated hence increased expression of low-density lipoprotein receptor and 3-hydroxy-3-methylglutaryl-coenzyme A reductase.[292] Conversely, excess of cytosolic cholesterol in hepatocytes inhibits sterol regulatory element-binding protein-2 and generate oxyterols, which activate liver X receptor and subsequent expression of ATP-binding cassette transporter A1.[292] The mechanism by which rohitukine decreases cholesterol in the liver of rodents is unknown.

3.58 *Walsura pinnata* Hassk.

Synonym: *Walsura cochinchinensis* (Baill.) Harrms, *Walsura yunnanensis* C.Y. Wu
Common name: yue nan ge she shu (China)
Subclass Rosidae, Superorder Rutanae, Order Rutales, Family Meliaceae
Pharmacological target: atherogenic hyperlipidemia

11β-hydroxysteroid dehydrogenase 1 catalyzes the conversion of cortisone to cortisol in human which at the hepatic level favors insulin resistance, triglyceride and cholesterol synthesis and plasma fatty acids uptake.[293] In obese patient with Metabolic Syndrome, human liver expresses 11β-hydroxysteroid dehydrogenase 1 is highly expressed.[294] Inhibition of 11β-hydroxysteroid dehydrogenase 1 improves insulin sensitivity, and in patients with type 2 diabetes it decreases glucose production rates, principally through decreased hepatic phosphoenolpyruvate carboxykinase and glucose-6-phosphatase expression.[295] In addition, inhibition of 11β-hydroxysteroid dehydrogenase 1 evokes a decreased in plasma cholesterol and low-density lipoprotein with no changes in high-density lipoprotein.[296] Conchinchinoid H, conchinchinoid K and mesendanin S isolated from the leaves and twigs of *Walsura pinnata* Hassk. (Figure 3.22) inhibited *in vitro* the enzymatic activity of human 11β-hydroxysteroid dehydrogenase 1 with IC_{50} values of 11.4, 3.2, and 3.7 μM, respectively.[297] Members of the family Meliaceae Juss. elaborate limonoids such as dysoxylumosin A, B and which inhibit the enzymatic activity of 11β-hydroxysteroid dehydrogenase 1 *in vitro*.[298]

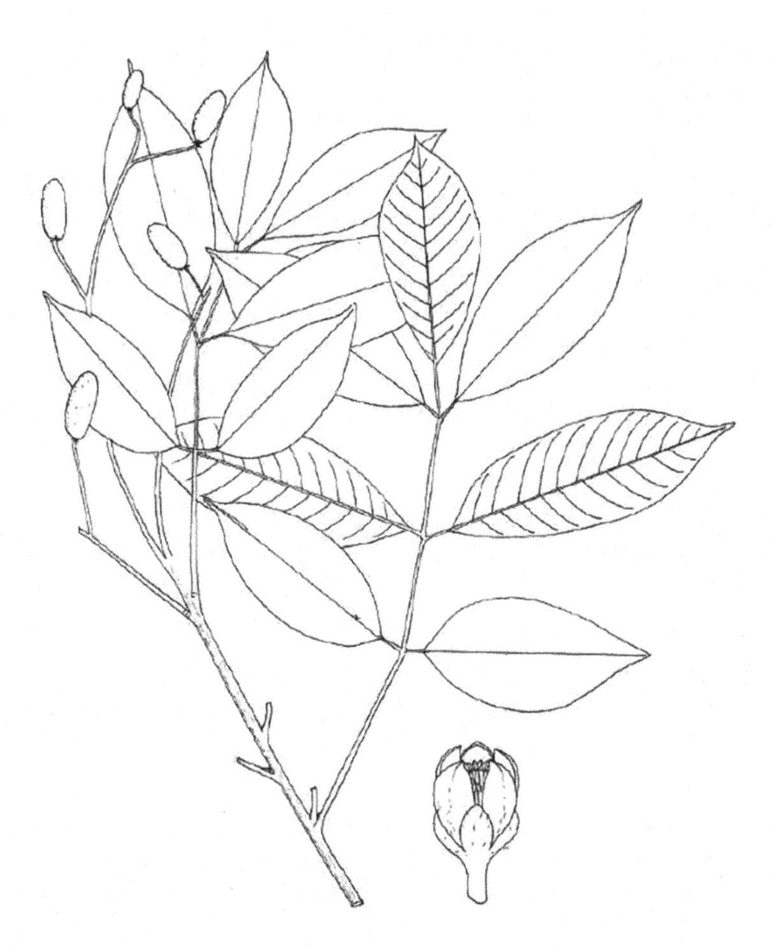

FIGURE 3.22 *Walsura pinnata* Hassk.

3.59 *Pistacia vera* L.

Common name: pistachio
Subclass Rosidae, Superorder Rutanae, Order Rutales, Family Anacardiaceae
Nutritional use: food (Turkey)
Pharmacological target: atherogenic hyperlipidemia

The seeds of *Pistacia vera* L. contain mainly oleic acid, lesser amounts of linoleic and palmitic acid, and traces of linolenic.[299] Kay et al. provided evidence that seeds of *Pistacia vera* L. administered at a dose of 63 g to 126 g/days for 4 weeks to volunteers with moderately elevated low-density lipoprotein–cholesterol (>2.8 mmol/L) evoked a decrease of low-density lipoprotein–cholesterol from 3.4 to 2.9 mmol/L and decreased of oxidized low-density lipoproteins from 51.2 to 43.4 U/L.[300] Elevation of oxidized low-density lipoprotein is associated with the increased risk of coronary heart disease.[301] Parthasarathy et al. provided evidence that low-density lipoprotein isolated from rabbits fed with a diet enriched in oleic acid were more resistant to copper-induced conversion of low-density lipoprotein into atherogenic oxidized low-density lipoprotein.[302] The exact hypolipidemic effect of oleic is not fully understood. It must be recalled that in hepatocytes, acyl coenzyme A:cholesterol acyl-transferase converts cholesterol into cholesteryl ester and the physiological substrate for this enzyme is oleic acid.[303] By increasing oleic acid supply by olive oil consumption, the activity of acyl coenzyme A:cholesterol acyl-transferase could be enhanced hence reduction of free unesterified cholesterol in the cytosol of hepatocytes translating into, at least, increased expression of low-density lipoprotein receptor and therefore increased plasmatic clearance of low-density lipoprotein.[304] Oleic acid inhibited the enzymatic activity of protein-tyrosine phosphatase-1B with an IC_{50} of 6.2 μM.[6] At a dose of 50 μM, this fatty acid boosted the insulin-induced phosphorylation of Akt in murine fibroblasts *in vitro*.[6] A bulk of evidence exists to demonstrate that olive oil consumption decrease atherogenic hyperlipidemia. Healthy volunteers fed with a diet enriched with olive oil experienced a decrease of plasma cholesterol by 9.5%, low-density lipoprotein–cholesterol by 12.2%, and triglycerides by 25.5%.[305] In a double-blind clinical trial, 30 g of olive oil given daily for 6 weeks to hypercholesterolemic subjects reduced serum total cholesterol from 6.4 to 6 mmol/L and low-density lipoprotein–cholesterol was lowered by 7.7%.[218] Serum total cholesterol above 6.2 mmol/L (240 mg/dL) and low-density lipoprotein–cholesterol above 4.1 mmol/L (190 mg/L) denote a state of hypercholesterolemia in human.[6] Of interest, one serving of pistachio per day to healthy volunteers evoked during exposure to acute stress a decrease in heart rate and a reduction of about 4.8 mmHg in systolic blood pressure.[306] Incorporation of pistachio and olive oil in diet could beneficial in metabolic syndrome.

3.60 *Rhus chinensis* Mill.

Synonym: *Rhus javanica* L.
Common name: yan fu mu (Chinese)
Subclass Rosidae, Superorder Rutanae, Order Rutales, Family Anacardiaceae
Medicinal use: tuberculosis (China)
Pharmacological target: insulin resistance

Morolic (Figure 3.23) and moronic acid (Figure 3.24) isolated from *Rhus chinensis* Mill. given to Wistar rats at a single oral dose of 50 mg/kg halved 30 minutes peak glycaemia in oral glucose tolerance tests.[307] In sucrose tolerance test morolic acid evoked a 20% reduction of glycaemia at 30 minutes whereby moronic acid reduced glycaemia by more than 50%.[307] These oleananes triterpenes given orally at a dose of 50 mg/kg/day 10 day to diabetic rats induced a 60% reduction of glycaemia, lowering of serum cholesterol, and triglycerides.[307] *In vitro*, morolic acid and moronic acid at a concentration of 10 μM inhibited the enzymatic activity of 11β-hydroxysteroid dehydrogenase 1 (11β-HSD 1) by 44% and 22%, respectively.[307] 11-β-hydroxysteroid dehydrogenase 1 catalyzes the conversion of cortisone to cortisol in human, mainly in the liver and adipose tissues and is related to Metabolic Syndrome and obesity-related disorders. Cortisol activates glucocorticoid receptor which is a cause for obesity.[308] Glucocorticoids in excess stimulate hepatic glucose production, decrease

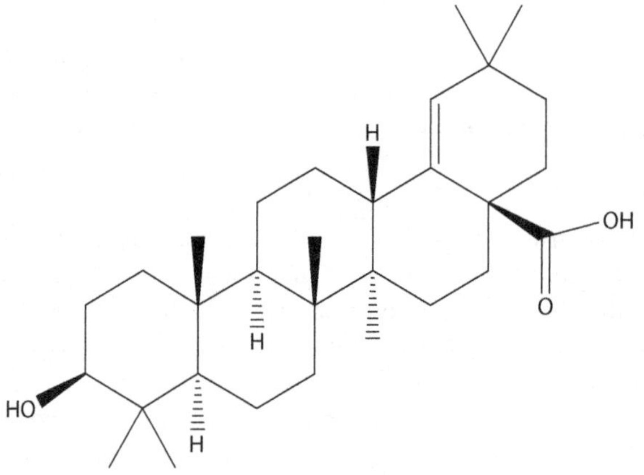

FIGURE 3.23 Morolic acid.

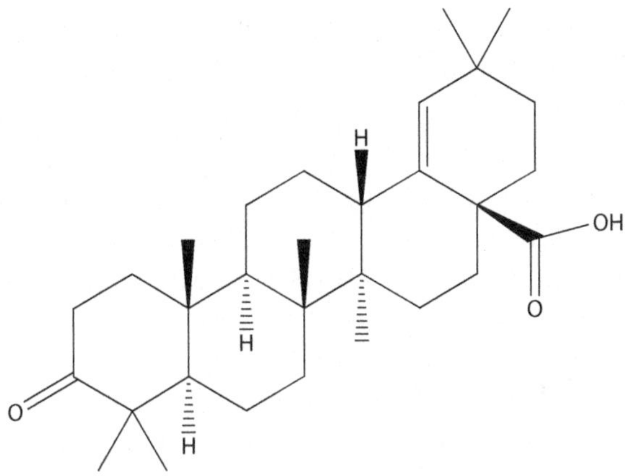

FIGURE 3.24 Moronic acid.

hepatic glucose utilization and decrease glycogen synthesis contributing to hyperglycemia, decrease insulin action and decrease glucose uptake by skeletal muscles.[309] However, 11-β-hydroxysteroid dehydrogenase-2 catalyzes the conversion of 11β-hydroxy ketoglucocorticoids to inactive metabolite in the kidneys and inhibition of this enzyme results in sodium retention and hypertension. Pentacyclic triterpenes inhibit protein-tyrosine phosphatase-1B and it would be interesting to examine the activity of morolic and moronic acid on this enzyme.

3.61 *Celastrus orbiculatus* Thunb.

Synonyms: *Celastrus articulatus* Thunb. *Celastrus insularis* Koidz. *Celastrus jeholensis* Nakai, *Celastrus lancifolius* Nakai, *Celastrus oblongifolius* Hayata, *Celastrus stephanotiifolius* (Makino) Makino, *Celastrus strigillosus* Nakai, *Celastrus tartarinowii* Rupr., *Celastrus versicolor* Nakai

Common names: nan she teng (China); oriental bittersweet

Subclass Rosidae, Superorder Celastranae, Order Celastrales, Family Celastraceae

Medicinal use: muscle pain (China)

Pharmacological target: atherogenic hyperlipidemia

Ethanol extract of stems of *Celastrus orbiculatus* Thunb. given at a dose of 5 g/kg/day for 8 weeks to guinea pigs on high-fat diet decreased body weight by 20% and liver weight by 34.3%, respectively.[310] This regimen lowered hepatic cholesterol and triglycerides and improved hepatic cytoarchitecture, and decreased alanine aminotransferase and serum aspartate aminotransferase activities by 53% and 26%.[310] The extract evoked an increased expression of 3-hydroxy-3-methylglutaryl-coenzyme A reductase and upregulated the expression of CYP7A1[310] as probable consequence of depleted hepatic cholesterol. In hepatocytes, high levels of free cholesterol translate into inhibition of 3-hydroxy-3-methylglutaryl-coenzyme A reductase via sterol regulatory element-binding protein-2. Ethanol extract of stem and roots of *Celastrus orbiculatus* Thunb. given orally at a dose of 10 g/kg/day for to C57BL/6J mice on high-fat diet for 6 weeks prevented body weight and body liver weight gain.[311] This regimen decreased plasma triglycerides, nonhigh-density lipoprotein–cholesterol and plasma cholesterol by 15.2%, 51.5%, and 21.7%, respectively, and increased high-density lipoprotein–cholesterol and apolipoprotein A-1 by 27.4% and 59.1%, respectively.[311] This treatment improved hepatic cytoarchitecture and decreased liver levels of cholesterol and triglycerides by 34.3% and 13.2%, respectively.[311] The treatment increased the expression of liver X receptor and subsequently ATB-binding cassette G5 and

CYP7A1.[311] From this extract, oleanolic acid at 150 μg/mL, ursolic acid at 20 μg/mL and celastrol at 0.4 μg/mL increased the expression of SRB1, ATB-binding cassette A1 and ATP-binding cassette G1 expression in Raw 264.7 macrophage.[310] Quercetin at 150 μg/mL from this extract increased the expression of SRB1, and ATP-binding cassette G1 expression in Raw 264.7 macrophage.[311] Members of the genus *Celastrus* have the ability to produce triterpenes such as 7β-hydroxy-3-oxo-D:A-friedooleanan-28-oic acid and 7β,29-dihydroxy-3-oxo-D:A-friedooleanan-28-oic acid which at a concentration of 100 μM induced the phosphorylation of insulin receptor in human hepatic Huh7 cells in absence of insulin.[312] This plant is toxic and has no place in therapeutic.[313]

3.62 *Hippophae rhamnoides* L.

Synonyms: *Elaeagnus rhamnoides* (L.) A. Nelson; *Rhamnoides hippophae* Moench
Common names: sha ji (Chinese); sea-buckthorn
Subclass Rosidae, Superorder Rhamnanae, Order Rhamnales, Family Eleagnaceae
Medicinal use: stomach pain (Tibet)
History: thirteenth century Mongol emperor Genghis Khan was feeding his armies horse with this plant before battles
Pharmacological target: atherogenic hyperlipidemia

The seeds of *Hippophae rhamnoides* L. contain linoleic acid and α-linolenic acid whereas the peel of the fruit contains palmitoleic acid.[314] Linoleic acid and linolenic acid, inhibited the enzymatic activity of protein-tyrosine phosphatase 1B *in vitro*.[315] Zhang et al. provided evidence that aqueous extract of deffated seeds of *Hippophae rhamnoides* L. given orally at a dose of 400 mg/kg/day to streptozotocin-induced diabetic type 2 Sprague–Dawley rats on high-fat diet for 6 weeks lowered glycaemia from about 10 to 9 mmol/L, reduced body weight, lowered insulinaemia and improved insulin sensitivity.[316] This extract reduced serum triglycerides, total cholesterol, and low-density lipoprotein–cholesterol.[316] In line, total flavonoid fraction of seeds given orally at a dose of 50 mg/kg/day for 12 weeks to ICR mice on high-fat diet evoked a decrease of liver weight and epididymal weight.[317] This treatment lowered plasma cholesterol by 13.5% and low-density lipoprotein–cholesterol by 26% and had no effect on plasma triglycerides.[317] Plasma glucose was lowered by 22.4% by the treatment, whereas plasma insulin was unchanged.[317] At the end of the treatment, area under the curve of glucose in oral glucose tolerance test was decreased by 8.3%.[317] At the hepatic level this treatment decreased hepatic triglycerides from 0.1 to 0.08 mg/g (normal 0.07 mg/g) and cholesterol from 0.07 to 0.05 mg/g (normal: 0.01 mg/g). This treatment improved hepatic cytoarchitecture with decrease in lipid droplets.[317] In a subsequent study, ethanol extract of leaves of *Hippophae rhamnoides* L. given to C57BL/6J mice on high-fat diet at a dose of 1 g/kg/day for 13 weeks lowered body weight gain from 18.4 to 11.2 g and epididymal fat weight from 2.6 to 1.5 g and improved glucose clearance.[318] This treatment evoked an increase of serum triglycerides from 127.2 to 141.5 mg/dL, a decrease of total cholesterol from 169.3 to 129.7 mg/dL, a decrease of high-density lipoprotein–cholesterol from 105.5 to 83.9 mg/dL.[317] At the hepatic level, the extract lowered triglycerides from 35.1 to 12.8 mg/g, total cholesterol from 1.6 to 0.8 mg/g, enhanced the expression of peroxisome proliferator-activated receptor-α and its target carnitine palmitoyltransferase-1, it also inhibited acetyl-CoA carboxylase and peroxisome proliferator-activated receptor-γ.[318] The precise mode of action and principle involved are not known but one could reasonably suggest, at least, the role of flavonoids. Indeed, the fruits of *Hippophae rhamnoides* L. contain flavonoids of which epicatechin, catechin, rutin, kaempferol, and quercetin as well as dimeric proanthocyanidins including catechin-(4α-8)-catechin, and catechin-(4α-8)-epicatechin as well as flavan-3-ols.[319,320] The leaves of this plant contain catechin, rutin, kaempferol, quercetin, and isorhamnetin.[321]

3.63 *Cornus officinalis* Siebold & Zucc.

Synonym: *Macrocarpium officinale* (Siebold & Zucc.) Nakai
Common name: shan zhu yu (Chinese)
Subclass Asteridae, Superorder Cornanae, Order Cornales, Family Cornaceae
Medicinal use: tonic (China)
Pharmacological target: atherogenic hyperlipidemia

Morroniside from the fruits of *Cornus officinalis* Siebold & Zucc. given to C57BLKS/J *db/db* orally at a dose of 100 mg/kg/day for 8 weeks had no effect on food intake, body weight but reduced water intake.[322] This secoiridoid glycoside had no effect on plasma glucose and cholesterol but decreased plasma triglycerides as well as serum alanine amino transferase.[322] At the hepatic level, morroniside treatment decreased glucose, triglycerides, and cholesterol.[322] Lipid peroxidation was significantly reduced in morroniside-treated *db/db* mice as well as reactive oxygen species. Besides, this regimen lowered the hepatic expression of nuclear factor-κB, and downstream targets cyclo-oxygenase 2 and nitric oxide.[322] In parallel, in liver, the regimen increased the expression of peroxisome proliferator-activated receptor-α and subsequently the decreased expression of sterol regulatory element-binding protein-1 and -2.[322] This iridoid glycoside given to obese *db/db* mice orally at a dose of 100 mg/kg/day for 8 weeks decreased water intake, had no significant effect on glycaemia but brought close to normal values serum aspartate aminotransferase and alanine aminotransferase and normalized hepatic lipid peroxidation.[322] At hepatic levels, this treatment normalized the expression of NAD(P)H oxidase 4 (NOX-4) and NAD(P)H oxidase component p22^{phox} and lowered the expression of Nrf2 and heme oxidase-1, nuclear factor-κB, cyclo-oxygenase 2, and inducible nitric oxide synthetase.[322] The treatment improved the histoarchitecture of livers of diabetic mice compared to vehicle treated animals.[322] Nuclear factor-erythroid 2-related factor 2 is a cellular a transcription factor activated by reactive oxygen species and its activation that protects hepatocytes against oxidative damage- by inducing the expression of antioxidant enzymes antioxidant/detoxification enzymes such as catalase, heme oxygenase 1, NAD(P)H: quinone oxidoreductase 1, catalase, and glutathione peroxidase.[323] The binding of insulin to its receptor in hepatocyte it activates NAD(P)H oxidase 4 (NOX-4) to transiently generate superoxides ions which inhibit protein-tyrosine phosphatase-1B, which dephosphorylates tyrosyl residues at the insulin receptor and its substrates 1 and 2.[325] In type 2 diabetes this enzyme is defectuous contributing to insulin resistance.[327] Morroside is not well absorbed orally[324] implying that the hepatoprotective effects of this iridoid glycoside are owed to bacterial metabolites including morroside aglycone, dehydroxylated aglycone, and methylated aglycone[327] and subsequent first pass metabolites. It would be of interest to decipher the exact pharmacological mechanism of this secoiridoid glycoside.

3.64 *Sambucus nigra* L.

Synonyms: *Sambucus graveolens* Willd.; *Sambucus peruviana* Kunth
Common names: u chu yu (Chinese); murver (Turkey); European elder
Subclass Asteridae, Superorder Cornanae, Order Dipsacales, Family Adoxaceae
Medicinal use: cough (Turkey)
History: the plant was known to Hippocrates Greek physician (460–370 BC)

Mice with targeted disruption of the apoliproptein-E gene ($ApoE^{-/-}$) develop atherosclerosis progressing from fatty streak to advanced lesions.[328] Anthocyanin-rich fraction of fruits of *Sambucus nigra* L. (containing mainly cyanidin 3-sambubioside and cyanidin 3-glucoside) given to $ApoE^{-/-}$ mice at 1.25% of AIN-93M diet for 6 weeks had no effect on body weight, had no effect on total cholesterol or triglycerides, lowered glycaemia, serum oxygen reactive species, and aspartate aminotransferase.[329] At the hepatic level, the fraction, increased lecithin-cholesterol acyltransferase, boosted low-density lipoprotein receptor and 3-hydroxy-3-methylglutaryl-coenzyme A reductase.[323] This fraction increased hepatic paraoxonase-1 expression.[329] The fraction lowered plasma monocyte chemoattractant peptide-1 and lowered aortic cholesterol contents.[329]

3.65 *Acanthopanax koreanum* Nakai

Subclass Cornanae, Superorder Cornanae, Order Apiales, Family Araliaceae
Medicinal use: inflammation (Korea)

The liver is the primary site of elimination of cholesterol in the body and this is done by direct secretion into bile through ATP-binding cassette G5 and G8 transporters or conversion into bile acids.[330] The rate-limiting enzyme for bile acids synthesis is cholesterol 7α-hydroxylase (CYP7A1), which catalyzes the conversion of cholesterol to 7α cholesterol.[331] The transcription of ABCG5 and ABCG8 transporters and CYP7A1 is induced, at least, by liver X receptors activation.[332] The root bark of *Acanthopanax koreanum* Nakai contains (−)-acanthoic acid (Figure 3.25) also known as (−)-pimara-9(11),15-dien-19-oic acid, which is an agonist for liver X receptor *in vitro* with IC_{50} below 10 μM.[333] Evidence indicates that liver X receptors not only induce genes involved in cholesterol efflux, but also repress inflammatory genes after tumor necrosis factor-α or interleukin-1β stimulation,[56] probably explaining why this diterpene given orally and prophylactically to Wistar rats at a dose of 100 mg/kg/day for 4 days attenuated carbon tetrachloride induced liver insults as evidenced by decreased plasma levels of aspartate aminotransferase.[334] Acanthoic acid given to C57BL/6 mice orally three times per weeks at a dose of 50 mg/kg for 8 weeks lowered plasma aspartate aminotransferase, tumor necrosis factor-α induced by carbon tetrachloride poisoning.[335] This diterpene improved hepatic cytoarchitecture as evidenced by decreased hepatocyte damage.[335] At the hepatic level, acanthoic acid upregulated liver X receptor and reduced the nuclear translocation of nuclear factor-κB.[335]

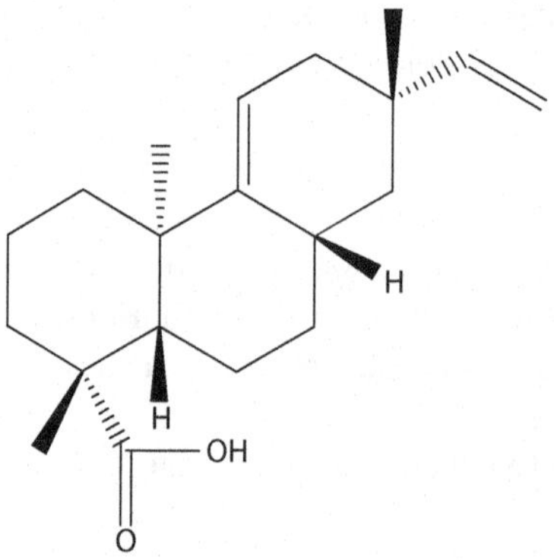

FIGURE 3.25 (−)-acanthoic acid.

3.66 *Aralia cordata* Thunb.

Synonyms: *Aralia schmidtii* Pojark; *Aralia taiwaniana* Y.C. Liu & F.Y. Lu
Common name: tu tang kuei (Chinese);
Subclass Cornanae, Superorder Cornanae, Order Apiales, Family Araliaceae
Medicinal use: indigestion (China)
Pharmacological target: atherogenic hyperlipidemia

In the liver, free fatty acids from adipose tissue are used for the synthesis of triglycerides, and a key an enzyme of triglyceride synthesis is glycerol-3-phosphate acyltransferase, which catalyzes the synthesis of 1-acyl-glycerol-3-phosphate from fatty acyl-CoA and glycerol-3-phosphate.[336] Methanol extract of roots of *Aralia cordata* Thunb. inhibited the enzymatic activity of glycerol-3-phosphate acyltransferase-1 with an IC_{50} value of 19.7 μg/mL.[337] From this extract *ent*-pimara-8(14),15-dien-19-oic acid inhibited this enzyme with an IC_{50} value of 60.5 μM.[337] In HepG2 cells, the diterpene at a concentration of 30 μM inhibited the synthesis of triglycerides by 80%.[337] Being able to inhibit the synthesis of triglycerides, this diterpenes could conceptually be of value to prevent hepatic steatosis. Roots of *Aralia cordata* Thunberg var *continentalis* (Kitagawa) Y. C. Zhu. contains continentalic acid, 7-oxo-*ent*-pimara-acid, 16α,17-dihydroxy-ent-kauran-19-oic acid and *ent*-thermarol, which inhibited the enzymatic activity of protein tyrosine phosphatase B with IC_{50} values below 2 μM (positive control ursolic acid: 1.2 μM).[338]

3.67 *Dendropanax morbiferum* H. Lév.

Synonyms: *Gilibertia morbifera* (H. Lév.) Nakai
Subclass Cornanae, Superorder Cornanae, Order Apiales, Family Araliaceae
Medicinal use: headache (Korea)

Essential oil containing mainly essential γ-elemene, β-selinene and β-zingiberene obtained from the flowers of *Dendropanax morbiferum* H. Lév. given orally to high cholesterol-fed Wistar rats for 2 weeks at a daily dose of 200 mg/kg normalized plasmal cholesterol, low-density lipoprotein–cholesterol and triglycerides and increased high-density lipoprotein–cholesterol and these effects were superior to clofibrate (100 mg/kg/day).[339] Zhong et al. provided evidence that oral administration of with 50 mg/kg/day for 5 weeks of β-elemene to rabbits on high-fat diet lowered plasma cholesterol and triglycerides as well as low-density lipoprotein–cholesterol and high-density lipoprotein–cholesterol.[340] β-Elemene injected intraperitoneally at a dose of 0.1 mL/100 g per day for 8 weeks protected Wistar rats against carbon tetrachloride-induced liver fibrosis.[341]

3.68 *Panax ginseng* C. A. Meyer

Synonyms: *Aralia ginseng* (C.A. Mey.) Baill.; *Panax chin-seng* Nees; *Panax quinquefolius* var. *ginseng* (C.A. Mey.) Regel & Maack
Subclass Cornanae, Superorder Cornanae, Order Apiales, Family Araliaceae
Medicinal use: tonic (China)
History: the plant was listed in Pen Ts'ao Kang Mu

Ethanol extract of *Panax ginseng* C. A. Meyer given orally at a dose of 10 g/kg/day to high-fat fed C57BL/6J mice for 13 weeks lowered weight gain from 27.6 to 22.5 g, and body fat from 5.8 to 4.9 g.[342] This extract lowered plasma insulin, cholesterol and normalized low-density lipoprotein–cholesterol resulting in a moderation of atherogenic index.[342] This treatment down-regulated CYP7A1, monoacyl glycerol *O*-acyltransferase 1, lysosomal lipase, low-density lipoprotein receptor, acyl-CoA thioesterase, and 3-hydroxyl-3-methylglutaryl-coenzyme.[342] In hepatocytes, low-density lipoproteins bind to their surface receptors and internalized cholesteryl esters are substrate for lysosomal acid lipase in lysosomes to release fatty acids and free cholesterol.[343] Acyl-CoA thioesterase cleave acyl-CoA to produce free fatty acid.[344] In the liver, free fatty acids are used for the synthesis of triglycerides, and a key an enzyme of triglyceride synthesis is monoacyl glycerol *O*-acyltransferase 1, which catalyzes the synthesis of diacyl-glycerol-3-phosphate from fatty acyl-CoA and monoacyl-glycerol-3-phosphate. The expression of this enzyme is increased by peroxisome proliferator-activated receptor-γ and is linked to triglyceride accumulation and steatosis. Wu et al. made the demonstration that in HepG2 cells, increased levels of cyclic adenosine monophosphate activates protein kinase A with subsequent of reduced expression of glycerol-3-phosphate acyltransferase, and decrease in triglyceride contents.[345]

3.69 *Panax notoginseng* (Burkill) F.H. Chen ex C.H. Chow

Synonyms: *Aralia quinquefolia* var. *notoginseng* Burkill; *Panax pseudoginseng* var. *notoginseng* (Burkill) C. Ho & C.J. Tseng
Common names: san qi (Chinese); san-chi ginseng
Subclass Cornanae, Superorder Cornanae, Order Apiales, Family Araliaceae
Medicinal use: bleeding (China)
Pharmacological target: atherogenic hyperlipidemia

Sprague–Dawley rats on high-fat diet treated orally with 100 mg/kg/day of saponin fraction (mainly notoginsenoside R1, ginsenosides Rg1, Re, Rb1, and Rd) of *Panax notoginseng* (Burkill) F.H. Chen ex C.H. Chow for 28 days had decreased plasma cholesterol from 5.5 to 3.9 mmol/L, triglycerides from 2.2 to 1 mmol/L, decreased low-density lipoprotein–cholesterol from 1.4 to 0.9 mmol/L, increased high-density lipoprotein–cholesterol from 1 to 1.4 mmol/L and decreased atherogenic index from 3.3 to 2 and these effects were similar with simvastatin at 3 mg/kg/day.[346] At the hepatic level the treatment increased the expression of superoxide dismutase, CYP7A1 and decreased the expression of peroxisome proliferator-activated receptor-α.[346] Butanol fraction of *Panax notoginseng* (Burkill) F.H. Chen ex C.H. Chow given orally to Sprague–Dawley rats on high-fat diet for 4 weeks of 100 mg/kg/day decreased plasma cholesterol, low-density lipoprotein–cholesterol, triglycerides and elevated high-density lipoprotein–cholesterol.[347] This treatment, brought down toward normal hepatic cholesterol and triglycerides.[347] The extract boosted the hepatic expression of liver X receptor target ATP-binding cassette A1, A5, A8 as well as farnesoid X receptor targets apolipoprotein C-II and SHP.[347] In line, the expression of CYP7A1, apolipoprotein C-III and sterol regulatory element-binding protein-1c and the expression level of low-density lipoprotein receptor was increased by the extract.[347] In a subsequent study, root powder of *Panax notoginseng* (Burkill) F.H. Chen ex C.H. Chow incorporated at 1% of high-fat diet given to Sprague–Dawley rats for 4 weeks lowered body weight whereby food intake was not significantly modified compared with untreated animals.[348] This supplementation lowered total cholesterol, low-density lipoprotein–cholesterol, triglycerides whereby high-density lipoprotein–cholesterol was brought above normal group values.[348] High-fat diet induced, at the hepatic level, the enzymatic of hepatic 3-hydroxy-3-methylglutaryl-coenzyme A reductase and the supplementation reduced that activity by 30.5%, as well as hepatic malondialdehyde by 35.4%, superoxide dismutase by 43.8%, and glutathione peroxidase by 15.9%.[348] The principle involved is apparently unknown.

3.70 *Ammi majus* L.

Synonyms: *Apium ammi* Crantz; *Carum majus* (L.) Koso-Pol.; *Selinum ammoides* E.H.L. Krause
Common names: da a mi qin (China); aatrilal (Pakistan); bishop's weed
Subclass Asteridae, Superorder Cornanae, Order Apiales, Family Apiaceae
Medicinal use: skin disease (Pakistan)
Pharmacological target: atherogenic hyperlipidemia

Ethanol extract of seeds *Ammi majus* L. given orally at a dose of 100 mg/kg/day for 2 months to Sprague–Dawley rats on high-fat diet lowered total cholesterol from 198.8 to 94.1 mg/dL (atorvastatin: 1 mg/kg twice per week: 89.5 mg/dL), triglycerides from 165.2 to 73.2 mg/dL (atorvastatin: 1 mg/kg twice per week: 71 mg/dL), low-density lipoprotein–cholesterol from 147.3 to 29.4 mg/dL (atorvastatin: 1 mg/kg twice per week: 28.7 mg/dL), and increased high-density lipoprotein–cholesterol from 17.9 to 46.8 mg/dL (atorvastatin: 1 mg/kg twice per week: 42.8 mg/dL).[349] The seeds of this plant contain oleanolic acid and series of coumarins of which ammirol, lomatin, 5-methoxy-S-hydroxy-psoralen, scopoletin[350] xanthotoxin, imperatorin, bergapten, isopimpinellin, and isoimperatorin and umbelliprenin.[351] Oral administration of imperatorin in rats orally for 12 weeks at a dose of 25 mg/kg/day has been reported to inhibit NADPH oxidase by approximately 25%.[352] NADPH oxidase in hepatocytes accounts for reactive oxygen species generation.[353] The hypocholesterolemic mechanism of this plant is apparently unknown. Clinical trials are warranted.

3.71 *Anethum graveolens* L.

Synonyms: *Anethum sowa* Roxb. ex Fleming *Ferula marathrophylla* Walp. *Peucedanum anethum* Baill. *Peucedanum graveolens* (L.) Hiern *Peucedanum sowa* (Roxb. ex Fleming) Kurz
Common names: sowa (India); dill
Subclass Asteridae, Superorder Cornanae, Order Apiales, Family Apiaceae
Medicinal use: lack of apetite (India)
Pharmacological target: atheogenic hyperlipidemia

Essential oil of aerial part of *Anethum graveolens* L. given to high-fat diet receiving Wistar rats orally at a daily dose of 180 mg/kg for 2 weeks lowered total cholesterol, low-density lipoprotein–cholesterol, triglycerides and atherogenic index and increased high-density lipoprotein–cholesterol more potently than clofibrate at 100 mg/kg/day orally.[354] Powder of dill given to Wistar rats at 10% of high-fat diet for 2 weeks lowered total cholesterol and triglycerides as potently as clofibrate at 100 mg/kg/day orally.[354] Powder of *Anethum graveolens* L. given to Wistar rats at 10% of high-fat diet for 2 weeks lowered low-density lipoprotein–cholesterol and increased high-density lipoprotein–cholesterol more potently than clofibrate at 100 mg/kg/day orally.[354] Hexane fraction of seeds as part of 0.05% on high-fat diet to KK-Ay for 4 weeks decreased plasma glucose, triglycerides and cholesterol, plasma fatty acid, and increased plasma adiponectin.[355] This treatment decreased liver cholesterol and triglycerides and increased the expression of peroxisome proliferator-activated receptor-α targets carnitine palmitoyltransferase-1, acyl-CoA synthetase and acyl-CoA carboxylase.[355] *In vitro*, this fraction at a concentration of 30 mg/mL induced the activation of peroxisome proliferator-activated receptor-α in transfected CV-1 cells.[355] A major constituent of essential oil of *Anethum graveolens* L. is carvone.[356] It is tempting to speculate that carvone which is an agonist of TRPV1 channels[357] increases cytosolic contents in calcium, activating thus adenosine monophosphate-activated protein kinase and downstream peroxisome proliferator-activated receptor-α. Clinical trials to ascertain the beneficial effect of *Anethum graveolens* L. in Metabolic Syndrome are warranted.

3.72 *Angelica acutiloba* (Siebold & Zucc.) Kitag.

Synonyms: *Ligusticum acutilobum* Siebold & Zucc.
Common names: dong dang gui (Chinese); Japanese angelica
Subclass Cornanae, Superorder Cornanae, Order Apiales, Family Apiaceae
Medicinal use: stimulate blood circulation (China)
Pharmacological target: atherogenic hyperlipidemia

Ethanol extract of roots of *Angelica acutiloba* (Siebold & Zucc.) Kitag. given to Wistar rats on high-fat diet at a daily dose of 300 mg/kg lowered body weight gain from 48.6 to 22.6 g/rats (normal: 16.9 g/rat) and lowered epididymal white adipose from 411.8 mg/100 g bw to 345.4 mg/100 g (normal 337.3 mg/100 g).[358] This extract lowered serum cholesterol by 28.3%, low-density lipoprotein–cholesterol by 51.4% and triglycerides by 25.3%, lowered free fatty acids and lowered atherogenic index from 2.9 to 1 (normal: 0.6).[358] This extract lowered hepatic cholesterol and triglycerides by 33% and 59%, respectively.[358] At the hepatic level, this extract increased the expression of peroxisome proliferator-activated receptor-α, as well as acyl-CoA oxidase and microsomal ω-oxidation (CYP4A) whereby the expression of sterol regulatory element-binding protein-1 and protein-2 were downregulated[358] implying peroxisome proliferator-activated receptor-α activation. Histological observation of liver tissues of treated rodent evidenced a decrease in average size of epididymal adipocytes and reduction of hepatic steatosis.[358] The roots of this plant contain the polyacetylenes falcarinol, falcarindiol, falcarinolone, choline, scopoletin, umbelliferone, and vanillic acid[359] as well as series of alkyl phthalide derivatives of which ligustilide.[360] Ligustilide at a concentration of 250 μM inhibited the production of tumor necrosis factor-α, prostaglandin E2, and nitric oxide as well as expression of inducible nitric oxide synthetase by Murine macrophage RAW 264.7 cells challenged with lipopolysaccharide via inhibition of nuclear factor-κB.[361] Ligustilide has a low oral bioavailability due to extensive first-pass metabolism in the liver.[362] Is metabolized in butylidenephthalide, senkyunolide I, and senkyunolide H.[362] Butylidenephthalide given orally at a dose of 80 mg/kg/day for 30 days to rats intoxicated with thioacetamide decrease the development of hepatic fibrosis.[363]

3.73 *Angelica keiskei* Koidz.

Synonym: *Archangelica keiskei* Miq.
Common names: ashitaba (Japanese); Japanese angelica
Subclass Cornanae, Superorder Cornanae, Order Apiales, Family Apiaceae
Nutritional use: vegetable (Japan)

Ethanol extract of aerial parts of *Angelica keiskei* Koidz. given for 11 weeks as 3% w/w of diet of to Wistar rats given 15% fructose solution as drinking water lowered liver weight compared to untreated animals.[364] This supplementation lowered blood glucose by 16.5%, insulinaemia by 47.3% and insulin resistance.[364] In regards to serum lipids, the treatment lowered triglycerides by 24.2%.[364] This extract had no effect on plasma cholesterol but increased high-density lipoprotein–cholesterol from 33.9 to 48.6 mg/dL.[364] Serum adiponectin was increased by the supplementation.[364] At the hepatic level, the supplementation increased the expression of acyl-CoA oxidase and medium chain acyl-CoA dehydrogenase which are involved in fatty acid β-oxidation in peroxisomes and mitochondria respectively, carnitine palmitoyltransferase-1, key enzyme transport of fatty acids in mitochondria, and apolipoprotein A1 and ATP-binding cassette A1 which are both involved in high-density lipoprotein production[364] which are downstream of peroxisome proliferator-activated receptor-α. This regimen had no effect on peroxisome proliferator-activated receptor-α expression in the liver[364] suggesting a possible agonistic activity on this nuclear receptor by a constituent of yet unidentified.

3.74 *Centella asiatica* (L.) Urb.

Synonyms: *Centella biflora* (P. Vell.) Nannf.; *Hydrocotyle asiatica* L.; *Hydrocotyle biflora* P. Vell.
Common names: pegaga (Malay/Indonesian); Asiatic pennywort
Subclass Cornanae, Superorder Cornanae, Order Apiales, Family Apiaceae
Nutritional use: vegetable (Malaysia)
Pharmacological target: insulin resistance

Centella asiatica (L.) Urb. contains asiatic acid which given orally at a daily dose of 20 mg/kg for 45 days to streptozotocin-induced diabetic Wistar rats (glycaemia >250 mg/dL) 45 days lowered fluid and food intake, plasma cholesterol, triglycerides, increased high-density lipoprotein–cholesterol, lowered low-density lipoprotein–cholesterol, brought unesterified fatty acids close to normal, and lowered atherogenic index.[365] At the hepatic level, this triterpene normalized cholesterol, triglycerides, and unesterified fatty acids.[365] Asiatic acid increased the activity of plasma lipoprotein lipase, which catalyze the endothelial release of fatty acids from very low-density lipoprotein triglycerides, lecithin cholesterol acyltransferase which esterifies cholesterol in high-density lipoprotein, and decreased hepatic 3-hydroxy-3-methylglutaryl-coenzyme A reductase.[365] In a parallel study, asiatic acid (Figure 3.26) given to C57BL/6 mice on high-fat diet at a dose of 20 mg/kg/day for 7 weeks decreased body weight gain, food intake, liver weight, and epididymal fats. The treatment decreased plasma triglycerides and cholesterol, had no effect on plasma glucose, decreased plasma insulin and decreased insulin resistance, and had no effect on plasma adiponectin nor leptin.[366] At the hepatic level, this triterpene decreased cholesterol from 7.6 to 6.4 mg/g and triglycerides from 64.5 to 38.2 mg/g.[366] This regimen increased fecal lipids from 13.6 to 15.7 mg/g feces.[366] At the hepatic level, asiatic acid decreased the expression of acetyl-CoA carboxylase fatty acid, synthetase stearoyl-CoA desaturase-1[366] which are targets of sterol regulatory element-binding protein-1c. This regimen decreased hepatic reactive oxygen species, increased glutathione as well as the activity of catalase and glutathione peroxidase.[366] This treatment decreased serum aspartate aminotransferase, improved hepatic cytoarchitecture with decreased hepatic triglycerides droplets.[366] This triterpene decreased hepatic expression of nuclear factor-κB and the levels of interleukin-6, interleukin-1β and tumor necrosis factor-α.[366] The precise hypolipidemic mode of action of asiatic acid is apparently unknown and could result, at least, from increased hepatic insulin sensitivity hence increased expression of sterol regulatory element-binding protein-1c and downstream targets.

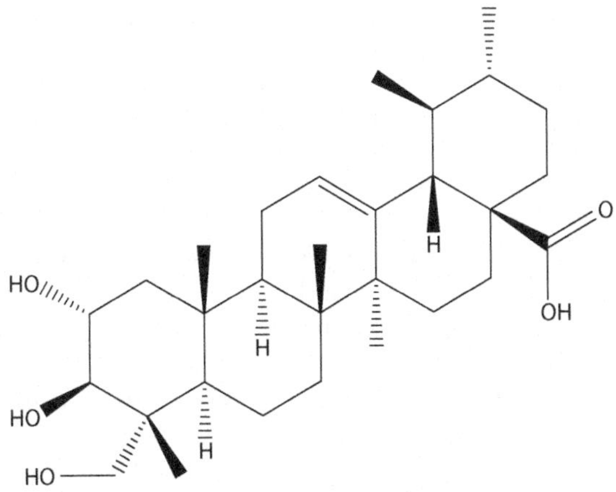

FIGURE 3.26 Asiatic acid.

3.75 *Cnidium monnieri* (L.) Cusson

Synonym: *Selinum monnieri* L.
Common name: she chuang (Chinese)
Subclass Cornanae, Superorder Cornanae, Order Apiales, Family Apiaceae
Medicinal use: tonic (China)
Pharmacological target: atherogenic hyperlipidemia

Osthol from the fruits of *Cnidium monnieri* (L.) Cusson given to spontaneously hypertensive rats at 0.05% of diet for 4 weeks lowered systolic blood pressure by approximately 15%.[367] This supplementation increased hepatic expression of 3-hydroxy-3-methylglutaryl-CoA reductase, acyl CoA oxidase, liver X receptor α and decreased hepatic apolipoprotein C-II and apolipoprotein C-III.[367] This prenylated coumarin from given to Kunming mice on high-fat diet, orally at a dose of 40 mg/kg/day for 6 weeks evoked a reduction of body weight from 47.1 to 45.2 g, lowered hepatic cholesterol, hepatic triglycerides, and hepatic free fatty acids.[368] In the serum of treated animals, osthol (Figure 3.27) reduced total cholesterol from 2.2 to 1.7 mmol/L (normal value: 1.8 mmol/L), triglycerides from 1.2 to 0.9 mmol/L, and free fatty acids from 6543 to 3373 μmol/L (below normal value: 5699 μmol/L). This coumarin evoked a reduction of hepatic expression of sterol regulatory element-binding protein-1c by about 45% and its target fatty acid synthetase.[368] This coumarin evoked a reduction of sterol regulatory element-binding protein-2 by 80% hence decreased low-density lipoprotein expression. This coumarin increased the expression of CYP7A1.[368] One could infer that osthol antagonizes liver X receptor hence repression of sterol regulatory element-binding protein-1c and CYP7A1.

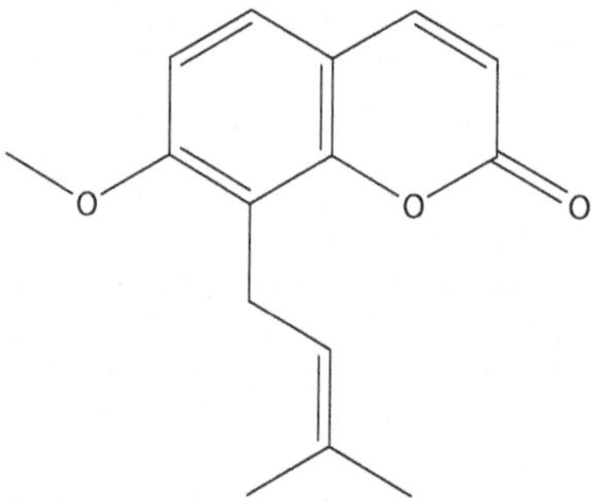

FIGURE 3.27 Osthol.

3.76 *Coriandrum sativum* L.

Common name: dhunia (India); coriander
Synonym: *Selinum coriandrum* Krause
Subclass Cornanae, Superorder Cornanae, Order Apiales, Family Apiaceae
Medicinal use: facilitates digestion (Bangladesh)
Pharmacological target: atherogenic hyperlipidemia

Sprague–Dawley rats on high-fat diet containing 10% of seeds of *Coriandrum sativum* L. for 75 days had serum cholesterol decreased from 160 to 89.1 mg/dL and triglycerides decreased from 14.8 to 7.3 mg/dL without reduction of food intake.[369] This regimen increased high-density lipoprotein–cholesterol, plasma lecithin:cholesterol acyltransferase and increased the activity of 3-hydroxy-3-methylglutaryl-coenzyme A CoA reductase.[369] Increased high-density lipoprotein–cholesterol and plasma lecithin:cholesterol acyltransferase indicates increase removal of cholesterol from extrahepatic tissues.[369] Besides, this supplementation increased hepatic and fecal bile acids and neutral sterols[369] suggesting, at least, an enhancement of cholesterol catabolism. Increased fecal excretion of bile acids and fecal sterol indicate an increased activity of cholesterol 7α-hydroxylase.[369] In a subsequent study coriander seeds given to Sprague–Dawley rats at part of 10% of a high-fat diet lowered serum cholesterol from 157.3 to 85.1 mg/dL and hepatic cholesterol from 1749.2 to 1048 mg/dL.[370] This supplementation lowered serum triglycerides from 12.4 to 7.2 mg/dL and liver triglycerides from 785.5 to 328.1 mg/dL.[370] The supplementation increased hepatic bile acids, hepatic neutral sterols, increased fecal bile acids and boosted fecal neutral sterols, and increased 3-hydroxy-3-methylglutaryl-coenzyme A CoA reductase activity.[370] With regard to serum lipids, this regimen increased high-density lipoprotein–cholesterol and lowered low-density lipoprotein–cholesterol and very low-density lipoprotein–cholesterol.[370] Aqueous decoction of seeds of *Coriandrum sativum* L. given orally at a single dose of 20 mg/kg to obese Meriones shawi rats lowered glycaemia from by 42% at 6 hours.[371] The extract given daily for 30 days lowered glycaemia by 58%, insulin by 54%, insulin resistance by 80%, total cholesterol by 80%, high-density lipoprotein–cholesterol by 28%, lowered low-density lipoprotein–cholesterol by 55%, triglycerides by 59%, and atherogenic index by 38%.[371] The seeds of *Coriandrum sativum* L. contain essential oil made principally of linalool (Figure 3.28). This monoterpene at a concentration of 100 μM evoked the activation of peroxisome proliferator-activated receptor-α in CHO-K1 cells.[372] In HepG2 cells, linalool at a concentration of 1 mM decreased triglyceride accumulation as a result of increased the expression of peroxisome proliferator-activated receptor-α.[372] In line, the expression of peroxisome proliferator-activated receptor-α targets including fatty acid transporter protein 4 (FATP4) and acyl-CoA synthetase 1 for fatty acid uptake; acyl-CoA oxidase and uncoupling protein 2 for fatty acid oxidation; and lipoprotein lipase and apolipoprotein C-III for triglyceride hydrolysis were increased.[372] Clinical trials to examine the beneficial effect of *Coriandrum sativum* L. for metabolic syndrome are needed.

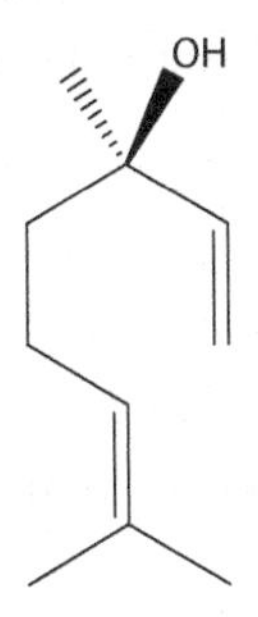

FIGURE 3.28 Linalool.

3.77 *Oenanthe javanica* (Blume) DC.

Synonyms: *Oenanthe stolonifera* Wall. ex DC.; *Sium graecium* Lour.
Common names: shui qin (Chinese), kankong cina (Malay); Japanese parsley (English)
Subclass Cornanae, Superorder Cornanae, Order Apiales, Family Apiaceae
Nutritional use: vegetable (Malaysia)
Pharmacological target: insulin resistance

Ethylacetate extract of *Oenanthe javanica* (Blume) DC. (Figure 3.29) given to spontaneous type 2 diabetic Goto–Kakizaki rats at a single oral dose improved glucose tolerance in oral glucose tolerance at 90 minutes.[373] From this extract, the polyacetylene falcarindiol inhibited the expression

FIGURE 3.29 *Oenanthe javanica* (Blume) DC.

of glucose-6 phosphatase, (the rate-controlling enzyme of gluconeogenesis), as well as glycogen synthetase kinase 3β in H4IIE rat hepatoma at a concentration of 50 μM.[373] It must be recalled that insulin binding to its hepatic receptor induced the inhibition of glucose-6 phosphatase and glycogen synthetase kinase-3β to command glycogen synthesis.[374] It would be of interest to examine the protective effect of falcarindiol on hepatic insulin signaling.

3.78 *Peucedanum japonicum* Thunb.

Common name: bin hai qian hu (Chinese)
Synonym: *Anethum japonicum* (Thunb.) Koso-Pol.
Subclass Asteridae, Superorder Cornanae, Order Apiales, Family Apiaceae
Medicinal use: cough (Japan)

Ethanol extract of leaves and stems of *Peucedanum japonicum* Thunb. given to C57BL/6 mice as part of 0.8% of diet for 4 weeks had no effect on food intake, lowered white adipose tissue weight, liver triglyceride from 34.9 to 21.4 mg/dL, and increased fecal triglycerides from 0.3 to 0.5 mg/day.[375,376] In the liver, the supplementation induced the expression of farnesoid X receptor and downstream target peroxisome proliferator-activated receptor-α.[375] In a subsequent study, hexane fraction of the plant decreased triglyceride accumulation in HepG2 cells with up-regulation of farnesoid X receptor and decreased expression of sterol regulatory element-binding protein-1c and downstream fatty acid synthetase.[377] From this plant, the pyranocoumarin pteryxin (Figure 3.30) at a concentration of 20 μg/mL lowered the content of triglycerides in hepatocytes by 27.4%.[377] This pyranocoumarin induced farnesoid X receptor, upregulated peroxisome proliferator-activated receptor-α, inhibited the expression of sterol regulatory element-binding protein-1c and downstream targets fatty acid synthetase, stearoyl-coenzyme A desaturase, and acetyl-CoA carboxylase.[377] The principle and hypolipidemic mechanism of action of *Peucedanum japonicum* Thunb. are apparently unknown.

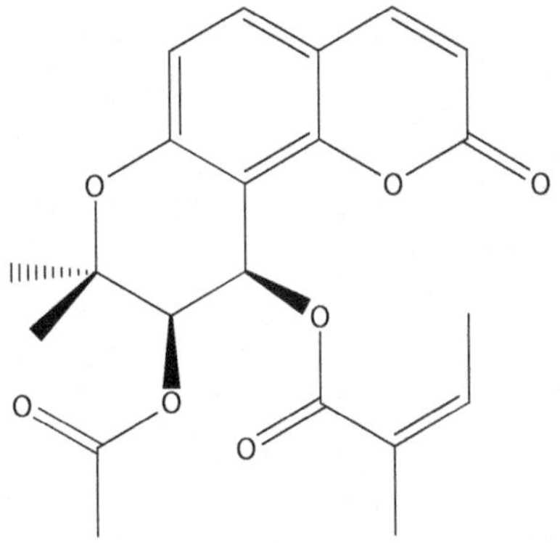

FIGURE 3.30 Pteryxin.

3.79 *Adenophora tetraphylla* (Thunberg) Fischer

Synonyms: *Adenophora obtusifolia* Merr.; *Adenophora triphylla* (Thunberg) A. DC.; *Campanula triphylla* Thunberg; *Campanula tetraphylla* Thunb.
Common names: nan sha shen (Chinese); giant bell flower
Subclass Asteridae, Superorder Asteranae, Order Campanulales, Family Campanulaceae
Medicinal use: cough (China)
Pharmacological target: atherogenic hyperlipidemia.

Ethylacetate fraction of roots of *Adenophora tetraphylla* (Thunberg) Fischer given orally at a dose of 75 mg/kg/day to C57BL/6J mice on high-fat diet for 4 weeks evoked a decrease of body weight compared with untreated rodent, whereas food intake was not changed.[378] This treatment decreased plasma cholesterol by 19.4%, reduced the low-density lipoprotein–cholesterol by 40.8%, reduced aspartate aminotransferase levels by 19.1%, and aspartate aminotransferase by 27.3%.[378] In metabolic syndrome, elevated low-density lipoprotein–cholesterol is a strong risk factor of cardiovascular diseases because it deposits cholesterol in vascular bed.[61] Conversely, elevated plasma high-density lipoprotein is a sign of lower risk of cardiovascular disease because it performs the transport of cholesterol from tissues to the liver.[379] At the hepatic level, the treatment increased the expression of low-density lipoprotein receptor as well as CYP7A1 and decreased the expression of 3-hydroxy-3-methylglutaryl-coenzyme A CoA reductase.[378] Ethanol extract of the plant given to C57BL/6 mice given as part of high-fat diet at 1% for 10 weeks had no effect of food intake and attenuated weight gain compared to untreated rodent.[380] This extract lowered liver weight, inhibited hepatic steatosis and decreased epididymal weight. The extract normalized plasma aspartate aminotransferase and alanine aminotransferase and alkaline phosphatase, glycaemia, triglycerides, cholesterol, high-density lipoprotein–cholesterol, low-density lipoprotein–cholesterol.[380] Plasma insulin was decreased from 42.6 to 22.5 μU/mL (normal: 20.7 μU/mL and insulin resistance was normalized).[380] In the liver of treated rodent, the extract increased the expression of adenosine monophosphate-activated protein kinase and consequently peroxisome proliferator-activated receptor-α.[380] In epididymal adipose tissue, the extract increased the expression of adiponectin,[380] a protein known to activate adenosine monophosphate-activated protein kinase in the liver as discussed elsewhere. The root contains phenylpropane glycosides such as siringinoside[381] and lupenone.[381–383] This triterpene inhibited protein-tyrosine phosphatase-1B with an IC_{50} value of 15.1 μM.[384] Lupenone (Figure 3.31) at a concentration of 200 μM inhibited the accumulation of 3T3-L1 mouse fibroblasts induced to differentiate by insulin-dexamethasone-3-isobutyl-1-methylxanthine via decreased expression of peroxisome proliferator-activated receptor-γ and CCAAT/enhancer-binding protein-α as well as adipogenic of fatty acid binding protein and resistin.[383]

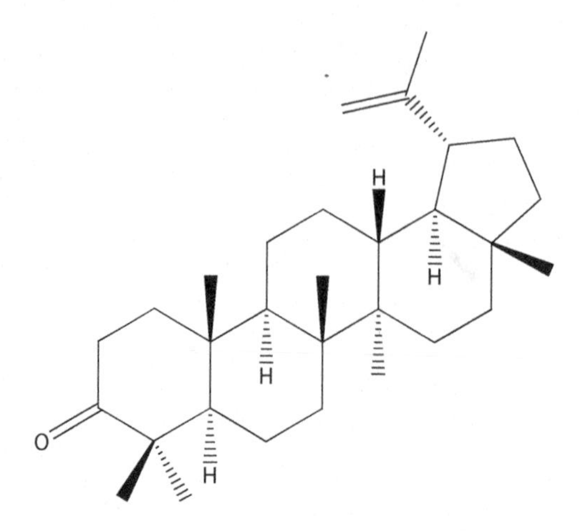

FIGURE 3.31 Lupenone.

3.80 *Codonopsis lanceolata* (Siebold & Zucc.) Trautv.

Synonyms: *Campanumoea japonica* Siebold ex Morren; *Campanumoea lanceolata* Siebold & Zucc.; *Codonopsis bodinieri* H. Lév.; *Glosocomia lanceolata* (Siebold & Zucc.) Maxim.
Common name: yang ru (Chinese)
Subclass Asteridae, Superorder Asteranae, Order Campanulales, Family Campanulaceae
Medicinal use: Tonic (China)
Pharmacological target: atherogenic hyperlipidemia

Aqueous extract of roots of *Codonopsis lanceolata* (Siebold & Zucc.) Trautv. given to C57BL/6 mice on high-fat diet at a dose of 360 mg/kg/day for 12 weeks had no effect on food consumption, lowered body weight by 26.1% and lowered weight of both subcutaneous and visceral fats by about 20% and 25%, respectively.[385] This regimen lowered plasma triglycerides, cholesterol and halved low-density cholesterol.[385] The extract lowered glycaemia and insulin levels.[385] The extract decreased liver weight and prevented hepatic steatosis induced by high-fat diet and lowered serum plasma aspartate aminotransferase and alanine aminotransferase.[385]

3.81 *Platycodon grandiflorus* (Jacq.) A. DC.

Synonyms: *Platycodon glaucum* (Thunb.) Nak.
Common name: jie geng (Chinese)
Subclass Asteridae, Superorder Asteranae, Order Campanulales, Family Campanulaceae
Medicinal use: cough (Korea)
Pharmacological target: insulin resistance

Saponin fraction of roots of *Platycodon grandiflorus* (Jacq.) A. DC. (containing platycodin D) at a concentration of 4 μg/mL inhibited triglyceride and cholesterol accumulation in HepG2 cells cultured with high-glucose concentration.[386] The fraction increased NAD^+/NADH ratio in HepG2 hence the activation of SIRT1 and downstream activation of adenosine monophosphate-activated protein kinase.[386] Activation of adenosine monophosphate-activated protein kinase by saponin, induced the phosphorylation of acetyl-CoA carboxylase and also decrease the expression sterol regulatory element-binding protein-1c and downstream target of fatty acid synthetase.[386] The fraction evoked an increase in cytoplasmic calcium.[386] It must be recalled that increased contents of calcium in the cytosol increase the activity of calcium/calmodulin-dependent protein-kinase kinase (CaMKK), which induce the activation of adenosine monophosphate-activated protein kinase.[387] Saponin fraction of roots of this plant (Platycodin D 25.1 mg/g) given as part of diet (0.5 g/100 g diet) for 6 week to C57BL/KsJ-*db/db* mice lowered food intake, prevented weight loss, lowered fasting plasma glucose by 37%, and improved glucose tolerance in intraperitoneal gluce tolerance test.[388] This regimen lowered plasma triglycerides from 1.6 to 0.9 mmol/L, free fatty acids from 1.1 to 0.9 mmol/L,[388] and had no effects on plasma cholesterol.[388] It increased fecal cholesterol from 3.2 to 4.6 mmol/g.[388] The fraction had no significant changes in the plasma leptin.[388] This fraction increased the activity of hepatic glucokinase and decreased the activity of glucose-6-phosphatase.[388] In physiological conditions, plasma glucose penetrates hepatocytes via glucose transporter-2, a membrane-bound transporter, the expression of which is independent of insulin.[49] Glucose is phosphorylated in the cytoplasm into glucose-6-phosphate by glucokinase, also termed hexokinase, which is induced by insulin and the first enzyme of glycolysis.[389] In diabetic patients decrease of insulin release reduces the activity of glucokinase favoring the release of glucose from the liver in the circulation. The supplementation lowered the activity of maltase and sucrase by 41% (as a result of insulin increase?). The fraction inhibited yeast α-glucosidase activity by 79% at concentrations of 10 mg/mL, respectively. In addition, the fraction was a more effective yeast α-glucosidase inhibitor than acarbose at 5 mg/mL.

3.82 *Artemisia scoparia* Maxim.

Synonym: *Artemisia scoparia* Waldst. & Kit.
Common name: zhu mao hao (Chinese)
Subclass Asteridae, Superorder Asteranae, Order Asterales, Family Asteraceae
Medicinal use: hepatoprotective (Pakistan)
Pharmacological target: insulin resistance

Ethanol extract of *Artemisia scoparia* Maxim. given orally to C57BL/6J mice on high-fat diet as part of 0.5% of diet for 4 weeks had no effect on fasting plasma glucose nor plasma cholesterol, attenuated plasma triglycerides, lowered fasting plasma insulin from about 10 to 6 ng/mL, potentiated the hypoglycemic effect of insulin in intraperitoneal insulin tolerance test.[390] This supplementation had no effect on plasma leptin but increased plasma adiponectin by about 2-folds.[390] At the hepatic level this extract prevented triglyceride and cholesterol accumulation, increased the expression of insulin receptor substrate-1/2, the phosphorylation of insulin receptor substrate-1, Akt and lowered protein-tyrosine phosphatase-1B evidencing increased insulin hepatic sensitivity.[390] In the liver, the extract activated adenosine monophosphate-activated protein kinase and downstream targets fatty acid synthetase, sterol regulatory element-binding protein-1c and of 3-hydroxy-3-methylglutaryl-coenzyme A CoA reductase.[390] In the liver, activation of adenosine monophosphate-activated protein kinase promotes insulin receptor sensitivity by inhibiting mTORC1 alleviating thus insulin receptor inhibition by S6K1.[391] Besides, S6K1 activates liver X receptor, which induces the expression of sterol regulatory element-binding protein-1c.[392] The plant contains rutin which was given at 4 doses of 20 mg/kg at 12 hours interval to mice prevented paracetamol-induced hepatic insults as evidenced by decreased plasma levels of aspartate aminotransferase.[393] Rutin given orally to Wister rats on high-fat diet at a dose of 100 mg/kg/day for 28 days decreased plasma cholesterol, triglycerides, and low-density lipoprotein–cholesterol and had no effect on high-density lipoprotein–cholesterol.[394] This flavonoid lowered aspartate aminotransferase, liver weight, and improved hepatic cytoarchitecture.[394]

3.83 *Artemisia sacrorum* Ledeb

Synonym: *Artemisia gmelinii* var. *gmelinii*
Common name: bai lian hao (Chinese)
Subclass Asteridae, Superorder Asteranae, Order Asterales, Family Asteraceae
Medicinal use: hepatoprotective (China)
Pharmacological target: insulin resistance

Petroleum ether fraction of *Artemisia sacrorum* Ledeb at a concentration of 100 μg/mL lowered glucose production by HepG2 cells by about 26.8%, and this effect was as efficient as insulin at 100 nM *in vitro*.[395] This fraction increased the activity of adenosine monophosphate-activated protein kinase and downstream phosphorylation of acetyl-CoA carboxylase as well as increased the phosphorylation of glycogen synthase kinase-3 and decreased the phosphorylation of CAMP-response binding element.[395] This fraction repressed the expression of gluconeogenic peroxisome proliferator-activated receptor coactivator-1α, phosphoenolpyruvate carboxykinase and glucose-6-phosphatase and this effect was abolished by cotreatment with adenosine monophosphate-activated protein kinase inhibitor compound C.[395] The fraction increased the expression of small heterodimer partner (SHP), and this effect was inhibited by adenosine monophosphate-activated protein kinase inhibitor Compound C.[395] Activation of AMP-activated protein kinase activates TORC2, which was associated with CAMP-response binding element, transcription of peroxisome proliferator-activated receptor coactivator-1α hence increased expression of gluconeogenic enzymes glucose-6-phosphatase and phosphoenolpyruvate carboxykinase.[396] Besides, AMP-activated protein kinase upregulates the nuclear receptor small heterodimer partner which inhibits the expression of glucose-6-phosphatase and phosphoenolpyruvate carboxykinase K.[397]

3.84 *Chrysanthemum coronarium* L.

Synonyms: *Glebionis coronaria* (L.) Cass. ex Spach; *Matricaria coronaria* (L.) Desr.
Common names: tung ho choi (Chinese); tango (Philippines); garland chrysanthemum
Subclass Asteridae, Superorder Asteranae, Order Asterales, Family Asteraceae
Nutritional use: vegetable (China)

Extract of flowering aerial parts of *Chrysanthemum coronarium* L. given orally to on high-fat diet streptozotocin diabetic Wistar rats at a dose of 500 mg/kg/day for 45 days lowered total cholesterol from 225 to 130 mg/dL (normal: 89.4 mg/dL), triglycerides from 156.9 to 60.3 mg/dL (normal: 56.7 mg/dL), low-density lipoprotein–cholesterol from 152.1 to 15.8 mg/dL (normal: 12.4 mg/dL), and increased high-density lipoprotein–cholesterol from 4.9 to 34.1 mg/dL (normal: 31.2 mg/dL).[398] This extract increased the hepatic glutathione contents and lowered both hepatic malondialdehyde and nitric oxide.[398] This extract lowered CAMP-response binding element interleukin-10 tumor necrosis factor-α.[398]

3.85 *Cirsium japonicum* DC.

Common names: ji (Chinese); Japanese thistle
Subclass Asteridae, Superorder Asteranae, Order Asterales, Family Asteraceae
Medicinal use: fever (China)

The flavones pectolinarin and 5,7-dihydroxy-6,4-dimethoxyflavone isolated from *Cirsium japonicum* DC. given orally at a dose of 50 mg/kg/day to streptozotocin-induced diabetic Sprague–Dawley rats (>230 mg/dL) on high-carbohydrate/high-fat diet lowered plasma glucose level by 24.5% and 19.6%, respectively, attenuated plasma cholesterol and triglycerides and had no effect on plasma insulin.[399] These flavones improved the activities of gluconeogenic enzyme glucose-6-phosphatase. Extract of *Cirsium japonicum* DC. at a concentration of 1 mg/mL inhibited the formation of lipid droplets in HepG2 cells exposed to high concentration of fatty acids.[400] This extract lowered triglycerides and cholesterol in HepG2 cells by about 20% and 50%, respectively.[400] The extract induced the phosphorylation of adenosine monophosphate-activated protein kinase and direct substrate acetyl-CoA carboxylase, lowered the expression of fatty acid synthetase and acetyl-CoA carboxylase in HepG2 cells challenged with fatty acids, and increased the expression of peroxisome proliferator-activated receptor-α and carnitine palmitoyltransferase-1.[400]

3.86 *Cynara scolymus* L.

Synonym: *Cynara cardunculus* L.
Common name: artichoke
Subclass Asteridae, Superorder Asteranae, Order Asterales, Family Asteraceae
Nutritional use: vegetable (Turkey)
Pharmacological target: atherogenic hyperlipidemia

Kirchhoff et al. provided evidence that an extract of *Cynara scolymus* L. (containing mainly 1,3-di-*O*-caffeoylquinic acid also known as cynarin) given orally at a single dose: 1.92 g to healthy volunteers increased bile secretion.[401] Aqueous extract of leaves of *Cynara scolymus* L. given orally to Wistar rats at a dose of 400 mg/kg increased bile flow from 0.1 to 0.2 mL/100 g animal/h, and this effect was similar to positive standard: dehydrocholic acid at 20 mg/kg.[402] The extract given for 7 days twice a day at a dose of 400 mg/kg increased bile flow from 0.1 to 0.2 mL/100 g animal/h (dehydrocholic acid at 20 mg/kg: 0.3 mL/100 g animal/h).[402] Ethanol extract for flowering head of *Cynara scolymus* L. given orally at a dose of 1500 mg/kg to Wistar rats before food pellet consumption lowered 120 minutes peak glycaemia from 130 to about 110 mg/dL. The extract given to obese Zucker obese rats orally at a dose of 1500 mg/kg before food pellet consumption lowered 60 minutes peak glycaemia from 150 to about 130 mg/dL.[403] Cynaropicrin, apigenin 7-*O*-glucoside, and cynarin at a concentration of 10 mg/mL increased the secretion of bile by perfused rat liver by 47%, 30%, and 5%, respectively.[404] Apigenin 7-*O*-glucoside is probably metabolized by bacterial intestinal flora and first-pass metabolism and it seems improbable to obtain hepatic pharmacological concentrations of this flavonoid upon oral administration. Cynaropicrin inhibited at a single dose of 100 mg/kg triglyceride absorption in rodents on partial account of gastric emptying delay.[405] The mode of action of cynaropicrin on bile secretion could be due to irritation of gut muscles. In a subsequent study, 200 mg of ethanol extract of flowering heads of *Cynara scolymus* L. given to overweight and obese volunteers with impaired fasting glycaemia (body mass index between 25 and 35 kg/m^2, fasting glycaemia between 6.1 and 7 mmol/L) three times per day orally before meals for 8 weeks evoked a decrease in fasting blood glucose by 9.6%, insulin resistance and had no effect on insulinemia.[406] This supplementation lowered plasma cholesterol by 6.3% and had no effect in triglycerides.[406] Aqueous extract of leaves given to Wistar rats on cholesterol-enriched diet orally at a dose of 300 mg/kg/day (simvastatin 4 mg/kg/day) for 30 days lowered total cholesterol by 51.9% (simvastatin 4 mg/kg/day: 41.9%) and low-density lipoprotein–cholesterol by 54.8% (simvastatin 4 mg/kg/day: 46.7%).[407] This extract had no effect on high-density lipoprotein–cholesterol and very low-density lipoprotein–cholesterol, lowered proinflammatory cytokines interleukin-1, interleukin-6, tumor necrosis factor-α, and INF-γ close to normal values.[407] Intake of artichoke could be beneficial in metabolic syndrome.

3.87 *Eclipta prostrata* (L.) L.

Synonyms: *Eclipta alba* (L.) Hassk.; *Eclipta erecta* L.; *Verbesina alba* L.; *Verbesina prostrata* L.
Common names: li chang (Chinese); bhrngaraja (India); yayaod (Philippines); eclipta
Subclass Asteridae, Superorder Asteranae, Order Asterales, Family Asteraceae
Medicinal use: hepatitis (Philippines)

Aqueous extract of leaves of *Eclipta prostrata* (L.) L. given orally to high-fat diet Wistar rats at a daily dose of 200 mg/kg prevented the increase of plasma cholesterol, triglycerides, and increased high-density lipoprotein above normal values.[408] This regimen lowered atherogenic index from 4.3 to 1.3 (normal: 3.1).[408] Ethanol fraction of this herb (containing 50.9 mg/g of wedelolactone) given orally at a dose of 250 mg/kg/day to Syrian golden hamsters on high-fat diet for 5 weeks prevented body weight gain, lowered total cholesterol from 9.1 to 5.3 mmol/L (normal: 2.1 mmol/L), triglycerides from 6.6 to 2.4 mmol/L (normal: 3.3 mmol/L), increased high-density lipoprotein–cholesterol above normal, lowered low-density lipoprotein–cholesterol below normal group, and decreased atherogenic index from 5.4 to 1.9 (normal: 0.3).[409] This fraction increased the hepatic level of antioxidative enzymes superoxide dismutase and glutathione peroxidase and lowered hepatic levels of malondialdehyde.[409] The treatment lowered plasma aspartate aminotransferase indicating hepatoprotective effect and reduced hepatosteatosis.[409] The fraction decreased the hepatic expression of 3-hydroxy-3-methyl-glutaryl-CoA reductase, increased the expression of peroxisome proliferator-activated receptor α, low-density lipoprotein receptor, lecithin-cholesterol transferase, and scavenger receptor class B type I (SR-BI).[409] Increased high-density lipoprotein receptor SR-B1 expression in the liver accounts for increased hepatic supply of arterial wall cholesterol decreasing thus atherogenesis in metabolic syndrome.[410]

3.88 *Smallanthus sonchifolius* (Poeppig) H. Robinson

Synonym: *Polymnia sonchifolia* Poeppig
Common names: ju shu (Chinese); yacon
Subclass Asteridae, Superorder Asteranae, Order Asterales, Family Asteraceae
Nutritional use: food (Japan)
Pharmacological target: insulin resistance

The tubers of *Smallanthus sonchifolius* (Poeppig) H. Robinson given to Zucker fa/fa rats with insulin resistance as part of 6.5% diet for 5 weeks had no effect on food intake nor weight gain, lowered plasma glucose from 184.1 to 167.9 mg/dL, plasma insulin from 13.2 to 10.4 ng/mL, and insulin resistance.[411] In addition, this supplementation lowered plasma triglycerides from 253.7 to 215.7 mg/dL and had no effect on plasma cholesterol, high-density lipoprotein–cholesterol and adiponectin.[411] This regimen improved insulin sensitivity in euglycemic-hyperinsulinemic clamp study as evidenced by a 12.3% increase in the glucose infusion rate necessary to maintain euglycaemia in rodents.[411] The supplementation increased Ser473 Akt phophorylation and lowered the expression of phosphoenolpyruvate carboxykinase and glucose 6-phosphatase by 49% and 64%, respectively, implying the increased hepatic insulin sensitivity.[411] Dual phosphorylation of Akt at Ser473/474 by phosphoinositide-dependent kinase 2 and Thr308/309 by phosphoinositide-dependent kinase 1 is required for complete activation following insulin binding to its receptor.[412] Besides, the liver of rats fed with yacon had expression of Trb3 reduced by 43%.[412] Trb3 contributes to insulin resistance by inhibiting Akt activation.[413] The plant contains smallanthaditerpenic acids A, B, C, and D that inhibited α-glucosidase with IC_{50} of 0.4, 0.5, 1, and 1.1 mg/mL, respectively.[414] The plant contains chlorogenic acid and 3,5-dicaffeoylquinic acid,[415] caffeic acid ferulic acid and protocatechuic acid,[416] as well as sequiterpene lactones such as sonchifolin, uvedalin, enhydrin, and fluctuanin.[417]

3.89 *Silybum marianum* (L.) Gaertn.

Synonyms: *Carduus marianus* L.; *Carthamus maculatum* (Scop.) Lam. *Cirsium maculatum* Scop.
Common name: milk thistle
Subclass Asteridae, Superorder Asteranae, Order Asterales, Family Asteraceae
Medicinal use: jaundice (India)

Silymarin given orally at a dose of 140 mg three times per day for 45 days to 20 type 2 diabetic patients had no effect on dietary intake, increased superoxide dismutase activity, glutathione peroxidase activity, and total antioxidant capacity by 12.8%, 30.3%, and 8.4%, respectively.[418] The treatment decreased malondialdehyde levels by 12%.[418]

3.90 *Taraxacum mongolicum* Hand.-Mazz.

Common names: pu gong yin; Mongolian dandelion
Subclass Asteridae, Superorder Asteranae, Order Asterales, Family Asteraceae
Medicinal use: jaundice (China)

Ethanol extract of *Taraxacum mongolicum* Hand.-Mazz. given to high-fat diet Sprague–Dawley rats at a dose of 900 mg/kg/day for 4 weeks lowered body weight gain from 94.7 to 35.7 g, normalized total cholesterol from 80 to 56.3 mg/dL (normal: 63.8 mg/dL), triglycerides from 64 to 46 mg/dL (normal: 41.3 mg/dL), low-density lipoprotein–cholesterol from 20.3 to 9.1 mg/dL (normal: 11.3 mg/dL), and increased high-density lipoprotein–cholesterol above normal values.[419] This regimen lowered adipose tissue and liver weight.[419] Hexane fraction from this extract (containing linoleic acid, phytol, and tetraconazole) *in vitro* at a concentration of 25 μg/mL HepG2 cells increased the phosphorylation of adenosine monophosphate-activated protein kinase and acetyl-CoA carboxylase and the expression of fatty acid synthetase.[419]

3.91 *Gardenia jasminoides* J. Ellis

Synonyms: *Gardenia augusta* Merr.; *Gardenia florida* L.; *Varneria augusta* L.
Common names: zhi zi (Chinese); karinga (India); cape jasmine
Subclass Lamiidae, Superorder Lamianae, Order Rubiales, Family Rubiaceae
Medicinal use: Jaundice (China)

In the postprandial state, the pancreatic secretion of insulin promotes the storage of glucose in the form of glycogen in the liver by via inhibition of glycogen phosphorylase, the rate-limiting enzyme of glycogenolysis and glucose-6-phosphatase.[420] *Gardenia jasminoides* J. Ellis contains geniposide, the oral absorption of which in mice improves insulin resistance. Geniposide given orally to streptozotocin high-fat diet-induced type 2 diabetic C57BL/6J mice (glycaemia between 10 and 20 mmol/L) at a daily dose of 400 mg/kg for 2 weeks lowered body weight, plasma glucose from 19.2 to 10.3 mmol/L (normal: 5.9 mmol/L), decreased insulinemia from 0.7 to 0.4 μU/mL (normal: 0.3 μU/mL), and plasma triglycerides from 164.6 to 120.9 mg/dL (normal: 15.4 mg/dL), but total cholesterol was unchanged.[421] This iridoid glycoside decreased the expression and activities of hepatic glycogen phosphorylase and glucose-6-phosphatase[421] implying increased hepatic sensitivity to insulin. Geniposide and genipin from the fruits of *Gardenia jasminoides* J. Ellis given to Tsumara Suzuki Obese Diabetes (TSOD) mice (model mice showing spontaneously multifactorial genetic type 2 diabetes based on visceral fat accumulation) at 0.3% of diet for 8 weeks had no effect on food intake and evoked a reduction of body weight gain as well as visceral and subcutaneous fat mass.[422] This iridoid glycoside improved glucose tolerance in oral glucose tolerance test with a decrease of 30 minutes postprandial glycaemia from about 550 to 425 mg/dL.[422] This regimen lowered plasma insulin from 15.3 to 2.5 ng/mL, lowered total cholesterol from 179.4 to 152 mg/dL, and increased the ratio high-density lipoprotein/total cholesterol.[422] Geniposide lowered liver total cholesterol from 8 to 6.9 mg/g liver.[422] After oral administration, geniposide is decomposed by intestinal bacteria into genipin.[423] Genipin at a concentration of 100 μM lowered triglyceride accumulation and cholesterol accumulation in HepG2 cells challenged with palmitic acid and increased peroxisome proliferator-activated receptor-α expression.[422]

3.92 *Mitragyna speciosa* (Korth.) Havil.

Synonym: *Nauclea speciosa* (Korth.) Miq.
Common name: kratom (Malay)
Subclass Lamiidae, Superorder Lamianae, Order Rubiales, Family Rubiaceae
Medicinal use: tonic (Malaysia)

Mitragyna speciosa (Korth.) Havil. elaborates indole alkaloids of which mitragynine, paynantheine, mitraphylline, and speciophylline.[424] Speciophylline induced an increase of biliary flow from 7.3 to 10.5 mg/100 g/min.[425] Biochemical analysis of the bile collected during biliary flow study revealed an increase in total bilirubin from 27.8 to 43.8 μmol/L and an increase in conjugated bilirubin from 12.1 to 19.2 μmol/L as well as absence of hepatotoxicity.[425] This plant is becoming popular but is in fact addictive and hepatotoxic. A 25-year-old man consuming 1 to 6 tea spoons of *Mitragyna speciosa* (Korth.) Havil. daily for 2 weeks developed severe intrahepatic cholestasis.[426] The mode of action of speciophylline is unknown. Evidence suggests that indole alkaloids have the tendency to activate farnesoid X receptor.[427,428] Mitragynine given orally to male rodent at a dose of 100 mg/kg/day for 28 days induced a weight gain by 130% whereby untreated animals had a weight gain of circa 140% and mitragynine at a dose of 1 mg/kg raised the weight gain by circa 160%.[429] The main food intake of rodents receiving 100 mg/kg/day for 28 days decreased from circa 300 to 220 g/kg/24 h/cage.[429] Besides, this alkaloid evoked in male rodents a decrease in triglycerides from 1.1 to 0.5 mmol/L and glycaemia from 6.2 to 5/8 mmol/L, whereby chesterolaemia was unchanged.[429]

3.93 *Neolamarckia cadamba* (Roxb.) Bosser

Synonyms: *Nauclea cadamba* Roxb.; *Anthocephalus cadamba* (Roxb.) Miq.; *Anthocephalus indicus* A. Richard; *Sarcocephalus cadamba* (Roxburgh) Kurz.
Common names: tuan hua (Chinese); kadambah (India)
Subclass Lamiidae, Superorder Lamianae, Order Rubiales, Family Rubiaceae
Medicinal use: dysentery (India)
Pharmacological target: insulin resistance

Ethanol extract of fruits of *Neolamarckia cadamba* (Roxb.) Bosser (Figure 3.32) given to Charles Foster rats on high-fat diet at a dose of 500 mg/kg/day for 30 days lowered plasma cholesterol by 22% and triglycerides by 25%.[430] This extract lowered hepatic total cholesterol and triglycerides by 21% and 27%, respectively.[430] Ethanol extract of flowering heads of this plant given orally to Swiss albino rats at a single oral dose of 200 mg/kg prior oral loading of glucose lowered postprandial glycaemia at 1 hour from 83.8 to 57.2 mg/dL.[431] This extract given orally to alloxan-induced diabetic Swiss albino rats at a dose of 400 mg/kg/day for 3 weeks lowered plasma glucose from 241.3 to 110.6 mg/dL.[431] This extract decreased plasma aspartate aminotransferase, alanine aminotransferase, and alkaline phosphatase to normal values and improved hepatic cytoarchitecture.[431] The leaves of *Neolamarckia cadamba* (Roxb.) Bosser contain chlorogenic acid that might be present in the flower heads.[432] Chlorogenic acid inhibited rat intestine maltase and sucrase with IC_{50} values of 2.9 and 2.1 mM, respectively, and given orally at a dose of 10 mg/kg to rats 10 minutes before oral administration of maltose or sucrose decreased postprandial glycaemia.[433] Chlorogenic acid to Sprague–Dawley rats on high-fat diet at a dose of 10 mg/kg/day for 28 days decreased plasma cholesterol and low-density lipoprotein–cholesterol and increased high-density lipoprotein–cholesterol, but no effect on plasma triglycerides was observed.[434] This regimen decreased atherogenic index.[434] At the hepatic level, this regimen decreased the expression of CYP7A1 and increased the expression of peroxisome proliferator-activated receptor-α[434] suggesting a depletion of unesterified hepatic cholesterol and subsequent inactivation of liver X receptor. Chlorogenic acid inhibited protein tyrosine phosphatase 1B with in vitro an IC_{50} value of 3.1 μg/mL.[435]

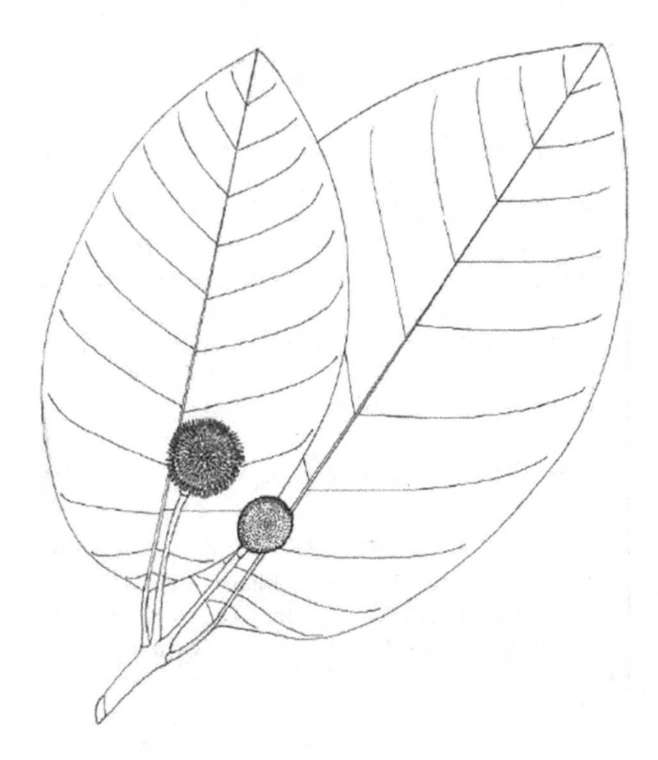

FIGURE 3.32 *Neolamarckia cadamba* (Roxb.) Bosser.

3.94 *Rubia yunnanensis* Diels

Synonym: *Rubia ustulata* Diels.
Common name: zi shen (Chinese)
Subclass Lamiidae, Superorder Lamianae, Order Rubiales, Family Rubiaceae
Medicinal use: rheumatism (China)

Methanol extract of roots of *Rubia yunnanensis* Diels at a concentration of 1 μg/mL decreased triglyceride contents by 36.1% in HepG2 cells cultured with high glucose concentration.[436] The extract given orally 30 minutes before oral loading with olive oil lowered plasma triglycerides at 4 hours from about 650 to 300 mg/dL (orlistat: 125 mg/kg; dose: about 150 mg/dL) evidencing pancreatic lipase inhibition.[436] From this extract, the triterpenes rubiarbonone C, rubiarbonol C and 2-methyl-1,3,6-trihydroxy-9,10-anthraquinone at a concentration of 1 μg/mL decreased triglyceride contents by 15.3%, 12%, and 23.6% in HepG2 cultured with high glucose concentration.[436] The arborinane-type triterpene rubianol and the lignan 5′-methoxylariciresinol isolated from this plant inhibited the production of nitric oxide from macrophages challenged with lipopolysacharide with IC_{50} of 70 and 78 μM, respectively,[437] supporting the idea of nuclear factor-κB inhibition.

3.95 *Rubia tinctorum* L.

Synonyms: *Galium rubia* (L.) E.H.L. Krause; *Rubia iberica* (Fisch. ex DC.) K. Koch
Common name: madder, ran se qian cao (Chinese)
Subclass Lamiidae, Superorder Lamianae, Order Rubiales, Family Rubiaceae
Medicinal use: inflammation (China)

The roots of *Rubia tinctorium* L. contain series of anthraquinones, some being carcinogenic, as well as mollugin.[438] Mollugin at a concentration of 50 μM inhibited the expression of fatty acid synthetase in SK-BR-3 cells via inhibition of Akt phosphorylation, downregulation of sterol regulatory element-binding protein-1c and suppressed nuclear factor-κB activation.[439] This naphthohydroquinone derivative *in vitro* at a concentration of 20 μM inhibited tumor necrosis factor-α induced nuclear factor-κB activation in HT-29 human colonic epithelial cells[440] and protected mouse hippocampal cells against glutamate toxicity via nuclear translocation of Nrf-2 and heme oxygenase-1.[441] Sterol regulatory element-binding protein-1c increases fatty acid synthesis by inducing the expression of fatty acid synthetase and increased levels of fatty acyl-CoA activates protein kinase C_θ to catalyze serine/threonine phosphorylation of the insulin receptor substrate-1/2 hence reduced insulin receptor sensitivity proteins and activate IKK-β to activate nuclear factor-κB hence interleukin-6 production and induction of SOCS-3.[172]

3.96 *Enicostemma littorale* Blume

Synonyms: *Enicostemma axillare* subsp. *Littorale* (Blume) A. Raynal; *Enicostemma hyssopifolium* (Willd.) I. Verd.; *Hippion littorale* (Blume) Miq.
Common names: Indian gentian (English); naahi (India)
Subclass Lamiidae, Superorder Lamianeae, Order Rubiales, Family Gentianaceae
Medicinal use: bitter stomachic (India)

Aqueous extract *Enicostemma littorale* Blume given to alloxan-induced diabetic Wistar rats (glycaemia 230–280 mg/dL) orally at a dose of 2 g/kg/day for 45 days evoked a decrease in glycaemia comparable with values obtained with daily intraperitoneal injection of protamine zinc insulin (6 units/kg).[442] The extract increased the activity of hepatic hexokinase, decreased the activity of neoglucogenic enzymes glucose-6-phosphatase and fructose-1,6-biphosphatase[442] suggesting the increased availability of insulin in the plasma of diabetic rodents or increased hepatic sensitivity to insulin. Swertiamarin from this plant given orally to Sprague–Dawley rats on high-cholesterol diet at a dose of 75 mg/kg/days for 7 days brought total serum cholesterol, low-density lipoprotein–cholesterol to level equivalent to atorvastatin given at a dose of 50 mg/kg/day.[443] The regimen was more active than atorvastatin (50 mg/kg/day) in reducing very low-density lipoprotein and serum triglycerides and had no effect on high-density lipoprotein.[443] This secoiridoid lowered atherogenic index similarly to atorvastatin (50 mg/kg/day) and inhibited of 3-hydroxy-3-methylglutaryl-coenzyme A CoA reductase activity more potently than atorvastatin (50 mg/kg/day).[443] Swertiamarin elevated the fecal secretion of bile and total fecal sterol more potently than atorvastatin (50 mg/kg/day).[443] In a subsequent study, the plant given to Charles Foster rats on fructose-enriched diet orally at a dose of 15 g/kg/day for 45 days lowered body mass from 307 to 271 g (normal: 329 g; rosiglitazone 10 mg/kg: 260 g), fasting glucose from 107 to 86 mg/dL (normal: 75 mg/dL; rosiglitazone 10 mg/kg: 84 mg/dL), fasting insulin from 41 to 22 mU/mL (normal: 13 mU/mL; rosiglitazone 10 mg/kg: 22 mU/mL), and fasting insulin resistance index from 176 to 76 (normal: 41; rosiglitazone 10 mg/kg: 76).[444] This regimen normalized glucose tolerance and lowered serum triglycerides and very low-density cholesterol by 43% and 54%, respectively.[444] The extract improved cardiac function.[444] The treatment lowered platelet hyperaggregability induced by fructose-enriched diet and lowered systolic blood pressure increased by fructose-enriched diet by 44%.[444] Swertimarin has anti-inflammatory activity.[445] It is metabolized by bacteria into 3 metabolites by intestinal bacteria of which gentianin that is pharmacologically active.[229]

3.97 *Gentiana olivieri* Griseb.

Synonyms: *Gentiana regeliana* Gand.; *Gentiana weschniakowii* Regel; *Gentianodes olivieri* (Griseb.) Omer, Ali & Qaiser
Common names: gul-khalle (India); bangera (Pakistan)
Subclass Lamiidae, Superorder Lamianeae, Order Rubiales, Family Gentianaceae
Medicinal use: hypertension (Turkey)

Methanol extract of aerial parts of *Gentiana oliveri* Griseb. given to Sprague–Dawley rats at a single oral dose of 300 mg/kg evoked a 36.5% fall of glycaemia at 1 hour in oral glucose tolerance test.[446] From this extract, isoorientin was given orally to streptozotocin-induced diabetic Sprague–Dawley rats (glycaemia >250 mg/dL) at a dose of 15 mg/kg/day for 30 days lowered glycaemia from 362 to 80.8 mg/dL, prevented weight loss lowered plasma triglycerides and cholesterol.[446] This C-glycosylflavones given orally at a daily dose of 40 mg/kg to Kunming mice on high-fructose diet for 56 days had no influence of food intake, attenuated body weight gain, and decreased body fat mass from 1.6 to 0.1 g (normal: 0.8 g).[447] This supplementation decreased plasma triglycerides from 3 to 1.6 mmol/L (normal: 1.7 mmol/L), cholesterol from 1.2 to 0.5 mmol/L (normal: 0.5 mmol/L), very low-density lipoprotein–cholesterol from 16 to 0.7 mmol/L (normal: 2.1 mmol/L), low-density lipoprotein–cholesterol from 1.4 to 0.9 mmol/L (normal: 0.9 mmol/L), and increased high-density lipoprotein–cholesterol from 1.3 to 1.8 mmol/L (normal: 1.7 mmol/L).[447] Isoorientin (Figure 3.33) supplementation decreased plasma glucose and insulin toward normal values evidencing improved glucose tolerance.[447] This regimen increased plasma apolipoprotein A-I, (major structural contents of high-density lipoprotein in human), decreased apolipoprotein B, (major structural contents of very low-density lipoprotein), and low-density lipoprotein and subsequently the ratio of Apo-B/ApoA-I evidencing a decreased risk of cardiovascular event.[447] In the liver of treated rodents, the activity of glutathione peroxidase and superoxide dismutase were increased whereas malondialdehyde, an indicator of oxidative insults, levels was decreased.[447] This regimen lowered hepatic tumor necrosis factor-α and interleukin-6 and increased interleukin-1.[447] This regimen decreased plasma alanine aminotransferase, aspartate aminotransferase, and alkaline phosphatase; improved hepatic cytoarchitecture; and decreased hepatic levels of fatty acid synthetase.[447] The precise mode hypolipidemic of action of this flavonoid is apparently unknown. The hypotensive use of the plant may account from gentinin within given intravenously at a dose of 10 mg/kg evoked within a minute a fall in blood pressure by 50.2% and a mild bradycardia that persisted for 2 minutes.[448]

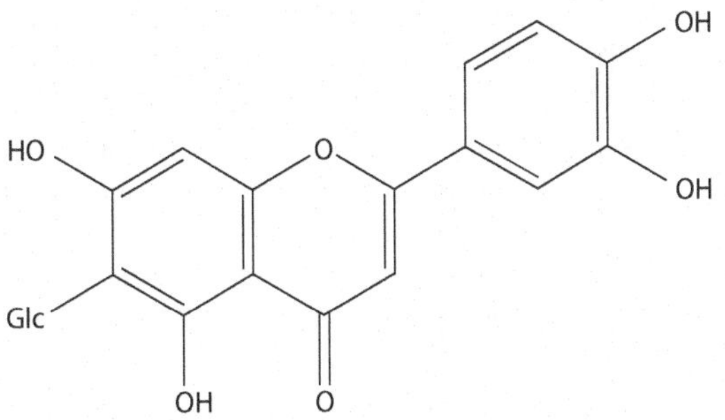

FIGURE 3.33 Isoorientin.

3.98 *Caralluma fimbriata* Wall.

Synonym: *Caralluma adscendens* var. *fimbriata* (Wall.) Gravely & Mayur.
Common names: monkey's horn (India); makur sing (India)
Subclass Lamiidae, Superorder Lamianeae, Order Rubiales, Family Apocynaceae
Nutritional use: vegetable (India)

Ethanol extract of *Caralluma fimbriata* Wall. given to healthy volunteers (body mass index greater than 25 kg/m^2) at a dose of 1 g/day for 60 days evoked a decrease of food intake and waist circumference from 95.1 to 93.9 cm, whereas plasma glucose, triglycerides, cholesterol, high-density lipoprotein, and low-density lipoproteins were not affected.[449] *Caralluma fimbriata* Wall. synthesizes the series of hydroxylated pregnane glycosides,[450] which are structurally close to P57AS3, oxypregnane steroidal glycoside isolated from members of the genus *Hoodia* with anorexigen properties.[451] Aqueous extract of *Caralluma fimbriata* Wall. given orally to high-fat diet Wistar rats at a dose of 200 mg/kg/day for 90 days lowered weight gain to values similar to the group of rodents fed with normal diet.[452] This regimen lowered glycaemia from about 140 to 70 mg/dL, plasma insulin, leptin, and lowered triglycerides from about 180 to 90 mg/dL similarly to metformin given orally at 20 mg/kg/day.[452]

3.99 *Rauvolfia serpentina* (L.) Benth. ex Kurz

Common names: sarpagandha (India); Indian snake root
Subclass Lamiidae, Superorder Lamianeae, Order Rubiales, Family Apocynaceae
Medicinal use: hypertension (India)
Pharmacological target: atherogenic hyperlipidemia
History: the plant was known to Sushruta

Rauvolfia serpentina (L.) Benth. ex Kurz contains the series of alkaloids of which reserpine that has been clinically used for the treatment of hypertension but owed to unpleasant side effects has been, perhaps hastily, sidelined.[453] Reserpine given on days 1, 3, and 5 during 5 days subcutaneously at a dose of 0.1 mg/kg to male New Zealand white rabbits fed with a standard chow diet evoked a decreased cholesterolaemia by 23%, whereas triglyceridemia was unchanged and evoked liver weight reduction by about 10%.[454] This treatment evoked a 3-fold increase in hepatic low-density lipoprotein receptor as a sign of cytolosolic deprivation in unesterified cholesterol.[454] In the same experiment, the amount of cholesterol present in arteries and heart was decreased, and a 50% decrease in aortic low-density lipoprotein was observed. Beside, reserpine induced a reduction of heart beat from 197 to 143 beats per minute.[455] Reserpine at given parenterally at a dose of 43 μg/kg for 6 weeks in rodents fed with a 0.2% cholesterol-enriched diet for reduced total cholesterol and aortic wall cholesterol by 42% and 73%, respectively, compared with control.[455] In the same experiment, aortic intimal-medial thickness ratio was reduced by 70% and heart rate was decreased by 28%.[455] Reserpine inhibits the secretion of norepinephrine by sympathetic neurons. In hepatocytes, the binding of norepinephrine to its receptor activates adenylate cyclase and increases cyclic adenosine monophosphate hence activation of inhibitor-1-phosphatase that inhibits protein phosphatase that dephosphorylates and therefore activates 3-hydroxy-3-methylglutaryl-coenzyme A CoA reductase and parallely maintains 3-hydroxy-3-methylglutaryl-coenzyme A CoA reductase in its active form by inhibiting protein phosphatase.[456]

3.100 *Wrightia tomentosa* (Roxb.) Roem. & Schult.

Synonym: *Wrightia arborea* (Dennst.) Mabb.
Common name: nelam pala (India)
Subclass Lamiidae, Superorder Lamianeae, Order Rubiales, Family Apocynaceae
Medicinal use: hypertension (India)

β-Amyrin acetate isolated from the leaves of *Wrightia tomentosa* (Roxb.) Roem. & Schult given to Golden-Syrian hamsters fed with high-fat diet orally at a dose of 10 mg/kg/day for 7 days lowered triglycerides by 35%, total cholesterol by 37%, low-density lipoprotein by 36%, and increased the ration high-density lipoprotein–cholesterol/total cholesterol by 49%.[457] From the same plant and at the same dose, β-amyrin palmitate lowered triglycerides by 24%, total cholesterol by 25%, low-density lipoprotein by 44%, and increased the ration high-density lipoprotein–cholesterol/total cholesterol by 28%.[457] *In vitro*, β-amyrin acetate and β-amyrin palmitate reduced by 53.1% and 56.8% 3-hydroxy-3-methylglutaryl-coenzyme A reductase at a concentration of 100 mM (Lovastatin at 100 μM: 86.6%).[457]

3.101 *Lycium barbarum* L.

Common names: ning xia gou qi (Chinese); goji
Subclass Lamiidae, Superorder Lamianae, Order Solanales, Family Solanaceae
Medicinal use: tonic (China)

Ethanol extract of root bark of *Lycium barbarum* L. given to streptozotocin high-fat diet–induced type 2 diabetic Sprague–Dawley rats (fasting glycaemia >11.1 mmol/L) orally at a dose of 30 g/kg/day for 7 weeks prevented weight gain and lowered serum glucose from 21.1 to 13.3 mmol/L (normal: 6.1 mmol/L; rosiglitazone 5 mg/kg/day: 14.6 mmol/L).[458] This regimen lowered insulin resistance as efficiently as rosiglitazone.[458] This extract reduced plasma triglycerides from 3.1 to 2.2 mmol/L (normal: 1.8 mmol/L), cholesterol from 3.6 to 2.7 mmol/L (normal: 2.1 mmol/L), lowered high-density lipoprotein–cholesterol, and lowered low-density lipoprotein–cholesterol.[458] At the hepatic level, the extract lowered aspartate aminotransferase and alanine aminotransferase indicating hepatoprotective effects. Histological observation of hepatic tissues revealed a decreased accumulation of fats in hepatocytes.[458] Aqueous extract of fruits of *Lycium barbatum* L. given orally to streptozotocin-induced diabetic Sprague–Dawley rats at a dose of 5 g/kg for 8 weeks prevented body weight loss and attenuates plasmatic glucotoxicity.[459] This regimen had no effect on plasma adiponectin, lowered tumor necrosis factor-α, and interleukin-6.[458] Polysaccharide fraction of fruits of *Lycium barbarum* L. given at a dose of 1 mg/kg/day to Sprague–Dawley rats on high-fat diet for 12 weeks attenuated body weight gain, had no effect on food intake, decreased plasma glucose and glucose tolerance as well as free fatty acds.[460] As a consequence of increased insulin receptor sensitivity, this regimen decreased phosphorylation level of glycogen synthase kinase-3 and phosphorylation of insulin receptor substrate-1.[460] This regimen improved hepatic cytoarchitecture with decrease of triglyceride accumulatiuon.[460] This fraction decreased expression of sterol regulatory element-binding protein-1c, peroxisome proliferator-activated receptor-γ, increased catalase and glutathione peroxidase, as well as adipocyte triglyceride lipase.[460] Also, at hepatic level, the fraction inhibited nuclear factor-κB translocation and subsequent expression of tumor necrosis factor-a, interleukin-1β, and cyclo-oxygenase-2.[460] Some preliminary clinical studies provide evidence of beneficial effect of this plant for metabolic syndrome.[461] Polysaccharides are not absorbed by the small intestine in any appreciable amount and must be hydrolyzed by digestive enzymes.[462] Furthermore, clinical trials are needed to examine the effects of the plant on metabolic syndrome.

3.102 *Solanum torvum* Swartz

Common names: shui qie (Chinese), kayangyin (Burma); sundai (India); terung pipit (Malay); barabihi (Nepal)
Subclass Lamiidae, Superorder Lamianae, Order Solanales, Family Solanaceae
Nutritional use: vegetable (Malaysia)
Pharmacological target: insulin resistance

FIGURE 3.34 *Solanum torvum* Swartz.

Ethanol extract of fruits of *Solanum torvum* Swartz (Figure 3.34) given to high-fructose diet-induced hypertensive Wistar rats orally at a dose of 300 mg/kg/day for 6 weeks of lowered mean systemic blood pressure from 141.3 to 101.4 mmHg, (nifedipine at 10 mg/kg/day: 99.7 mmHg).[463] This regimen lowered serum glucose from 13.9 to 6.7 mM (normal: 4 mM), cholesterol from 3.1 to 1.8 mM (normal 1.5 mM), serum triglycerides from 4.3 to 1 mM (normal: 0.9 mM), insulinaemia from 96.6 to 64.3 pM/L (normal: 50.3 pM/L), and reduced insulin resistance.[463] Gandhi et al. administered methanol extract of fruits containing rutin, caffeic acid, gallic acid, and catechin to streptozotocin-induced diabetic Wistar rats at a dose of 400 mg/kg/day for 28 days and observed a decrease of plasma glucose and increase in plasma insulin equivalent to glibenclamide (10 mg/kg/day).[464] As a consequence of increased insulinemia, the activities of plasma alanine aminotransferase, aspartate aminotransferase, alkaline phosphatase, and hepatic gluconeogenic enzymes glucose-6-phosphatase and fructose-1,6-bisphosphatase were decreased, whereas hexokinase and glycogen content were elevated.[464] This regimen improved hepatic and pancreatic cytoarchitecture and increased the hepatic activity of superoxide dismutase, catalase, and glutathione peroxidase.[464] Caffeic acid (Figure 3.35) given orally to spontaneously hypertensive rats at a dose of 0.1 g/kg/day for 6 weeks

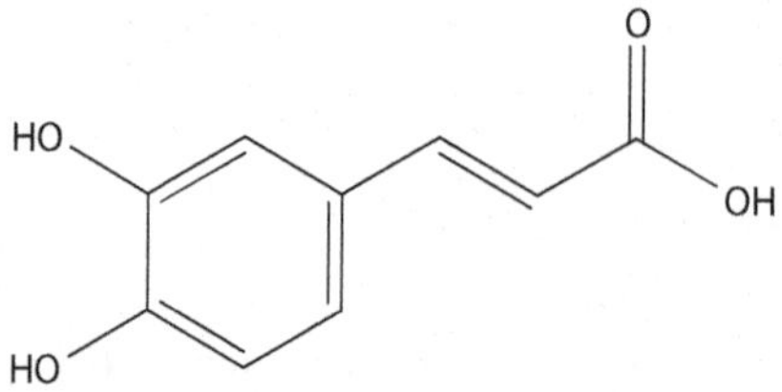

O
HO
OH
HO

FIGURE 3.35 Caffeic acid.

evoked a reduction of systolic blood pressure by 23.4% and diatolic blood pressure by 15%.[465] This extract lowered plasma and hepatic lipid peroxidation by 30% and 27%, respectively.[465] In the liver and kidneys, caffeic acid increased the activity of superoxide dismutase and glutathione peroxidase.[465] Caffeic acid was found intact in the plasma with a half-life of about 2 hours.[465]

3.103 *Capsicum annuum* L.

Common names: pepper (English); la jiao (Chinese); cili padi (Malay)
Subclass Lamiidae, Superorder Lamianae, Order Solanales, Family Solanaceae
Medicinal use: facilitate digestion (Malaysia)
Pharmacological target: atherogenic hyperlipidemia

Powder of fruits of *Capsicum annum* L. given to New Zealand white rabbits given as 1% of a high-cholesterol diet for 12 weeks had no effect on body weight and lowered the activity of plasmatic cholesteryl ester transfer protein activity by 11%.[466] This regimen lowered plasma cholesterol and triglyceride concentrations by 15% and 52%, respectively.[466] This supplementation lowered low-density lipoprotein–cholesterol, very low-density lipoprotein–cholesterol, whereas high-density lipoprotein–cholesterol was elevated by 34%.[466] The supplementation increased fecal triglycerides[466] suggesting, at least, the inhibition of cholesterol absorption by intestine. Li et al. provided evidence that C57BL/6J on high-fat diet containing 0.01% of capsaicin for 24 weeks of decreased triglycerides content in the liver from 38.7 to 21.4 mg/g liver tissue. This supplementation upregulated fatty acids mitochondrial membrane transporter uncoupling protein-2 expression in the liver implying increased hepatic β-oxidation.[467] It must be recalled that in hepatocytes uncoupling protein-2 is induced by peroxisome proliferator-activated receptor-α. This regimen lowered plasma glucose from 5.3 to 3.5 mmol/L, triglycerides from 1.2 to 0.7 mmol/L, cholesterol from 2.3 to 1.9 mmol/L, and high-density lipoprotein–cholesterol from 1.4 to 1.2 mmol/L.[467] This supplementation improved hepatic cytoarchitecture with respect to triglyceride accumulation.[467] These beneficial effects were not observed in C57BL/6J TRPV1−/− mice on high-fat diet.[467] A mixture of capsaicin and dihydrocapsaicin from *Capsicum annuum* L. given orally at a dose of 10 mg/kg/day for 28 days to rodent on cholesterol-enriched diet reduced plasma cholesterol from 4 to 2.6 mmol/L and triglycerides from 0.7 to 0.5 mmol/L, respectively, together with a reduction of atherogenic index by 51.1%.[468] This regimen evoked a reduction of hepatic cholesterol, hepatic triglycerides, and total hepatic lipids from 110.4 to 91 μmol/liver and 176 to 159.7 μmol/liver.[468] Furthermore, this alkaloidal mixture evoked an increase in expression of hepatic CYP7A1 and TRPV1.[468] At intestinal level, this mixture increased the expression of apical sodium-bile acid transporter that reabsorbs luminal bile acids into enterocytes during enterohepatic circulation and TRPV1 in rats.[468] The amounts of bile acids in the excreted feces were increased from 15.1 to 17.1 μmol/d fecal.[468] In a previous study, however, capsaicin given orally to Sprague–Dawley rats at a dose of 1 mg/kg evoked a decrease in bile flow bile flow by about 30%.[469] Capsaicin at a concentration of 10 μM increased cytosolic concentration of calcium in hepatocytes *in vitro* as a consequence of TRVP1 activation.[467] It must be recalled in this context that increased contents of calcium in the cytosol of liver cells increases the activity of calcium/calmodulin-dependent protein-kinase kinase (CaMKK), which induces the activation of adenosine monophosphate-activated protein kinase.[387] The stimulation of adenosine monophosphate-activated protein kinase inhibits liver X receptor, which normally activates both CYP7A1 and ATP-binding cassette G5 and G8.[470]

3.104 *Tecoma stans* (L) Juss. ex Kunth

Synonyms: *Bignonia stans* L.; *Stenolobium stans* (L.) Seem.
Common name: yellow bells
Subclass Lamiidae, Superorder Lamianae, Order Lamiales, Family Bignoniaceae
Medicinal use: diuretic (India)

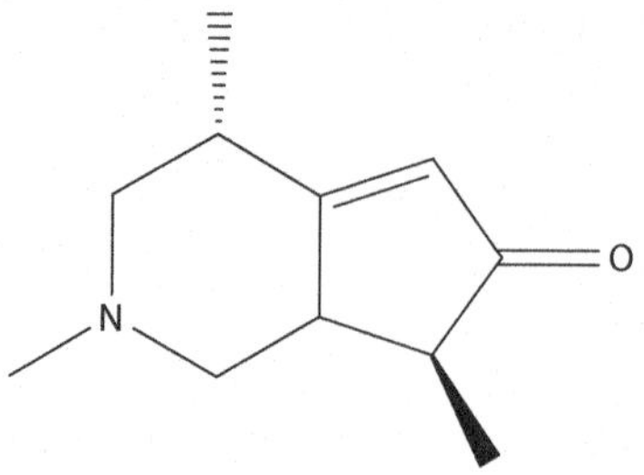

FIGURE 3.36 Tecomine.

Tecomine (Figure 3.36) from *Tecoma stans* (L) Juss. ex Kunth administered parenterally induced hypoglycaemia in rodents.[471,472] This monoterpene alkaloid given orally at a dose of 50 mg/kg reduced the cholesterolaemia of a model of type 2 diabetes rodent by 20%.[471–473] Tecomine hydrochloride at 50 mg/kg or 5β-hydroxyskitanthine at 63.4 mg/kg isolated from the leaves given orally to C57BL/KsJ *db/db* mouse (genetic model of type 2 diabetes characterized by insulin resistance and obesity) twice a day for 7 days had no effect on glycaemia suggesting unfavorable oral bioavailability.[473] However tecomine hydrochloride reduced plasma cholesterol levels from 94.4 to 72.96 mg/dL while body weight, free fatty acids, and triglycerides did not change following the treatment.[473] The first-pass metabolite(s) of tecomine and precise hypocholesterolemic mode of action remain unknown.

3.105 *Andrographis paniculata* (Burm. f.) Wall. ex Nees

Synonym: *Justicia paniculata* Burm. f.
Common names: chuan xin lian (Chinese); hempedu bumi (Malay), kalmegh (India)
Subclass Lamiidae, Superorder Lamianae, Order Lamiales, Family Acanthaceae
Medicinal use: diabetes (Malaysia)
Pharmacological target: atherogenic hyperlipidemia

Phunikhom et al. administered a fraction of *Andrographis paniculata* (Burm. f.) Wall. ex Nees (containing 24 mg of andrographolide) 5 times per day orally to 20 volunteers (plasma triglycerides >150 mg/dL) for 8 weeks and observed a decrease of plasma triglycerides from 270.1 to 228.5 mg/dL (gemfibrozil 300 mg/kg/day: 237.6–180.5 mg/dL). In the same experiment, this fraction did not affect high-density lipoprotein–cholesterol similarly to gemfibrozil.[474] Gemfibrozil is a peroxisome proliferator-activated receptor-α agonist and it tempting to speculate that andrographolide could have similar target. In a subsequent study, andrographolide (Figure 3.37) given at a dose of 100 mg/kg/day for 12 weeks to C57BL/6 mice on high-fat diet had no effect on food intake, suppressed body weight gain by 20%.[475] This regimen decreased plasma glucose and insulin, serum cholesterol, triglycerides, and low-density lipoprotein–cholesterol and did not affect serum high-density lipoprotein–cholesterol.[475] This treatment decreased hepatic triglycerides.[475] At the hepatic level, andrographolide decreased the expression of sterol regulatory element-binding protein-1c and targets fatty acid synthetase and stearoyl-coenzyme A desaturase-1. In parallel, this regimen decreased the expression sterol regulatory element-binding protein-2 and targets 3-hydroxy-3-methylglutaryl-coenzyme A reductase and low-density lipoprotein receptor.[475] It must be recalled that activation of peroxisome proliferator-activated receptor-α in hepatocytes translates into repression of liver X receptor, induction of regulatory element-binding protein-1c sterol regulatory element-binding protein-1c[476] and inhibition.[477]

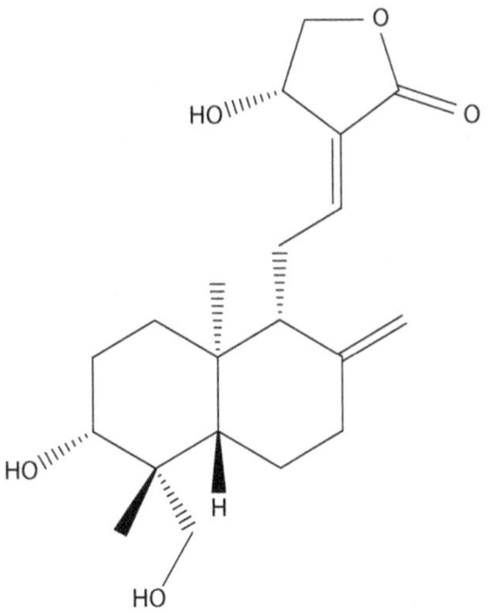

FIGURE 3.37 Andrographolide.

3.106 *Hygrophila auriculata* Heine

Synonyms: *Asteracantha longifolia* (L.) Nees; *Barleria auriculata* Schumach.; *Barleria longifolia* L.; *Hygrophila spinosa* T. Anderson
Common names: ikshura (India); yalmakhana (Pakistan)
Subclass Lamiidae, Superorder Lamianae, Order Lamiales, Family Acanthaceae
Medicinal use: jaundice (India)
Pharmacological profile: atherogenic hyperlipidemia

Aqueous extract of roots given at a dose of 150 mg/kg/day for 5 days protected rats against carbon tetrachloride liver insults as evidenced by regeneration of hepatocytes, normalization of necrosis of the liver and decrease of plasma activities of alanine aminotransferase, aspartate aminotransferase, and alkaline phosphatase.[478] In a subsequent study, total alkaloid fraction of leaves at a dose of 80 mg/kg/day for 7 days protected Wistar albino rats against liver damage induced by carbon tetrachloride.[479] Methanol extract of aerial parts of *Hygrophila auriculata* Heine given to streptozotocin-induced diabetic Sprague–Dawley rats (glycaemia >200 mg/dL) at a dose of 250 mg/kg/day for 3 weeks lowered glycaemia from 240 to 119.7 mg/dL (normal: 93.6 μg/dL; glibenclamide 600 μg/kg/day: 94.3 μg/dL).[480] In untreated rats, glycaemia decreased from 240.7 to 228.3 mg/dL[480] evidencing the fact that β-cells[480] are able to regenerate after streptozotocin intoxication in rodents (not in humans). This regimen lowered reactive oxygen species and hydroperoxide in the liver and kidneys.[480] The extract increased glutathione, the enzymatic activity of glutathione S-transferase, glutathione peroxidase, and catalase in liver and kidneys and increased superoxide dismutase activity in erythrocytes.[480] The hepatoprotective principles are apparently unknown as well as precise mechanism. The plant contains lupeol, betulin, and lupenone as well as alkaloids asteracanthine and asteracanthicine.[481] Unesterified cholesterol is converted to cholesteryl ester by acyl-CoA:cholesterol acyltransferase.[482] Acyl-CoA:cholesterol acyltransferase-1 is ubiquitously expressed whereas acyl-CoA:cholesterol acyltransferase-2 is restricted to enterocytes for chylomicron formation and hepatocytes for very low-density lipoprotein formation.[482] Lupeol and betulin inhibited at a concentration of 100 μM the enzymatic activity of acyl-CoA:cholesterol acyltransferase-1 by 72.3% and 52.4%, respectively.[483] Betulin given orally at a dose of 30 mg/kg/day for 6 weeks to C57BL/6J mice on high-fat diet had no effect on food intake, decreased body weight, decreased insulin resistance, cholesterol and triglycerides, decreased low-density lipoprotein–cholesterol, and increased high-density lipoprotein–cholesterol as potently as lovastatin.[484] This triterpene decreased hepatic contents in cholesterol and triglycerides and improved hepatic structure.[484] At the hepatic level, betulin treatment decreased expression of sterol regulatory element-binding protein-2 and targets 3-hydroxy-3-methylglutaryl-coenzyme A reductase.[484] Betulin decreased the expression of sterol regulatory element-binding protein-1c and its target fatty acid synthetase and stearoyl-coenzyme A desaturase-1. This triterpene increased the expression of peroxisome proliferator-activated receptor-α as well as of hepatic lipase and apolipoprotein E.[484] This triterpene increased the glucose-6-phosphate dehydrogenase.[484] Liver X receptor in hepatocytes is activated by oxysterols, which share somewhat some structural similarities triterpenes in general and it is tempting to speculate, although highly speculative, that betulin could be a liver X receptor antagonist. Note that the plant contains the phytosterol stigmasterol, which is structurally not so far from cholesterol.[485]

3.107 *Ocimum basilicum* L.

Synonym: *Ocimum thyrsiflorum* L.
Common names: luo le (China); kali tulasi (India); basil
Subclass Lamiidae, Superorder Lamiidae, Order Lamiales, Family Lamiaceae
Medicinal use: diuretic (India)
History: this plant was known of Pliny the elder

Aqueous fraction of *Ocimum basilicum* L. given orally at a daily dose of 200 mg/kg for 5 weeks to mice on hypercholesterolemic diet lowered plasma cholesterol by 42%, triglycerides by 42%, and lowered low-density lipoprotein–cholesterol by 88% and increased high-density lipoprotein by 80%.[486] This treatment reduced atherogenic index by 88%.[486] Fenofibrate given orally at a dose of 200 mg/kg lowered plasma cholesterol by 25%, triglycerides by 52%, low-density lipoprotein–cholesterol by 84%.[486] The aqueous fraction lowered hepatic cholesterol by 52% and hepatic triglycerides by 58%.[486] Fenofibrate lowered hepatic cholesterol and triglyceride by 59% and 71%, respectively.[486] Clinical trials are warranted.

3.108 *Salvia miltiorrhiza* Bunge

Common name: dan shen (Chinese)
Subclass Lamiidae, Superorder Lamiidae, Order Lamiales, Family Lamiaceae
Medicinal use: heart diseases (China)
Pharmacological target: insulin resistance

Abietane diterpenes isotanshinone IIA, dihydroisotanshinone I and isocryptotanshinone isolated from *Salvia miltiorrhiza* Bunge inhibited protein-tyrosine phosphatase1B activity with IC_{50} values of 11.4, 22.4, and 56.1 μM, respectively.[487] Cryptotanshinone and 15,16-dihydrotanshinone I isolated from the roots inhibited rat liver diacylglycerol acyltransferase activity with IC_{50} values of 10.5 and 11.1 μg/mL, respectively.[488] Phenolic fraction of roots of given orally at a dose of 187 mg/kg/day for 28 days to high-fat diet-streptozotocin-induced type 2 diabetes Sprague–Dawley rats (glycaemia between 11.1 and 33.3 mM) attenuated weight loss, lowered fasting blood glucose from 23.1 to 16.2 mM (normal: 3.5 mM; gliclazide 26.7 mg/kg/day: 18 mM).[489] This regimen lowered plasma insulin from 20 to 15.2 mIU/L (normal: 7.6 mIU/L; gliclazide 26.7 mg/kg/day: 24.9 mIU/L) and increased insulin sensitivity index.[489] The fraction lowered plasma cholesterol and triglycerides.[489]

3.109 *Salvia plebeia* R. Br.

Synonyms: *Lumnitzera fastigiata* (Roth) Spreng.; *Ocimum fastigiatum* Roth; *Ocimum virgatum* Thunb.; *Salvia brachiata* Roxb.; *Salvia minutiflora* Bunge
Common names: li zhi cao (Chinese); Australian sage (English)
Subclass Lamiidae, Superorder Lamiidae, Order Lamiales, Family Lamiaceae
Medicinal use: hepatitis (Taiwan)
Pharmacological target: insulin resistance

Oshima et al. provided evidence that eupafolin, hispidulin, and hispidulin-7 glucoside from *Salvia plebeia* R. Br. could prevent carbon tetrachloride insults in cultured rat hepatocytes substantiating the hepatoprotective use of the plant.[490] The plant elaborates ursolic acid and luteolin-7-*O*-glucoside,[491] which given orally at a daily dose of 2 mg/day for 7 days to Wistar rats lowered glycaemia from 176.1 to 158.1 mg/dL and from 176.1 to 157.6 mg/dL, respectively, and increased glycogen contents in the liver.[492] Ursolic acid increased the hepatic expression of phosphoglycogen synthetase kinase-3.[492] In the postprandial state, insulin promotes the storage of glucose in the form of glycogen in the liver by activating glycogen synthetase, protein phosphatase 1, inhibition of glycogen phosphorylase (rate-limiting enzyme of glycogenolysis), and inhibiting glycogen synthetase kinase 3 hence activation of glycogen synthetase.[493] Both ursolic acid and luteolin-7-*O*-glucoside decreased plasma cholesterol, low-density lipoprotein, and increased high-density lipoprotein.[492] Ursolic acid in CHO/IR cells increased insulin receptor sensitivity with in insulin, increased receptor autophosphorylation and a subsequent activation of downstream phosphatidylinositol 3-kinase resulting in phosphorylation and inactivation of glycogen synthase kinase-3, leading to glycogen synthase activation and glycogen synthesis.[491] Glycogen synthase kinase-3 is a key enzyme that suppresses the production of liver glycogen and its activation increases plasma glucose concentration.[494] Ursolic acid at a dose of 25 μg/mL repressed the expression of phosphoenolpyruvate carboxykinase in H4IIE cells induced by dexamethasone and 8-bromo-cyclic adenosine monophosphate[495] supporting further, at least, the notion of hepatic insulin sensitizing effect of pentacyclic triterpenes.[496,497]

3.110 *Scutellaria baicalensis* Georgi

Synonyms: *Scutellaria lanceolaria* Miq.; *Scutellaria macrantha* Fisch.
Common name: huang qin (Chinese)
Subclass Lamiidae, Superorder Lamiidae, Order Lamiales, Family Lamiaceae
Medicinal use: fever (China)

Aqueous extract of *Scutellaria baicalensis* Georgi given orally to C57BLKS/J-db/db mice at 100 mg/kg/day for 4 weeks decreased body weight gain and had no effect on food intake.[498] This supplementation lowered serum alanine aminotransferase by 29%, plasma triglycerides by about 20% and plasma insulin by about 35%[498] implying, at least, a decrease in insulin resistance. This extract had no effect on plasma cholesterol or glycaemia and increased the phosphorylation of adenosine monophosphate-activated protein kinase in the liver of C57BLKS/J-*db/db* mice.[498]

3.111 *Thymbra spicata* L.

Synonym: *Satureja spicata* (L.) Garsault
Common name: spiked thyme
Medicinal use: high cholesterol (Turkey)
Pharmacological targets: atherogenic hyperlipidemia; insulin resistance

Thymbra spicata L. given orally at a dose of 100 mg/kg/day for 30 days to Male Swiss albino mice feeding on diet containing 1% cholesterol, reduced cholesterolaemia from 218.4 to 93.6 mg/dL, plasma triglycerides and glucose.[77] Serum high-density lipoprotein–cholesterol was increased by from 25.8 to 43.6 mg/dL and low-density lipoprotein–cholesterol was reduced from 143.0 to 42.0 mg/L.[77] Diethyl ether fraction of *Thymbra spicata* L. containing mainly carvacrol given orally at a dose of 100 mg/kg/day to Swiss albino mice on high-fat diet for 4 weeks decreased plasma cholesterol from 141.4 to 76.9 mg/dL (normal: 65 mg/dL), increased high-density lipoprotein from 23.6 to 27.4 mg/dL, lowered low-density lipoprotein from 84.2 to 64.9 mg/dL (normal: 24.5 mg/dL), lowered triglycerides from 167.7 to 75.1 mg/dL (normal: 120.3 mg/dL), and lowered glycaemia from 125.2 to 95.2 mg/dL (normal: 84.3 mg/dL).[499] This fraction lowered plasma malondialdehyde, increased plasma glutathione, lowered red blood cell superoxide dismutase, and increased red blood cell catalase.[499] The fraction normalized hepatic cytoarchitecture as evidence as a reduction in fat droplets in hepatocytes.[499] Carvacrol given to C57BL/6N mice at 0.1% of high-fat diet for 10 weeks had no effect on food intake, decreased body weight gain, decreased liver cholesterol, triglycerides, and fatty acids toward normal values.[500] This supplementation improved hepatic cytoarchitecture with decrease in triglyceride droplets and decreased plasma aspartate aminotransferase and alanine aminotransferase.[500] At the hepatic level, carvacrol increased the expression of SIRT1, hence increased phosphorylation of adenosine monophosphate-activated protein kinase with subsequent decreased expression of liver X receptor, sterol regulatory element-binding protein-1c, fatty acid synthetase, fatty acid translocase, and increased expression of carnitine palmitoyltransferase-1.[500] In parallel, this supplementation increased the expression of sterol regulatory element-binding protein-2, 3-hydroxy-3-methylglutaryl-CoA reductase, low-density lipoprotein receptor, ATP-binding cassette G5 and G8 and CYP7A1, and decreased the expression of acyl-CoA:cholesterol acyltransferase-1.[500] It must be recalled that in hepatocytes, activated adenosine monophosphate-activated protein kinase stimulates TSC1/2 which in turn inhibits mTOR and downstream liver X receptor via S6K1.[501] The hypolipidemic mode of action of carvacrol, which is not uncommon in members of the family Lamiaceae, is apparently unknown. This monoterpene could potentially increase NAD^+ contents in hepatocytes.[151] Carvacrol given intraperitonally to anesthetized Sprague–Dawley rats at dose of 100 μg/kg evoked a decrease of systolic and diastolic blood from pressure from 99 mmHg to about 65 mmHg and 50 mmHg, respectively at 90 minutes and lowered heart rate at 30 minutes.[502] Carvacrol was inactive on phenylephrine, calcium, or potassium-induced contraction of isolated rat aorta.[152]

3.112 *Alisma orientale* (Sam.) Juz.

Synonym: *Alisma plantago-aquatica* var. *orientale* Sam.
Common name: dong fang ze xie (Chinese)
Class Liliopsida, Subclass Alismatidae, Superorder Alismatanae, Order Alismatales, Family Alismataceae
Medicinal use: diabetes (China)

The protostane-type triterpene alisol M 23-acetate and alisol A 23-acetate (Figure 3.38) isolated from *Alisma orientale* (Sam.) Juz. at a concentration of 10 μM evoked farnesoid X receptor agonistic properties in HepG2 cells.[503]

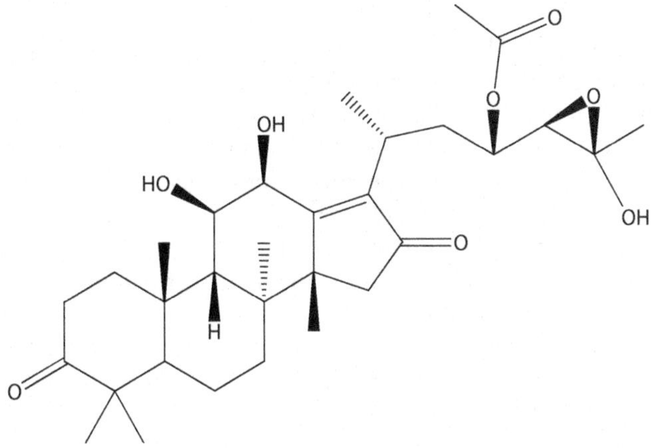

FIGURE 3.38 Alisol M 23-acetate.

3.113 *Asparagus officinalis* L.

Synonym: *Asparagus polyphyllus* Steven
Common names: shi diao bai (China); halgun (India); asparagus
Class Liliopsida; Subclass Lillidae, Superorder Lilianae, Order Asparagales, Family Asparagaceae
Medicinal use: diuretic (India)

Butanol fraction (enriched with steroidal saponins) of *Asparagus officinale* L. given to ICR mice on high-fat diet at a dose of 160 mg/kg/day for 8 weeks had no effect on food intake, lowered body weight gain by 31% and decreased plasma cholesterol by 21%.[504] This extract increased high-density lipoprotein–cholesterol by 34%, decreased low-density lipoprotein–cholesterol by 30% whereby plasma triglycerides were unchanged.[504] This regimen lowered plasma aspartate aminotransferase, alanine aminotransferase and alkaline phosphatase.[504] The extract attenuated hepatic malondialdehyde and increased superoxide dismutase.[504]

3.114 *Polygonatum odoratum* (Mill.) Druce

Synonyms: *Convallaria odorata* Mill.; *Polygonatum japonicum* C. Morren & Decne.; *Polygonatum maximowiczii* F. Schmidt; *Polygonatum officinale* All; *Polygonatum vulgare* Desf.
Common name: yu zhu (Chinese)
Subclass Liliidae, Superorder Lilianae, Order Asparagales, Family Asparagaceae
Medicinal use: cough (China)
Pharmacological target: insulin resistance

Saponin fraction of rhizome at a concentration of 0.1 mg/mL boosted glucose uptake by HepG2 cells.[505] The fraction given orally at a single dose of 500 mg/kg to streptozotocin-induced diabetic Sprague–Dawley rats lowered 30 minutes peak glycaemia from 218 mg/dL to about 160 mg/dL.[505]

In oral glucose tolerance test, it inhibited α-glucosidase activity with an IC_{50} of 2 mg/mL *in vitro* (acarbose: 0.1 mg/mL).[505] The fraction given orally at a dose of 500 mg/kg/day for 60 days to diabetic rodents lowered glycaemia from 405.1 to 218.9 mg/dL (normal: 80.2 mg/dL; metformin: 300 mg/kg/day 154.4 mg/dL).[505] The regimen increased superoxide dismutase activity and decreased malondialdehyde level in plasma.[505] Flavonoid fraction of rhizome of *Polygonatum odoratum* (Mill.) Druce given orally to streptozotocin-induced type 1 diabetic Kunming mice (glycaemia >11 mmol/L) at a dose of 200 mg/kg/day for 9 days (acarbose: 20 mg/kg/day) lowered glycaemia from 27.6 to 14.1 mmol/L (normal: 5.3 mmol/L; acarbose: 20 mg/kg/day: 13.6 mmol/L), improved oral glucose tolerance, and had no effect on insulin levels.[506] Given to high-fat diet alloxan-induced type 2 diabetic Sprague–Dawley rats (glycaemia >11 mmol/L) for 10 days at a dose of 200 mg/kg/day the extract lowered glycaemia from 24.2 to 15.6 mmol/L (normal: 5.9 mmol/L; gliclazide: 15 mg/kg/day: 17.5 mmol/L) and increased plasma insulin as efficiently as gliclazide after 30 days of treatment.[506] The extract had some levels of inhibition on porcine pancreatic α-amylase.[506] (3R)-5,7-Dihydroxyl-6-methyl-8-methoxy-3-(4′-hydroxylbenzyl)-chroman-4-one, (3R)-5,7-dihydroxy-6,8-deimethyl-3-(4′-hydroxybenzyl)-chroman-4-one, (3R)-5,7-dihydroxyl-6-methyl-3-(4′-hydroxylbenzyl)-chroman-4-one and polygonatone D isolated from the rhizome induced at a concentration of 10 μM adenosine monophosphate-activated protein kinase phosphorylation and the phosphorylation and inhibition of acetyl-CoA carboxylase in rat liver epithelial cell line IAR-20.[507]

3.115 *Dracaena cochinchinensis* (Lour.) S.C. Chen

Synonyms: *Aletris cochinchinensis* Lour.; *Dracaena loureiroi* (Lour.) Gagnep.; *Pleomele cochinchinensis* Merr. ex Gagnep.
Common name: jian ye long xue shu (Chinese)
Subclass Liliidae, Superorder Lilianae, Order Asparagales, Family Asparagaceae
Medicinal use: dysentery (Laos)

Total flavonoid fraction wood of *Dracaena cochinchinensis* given to high-fat diet-streptozotocin-induced type 2 diabetic Sprague–Dawley rats (glycaemia ≥11.1 mmol/L) orally at a dose of 200 mg/kg/day for 21 days lowered plasma glucose from about 16 to 10 mmol/L and lowered plasma insulin.[508] This regimen improved glucose tolerance in oral glucose tolerance test and intraperitoneal glucose test.[508] This fraction lowered triglycerides from 197.8 to 112.1 mg/dL, normalized cholesterol from 152.2 to 75.7 mg/dL, normalized free fatty acids from 182.2 to 76.6 mg/dL, lowered low-density lipoprotein–cholesterol and increased high-density lipoprotein–cholesterol.[508] The fraction prevented hepatic steatosis in diabetic rodents.[508] The fraction lowered serum malondialdehyde and increased glutathione and superoxide dismutase.[508] The fraction lowered serum interleukin-6, tumor necrosis factor-α.[508] The treatment alleviated streptozotocin-induced pancreatic and insults.[508]

3.116 *Dioscorea nipponica* Makino

Common names: chuan long shu yu (Chinese); Japanese yam
Subclass Liliidae, Superorder Dioscorenae, Order Dioscoreales, Family Dioscoreaceae
Medicinal use: rheumatisms (China)
Pharmacological targets: atherogenic hyperlipidemia; insulin resistance

Butanol fraction of rhizomes of *Dioscorea nipponica* Makino given orally to Sprague–Dawley rats at a dose of 100 mg/kg for 28 days lowered plasma cholesterol and decreased triglycerides and decreased high-density lipoprotein and low-density lipoprotein.[509] From this extract trillin injected intraperitoneally at a dose of 500 mg/kg/day for 2 weeks on high-fat diet rats increased tail bleeding time, lowered plasma triglycerides and cholesterol near to normal values.[509] This isospirostanol saponin lowered low-density lipoprotein to normal values and increased high-density lipoprotein, lowered serum lipid peroxydes to normal and increased superoxide dismutase activity.[509] A Total saponin fraction of *Dioscorea nipponica* Makino (containing protodioscin, methylprotodioscin and dioscin) given to high-fat diet-streptozotocin-induced type 2 diabetic Wistar rats (glycaemia >16.7 mmol/L) orally at a dose of 200 mg/kg/day for 12 weeks lowered food and water intake, prevented body weight loss, decreased glycaemia and increased plasma insulin close to normal.[510] As a consequence of increased insulin sensitivity and availability, at the hepatic level, the regimen increased glycogen contents and improved hepatic cytoarchitecture with notably reduction of lipid droplets and reduced plasma aspartate aminotransferase and alanine aminotransferase.[510] This regimen decreased plasma cholesterol by 47.8%, triglycerides by 24.6%, free fatty acids by 29.4%, lowered low density lipoporotein by 66.1%.[510] The extract increased plasma glutathione, glutathione peroxidase, catalase and lowered malondialdehyde as well as nitric oxide.[510] At the hepatic level, the extract increased the expression of insulin receptor substrate-2, glucose transporter-4 and phosphorylated-Akt, increased insulin receptor expression as well as phosphofructokinase, glucokinase, and pyruvate kinase and decreased the expression phophoenolpyruvate carboxykinase, glycogen synthase kinase-3 and glucose-6-phosphatase with a decrease in activity of glycogen synthetase kinase- 3.[510] In addition, in the liver the extract decreased the expression of phosphorylated adenosine monophosphate-activated protein kinase, increased the expression of peroxisome proliferator-activated receptor-γ, carnitine palmitoyltranferase-1 and acyl-CoA oxidase while sterol regulatory element-binding protein-1c, and downstream targets acyl-CoA carboxylase and stearoyl CoA desaturase-1 and fatty acid synthetase expression were lowered.[510] The fraction lowered the expression of hepatic tumor necrosis factor-α, interleukin 6, Nuclear factor-κB and cyclooxygenase-2.[510] Insulin resistance results in increased release of glucose by the liver by decreasing the activities of glucokinase, phosphofructokinase and pyruvate kinase.[511] Protodioscin given orally to streptozotocin-induced diabetic Sprague–Dawley rats at a dose of 40 mg/kg/day for 12 weeks decreased food and water intake, improved glucose tolerance in oral glucose tolerance test, had no effect on plasma insulin, decreased plasma glucose from 26.6 to 15.4 mmol/L (6.7 mmol/L) and reduced insulin resistance. This regimen increased plasma adiponectin and glucose transporter-4 contents in muscles.[512] The mechanism by which saponins in this plant ameliorate hepatic insulin sensitivity could involve at least some anti-inflammatory effects of first pass metabolites. Methyl-protodioscin isolated from this plant at a concentration of 100 μM increased the expression of ATP-binding cassette A1 in macrophages *in vitro* as well as apolipoprotein A1-mediated cholesterol efflux.[513] This steroidal saponin decreased.[513] At a concentration of 60 μM, methyl-protodisocin increased the expression of ATP-binding cassette A1 and apolipoprotein A1-mediated cholesterol efflux in HepG2 cells, lowered the expression of 3-hydroxy-3-methylglutaryl-coenzyme A reductase, liver X receptor-α, acetyl CoA carboxylase and fatty acid synthetase and increased the expression of low-density lipoprotein receptor.[513]

3.117 *Dioscorea oppositifolia* L.

Common name: sarpakhya (India)
Subclass Dilleniidae, Superorder Dioscoreanae, Order Dioscoreales, Family Dioscoreaceae
Medicinal use: swelling (India)

Butanol fraction of rhizomes of *Dioscorea oppositifolia* L. (containing 21.2% of (3*R*,5*R*)-3,5-dihydroxy-1,7-bis(4-hydroxyphenyl)-3,5-heptanediol) given to ICR mice on high-fat diet at a dose of 100 mg/kg/day for 8 weeks had no effect on food intake, and prevented increase in weight gain by 45.2% (orlistat 15mg/kg/day: 47.3%).[514] This regimen prevented parametrila adipose tissue weight increase by 22.5%.[514] The fraction intake lowered plasma triglycerides from 147.4 to 119.7 mg/dL (normal: 87.4 mg/dL), cholesterol from 130.7 to 110.6 mg/dL (normal: 86.5 mg/dL), increased high-density lipoprotein–cholesterol from 30.3 to 40.6 mg/dL (normal: 39.3 mg/dL), lowered low-density lipoprotein–cholesterol from 77.6 to 50.6 mg/dL (normal: 38.9 mg/dL) and reduced atherogenic index from 3.5 to 1.8 (normal: 1.2).[514] At the hepatic level, the extract lowered triglycerides, cholesterol and afforded hepatoprotection as evidenced by a reduction of plasmatic aspartate aminotransferase and normalization of alanine aminotransferase.[514] The treatment increased fecal triglycerides from 43.2 to 72.2 mg/g feces/day (orlistat: 90.5 mg/g feces/day).[514]

3.118 *Alpinia katsumadae* Hayata

Synonyms: *Alpinia hainanensis* K. Schum.; *Languas katsumadae* (Hayata) Merr.
Common name: cao dou kou (Chinese)
Subclass Commelinidae, Superorder Zingiberanae, Order Zingiberales, Family Zingiberaceae
Medicinal use: facilitates digestion (China)
Pharmacological target: atherogenic hyperlipidemia

Ethanol extract of seeds of *Alpinia katsumadai* Hayata given to ICR mice on high-fat diet at a dose of 200 mg/kg/day for 12 weeks lowered plasma cholesterol from 163 to 117.4 mg/dL and triglycerides from 150.6 to 111.2 mg/dL.[515] This extract mildly increased high-density lipoprotein and lowered low-density lipoprotein from 24.2 to 11.4 mg/dL.[515] From this extract, 2,3,22,23-tetrahydroxyl-2,6,10,15,19,23-hexamethyl-6,10,14,18-tetracosatetraene inhibited rat liver microsomal activity acyl-CoA: cholesterol acyltransferase activity of rat liver microsomes with an IC_{50} value of 47.9 μM.[515] In hepG2 cells, this linear triterpene inhibited cholesteryl oleate synthesis with an IC_{50} value of 26 μM.[515] Inhibition of acyl-CoA: cholesterol acyltransferase implies an increase of unesterified cholesterol in hepatocytes hence inhibition of sterol regulatory element-binding protein-2[516] and its target 3-hydroxy-3-methylglutaryl-coenzyme A reductase.[517]

3.119 *Alpinia pricei* Hayata

Synonyms: *Alpinia sasakii* Hayata; *Alpinia tarokoensis* (Sasaki) Hayata; *Languas pricei* (Hayata) Sasaki; *Languas sasakii* (Hayata) Sasaki; *Languas tarokoensis* Sasaki
Common name: duan sui shan jiang (Chinese)
Subclass Commelinidae, Superorder Zingiberanae, Order Zingiberales, Family Zingiberaceae
Nutritional use: spice (Taiwan)

Ethanol extract (containing desmethyoxyyangonin, cardamonin and flavokawain B) of rhizome of *Alpinia pricei* Hayata given to Syrian hamsters at 0.8% of high-fat diet fed for 4 weeks had no effect on food intake and body weight, lowered plasma triglycerides, cholesterol, low-density lipoprotein–cholesterol and had no effect on high-density lipoprotein–cholesterol.[518] This regimen lowered plasma reactive oxygen species, oxidized low-density lipoprotein, increased alanine aminotransferase and had no effect on lipoprotein lipase.[518] The supplementation had no effect on cholesterol or triglyceride and decreased hepatic expression of peroxisome proliferator-activated receptor-γ.[518]

3.120 *Curcuma comosa* Roxb.

Common name: wan chak motluk (Thailand)
Subclass Commelinidae, Superorder Zingiberanae, Order Zingiberales, Family Zingiberaceae
Medicinal use: menopause (Thailand)
Pharmacological target: atherogenic hyperlipidemia

Curcuma comosa Roxb. extract given orally at a dose of 400 mg/kg/day orally to New Zealand White rabbits receiving 0.5% cholesterol for 3 months lowered plasma cholesterol, triglycerides and low-density lipoprotein by about 31%, 30%, and 27% and these effects were comparable with simvastatin at 5 mg/kg/day.[519] This regimen decreased expression of interleukin-1β, monocyte chemoattractant peptide-1 and tumor necrosis factor-α by abdominal aorta and live.[519] Clinical trials are warranted.

3.121 *Curcuma longa* L.

Synonym: *Curcuma domestica* Valeton
Common names: jiang huang (Chinese); kunyit (Malay); dilau (Philippines); turmeric
Subclass Commelinidae, Superorder Zingiberanae, Order Zingiberales, Family Zingiberaceae
Medicinal use: diabetes (Philippines)
Pharmacological target: atherogenic hyperlipidemia

Aqueous extract of rhizome of *Curcuma longa* (enriched with curcumin) given to Sprague–Dawley rats on high-cholesterol diet for 28 days attenuated liver weight increase, lowered plasma cholesterol from 4.4 to 3.1 mmol/L (normal: 2.3 mmol/L), lowered low-density lipoprotein–cholesterol from 1.8 to 0.4 mmol/L (normal: 0.4 mmol/L), increased high-density lipoprotein–cholesterol above normal group value.[520] The regimen improved hepatic cytoarchitecture with decrease of lipid accumulation by about 30%.[520] At the hepatic level, the extract repressed 3-hydroxy-3-methylglutaryl-coenzyme A reductase expression, increased the expression of CYP7A1, low-density lipoprotein receptor and heme oxygenase-1.[520]

3.122 *Elettaria cardamomum* (L.) Matton

Synonyms: *Amomum cardamomum* L.; *Cardamomum officinale* (L.) Salisb.
Common name: ela (India); small cardamom
Subclass Commelinidae, Superorder Zingiberanae, Order Zingiberales, Family Zingiberaceae
Medicinal use: bilious affections (India)
History: the plant was known of Theophrastus

Powder of fruits of *Elettaria cardamomum* (L.) Matton given to hypertensive patients taken at a dose of 3 g/day for 12 weeks lowered mean blood pressure from 112.5 to 97.9 mmHg.[521] This regimen lowered plasma cholesterol, triglycerides, very low-density lipoprotein–cholesterol and low-density lipoprotein–cholesterol by 19%, 15%, 15%, and 25%, respectively and increased plasma antioxidant status by 90%.[521] A decrease of 5 mmHg of blood pressure in a hypertensive individual can decrease the mortality due to stroke and coronary events by 14% and 9%, respectively.[522] Plasma triglycerides above 200 mg/dL denotes a state of hypertriglyceridemia which is associated with increased risk of cardiovascular disease.[6] The beneficial effect of small cardamom intake in Metabolic Syndrome need to be examined further.

3.123 *Zingiber mioga* (Thunb.) Roscoe

Synonyms: *Amomum mioga* Thunb.; *Zingiber echuanense* Y.K. Yang; *Zingiber oligophyllum* K. Schum.
Common names: jang ho (Chinese); myoga ginger (English); yang ha (Korean)
Subclass Commelinidae, Superorder Zingiberanae, Order Zingiberales, Family Zingiberaceae
Medicinal use: malaria (China)
Pharmacological targets: atherogenic hyperlipidemia; insulin resistance

Zingiber mioga (Thunb.) Roscoe (Figure 3.39) contains miogadial (Figure 3.40) also known as (E)-8β,17-epoxylabd-12-ene-15,16-dial,[523] which at a concentration of 10^{-2} M inhibited the

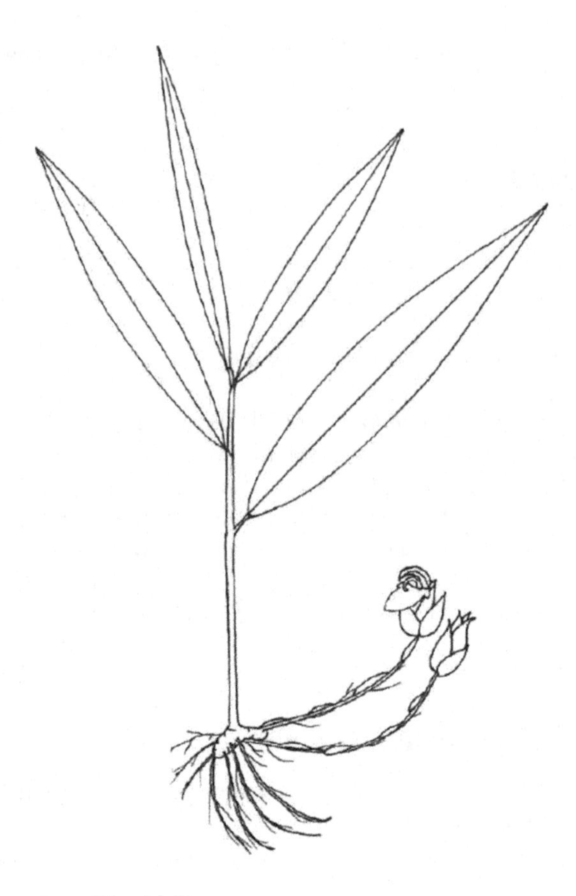

FIGURE 3.39 *Zingiber mioga* (Thunb.) Roscoe.

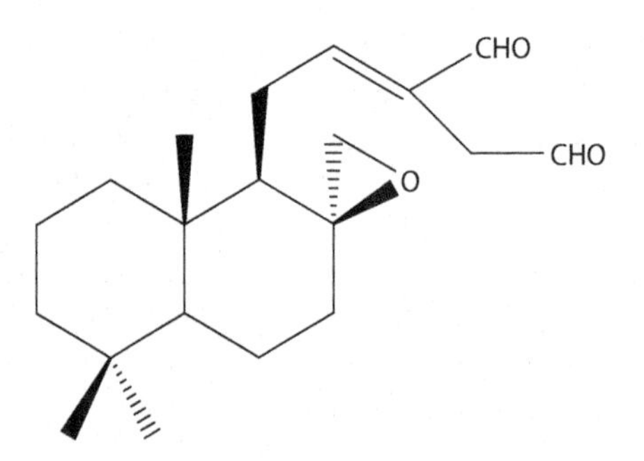

FIGURE 3.40 Miogadial.

synthesis of cholesterol in rat liver homogenate *in vitro* by 41.6%.[524] This diterpene could at least partially account to the fact that aqueous extract flower buds of *Zingiber mioga* (Thunb.) Roscoe given to C57BL/6J on high-fat diet at 0.5% of diet for 8 weeks lowered body weight and liver weight compared to untreated animals.[525] This regimen decreased plasma triglycerides from 62.3 to 48.3 mg/dL (normal: 37.5 mg/dL) and cholesterol from 243.2 to 208.4 mg/dL (normal: 162 mg/dL) and had no effect on high-density lipoprotein–cholesterol.[525] The extract lowered liver cholesterol and triglycerides.[525] The extract decreased glucose plasma area under the curve in oral glucose tolerance test indicating improved insulin sensitivitiy.[525] The supplementation reduced hepatic expression of gluconeogenic enzymes phosphoenolpyruvate carboxykinase and glucose-6-phosphatase indicating increased hepatic sensitivity to insulin. This could be owed to adenosine monophosphate-activated protein kinase since the extract induced the activation of adenosine monophosphate-activated protein kinase and decreased sterol regulatory element-binding protein-1c in the liver.[525] At the hepatic level, the extract reduced nuclear factor-κB phosphorylation and interleukin-6 expression.[525] Clinical trails are warranted.

3.124 *Zingiber officinale* Roscoe

Synonyms: *Amomum zingiber* L.; *Zingiber aromaticum* Noronha; *Zingiber sichuanense* Z.Y. Zhu, S.L. Zhang & S.X. Chen
Common names: jiang (Chinese); aderuck (India); halia (Malay); ginger
Subclass Commelinidae, Superorder Zingiberanae, Order Zingiberales, Family Zingiberaceae
Medicinal use: indigestion (India)
History: the plant was known to Dioscorides
Pharmacological targets: atherogenic hyperlipidemia; insulin resistance

A fraction enriched with gingerol and shoagol from *Zingiber officinale* Roscoe given to Syrian hamsters as part of 1% of cholesterol-enriched diet for 6 weeks lowered plasma cholesterol from 259.4 to 209.1 mg/dL, attenuated triglycerides, lowered high density-cholesterol, liver cholesterol and decreased atherogenic plaque from 17.1% to 7.2%.[526] This regimen increased total fecal fatty acids from 6.7 to 9.9 mg/g, fecal bile acids, fecal cholesterol from 0.8 to 1.4 mg/day and fecal neutral sterols from 2 to 4 mg/day.[526] This supplementation decreased the expression of intestinal acyl coenzyme A:cholesterol acyltransferase 2, triacylglycerol transport protein and ATP-binding cassette G5 and G8.[526] The fraction at the hepatic level increased the expression of sterol regulatory element-binding protein-2 and cholesterol 7α-hydroxylase (CYP7A1) and decreased liver X receptor-α expression.[526] Cholesteryl ester transfer protein is structurally associated with high-density lipoprotein and account for the delivery of cholesteryl ester to very low-density lipoprotein and low-density lipoprotein in exchange for triglycerides.[527] In obese patients with insulin resistance, increased circulating very low-density lipoprotein in the plasma favors the transfer of triglycerides from very low-density lipoprotein to high-density lipoprotein by cholesteryl ester transfer protein leading to the formation of triglyceride-rich high-density lipoprotein which become substrate for hepatic lipoprotein lipase resulting in a reduction of circulating high-density lipoprotein.[528] It can be said that upon cholesteryl ester transfer protein low-density lipoproteins are enriched with cholesterol and become more atherogenic.[529] Cholesteryl ester transfer protein activity is elevated in dyslipidaemia and early onset coronary heart disease.[45] [10]-dehydrogingeridione isolated from ginger inhibited human cholesteryl ester transfer protein *in vitro* with IC_{50} of 35 μM.[530] 6-Gingerol (Figure 3.41) given orally to Wistar rats on high-fat diet at a dose of 75 mg/kg/day for 30 days

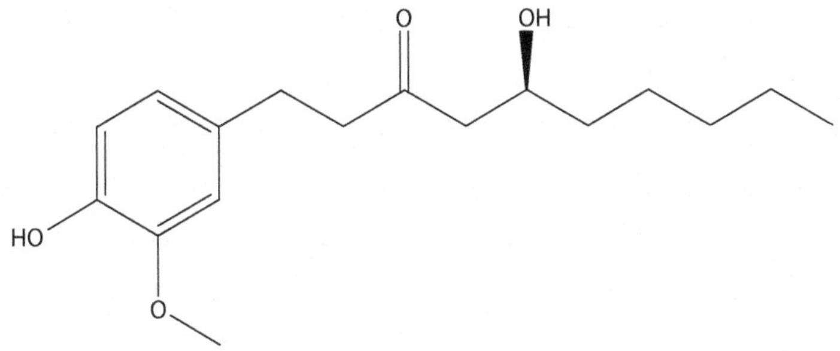

O OH HO O

FIGURE 3.41 6-Gingerol.

decreased body weight gain, plasma glucose, insulin and leptin and decreased insulin resistance.[531] This regimen decreased plasma cholesterol, triglycerides, low-density lipoprotein, very low-density lipoprotein–cholesterol and free fatty acids.[531] In a subsequent study, 6-gingerol at a concentration of 100 μmol/L prevented triglyceride droplet formation on HepG2 cells challenged with free fatty acids and this effect was comparable with peroxisome proliferator-activated receptor-α agonist ciprofibrate at 100 μmol/L. 6-Gingerol treatment attenuated free fatty acid-induced production of tumor necrosis factor-α and interleukin-by HepG2 cells.[532] This alkylphenol given orally at a dose of 100 mg/kg/day to methionine-choline- diet-induced hepatosteatosis in C57BL/6 mice for 4 weeks lowered hepatic cholesterol and triglycerides 26.3% and 36.8% (ciprofibrate 10 mg/kg/day: 29.5% and 40.6%, respectively).[532] This phenol lowered plasma alanine aminotransferase and aspartate aminotransferase and improved liver cytoarchitecture with attenuation of steatosis and necro-inflammation.[532] The treatment lowered hepatic expression of tumor necrosis factor-α, and interleukin-6, inhibited nuclear factor-κB.[532] 6-Gingerol downregulated the expression of diacylglycerol acyltransferase-2 and up-regulated peroxisome proliferator-activated receptor-α expression.[532] Rhizomes of *Zingiber officinale* (thumb) Roscoe given to obese volunteers at a dose of 2 g for 12 weeks lowered body weight from 89.5 to 87.6 kg, body mass index from 35.4 to 33.9 kg/m^2.[533] Methanol extract of rhizome of *Zingiber officinale* Roscoe inhibited α-amylase and inhibited porcine pancreatic lipase activity *in vitro*.[534] Clinical trials are warranted.

3.125 *Zingiber zerumbet* (L.) Roscoe ex Sm.

Synonyms: *Amomum zerumbet* L.; *Zingiber sylvestre* (L.) Garsault
Common names: hong qiu jiang (Chinese); lempoyang (Malay); shampoo ginger
Medicinal use: fatigue (China)
Pharmacological targets: insulin resistance; atherogenic hyperlipidemia

Ethanol extract of rhizomes of *Zingiber zerumbet* (L.) Roscoe ex Sm. (containing kaempferol) given orally at a dose of 300 mg/kg/day for days to Wistar rats on high-fructose diet for 8 weeks had no effect on food intake and decreased body weight gain, decreased plasma glucose by 14%, insulinemia by 8% and insulin resistance.[535] This regimen increased hepatic glycogen contents and decreased the expression of phosphoenolpyruvate carboxykinase[535] implying, at least, increased insulin sensitivity. In the soleus muscle of treated rats, this regimen increased glucose transporter-4 cytoplasmic membrane translocation compared to untreated rodents after a single injection of insulin.[535] Wistar rats on high-fat diet given kaempferol (Figure 3.42) daily and orally at a dose of 300 mg/kg had no effect on food intake, lowered body weight gain from 53.4 to 24.9 g (normal: 18.6 g; fenofibrate 100 mg/kg/day: 20.4 g), lowered epididymal white adipose tissue mass from 422.1 to 328.1 mg (normal: 303.6 mg; fenofibrate 100 mg/kg/day: 312.7 mg).[536] This regimen lowered plasma cholesterol from 130.2 to 89.1 mg/dL (normal: 72.9 mg/dL; fenofibrate: 80.2 mg/dL); triglycerides from

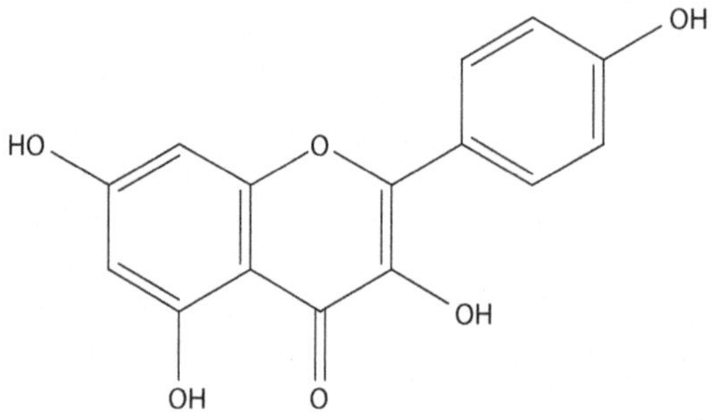

OH
HO
O
OH
OH
O

FIGURE 3.42 Kaempferol.

124.5 to 80.3 mg/dL (normal: 56.8 mg/dL; fenofibrate: 62.5 mg/dL); lowered low-density lipoprotein–cholesterol from 100.2 to 48.1 mg/dL (normal: 31.8 mg/dL; fenofibrate: 39.2 mg/dL), normalized high-density lipoprotein–cholesterol, lowered plasma free fatty acids from 61.9 to 39.5 mg/dL (normal: 28.2 mg/dL; fenofibrate: 33.2 mg/dL) and decreased atherogenic index from 3.6 to 1 (normal: 0.7; fenofibrate: 0.9).[536] This flavonol lowered hepatic triglycerides and cholesterol as efficiently as fenofibrate and normalized hepatic cytoarchitecture.[536] In the liver of treated rodents, kaempferol increased the expression of peroxisome proliferator-activated receptor-α, acetyl-CoA oxidase, and decreased the expression of sterol regulatory element-binding protein-1 and -2.[536] In fact, kaempferol at a concentration of 30 μM induced glucose uptake by L6 myotubes *in vitro* as efficiently as insulin at 0.1 μM.[536] Zerumbone from *Zingiber zerumbet* (L.) Roscoe ex Sm. inhibited porcine pancreatic lipase.[537] This sesquiterpene given orally at 100 mg/kg/day for 8 weeks Syrian hamsters on high-fat diet had no effect on food intake, attenuated body weight, lowered plasma cholesterol from 3.3 to 2.3 mmol/L (normal: 1.8 mmol/L), triglycerides from 1.4 to 0.9 mmol/L (normal: 0.6 mmol/L), lowered low-density lipoprotein and increased high-density lipoprotein.[537] This sesquiterpene increased fecal cholesterol from 3.8 to 5.3 μmol/g feces and increased fecal triglycerides from 2.5 to 6.4 μmol/g feces.[537] Zerumbone (Figure 3.43) elevated peroxisome proliferator-activated receptor-α expression, increased the expression of target carnitine palmitoyltransferase-1 and acyl-CoA oxidase.[537] At the hepatic level, this regimen lowered the expression of peroxisome proliferator-activated receptor-α and targets sterol regulatory element-binding protein-1c and sterol regulatory element-binding protein-2, fatty acid synthetase and 3-hydroxy-3-methyl-glutaryl-CoA reductase.[537] The treatment with zerumbone improved hepatic cytoarchitecture with lower accumulation of lipids and decreased aspartate aminotransferase and alanine aminotransferase.[537] Clinical trails are warranted.

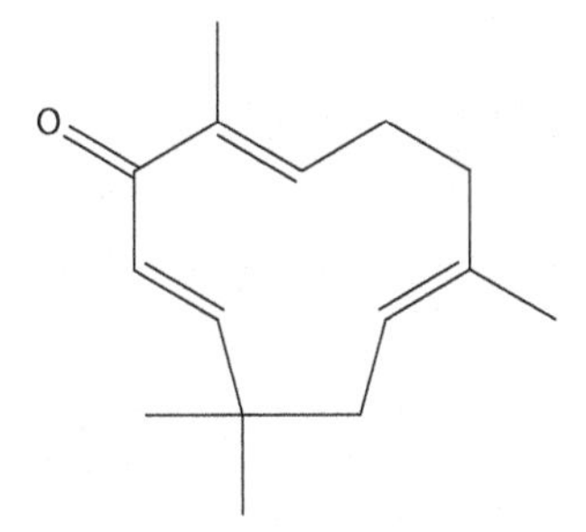

FIGURE 3.43 Zerumbone.

3.126 *Stemona sessilifolia* (Miq.) Miq.

Common name: bai bu (Chinese)
Subclass Liliidae, Superorder Pandadanae, Order Stemonales, Family Stemonaceae
Medicinal use: tuberculosis (China)
Pharmacological target: atherogenic hyperlipidemia

Plasma high-density lipoproteins release their triglyceride contents in the liver via hepatic lipase.[538] Cholesterol collected from arteries by high-density lipoproteins is released in hepatocytes upon binding with by scavenger receptor class B type I.[538] Cholesterol in the liver is either excreted in the form of biliary cholesterol or bile acids or recylcled.[45] Isooxymaistemonine isolated from the roots of *Stemona sessifolia* (Miq.) Miq. upregulated human high-density lipoprotein receptor SRB-1 by transfected human hepatocellular carcinoma BEL-7402 cells.[539] The mode of action of this stemona-type alkaloids is unknown.

APPENDIX

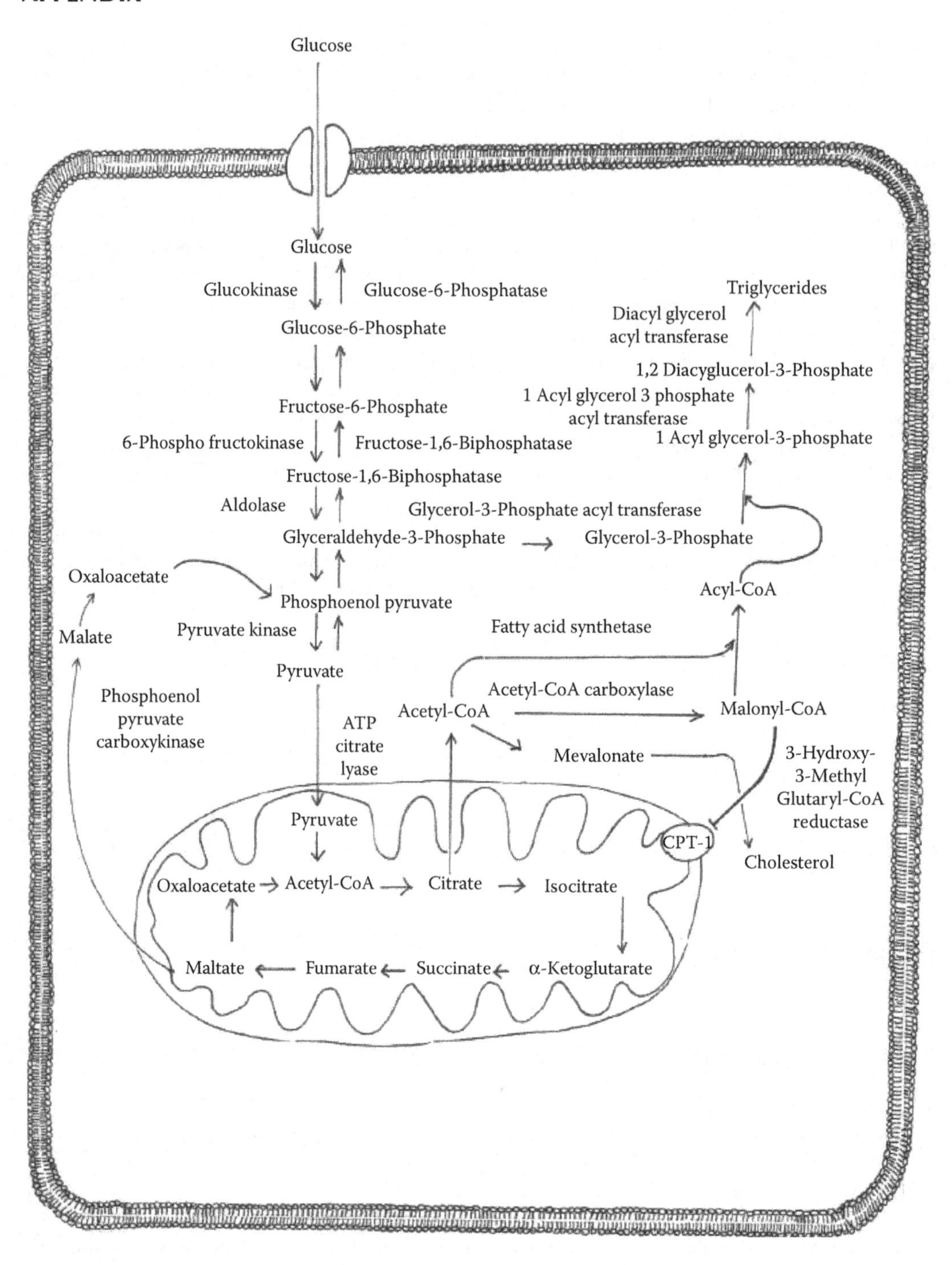

APPENDIX 3.1 Glucose metabolism in hepatocyte.

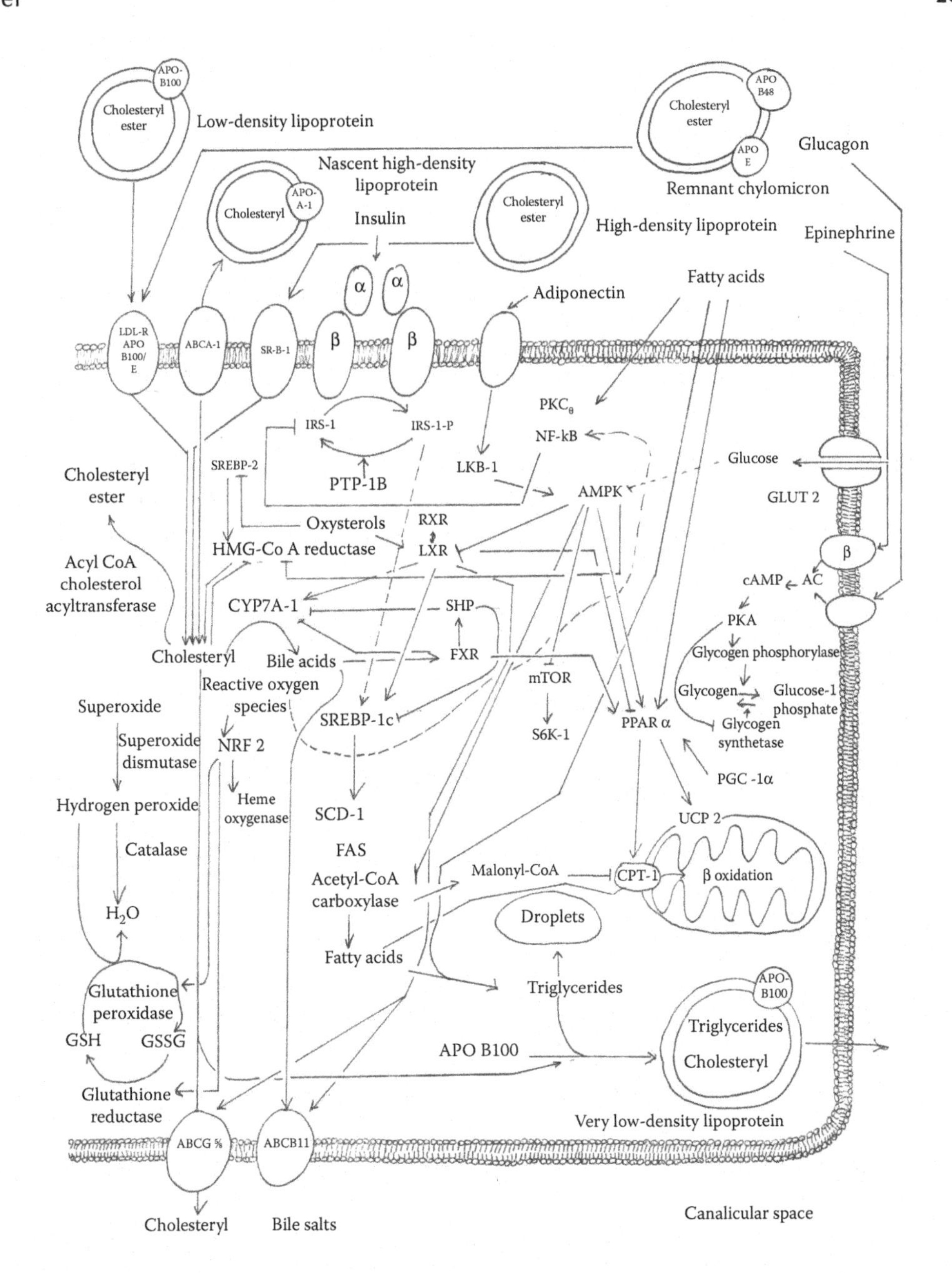

APPENDIX 3.2 Metabolism of cholesterol and fatty acids in hepatocyte.

ABCA1: ATP binding cassette transporter-1: ABCB11: ATP-binding cassette, subfamily B, member 11; ABCG5/8: ATP binding cassette transporter G5 and G8; AC: adenylate cyclase; Akt: protein kinase B; AMPK: adenosine monophosphate-activated protein kinase; ApoB48: apolipoprotein-B48; ApoB100: apoliporptein-B100; ApoE: apolipoprotein-E; CPT-1: carnitine palmitoyl transferase I; CYP7A1: cholesterol 7a-hydroxylase; FAS: fatty acid synthetase; FXR: farnesoid X receptor; GLUT-2: glucose transporter-2; GSH: glutathione; GSSG: glutathione disulfide; HMG-CoA reductase: 3-hydroxy-3-methylglutaryl coenzyme A reductase; IRS-1: Insulin receptor substrate-1; LDL-R: Low-density lipoprotein receptor; NRF-2: nuclear factor-erythroid 2-related factor 2; LXR: Liver X receptor; PGC-1a: peroxisome proliferator-activated receptor-gamma coactivator-1 alpha; PKA: protein kinase A; PKCq: protein kinase Cq; RXR: retinoid X receptor ; SCD-1: stearoyl-CoA desaturase-1; SHP: small heterodimer partner ; UCP2: uncoupling protein-2; SR-B1: scavenger receptor class B type 1; SREBP-1c: sterol regulatory element-binding protein-1c; SREBP-2: sterol regulatory element-binding protein-2;

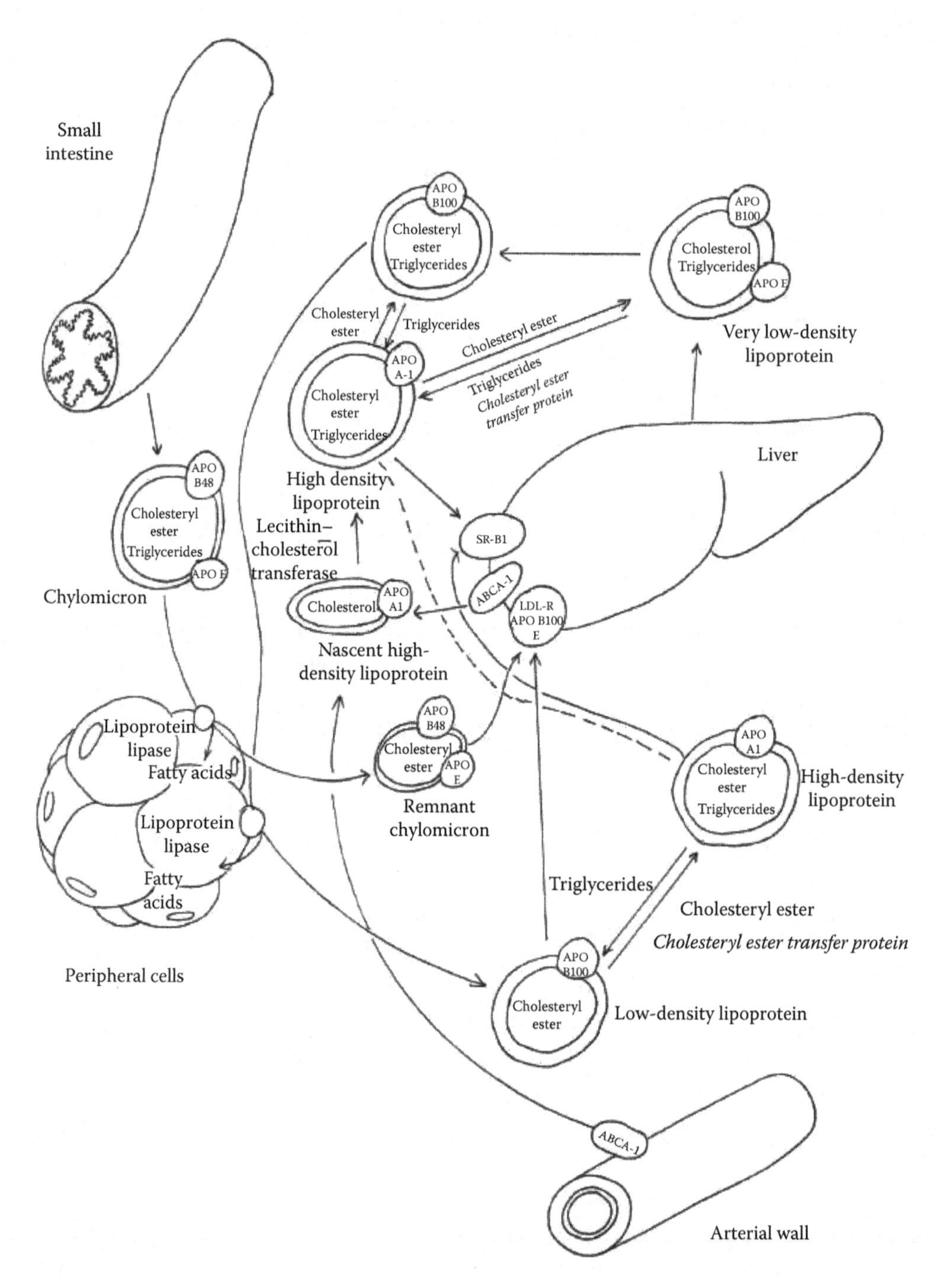

APPENDIX 3.3 Plasma lipoproteins.

REFERENCES

1. Hotamisligil, G.S., Shargill, N.S. and Spiegelman, B.M., 1993. Adipose expression of tumor necrosis factor-alpha: Direct role in obesity-linked insulin resistance. *Science*, 259, 87.
2. W.H.O., 2006. Definition and diagnosis of diabetes mellitus and intermediate hyperglycemia: Report of a WHO/IDF consultation. Geneva, Switzerland: WHO, p. 50.
3. Neuschwander-Tetri, B.A., 2010. Hepatic lipotoxicity and the pathogenesis of nonalcoholic steatohepatitis: The central role of nontriglyceride fatty acid metabolites. *Hepatology*, 52(2), 774–788.
4. Grundy, S.M., 2008. Metabolic syndrome pandemic. *Arteriosclerosis, Thrombosis, and Vascular Biology*, 28(4), 629–636.
5. Fong, D.G., Nehra, V., Lindor, K.D. and Buchman, A.L., 2000. Metabolic and nutritional considerations in nonalcoholic fatty liver. *Hepatology*, 32(1), 3–10.
6. Williams, L., 2002. Third report of the National Cholesterol Education Program (NCEP) expert panel on detection, evaluation, and treatment of high blood cholesterol in adults (Adult Treatment Panel III) final report. *Circulation*, 106(25), 3143.
7. Jansen, P.L., 2004. Non-alcoholic steatohepatitis. *European Journal of Gastroenterology & Hepatology*, 16(11), 1079–1085.
8. Sohn, J.H., Han, K.L., Choo, J.H. and Hwang, J.K., 2007. Macelignan protects HepG2 cells against tert-butylhydroperoxide-induced oxidative damage. *Biofactors*, 29(1), 1–10.
9. Horton, J.D., Goldstein, J.L. and Brown, M.S., 2002. SREBPs: Activators of the complete program of cholesterol and fatty acid synthesis in the liver. *The Journal of Clinical Investigation*, 109(9), 1125–1131.
10. Lee, M.S., Kim, K.J., Kim, D., Lee, K.E. and Hwang, J.K., 2011. Meso-Dihydroguaiaretic acid inhibits hepatic lipid accumulation by activating AMP-activated protein kinase in human HepG2 cells. *Biological and Pharmaceutical Bulletin*, 34(10), 1628–1630.
11. Mandard, S., Müller, M. and Kersten, S., 2004. Peroxisome proliferator-activated receptor α target genes. *Cellular and Molecular Life Sciences CMLS*, 61(4), 393–416.
12. González-Rodríguez, Á., Gutierrez, J.A.M., Sanz-González, S., Ros, M., Burks, D.J. and Valverde, Á.M., 2010. Inhibition of PTP1B restores IRS1-mediated hepatic insulin signaling in IRS2-deficient mice. *Diabetes*, 59(3), 588–599.
13. Yang S., Na, M.K., Jang, J.P., Kim, K.A., Kim, B.Y., Sung, N.J., Oh, W.K. and Ahn, J.S., 2006. Inhibition of protein tyrosine phosphatase 1B by lignans from *Myristica fragrans*. *Phytotherapy Research*, 20(8), 680–682.
14. Wilson, P.W., D'Agostino, R.B., Levy, D., Belanger, A.M., Silbershatz, H. and Kannel, W.B., 1998. Prediction of coronary heart disease using risk factor categories. *Circulation*, 97(18), 1837–1847.
15. Nguyen, P.H., Le, T.V.T., Kang, H.W., Chae, J., Kim, S.K., Kwon, K.I., Seo, D.B., Lee, S.J. and Oh, W.K., 2010. AMP-activated protein kinase (AMPK) activators from Myristica fragrans (nutmeg) and their anti-obesity effect. *Bioorganic & Medicinal Chemistry Letters*, 20(14), 4128–4131.
16. Stapleton, D., Mitchelhill, K.I., Gao, G., Widmer, J., Michell, B.J., Teh, T., House, C.M., et al., 1996. Mammalian AMP-activated protein kinase subfamily. *Journal of Biological Chemistry*, *271*(2), 611–614.
17. Park, H., Kaushik, V.K., Constant, S., Prentki, M., Przybytkowski, E., Ruderman, N.B. and Saha, A.K., 2002. Coordinate regulation of malonyl-CoA decarboxylase, sn-glycerol-3-phosphate acyltransferase, and acetyl-CoA carboxylase by AMP-activated protein kinase in rat tissues in response to exercise. *Journal of Biological Chemistry*, 277(36), 32571–32577.
18. Saha, A.K., Avilucea, P.R., Ye, J.M., Assifi, M.M., Kraegen, E.W. and Ruderman, N.B., 2004. Pioglitazone treatment activates AMP-activated protein kinase in rat liver and adipose tissue in vivo. *Biochemical and Biophysical Research Communications*, 314(2), 580–585.
19. Towler, M.C. and Hardie, D.G., 2007. AMP-activated protein kinase in metabolic control and insulin signaling. *Circulation Research*, 100(3), 328–341.
20. Viollet, B., Mounier, R., Leclerc, J., Yazigi, A., Foretz, M. and Andreelli, F., 2007. Targeting AMP-activated protein kinase as a novel therapeutic approach for the treatment of metabolic disorders. *Diabetes & Metabolism*, 33(6), 395–402.
21. Leclercq, I.A., Morais, A.D.S., Schroyen, B., Van Hul, N. and Geerts, A., 2007. Insulin resistance in hepatocytes and sinusoidal liver cells: Mechanisms and consequences. *Journal of Hepatology*, 47(1), 142–156.
22. Saltiel, A.R. and Kahn, C.R., 2001. Insulin signalling and the regulation of glucose and lipid metabolism. *Nature*, 414(6865), 799–806.
23. Anand, P., Murali, K.Y., Tandon, V., Murthy, P.S. and Chandra, R., 2010. Insulinotropic effect of cinnamaldehyde on transcriptional regulation of pyruvate kinase, phosphoenolpyruvate carboxykinase, and GLUT4 translocation in experimental diabetic rats. *Chemico-Biological Interactions*, 186(1), 72–81.

24. Hanson, R.W. and Patel, Y.M., 1994. Phosphoenolpyruvate carboxykinase (GTP): the gene and the enzyme. *Advances in Enzymology and Related Areas of Molecular Biology*, 69, 203–281.
25. Cheng, D.M., Kuhn, P., Poulev, A., Rojo, L.E., Lila, M.A. and Raskin, I., 2012. In vivo and in vitro antidiabetic effects of aqueous cinnamon extract and cinnamon polyphenol-enhanced food matrix. *Food Chemistry*, 135(4), 2994–3002.
26. Saifudin, A., Kadota, S. and Tezuka, Y., 2013. Protein tyrosine phosphatase 1B inhibitory activity of Indonesian herbal medicines and constituents of Cinnamomum burmannii and Zingiber aromaticum. *Journal of Natural Medicines*, 67(2), 264–270.
27. Goldstein, B.J., Ahmad, F., Ding, W., Li, P.M., and Zhang, W.R., 1998. Regulation of the insulin signalling pathway by cellular protein-tyrosine phosphatases. *Molecular and Cellular Biochemistry*. 182 (1–2), 91–99.
28. Kim, S.H. and Choung, S.Y., 2010. Antihyperglycemic and antihyperlipidemic action of *Cinnamomi Cassiae* (Cinnamon bark) extract in C57BL/Ks db/db mice. *Archives of Pharmacal Research*, 33(2), 325–333.
29. Gervois, P. and Mansouri, R.M., 2012. PPARα as a therapeutic target in inflammation-associated diseases. *Expert Opinion on Therapeutic Targets*, 16(11), 1113–1125.
30. Ranasinghe, P., Perera, S., Gunatilake, M., Abeywardene, E., Gunapala, N., Premakumara, S., Perera, K., Lokuhetty, D. and Katulanda, P., 2012. Effects of *Cinnamomum zeylanicum* (Ceylon cinnamon) on blood glucose and lipids in a diabetic and healthy rat model. *Pharmacognosy Research*, 4(2), 73.
31. Gao, Z., Zhang, X., Zuberi, A., Hwang, D., Quon, M.J., Lefevre, M. and Ye, J., 2004. Inhibition of insulin sensitivity by free fatty acids requires activation of multiple serine kinases in 3T3-L1 adipocytes. *Molecular Endocrinology*, 18(8), 2024–2034.
32. Ferré, P., 2004. The biology of peroxisome proliferator-activated receptors. *Diabetes*, 53(suppl 1), S43–S50.
33. Xi Li, J., Liu, T., Wang, L., Guo, X., Xu, T., Wu, L., Qin, L. and Sun, W., 2012. Antihyperglycemic and antihyperlipidemic action of cinnamaldehyde in C57BLKS/J db/db mice. *Journal of Traditional Chinese Medicine*, 32(3), 446–452.
34. Gordon, T., Castelli, W.P., Hjortland, M.C., Kannel, W.B. and Dawber, T.R., 1977. High density lipoprotein as a protective factor against coronary heart disease: The Framingham study. *The American Journal of Medicine*, 62(5), 707–714.
35. Chawla, A., Repa, J.J., Evans, R.M. and Mangelsdorf, D.J., 2001. Nuclear receptors and lipid physiology: Opening the X-files. *Science*, 294(5548), 1866–1870.
36. Cuperus, F.J., Claudel, T., Gautherot, J., Halilbasic, E. and Trauner, M., 2014. The role of canalicular ABC transporters in cholestasis. *Drug Metabolism and Disposition*, 42(4), 546–560.
37. Yu, L., Li-Hawkins, J., Hammer, R.E., Berge, K.E., Horton, J.D., Cohen, J.C. and Hobbs, H.H., 2002. Overexpression of ABCG5 and ABCG8 promotes biliary cholesterol secretion and reduces fractional absorption of dietary cholesterol. *The Journal of Clinical Investigation*, 110(5), 671–680.
38. Wang, Z. and Xie, P. (Eds.). 2015. *Monographs for Quality Evaluation of Chinese Crude Drugs*. World Scientific: Singapore.
39. Jang, Y.Y., Song, J.H., Shin, Y.K., Han, E.S. and Lee, C.S., 2000. Protective effect of boldine on oxidative mitochondrial damage in streptozotocin-induced diabetic rats. *Pharmacological Research*, 42(4), 361–371.
40. Zagorova, M., Prasnicka, A., Kadova, Z., Dolezelova, E., Kazdova, L., Cermanova, J., Rozkydalova, L., Hroch, M., Mokry, J. and Micuda, S., 2015. Boldine attenuates cholestasis associated with nonalcoholic fatty liver disease in hereditary hypertriglyceridemic rats fed by high-sucrose diet. *Physiological Research*, 64, S467.
41. Cermanova, J., Kadova, Z., Zagorova, M., Hroch, M., Tomsik, P., Nachtigal, P., Kudlackova, Z. et al., 2015. Boldine enhances bile production in rats via osmotic and Farnesoid X receptor dependent mechanisms. *Toxicology and Applied Pharmacology*, 285(1), 12–22.
42. Yamahara, J., Matsuda, H., Sawada, T. and Kushida, H., 1983. Biologically active principles of crude drugs preventive effect of sesquiterpenoid components of the root of *Lindera strychinifolia* on experimental liver damage. 生薬学雑誌, 37(1), 84–86.
43. Wang, F., Gao, Y., Zhang, L. and Liu, J.K., 2010. Bi-linderone, a highly modified methyl-linderone dimer from *Lindera aggregata* with activity toward improvement of insulin sensitivity in vitro. *Organic Letters*, 12(10), 2354–2357.
44. Hashimura, H., Ueda, C., Kawabata, J. and Kasai, T., 2001. Acetyl-CoA carboxylase inhibitors from avocado (*Persea americana* Mill.) fruits. *Bioscience, Biotechnology, and Biochemistry*, 65(7), 1656–1658.
45. Durrington, P., 2007. *Hyperlipidaemia 3Ed: Diagnosis and Management*. Boca Raton, FL: CRC Press.
46. Brai, B.I., Odetola, A.A. and Agomo, P.U., 2007. Hypoglycemic and hypocholesterolemic potential of *Persea americana* leaf extracts. *Journal of Medicinal Food*, 10(2), 356–360.

47. Coleman, R.A. and Lee, D.P., 2004. Enzymes of triacylglycerol synthesis and their regulation. *Progress in Lipid Research*, 43, 134–176.
48. Kawagishi, H., Fukumoto, Y., Hatakeyama, M., He, P., Arimoto, H., Matsuzawa, T., Arimoto, Y., Suganuma, H., Inakuma, T. and Sugiyama, K., 2001. Liver injury suppressing compounds from avocado (*Persea americana*). *Journal of Agricultural and Food Chemistry*, 49(5), 2215–2221.
49. Bechmann, L.P., Hannivoort, R.A., Gerken, G., Hotamisligil, G.S., Trauner, M. and Canbay, A., 2012. The interaction of hepatic lipid and glucose metabolism in liver diseases. *Journal of Hepatology*, 56(4), 952–964.
50. Savage, D.B., Choi, C.S., Samuel, V.T., Liu, Z.X., Zhang, D., Wang, A., Zhang, X.M. et al., 2006. Reversal of diet-induced hepatic steatosis and hepatic insulin resistance by antisense oligonucleotide inhibitors of acetyl-CoA carboxylases 1 and 2. *The Journal of Clinical Investigation*, 116(3), 817–824.
51. Lima, C.R., Vasconcelos, C.F.B., Costa-Silva, J.H., Maranhão, C.A., Costa, J., Batista, T.M., Carneiro, E.M., Soares, L.A.L., Ferreira, F. and Wanderley, A.G., 2012. Anti-diabetic activity of extract from *Persea americana* Mill. leaf via the activation of protein kinase B (PKB/Akt) in streptozotocin-induced diabetic rats. *Journal of Ethnopharmacology*, 141(1), 517–525.
52. Li, Z., Wong, A., Henning, S.M., Zhang, Y., Jones, A., Zerlin, A., Thames, G., Bowerman, S., Tseng, C.H. and Heber, D., 2013. Hass avocado modulates postprandial vascular reactivity and postprandial inflammatory responses to a hamburger meal in healthy volunteers. *Food & Function*, 4(3), 384–391.
53. Lee, S.W., Rho, M.C., Park, H.R., Choi, J.H., Kang, J.Y., Lee, J.W., Kim, K., Lee, H.S. and Kim, Y.K., 2006. Inhibition of diacylglycerol acyltransferase by alkamides isolated from the fruits of *Piper longum* and *Piper nigrum*. *Journal of Agricultural and Food Chemistry*, 54(26), 9759–9763.
54. Jin, Z., Borjihan, G., Zhao, R., Sun, Z., Hammond, G.B. and Uryu, T., 2009. Antihyperlipidemic compounds from the fruit of Piper longum L. *Phytotherapy Research*, 23(8), 1194–1196.
55. Jwa, H., Choi, Y., Park, U.H., Um, S.J., Yoon, S.K. and Park, T., 2012. Piperine, an LXRα antagonist, protects against hepatic steatosis and improves insulin signaling in mice fed a high-fat diet. *Biochemical Pharmacology*, 84(11), 1501–1510.
56. Zhu, R., Ou, Z., Ruan, X. and Gong, J., 2012. Role of liver X receptors in cholesterol efflux and inflammatory signaling (Review). *Molecular Medicine Reports*, 5(4), 895–900.
57. Shaw, R.J., Lamia, K.A., Vasquez, D., Koo, S.H., Bardeesy, N., DePinho, R.A., Montminy, M. and Cantley, L.C., 2005. The kinase LKB1 mediates glucose homeostasis in liver and therapeutic effects of metformin. *Science*, 310(5754), 1642–1646.
58. Sakamoto, K., McCarthy, A., Smith, D., Green, K.A., Hardie, D.G., Ashworth, A. and Alessi, D.R., 2005. Deficiency of LKB1 in skeletal muscle prevents AMPK activation and glucose uptake during contraction. *The EMBO Journal*, 24(10), 1810–1820.
59. Choi, S., Choi, Y., Choi, Y., Kim, S., Jang, J. and Park, T., 2013. Piperine reverses high fat diet-induced hepatic steatosis and insulin resistance in mice. *Food Chemistry*, 141(4), 3627–3635.
60. Howard, B.V., Robbins, D.C., Sievers, M.L., Lee, E.T., Rhoades, D., Devereux R.B., Cowan, L.D. et al., 2000. LDL cholesterol as a strong predictor of coronary heart disease in diabetic individuals with insulin resistanceand low LDL: The Strong Heart Study. *Arteriosclerosis, Thrombosis and Vascular Biology*, 20(3), 830–835.
61. Steinberg, D., Parthasarathy, S., Carew, T.E., Khoo, J.C. and Witztum, J.L., 1989. Beyond cholesterol. Modification of low-density lipoprotein that increase atherogenicity. *The New England Journal of Medicine*, 320, 915–924.
62. Kim, K.J., Lee, M.S., Jo, K. and Hwang, J.K., 2011. Piperidine alkaloids from Piperretrofractum Vahl. protect against high-fat diet-induced obesity by regulating lipid metabolism and activating AMP-activated protein kinase. *Biochemical and Biophysical Research Communications*, 411(1), 219–225.
63. Finck, B.N. and Kelly, D.P., 2006. PGC-1 coactivators: Inducible regulators of energy metabolism in health and disease. *The Journal of Clinical Investigation*, 116(3), 615–622.
64. Pouliot, M.C., Després, J.P., Nadeau, A., Moorjani, S., Prud'Homme, D., Lupien, P.J., Tremblay, A. and Bouchard, C., 1992. Visceral obesity in men: Associations with glucose tolerance, plasma insulin, and lipoprotein levels. *Diabetes*, 41(7), 826–834.
65. Guo, F., Yang, X., Li, X., Feng, R., Guan, C., Wang, Y., Li, Y. and Sun, C., 2013. Nuciferine prevents hepatic steatosis and injury induced by a high-fat diet in hamsters. *PLoS ONE*, 8(5), e63770.
66. Ye, L.H., Xiao, B.X., Liao, Y.H., Liu, X.M., Pan, R.L. and Chang, Q., 2016. Metabolism profiles of nuciferine in rats using ultrafast liquid chromatography with tandem mass spectrometry. *Biomedical Chromatography*, 30, 1216–1222.
67. Cordell, G.A., 1993. *The Alkaloids: Chemistry and Pharmacology*. San Diego, CA: Academic Press.

68. Macko, E., Douglas, B., Weisbach, J.A. and Waltz, D.T., 1972. Studies on the pharmacology of nuciferine and related aporphines. *Archives internationales de pharmacodynamie et de therapie*, 197(2), 261.
69. Brusq, J.M., Ancellin, N., Grondin, P., Guillard, R., Martin, S., Saintillan, Y. and Issandou, M., 2006. Inhibition of lipid synthesis through activation of AMP kinase: An additional mechanism for the hypolipidemic effects of berberine. *Journal of Lipid Research*, 47(6), 1281–1288.
70. Fan, H., Chen, Y.Y., Bei, W.J., Wang, L.Y., Chen, B.T. and Guo, J., 2013. In vitro screening for antihepatic steatosis active components within Coptidis Rhizoma alkaloids extract using liver cell extraction with hplc analysis and a free fatty acid-induced hepatic steatosis hepg2 cell assay. *Evidence-Based Complementary and Alternative Medicine*, 2013.
71. Cao, S., Zhou, Y., Xu, P., Wang, Y., Yan, J., Bin, W., Qiu, F. and Kang, N., 2013. Berberine metabolites exhibit triglyceride-lowering effects via activation of AMP-activated protein kinase in Hep G2 cells. *Journal of Ethnopharmacology*, 149(2), 576–582.
72. Abidi, P., Chen, W., Kraemer, F.B., Li, H. and Liu, J., 2006. The medicinal plant goldenseal is a natural LDL-lowering agent with multiple bioactive components and new action mechanisms. *Journal of Lipid Research*, 47(10), 2134–2147.
73. Cao, Y., Bei, W., Hu, Y., Cao, L., Huang, L., Wang, L., Luo, D. et al., 2012. Hypocholesterolemia of *Rhizoma Coptidis* alkaloids is related to the bile acid by up-regulated CYP7A1 in hyperlipidemic rats. *Phytomedicine*, 19(8), 686–692.
74. Yan, F., Benrong, H., Qiang, T., Qin, F. and Jizhou, X., 2005. Hypoglycemic activity of jatrorrhizine. *Journal of Huazhong University of Science and Technology* [*Medical Sciences*], 25(5), 491–493.
75. Wu, H., He, K., Wang, Y., Xue, D., Ning, N., Zou, Z., Ye, X., Li, X., Wang, D. and Pang, J., 2014. The antihypercholesterolemic effect of jatrorrhizine isolated from *Rhizoma Coptidis*. *Phytomedicine*, 21(11), 1373–1381.
76. Yang, Y., Kang, N., Xia, H., Li, J., Chen, L. and Qiu, F., 2010. Metabolites of protoberberine alkaloids in human urine following oral administration of Coptidis Rhizoma decoction. *Planta medica*, 76(16), 1859–1863.
77. Avcı, G., Kupeli, E., Eryavuz, A., Yesilada, E. and Kucukkurt, I., 2006. Antihypercholesterolaemic and antioxidant activity assessment of some plants used as remedy in Turkish folk medicine. *Journal of Ethnopharmacology*, 107(3), 418–423.
78. Benhaddou-Andaloussi, A., Martineau, L.C., Vallerand, D., Haddad, Y., Afshar, A., Settaf, A. and Haddad, P.S., 2010. Multiple molecular targets underlie the antidiabetic effect of *Nigella sativa* seed extract in skeletal muscle, adipocyte and liver cells. *Diabetes, Obesity and Metabolism*, 12(2), 148–157.
79. Morikawa, T., Ninomiya, K., Xu, F., Okumura, N., Matsuda, H., Muraoka, O., Hayakawa, T. and Yoshikawa, M., 2013. Acylated dolabellane-type diterpenes from Nigella sativa seeds with triglyceride metabolism-promoting activity in high glucose-pretreated HepG2 cells. *Phytochemistry Letters*, 6(2), 198–204.
80. Kaleem, M., Kirmani, D., Asif, M., Ahmed, Q. and Bano, B., 2006. Biochemical effects of *Nigella sativa* L seeds in diabetic rats. *Indian Journal of Experimental Biology*, 44(9), 745.
81. Al-Naqeep, G., Ismail, M. and Yazan, L.S., 2009. Effects of thymoquinone rich fraction and thymoquinone on plasma lipoprotein levels and hepatic low density lipoprotein receptor and 3-hydroxy-3-methylglutaryl coenzyme A reductase genes expression. *Journal of Functional Foods*, 1(3), 298–303.
82. Awad, A.S., Al Haleem, E.N.A., El-Bakly, W.M. and Sherief, M.A., 2016. Thymoquinone alleviates nonalcoholic fatty liver disease in rats via suppression of oxidative stress, inflammation, apoptosis. *Naunyn-Schmiedeberg's Archives of Pharmacology*, 389(4), 381–391.
83. Dehkordi, F.R. and Kamkhah, A.F., 2008. Antihypertensive effect of *Nigella sativa* seed extract in patients with mild hypertension. *Fundamental & Clinical Pharmacology*, 22(4), 447–452.
84. Koruk, M., Taysi, S., Savas, M.C., Yilmaz, O., Akcay, F. and Karakok, M., 2004. Oxidative stress and enzymatic antioxidant status in patients with nonalcoholic steatohepatitis. *Annals of Clinical & Laboratory Science*, 34(1), 57–62.
85. Harris, E.D. 1992. Regulation of antioxidant enzymes. *FASEB Journal*, 6, 2675–8334.
86. Min, Q., Bai, Y.T., Shu, S.J. and Ren, P., 2006. [Protective effect of dl-tetrahydropalmatine on liver injury induced by carbon tetrachloride in mice]. *Zhongguo Zhong yao za zhi= Zhongguo zhongyao zazhi= China journal of Chinese materia medica*, 31(6), 483–484.
87. Lee, W.C., Kim, J.K., Kang, J.W., Oh, W.Y., Jung, J.Y., Kim, Y.S., Jung, H.A., Choi, J.S. and Lee, S.M., 2010. Palmatine attenuates D-galactosamine/lipopolysaccharide-induced fulminant hepatic failure in mice. *Food and Chemical Toxicology*, 48(1), 222–228.
88. Wei, H.L. and Liu, G.T., 1997. [Protective action of corynoline, acetylcorynoline and protopine against experimental liver injury in mice]. *Yao xue xue bao= Acta pharmaceutica Sinica*, 32(5), 331–336.

89. Wu, Y.R., Ma, Y.B., Zhao, Y.X., Yao, S.Y., Zhou, J., Zhou, Y. and Chen, J.J., 2007. Two new quaternary alkaloids and anti-hepatitis B virus active constituents from *Corydalis saxicola*. *Planta Medica*, 73(8), 787–791.
90. Wang, T., Sun, N.L., Zhang, W.D., Li, H.L., Lu, G.C., Yuan, B.J., Jiang, H., She, J.H. and Zhang, C., 2008. Protective effects of dehydrocavidine on carbon tetrachloride-induced acute hepatotoxicity in rats. *Journal of Ethnopharmacology*, 117(2), 300–308.
91. Slater, T.F., 1984. Overview of methods used for detecting lipid peroxidation. *Methods in Enzymology*, 105, 283–293.
92. Domitrović, R., Jakovac, H. and Blagojević, G., 2011. Hepatoprotective activity of berberine is mediated by inhibition of TNF-α, COX-2, and iNOS expression in CCl 4-intoxicated mice. *Toxicology*, 280(1), 33–43.
93. Gilani, A.H., Janbaz, K.H. and Akhtar, M.S., 1996. Selective protective effect of an extract from Fumaria parviflora on paracetamol-induced hepatotoxicity. *General Pharmacology: The Vascular System*, 27(6), 979–983.
94. Rathi, A., Srivastava, A.K., Shirwaikar, A., Rawat, A.K.S. and Mehrotra, S., 2008. Hepatoprotective potential of Fumaria indica Pugsley whole plant extracts, fractions and an isolated alkaloid protopine. *Phytomedicine*, 15(6), 470–477.
95. Beynen, A.C. and Katan, M.B., 1985. Why do polyunsaturated fatty acids lower serum cholesterol? *The American Journal of Clinical Nutrition*, 42(3), 560–563.
96. Amaral, J.S., Casal, S., Pereira, J.A., Seabra, R.M. and Oliveira, B.P., 2003. Determination of sterol and fatty acid compositions, oxidative stability, and nutritional value of six walnut (*Juglans regia* L.) cultivars grown in Portugal. *Journal of Agricultural and Food Chemistry*, 51(26), 7698–7702.
97. Zibaeenezhad, M.J., Rezaiezadeh, M., Mowla, A., Ayatollahi, S.M.T. and Panjehshahin, M.R., 2003. Antihypertriglyceridemic effect of walnut oil. *Angiology*, 54(4), 411–414.
98. Wu, L., Piotrowski, K., Rau, T., Waldmann, E., Broedl, U.C., Demmelmair, H., Koletzko, B. et al., 2014. Walnut-enriched diet reduces fasting non-HDL-cholesterol and apolipoprotein B in healthy Caucasian subjects: A randomized controlled cross-over clinical trial. *Metabolism*, 63(3), 382–391.
99. Shimoda, H., Tanaka, J., Kikuchi, M., Fukuda, T., Ito, H., Hatano, T. and Yoshida, T., 2009. Effect of polyphenol-rich extract from walnut on diet-induced hypertriglyceridemia in mice via enhancement of fatty acid oxidation in the liver. *Journal of Agricultural and Food Chemistry*, 57(5), 1786–1792.
100. Yoshimura, Y., Nishii, S., Zaima, N., Moriyama, T. and Kawamura, Y., 2013. Ellagic acid improves hepatic steatosis and serum lipid composition through reduction of serum resistin levels and transcriptional activation of hepatic ppara in obese, diabetic KK-Ay mice. *Biochemical and Biophysical Research Communications*, 434(3), 486–491.
101. Poulose, N., Prasad, C.V., Haridas, P.N. and Anilkumar, G., 2012. Ellagic acid stimulates glucose transport in adipocytes and muscles through AMPK mediated pathway. *Journal of Diabetes & Metabolism*, 2011, 2:7.
102. Harano, Y., Yasui, K., Toyama, T., Nakajima, T., Mitsuyoshi, H., Mimani, M., Hirasawa, T., Itoh, Y. and Okanoue, T., 2006. Fenofibrate, a peroxisome proliferator-activated receptor α agonist, reduces hepatic steatosis and lipid peroxidation in fatty liver Shionogi mice with hereditary fatty liver. *Liver International*, 26(5), 613–620.
103. Kannel, W.B. and Belanger, A.J., 1991. Epidemiology of heart failure. *American Heart Journal*, 121(3), 951–957.
104. Fruchart, J.C., 2009. Peroxisome proliferator-activated receptor-alpha (PPARα): At the crossroads of obesity, diabetes and cardiovascular disease. *Atherosclerosis*, 205(1), 1–8.
105. Watanabe, M. and Ayugase, J., 2009. Chiral separation of catechins in buckwheat groats and the effects of phenolic compounds in mice subjected to restraint stress. *Journal of Agricultural and Food Chemistry*, 57(14), 6438–6442.
106. Lee, J.S., Bok, S.H., Jeon, S.M., Kim, H.J., Do, K.M., Park, Y.B. and Choi, M.S., 2010. Antihyperlipidemic effects of buckwheat leaf and flower in rats fed a high-fat diet. *Food Chemistry*, 119(1), 235–240.
107. Wang, M., Liu, J.R., Gao, J.M., Parry, J.W. and Wei, Y.M., 2009. Antioxidant activity of Tartary buckwheat bran extract and its effect on the lipid profile of hyperlipidemic rats. *Journal of Agricultural and Food Chemistry*, 57(11), 5106–5112.
108. Walldius, G. and Jungner, I., 2004. Apolipoprotein B and apolipoprotein A-I: Risk indicators of coronary heart disease and targets for lipid-modifying therapy. *Journal of Internal Medicine*, 255(2), 188–205.
109. Lamarche, B., Moorjani, S., Lupien, P.J., Cantin, B., Bernard, P.M., Dagenais, G.R. and Després, J.P., 1996. Apolipoprotein A-1 and B levels and the risk of ischemic heart disease during a five-year follow-up of men in the Québec Cardiovascular Study. *Circulation*, 94, 273–278.

110. Lee, C.C., Shen, S.R., Lai, Y.J. and Wu, S.C., 2013. Rutin and quercetin, bioactive compounds from tartary buckwheat, prevent liver inflammatory injury. *Food & Function*, 4(5), 794–802.
111. Chen, J., Ma, M., Lu, Y., Wang, L., Wu, C. and Duan, H., 2009. Rhaponticin from rhubarb rhizomes alleviates liver steatosis and improves blood glucose and lipid profiles in KK/Ay diabetic mice. *Planta medica*, 75(5), 472–477.
112. De Schrijver, E., 2005. *Fatty Acid Synthase: A Novel Target for Antineoplastic Therapy?* (Vol. 339). Leuven University Press: Belgium.
113. Li, P., Tian, W., Wang, X. and Ma, X., 2014. Inhibitory effect of desoxyrhaponticin and rhaponticin, two natural stilbene glycosides from the Tibetan nutritional food *Rheum tanguticum* Maxim. ex Balf., on fatty acid synthase and human breast cancer cells. *Food & Function*, 5(2), 251–256.
114. Janowski, B.A., Willy, P.J., Devi, T.R., Falck, J.R. and Mangelsdorf, D.J., 1996. An oxysterol signalling pathway mediated by the nuclear receptor LXR alpha. *Nature*, 383(6602), 728.
115. Gupta, S., Pandak, W.M. and Hylemon, P.B., 2002. LXRα is the dominant regulator of CYP7A1 transcription. *Biochemical and Biophysical Research Communications*, 293(1), 338–343.
116. Grempler, R., Günther, S., Steffensen, K.R., Nilsson, M., Barthel, A., Schmoll, D. and Walther, R., 2005. Evidence for an indirect transcriptional regulation of glucose-6-phosphatase gene expression by liver X receptors. *Biochemical and Biophysical Research Communications*, 338(2), 981–986.
117. Dalen, K.T., Ulven, S.M., Bamberg, K., Gustafsson, J.Å. and Nebb, H.I., 2003. Expression of the insulin-responsive glucose transporter GLUT4 in adipocytes is dependent on liver X receptor α. *Journal of Biological Chemistry*, 278(48), 48283–48291.
118. Sheng, X., Zhu, X., Zhang, Y., Cui, G., Peng, L., Lu, X. and Zang, Y.Q., 2012. Rhein protects against obesity and related metabolic disorders through liver X receptor-mediated uncoupling protein 1 upregulation in brown adipose tissue. *International Journal of Biological Sciences*, 8(10), 1375–1384.
119. Jo, S.P., Kim, J.K. and Lim, Y.H., 2014. Antihyperlipidemic effects of rhapontin and rhapontigenin from *Rheum undulatum* in rats fed a high-cholesterol diet. *Planta medica*, 80(13), 1067–1071.
120. Garcia-Diez, F. and Garcia-Mediavilla, V., 1996. Pectin feeding influences fecal bile acid excretion, hepatic bile acid and cholesterol synthesis and serum cholesterol in rats. *The Journal of Nutrition*, 126(7), 1766.
121. Shepherd, J., Packard, C.J., Bicker, S., Lawrie, T.V. and Morgan, H.G., 1980. Cholestyramine promotes receptor-mediated low-density-lipoprotein catabolism. *New England Journal of Medicine*, 302(22), 1219–1222.
122. Kritchevsky, D. and Story, J.A., 1978. Fiber, hypercholesteremia, and atherosclerosis. *Lipids*, 13(5), 366–369.
123. Arguello, G., Balboa, E., Arrese, M. and Zanlungo, S., 2015. Recent insights on the role of cholesterol in non-alcoholic fatty liver disease. *Biochimica et Biophysica Acta (BBA)-Molecular Basis of Disease*, 1852(9), 1765–1778.
124. Qureshi, A.A., Lehmann, J.W. and Peterson, D.M., 1996. Amaranth and its oil inhibit cholesterol biosynthesis in 6-week-old female chickens. *The Journal of Nutrition*, 126(8), 1972.
125. Park, S.H., Ko, S.K., Choi, J.G. and Chung, S.H., 2006. *Salicornia herbacea* prevents high fat diet-induced hyperglycemia and hyperlipidemia in ICR mice. *Archives of Pharmacal Research*, 29(3), 256–264.
126. Pil Hwang, Y., Gyun Kim, H., Choi, J.H., Truong Do, M., Tran, T.P., Chun, H.K., Chung, Y.C., Jeong, T.C. and Jeong, H.G., 2013. 3-Caffeoyl, 4-dihydrocaffeoylquinic acid from *Salicornia herbacea* attenuates high glucose-induced hepatic lipogenesis in human HepG2 cells through activation of the liver kinase B1 and silent information regulator T1/AMPK-dependent pathway. *Molecular Nutrition & Food Research*, 57(3), 471–482.
127. Li, Y., Xu, S., Mihaylova, M.M., Zheng, B., Hou, X., Jiang, B., Park, O. et al., 2011. AMPK phosphorylates and inhibits SREBP activity to attenuate hepatic steatosis and atherosclerosis in diet-induced insulin-resistant mice. *Cell Metabolism*, 13(4), 376–388.
128. Lage, R., Diéguez, C., Vidal-Puig, A. and López, M., 2008. AMPK: A metabolic gauge regulating whole-body energy homeostasis. *Trends in Molecular Medicine*, 14(12), 539–549.
129. Ye, C., Lin, Y., Su, J., Song, X. and Zhang, H., 1998. Purine alkaloids in *Camellia assamica* var. kucha Chang et Wang. *Acta Scientiarum Naturalium Universitatis Sunyatseni*, 38(5), 82–86.
130. Sawanishi, H., Suzuki, H., Yamamoto, S., Waki, Y., Kasugai, S., Ohya, K., Suzuki, N., Miyamoto, K.I. and Takagi, K., 1997. Selective inhibitors of cyclic AMP-specific phosphodiesterase: Heterocycle-condensed purines. *Journal of Medicinal Chemistry*, 40(20), 3248–3253.
131. Trauner, M., Meier, P.J. and Boyer, J.L., 1998. Molecular pathogenesis of cholestasis. *New England Journal of Medicine*, 339(17), 1217–1227.

132. Kortz, W.J., Meyers, W.C., Schirmer, B.D. and Jones, R.S., 1984. Effects of dibutyryl cyclic AMP and theophylline on biliary cholesterol secretion. *Journal of Surgical Research*, 36(1), 62–70.
133. Obel, L.F., Müller, M.S., Walls, A.B., Sickmann, H.M., Bak, L.K., Waagepetersen, H.S. and Schousboe, A., 2012. Brain glycogen—new perspectives on its metabolic function and regulation at the subcellular level. *Frontiers in Neuroenergetics*, 4, 3.
134. Hvidberg, A., Rasmussen, M.H., Christensen, N.J. and Hilsted, J., 1994. Theophylline enhances glucose recovery after hypoglycemia in healthy man and in type I diabetic patients. *Metabolism*, 43(6), 776–781.
135. Emerling, B.M., Weinberg, F., Snyder, C., Burgess, Z., Mutlu, G.M., Viollet, B., Budinger, G.S. and Chandel, N.S., 2009. Hypoxic activation of AMPK is dependent on mitochondrial ROS but independent of an increase in AMP/ATP ratio. *Free Radical Biology and Medicine*, 46(10), 1386–1391.
136. Fu, D., Wakabayashi, Y., Lippincott-Schwartz, J. and Arias, I.M., 2011. Bile acid stimulates hepatocyte polarization through a cAMP-Epac-MEK-LKB1-AMPK pathway. *Proceedings of the National Academy of Sciences*, 108(4), 1403–1408.
137. Steinberg, G.R. and Kemp, B.E., 2009. AMPK in health and disease. *Physiological Reviews*, 89(3), 1025–1078.
138. Eteng, M.U. and Ettarh, R.R., 2000. Comparative effects of theobromine and cocoa extract on lipid profile in rats. *Nutrition Research*, 20(10), 1513–1517.
139. Fears, R., 1978. The hypercholesterolaemic effect of caffeine in rats fed on diets with and without supplementary cholesterol. *British Journal of Nutrition*, 39(2), 363–374.
140. Sachs, M. and Förster, H., 1984. Effect of caffeine on various metabolic parameters in vivo. *Zeitschrift fur Ernahrungswissenschaft*, 23(3), 181–205.
141. Curb, J.D., Reed, D.M., Kautz, J.A. and Yano, K., 1986. Coffee, caffeine, and serum cholesterol in Japanese men in Hawaii. *American Journal of Epidemiology*, 123(4), 648–655.
142. Li, W.X., Li, Y.F., Zhai, Y.J., Chen, W.M., Kurihara, H. and He, R.R., 2013. Theacrine, a purine alkaloid obtained from *Camellia assamica* var. kucha, attenuates restraint stress-provoked liver damage in mice. *Journal of Agricultural and Food Chemistry*, 61(26), 6328–6335.
143. Zaidi, N., Swinnen, J.V. and Smans, K., 2012. ATP-citrate lyase: A key player in cancer metabolism. *Cancer Research*, 72(15), 3709–3714.
144. Watson, J.A., Fang, M. and Lowenstein, J.M., 1969. Tricarballylate and hydroxycitrate: Substrate and inhibitor of ATP: Citrate oxaloacetate lyase. *Archives of Biochemistry and Biophysics*, 135, 209–217.
145. Rao, R.N. and Sakariah, K.K., 1988. Lipid-lowering and antiobesity effect of (–) hydroxycitric acid. *Nutrition Research*, 8(2), 209–212.
146. Thom, E., 1996. Hydroxycitrate (HCA) in the treatment of obesity. *International Journal of Obesity*, 20(suppl. 4), 48.
147. Tuansulong, K.A., Hutadilok-Towatana, N., Mahabusarakam, W., Pinkaew, D. and Fujise, K., 2011. Morelloflavone from *Garcinia dulcis* as a novel biflavonoid inhibitor of HMG-CoA reductase. *Phytotherapy Research*, 25(3), 424–428.
148. Deachathai, S., Mahabusarakam, W., Phongpaichit, S., Taylor, W.C., Zhang, Y.J. and Yang, C.R. (2006). Phenolic compounds from the flowers of *Garcinia dulcis. Phytochemistry*, 67(5), 464–469.
149. Herath, K., Jayasuriya, H., Ondeyka, J.G., Guan, Z., Borris, R.P., Stijfhoorn, E., Stevenson, D. et al., 2005. Guttiferone I, a new prenylated benzophenone from *Garcinia humilis* as a liver X receptor ligand. *Journal of Natural Products*, 68(4), 617–619.
150. Choi, Y.H., Bae, J.K., Chae, H.S., Kim, Y.M., Sreymom, Y., Han, L., Jang, H.Y. and Chin, Y.W., 2015. α-Mangostin regulates hepatic steatosis and obesity through SirT1-AMPK and PPARγ pathways in high-fat diet-induced obese mice. *Journal of Agricultural and Food Chemistry*, 63(38), 8399–8406.
151. Cantó, C. and Auwerx, J., 2012. Targeting sirtuin 1 to improve metabolism: All you need is NAD(+)? *Pharmacological Reviews*, 64, 166–187.
152. Han, S.Y., You, B.H., Kim, Y.C., Chin, Y.W. and Choi, Y.H., 2015. Dose-independent ADME properties and tentative identification of metabolites of α-mangostin from *Garcinia mangostana* in mice by automated microsampling and UPLC-MS/MS methods. *PLoS ONE*, 10(7), e0131587.
153. Jiang, H.Z., Quan, X.F., Tian, W.X., Hu, J.M., Wang, P.C., Huang, S.Z., Cheng, Z.Q. et al., 2010. Fatty acid synthase inhibitors of phenolic constituents isolated from *Garcinia mangostana. Bioorganic & Medicinal Chemistry Letters*, 20(20), 6045–6047.
154. Matsuura, N., Kanae, G.A.M.O., Miyachi, H., Iinuma, M., Kawada, T., Takahashi, N. and Yukihiro, A.K.A.O., 2013. γ-Mangostin from *Garcinia mangostana* pericarps as a dual agonist that activates both PPARα and PPARδ. *Bioscience, Biotechnology, and Biochemistry*, 77(12), 2430–2435.
155. Sweetser, D.A., Heuckeroth, R.O. and Gordon, J.I., 1987. The metabolic significance of mammalian fatty-acid-binding proteins: Abundant proteins in search of a function. *Annual Review of Nutrition*, 7(1), 337–359.

156. Zou, Y., Lu, Y. and Wei, D., 2005. Hypocholesterolemic effects of a flavonoid-rich extract of *Hypericum perforatum* L. in rats fed a cholesterol-rich diet. *Journal of Agricultural and Food Chemistry*, 53(7), 2462–2466.
157. Kliewer, S.A. and Willson, T.M., 2002. Regulation of xenobiotic and bile acid metabolism by the nuclear pregnane X receptor. *Journal of Lipid Research*, 43(3), 359–364.
158. Moore, L.B., Goodwin, B., Jones, S.A., Wisely, G.B., Serabjit-Singh, C.J., Willson, T.M., Collins, J.L. and Kliewer, S.A., 2000. St. John's wort induces hepatic drug metabolism through activation of the pregnane X receptor. *Proceedings of the National Academy of Sciences*, 97(13), 7500–7502.
159. Biheimer, D.W., Grundy, S.M., Brown, M.S. and Goldstein, J.L., 1983. Mevilonin and cholestepol stimulate receptor-mediated clearance of low density lipoprotein from plasma in familial hypercholesterolemia heterozygotes. *Proceedings of the National Academy of Sciences USA*, 80, 4124–4128.
160. Durkar, A.M., Patil, R.R. and Naik, S.R., 2014. Hypolipidemic and antioxidant activity of ethanolic extract of *Symplocos racemosa* Roxb. in hyperlipidemic rats: an evidence of participation of oxidative stress in hyperlipidemia. *Indian Journal of Experimental Biology*, 52(1), 36–45.
161. Na, M.K., Yang, S., He, L., Oh, H., Kim, B.S., Oh, W.K., Kim, B.Y. and Ahn, J.S., 2006. Inhibition of protein tyrosine phosphatase 1B by ursane-type triterpenes isolated from *Symplocos paniculata*. *Planta medica*, 72(3), 261–263.
162. Rao, V.S., de Melo, C.L., Queiroz, M.G.R., Lemos, T.L., Menezes, D.B., Melo, T.S. and Santos, F.A., 2011. Ursolic acid, a pentacyclic triterpene from Sambucus australis, prevents abdominal adiposity in mice fed a high-fat diet. *Journal of Medicinal Food*, 14(11), 1375–1382.
163. Matsumoto, K., Watanabe, Y., Ohya, M.A. and Yokoyama, S.I., 2006. Young persimmon fruits prevent the rise in plasma lipids in a diet-induced murine obesity model. *Biological and Pharmaceutical Bulletin*, 29(12), 2532–2535.
164. Tammela, T., Enholm, B., Alitalo, K. and Paavonen, K., 2005. The biology of vascular endothelial growth factors. *Cardiovascular Research*, 65, 550–563.
165. Desvergne, B., Michalik, L. and Wahli, W., 2006. Transcriptional regulation of metabolism. *Physiological Reviews*, 86(2), 465–514.
166. Nakatsubo, F., Enokita, K., Murakami, K., Yonemori, K., Sugiura, A., Utsunomiya, N. and Subhadrabandhu, S., 2002. Chemical structures of the condensed tannins in the fruits ofDiospyros species. *Journal of Wood Science*, 48(5), 414–418.
167. Zou, B., Li, C.M., Chen, J.Y., Dong, X.Q., Zhang, Y. and Du, J., 2012. High molecular weight persimmon tannin is a potent hypolipidemic in high-cholesterol diet fed rats. *Food Research International*, 48(2), 970–977.
168. Thuong, P.T., Lee, C.H., Dao, T.T., Nguyen, P.H., Kim, W.G., Lee, S.J. and Oh, W.K., 2008. Triterpenoids from the leaves of *Diospyros kaki* (persimmon) and their inhibitory effects on protein tyrosine phosphatase 1B. *Journal of Natural Products*, 71(10), 1775–1778.
169. Dewanjee, S., Das, A.K., Sahu, R. and Gangopadhyay, M., 2009. Antidiabetic activity of *Diospyros peregrina* fruit: effect on hyperglycemia, hyperlipidemia and augmented oxidative stress in experimental type 2 diabetes. *Food and Chemical Toxicology*, 47(10), 2679–2685.
170. Bremer, A.A., 2012. Insulin resistance in pediatric disease. *Pediatric Annals*, 41(2), e18–e24.
171. Grimsby, J., Sarabu, R., Corbett, W.L., Haynes, N.E., Bizzarro, F.T., Coffey, J.W., Guertin, K.R. et al., 2003. Allosteric activators of glucokinase: Potential role in diabetes therapy. *Science*, 301(5631), 370–373.
172. Larter, C.Z. and Farrell, G.C., 2006. Insulin resistance, adiponectin, cytokines in NASH: Which is the best target to treat? *Journal of Hepatology*, 44(2), 253–261.
173. Poduri, A., Rateri, D.L., Saha, S.K., Saha, S. and Daugherty, A., 2013. *Citrullus lanatus* "sentinel"(watermelon) extract reduces atherosclerosis in LDL receptor-deficient mice. *The Journal of Nutritional Biochemistry*, 24(5), 882–886.
174. Hong, M.Y., Hartig, N., Kaufman, K., Hooshmand, S., Figueroa, A. and Kern, M., 2015. Watermelon consumption improves inflammation and antioxidant capacity in rats fed an atherogenic diet. *Nutrition Research*, 35(3), 251–258.
175. Jegatheesan, P. and De Bandt, J.P., 2016. Hepatic steatosis: a role for citrulline. *Current Opinion in Clinical Nutrition & Metabolic Care*, 19(5), 360–365.
176. Sellmann, C., Jin, C.J., Engstler, A.J., De Bandt, J.P. and Bergheim, I., 2016. Oral citrulline supplementation protects female mice from the development of non-alcoholic fatty liver disease (NAFLD). *European Journal of Nutrition*, 1–9.
177. Jegatheesan, P., Beutheu, S., Freese, K., Waligora-Dupriet, A.J., Nubret, E., Butel, M.J., Bergheim, I. and De Bandt, J.P., 2016. Preventive effects of citrulline on Western diet-induced non-alcoholic fatty liver disease in rats. *The British Journal of Nutrition*, 116, 1–13.

178. Makni, M., Fetoui, H., Gargouri, N.K., Garoui, E.M., Jaber, H., Makni, J., Boudawara, T. and Zeghal, N., 2008. Hypolipidemic and hepatoprotective effects of flax and pumpkin seed mixture rich in ω-3 and ω-6 fatty acids in hypercholesterolemic rats. *Food and Chemical Toxicology*, 46(12), 3714–3720.
179. Yoshinari, O., Sato, H. and Igarashi, K., 2009. Anti-diabetic effects of pumpkin and its components, trigonelline and nicotinic acid, on Goto-Kakizaki rats. *Bioscience, biotechnology, and biochemistry*, 73(5), 1033–1041.
180. Zhang, D.F., Zhang, F., Zhang, J., Zhang, R.M. and Li, R., 2015. Protection effect of trigonelline on liver of rats with non-alcoholic fatty liver diseases. *Asian Pacific Journal of Tropical Medicine*, 8(8), 651–654.
181. Clark, J.M. and Diehl, A.M., 2002. Hepatic steatosis and type 2 diabetes mellitus. *Current Diabetes Reports*, 2(3), 210–215.
182. Lukasova, M., Hanson, J., Tunaru, S. and Offermanns, S. (2011). Nicotinic acid (niacin): New lipid-independent mechanisms of action and therapeutic potentials. *Trends in Pharmacological Sciences*, 32(12), 700–707.
183. Creider, J.C., Hegele, R.A. and Joy, T.R., 2012. Niacin: Another look at an underutilized lipid-lowering medication. *Nature Reviews Endocrinology*, 8(9), 517–528.
184. Kamanna, V.S. and Kashyap, M.L., 2008. Mechanism of action of niacin. *The American Journal of Cardiology*, 101(8), S20–S26.
185. Bender, D.A., 2003. *Nutritional Biochemistry of the Vitamins*. Cambridge: Cambridge University Press.
186. Ghule, B.V., Ghante, M.H., Saoji, A.N. and Yeole, P.G., 2009. Antihyperlipidemic effect of the methanolic extract from *Lagenaria siceraria* Stand. fruit in hyperlipidemic rats. *Journal of Ethnopharmacology*, 124(2), 333–337.
187. Nadeem, S., Dhore, P., Quazi, M., Pawar, S. and Raj, N., 2012. *Lagenaria siceraria* fruit extract ameliorate fat amassment and serum TNF–in high–fat diet–induced obese rats. *Asian Pacific Journal of Tropical Medicine*, 5(9), 698–702.
188. Enslin, P.R., 1954. Bitter principles of the cucurbitaceae. I.—Observations on the chemistry of cucurbitacin A. *Journal of the Science of Food and Agriculture*, 5(9), 410–416.
189. Agil, A., Miró, M., Jimenez, J., Aneiros, J., Caracuel, M.D., García-Granados, A. and Navarro, M.C., 1999. Isolation of an anti-hepatotoxic principle from the juice of *Ecballium elaterium*. *Planta medica*, 65(7), 673–675.
190. Yang, S.J., Chang, Y.Q., Zheng, L.H., Wei, Z.R., Qu, H.G. and Cao, S.G., 2005. Protective effects of cucurbitacin B on the acute liver injury induced by CCl4. *Food Science*, 26(9), 524–526.
191. Klover, P.J., Zimmers, T.A., Koniaris, L.G. and Mooney, R.A., 2003. Chronic exposure to interleukin-6 causes hepatic insulin resistance in mice. *Diabetes*, 52(11), 2784–2789.
192. Esmaili, S., Xu, A. and George, J., 2014. The multifaceted and controversial immunometabolic actions of adiponectin. *Trends in Endocrinology & Metabolism*, 25(9), 444–451.
193. Seo, C.R., Yang, D.K., Song, N.J., Yun, U.J., Gwon, A.R., Jo, D.G., Cho, J.Y. et al., 2014. Cucurbitacin B and cucurbitacin I suppress adipocyte differentiation through inhibition of STAT3 signaling. *Food and Chemical Toxicology*, 64, 217–224.
194. Verma, A. and Jaiswal, S., 2015. Bottle gourd (*Lagenaria siceraria*) juice poisoning. *World Journal of Emergency Medicine*, 6(4), 308.
195. Chait, A., Bierman, E.L. and Albers, J.J., 1979. Low-density lipoprotein receptor activity in cultured human skin fibroblasts. Mechanism of insulin-induced stimulation. *Journal of Clinical Investigation*, 64(5), 1309.
196. Harinantenaina, L., Tanaka, M., Takaoka, S., Oda, M., Mogami, O., Uchida, M. and Asakawa, Y., 2006. *Momordica charantia* constituents and antidiabetic screening of the isolated major compounds. *Chemical and Pharmaceutical Bulletin*, 54(7), 1017–1021.
197. Cheng, H.L., Kuo, C.Y., Liao, Y.W. and Lin, C.C., 2012. EMCD, a hypoglycemic triterpene isolated from *Momordica charantia* wild variant, attenuates TNF-α-induced inflammation in FL83B cells in an AMP-activated protein kinase-independent manner. *European Journal of Pharmacology*, 689(1), 241–248.
198. Xu, J., Cao, K., Li, Y., Zou, X., Chen, C., Szeto, I.M.Y., Dong, Z. et al., 2014. Bitter gourd inhibits the development of obesity-associated fatty liver in C57BL/6 mice fed a high-fat diet. *The Journal of Nutrition*, 144(4), 475–483.
199. Watanabe, M., Houten, S.M., Wang, L., Moschetta, A., Mangelsdorf, D.J., Heyman, R.A., Moore, D.D. and Auwerx, J., 2004. Bile acids lower triglyceride levels via a pathway involving FXR, SHP, and SREBP-1c. *The Journal of Clinical Investigation*, 113(10), 1408–1418.

200. Shih, C.C., Shlau, M.T., Lin, C.H. and Wu, J.B., 2014. *Momordica charantia* ameliorates insulin resistance and dyslipidemia with altered hepatic glucose production and fatty acid synthesis and AMPK phosphorylation in high-fat-fed mice. *Phytotherapy Research*, 28(3), 363–371.
201. Teodoro, J.S., Rolo, A.P. and Palmeira, C.M., 2011. Hepatic FXR: Key regulator of whole-body energy metabolism. *Trends in Endocrinology & Metabolism*, 22(11), 458–466.
202. Takahashi, H., Hara, H., Goto, T., Kamakari, K., Wataru, N., Mohri, S., Takahashi, N., Suzuki, H., Shibata, D. and Kawada, T., 2015. 13-Oxo-9(Z),11(E),15(Z)-octadecatrienoic acid activates peroxisome proliferator-activated receptor γ in adipocytes. *Lipids*, 50(1), 3–12.
203. Iwabu, M., Yamauchi, T., Okada-Iwabu, M., Sato, K., Nakagawa, T., Funata, M., Yamaguchi, M. et al., 2010. Adiponectin and AdipoR1 regulate PGC-1 [agr] and mitochondria by Ca2+ and AMPK/SIRT1. *Nature*, 464(7293), 1313–1319.
204. Basch, E., Gabardi, S. and Ulbricht, C., 2003. Bitter melon (Momordica charantia): A review of efficacy and safety. *American Journal of Health and Systemic Pharmacology*, 65, 356–359.
205. Wu, C.H., Ou, T.T., Chang, C.H., Chang, X.Z., Yang, M.Y. and Wang, C.J., 2014. The polyphenol extract from *Sechium edule* shoots inhibits lipogenesis and stimulates lipolysis via activation of AMPK signals in HepG2 cells. *Journal of Agricultural and Food Chemistry*, 62(3), 750–759.
206. Fuchs, M., 2003. III. Regulation of bile acid synthesis: Past progress and future challenges. *American Journal of Physiology-Gastrointestinal and Liver Physiology*, 284(4), G551–G557.
207. Azahar, M.A., Al-Naqeb, G., Hasan, M. and Adam, A., 2012. Hypoglycemic effect of *Octomeles sumatrana* aqueous extract in streptozotocin–induced diabetic rats and its molecular mechanisms. *Asian Pacific Journal of Tropical Medicine*, 5(11), 875–881.
208. Melega, S., Canistro, D., De Nicola, G.R., Lazzeri, L., Sapone, A. and Paolini, M., 2013. Protective effect of Tuscan black cabbage sprout extract against serum lipid increase and perturbations of liver antioxidant and detoxifying enzymes in rats fed a high-fat diet. *British Journal of Nutrition*, 110(6), 988–997.
209. Stoewsand, G.S., 1995. Bioactive organosulfur phytochemicals in *Brassica oleracea* vegetables—A review. *Food and Chemical Toxicology*, 33(6), 537–543.
210. Ernst, I.M., Wagner, A.E., Schuemann, C., Storm, N., Höppner, W., Döring, F., Stocker, A. and Rimbach, G., 2011. Allyl-, butyl-and phenylethyl-isothiocyanate activate Nrf2 in cultured fibroblasts. *Pharmacological Research*, 63(3), 233–240.
211. Braun, S., Hanselmann, C., Gassmann, M.G., auf dem Keller, U., Born-Berclaz, C., Chan, K., Kan, Y.W. and Werner, S., 2002. Nrf2 transcription factor, a novel target of keratinocyte growth factor action which regulates gene expression and inflammation in the healing skin wound. *Molecular and Cellular Biology*, 22(15), 5492–5505.
212. Sahu, R.P., Zhang, R., Batra, S., Shi, Y. and Srivastava, S.K., 2009. Benzyl isothiocyanate-mediated generation of reactive oxygen species causes cell cycle arrest and induces apoptosis via activation of MAPK in human pancreatic cancer cells. *Carcinogenesis*, 30(10), 1744–1753.
213. Kim, S.Y., Sun, Y.O.O.N., Kwon, S.M., Park, K.S. and Lee-Kim, Y.C., 2008. Kale juice improves coronary artery disease risk factors in hypercholesterolemic Men11. This research was supported by the Brain Korea 21 Project from the Korea Research Foundation. *Biomedical and Environmental Sciences*, 21(2), 91–97.
214. Bahadoran, Z., Mirmiran, P., Hosseinpanah, F., Rajab, A., Asghari, G. and Azizi, F., 2012. Broccoli sprouts powder could improve serum triglyceride and oxidized LDL/Low-density lipoprotein–cholesterol ratio in type 2 diabetic patients: A randomized double-blind placebo-controlled clinical trial. *Diabetes Research and Clinical Practice*, 96(3), 348–354.
215. Jung, U.J., Baek, N.I., Chung, H.G., Bang, M.H., Jeong, T.S., Lee, K.T., Kang, Y.J. et al., 2008. Effects of the ethanol extract of the roots of Brassica rapa on glucose and lipid metabolism in C57BL/KsJ-db/db mice. *Clinical Nutrition*, 27(1), 158–167.
216. Padilla, G., Cartea, M.E., Velasco, P., de Haro, A. and Ordás, A., 2007. Variation of glucosinolates in vegetable crops of Brassica rapa. *Phytochemistry*, 68(4), 536–545.
217. Choi, K.M., Lee, Y.S., Kim, W., Kim, S.J., Shin, K.O., Yu, J.Y., Lee, M.K. et al., 2014. Sulforaphane attenuates obesity by inhibiting adipogenesis and activating the AMPK pathway in obese mice. *The Journal of Nutritional Biochemistry*, 25(2), 201–207.
218. Karvonen, H.M., Aro, A., Tapola, N.S., Salminen, I., Uusitupa, M.I. and Sarkkinen, E.S., 2002. Effect of alpha-linolenic acid-rich *Camelina sativa* oil on serum fatty acid composition and serum lipids in hypercholesterolemic subjects. *Metabolism*, 51(10), 1253–1260.
219. Omidi, H., Tahmasebi, Z., Badi, H.A.N., Torabi, H. and Miransari, M., 2010. Fatty acid composition of canola (Brassica napus L.), as affected by agronomical, genotypic and environmental parameters. *Comptes Rendus Biologies*, 333(3), 248–254.

220. Fuster, V., Topol, E.J. and Nabel, E.G. (Eds.), 2005. *Atherothrombosis and Coronary Artery Disease*. Philadelphia, PA: Lippincott Williams & Wilkins.
221. Lang, C.A. and Davis, R.A., 1990. Fish oil fatty acids impair VLDL assembly and/or secretion by cultured rat hepatocytes. *Journal of Lipid Research*, 31(11), 2079–2086.
222. Maiyoh, G.K., Kuh, J.E., Casaschi, A. and Theriault, A.G., 2007. Cruciferous indole-3-carbinol inhibits apolipoprotein B secretion in HepG2 cells. *Journal of Nutrition*, 137(10), 2185–2189.
223. Yazdanparast, R., Bahramikia, S. and Ardestani, A., 2008. Nasturtium officinale reduces oxidative stress and enhances antioxidant capacity in hypercholesterolaemic rats. *Chemico-Biological Interactions*, 172(3), 176–184.
224. Lugasi, A., Blázovics, A., Hagymási, K., Kocsis, I. and Kéry, Á., 2005. Antioxidant effect of squeezed juice from black radish (*Raphanus sativus* L. var. niger) in alimentary hyperlipidaemia in rats. *Phytotherapy Research*, 19(7), 587–591.
225. Barillari, J., Cervellati, R., Costa, S., Guerra, M.C., Speroni, E., Utan, A. and Iori, R., 2006. Antioxidant and choleretic properties of *Raphanus sativus* L. sprout (Kaiware Daikon) extract. *Journal of Agricultural and Food Chemistry*, 54(26), 9773–9778.
226. Castro-Torres, I.G., Naranjo-Rodríguez, E.B., Domínguez-Ortíz, M.Á., Gallegos-Estudillo, J. and Saavedra-Vélez, M.V., 2012. Antilithiasic and hypolipidaemic effects of *Raphanus sativus* L. var. niger on mice fed witha lithogenic diet. *Journal of Biomedicine and Biotechnology*, 2012, 161205.
227. Manley, S. and Ding, W., 2015. Role of farnesoid X receptor and bile acids in alcoholic liver disease. *Acta Pharmaceutica Sinica B*, 5(2), 158–167.
228. Scott, L.J. and Spencer, C.M., 2000. Miglitol. *Drugs*, 59(3), 521–549.
229. Tsuduki, T., Nakamura, Y., Honma, T., Nakagawa, K., Kimura, T., Ikeda, I. and Miyazawa, T., 2009. Intake of 1-deoxynojirimycin suppresses lipid accumulation through activation of the β-oxidation system in rat liver. *Journal of Agricultural and Food Chemistry*, 57(22), 11024–11029.
230. Tsuduki, T., Kikuchi, I., Kimura, T., Nakagawa, K. and Miyazawa, T., 2013. Intake of mulberry 1-deoxynojirimycin prevents diet-induced obesity through increases in adiponectin in mice. *Food Chemistry*, 139(1), 16–23.
231. Lee, W.J., Kim, M., Park, H.S., Kim, H.S., Jeon, M.J., Oh, K.S., Koh, E.H. et al., 2006. AMPK activation increases fatty acid oxidation in skeletal muscle by activating PPARα and PGC-1. *Biochemical and Biophysical Research Communications*, 340(1), 291–295.
232. Bnouham, M., Merhfour, F.Z., Ziyyat, A., Aziz, M., Legssyer, A. and Mekhfi, H., 2010. Antidiabetic effect of some medicinal plants of Oriental Morocco in neonatal non-insulin-dependent diabetes mellitus rats. *Human & Experimental Toxicology*, 29(10), 865–871.
233. Daher, C.F., Baroody, K.G. and Baroody, G.M., 2006. Effect of *Urtica dioica* extract intake upon blood lipid profile in the rats. *Fitoterapia*, 77(3), 183–188.
234. Kavtaradze, N.S., Alaniya, M.D. and Aneli, J.N., 2001. Chemical components of *Urtica dioica* growing in Georgia. *Chemistry of Natural Compounds*, 37(3), 287–287.
235. Pinelli, P., Ieri, F., Vignolini, P., Bacci, L., Baronti, S. and Romani, A., 2008. Extraction and HPLC analysis of phenolic compounds in leaves, stalks, and textile fibers of *Urtica dioica* L. *Journal of Agricultural and Food Chemistry*, 56(19), 9127–9132.
236. Budzianowski, J., 1991. Caffeic acid esters from *Urtica dioica* and U. urens. *Planta medica*, 57(5), 507.
237. Li, X.C., Joshi, A.S., ElSohly, H.N., Khan, S.I., Jacob, M.R., Zhang, Z., Khan, I.A. et al., 2002. Fatty acid synthase inhibitors from plants: isolation, structure elucidation, and SAR studies. *Journal of Natural Products*, 65(12), 1909–1914.
238. Adeneye, A.A., 2012. The leaf and seed aqueous extract of *Phyllanthus amarus* improves insulin resistance diabetes in experimental animal studies. *Journal of Ethnopharmacology*, 144(3), 705–711.
239. Chang, C.C., Lien, Y.C., Liu, K.C.C. and Lee, S.S., 2003. Lignans from *Phyllanthus urinaria*. *Phytochemistry*, 63(7), 825–833.
240. Jagtap, S., Khare, P., Mangal, P., Bishnoi, M., Kondepudi, K.K. and Bhutani, K.K., 2016. Protective effects of phyllanthin, a lignan from *Phyllanthus amarus*, against progression of high fat diet induced metabolic disturbances in mice. *RSC Advances*, 6, 58343–58353.
241. Fast, D.G. and Vance, D.E., 1995. Nascent VLDL phospholipid composition is altered when phosphatidylcholine biosynthesis is inhibited: Evidence for a novel mechanism that regulates VLDL secretion. *Biochimica et Biophysica Acta (BBA)-Lipids and Lipid Metabolism*, 1258(2), 159–168.
242. Shen, B., Yu, J., Wang, S., Chu, E.S., Wong, V.W., Zhou, X., Lin, G., Sung, J.J. and Chan, H.L., 2008. *Phyllanthus urinaria* ameliorates the severity of nutritional steatohepatitis both in vitro and in vivo. *Hepatology*, 47(2), 473–483.

243. Tomczyk, M. and Latté, K.P., 2009. Potentilla—A review of its phytochemical and pharmacological profile. *Journal of Ethnopharmacology*, 122(2), 184–204.
244. Hsu, C.L. and Yen, G.C., 2007. Effect of gallic acid on high fat diet-induced dyslipidaemia, hepatosteatosis and oxidative stress in rats. *British Journal of Nutrition*, 98(4), 727–735.
245. Klok, M.D., Jakobsdottir, S. and Drent, M.L., 2007. The role of leptin and ghrelin in the regulation of food intake and body weight in humans: A review. *Obesity Review*, 8, 21–34.
246. Zhao, J., Zheng, H., Liu, Y., Lin, J., Zhong, X., Xu, W., Hong, Z. and Peng, J., 2013. Anti-inflammatory effects of total alkaloids from *Rubus aleaefolius* Poir. on non-alcoholic fatty liver disease through regulation of the NF-κB pathway. *International Journal of Molecular Medicine*, 31(4), 931–937.
247. Lin, J., Zhao, J., Li, T., Zhou, J., Hu, J. and Hong, Z., 2011. Hepatoprotection in a rat model of acute liver damage through inhibition of CY2E1 activity by total alkaloids extracted from *Rubus alceifolius* Poir. *International Journal of Toxicology*, 30(2), 237–243.
248. Ludovico, A., Gabriella, A., Raffaele, C., Natasa, M. and Francesco, C., 2011. Milk thistle for treatment of nonalcoholic fatty liver disease. *Hepatitis Monthly*, 2011(3), 173–177.
249. Ramgopal, M., Kruthika, B.S., Surekha, D. and Meriga, B., 2014. *Terminalia paniculata* bark extract attenuates non-alcoholic fatty liver via down regulation of fatty acid synthase in high fat diet-fed obese rats. *Lipids in Health and Disease*, 13, 58.
250. Ramachandran, S., Rajasekaran, A. and Adhirajan, N., 2013. In vivo and in vitro antidiabetic activity of *Terminalia paniculata* bark: an evaluation of possible phytoconstituents and mechanisms for blood glucose control in diabetes. *ISRN Pharmacology*, 2013, 1–10.
251. Kannur, D.M., Hukkeri, V.I. and Akki, K.S., 2006. Antidiabetic activity of *Caesalpinia bonducella* seed extracts in rats. *Fitoterapia*, 77(7), 546–549.
252. Chakrabarti, S., Biswas, T.K., Rokeya, B., Ali, L., Mosihuzzaman, M., Nahar, N., Khan, A.A. and Mukherjee, B., 2003. Advanced studies on the hypoglycemic effect of *Caesalpinia bonducella* F. in type 1 and 2 diabetes in Long Evans rats. *Journal of Ethnopharmacology*, 84(1), 41–46.
253. Poretsky, L., ed, 2010. *Principles of diabetes mellitus*. 21. Springer, New York.
254. Chakrabarti, S., Biswas, T.K., Seal, T., Rokeya, B., Ali, L., Khan, A.A., Nahar, N., Mosihuzzaman, M. and Mukherjee, B., 2005. Antidiabetic activity of *Caesalpinia bonducella* F. in chronic type 2 diabetic model in Long-Evans rats and evaluation of insulin secretagogue property of its fractions on isolated islets. *Journal of Ethnopharmacology*, 97(1), 117–122.
255. Peter, S., Tinto, W.F., McLean, S., Reynolds, W.F. and Yu, M., 1998. *Cassane diterpenes* from *Caesalpinia bonducella*. *Phytochemistry*, 47(6), 1153–1155.
256. Ruotolo, G., Parlavecchia, M., Taskinen, M.R., Galimberti, G., Zoppo, A., Le, N.A., Ragogna, F., Micossi, P. and Pozza, G., 1994. Normalization of lipoprotein composition by intraperitoneal insulin in IDDM: Role of increased hepatic lipase activity. *Diabetes Care*, 17(1), 6–12.
257. Rader, D.J., Alexander, E.T., Weibel, G.L., Bilheimer, J. and Rothblat, G.H., 2009. The role of reverse cholesterol transport in animals and humans ans relationship to atherosclerosis. *The Journal of Lipid Research*, 50(Suppl.), S189–S194.
258. Zheng, T., Shu, G., Yang, Z., Mo, S., Zhao, Y. and Mei, Z., 2012. Antidiabetic effect of total saponins from *Entada phaseoloides* (L.) Merr. in type 2 diabetic rats. *Journal of Ethnopharmacology*, 139(3), 814–821.
259. Liu, X., Yuan, H., Niu, Y., Niu, W. and Fu, L., 2012. The role of AMPK/mTOR/S6K1 signaling axis in mediating the physiological process of exercise-induced insulin sensitization in skeletal muscle of C57BL/6 mice. *Biochimica et Biophysica Acta (BBA)-Molecular Basis of Disease*, 1822(11), 1716–1726.
260. Xiong, H., Zheng, Y., Yang, G., Wang, H. and Mei, Z., 2015. *Triterpene saponins* with anti-inflammatory activity from the stems of *Entada phaseoloides*. *Fitoterapia*, 103, 33–45.
261. Rebhun, J.F., Glynn, K.M. and Missler, S.R., 2015. Identification of glabridin as a bioactive compound in licorice (*Glycyrrhiza glabra* L.) extract that activates human peroxisome proliferator-activated receptor gamma (PPARγ). *Fitoterapia*, 106, 55–61.
262. Tachibana, K., Kobayashi, Y., Tanaka, T., Tagami, M., Sugiyama, A., Katayama, T., Ueda, C. et al., 2005. Gene expression profiling of potential peroxisome proliferator-activated receptor (PPAR) target genes in human hepatoblastoma cell lines inducibly expressing different PPAR isoforms. *Nuclear Receptor*, 3(1), 1.
263. Choi, J.H., Choi, J.N., Lee, S.Y., Lee, S.J., Kim, K. and Kim, Y.K., 2010. Inhibitory activity of diacylglycerol acyltransferase by glabrol isolated from the roots of licorice. *Archives of Pharmacal Research*, 33(2), 237–242.
264. Gupta, S.C., Sundaram, C., Reuter, S. and Aggarwal, B.B., 2010. Inhibiting NF-κB activation by small molecules as a therapeutic strategy. *Biochimica et Biophysica Acta (BBA)-Gene Regulatory Mechanisms*, 1799(10), 775–787.

265. Sil, R., Ray, D. and Chakraborti, A.S., 2013. Glycyrrhizin ameliorates insulin resistance, hyperglycemia, dyslipidemia and oxidative stress in fructose-induced metabolic syndrome-X in rat model. *Indian Journal of Experimental Biology*, 51(2), 129–138.
266. Osno, Y., Nakajima, K. and Hata, Y., 1995. Hypertriglyceridemia and fatty lover: Clinical diagnosis of fatty liver and lipoprotein profiles in hypertriglyceridemic patients with fatty liver. *Journal of Atherosclerosis and Thrombosis*, 2, S47–S52.
267. Kondeti, V.,K., Badri, K.,R., Maddirala, D.,R., Thur, S.,K., Fatima, S.,S., Kasetti, R.,B. and Rao, C.A., 2010. Effect of *Pterocarpus santalinus* bark, on blood glucose, serum lipids, plasma insulin and hepatic carbohydrate metabolic enzymes in streptozotocin-induced diabetic rats. *Food and Chemical Toxicology*, 48(5), 1281–1287.
268. Zhang, H.F., Shi, L.J., Song, G.Y., Cai, Z.G., Wang, C. and An, R.J., 2013. Protective effects of matrine against progression of high-fructose diet-induced steatohepatitis by enhancing antioxidant and anti-inflammatory defences involving Nrf2 translocation. *Food and Chemical Toxicology*, 55, 70–77.
269. Guo, C., Zhang, C., Li, L., Wang, Z., Xiao, W. and Yang, Z., 2014. Hypoglycemic and hypolipidemic effects of oxymatrine in high-fat diet and streptozotocin-induced diabetic rats. *Phytomedicine*, 21(6), 807–814.
270. Sasaki, T., Li, W., Higai, K., Quang, T.,H., Kim, Y.,H., Koike, K., 2014. Protein tyrosine phosphatase 1B inhibitory activity of lavandulyl flavonoids from roots of *Sophora flavescens*. *Planta Medica*, 80(7), 557–560.
271. Song, C.Y., Zeng, X., Chen, S.W., Hu, P.F., Zheng, Z.W., Ning, B.F., Shi, J., Xie, W.F. and Chen, Y.X., 2011. Sophocarpine alleviates non-alcoholic steatohepatitis in rats. *Journal of Gastroenterology and Hepatology*, 26(4), 765–774.
272. Song, C.Y., Shi, J., Zeng, X., Zhang, Y., Xie, W.F. and Chen, Y.X., 2013. Sophocarpine alleviates hepatocyte steatosis through activating AMPK signaling pathway. *Toxicology In Vitro*, 27(3), 1065–1071.
273. Park, K.W., Lee, J.E. and Park, K.M., 2009. Diets containing *Sophora japonica* L. prevent weight gain in high-fat diet-induced obese mice. *Nutrition Research*, 29(11), 819–824.
274. Pavana, P., Manoharan, S., Renju, G.L. and Sethupathy, S., 2007. Antihyperglycemic and antihyperlipidemic effects of *Tephrosia purpurea* leaf extract in streptozotocin induced diabetic rats. *Journal of Environmental Biology*, 28(4), 833–837.
275. Stampfer, M.J., Sacks, F.M., Salvini, S., Willnett, W.C. and Hennekens, C.H., 1991. A prospective study of cholesterol, apolipoproteins, and the risk of myocardial infarction. *The New England Journal of Medicine*, 325, 373–381.
276. Reichi, D. and Miller, N.E., 1989. Pathophysiology of reverse cholesterol transport: Insights from inherited disorders of lipoprotein metabolism. *Arteriosclerosis*, 9, 785–797.
277. Maciejko, J.J., Holmes, D.R., Kottke, B.A., Zinsmeister, A.R., Dinh, D.M. and Mao, S.J.T., 1983. Apoliporptein A-I as a marker of angiographycally assessed coronary-artery disease. *The New England Journal of Medicine*, 309, 309–389.
278. Gross, D., 2009. *Animal Models in Cardiovascular Research*. New York: Springer Science & Business Media.
279. Chang, L.C., Gerhäuser, C., Song, L., Farnsworth, N.R., Pezzuto, J.M. and Kinghorn, A.D., 1997. Activity-guided isolation of constituents of Tephrosia purpurea with the potential to induce the phase II enzyme, quinone reductase. *Journal of Natural Products*, 60(9), 869–873.
280. Erlund, I., Meririnne, E., Alfthan, G. and Aro, A., 2001. Plasma kinetics and urinary excretion of the flavanones naringenin and hesperetin in humans after ingestion of orange juice and grapefruit juice. *The Journal of Nutrition*, 131(2), 235–241.
281. Cho, K.W., Kim, Y.O., Andrade, J.E., Burgess, J.R. and Kim, Y.C., 2011. Dietary naringenin increases hepatic peroxisome proliferators-activated receptor α protein expression and decreases plasma triglyceride and adiposity in rats. *European Journal of Nutrition*, 50(2), 81–88.
282. Constantin, R.P., Constantin, R.P., Bracht, A., Yamamoto, N.S., Ishii-Iwamoto, E.L. and Constantin, J., 2014. Molecular mechanisms of citrus flavanones on hepatic gluconeogenesis. *Fitoterapia*, 92, 148–162.
283. de la Garza, A.L., Etxeberria, U., Palacios-Ortega, S., Haslberger, A.G., Aumueller, E., Milagro, F.I. and Martínez, J.A., 2014. Modulation of hyperglycemia and TNFα-mediated inflammation by helichrysum and grapefruit extracts in diabetic db/db mice. *Food & Function*, 5(9), 2120–2128.
284. Chudnovskiy, R., Thompson, A., Tharp, K., Hellerstein, M., Napoli, J.L. and Stahl, A., 2014. Consumption of clarified grapefruit juice ameliorates high-fat diet induced insulin resistance and weight gain in mice. *PLoS ONE*, 9(10), e108408.
285. Tan, S., Li, M., Ding, X., Fan, S., Guo, L., Gu, M., Zhang, Y. et al., 2014. Effects of *Fortunella margarita* fruit extract on metabolic disorders in high-fat diet-induced obese C57BL/6 mice. *PLoS ONE*, 9(4), e93510.

286. Luis, A. and Río, D. (Eds.), 2013. *Peroxisomes and Their Key Role in Cellular Signaling and Metabolism*. Dordrecht, the Netherlands: Springer.
287. Kumamoto, H., Matsubara, Y., Iizuka, Y., Okamoto, K. and Yokoi, K., 1985. Structure and hypotensive effect of flavonoid glycosides in kinkan (*Fortunella japonica*) peelings. *Agricultural and Biological Chemistry*, 49(9), 2613–2618.
288. Assini, J.M., Mulvihill, E.E. and Huff, M.W., 2013. Citrus flavonoids and lipid metabolism. *Current Opinion in Lipidology*, 24(1), 34–40.
289. Mishra, S.K., Tiwari, S., Shrivastava, S., Sonkar, R., Mishra, V., Nigam, S.K., Saxena, A.K., Bhatia, G. and Mir, S.S., 2014. Pharmacological evaluation of the efficacy of *Dysoxylum binectariferum* stem bark and its active constituent rohitukine in regulation of dyslipidemia in rats. *Journal of Natural Medicines*, 28:1–9.
290. Varshney, S., Shankar, K., Beg, M., Balaramnavar, V.M., Mishra, S.K., Jagdale, P., Srivastava, S. et al., 2014. Rohitukine inhibits in vitro adipogenesis arresting mitotic clonal expansion and improves dyslipidemia in vivo. *Journal of Lipid Research*, 55(6), 1019–1032.
291. Lund, E.G., Menke, J.G. and Sparrow, C.P., 2003. Liver X receptor agonists as potential therapeutic agents for dyslipidemia and atherosclerosis. *Arteriosclerosis, Thrombosis, and Vascular Biology*, 23(7), 1169–1177.
292. Gathercole, L.L., Lavery, G.G., Morgan, S.A., Cooper, M.S., Sinclair, A.J., Tomlinson, J.W. and Stewart, P.M., 2013. 11β-Hydroxysteroid dehydrogenase 1: Translational and therapeutic aspects. *Endocrine Reviews*, 34(4), 525–555.
293. Torrecilla, E., Fernandez-Vazquez, G., Vicent, D., Sánchez-Franco, F., Barabash, A., Cabrerizo, L., Sánchez-Pernaute, A., Torres, A.J. and Rubio, M.A., 2012. Liver upregulation of genes involved in cortisol production and action is associated with metabolic syndrome in morbidly obese patients. *Obesity Surgery*, 22, 478–486.
294. Andrews, R.C., Rooyackers, O., Walker, B.R., 2003. Effects of the inhibitor carbenoxolone on insulin sensitivity in men with type 2 diabetes. *Journal of Clinical Endocrinology and Metabolism*, 88, 285–291.
295. Rosenstock, J., Banarer, S., Fonseca, V.A., Inzucchi, S.E., Sun, W., Yao, W., Hollis, G. et al., 2010. The 11-beta-hydroxysteroid dehydrogenase type 1 inhibitor INCB13739 improves hyperglycemia in patientswith type2diabetesinadequatelycontrolled by metformin monotherapy. *Diabetes Care*, 33, 1516–1522.
296. Han, M.L., Shen, Y., Wang, G.C., Leng, Y., Zhang, H. and Yue, J.M., 2013. 11β-HSD1 inhibitors from Walsura cochinchinensis. *Journal of Natural Products*, 76(7), 1319–1327.
297. Zhou, B., Shen, Y., Wu, Y., Leng, Y. and Yue, J.M., 2015. Limonoids with 11β-hydroxysteroid dehydrogenase type 1 inhibitory activities from dysoxylum mollissimum. *Journal of Natural Products*, 78(8), 2116–2122.
298. Satil, F., Azcan, N. and Baser, K.H.C., 2003. Fatty acid composition of pistachio nuts in Turkey. *Chemistry of Natural Compounds*, 39(4), 322–324.
299. Kay, C.D., Gebauer, S.K., West, S.G. and Kris-Etherton, P.M., 2010. Pistachios increase serum antioxidants and lower serum oxidized-LDL in hypercholesterolemic adults. *Journal of Nutrition*, 140(6), 1093–1098.
300. Toshima, S., Hasegawa, A., Kurabayashi, M., Itabe, H., Takano, T., Sugano, J., Shimamura, K. et al., 2000. Circulating oxidized low density lipoprotein levels. A biochemical risk marker for coronary heart disease. *Arteriosclerosis, Thrombosis, and Vascular Biology*, 20, 2243–2247.
301. Parthasarathy, S., Khoo, J.C., Miller, E., Barnett, J., Witztum, J.L. and Steinberg, D., 1990. Low density lipoprotein rich in oleic acid is protected against oxidative modification: Implications for dietary prevention of atherosclerosis. *Proceedings of the National Academy of Sciences*, 87(10), 3894–3898.
302. Chang, T.Y., Chang, C.C. and Cheng, D., 1997. Acyl-coenzyme A: Cholesterol acyltransferase. *Annual Review of Biochemistry*, 66(1), 613–638.
303. Bendich, A. and Deckelbaum, R.J. (Eds.), 2016. *Preventive Nutrition: The Comprehensive Guide for Health Professionals*. London: Springer.
304. Baggio, G., Pagnan, A., Muraca, M., Martini, S., Opportuno, A., Bonanome, A., Ambrosio, G.B., Ferrari, S., Guarini, P. and Piccolo, D., 1988. Olive-oil-enriched diet: Effect on serum lipoprotein levels and biliary cholesterol saturation. *The American Journal of Clinical Nutrition*, 47(6), 960–964.
305. West, S.G., Gebauer, S.K., Kay, C.D., Bagshaw, D.M., Savastano, D.M., Diefenbach, C. and Kris-Etherton, P.M., 2012. Diets containing pistachios reduce systolic blood pressure and peripheral vascular responses to stress in adults with dyslipidemia. *Hypertension*, 60(1), 58–63.
306. Newsholme, P., E.P. Haber, S.M. Hirabara, E.L.O. Rebelato, J. Procopio, D. Morgan, H.C. Oliveira-Emilio, A.R. Carpinelli, and Curi, R., 2007. Diabetes associated cell stress and dysfunction: role of mitochondrial and non-mitochondrial ROS production and activity. *The Journal of physiology* 583(1): 9–24.

307. Anagnostis, P., Athyros, V.G., Tziomalos, K., Karagiannis, A. and Mikhailidis, D.P., 2009. The pathogenetic role of cortisol in the metabolic syndrome: A hypothesis. *The Journal of Clinical Endocrinology & Metabolism*, 94(8), 2692–2701.
308. McCowen, K.C., Malhotra, A. and Bistrian, B.R., 2001. Stress-induced hyperglycemia. *Critical Care Clinics*, 17(1), 107–124.
309. Zhang, Y., Si, Y., Zhai, L., Yang, N., Yao, S., Sang, H., Zu, D., Xu, X., Qin, S. and Wang, J., 2013. *Celastrus orbiculatus* Thunb. ameliorates high-fat diet-induced non-alcoholic fatty liver disease in guinea pigs. *Pharmazie*, 68(10), 850–854.
310. Zhang, Y., Si, Y., Zhai, L., Guo, S., Zhao, J., Sang, H., Pang, X., Zhang, X., Chen, A. and Qin, S., 2016. *Celastrus orbiculatus* Thunb. reduces lipid accumulation by promoting reverse cholesterol transport in hyperlipidemic mice. *Lipids*, 51(6), 677–692.
311. Ardiles, A.E., González-Rodríguez, A., Núñez, M.J., Perestelo, N.R., Pardo, V., Jiménez, I.A., Valverde, A.M. and Bazzocchi, I.L., 2012. Studies of naturally occurring friedelane triterpenoids as insulin sensitizers in the treatment type 2 diabetes mellitus. *Phytochemistry*, 84, 116–124.
312. Nelson, L.S., Shih, R.D., Balick, M.J. and Lampe, K.F., 2007. *Handbook of Poisonous and Injurious Plants*. New York: New York Botanical Garden.
313. Yang, B., Kalimo, K.O., Mattila, L.M., Kallio, S.E., Katajisto, J.K., Peltola, O.J. and Kallio, H.P., 1999. Effects of dietary supplementation with sea buckthorn (*Hippophae rhamnoides*) seed and pulp oils on atopic dermatitis. *The Journal of Nutritional Biochemistry*, 10(11), 622–630.
314. Steinmann, D., Baumgartner, R.R., Heiss, E.H., Bartenstein, S., Atanasov, A.G., Dirsch, V.M., Ganzera, M. and Stuppner, H., 2012. Bioguided isolation of (9Z)-octadec-9-enoic acid from *Phellodendron amurense* Rupr. and identification of fatty acids as PTP1B inhibitors. *Planta Medica*, 78(3), 219–224.
315. Zhang, W., Zhao, J., Zhu, X., Zhuang, X., Pang, X., Wang, J. and Qu, W., 2010. Antihyperglycemic effect of aqueous extract of sea buckthorn (*Hippophae Rhamnoides* L.) seed residues in streptozotocin-treated and high fat-diet-fed rats. *Journal of Food Biochemistry*, 34(4), 856–868.
316. Wang, J., Zhang, W., Zhu, D., Zhu, X., Pang, X. and Qu, W., 2011. Hypolipidaemic and hypoglycaemic effects of total flavonoids from seed residues of Hippophae rhamnoides L. in mice fed a high-fat diet. *Journal of the Science of Food and Agriculture*, 91(8), 1446–1451.
317. Pichiah, P.B., Moon, H.J., Park, J.E., Moon, Y.J. and Cha, Y.S., 2012. Ethanolic extract of seabuckthorn (*Hippophae rhamnoides* L.) prevents high-fat diet-induced obesity in mice through down-regulation of adipogenic and lipogenic gene expression. *Nutrition Research*, 32(11), 856–864.
318. Chu, Q.C., Qu, W.Q., Peng, Y.Y., Cao, Q.H. and Ye, J.N., 2003. Determination of flavonoids in *Hippophae rhamnoides* L. and its phytopharmaceuticals by capillary electrophoresis with electrochemical detection. *Chromatographia*, 58(1–2), 67–71.
319. Fan, J., Ding, X. and Gu, W., 2007. Radical-scavenging proanthocyanidins from sea buckthorn seed. *Food Chemistry*, 102, 168–177.
320. Zu, Y., Li, C., Fu, Y. and Zhao, C., 2006. Simultaneous determination of catechin, rutin, quercetin kaempferol and isorhamnetin in the extract of sea buckthorn (*Hippophae rhamnoides* L.) leaves by RP-HPLC with DAD. *Journal of Pharmaceutical and Biomedical Analysis*, 41(3), 714–719.
321. Park, C.H., Noh, J.S., Kim, J.H., Tanaka, T., Zhao, Q., Matsumoto, K., Shibahara, N. and Yokozawa, T., 2011. Evaluation of morroniside, iridoid glycoside from *Corni Fructus*, on diabetes-induced alterations such as oxidative stress, inflammation, and apoptosis in the liver of type 2 diabetic db/db mice. *Biological and Pharmaceutical Bulletin*, 34(10), 1559–1565.
322. Park, C.H., Yamabe, N., Noh, J.S., Kang, K.S., Tanaka, T. and Yokozawa, T., 2009. The beneficial effects of morroniside on the inflammatory response and lipid metabolism in the liver of db/db mice. *Biological and Pharmaceutical Bulletin*, 32(10), 1734–1740.
323. Aleksunes, L.M., Reisman, S.A., Yeager, R.L., Goedken, M.J. and Klaassen, C.D., 2010. Nuclear factor erythroid 2-related factor 2 deletion impairs glucose tolerance and exacerbates hyperglycemia in type 1 diabetic mice. *Journal of Pharmacology and Experimental Therapeutics*, 333(1), 140–151.
324. Wu, X. and Williams, K.J., 2012. NOX4 pathway as a source of selective insulin resistance and responsiveness. *Arteriosclerosis, Thrombosis, and Vascular Biology*, 32(5), 1236–1245.
325. Li, X., Wang, Q., Zhang, X., Sheng, X., Zhou, Y., Li, M., Jing, X., Li, D. and Zhang, L., 2007. HPLC study of pharmacokinetics and tissue distribution of morroniside in rats. *Journal of Pharmaceutical and Biomedical Analysis*, 45(2), 349–355.
326. Zhao, M., Du, L., Tao, J., Qian, D., Guo, J., Jiang, S., Shang, E.X., Duan, J.A. and Wu, C., 2015. Ultra-performance liquid chromatography coupled with quadrupole time-of-flight mass spectrometry for rapid analysis of the metabolites of morroniside produced by human intestinal bacteria. *Journal of Chromatography B*, 976, 61–67.

327. Owen, J.B., Treasure, J.L. and Collier, D.A. (Eds.), 2001. *Animal Models-Disorders of Eating Behaviour and Body Composition*. Dordrecht, the Netherlands: Kluwer Academic Publishers.
328. Farrell, N., Norris, G., Lee, S.G., Chun, O.K. and Blesso, C.N., 2015. Anthocyanin-rich black elderberry extract improves markers of HDL function and reduces aortic cholesterol in hyperlipidemic mice. *Food & Function*, 6(4), 1278–1287.
329. Graf, G.A., Li, W.P., Gerard, R.D., Gelissen, I., White, A., Cohen, J.C. and Hobbs, H.H., 2002. Coexpression of ATP-binding cassette proteins ABCG5 and ABCG8 permits their transport to the apical surface. *The Journal of Clinical Investigation*, 110(5), 659–669.
330. Shimano, H., 2001. Sterol regulatory element-binding proteins (SREBPs): Transcriptional regulators of lipid synthetic genes. *Progress in Lipid Research*, 40(6), 439–452.
331. Lehmann, J.M., Lenhard, J.M., Oliver, B.B., Ringold, G.M. and Kliewer, S.A., 1997. Peroxisome proliferator-activated receptors α and γ are activated by indomethacin and other non-steroidal anti-inflammatory drugs. *Journal of Biological Chemistry*, 272(6), 3406–3410.
332. Jayasuriya, H., Herath, K.B., Ondeyka, J.G., Guan, Z., Borris, R.P., Tiwari, S., de Jong, W. et al., 2005. Diterpenoid, steroid, and triterpenoid agonists of liver X receptors from diversified terrestrial plants and marine sources. *Journal of Natural Products*, 68(8), 1247–1252.
333. Park, E.J., Zhao, Y.Z., Kim, Y.H., Lee, J.J. and Sohn, D.H., 2004. Acanthoic acid from Acanthopanax koreanum protects against liver injury induced by tert-butyl hydroperoxide or carbon tetrachloride in vitro and in vivo. *Planta medica*, 70(4), 321–327.
334. Bai, T., Yao, Y.-L., Jin, X.-J., Lian, L.-H., Li, Q., Yang, N., Jin, Q., Wu, Y.-L. and Nan, J.-X., 2014. Acanthoic acid, a diterpene in *Acanthopanax koreanum*, ameliorates the development of liver fibrosis via LXRs signals. *Chemico-Biological Interactions*, 218, 63–70.
335. Lee, Y.J., Ko, E.H., Kim, J.E., Kim, E., Lee, H., Choi, H., Yu, J.H. et al., 2012. Nuclear receptor PPARγ-regulated monoacylglycerol O-acyltransferase 1 (MGAT1) expression is responsible for the lipid accumulation in diet-induced hepatic steatosis. *Proceedings of the National Academy of Sciences*, 109(34), 13656–13661.
336. Kim, M.O., Lee, S.H., Seo, J.H., Kim, I.S., Han, A.R., Moon, D.O., Cho, S., Cui, L., Kim, J. and Lee, H.S., 2013. *Aralia cordata* inhibits triacylglycerol biosynthesis in HepG2 cells. *Journal of Medicinal Food*, 16(12), 1108–1114.
337. Jung, H.J., Jung, H.A., Kang, S.S., Lee, J.H., Cho, Y.S., Moon, K.H. and Choi, J.S., 2012. Inhibitory activity of *Aralia continentalis* roots on protein tyrosine phosphatase 1B and rat lens aldose reductase. *Archives of Pharmacal Research*, 35(10), 1771–1777.
338. Chung, I.M., Kim, M.Y., Park, W.H. and Moon, H.I., 2006. Antiatherogenic activity of *Dendropanax morbifera* essential oil in rats. *Pharmazie*, 64(8), 547–549.
339. Zhong, Y., Jun, L., Wei-Min, H., Wen-Li, D., Xue, W. and Shang, J., 2015. β-Elemene reduces the progression of atherosclerosis in rabbits. *Chinese Journal of Natural Medicines*, 13(6), 415–420.
340. Liu, J., Zhang, Z., Gao, J., Xie, J., Yang, L. and Hu, S., 2011. Downregulation effects of beta-elemene on the levels of plasma endotoxin, serum TNF-alpha, and hepatic CD14 expression in rats with liver fibrosis. *Frontiers of Medicine*, 5(1), 101–105.
341. Song, Y.B., An, Y.R., Kim, S.J., Park, H.W., Jung, J.W., Kyung, J.S., Hwang, S.Y. and Kim, Y.S., 2012. Lipid metabolic effect of Korean red ginseng extract in mice fed on a high-fat diet. *Journal of the Science of Food and Agriculture*, 92(2), 388–396.
342. Benlian, P., 2001. *Genetics of Dyslipidemia* (Vol. 7). New York: Springer Science & Business Media.
343. Ziboh, V.A., 2001. *Gamma Linolenic Acid: Recent Advances in Biotechnology and Clinical Applications*. Champaign, IL: The American Oil Chemists Society.
344. Wu, C., Jia, Y., Lee, J.H., Kim, Y., Sekharan, S., Batista, V.S. and Lee, S.J., 2015. Activation of OR1A1 suppresses PPAR-γ expression by inducing HES-1 in cultured hepatocytes. *The International Journal of Biochemistry & Cell Biology*, 64, 75–80.
345. Wu, J.H., Leung, G.P.H., Kwan, Y.W., Sham, T.T., Tang, J.Y., Wang, Y.H., Wan, J.B., Lee, S.M.Y. and Chan, S.W., 2013. Suppression of diet-induced hypercholesterolaemia by saponins from *Panax notoginseng* in rats. *Journal of Functional Foods*, 5(3), 1159–1169.
346. Ji, W. and Gong B.Q., 2007. Hypolipidemic effects and mechanisms of *Panax notoginseng* on lipid profile in hyperlipidemic rats. *Journal of Ethnopharmacology*, 113(2), 318–324.
347. Xia, W., Sun, C., Zhao, Y. and Wu, L., 2011. Hypolipidemic and antioxidant activities of sanchi (*Radix notoginseng*) in rats fed with a high fat diet. *Phytomedicine*, 18(6), 516–520.
348. Koriem, K.M., Asaad, G.F., Megahed, H.A., Zahran, H. and Arbid, M.S., 2012. Evaluation of the antihyperlipidemic, anti-inflammatory, analgesic, and antipyretic activities of ethanolic extractof *Ammi majus* seeds in albino rats and mice. *International Journal of Toxicology*, 31(3), 294–300.

349. Elgamal, M.H.A., Shalaby, N.M., Duddeck, H. and Hiegemann, M., 1993. Coumarins and coumarin glucosides from the fruits of *Ammi majus. Phytochemistry*, 34(3), 819–823.
350. Abu-Mustafa, E.A., El-Bay, F.K.A. and Fayez, M.B.E., 1971. Natural coumarins XII: Umbelliprenin, a constituent of *Ammi majus* L. fruits. *Journal of Pharmaceutical Sciences*, 60(5), 788–789.
351. Cao, Y., Zhang, Y., Wang, N. and He, L., 2014. Antioxidant effect of imperatorin from *Angelica dahurica* in hypertension via inhibiting NADPH oxidase activation and MAPK pathway. *Journal of the American Society of Hypertension*, 8(8), 527–536.
352. Guichard, C., Moreau, R., Pessayre, D., Epperson, T.K. and Krause, K.H., 2008. NOX family NADPH oxidases in liver and in pancreatic islets: A role in the metabolic syndrome and diabetes? *Biochemical Society Transactions*, 36(5), 920–929.
353. Hajhashemi, V. and Abbasi, N., 2008. Hypolipidemic activity of *Anethum graveolens* in rats. *Phytotherapy Research*, 22(3), 372–375.
354. Takahashi, N., Yao, L., Kim, M., Sasako, H., Aoyagi, M., Shono, J., Tsuge, N., Goto, T. and Kawada, T., 2013. Dill seed extract improves abnormalities in lipid metabolism through peroxisome proliferator-activated receptor-α (PPAR-α) activation in diabetic obese mice. *Molecular Nutrition & Food Research*, 57(7), 1295–1299.
355. Singh, G., Maurya, S., Lampasona, M.P. and Catalan, C., 2005. Chemical constituents, antimicrobial investigations, and antioxidative potentials of *Anethum graveolens* L. essential oil and acetone extract: Part 52. *Journal of Food Science*, 70(4), M208–M215.
356. Gonçalves, J.C., Silveira, A.L., de Souza, H.D., Nery, A.A., Prado, V.F., Prado, M.A., Ulrich, H. and Araújo, D.A., 2013. The monoterpene (-)-carvone: A novel agonist of TRPV1 channels. *Cytometry A*, 83(2), 212–219.
357. Liu, I.M., Tzeng, T.F., Liou, S.S. and Chang, C.J., 2012. Regulation of obesity and lipid disorders by extracts from *Angelica acutiloba* root in high-fat diet-induced obese rats. *Phytotherapy Research*, 26(2), 223–230.
358. Tanaka, S., Ikeshiro, Y., Tabata, M. and Konoshima, M., 1976. Anti-nociceptive substances from the roots of *Angelica acutiloba. Arzneimittel-Forschung*, 27(11), 2039–2045.
359. Tsuchida, T., Kobayashi, M., Kaneko, K. and Mitsuhashi, H., 1987. Studies on the constituents of Umbelliferae plants. XVI. Isolation and structures of three new ligustilide derivatives from *Angelica acutiloba. Chemical and Pharmaceutical Bulletin*, 35(11), 4460–4464.
360. Su, Y.W., Chiou, W.F., Chao, S.H., Lee, M.H., Chen, C.C. and Tsai, Y.C., 2011. Ligustilide prevents LPS-induced iNOS expression in RAW 264.7 macrophages by preventing ROS production and down-regulating the MAPK, NF-κB and AP-1 signaling pathways. *International Immunopharmacology*, 11(9), 1166–1172.
361. Yan, R., Ko, N.L., Li, S.L., Tam, Y.K. and Lin, G., 2008. Pharmacokinetics and metabolism of ligustilide, a major bioactive component in Rhizoma Chuanxiong, in the rat. *Drug Metabolism and Disposition*, 36(2), 400–408.
362. Chuang, H.M., Su, H.L., Li, C., Lin, S.Z., Yen, S.Y., Huang, M.H., Ho, L.I., Chiou, T.W. and Harn, H.J., 2016. The role of butylidenephthalide in targeting the microenvironment which contributes to liver fibrosis amelioration. *Frontiers in Pharmacology*, 7, 112.
363. Ohnogi, H., Hayami, S., Kudo, Y., Deguchi, S., Mizutani, S., Enoki, T., Tanimura, Y. et al., 2012. *Angelica keiskei* extract improves insulin resistance and hypertriglyceridemia in rats fed a high-fructose drink. *Bioscience, Biotechnology, and Biochemistry*, 76(5), 928–932.
364. Ramachandran, V., Saravanan, R. and Senthilraja, P., 2014. Antidiabetic and antihyperlipidemic activity of asiatic acid in diabetic rats, role of HMG CoA: In vivo and in silico approaches. *Phytomedicine*, 21(3), 225–232.
365. Yan, S.L., Yang, H.T., Lee, Y.J., Lin, C.C., Chang, M.H. and Yin, M.C., 2014. Asiatic acid ameliorates hepatic lipid accumulation and insulin resistance in mice consuming a high-fat diet. *Journal of Agricultural and Food Chemistry*, 62(20), 4625–4631.
366. Ogawa, H., Sasai, N., Kamisako, T. and Baba, K., 2007. Effects of osthol on blood pressure and lipid metabolism in stroke-prone spontaneously hypertensive rats. *Journal of Ethnopharmacology*, 112(1), 26–31.
367. Du, R., Xue, J., Wang, H.B., Zhang, Y. and Xie, M.L., 2011. Osthol ameliorates fat milk-induced fatty liver in mice by regulation of hepatic sterol regulatory element-bindingprotein-1c/2-mediated target gene expression. *European Journal of Pharmacology*, 666(1–3), 183–188.
368. Chithra, V. and Leelamma, S., 1997. Hypolipidemic effect of coriander seeds (*Coriandrum sativum*): Mechanism of action. *Plant Foods for Human Nutrition*, 51(2), 167–172.
369. Dhanapakiam, P., Joseph, J.M., Ramaswamy, V.K., Moorthi, M. and Kumar, A.S., 2008. The cholesterol lowering property of coriander seeds (Coriandrum sativum): Mechanism of action. *Journal of Environmental Biology*, 29(1), 53–56.

370. Aissaoui, A., Zizi, S., Israili, Z.H. and Lyoussi, B., 2011. Hypoglycemic and hypolipidemic effects of *Coriandrum sativum* L. in Meriones shawi rats. *Journal of Ethnopharmacology*, 137(1), 652–661.
371. Jun, H.J., Lee, J.H., Kim, J., Jia, Y., Kim, K.H., Hwang, K.Y., Yun, E.J., Do, K.R. and Lee, S.J., 2014. Linalool is a PPARα ligand that reduces plasma TG levels and rewires the hepatic transcriptome and plasma metabolome. *Journal of Lipid Research*, 55(6), 1098–1110.
372. Yoshida, J., Seino, H., Ito, Y., Nakano, T., Satoh, T., Ogane, Y., Suwa, S., Koshino, H. and Kimura, K., 2013. Inhibition of glycogen synthase kinase-3β by falcarindiol isolated from Japanese Parsley (*Oenanthe javanica*). *Journal of Agriculture and Food Chemistry*, 61(31), 7515–7521.
373. Lawrence, J.C. and Roach, P.J., 1997. New insights into the role and mechanism of glycogen synthase activation by insulin. *Diabetes*, 46(4), 541–547.
374. Nukitrangsan, N., Okabe, T., Toda, T., Inafuku, M., Iwasaki, H. and Oku, H., 2012. Effect of *Peucedanum japonicum* Thunb extract on high-fat diet-induced obesity and gene expression in mice. *Journal of Oleo Science*, 61(2), 89–101.
375. Nugara, R.N., Inafuku, M., Takara, K., Iwasaki, H. and Oku, H., 2014. Pteryxin: A coumarin in *Peucedanum japonicum* Thunb leaves exerts antiobesity activity through modulation of adipogenic gene network. *Nutrition*, 30(10), 1177–1184.
376. Nugara, R.N., Inafuku, M., Iwasaki, H. and Oku, H., 2014. Partially purified *Peucedanum japonicum* Thunb extracts exert anti-obesity effects in vitro. *Nutrition*, 30(5), 575–583.
377. Choi, H.J., Chung, M.J. and Ham, S.S., 2010. Antiobese and hypocholesterolaemic effects of an *Adenophora triphylla* extract in HepG2 cells and high fat diet-induced obese mice. *Food Chemistry*, 119(2), 437–444.
378. Briel, M., Ferreira-Gonzalez, I., You, J.J., Karanicolas, P.J., Akl, E.A., Wu, P., Blechacz, B. et al., 2009. Association between change in high density lipoprotein–cholesterol and cardiovascular disease morbidity and mortality: Systematic review and meta-regression analysis. *BMJ*, 338, b92.
379. Lee, S.E., Lee, E.H., Lee, T.J., Kim, S.W. and Kim, B.H., 2013. Anti-obesity effect and action mechanism of *Adenophora triphylla* root ethanol extract in C57BL/6 obese mice fed a high-fat diet. *Bioscience, Biotechnology, and Biochemistry*, 77(3), 544–550.
380. Kuang, H.-X., Shao, C.-J., Kasai, R., Ohtani, K., Tian, C.-K., Xuh, J.-D. and Tanaka, O., 1991. Phenolic glycosides from roots of *Adenophora tetraphylla* collected in Heilongjiang, China. *Chemical and Pharmaceutical Bulletin*, 39(9), 2440–2442.
381. Yao, S., Liu, R., Huang, X. and Kong, L., 2007. Preparative isolation and purification of chemical constituents from the root of *Adenophora tetraphlla* by high-speed counter-current chromatography with evaporative light scattering detection. *Journal of Chromatography A*, 1139(2), 254–262.
382. Ahn, E.K. and Oh, J.S., 2013. Lupenone Isolated from Adenophora triphylla var. japonica extract inhibits adipogenic differentiation through the downregulation of PPARγ in 3T3-L1 cells. *Phytotherapy Research*, 27(5), 761–766.
383. Seong, S.H., Roy, A., Jung, H.A., Jung, H.J. and Choi, J.S., 2016. Protein tyrosine phosphatase 1B and α-glucosidase inhibitory activities of *Pueraria lobata* root and its constituents. *Journal of Ethnopharmacology*, 194, 706–716.
384. Lee, J.S., Kim, K.J., Kim, Y.H., Kim, D.B., Shin, G.H., Cho, J.H., Kim, B.K., Lee, B.Y. and Lee, O.H., 2014. *Codonopsis lanceolata* extract prevents diet-induced obesity in C57BL/6 mice. *Nutrients*, 6(11), 4663–4677.
385. Hwang, Y.P., Choi, J.H., Kim, H.G., Lee, H.S., Chung, Y.C. and Jeong, H.G., 2013. Saponins from *Platycodon grandiflorus* inhibit hepatic lipogenesis through induction of SIRT1 and activation of AMP-activated protein kinase in high-glucose-induced HepG2 cells. *Food Chemistry*, 140(1–2), 115–123.
386. Carling, D., Sanders, M.J. and Woods, A., 2008. The regulation of AMP-activated protein kinase by upstream kinases. *International Journal of Obesity*, 32, S55–S59.
387. Lee, J.S., Choi, M.S., Seo, K.I., Lee, J., Lee, H.I., Lee, J.H., Kim, M.J. and Lee, M.K., 2014. *Platycodi radix* saponin inhibits α-glucosidase in vitro and modulates hepatic glucose-regulating enzyme activities in C57BL/KsJ-db/db mice. *Archives of Pharmacal Research*, 37(6), 773–782.
388. Laakso, M., Malkki, M. and Deeb, S.S., 1995. Amino acid substitutions in hexokinase II among patients with NIDDM. *Diabetes*, 44(3), 330–334.
390. Storey, J. M., and Bailey, E., 1978. Effect of streptozotocin diabetes and insulin administration on some liver enzyme activities in the post-weaning rat. Enzyme 23(6): 389.
390. Wang, Z.Q., Zhang, X.H., Yu, Y., Tipton, R.C., Raskin, I., Ribnicky, D., Johnson, W. and Cefalu, W.T., 2013. *Artemisia scoparia* extract attenuates non-alcoholic fatty liver disease in diet-induced obesity mice by enhancing hepatic insulin and AMPK signaling independently of FGF21 pathway. *Metabolism*, 62(9), 1239–1249.

391. Ginion, A., Auquier, J., Benton, C.R., Mouton, C., Vanoverschelde, J.L., Hue, L., Horman, S., Beauloye, C. and Bertrand, L., 2011. Inhibition of the mTOR/p70S6K pathway is not involved in the insulin-sensitizing effect of AMPK on cardiac glucose uptake. *American Journal of Physiology-Heart and Circulatory Physiology*, 301(2), H469–H477.
392. Bakan, I. and Laplante, M., 2012. Connecting mTORC1 signaling to SREBP-1 activation. *Current Opinion in Lipidology*, 23(3), 226–234.
393. Janbaz, K.H., Saeed, S.A. and Gilani, A.H., 2002. Protective effect of rutin on paracetamol-and CCl 4-induced hepatotoxicity in rodents. *Fitoterapia*, 73(7), 557–563.
394. Ziaee, A., Zamansoltani, F., Nassiri-Asl, M. and Abbasi, E., 2009. Effects of rutin on lipid profile in hypercholesterolaemic rats. *Basic & Clinical Pharmacology & Toxicology*, 104(3), 253–258.
395. Yuan, H.D. and Piao, G.C., 2011. An active part of Artemisia sacrorum Ledeb. suppresses gluconeogenesis through AMPK mediated GSK3β and CREB phosphorylation in human HepG2 cells. *Bioscience, Biotechnology, and Biochemistry*, 75(6), 1079–1084.
396. Mantovani, J. and Roy, R., 2011. Re-evaluating the general (ized) roles of AMPK in cellular metabolism. *FEBS Letters*, 585(7), 967–972.
397. Wu, H., Deng, X., Shi, Y., Su, Y., Wei, J. and Duan, H., 2016. PGC-1α, glucose metabolism and type 2 diabetes mellitus. *Journal of Endocrinology*, 229, R99–R115.
398. Abd-Alla, H.I., Albalawy, M.A., Aly, H.F., Shalaby, N.M. and Shaker, K.H., 2014. Flavone composition and antihypercholesterolemic and antihyperglycemic activities of *Chrysanthemum coronarium* L. *Zeitschrift fur Naturforschung C*, 69(5–6), 199–208.
399. Liao, Z., Chen, X. and Wu, M., 2010. Antidiabetic effect of flavones from *Cirsium japonicum* DC in diabetic rats. *Archives of Pharmacal Research*, 33(3), 353–362.
400. Wan, Y., Liu, L.Y., Hong, Z.F. and Peng, J., 2014. Ethanol extract of *Cirsium japonicum* attenuates hepatic lipid accumulation via AMPK activation in human HepG2 cells. *Experimental and Therapeutic Medicine*, 8(1), 79–84.
401. Kirchhoff, R., Beckers, C.H., Kirchhoff, G.M., Trinczek-Gärtner, H., Petrowicz, O. and Reimann, H.J., 1994. Increase in choleresis by means of artichoke extract. *Phytomedicine*, 1(2), 107–115.
402. Rodriguez, T.S., Gimenez, D.G. and de la Puerta Vazquez, R., 2002. Choleretic activity and biliary elimination of lipids and bile acids induced by an artichoke leaf extract in rats. *Phytomedicine*, 9, 687–693.
403. Fantini, N., Colombo, G., Giori, A., Riva, A., Morazzoni, P., Bombardelli, E. and Carai, M.A., 2011. Evidence of glycemia-lowering effect by a *Cynara scolymus* L. extract in normal and obese rats. *Phytotherapy Research*, 25(3), 463–466.
404. Glasl, S., Tsendayush, D., Batchimeg, U., Holec, N., Wurm, E., Kletter, C., Gunbilig, D., Daariimaa, K., Narantuya, S. and Thalhammer, T., 2007. Choleretic effects of the Mongolian medicinal plant Saussurea amara in the isolated perfused rat liver. *Planta medica*, 73(1), 59–66.
405. Shimoda, H., Ninomiya K., Nishida, N., Yoshino, T., Morikawa, T., Matsuda, H., and Yoshikawa, M., 2003. Anti-hyperlipidemic sesquiterpenes and new sesquiterpene glycosides from the leaves of artichoke (Cynara scolymus L.): structure requirement and mode of action. *Bioorganic & Medicinal Chemistry letters*, 13(8): 223-228.
406. Rondanelli, M., Opizzi, A., Faliva, M., Sala, P., Perna, S., Riva, A., Morazzoni, P., Bombardelli, E. and Giacosa, A., 2014. Metabolic management in overweight subjects with naive impaired fasting glycaemia by means of a highly standardized extract from *Cynara scolymus*: A double-blind, placebo-controlled, randomized clinical trial. *Phytotherapy Research*, 28(1), 33–41.
407. Mocelin, R., Marcon, M., Santo, G.D., Zanatta, L., Sachett, A., Schönell, A.P., Bevilaqua, F. et al., 2016. Hypolipidemic and antiatherogenic effects of *Cynara scolymus* in cholesterol-fed rats. *Revista Brasileira de Farmacognosia*, 26(2), 233–239.
408. Dhandapani, R., 2007. Hypolipidemic activity of *Eclipta prostrata* (L.) L. leaf extract in atherogenic diet induced hyperlipidemic rats. *Indian Journal of Experimental Biology*, 45(7), 617–619.
409. Zhao, Y., Peng, L., Lu, W., Wang, Y., Huang, X., Gong, C., He, L., Hong, J., Wu, S. and Jin, X., 2015. Effect of *Eclipta prostrata* on lipid metabolism in hyperlipidemic animals. *Experimental Gerontology*, 62, 37–44.
410. Kozarsky, K.F., Donahee, M.H., Rigotti, A., Iqbal, S.N., Edelman, E.R. and Kriger, M., 1997. Overexpression of the HDL receptor SR-B1 alters plasma HDL and bile cholesterol levels. *Nature*, 387, 414–417.
411. Satoh, H., Audrey Nguyen, M.T., Kudoh, A. and Watanabe, T., 2013. Yacon diet (*Smallanthus sonchifolius*, Asteraceae) improves hepatic insulin resistance via reducing Trb3 expression in Zucker fa/fa rats. *Nutrition & Diabetes*, 3, e70.

412. Thing, F.S.L., Dunagi, C.B. and Klip, A., 2005. Turning signals on and off: GLUT4 traffic in the insulin-signaling highway. *Physiology*, 20, 271–284.
413. Du, K., Herxing, S., Kulkarni, R.N. and Montminy, M., 2003. TRB3: A tribbles homolog that inhibits Akt/PKB activation by insulin in liver. *Science*, 300, 1574–1577.
414. Xiang, Z., He, F., Kang, T.G., Dou, D.Q., Gai, K., Shi, Y.Y., Kim, Y.H. and Dong, F., 2010. Anti-diabetes constituents in leaves of *Smallanthus sonchifolius. Natural Product Communications*, 5(1), 95–98.
415. Valentova, K., Cvak, L., Muck, A., Ulrichova, J. and Simanek, V., 2003. Antioxidant activity of extracts from the leaves of *Smallanthus sonchifolius. European Journal of Nutrition*, 42(1), 61–66.
416. Takenaka, M., Yan, X., Ono, H., Yoshida, M., Nagata, T. and Nakanishi, T., 2003. Caffeic acid derivatives in the roots of yacon (*Smallanthus sonchifolius*). *Journal of Agricultural and Food Chemistry*, 51(3), 793–796.
417. Inoue, A., Tamogami, S., Kato, H., Nakazato, Y., Akiyama, M., Kodama, O., Akatsuka, T. and Hashidoko, Y., 1995. Antifungal melampolides from leaf extracts of *Smallanthus sonchifolius. Phytochemistry*, 39(4), 845–848.
418. Ebrahimpour Koujan, S., Gargari, B.P., Mobasseri, M., Valizadeh, H. and Asghari-Jafarabadi, M., 2015. Effects of *Silybum marianum* (L.) Gaertn. (silymarin) extract supplementation on antioxidant status and hs-CRP in patients with type 2 diabetes mellitus: A randomized, triple-blind, placebo-controlled clinical trial. *Phytomedicine*, 22(2), 290–296.
419. Liu, Y.J., Shieh, P.C., Lee, J.C., Chen, F.A., Lee, C.H., Kuo, S.C., Ho, C.T., Kuo, D.H., Huang, L.J. and Way, T.D., 2014. Hypolipidemic activity of Taraxacum mongolicum associated with the activation of AMP-activated protein kinase in human HepG2 cells. *Food & Function*, 5(8), 1755–1762.
420. Lieberman, M., Marks, A.D., Smith, C.M. and Marks, D.B., 2007. *Marks' Essential Medical Biochemistry*. Philadelphia, PA: Lippincott Williams & Wilkins.
421. Wu, S.Y., Wang, G.F., Liu, Z.Q., Rao, J.J., Lü, L., Xu, W., Wu, S.G., Zhang, J.J., 2009. Effect of geniposide, a hypoglycemic glucoside, on hepatic regulating enzymes in diabetic mice induced by a high-fat diet and streptozotocin. *Acta Pharmacologica Sinica*, 30(2), 202–208.
422. Kojima, K., Shimada, T., Nagareda, Y., Watanabe, M., Ishizaki, J., Sai, Y., Miyamoto, K. and Aburada, M., 2011. Preventive effect of geniposide on metabolic disease status in spontaneously obese type 2 diabetic mice and free fatty acid-treated HepG2 cells. *Biological and Pharmaceutical Bulletin*, 34(10), 1613–1618.
423. Aburada, M., Takeda, S., Shibata, Y. and Harada, M., 1978. Pharmacological studies of gardenia fruit. III. Relationship between in vivo hydrolysis of geniposide and its choleretic effect in rats. *Journal of Pharmacobio-Dynamics*, 1, 81–88.
424. Raffa, R.B. (Ed.), 2014. *Kratom and Other Mitragynines: The Chemistry and Pharmacology of Opioids from a Non-Opium Source*. Boca Raton, FL: CRC Press.
425. Toure, H., Balansard, G., Pauli, A.M. and Scotto, A.M., 1996. Pharmacological investigation of alkaloids from leaves of *Mitragyna inermis* (Rubiaceae). *Journal of Ethnopharmacology*, 54(1), 59–62.
426. Kapp, F.G., Maurer, H.H., Auwärter, V., Winkelmann, M. and Hermanns-Clausen, M., 2011. Intrahepatic cholestasis following abuse of powdered kratom (*Mitragyna speciosa*). *Journal of Medical Toxicology*, 7(3), 227–231.
427. Evans, M.J., Mahaney, P.E., Borges-Marcucci, L., Lai, K., Wang, S., Krueger, J.A., Gardell, S.J. et al., 2009. A synthetic farnesoid X receptor (FXR) agonist promotes cholesterol lowering in models of dyslipidemia. *American Journal of Physiology-Gastrointestinal and Liver Physiology*, 296(3), G543–G552.
428. Flatt, B., Martin, R., Wang, T.L., Mahaney, P., Murphy, B., Gu, X.H., Foster, P. et al., 2009. Discovery of XL335 (WAY-362450), a highly potent, selective, and orally active agonist of the farnesoid X receptor (FXR). *Journal of Medicinal Chemistry*, 52(4), 904–907.
429. Sabetghadam, A., Ramanathan, S., Sasidharan, S. and Mansor, S.M., 2013. Subchronic exposure to mitragynine, the principal alkaloid of *Mitragyna speciosa*, in rats. *Journal of Ethnopharmacology*, 146(3), 815–823.
430. Kumar, V., Singh, S., Khanna, A.K., Khan, M.M., Chander, R., Mahdi, F., Saxena, J.K., Singh, R. and Singh, R.K., 2008. Hypolipidemic activity of *Anthocephalus indicus* (kadam) in hyperlipidemic rats. *Medicinal Chemistry Research*, 17(2–7), 152–158.
431. Alam, M.A., Subhan, N., Chowdhury, S.A., Awal, M.A., Mostofa, M., Rashid, M.A., Hasan, C.M., Nahar, L. and Sarker, S.D., 2011. *Anthocephalus cadamba* (Roxb.) Miq., Rubiaceae, extract shows hypoglycemic effect and eases oxidative stress in alloxan-induced diabetic rats. *Revista Brasileira de Farmacognosia*, 21(1), 155–164.
432. Pandey, A. and Negi, P.S., 2016. Traditional uses, phytochemistry and pharmacological properties of *Neolamarckia cadamba*: A review. *Journal of Ethnopharmacology*, 181, 118–135.

433. Ishikawa, A., Yamashita, H., Hiemori, M., Inagaki, E., Kimoto, M., Okamoto, M., Tsuji, H., Memon, A.N., Mohammadi, A. and Natori, Y., 2007. Characterization of inhibitors of postprandial hyperglycemia from the leaves of *Nerium indicum*. *Journal of Nutritional Science and Vitaminology*, 53(2), 166–173.
434. Wan, C.W., Wong, C.N.Y., Pin, W.K., Wong, M.H.Y., Kwok, C.Y., Chan, R.Y.K., Yu, P.H.F. and Chan, S.W., 2013. Chlorogenic acid exhibits cholesterol lowering and fatty liver attenuating properties by up-regulating the gene expression of PPAR-α in hypercholesterolemic rats induced with a high-cholesterol diet. *Phytotherapy Research*, 27(4), 545–551.
435. Muthusamy, V.S., Saravanababu, C., Ramanathan, M., Bharathi Raja, R., Sudhagar, S., Anand, S. and Lakshmi, B.S., 2010. Inhibition of protein tyrosine phosphatase 1B and regulation of insulin signalling markers by caffeoyl derivatives of chicory (Cichorium intybus) salad leaves. *British Journal of Nutrition*, 104(6), 813–823.
436. Gao, Y., Su, Y., Huo, Y., Mi, J., Wang, X., Wang, Z., Liu, Y. and Zhang, H., 2014. Identification of antihyperlipidemic constituents from the roots of *Rubia yunnanensis* Diels. *Journal of Ethnopharmacology*, 155(2), 1315–1321.
437. Tao, J., Morikawa, T., Ando, S., Matsuda, H. and Yoshikawa, M., 2003. Bioactive constituents from Chinese natural medicines. XI. Inhibitors on NO production and degranulation in RBL-2H3 from *Rubia yunnanensis*: Structures of rubianosides II, III, and IV, rubianol-g, and rubianthraquinone. *Chemical and Pharmaceutical Bulletin*, 51(6), 654–662.
438. Kawasaki, Y., Goda, Y. and Yoshihira, K., 1992. The mutagenic constituents of *Rubia tinctorum*. *Chemical & Pharmaceutical Bulletin*, 40(6), 1504–1509.
439. Do, M.T., Hwang, Y.P., Kim, H.G., Na, M. and Jeong, H.G., 2013. Mollugin inhibits proliferation and induces apoptosis by suppressing fatty acid synthase in HER2-overexpressing cancer cells. *Journal of Cellular Physiology*, 228(5), 1087–1097.
440. Kim, K.J., Lee, J.S., Kwak, M.K., Choi, H.G., Yong, C.S., Kim, J.A., Lee, Y.R., Lyoo, W.S. and Park, Y.J., 2009. Anti-inflammatory action of mollugin and its synthetic derivatives in HT-29 human colonic epithelial cells is mediated through inhibition of NF-κB activation. *European Journal of Pharmacology*, 622(1), 52–57.
441. Jeong, G.S., Lee, D.S., Kim, D.C., Jahng, Y., Son, J.K., Lee, S.H. and Kim, Y.C., 2011. Neuroprotective and anti-inflammatory effects of mollugin via up-regulation of heme oxygenase-1 in mouse hippocampal and microglial cells. *European Journal of Pharmacology*, 654(3), 226–234.
442. Srinivasan, M., Padmanabhan, M. and Prince, P.S., 2005. Effect of aqueous *Enicostemma littorale* Blume extract on key carbohydrate metabolic enzymes, lipid peroxidesand antioxidants in alloxan-induced diabetic rats. *Journal of Pharmacy and Pharmacology*, 57(4), 497–503.
443. Vaidya, H., Rajani, M., Sudarsanam, V., Padh, H. and Goyal, R., 2009. Swertiamarin: A lead from *Enicostemma littorale* Blume. for anti-hyperlipidaemic effect. *European Journal of Pharmacology*, 617(1–3), 108–112.
444. Bhatt, N.M., Chavda, M., Desai, D., Zalawadia, R., Patel, V.B., Burade, V., Sharma, A.K., Singal, P.K. and Gupta, S., 2012. Cardioprotective and antihypertensive effects of *Enicostemma littorale* Blume extract in fructose-fed rats. *Canadian Journal of Physiology and Pharmacology*, 90(8), 1065–1073.
445. Saravanan, S., Islam, V.I., Thirugnanasambantham, K., Pazhanivel, N., Raghuraman, N., Paulraj, M.G. and Ignacimuthu, S., 2014. Swertiamarin ameliorates inflammation and osteoclastogenesis intermediates in IL-1β induced rat fibroblast-like synoviocytes. *Inflammation Research*, 63(6), 451–462.
446. Sezik, E., Aslan, M., Yesilada, E. and Ito, S., 2005. Hypoglycaemic activity of *Gentiana olivieri* and isolation of the active constituent through bioassay-directedfractionation techniques. *Life Sciences*, 76(11), 1223–1238.
447. Yuan, L., Han, X., Li, W., Ren, D. and Yang, X., 2016. Isoorientin prevents hyperlipidemia and liver injury by regulating lipid metabolism, antioxidant capability, and inflammatory cytokine release in high-fructose-fed mice. *Journal of Agricultural and Food Chemistry*, 64(13), 2682–2689.
448. Mansoor, A., Samad, A., Zaidi, M.I. and Aftab, K., 1998. Hypotensive effect of *Gentiana olivieri* and its alkaloid gentianine in rats. *Pharmacy and Pharmacology Communications*, 4(4), 229–230.
449. Kuriyan, R., Raj, T., Srinivas, S.K., Vaz, M., Rajendran, R. and Kurpad, A.V., 2007. Effect of *Caralluma fimbriata* extract on appetite, food intake and anthropometry in adult Indian men and women. *Appetite*, 48(3), 338–344.
450. Kunert, O., Rao, V.G., Babu, G.S., Sujatha, P., Sivagamy, M., Anuradha, S., Rao, B.V.A. et al., 2008. Pregnane glycosides from *Caralluma adscendens* var. fimbriata. *Chemistry & Biodiversity*, 5(2), 239–250.
451. Van Heerden, F.R., Horak, R.M., Maharaj, V.J., Vleggaar, R., Senabe, J.V. and Gunning, P.J., 2007. An appetite suppressant from *Hoodia* species. *Phytochemistry*, 68, 2545–2553.

452. Sudhakara, G., Mallaiah, P., Sreenivasulu, N., Sasi Bhusana Rao, B., Rajendran, R. and Saralakumari, D., 2014. Beneficial effects of hydro-alcoholic extract of Caralluma fimbriata against high-fat diet-induced insulin resistance and oxidative stress in Wistar male rats. *Journal of Physiology and Biochemistry*, 70(2), 311–320.
453. Ghosh, A., 2010. *Mayo Clinic Internal Medicine Board Review*. Oxford: Oxford University Press.
454. Shafi, S., Stepanova, I.P., Fitzsimmons, C., Bowyer, D.E., Welzel, D. and Born, G.V., 2000. Effects of reserpine on expression of the LDL receptor in liver and on plasma and tissue lipids, low density lipoprotein and fibrinogen in rabbits in vivo. *Atherosclerosis*, 149(2), 267–275.
455. Shafi, S., Stepanova, I.P., Fitzsimmons, C., Bowyer, D.E. and Born, G.V., 2002. Long-term low-dose treatment with reserpine of cholesterol-fed rabbits reduces cholesterol in plasma, non-high density lipoproteins and arterial walls. *Journal of Cardiovascular Pharmacology*, 40(1), 67–79.
456. Bhagavan, N.V. and Ha, C.E., 2015. *Essentials of Medical Biochemistry: With Clinical Cases*. San Diego, CA: Academic Press.
457. Maurya, R., Srivastava, A., Shah, P., Siddiqi, M.I., Rajendran, S.M., Puri, A. and Yadav, P.P., 2012. β-Amyrin acetate and β-amyrin palmitate as antidyslipidemic agents from *Wrightia tomentosa* leaves. *Phytomedicine*, 19(8–9), 682–685.
458. Ye, Z., Huang, Q., Ni, H.X. and Wang, D., 2008. Cortex Lycii Radicis extracts improve insulin resistance and lipid metabolism in obese-diabetic rats. *Phytotherapy Research*, 22(12), 1665–1670.
459. Hu, C.K., Lee, Y.J., Colitz, C.M., Chang, C.J. and Lin, C.T., 2012. The protective effects of *Lycium barbarum* and *Chrysanthemum morifolum* on diabetic retinopathies in rats. *Veterinary Ophthalmology*, 15(Suppl 2), 65–71.
460. Xiao, J., Xing, F., Huo, J., Fung, M.L., Liong, E.C., Ching, Y.P., Xu, A., Chang, R.C.C., So, K.F. and Tipoe, G.L., 2014. *Lycium barbarum* polysaccharides therapeutically improve hepatic functions in non-alcoholic steatohepatitis rats and cellular steatosis model. *Scientific Reports*, 4, 5587.
461. Potterat, O., 2010. Goji (*Lycium barbarum* and L. chinense): Phytochemistry, pharmacology and safety in the perspective of traditional uses and recent popularity. *Planta medica*, 76(1), 7–19.
462. Guslandi, M. and Braga, P.C., 1993. *Drug-Induced Injury to the Digestive System*. New York: Springer Science & Business Media.
463. Mohan, M., Jaiswal, B.S. and Kasture, S., 2009. Effect of Solanum torvum on blood pressure and metabolic alterations in fructose hypertensive rats. *Journal of Ethnopharmacology*, 126(1), 86–89.
464. Gandhi, G.R., Ignacimuthu, S. and Paulraj, M.G., 2011. *Solanum torvum* Swartz. fruit containing phenolic compounds shows antidiabetic and antioxidant effects in streptozotocin induced diabetic rats. *Food and Chemical Toxicology*, 49(11), 2725–2733.
465. Yeh, C.T., Huang, W.H. and Yen, G.C., 2009. Antihypertensive effects of Hsian-tsao and its active compound in spontaneously hypertensive rats. *Journal of Nutritional Biochemistry*, 20(11), 866–875.
466. Kwon, M.J., Song, Y.S., Choi, M.S. and Song, Y.O., 2003. Red pepper attenuates cholesteryl ester transfer protein activity and atherosclerosis in cholesterol-fed rabbits. *Clinica Chimica Acta*, 332(1–2), 37–44.
467. Li, L., Chen, J., Ni, Y., Feng, X., Zhao, Z., Wang, P., Sun, J. et al., 2012. TRPV1 activation prevents nonalcoholic fatty liver through UCP2 upregulation in mice. *Pflügers Archiv-European Journal of Physiology*, 463(5), 727–732.
468. Zhang, L., Fang, G., Zheng, L., Chen, Z. and Liu, X., 2013. Hypocholesterolemic effect of capsaicinoids in rats fed diets with or without cholesterol. *Journal of Agricutural and Food Chemistry*, 61(18), 4287–4293.
469. Abdel Salam, O.M., Heikal, O.A. and El-Shenawy, S.M., 2005. Effect of capsaicin on bile secretion in the rat. *Pharmacology*, 73(3), 121–128.
470. Zhao, C. and Dahlman-Wright, K., 2010. Liver X receptor in cholesterol metabolism. *Journal of Endocrinology*, 204(3), 233–240.
471. Hammouda, Y. and Amer, M.S., 1966. Antidiabetic effect of tecomine and tecostanine. *Journal of Pharmaceutical Sciences*, 55(12), 1452–1454.
472. Hammouda, Y., Rashid, A.K. and Amer, M.S., 1964. Hypoglycaemic properties of tecomine and tecostanine. *Journal of Pharmacy and Pharmacology*, 16(12), 833–834.
473. Costantino, L., Raimondi, L., Pirisino, R., Brunetti, T., Pessotto, P., Giannessi, F., Lins, A.P., Barlocco, D., Antolini, L. and El-Abady, S.A., 2003. Isolation and pharmacological activities of the *Tecoma stans* alkaloids. *Il Farmaco*, 58(9), 781–785.
474. Phunikhom, K., Khampitak, K., Aromdee, C., Arkaravichien, T. and Sattayasai, J., 2015. Effect of *Andrographis paniculata* extract on triglyceride levels of the patients with hypertriglyceridemia: A randomized controlled trial. *Journal of the Medical Association of Thailand*, 98(Suppl 6), S41–S47.

475. Ding, L., Li, J., Song, B., Xiao, X., Huang, W., Zhang, B., Tang, X. et al., 2014. Andrographolide prevents high-fat diet–induced obesity in C57BL/6 mice by suppressing the sterol regulatory element-binding protein pathway. *Journal of Pharmacology and Experimental Therapeutics*, 351(2), 474–483.
476. Yoshikawa, T., Ide, T., Shimano, H., Yahagi, N., Amemiya-Kudo, M., Matsuzaka, T., Yatoh, S. et al., 2003. Cross-talk between peroxisome proliferator-activated receptor (PPAR) α and liver X receptor (LXR) in nutritional regulation of fatty acid metabolism. I. PPARs suppress sterol regulatory element binding protein-1c promoter through inhibition of LXR signaling. *Molecular Endocrinology*, 17(7), 1240–1254.
477. König, B., Koch, A., Spielmann, J., Hilgenfeld, C., Stangl, G.I. and Eder, K., 2007. Activation of PPARα lowers synthesis and concentration of cholesterol by reduction of nuclear SREBP-2. *Biochemical Pharmacology*, 73(4), 574–585.
478. Shanmugasundaram, P. and Venkataraman, S., 2006. Hepatoprotective and antioxidant effects of Hygrophila auriculata (K. Schum) Heine Acanthaceae root extract. *Journal of Ethnopharmacology*, 104(1), 124–128.
479. Raj, V.P., Chandrasekhar, R.H., Vijayan, P., Dhanaraj, S.A., Rao, M.C., Rao, V.J. and Nitesh, K., 2010. In vitro and in vivo hepatoprotective effects of the total alkaloid fraction of *Hygrophila auriculata* leaves. *Indian Journal of Pharmacology*, 42(2), 99.
480. Vijayakumar, M., Govindarajan, R., Rao, G.M., Rao, C.V., Shirwaikar, A., Mehrotra, S. and Pushpangadan, P., 2006. Action of *Hygrophila auriculata* against streptozotocin-induced oxidative stress. *Journal of Ethnopharmacology*, 104(3), 356–361.
481. Chauhan, N.S. and Dixit, V.K., 2010. *Asteracantha longifolia* (L.) Nees, Acanthaceae: chemistry, traditional, medicinal uses and its pharmacological activities-a review. *Revista Brasileira de Farmacognosia*, 20(5), 812–817.
482. Feldman, M., Friedman, L.S. and Brandt, L.J., 2010. *Sleisenger and Fordtran's Gastrointestinal and Liver Disease: Pathophysiology, Diagnosis, Management, Expert Consult Premium Edition-Enhanced Online Features* (Vol. 1). Elsevier Health Sciences: Philadelphia.
483. Im, K.R., Jeong, T.S., Kwon, B.M., Baek, N.I., Kim, S.H. and Kim, D.K., 2006. Acyl-CoA: Cholesterol acyltransferase inhibitors from *Ilex macropoda*. *Archives of Pharmacal Research*, 29(3), 191–194.
484. Tang, J.J., Li, J.G., Qi, W., Qiu, W.W., Li, P.S., Li, B.L. and Song, B.L., 2011. Inhibition of SREBP by a small molecule, betulin, improves hyperlipidemia and insulin resistance and reduces atherosclerotic plaques. *Cell Metabolism*, 13(1), 44–56.
485. Hussain, M.S., Fareed, S. and Ali, M., 2012. Simultaneous HPTLC-UV530 nm analysis and validation of bioactive lupeol and stigmasterol in *Hygrophila auriculata* (K. Schum) Heine. *Asian Pacific Journal of Tropical Biomedicine*, 2(2), S612–S617.
486. Harnafi, H., Hennebelle, T., Martin-Nizard, F. and Amrani, S., 2010. Hypolipidemic effect of methanol and aqueous fractions from sweet basil in hyperlipidemic mice. *Phytotherapie*, 8(1), 9–15.
487. Han, Y.M., Oh, H., Na, M., Kim, B.S., Oh, W.K., Kim, B.Y., Jeong, D.G., Ryu, S.E., Sok, D.E. and Ahn, J.S., 2005. PTP1B inhibitory effect of abietane diterpenes isolated from *Salvia miltiorrhiza*. *Biological and Pharmaceutical Bulletin*, 28(9), 1795–1797.
488. Ko, J.S., Ryu, S.Y., Kim, Y.S., Chung, M.Y., Kang, J.S., Rho, M.C., Lee, H.S. and Kim, Y.K., 2002. Inhibitory activity of diacylglycerol acyltransferase by tanshinones from the root of *Salvia miltiorrhiza*. *Archives of Pharmacal Research*, 25(4), 446–448.
489. Huang, M., Xie, Y., Chen, L., Chu, K., Wu, S., Lu, J., Chen, X., Wang, Y. and Lai, X., 2012. Antidiabetic effect of the total polyphenolic acids fraction from *Salvia miltiorrhiza* Bunge in diabetic rats. *Phytotherapy Research*, 26(6), 944–948.
490. Oshima, Y., Kawakami, Y., Kiso, Y., Hikino, H., Yang, L.L. and Yen, K.Y., 1984. Antihepatotoxic principles of *Salvia plebeia* herbs. 生薬学雑誌, 38(2), 201–202.
491. Lee, J. and Kim, M.S., 2007. The role of GSK3 in glucose homeostasis and the development of insulin resistance. *Diabetes Research and Clinical Practice*, 77, S49–S57.
492. Azevedo, M.F., Camsari, C., Sá, C.M., Lima, C.F., Fernandes-Ferreira, M. and Pereira-Wilson, C., 2010. Ursolic acid and luteolin-7 glucoside improve lipid profiles and increase liver glycogen content through glycogensynthase kinase-3. *Phytotherapy Research*, 24(Suppl 2), S220–S224.
493. Oikonomakos, N.G. and Somsak, L., 2008. Advances in glycogen phosphorylase inhibitor design. *Current Opinion in Investigational Drugs (London, England: 2000)*, 9(4), 379–395.
494. Weston, C.R. and Davis, R.J., 2001. Signaling specificity a complex affair. *Science*, 292, 2439–2440.
495. Chen, C.C., Hsu, C.Y., Chen, C.Y. and Liu, H.K., 2008. Fructus Corni suppresses hepatic gluconeogenesis related gene transcription, enhances glucose responsiveness of pancreatic beta-cells, and prevents toxin induced beta-cell death. *Journal of Ethnopharmacology*, 117(3), 483–490.

496. Jung, S.H., Ha, Y.J., Shim, E.K., Choi, S.Y., Jin, J.L., Yun-Choi, H.S. and Lee, J.R., 2007. Insulin-mimetic and insulin-sensitizing activities of a pentacyclic triterpenoid insulin receptor activator. *Biochemical Journal*, 403, 243–250.
497. Lee, S.J., Jang, H.J., Kim, Y., Oh, H.M., Lee, S., Jung, K., Kim, Y.H., Lee, W.S., Lee, S.W. and Rho, M.C., 2016. Inhibitory effects of IL-6-induced STAT3 activation of bio-active compounds derived from Salvia plebeia R. Br. *Process Biochemistry*, 51(12), 2222–2229.
498. Song, K.H., Lee, S.H., Kim, B.Y., Park, A.Y. and Kim, J.Y., 2013. Extracts of *Scutellaria baicalensis* reduced body weight and blood triglyceride in db/db Mice. *Phytotherapy Research*, 27(2), 244–250.
499. Akkol, E.K., Avci, G., Küçükkurt, I., Keleş, H., Tamer, U., Ince, S. and Yesilada, E., 2009. Cholesterol-reducer, antioxidant and liver protective effects of *Thymbra spicata* L. var. spicata. *Journal of Ethnopharmacology*, 126(2), 314–319.
500. Kim, E., Choi, Y., Jang, J. and Park, T., 2013. Carvacrol protects against hepatic steatosis in mice fed a high-fat diet by enhancing SIRT1-AMPK signaling. *Evidence-Based Complementary and Alternative Medicine*, 2013, 1–10.
501. Yang, Y.M., Han, C.Y., Kim, Y.J. and Kim, S.G., 2010. AMPK-associated signaling to bridge the gap between fuel metabolism and hepatocyte viability. *World Journal of Gastroenterology*, 16(30), 3731–3742.
502. Aydin, Y., Kutlay, O., Ari, S., Duman, S., Uzuner, K. and Aydin, S., 2007. Hypotensive effects of carvacrol on the blood pressure of normotensive rats. *Planta Medica*, 73(13), 1365–1371.
503. Lin, H.R., 2012. Triterpenes from *Alisma orientalis* act as farnesoid X receptor agonists. *Bioorganic & Medicinal Chemistry Letters*, 22(14), 4787–4792.
504. Zhu, X., Zhang, W., Pang, X., Wang, J., Zhao, J. and Qu, W., 2011. Hypolipidemic effect of n-butanol extract from *Asparagus officinalis* L. in mice fed a high-fat diet. *Phytotherapy Research*, 25(8), 1119–1124.
505. Deng, Y., He, K., Ye, X., Chen, X., Huang, J., Li, X., Yuan, L., Jin, Y., Jin, Q. & Li, P., 2012. Saponin rich fractions from *Polygonatum odoratum* (Mill.) Druce with more potential hypoglycemic effects. *Journal of Ethnopharmacology*, 141(1), 228–233.
506. Shu, X.S., Lv, J.H., Tao, J., Li, G.M., Li, H.D. and Ma, N., 2009. Antihyperglycemic effects of total flavonoids from Polygonatum odoratum in STZ and alloxan-induced diabetic rats. *Journal of Ethnopharmacology*, 124(3), 539–543.
507. Guo, H., Zhao, H., Kanno, Y., Li, W., Mu, Y., Kuang, X., Inouye, Y., Koike, K., Jiang, H. and Bai, H., 2013. A dihydrochalcone and several homoisoflavonoids from Polygonatum odoratum are activators of adenosine monophosphate-activated protein kinase. *Bioorganic & Medicinal Chemistry Letters*, 23(11), 3137–3139.
508. Chen, F., Xiong, H., Wang, J., Ding, X., Shu, G. and Mei, Z., 2013. Antidiabetic effect of total flavonoids from Sanguis draxonis in type 2 diabetic rats. *Journal of Ethnopharmacology*, 149(3), 729–736.
509. Wang, T., Choi, R.C., Li, J., Bi, C.W., Ran, W., Chen, X., Dong, T.T., Bi, K. and Tsim, K.W., 2012. Trillin, a steroidal saponin isolated from the rhizomes of *Dioscorea nipponica*, exerts protective effects against hyperlipidemia and oxidative stress. *Journal of Ethnopharmacology*, 139(1), 214–220.
510. Yu, H., Zheng, L., Xu, L., Yin, L., Lin, Y., Li, H., Liu, K. and Peng, J., 2015. Potent effects of the total saponins from *Dioscorea nipponica* Makino against streptozotocin-induced type 2 diabetes mellitus in rats. *Phytotherapy Research*, 29(2), 228–240.
511. Hiroshi, H., Masako, K., Yutaka, S. and Chohachi, K., 1989. Mechanisms of hypoglycemic activity of aconitan A, a glycan from *Aconitum carmichaeli* roots. *Journal of Ethnopharmacology*, 25(3), 295–304.
512. Guo, C., Li, C., Yu, Y., Chen, W., Ma, T. and Zhou, Z., 2016. Antihyperglycemic and antihyperlipidemic activities of protodioscin in a high-fat diet and streptozotocin-induced diabetic rats. *RSC Advances*, 6(91), 88640.
513. Ma, W., Ding, H., Gong, X., Liu, Z., Lin, Y., Zhang, Z. and Lin, G., 2015. Methyl protodioscin increases ABCA1 expression and cholesterol efflux while inhibiting gene expressions forsynthesis of cholesterol and triglycerides by suppressing SREBP transcription and microRNA 33a/b levels. *Atherosclerosis*, 239(2), 566–570.
514. Jeong, E.J., Jegal, J., Ahn, J., Kim, J. and Yang, M.H., 2016. Anti-obesity effect of *Dioscorea oppositifolia* extract in high-fat diet-induced obese mice and its chemical characterization. *Biological and Pharmaceutical Bulletin*, 39(3), 409–414.
515. Choi, S.Y., Lee, M.H., Choi, J.H. and Kim, Y.K., 2012. 2,3,22,23-tetrahydroxyl-2,6,10,15,19,23-hexamethyl-6,10,14,18-tetracosatetraene, an acyclic triterpenoid isolated from the seeds of *Alpinia katsumadai*, Inhibits acyl-CoA: cholesterol acyltransferase activity. *Biological and Pharmaceutical Bulletin*, 35(11), 2092–2096.

516. Li, T., Francl, J.M., Boehme, S. and Chiang, J.Y., 2013. Regulation of cholesterol and bile acid homeostasis by the cholesterol 7α-hydroxylase/steroid response element-binding protein 2/microRNA-33a axis in mice. *Hepatology*, 58(3), 1111–1121.
517. Enjoji, M., Kohjima, M., Kotoh, K. and Nakamuta, M., 2012. Metabolic disorders and steatosis in patients with chronic hepatitis C: Metabolic strategies for antiviral treatments. *International Journal of Hepatology*, 2012, 1–7.
518. Chang, N.W., Wu, C.T., Wang, S.Y., Pei, R.J. and Lin, C.F., 2010. *Alpinia pricei* Hayata rhizome extracts have suppressive and preventive potencies against hypercholesterolemia. *Food and Chemical Toxicology*, 48(8–9), 2350–2356.
519. Charoenwanthanang, P., Lawanprasert, S., Phivthong-Ngam, L., Piyachaturawat, P., Sanvarinda, Y. and Porntadavity, S., 2011. Effects of *Curcuma comosa* on the expression of atherosclerosis-related cytokine genes in rabbits fed a high-cholesterol diet. *Journal of Ethnopharmacology*, 134(3), 608–613.
520. Yiu, W.F., Kwan, P.L., Wong, C.Y., Kam, T.S., Chiu, S.M., Chan, S.W. and Chan, R., 2011. Attenuation of fatty liver and prevention of hypercholesterolemia by extract of *Curcuma longa* through regulating the expression of CYP7A1, LDL-receptor, heme oxygenase-1, and HMG-CoA reductase. *Journal of Food Science*, 76(3), H80–H89.
521. Verma, S.K., Jain, V. and Katewa, S.S., 2009. Blood pressure lowering, fibrinolysis enhancing and antioxidant activities of cardamom (*Elettaria cardamomum*). *Indian Journal of Biochemistry and Biophysics*, 46(6), 503–506.
522. Chobanian, A.V., Bakris, G.L., Black, H.R., Cushman, W.C., Green, L.A., Izzo, J.L., Jones, D.W. et al., 2003. Seventh report of the jin National Committee on prevention, detection, evaluation, and treatment of high blood pressure. *Hypertension*, 42, 1206–1252.
523. Abe, M., Ozawa, Y., Uda, Y., Yamada, Y., Morimitsu, Y., Nakamura, Y. and Osawa, T., 2002. Labdane-type diterpene dialdehyde, pungent principle of myoga, *Zingiber mioga* Roscoe. *Bioscience, Biotechnology, and Biochemistry*, 66(12), 2698–2700.
524. Tanabe, M., Chen, Y.-D., Saito, K.-I. and Kano, Y., 1993. Cholesterol biosynthesis inhibitory component from *Zingiber officinale* Roscoe. *Chemical and Pharmaceutical Bulletin*, 41(4), 710–713.
525. Lee, D.H., Ahn, J., Jang, Y.J., Ha, T.Y. and Jung, C.H., 2016. *Zingiber mioga* reduces weight gain, insulin resistance and hepatic gluconeogenesis in diet-induced obese mice. *Experimental and Therapeutic Medicine*, 12(1), 369–376.
526. Lei, L., Liu, Y., Wang, X., Jiao, R., Ma, K.Y., Li, Y.M., Wang, L. et al., 2014. Plasma cholesterol-lowering activity of gingerol- and shogaol-enriched extract is mediated by increasing sterol excretion. *Journal of Agricutural and Food Chemistry*, 62(43), 10515–10521.
527. Le Goff, W., Guerin, M. and Chapman, M.J., 2004. Pharmacological modulation of cholesteryl ester transfer protein, a new therapeutic target in atherogenic dyslipidemia. *Pharmacology & Therapeutics*, 101(1), 17–38.
528. Gerber, P.A., Spinas, G.A. and Berneis, K., 2012. Small dense low-density lipoprotein particles: Priority as a treatment target in Type 2 diabetes? *Diabetes Management*, 2(1), 65–74.
529. Williams, K.J. and Tabas, I., 1995. The response-to-retention hypothesis of early atherogenesis. *Arteriosclerosis, Thrombosis, and Vascular Biology*, 15(5), 551–561.
530. Choi, S.Y., Park, G.S., Lee, S.Y., Kim, J.Y. and Kim, Y.K., 2011. The conformation and CETP inhibitory activity of [10]-dehydrogingerdione isolated from *Zingiber officinale. Archives of Pharmacal Research*, 34(5), 727–731.
531. Saravanan, G., Ponmurugan, P., Deepa, M.A. and Senthilkumar, B., 2014. Anti-obesity action of gingerol: Effect on lipid profile, insulin, leptin, amylase and lipase in male obese rats induced by a high-fat diet. *Journal of the Science of Food and Agriculture*, 94(14), 2972–2977.
532. Tzeng, T.F., Liou, S.S., Chang, C.J. and Liu, I.M., 2015. 6-gingerol protects against nutritional steatohepatitis by regulating key genes related to inflammation and lipid metabolism. *Nutrients*, 7(2), 999–1020.
533. Ebrahimzadeh Attari, V., Asghari Jafarabadi, M., Zemestani, M. and Ostadrahimi, A., 2015. Effect of *Zingiber officinale* supplementation on obesity management with respect to the uncoupling protein 1 -3826A>G and ß3-adrenergic receptor Trp64Arg polymorphism. *Phytotherapy Research*, 29(7), 1032–1039.
534. Buchholz, T. and Melzig, M.K., 2016. Medicinal plants traditionally used for the treatment of obesity and diabetes mellitus-Screening for pancreatic lipase and a-amylase inhibition. *Phytotherapy Research*, 30, 260–266.
535. Chang, C.J., Tzeng T.F., Chang, Y.S. and Liu, I.M., 2012. Beneficial impact of *Zingiber zerumbet* on insulin sensitivity in fructose-fed rats. *Planta Medica*, 78(4), 317–325.

536. Chang, C.J., Tzeng, T.F., Liou, S.S., Chang, Y.S. and Liu, I.M., 2011. Kaempferol regulates the lipid-profile in high-fat diet-fed rats through an increase in hepatic PPARα levels. *Planta Medica*, 77(17), 1876–1882.
537. Tzeng, T.F., Lu, H.J., Liou, S.S., Chang, C.J. and Liu, I.M., 2014. Lipid-lowering effects of zerumbone, a natural cyclic sesquiterpene of *Zingiber zerumbet* Smith, in high-fat diet-induced hyperlipidemic hamsters. *Food and Chemical Toxicology*, 69, 132–139.
538. Azzam, K.M. and Fessler, M.B., 2012. Crosstalk between reverse cholesterol transport and innate immunity. *Trends in Endocrinology & Metabolism*, 23(4), 169–178.
539. Guo, A., Jin, L., Deng, Z., Cai, S., Guo, S. and Lin, W., 2008. New Stemona alkaloids from the roots of *Stemona sessilifolia*. *Chemistry and Biodiversity*, 5(4), 598–605.

4 Increasing the Sensitivity of Adipocytes and Skeletal Muscle Cells to Insulin

Visceral adiposity is the first step of a spiral of metabolic events leading to type 2 diabetes and cardiovascular disease in metabolic syndrome. The adipose mass in obese patients is responsible for the secretion of proinflammatory cytokines including tumor necrosis factor-α and interleukin-6 as well as unesterified fatty acids, which favor insulin resistance.[1] Insulin resistance results in impaired glucose tolerance, which is defined as a 2-hour post 75 g of glucose drink glycemia between 7.8 mmol/L (140 mg/dL) and 11.1 mmol/L (199 mg/dL).[2] Postprandial-elevated glycemia generates reactive oxygen species and cellular inflammatory response leading notably to the phosphorylation of insulin receptor substrate and subsequent insulin resistance.[3] In adipose tissue, insulin inhibits lipolysis of stored triglycerides.[4] In a state of insulin resistance, impaired insulin-dependent inhibition of lipolysis evokes an increase in plasma unesterified fatty acids leading to an increase of triglycerides in skeletal muscles which in turn further aggravates insulin resistance.[5,6] In this context, preventing the accumulation of triglycerides in adipose tissues and increasing the sensitivity of adipocytes and skeletal muscle cells to insulin with natural products constitutes one therapeutic strategy to prevent or manage insulin resistance in metabolic syndrome.

4.1 *Lindera erythrocarpa* Makino

Synonyms: *Lindera funiushanensis* C.S. Zhu; *Lindera henanensis* H.B. Cui
Common names: hong guo shan hu jiao (Chinese); kanakugi-no-ki (Japanese)
Subclass Magnoliidae, Superorder Lauranae, Order Laurales, Family Lauraceae
Medicinal use: indigestion (Japan)

Swiss mouse 3T3-L1 fibroblasts slowly differentiate functionally and morphologically into adipocytes and are used extensively to assess the adipogenic or antiadipogenic properties of natural products *in vitro*.[7] Lucidone from *Lindera erythrocarpa* Makino at a concentration of 80 μmol/L inhibited the accumulation of lipids in 3T3-L1 cells by 40%.[8] This antiadipogenic effect was accompanied with a mild increase in CCAAT enhancer binding protein-δ and CCAAT/enhancer-binding protein-β expression and a downregulation of peroxisome proliferator-activated receptor-γ, CCAAT enhancer binding protein-α, liver X receptor, lipoprotein lipase, adipocyte fatty acid-binding protein, adiponectin, and glucose transporter-4.[8] In adipocytes, peroxisome proliferator-activated receptor-γ is the master adipogenic regulator.[9] Activation of peroxisome proliferator-activated receptor-γ induces heterodimerization with retinoid X receptor and binding to peroxisome proliferator-activated receptor response element in the promoters of target genes such as adiponectin.[10] Adipose tissue expression of adiponectin is lower in visceral adiposity, insulin-resistant state, and diabetes.[11] Adiponectin is a protein secreted by fully differentiated adipocytes, and lucidone decreased adiponectin expression.[8] This cyclopentenedione given in at 0.125% of high-fat diet to C57BL/6 mice for 12 weeks decreased body weight gain, food intake, fasting blood glucose, insulinemia, triglycerides, cholesterol, and total fat pad mass.[8] CCAAT/enhancer-binding protein-β and protein-δ are expressed transiently at the early stage of adipocyte differentiation and drive the subsequent expression of CCAAT/enhancer-binding protein-α and peroxisome proliferator-activated receptor-γ.[12]

4.2 *Lindera obtusiloba* Blume

Synonym: *Benzoin obtusilobum* (Blume) Kuntze
Common name: san ya wu yao (Chinese)
Subclass Magnoliidae, Superorder Lauranae, Order Laurales, Family Lauraceae Juss.
Medicinal use: promotes liver function (Korea)

Mixture of methylisobutylxanthine, dexamethasone and insulin (sMDI) is used to induce adipocyte formation. Murine 3T3-L1 preadipocytes differentiation induced by MDI was abrogated by episesamin at 10 μM together with the reduced expression of glucose transporter-4, a decrease in β-catenin and phosphorylated catenin, phosphorylated extracellular signal-regulated kinase-1/2, as well as peroxisome proliferator-activated receptor-γ.[13] This transcription factor upregulates fatty acid synthetase and downregulates carnitine palmitoyltransferase-1 in adipocytes to induce a shift from lipid metabolism toward lipid accumulation.[14] This lignan (Figure 4.1) at 10 μM reduced fat contents in mature adipocytes by 33%.[13] Preadipocytes challenged with tumor necrosis factor-α produced interleukin-6, and that effect was inhibited by this lignan.[13]

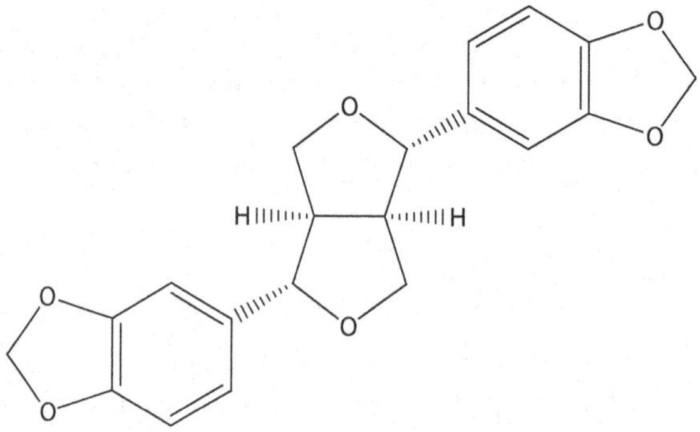

FIGURE 4.1 Episesamin.

4.3 *Piper retrofractum*

Synonyms: *Chavica officinarum* Miq.; *Piper chaba* Hunter; *Piper officinarum* (Miq.) C. DC.
Common names: jia bi ba (Chinese); litlit (Philippines)
Subclass Magnoliidae, Superorder Piperanae, Order Piperales, Family Piperaceae
Medicinal use: indigestion (Philippines)

Pipernonaline and dehydropipernonaline from this plant inhibited the enzymatic activity of acyl CoA:diacylglycerol acyltransferase with IC_{50} values equal to 37.2 and 21.2 μM, respectively, whereby refractamide was inactive.[15] In C57BL/6J mice fed with high-fat diet, piperidine alkaloids of which piperidine, dehydropipernonaline, and pipernonaline at a dose of 300 mg/kg/day for 8 weeks decreased body weight by 29.2% with the reduction of perirenal, epididymal, and subcutaneous adipose tissue.[16] This regimen reduced the total cholesterol by 44.3%, the low-density lipoprotein–cholesterol by 57.6%, and the serum total lipid concentration by 49.1%.[16] In L6 myocytes, which are commonly used in metabolic studies, piperine, pipernonaline, and dehydropipernonaline at a concentration of 30 μM induced the expression of peroxisome proliferator-activated receptor-δ and downstream target uncoupling protein-3, which catalyses fatty acids oxidation in skeletal muscles.[16] Piperine at a concentration of

50 μg/mL inhibited the accumulation of lipid in 3T3-L1 cells by 44.9% via reduction of peroxisome proliferator-activated receptor-γ, sterol regulatory element-binding protein-1c, and CCAAT/enhancer-binding protein-β expression as well as inhibition of peroxisome proliferator-activated receptor-γ and CREB-binding protein.[17]

4.4 *Nelumbo nucifera* Gaertn.

Synonyms: *Nelumbium nuciferum* Gaertn.; *Nelumbo speciosa* Willd.; *Nymphaea nelumbo* L.
Common names: lian (Chinese); sacred lotus
Subclass Ranunculidae, Superorder Proteanae, Order Nelumbonales, Family Nelumbonaceae
Medicinal use: anxiety (China)

Ethanol extract of *Nelumbo nucifera* Gaertn. and physical exercise protected rodents against high-fat diet-induced weight gain, whereby physical exercise per se mitigated weight gain and the extract itself had no activity implying an effect of increasing ability to burn calories.[18] In fact, the extract doubled the expression of uncoupling protein-3 in C2C12 myocytes at 50 μg/mL.[18] In a parallel study, at a concentration of 500 μg/mL, ethanol extract of this plant increased the production of glycerol by 8-fold, and this effect was nullified by propranolol suggesting a lipolytic effect mediated via β-receptors.[18] Boldine at 25 μM, at the inductive phase, evoked the expression of peroxisome proliferator-activated receptor-γ, CCAAT/enhancer-binding protein-α, and adiponectin by 3T3-L1 cells challenged with hydrogen peroxide or tumor necrosis factor-α.[19] In the same experiment, this aporphine alkaloid at 25 μM induced the production of adiponectin by 3T3-L1 cells without oxidative challenge.[19] In adipocytes, CCAAT/enhancer-binding protein-α induces adipogenesis through peroxisome proliferator-activated receptor-γ.[20] The leaves of *Nelumbo nucifera* Gaertn shelter 6*R*,6aR-roemoerine-N_β-oxide, liriodenine, and pronuciferine, which inhibited the differentiation of 3T3-L1 preadipocytes into adipocytes *in vitro*.[21]

4.5 *Coptis chinensis* Franch.

Common name: huang lian (Chinese)
Subclass Ranunculidae, Superorder Ranunculanae, Order Ranunculales, Family Ranunculaceae
Medicinal use: fever (China)

Berberine given orally at a dose of 150 mg/kg/day lowered the glycemia of healthy rodents to 52.5 mg/dL after 12 days.[22] In the same experiment, berberine given orally at a dose of 150 mg/kg/day lowered the glycemia of streptozotocin-nicotinamide-induced type 2 diabetic rodents model.[22] Berberine inhibited the oxygen consumption of L6 myotubes by 50% at a concentration of 15 μmol/L.[23] In isolated mitochondria, berberine reduced oxygen consumption via mitochondrial respiration blockade activating thus adenosine monophosphate-activated protein kinase.[23] The derivative dihydroberberine inhibiting cellular respiration on account of mitochondrial consumption of oxygen given to rodent poisoned with high-fat diet at a dose of 100 mg/kg/day reduced epididymal fat and inguinal fat and improved glucose tolerance, an effect obtained with 560 mg/kg/day of berberine.[23] Furthermore, dihydroberberine increased but did not normalized insulin sensitivity in treated animals with improved uptake of glucose into peripheral tissues.[23] Berberine at 5 μmol/L boosted the uptake of glucose induced by insulin in L6 myoblasts cultured *in vitro* and normalized the uptake of glucose induced by insulin despite cotreatment with palmitic acid.[24] In myotubes exposed to insulin and palmitic acid, berberine at a concentration of 5 μmol/L decreased peroxisome proliferator-activated receptor-γ, whereby the expression of peroxisome proliferator-activated receptor-α was unchanged.[24] Furthermore, this alkaloid at a concentration of 5 μmol/L reduced the expression of fatty acid translocase in both insulin-sensitive and palmitic acid-induced insulin resistant myoblasts.[24] Protein tyrosine phosphatase-1B catalyzes the dephosphorylation of insulin receptor substrate and natural products inhibiting this enzyme increase insulin receptor sensitivity. Berberine inhibited protein tyrosine phosphatase-1B activities by 60% at a concentration of 100 μM.[25] Ethanol extract of roots of *Coptis chinensis* Franch. increased the intake of glucose by C2C12 myocytes *in vitro*.[26] From this extract, coptisonine at a concentration of 50 μg/mL and berberine increased glucose uptake by C2C12 myocytes.[26] *In vitro*, berberine at 20 μmol/mL increased the absorption of glucose by 3T3-L1 adipocytes and C2C12 cells by 113.7% and 74%, respectively, via the reduction of aerobic mitochondrial respiration, increase of AMP/ATP ration and therefore phosphorylation of adenosine monophosphate-activated protein kinase.[27] Berberine at a concentration of 4 μM halved lipid accumulation in 3T3-L1 cells together with an increase in peroxisome proliferator-activated receptor-α, receptor-δ, and receptor-γ; fatty acid-binding protein; CDK9; and cyclin T1 expression. In diabetic rodents, 300 mg/kg of berberine increased the expression of peroxisome proliferator-activated receptor-α, peroxisome proliferator-activated receptor-δ, and peroxisome proliferator-activated receptor-γ, CDK9 cyclin T1 in adipose tissue as well as reduction of tumor necrosis factor-α.[28] Berberine at a concentration of 8 μM abrogated adipogenesis in 3T3-L1 cells with a reduction of fat amount by 72.6% with decrease in expression of peroxisome proliferator-activated receptor-γ and CCAAT/enhancer-binding protein-α and increase in GATA binding protein-2 and protein-3.[29] 3T3-L1 preadipocytes exposed to 4 μM of berberine for 8 days developed lower lipid content, together with the reduction in adiponectin, fatty acid-binding protein, and peroxisome proliferator-activated receptor-γ.[30] At the same concentration, berberine reduced the production of adiponectin in mature 3T3-L1.[30] It must be recalled that antidiabetic agent thiazolidinediones in adipocytes activate peroxisome proliferator-activated receptor-γ and decrease the secretion of tumor necrosis-α, interleukin-6, resistin, and augment adiponectin production.[31] Resistin secreted by adipocytes is responsible for insulin resistance.[31] In the same experiment, berberine at a concentration of 4 μM induced after 1 hour exposure, the phosphorylation of adenosine monophosphate-activated protein kinase hence phosphorylation of peroxisome proliferator-activated receptor-γ and inhibiting its transcriptional activity.[32] In 3T3-L1 preadipocytes induced to differentiate for 7 days berberine attenuated lipid accumulation at a concentration of 3 μM with the reduction of CCAAT/enhancer-binding protein-α and peroxisome proliferator-activated receptor-γ, whereas in L929 fibroblast, 40 μM of berberine boosted the intake of glucose as a result of glucose transporter-1 activation.[27,31–34]

4.6 *Tinospora cordifolia* (Willd.) Miers ex Hook. f. & Thomson

Synonym: *Menispermum cordifolium* Willd.
Common name: goorcha (India)
Subclass Ranunculidae, Superorder Ranunculanae, Order Menispermales, Family Menispermaceae
Medicinal use: jaundice (India)

The skeletal muscle mass account for the absorption of postprandial glucose by 50%–60% and muscle insulin resistance participates in postprandial hyperglycemia in obese patients and type 2 diabetic patients.[35] Physiologically, the binding of insulin to its receptors on skeletal muscle cells activates insulin receptor tyrosine kinase, which phosphorylates insulin receptor substrate.[36] The phosphorylation of insulin receptor substrate propulses the activation of phosphoinositide-3-kinase, activation of phosphoinositide-dependent kinase-1, protein kinase B (also termed Akt), phosphorylation of AS160, translocation of glucose transporter-4, and engulfment of plasma glucose into skeletal muscle cells.[36] L6 myotubes exposed to extract of stems of *Tinospora cordifolia* (Willd.) Miers ex Hook. f. & Thomson given at a concentration of 10 μg/mL or palmatine at a concentration of 625 nM induced about a 2.5-fold in glucose intake dependently of insulin receptor tyrosine kinase and phosphatidylinositol-3-kinase, which are part of the insulin pathway.[37] In line, both the extract and the alkaloid increased by about 4 folds the expression of glucose transporter-4 and commanded the expression of peroxisome proliferator-activated receptor-α by about 2 folds whereby the expression of peroxisome proliferator-activated receptor-γ was halved.[37] In skeletal muscles, peroxisome proliferator-activated receptor-γ activation increases insulin sensitivity.[38] Agonists are used for the treatment of type 2 diabetes.[38]

4.7 *Juglans regia* L.

Synonyms: *Juglans duclouxiana* Dode; *Juglans fallax* Dode; *Juglans kamaonia* (C. DC.) Dode; *Juglans orientis* Dode; *Juglans sinensis* (C. DC.) Dode
Common names: hu tao (Chinese); walnut
Subclass Hamamelidae, Superorder Juglandanae, Order Juglandales, Family Juglandaceae
Medicinal use: abscesses (China)

Ethanol extract of leaves of *Juglans regia* L. at a dose of 100 mg twice a day for 3 months given to type 2 diabetic patients (fasting blood glucose between 150 and 200 mg/dL, taking no more than two 5 mg glibenclamide and two 500 mg metformin tablets per day) lowered fasting glycemia from 165 to 143 mg/dL, cholesterol from 192.1 to 179.6 mg/dL, and triglycerides from 162.5 to 146.3 mg/dL.[39] The supplementation had no effect on serum insulin[39] suggesting a possible effect on glucose absorption by skeletal muscles. Methanol extract of leaves of *Juglans regia* (containing flavonoid glycosides) at a concentration of 25 μg/mL increased the glucose uptake by C2C12 myocytes without or with insulin by about 1.6 and 2.5 folds, respectively.[40] This extract inhibited protein-tyrosine phosphatase 1B activity by about 87.5% at a concentration of 30 μg/mL.[40] In type 2 diabetic patients, adenosine monophosphate-activated protein kinase activators such as metformin decrease glycemia and insulinemia, tend to lower body weight loss (unlike insulin and thiazolidinedione), decrease plasma triglycerides and fatty acids, attenuate total cholesterol and low-density lipoprotein–cholesterol, and slightly increase high-density lipoprotein–cholesterol.[41,42]

4.8 *Aerva lanata* (L.) Juss. ex Schult.

Synonyms: *Achyranthes lanata* L.; *Achyranthes lanata* Roxb.; *Aerva elegans* Moq.
Common names: astmabayada (India); tabang ahas (Philippines); wool plant
Subclass Caryophyllidae, Superorder Caryophyllanae, Order Caryophyllales, Family Amaranthaceae
Medicinal use: diuretic (Philippines)

The glycemia of Swiss albino male rats poisoned with alloxan was reduced by 48%, 6 hours after a single oral administration of 500 mg/kg of an ethanol extract of *Aerva lanata* (L.) Juss. ex Schult.[43] The administration of this extract at the same dose for 15 days to diabetic animals lowered glycemia by 64% and sustained body weight.[43] Furthermore, blood urea was reduced from 70 mg/100 mL to 32.1 mg/100 mL and lowered serum peroxide.[43] Agrawal et al. provided evidence that a single oral administration of an alkaloid fraction containing canthin-6-one derivatives to streptozotocin–nicotinamide-induced diabetic rats at a dose of 20 mg/kg evoked a reduction of glycemia after 20 hours.[44] In a subsequent study, aqueous extract of leaves (containing α-amyrin, betulin, and β-sitosterol) inhibited yeast and rat intestinal α-glucosidase with IC_{50} values of 81.7 and 108.7 μg/mL, respectively. The extract inhibited dipeptidyl peptidase-IV activity with an IC_{50} value of 118.62 μg/mL and protein-tyrosine phosphatase-1B with an IC_{50} value of 94.6 μg/mL.[45] This extract at enhanced insulin mediated glucose uptake by 3-fold of basal at 100 μg/mL in L6 myotubes.[45] The precise mode of action of this plant is yet unknown but one could make the inference that, at least, it promotes glucose uptake by peripheral tissues. In this context, it must be noted that canthin-6-one derivatives in *Aerva lanata* (L.) Juss. ex Schult. may possess protein-tyrosine phosphatase-1B inhibitory activities[46] improving therefore plasmatic clearance of glucose by skeletal and adipose tissues under insulin influence. Betulin at a concentration of 50 μg/mL increased by 21.4% uptake of glucose by C2C12 myocytes.[47]

4.9 *Persicaria hydropiper* (L.) Delarbre

Synonym: *Polygonum hydropiper* L.
Subclass Caryophyllidae, Superorder Polygonanae, Order Polygonales, Family Polygonaceae
Medicinal use: intestinal worms (Korea)

Persicaria hydropiper (L.) Delarbre methanol extract at concentration of 1 μg/mL inhibited the differentiation of 3T3-L1 cells.[48] At a concentration of 50 μM isoquercitrin (Figure 4.2) from this extract increased Wnt/β-catenin by approximately 4-fold.[48]

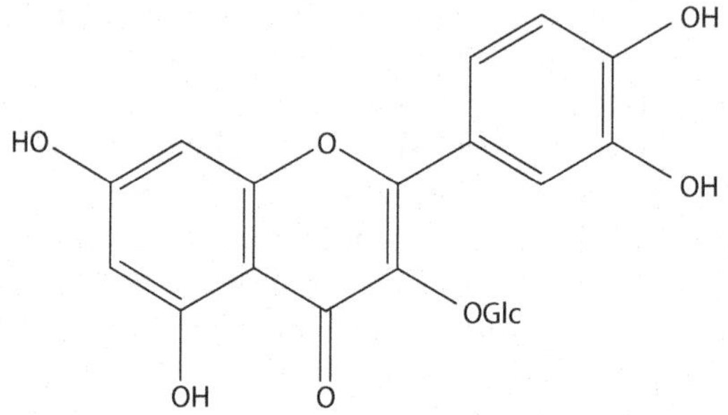

FIGURE 4.2 Isoquercitrin.

4.10 *Polygonum cuspidatum* Siebold & Zucc.

Synonyms: *Pleuropterus cuspidatus* (Siebold & Zucc.) H. Gross; *Polygonum pictum* Siebold; *Polygonum reynoutria* Makino
Common names: hu zhang (Chinese); giant knotgrass
Subclass Caryophyllidae, Superorder Polygonanae, Order Polygonales, Family Polygonaceae
Medicinal use: jaundice (China)

Butanol extract of rhizomes *Polygonum cuspidatum* Siebold & Zucc. inhibited the enzymatic activity of pancreatic lipase with an IC_{50} value equal to 15.8 μg/mL.[49] This extract at 25 μg/mL inhibited 3T3-L1 preadipocyte differentiation into adipocyte concomitantly with a decrease in glycerol-3-phosphate dehydrogenase, decrease in expression adipocyte differentiation-related protein, perilipin, adipogenic transcription factors peroxisome proliferator-activated receptor-γ, and CCAAT/enhancer-binding protein-α whereby phosphorylated adenosine monophosphate-activated protein kinase was boosted.[49] In adipocytes, the activation of peroxisome proliferator-activated receptor-γ activates perilipin.[50]

4.11 *Rheum rhabarbarum* L.

Synonyms: *Rheum franzenbachii* Münter; *Rheum undulatum* L.
Common names: bo ye da huang (Chinese); rhubarb
Subclass Caryophyllidae, Superorder Polygonanae, Order Polygonales, Family Polygonaceae
Medicinal use: stop bleeding (China)

Rheum rhabarbarum L. contains desoxyrhapontigenin, emodin, and chrysophanol, which at a single oral dose of 0.2, 0.4, and 0.1 mg/kg, respectively, decreased postprandial glycemia at 30 minutes by about 30% in mice.[51] This plant contains emodin that given orally at a dose of 80 mg/kg/day for 8 weeks to Wistar rats on high-fat diet decreased body weight gain from 49.5 to 24.5 g/rats.[52] This anthraquinone brought plasma cholesterol from 228.4 to 123.7 mg/dL, whereby atherogenic index was decreased from 6.2 to 1.3.[52] Emodin treatment increased the expression of adenosine monophosphate-activated protein kinase, phosphorylated adenosine monophosphate-activated protein kinase, acetyl-CoA carboxylase, phosphorylated-acetyl-CoA carboxylase, and carnitine palmitoyltransferase-1 in epididymal adipocytes of high-fat diet fed rats, sterol regulatory element-binding protein 1 and fatty acid synthetase were decreased.[52] *In vitro*, this anthraquinone at a concentration of 1 μmol/L inhibited the accumulation of triglycerides in 3T3-L1 preadipocytes by about 30%.[52] In a parallel study, ethanol extract of rhizomes given orally to high-fat diet male C57BL/6 mice for 8 weeks at a dose 100 mg/kg/day reduced weight gain, fasting blood glucose, cholesterol, low-density lipoprotein–cholesterol.[53] This regimen induced the expression adiponectin, fatty acid-binding protein, and uncoupling protein-3 in epididymal adipocytes.[53] From the extract, chrysophanol and physcion inhibited the enzymatic activity of protein-tyrosine phosphatase-1B. Chrysophanol or physcion, at 30 μM boosted the phosphorylation of insulin receptor and Akt in 3T3L1 cells challenged with insulin.[53]

4.12 *Tetracera scandens* (L.) Merr.

Synonym: *Delima scandens* Burk.
Subclass Dilleniidae, Superorder Dillenianae, Order Dilleniales, Family Dilleniaceae
Medicinal use: dysentery (Malaysia)

Aqueous extract of leaves of *Tetracera scandens* (L.) Merr. given orally at a single oral dose of 0.2 g/kg decreased fasting blood glucose glycemia of Wistar rats alloxan by 62.5%, whereby glibenclamide at a dose of 0.2 mg/kg evoked a 40% fall.[54] In healthy rays, the extract did not affect fasting blood glucose glycemia[54] suggesting, at least, an hypoglycemic effect dependent on peripheral uptake of glucose.

4.13 *Camellia japonica* L.

Synonym: *Thea japonica* (L.) Baill.
Common names: cha hua (Chinese); camellia
Subclass Dilleniidae, Superorder Ericanae, Order Theales, Family Theaceae
Medicinal use: intestinal bleeding (China)

In diabetic rodents, theophylline, which is not uncommon in members of the genus *Camellia*, boosted the lipolytic effects of epinephrine.[55] This purine alkaloid at a dose of 100 mg/kg/day given subchronically to rodent induced the enzymatic activity of carnitine palmitoyltransferase in skeletal muscle but not in the liver.[56] The plant contains 3β,16α,17β-trihydroxy-olean-12-ene, 3β-hydroxy-olean-11,13(18)-diene-28-oic acid, 3β-acetoxy-olean-12-ene-28-oic acid, camellenodiol, 3β-hydroxy-16-oxo-olean-11,13(18)-diene, and oleanolic acid inhibited the enzymatic activity of protein-tyrosine phosphatase-1B with IC_{50} values below 10 μM.[57] Aqueous extract of green tea, polyphenol given orally at a dose of 400 mg/kg/day to high-fat fed rodent for 6 weeks reduced body weight from 384.9 to 319.8 g, reduced fat weight from 10.9 to 4.7 g, lowered total cholesterol from 4.3 to 2.8 mmol/L, triglycerides from 1.5 to 0.9 mmol/L, low-density lipoprotein–cholesterol from 3 to 1.5 mmol/L, increased fecal fatty acids, and prevented hepatic steatosis.[58] In the white adipose tissue of treated animals, the expression of TNFα, leptin and interleukin-6 was reduced.[58] Leptin levels are elevated in obese patients[59] and accelerate visceral adiposity-related pathologies such as insulin resistance and cardiovascular diseases.[60] The secretion of leptin by adipocytes increases during triglyceride accumulation.[61] Ethanol extract of *Camellia japonica* L. inhibited the enzymatic activity of lipase[62] on probable account of tannins.

4.14 *Arbutus unedo* L.

Common names: kocayemiş yaprağı (Turkish); strawberry tree
Subclass Dillenidae, Superorder Ericanae, Order Ericales, Family Ericaceae
Medicinal use: kidney stones (Turkey)

Aqueous extract of *Arbutus unedo* L. given for 5 weeks in the drinking water of streptozotocin-diabetic Wistar rats lowered fasting glucose by 31%.[63] In isolated hemidiaphragms of healthy rats, the extracts of *Arbutus unedo* L. at a dose of 1 mg/mL increased glucose consumption.[63]

4.15 *Rhododendron brachycarpum* G. Don

Synonym: *Rhododendron faurieri* var *rufescens* Nakai
Subclass Dillenidae, Superorder Ericanae, Order Ericales, Family Ericaceae
Medicinal use: hypertension (Korea)

Pentacyclic triterpenes are often able to inhibit protein-tyrosine phosphatase-1B perhaps of account of the similarity of their chemical structure with an endogenous ligand. Rhododendric acid, ursolic acid, corosolic acid, rotundic acid, 23-hydroxyursolic acid, and actinidic acid from *Rhododendron brachycarpum* G. Don (Ericaceae) inhibited the enzymatic activity of protein-tyrosine phosphatase-1B with IC_{50} below 20 μM, respectively.[64]

4.16 *Styrax japonicus* Siebold & Zucc.

Synonym: *Cyrta japonica* (Siebold & Zucc.) Miers
Common names: ye mo li (Chinese); chishanoki (Japanese)
Subclass Dilleniidae, Superorder Primulanae, Order Styracales, Family Styracaceae
Medicinal use: cough (Japan)

3β-Acetoxy-28-hydroxyolean-12-ene, 3β-acetoxyolean-12-ene-28-acid, and 3β-acetoxyolean-12-en-28-aldehyde isolated from the stem bark of *Styrax japonicus* Siebold & Zucc. inhibited the enzymatic activity of protein-tyrosine phosphatase-1B with IC_{50} below 50 μM.[65]

4.17 *Ardisia japonica* (Thunb.) Blume

Synonym: *Tinus japonica* (Thunb.) Kuntze
Common name: zi jin niu (Chinese)
Subclass Dilleniidae, Superorder Primulanae, Order Primulales, Family Myrsinaceae
Medicinal use: cough (China)

Ardisia japonica (Thunb.) Blume produces maesanin, 2,5-dihydroxy-3-[(10Z)-pentadec-10-en-1-yl][1,4]benzoquinone), 5-ethoxy-2-hydroxy-3-[(10Z)-pentadec-10-en-1-yl][1,4]benzoquinone, and 5-ethoxy-2-hydroxy-3-[(8Z)-tridec-8-en-1-yl][1,4]benzoquinone inhibited the enzymatic activity of protein tyrosine phosphatase-1B *in vitro* with IC_{50} values below 20 μM.[66]

4.18 *Embelia ribes* Burm.f.

Common names: bai hua suan teng guo (Chinese); vidanga (India)
Subclass Dilleniidae, Superorder Primulanae, Order Primulales, Family Myrsinaceae
Medicinal use: jaundice (India)
History: the plant was known to Susruta

The seeds of *Embelia ribes* Burm.f. contain embelin (Figure 4.3), which given orally at a dose of 25 mg/kg/day for 15 days to Wistar rats intoxicated with carbon tetrachloride lowered serum aspartate aminotransferase, alanine aminotransferase, alkaline phosphatase activities, and total bilirubin.[67] At the hepatic level, this benzoquinone increased the activity of superoxide dismutase, catalase, and glutathione peroxidase. It also increased glutathione content; lowered hepatic lipoperoxydation; and improved hepatic cytoarchitecture.[67] Naik et al. provided evidence that embelin given at a dose of 50 mg/kg/day orally for 3 days to Wistar rats poisoned, with streptozotocin reduced glycemia by 42%.[68] Embelin reduced serum tumor necrosis-α and interleukin-6 by 60% and 40%, respectively, and improved hepatic glutathione, superoxide dismutase, and catalase.[68] The plasma lipid profile of rodents treated by embelin evidenced a decrease of cholesterol from 187.7 to 111 mg/dL, triglycerides from 173.1 to 108 mg/dL, very low-density lipoprotein from 34.6 to 21.3 mg/dL and low-density lipoprotein from 126.9 to 59.7 mg/dL.[68] This alkylresorcinol had no effect on the glycemia of normoglycemic rodents fed with normal diet implying an increase of peripheral glucose uptake.[68] In a parallel study, embelin given orally to streptozotocin high fat-induced type 2 diabetic Wistar rats (fasting blood glucose >250 mg/dL) at a dose of 50 mg/kg/day for 28 days had no effect on body weight and lowered glycemia from 305.1 to 138.6 mg/dL (normal: 139.6 mg/dL; rosiglitazone 10 mg/kg/day: 134.2 mg/dL).[69] This aryl-1,4-benzoquinone lowered plasma insulin from 28.5 to 15.6 μU/mL (normal: 16.7 μU/mL; rosiglitazone 10 mg/kg/day: 16.8 μU/mL) evidencing a decrease in insulin resistance.[69] This treatment lowered plasma cholesterol by 23.2%, triglycerides by 53.3%, and free fatty acids by 41.3%.[69] In oral glucose tolerance test on day 25, embelin lowered 60 minutes peak glycemia at 356.1 to 141.2 mg/dL (normal: 158.6; rosiglitazone 10 mg/kg/day: 143.1 mg/dL).[69] Embelin evoked an increased expression of peroxisome proliferator-activated receptor-γ in epididymal adipose tissue and to a much lesser extend in liver and skeletal muscle.[69] Embelin in epididymal adipose tissue upregulated phosphoinositide 3-kinase, phosphorylated Akt and glucose transporter-4.[69] The glucose lowering effects of thiazolidinedione antidiabetic agents is based on agonist effect on peroxisome proliferator-activated receptor-γ, which induces the expression of glucose transporter-4, the secretion of leptin, adiponectin and tumor necrosis-α, fatty acid-binding protein, fatty acid transport protein, and acyl-CoA-oxidase.[70] Furthermore, agonists of peroxisome proliferator-activated receptor-γ improve the sensitivity of adipocyte to insulin resulting in decreased release of nonesterified free fatty acids in the circulation.[71]

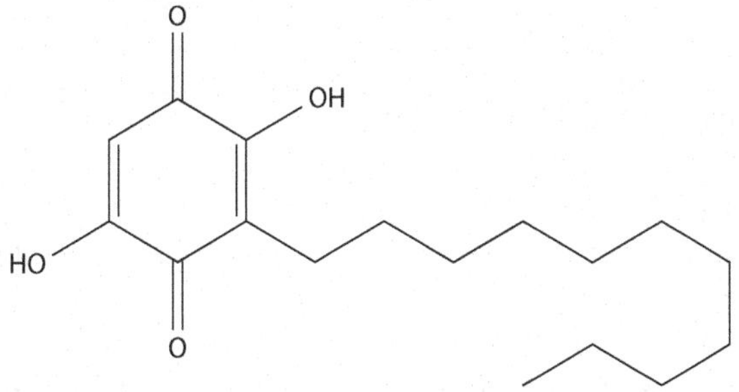

FIGURE 4.3 Embelin.

4.19 *Coccinia grandis* (L.) Voigt.

Synonyms: *Bryonia grandis* L.; *Cephalandra indica* Naudin; *Coccinia cordifolia* (L.) Cogn.; *Coccinia indica* Wight & Arn.
Common names: hong gua (China); bimbi (India); kanduri (Pakistan); ivy gourd
Subclass Dilleniidae, Superorder Violanae, Order Cucurbitales, Family Cucurbitaceae
Medicinal use: jaundice (India)

Chloroform extract of leaves of *Coccinia grandis* (L.) Voigt. given orally to hamsters on high-fat diet at a dose of 250 mg/kg/day for 7 days reduced plasma triglycerides and cholesterol by about 30%, as well as high-density lipoprotein–cholesterol and free fatty acids.[72] From this extract was isolated the long chain isoprenoid alcohol polyprenol that reduced at a dose of 50 mg/kg triglycerides and cholesterol by 42% and 25%, respectively, and free fatty acid by 9%, whereby high-density lipoprotein–cholesterol was unchanged.[72] Ethanol extract of roots at a concentration of 500 μg/mL reduced lipid accumulation in differentiating 3T3-L1 adipocytes *in vitro* by about 45%.[73] This extract inhibited the expression of peroxisome proliferator-activated receptor-γ and CCAAT/enhancer-binding protein-α in 3T3-L1 adipocytes as well as glucose transporter-4 and adiponectin.[73] Hexane fraction from this extract inhibited the expression of peroxisome proliferator-activated receptor-γ and CCAAT/enhancer-binding protein-α in 3T3-L1 adipocytes as well as fatty acid-binding protein, lipoprotein lipase, and induced acetyl-CoA carboxylase-1, pyruvate dehydrogenase kinase-4, and adiponectin receptor-1.[73] Peroxisome proliferator-activated receptor-γ promotes lipid storage by increasing the expression of fatty acid-binding protein.[74]

4.20 *Cucurbita ficifolia* Bouché

Synonym: *Cucurbita melanosperma* Gasp. *Pepo ficifolius* (Bouché) Britton
Common name: fig leaf gourd
Subclass Dilleniidae, Superorder Violanae, Order Cucurbitales, Family Cucurbitaceae
Nutritional use: vegetable (Philippines)

Aqueous extract of fruits of *Cucurbita ficifolia* Bouché at a concentration of 0.1 mM protected 3T3-L1 murine fibroblasts against the oxidative stress induced by high-glucose concentration as evidenced by an increase in glutathione, glutathione disulfide, glutathione peroxidase, and glutathione reductase activities.[75] This extract inhibited tumor necrosis-α and resistin expression and increased the expression of interleukin-6.[75]

4.21 *Cucurbita moschata* Duschesne

Synonym: *Cucurbita pepo* var. *moschata* (Duchesne) Duchesne
Common names: nan kua (Chinese); squash
Subclass Dilleniidae, Superorder Violanae, Order Cucurbitales, Family Cucurbitaceae
Medicinal use: jaundice (China)

Dehydrodiconifery-l alcohol (Figure 4.4) isolated from the stems of *Cucurbita moschata* Duschesne at a concentration of 70 μM inhibited triglyceride accumulation, adipogenic transcription factors peroxisome proliferator-activated receptor-γ and CCAAT/enhancer-binding protein-α expression and abrogated lipogenic sterol regulatory element-binding protein-1c and stearoyl-CoA desaturase-1 expression in differentiating 3T3-L1 adipocytes.[76] This lignan inhibited the phosphorylation and DNA binding of CCAAT/enhancer-binding protein-β.[76] Peroxisome proliferator-activated receptor-γ promotes adipocyte differentiation by inducing the expression of CAAT-enhancer binding protein-α.[77]

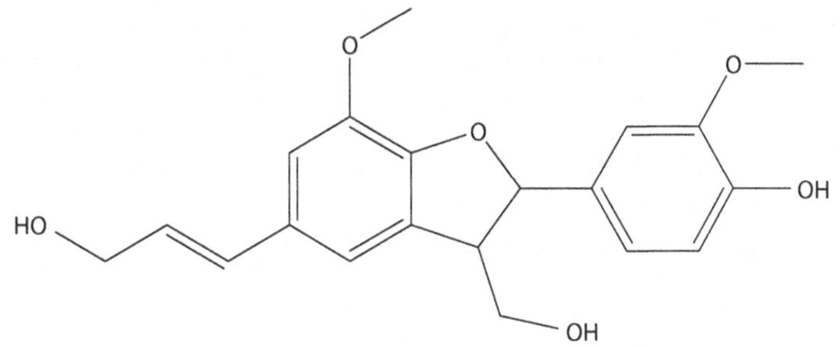

FIGURE 4.4 Dehydrodiconiferyl alcohol.

4.22 *Lagenaria siceraria* (Mol.) Standl.

Synonyms: *Cucumis mairei* H. Lév.; *Cucurbita lagenaria* L.; *Cucurbita leucantha* Duchesne; *Cucurbita siceraria* Molina; *Lagenaria vulgaris* Ser.
Common names: hu lu (Chinese); kalubay (Philippines); bottle gourd
Subclass Dilleniidae, Superorder Violanae, Order Cucurbitales, Family Cucurbitaceae
Medicinal use: cough (Philippines)

In adipocytes, peroxisome proliferator-activated receptor-γ promotes triglycerides storage by increasing the expression of fatty acid-binding protein, acyl-CoA-binding protein, lipoprotein lipase and fatty acid translocase.[78,79] Cucurbitacin B at concentration of 300 nM abrogated accumulation in 3T3-L1 preadipocytes *in vitro* together with reduced expression of peroxisome proliferator-activated receptor-γ, fatty acid-binding protein, CCAAT/enhancer-binding protein-α, adiponectin, and fatty acid translocase.[80] This cucurbitane-type steroid at a concentration of 300 nM in 3T3-L1 adipocytes halved the phosphorylation of STAT3 repressed CCAAT/enhancer-binding protein-δ and cyclin D1 and increased KLF5.[80] In multipotent C3H10T1/2 cells, this cucurbitane at a dose of 200 nM lowered CCAAT/enhancer-binding protein-δ, mitigated cyclin D1, and boosted KLF5.[80] Likewise, cucurbitacin I at a concentration of 300 nM lowered peroxisome proliferator-activated receptor-γ, CCAAT/enhancer-binding protein-α, and fatty acid translocase in 3T3-L1 preadipocytes.[80] Adipogenesis is regulated by a balance of positive and negative regulators.[81] In preadipocytes, the negative regulators of adipogenesis such as CCAAT/enhancer-binding protein-γ, C/EBP homologous protein, and Kruppel-like factor are predominantly expressed. Positive regulators such as CCAAT/enhancer-binding protein-δ, sterol regulatory element binding protein-1c, and KLF5 have low expression levels.[81,82]

4.23 *Momordica charantia* L.

Common names: ku gua (Chinese); periah (Malay); karela (Pakistan); balsam-apple
Subclass Dilleniidae, Superorder Violanae, Order Cucurbitales, Family Cucurbitaceae
Medicinal use: diabetes (Sri Lanka)

Butanol extract of fruits of *Momordica charantia* L. given at a dose of 0.2 g/kg/day for 2 weeks to Sprague–Dawley rats fed high-fructose diet attenuated body weight gain, glycemia from 145.2 to 103.6 mg/dL, insulinemia from 3.6 to 0.9 μg/L, leptin from 5.7 to 2 μg/L, triglycerides from 387.3 to 219.1 mg/dL, and cholesterol from 56.4 and 46.8 mg/dL, whereby adiponectin was increased from 0.2 to 0.4 ng/mL.[83] This extract evoked a mild increase in peroxisome proliferator-activated receptor-γ in the white adipose tissue of treated rats and induced a 3-fold increase in glucose transporter-4 expression in skeletal muscle.[83] In physiological conditions, adiponectin increases insulin sensitivity and its secretion decreases with visceral adiposity.[84] This secretory protein activates adenosine monophosphate-activated kinase in liver and skeletal muscles[85] leading to inhibition of gluconeogenesis in the liver, induction of fatty acid oxidation, and glucose uptake by skeletal muscles.[86]

4.24 *Capparis decidua* (Forssk.) Edgew.

Synonym: *Sodada decidua* Forssk.
Common names: karimulli (India); wild caper
Medicinal use: colic (India)

Ethanol extract of fruits of *Capparis decidua* (Forssk.) Edgew. given at a dose of 500 mg/kg/day orally to rabbits fed with cholesterol-enriched diet for 60 days reduced total cholesterol by 60%, low-density lipoprotein–cholesterol by 70%, triglycerides, high-density lipoprotein–cholesterol, and very low-density lipoprotein–cholesterol from by 30% and brought the atherogenic index close to normal.[87] This treatment lowered hepatic, cardiac, and arterial cholesterol by 40%, 70%, and 45%, respectively.[87] Alkaloidal extract of fruits of *Capparis decidua* (Forssk.) Edgew. given at a dose of 50 mg/kg for 28 days lowered the fasting blood glucose value of streptozotocin-diabetic rodents from 162 to 147 mg/dL and decreased insulin resistance as evidenced by a decrease of postprandia glycemia from 232 to 125 mg/dL after 120 minutes in oral glucose tolerance test.[88] Likewise, this treatment lowered total cholesterol from 175.6 to 132.6 mg/dL, low-density lipoprotein–cholesterol from 130.4 to 89.3 mg/dL, triglycerides from 132.4 to 97.2 mg/dL, very low-density lipoprotein–cholesterol from 26.5 to 19.4 mg/dL and increased high-density lipoprotein–cholesterol from 18.7 to 23.9 mg/dL.[88] The extract increased the glycogen contents of liver and muscles by 28% and 33%, respectively, and halved hepatic glucose-6-phosphatase activity.[88] The expression of glucose-6-phosphatase and phosphoenolpyruvate carboxy kinase was decreased in hepatocytes, whereby the expression of glucokinase and glucose transporter-4 was increased.[88] Increased expression of peroxisome proliferator-activated receptor-γ and decreased expression of tumor necrosis-α in adipose tissues were noted.[88]

4.25 *Capparis moonii* Wight

Common name: rudanti (India)
Subclass Dillenidae, Superorder Capparanae, Order Capparales, Family Capparaceae
Medicinal use: cough (India)

The gallotannins 1,3,6-tri-*O*-galloyl-2-chebuloyl-β-D-glucopyranoside at 10 ng/mL and 1,3,6-tri-*O*-galloyl-2-chebuloyl ester-β-D-glucopyranoside at 100 ng/mL isolated from the fruits of *Capparis moonii* Wight induced glucose uptake by L6 myotubes *in vitro* by 223% and 219%, respectively, by increasing insulin receptor-β subunit phosphorylation and insulin receptor substrate-1 phosphorylation, expression of phosphoinositide 3-kinase and glucose transporter-4.[89] It must be recalled that ellagitannins and gallotannins are decomposed by commendal bacteria in the guts in gallic acid and ellagic acid.

4.26 *Capparis spinosa* L.

Synonym: *Capparis murrayana* J. Graham
Common names: shan ga (Chinese); Kebere (Turkey); caper
Subclass Dillenidae, Superorder Capparanae, Order Capparales, Family Capparaceae
Medicinal use: hepatitis (Turkey)

Ethanol extract of roots of *Capparis spinosa* L. given to streptozotocin-induced diabetic Wistar rats at a dose of 200 mg/kg/day for 28 evoked a decrease in glycemia and had no effect on plasma insulin and improved plasma lipids implying, at least, increased insulin sensitivity.[90] One could speculate that principles of *Capparis spinosa* and/or metabolites may lower plasma glucose by increasing glucose absorption by skeletal muscles and adipocytes via probable adenosine monophosphate-activated protein kinase activation. Of note, ethanol extract of fruits of *Capparis spinosa* L. given at a dose of 400 mg, 3 times per day before meals to type 2 diabetic patients for 2 months reduced fasting glycemia and plasma, triglycerides.[91] Intake of caper could be of value in metabolic syndrome.

4.27 *Sida cordifolia* L.

Synonyms: *Sida herbacea* Cav.; *Sida hongkongensis* Gand.; *Sida rotundifolia* Lam. ex Cav.; *Sida holosericea* Willd. ex Spreng.
Common names: xin ye huang hua ren (Chinese); gulipas (Philippines)
Subclass Dillenidae, Superorder Malvanae, Order Malvales, Family Malvacae
Medicinal use: fever (Philippines)

Sida cordifolia L. elaborates a tryptophan derivative known as hypaphorine that given to streptozotocin-induced diabetic Sprague–Dawley rats (fasting blood glucose >300 mg/dL) at a single oral dose of 50 mg/kg lowered glycemia by 27.8%.[92]

4.28 *Ficus carica* L.

Synonym: *Ficus kopetdagensis* Pachom.
Common names: İncir ağacı (Turkey); common fig-tree
Subclass Dillenidae, Superorder Malvanae, Order Urticales, Family Moraceae
Medicinal use: hemorrhoids (Turkey)

Ficusin, also known as psoralen, isolated from the leaves of *Ficus carica* L. given orally at a dose of 40 mg/kg/day for 28 days to streptozotocin high-fat diet-induced type 2 diabetic Wistar decreased glycemia by 66% and improved glucose tolerance.[93] This furanocoumarin increased body weight, plasma insulin by 67.8%, and glycogen contents. This regimen lowered plasma cholesterol, triglycerides, fatty acids, alanine aminotransferase, and aspartate aminotransferase.[93] In adipose tissues, ficusin increased the expression peroxisome proliferator-activated receptor-γ and of glucose transporter-4.[93] Being toxic, this natural product has no place in therapeutic.[94]

4.29 *Broussonetia kazinoki* Siebold & Zucc.

Synonym: *Broussonetia monoica* Hance
Common names: xia gu shu (Chinese); kuzo (Japanese)
Subclass Dillenidae, Superorder Malvanae, Order Urticales, Family Moraceae
Medicinal use: abdominal pain (China)

The stem bark of *Broussonetia kazinoki* Sieb. contains series of flavonoids including broussonone A, broussonin A, broussonin B, 7,4′-dihydroxyflavan, and 3′,7-dihydroxy-4′-methoxyflavan at which 100 μM inhibited the accumulation of triglycerides in 3T3-L1 mouse embryo fibroblasts.[95]

4.30 *Morus alba* L.

Synonyms: *Morus atropurpurea* Roxb.; *Morus australis* Poir.; *Morus indica* L.
Common names: sang (Chinese); white mulberry
Subclass Dillenidae, Superorder Malvanae, Order Urticales, Family Moraceae
Medicinal use: diuretic (China)

The major enzymes involved in triglyceride homeostasis in adipocytes are endothelial lipoprotein lipase and hormone-sensitive lipase that have concerted opposite effects. In the postprandial state, insulin activates endothelial lipoprotein lipase and blocks simultaneously hormone-sensitive lipase.[96] The activation of endothelial lipoprotein lipase commands the hydrolysis of triglycerides brought by very low-density lipoprotein and chylomicron to release fatty acids in adipocytes that are used to synthetise triglycerides.[43] Ethanol extract of roots of *Morus alba* L. inhibited the enzymatic activity of lipase *in vitro* with an IC_{50} value equal to 2 μg/mL and at 100 μg/mL abrogated the enzymatic activity of phosphodiesterase *in vitro*.[97] At 100 μg/mL, this extract reduced triglyceride accumulation in 3T3-L1 adipocytes by about 15%.[97] In epididymal adipose from male Wister rats, the extract doubled the secretion of glycerol.[97] Hydroalcoholic extract of leaves of *Morus alba* L. at a concentration of 45 μg/mL induced lipid accumulation in 3T3-L1 preadipocytes by 19.2% with parallel increased expression in CCAAT/enhancer-binding protein-α, peroxisome proliferator-activated receptor-γ, and fatty acid-binding protein by 2.2-, 6-, and 2.3-fold, respectively.[98] Furthermore, this extract increased adiponectin secretion by 3T3-L1 preadipocytes from 70 to 176 ng/mL.[98] In a subsequent study, kuwanon A, kuwanon C, kuwanon T, morusin, kuwanon E, sanggenon F, betulinic acid, uvaol, and β-sitosterol at a concentration of 20 μM inhibited adipogenesis in 3T3-L1 adipocytes, with a decrease in triglyceride accumulation.[99] Kuwanon C, morusin and betulinic acid, at a concentration of 20 μM inhibited glycerophosphate dehydrogenase activity by 60.1%, 70%, and 44.1%, respectively, in 3T3-L1 adipocytes.[99]

4.31 *Morus notabilis* C.K. Schneid.

Common name: chuan sang (Chinese)
Subclass Dillenidae, Superorder Malvanae, Order Urticales, Family Moraceae

Morus notabilis C.K. Schneid. elaborates isoprenylated flavonoids notabilisins A and C, which at a concentration of 10 and 15 μM, respectively, induced triglyceride accumulation in 3T3L1 cells with concomitant increased the gene expression of fatty acid-binding protein by 60% and 100% and glucose transporter-4 by 140% and 70% in 3T3L1 cells suggesting, at least, the activation of peroxisome proliferator-activated receptor-γ.[100]

4.32 *Phyllanthus acidus* (L.) Skeels

Synonyms: *Averrhoa acida* L. *Cicca acida* (L.) Merr.; *Phyllanthus acidissimus* (Blanco) Müll. Arg.; *Phyllanthus cicca* Müll. Arg.; *Phyllanthus distichus* Hook. & Arn.
Common names: bangkiling (Philippines); ma yom (Thai) Malay gooseberry
Subclass Dillenidae, Superorder Euphorbianae, Order Euphorbiales, Family Phyllanthaceae
Medicinal use: urticaria (Philippines)

Aqueous extract of leaves of *Phyllanthus acidus* (L.) Skeels given orally and daily at a dose of 1 g/kg to Wistar rats for 6 weeks evoked a reduction of body weight from 430.2 to 416 g, as well as reduction of adipose tissues weight epididymis, retroperitoneal, mesentery, and subcutaneous area.[101] Besides, this treatment lowered glycemia, serum triglycerides, cholesterolemia, high-density lipoprotein–cholesterol, and reduced by more than 50% low-density lipoprotein–cholesterol.[101] This regimen had no effect on basal blood pressure and heart rate.[101]

4.33 *Euphorbia lathyris* L.

Common names: xu sui zi (Chinese); burg sadab (India); caper spurge
Subclass Dillenidae, Superorder Euphorbianae, Order Euphorbiales, Family Phyllanthaceae
Medicinal use: diuretic (China)

Euphorbiasteroid from *Euphorbia lathyrus* L. at a concentration of 25 μM abrogated the accumulation of triglycerides in 3T3-L1 via decreased expression in fatty acid synthetase, and upstream regulators CCAAT/enhancer-binding protein-α and protein-β, peroxisome proliferator-activated receptor-γ, and sterol regulatory element-binding protein-1c.[102] This tricyclic diterpene induced the phosphorylation and therefore activated the adenosine monophosphate-activated protein kinase and induced the phosphorylation of acetyl-CoA carboxylase leading to inhibition of fatty acid synthesis.[102] This plant is poisonous.[103]

4.34 *Rubus fruticosus* L.

Common names: bogurtlen (Turkey); European blackberry
Subclass Rosiidae, Superorder Rosanae, Order Rosales, Family Rosaceae
Medicinal use: constipations (Turkey)

Aqueous extract of leaves of *Rubus fruticosus* L. given orally to Wistar rats at a single dose of 100 mg/kg evoked a reduction of plasma glucose after 2 hours.[104] The extract given orally to streptozotocin-induced diabetic Wistar rats at a dose of 100 mg/kg/day for 9 days evoked a reduction of plasma glucose by about 75% similarly to metformin at 500 mg/kg/day.[104] This extract had no effect on plasma insulin[104] implying, at least, increased glucose uptake by skeletal muscles and adipose tissues.

4.35 *Terminalia pallida* Brandis

Common name: haritaki (India)
Subclass Rosidae, Superorder Myrtanae, Order Myrtales, Family Combretaceae
Medicinal use: diabetes (India)

Ethanol extract of fruits of *Terminalia pallida* Brandis given at a single oral dose of 0.5 g/kg to alloxan-induced diabetic Wistar rats reduced fasting glycemia by 18.6% after 5 hours of treatment.[105] In normal rats, no decrease in glycemic was observed.[105] Adenosine monophosphate-activated protein kinase is activated by the antidiabetic drug metformin via LKB1.[106] This biguanide developed from *Galega officinalis* L. (family Fabaceae) activates adenosine monophosphate-activated protein kinase. It does not affect glycemia of normal subjects because increase in peripheral glucose utilization is compensated by an increase in hepatic glucose output.[107] One could suggest that *Terminallia pallida* Brandis contains some tannins which upon gut bacterial flora metabolism and first-pass metabolism release phenolic entities with activities related to increased uptake of glucose by skeletal muscles.

4.36 *Punica granatum* L.

Common names: shi liu (Chinese); dhalim (India); pomegranate
Subclass Rosidae, Superorder Myrtanae, Order Myrtales, Family Lythraceae
Medicinal use: diabetes (India)
History: the plant was known to Serapion

Type 2 diabetic patients (fasting glucose >7 mmol/L) taking the juice of fruits of *Punica granatum* L. at a single dose of 1.5 mL/kg experienced a decrease in glycemia from 9.4 to 8.5 mmol/L after 3 hours, whereas in normal subject, the juice intake had no effect.[108] This supplementation decreased plasma insulin from 11 to 9.8 μIU/mL, increased β-cell function and lowered insulin resistance.[108] The juice of fruit of *Punica granatum* L. accumulates ellagitannins (between about 10 and 150 mg/mL for Spanish cultivars) and accumulates ascorbic acid (between about 80 and 200 mg/L for Spanish cultivars).[109] Ellagitannins are not absorbed but converted in the small intestine to ellagic acid and by intestinal bacteria leads urolithins that are glucuronidated and excreted in the urine. Mason et al. provided evidence that intake of ascorbic acid at a dose of 500 mg/kg twice a day for 4 months improved insulin-mediated plasmatic clearance of glucose in type 2 diabetic patients.[110] Besides, gallic acid is known to promote intake of glucose by rats myotubes *in vitro*.[111] Therefore, it is reasonable to suggest that intake of *Punica granatum* L. fruit juice attenuates the glycemia of type 2 diabetic patients by, at least, increasing the sensitivity of skeletal muscles and adipocytes to insulin. Decreased plasma insulin recorded in type 2 diabetic patient support this contention. More clinical studies are required to ascertain the value of pomegranate for the management of metabolic syndrome.

4.37 *Cyamopsis tetragonoloba* (L.) Taub.

Synonyms: *Cyamopsis psoraloides* DC.; *Dolichos fabiformis* L'Hér.; *Dolichos psoraloides* Lam.; *Lupinus trifoliatus* Cav.; *Psoralea tetragonoloba* L.
Common names: gua er dou (Chinese); govar (India); cluster bean
Subclass Rosidae, Superorder Myrtanae, Order Myrtales, Family Lythraceae
Medicinal use: tonic (India)

Gallic acid isolated from beans of *Cyamopsis tetragonoloba* (L.) Taub. at a dose of 20 mg/kg/day given orally for 28 days normalized fasting blood glucose of streptozotocin-induced diabetic Wistar rats on high-fat diet.[112] In addition, this simple phenolic acid, brought close to normal insulinemia, and at the hepatic level normalized glucose-6-phosphatase, fructose-1,6-biphosphatase and brought close to normal values hexokinase and hepatic glycogen[112] suggesting the improved insulin sensitivity. The treatment normalized the expression of peroxisome proliferator-activated receptor-γ in adipocytes of diabetic rodents.[112] In adipose tissue, the treatment normalized the expression of glucose transporter-4 and boosted phosphoinositide 3-kinase and p-Akt expression.[112] Gallic acid is known to promote the intake of glucose by rats myotubes. The chronic toxicity of gallic acid needs to be examined.

4.38 *Sophora flavescens* Aiton

Synonyms: *Sophora angustifolia* Siebold & Zucc.; *Sophora macrosperma* DC.; *Sophora tetragonocarpa* Hayata
Common name: ku shen (Chinese)
Subclass Rosidae, Superorder Fabanae, Order Rosiidae, Family Fabaceae
Medicinal use: jaundice (China)
Pharmacological target: insulin resistance

Sophora flavescens Aiton produces matrine which at a concentration of 500 μg/mL reduced after 6 days the accumulation of triglycerides in 3T3-L1 preadipocytes by 60% with decrease in peroxisome proliferator-activated receptor-γ and CCAAT/enhancer-binding protein-α, protein-β, and protein-δ on account of extracellular signal-regulated kinase-1/2 hypophosphorylation.[113] It must be recalled that insulin is involved in the early stage of adipogenesis throughout the activation of phosphoinositide 3-kinase and extracellular signal-regulated kinase-1/2.[114] The extracellular signal-regulated kinases 1 and 2 are involved in the early stage of adipocytes differentiation as its phosphorylation is required for the expression of adipogenic transcriptional factors peroxisome proliferator-activated receptor-γ and CCAAT/enhancer-binding protein-α.[115]

4.39 *Amorpha fruticosa* L.

Common names: zi sui huai (Chinese); bastard indigo
Subclass Rosidae, Superorder Fabanae, Order Rosiidae, Family Fabaceae
Medicinal use: eczema (China)

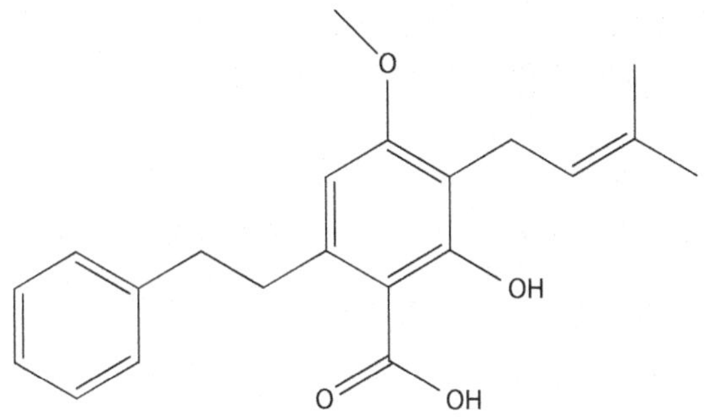

FIGURE 4.5 Amorfrutin-1.

Amorfrutin-1 (Figure 4.5) from *Amorpha fruticosa* L. activates peroxisome proliferator-activated receptor-γ with EC_{50} values equal to 0.4 μM *in vitro*.[116] This isoprenoid-substituted benzoic acid given orally C57BL/6 mice on high-fat diet for 23 days at a dose of 100 mg/kg/day improved insulin sensibility, glucose tolerance, lowered plasma triglycerides, and free fatty acid together with an increase in food intake and reduced body weight gain.[116] Amorfrutin-1 blocked peroxisome proliferator-activated receptor-γ phosphorylation in visceral white adipose tissue[116] enhancing thereby peroxisome proliferator-activated receptor-γ activity.[117] Phenolics from medicinal plants have the tendency to activate peroxisome proliferator-activated receptor-γ.

4.40 *Macrotyloma uniflorum* (Lam.) Verdc.

Synonyms: *Dolichos uniflorus* Lam.; *Kerstingiella uniflora* (Lam.) Lackey
Common names: ying pi dou (Chinese); horse-gram
Subclass Rosidae, Superorder Fabanae, Order Fabales, Family Fabaceae
Nutritional use: food (India)

Ethanol extract of seeds of *Macrotyloma uniflorum* (Lam.) Verdc. given orally for 42 days to Sprague–Dawley rats on high-fat diet lowered body weights from 307 to 252 g.[118] Food intake was reduced from 6 week of treatment onward. This treatment lowered plasma cholesterol, triglycerides, very low-density lipoprotein, very low-density lipoprotein, and increased high-density lipoprotein, whereas increased fecal cholesterol was observed.[118] The seeds of this plant contains phenolic acids of which p-coumaric acid and ferulic acid.[119] Konishi et al. provided evidence that p-coumaric is absorbed orally in rats.[120] This phenolic acid at a concentration of 50 μM increased the phosphorylation of adenosine monophosphate-activated protein kinase in L6 skeletal muscle, phosphorylation of acetyl-CoA carboxylase, and the expression of carnitine palmitoyltransferase-1.[121] It induced peroxisome proliferator-activated receptor-α, promoting thus the β-oxidation of fatty acids.[121] In skeletal muscles, increased AMP, low ATP levels, hypoxia, glucose deprivation induced by muscular contraction or mitochondrial poisons activate allosterically activate adenosine monophosphate-activated protein kinase.[107] Besides, this heterotrimeric protein is directly activated by LKB1 which is a target for metformin and CaMKK in response to increased cytoplasmic calcium concentration.[122] This energy sensor is also activated by silent information regulator T1 (SIRT1) in response to increased NAD^+.[107] Activation of adenosine monophosphate-activated protein kinase commands the phosphorylation of TBC1D1 (also known as AS160) and subsequent translocation of glucose transporter-4 allowing glucose uptake by skeletal muscle.[123] In skeletal muscles, adenosine monophosphate-activated protein kinase induces fatty acids uptake by fatty acid translocase, and phosphorylates and inhibits acetyl-CoA carboxylase and therefore the synthesis of malonyl-CoA from acetyl-CoA decreasing thus the rate of fatty acid synthesis.[124] Malonyl-CoA is the precursor for fatty acid synthesis and an inhibitor of carnitine palmitoyltransferase-1 and therefore adenosine monophosphate-activated protein kinase activation stimulates mitochondrial β-oxidation of fatty acids.[125]

4.41 *Pterocarpus marsupium* Roxb.

Synonyms: *Lingoum marsupium* (Roxb.) Kuntze; *Pterocarpus bilobus* Roxb. ex G. Don
Common names: ma la ba zi tan (Chinese); kum kusrala (India); kino
Subclass Rosidae, Superorder Fabanae, Order Rosiidae, Family Fabaceae
Medicinal use: diabetes (India)

The plant accumulates series of flavonoids of which (–)-epicatechin which is absorbed orally in human.[126] This flavanol at a concentration of 25 μg/mL increased the uptake of glucose by C2C12 by 166.3%.[47] *Pterocarpus marsupium* Roxb. also elaborates 7-*O*-α-L-rhamnopyranosyl-oxy-4′-methoxy-5-hydroxy isoflavone that enhanced *in vitro* glucose uptake by L6 myotubes with elevation in peroxisome proliferator-activated receptor-γ and glucose transporter-4 expression.[127] Furthermore, the wood contains series of flavonoid C-glucosides[128] with the ability to enhance glucose absorption by rat L6 myotubes *in vitro*.[129] Aqueous extract of this plant given orally for 30 days at a dose of 1 g/kg/day to rats on high-fructose diet reduced body weight from 180 to 173.8 g (normal: 171.7 g), lowered serum glucose from 118 to 89.7 mg/dL (normal: 84.3 mg/dL), and lowered plasma triglycerides from 90.9 to 80.4 mg/dL (normal: 66.3 mg/dL).[130] This extract lowered insulinemia from 28.5 to 18.4 ng/dL (normal: 17.5 ng/dL)[130] suggesting, at least, increased plasmatic glucose clearance by skeletal muscles. Aqueous extract of wood of this tree enhanced glucose absorption by mouse skeletal muscles. Epicatechin at a concentration of 25 μg/mL increased the uptake of glucose by C2C12 myocytes by 166.3%.[131]

4.42 *Trigonella foenum-graecum* L.

Synonym: *Trigonella tibetana* (Alef.) Vassilcz.
Common names: hu lu ba (Chinese); fenugreek
Subclass Rosidae, Superorder Fabanae, Order Rosiidae, Family Fabaceae
Medicinal use: promote urination (Malaysia)

Ethanol extract of seeds of *Trigonella foenum-graecum* L. given orally to C57BL/6J mice on high-fat diet at a dose of 2 g/kg/day for 20 weeks had no effect on food intake, decreased body weight close to normal, lowered plasma glucose from 183.1 to 129.3 mg/dL, decreased plasma triglycerides from 48.9 to 18.9 mg/dL (normal: 29 mg/dL), and prevented hepatic steatosis.[132] The extract given from 17th to 35th week at the same dose to the rodents on high-fat diet from the 1st week, decreased plasma glucose from 229 to 170 mg/dL, lowered plasma insulin, insulin resistance, plasma cholesterol, and triglycerides as well as low-density lipoprotein–cholesterol and increased high-density lipoprotein–cholesterol.[132] *Trigonella foenum-graecum* L. contains trigonelline that incorporated into the diet of Goto–Kakizaki rats at 0.05% for 43 days had no effect on glycemia but improved glucose tolerance in oral glucose tolerance tests, decreased epididymal adipose tissue weigh, attenuated plasma insulin and decreased circulating tumor necrosis factor-α from 123 to 111 pg/mL (normal: 121 pg/mL).[133] Trigonelline supplementation lowered plasma cholesterol, high-density lipoprotein–cholesterol, brought plasma triglycerides below normal values as well as plasma fatty acids, and decreased total bile acid.[133] At the hepatic level, trigonelline treatment lowered triglycerides, cholesterol, and fatty acids to normal value.[133] Besides, this alkaloid decreased fatty acid synthetase activity and increased carnitine palmitoyltransferase-1.[133] This alkaloid decreased hepatic activity of glucose-6-phosphate dehydrogenase and 6-phosphogluconate dehydrogenase activity and increased the activity of rate-limiting enzyme on the glycolytic pathway glucokinase.[133] In addition, trigonelline treatment decreased hepatic lipoperoxidation to normal.[133] *In vitro*, trigonelline at a concentration of 300 μM inhibited the accumulation of triglycerides in 3T3-L1 adipocytes induced by MDI by 23.4%.[134] In a parallel study, trigonelline at 100 μM mitigated the proliferation of adipocytes after 48 hours and inhibited the differentiation of 3T3-L1 preadipocytes into adipocytes after 10 days via repression of peroxisome proliferator-activated receptor-γ and CCAAT/enhancer-binding protein-α, adiponectin, leptin, resistin, and adipocyte fatty acid-binding protein.[135] Fatty acid-binding protein is highly expressed during the differentiation of adipocytes. In preadipocytes treated with 100 μM of trigonelline and 10 μM of isoproterenol for 48 hours, the expression of fatty acid synthase and glucose transporter-4 were inhibited.[135]

4.43 *Vigna nakashimae* (Ohwi) Ohwi & H. Ohashi

Synonyms: *Azukia nakashimae* (Ohwi) Ohwi; *Phaseolus nakashimae* Ohwi
Common name: jom dol pat (Korean)
Subclass Rosidae, Superorder Fabanae, Order Rosiidae, Family Fabaceae
Nutritional use: food (Korea)

Ethanol extract of seeds of *Vigna nakashimae* (Ohwi) Ohwi & H. Ohashi at a concentration of 100 μg/mL inhibited 3T3-L1 differentiation, triglyceride accumulation, and decreased the expression of peroxisome proliferator-activated receptor-γ and targets fatty acid synthetase and fatty acid-binding protein-2.[136] In differentiated 3T3-L1 cells, this extract lowered the expression of peroxisome proliferator-activated receptor-γ, fatty acid synthetase, and fatty acid-binding protein-2 expression; induced adenosine monophosphate-activated protein kinase phosphorylation; and consequently induced acetyl-CoA carboxylase phosphorylation, carnitine palmitoyltransferase-1, and acyl-CoA-oxidase.[136] Given orally at a dose of 500 mg/kg/day for 40 days to C57BL/6J mice on high-fat diet. It decreased a weight gain by 129% and decreased epididymal fat without changes in food intake.[136] Expression of peroxisome proliferator-activated receptor-γ, CCAAT/enhancer-binding protein-α, fatty acid synthetase, and fatty acid-binding protein-2 was decreased in epididymal fat upon extract treatment. The phosphorylation of adenosine monophosphate-activated protein kinase and acetyl-CoA carboxylase was promoted as well as expression of fatty acid oxidation carnitine palmitoyltransferase-1 and acyl-CoA-oxidase.[136] Serum total cholesterol, tumor necrosis-α, interleukin-6, nonesterified fatty acids were reduced, triglycerides were normalized, and adiponectin was increased.[136] In liver tissue, the extract inhibited peroxisome proliferator-activated receptor-γ, CCAAT/enhancer-binding protein-α, fatty acid-binding protein-2, and fatty acid synthetase.[136]

4.44 *Aegle marmelos* (L.) Corrêa

Synonym: *Crataeva marmelos* L.
Common names: mu ju (Chinese); bael (India); bengal quince
Subclass Rosidae, Superorder Rutanae, Order Rutales, Family Rutaceae
Medicinal use: constipation (India)
History: sacred tree of the Hindus, known to Sushruta

Aegeline from the leaves of *Aegle marmelos* (L.) Corrêa given at a single oral dose of 100 mg/kg 30 minutes before sucrose loading lowered glycemia by 12.9% at 5 hours in streptozotocin-induced diabetic Sprague–Dawley rats.[137] In the same experiment, the adenosine monosphosphate-activated protein kinase activator metformin at same dose decreased glycemia by 23.5% at 5 hours.[137] This phenylethylamide given orally to high-fat diet-induced hyperlipidemic hamster for 7 days at a dose of 50 mg/kg/day lowered serum cholesterol and triglycerides by 24% and 55%, respectively, increased high-density lipoprotein–cholesterol by 28%, decreased plasma fatty acids, and increased high-density lipoprotein–cholesterol ratio by 66%.[137] Fenofibrate at a dose of 108 mg/kg/day lowered triglycerides and cholesterol by 42% and 18%, respectively, lowered fatty acids and increased high-density lipoprotein-cholesterol/cholesterol ratio.[137] Aqueous extract of fruits of *Aegle marmelos* (L.) Corrêa given orally at a dose of 1 g/kg/day for 21 days to high-fat diet streptozotocin-induced type 2 diabetes (fasting glycemia >13.89 mmol/L) lowered fasting glycemia from 16.7 to 8.8 mmol/L (normal: 4.9 mmol/L; metformin 100 mg/kg/day: 6.8 mmol/L), decreased insulinemia from 117.1 to 94.6 pmol/L (normal: 77.7 pmol/L; metformin at 100 mg/kg/day: 89.3 pmol/L), and decreased insulin resistance by 57.2%.[138] This extract lowered serum tumor necrosis factor-α and interleukin-6 increased superoxide dismutase activity in pancreas and improved β-cell cytoarchitecture.[138] The extract decreased plasma cholesterol by 47.5%, triglycerides by 51.6%, low-density lipoprotein–cholesterol by 37.9% and increased high-density lipoprotein–cholesterol by 25.8%.[138] This treatment increased the expression of hepatic peroxisome proliferator-activated receptor-γ.[138] Peroxisome proliferator-activated receptor-γ agonists have insulin-sensitizing activity. A dichloromethane extract of leaves decreased lipid accumulation in 3T3-L1 adipocytes by 33.9% at a dose of 100 μg/mL.[139] From this extract, halfordinol, ethyl ether aegeline, methyl ether aegeline reduced lipid content from 3T3-L1 adipocytes by 90%, 80%, and 40%, respectively, at a dose of 100 μM.[139] (3,3-dimethylallyl) halfordinol isolated from the leaves given orally at a daily dose of 50 mg/kg/day for 4 weeks to high-fat diet C57BL/6 mice lowered glycemia from 228.5 to 187.5 mg/dL (normal: 115.2 mg/dL), insulinemia from 7.8 to 5.6 ng/mL (normal: 2.3 ng/mL), total cholesterol from 98.5 to 82.2 mg/dL (normal: 64.3 mg/dL), and triglycerides from 132.2 to 100.2 mg/dL (normal: 91.6 mg/dL).[140] This treatment decreased hepatic triglycerides from 6.5 mg/g to 3.5 mg/dL (normal: 0.7 mg/dL), liver lipid peroxidation, and increased hepatic glycogen contents.[140] This treatment inhibited the expression of CCAAT/enhancer-binding proteins-α, peroxisome proliferator-activated receptor-γ, and increased the expression of sterol regulatory element-binding protein-1c in adipose tissues.[140] *In vitro*, this compound at 20 μg/mL inhibited the accumulation of triglycerides in differentiating 3T3-L1 preadipocytes by 44% and induced lipolysis as evidenced by an increase in glycerol production.[140] Aqueous extract of leaves given orally at a daily dose of 200 mg/kg for 35 days protected Wistar rats against isoprenaline-induced myocardial infarction.[141]

4.45 *Citrus reticulata* Blanco

Synonyms: *Citrus chrysocarpa* Lush.; *Citrus daoxianensis* S.W. He & G.F. Liu; *Citrus erythrosa* Yu. Tanaka; *Citrus nobilis* Lour.; *Citrus sunki* hort. ex Tanaka; *Citrus unshiu* Marcov.
Common names: narangah (India); loose-skinned orange
Subclass Rosidae, Superorder Rutanae, Order Rutales, Family Rutaceae
Medicinal use: indigestion (India)

Limonene, naringenin, and auraptene are common in members of the genus *Citrus* L. At a concentration of 50 μM, this natural products inhibited the differentiation of preadipocyte 3T3-L1 cells into adipocytes 3T3-L1 cells.[142] *Citrus reticulata* Blanco shelters limonene, which added to high-fat diet at 0.5% given for 4 weeks to C57BL/6 mice did not prevent weight gain, had no effect on total cholesterol nor low-density lipoprotein–cholesterol but lowered serum triglycerides by 47%.[142] This regimen decreased fasting blood glucose by 29.1%.[142] This monoterpene reduced the size of white adipose tissue and prevented hepatic lipid accumulation.[142] Limonene added to high-fat diet at 0.5% given for 2 weeks to C57BL/6 mice undergoing 12 weeks of high-fat diet did not prevent weight gain, lowered low-density lipoprotein–cholesterol by 20.4%, lowered serum triglycerides by 36.1%, and increased high-density lipoprotein–cholesterol by 18.3%.[142] This regimen lowered fasting blood glucose by 30.3% and improved glucose tolerance in obese mice.[142] In the 293T cells, limonene increased the transactivity of peroxisome proliferator-activated receptor-α.[142] In the white adipose tissue of preventively treated animals, limonene induced the expression of peroxisome proliferator-activated receptor-α uncoupling protein-2, acetyl-CoA carboxylase, and peroxisome proliferator-activated receptor coactivator-1α.[142] In the liver of preventively treated rodents, this monoterpenes repressed liver X receptor targets sterol regulatory element-binding protein-1 and apolipoprotein-E.[142] Ethanol extract of peel given orally to C57BL/6 mice on high-fat diet at a dose of 150 mg/kg/day for 70 days lowered body weight gain from 38.2 to 9.2 g (normal diet: 5.6 g), total cholesterol from 179.1 to 147.5 mg/dL (normal diet: 119.7 mg/dL), triglycerides, and liver weight.[143] This flavonoid-rich extract reduced epididymal adipose tissue weight and perirenal adipose tissue weight by about 35%. It induced the expression of phosphorylated adenosine monophosphate-activated protein kinase and subsequently the phosphorylated acetyl-CoA carboxylase as well as adiponectin expression in epididymal adipose tissue.[143] This regimen lowered liver weight, serum and histological observation of liver tissues revealed decreased accumulation of triglycerides.[143] At a concentration of 200 μM, this extract induced the activation of LKB1, adenosine monophosphate-activated protein kinase, acetyl-CoA carboxylase, and carnitine palmitoyltransferase-1, as well as protein kinase A and hormone sensitive lipase in 3T3-L1 cells.[143] The flavanone naringenin, which at a concentration of 100 μM inhibited TLR2 expression during 3T3-L1 cells differentiation and increased lipid content by activating peroxisome proliferator-activated receptor-γ.[144] In differentiated adipocytes, this flavonoid at 100 μM suppressed tumor necrosis factor-α induced expression of TRL2 by differentiated adipocytes by inhibiting nuclear factor-κB and c-jun terminal kinase.[144] In C57BL/6J mice high-fat diet containing 1% of this flavanone for 16 weeks glycemia was lowered by about 30%, body weight was not affected whereby epididymal fat tumor necrosis factor-α, monocyte chemoattractant peptide-1, and TLR2 expression (proinflammatory) adipocytic expressions were decreased compared with untreated rodents.[144] Growing evidence suggests that insulin resistance in obese patients is a chronic inflammatory disease initiated in adipose tissues and liver.[145] Reactive oxygen species production in adipocytes increases the expression of monocyte chemoattractant peptide-1, which is a chemoattractant for monocytes and macrophages.[146] Visceral adiposity encompasses the infiltration of macrophages in adipose tissue via monocyte chemoattractant protein-1 resulting in tumor necrosis factor-α secretion and subsequent adipocytes insulin resistance, release of unesterified fatty acids, and decrease of adiponectin secretion.[145] In a subsequent study, Takahashi et al. provided evidence that 13-oxo-9(Z),11(E),15(Z)-octadecatrienoic acid from *Citrus reticulata* Blanco at concentration of 10 μM induced triglyceride accumulation in 3T3-L1 adipocytes and induced adiponectin production. 13-oxo-9(Z),11(E),15(Z)-octadecatrienoic acid induced peroxisome proliferator-activated receptor-γ activity and expression.[147] Induction of peroxisome proliferator-activated receptor-γ activity by 13-oxo-9(Z),11(E),15(Z)-octadecatrienoic acid was abrogated by peroxisome proliferator-activated receptor-γ antagonist GW9662.[147] This oxidized fatty acid evoked the expression of fatty acid-binding protein-2, lipoprotein lipase, and CCAAT/enhancer-binding proteins-α.[147] In differentiated 3T3-L1 adipocytes treated with or without insulin, this fatty acid induced glucose uptake.[147] It must be recalled here that, in adipocytes peroxisome proliferator-activated receptor-γ is endogenously activated by fatty acids.[148]

4.46 *Euodia rutaecarpa* (Juss.) Hook. f. & Thoms.

Synonym: *Boymia rutaecarpa* Juss.
Subclass Rosidae, Superorder Rutanae, Order Rut
Medicinal use: carminative (China)

Euodia rutaecarpa (Juss.) Hook. f. & Thoms. produces evodiamine, a vanilloid receptor agonist.[149] This alkaloid given at 0.02% in the high-fat diet of Sprague–Dawley rats evoked a reduction of adipose tissues, serum unesterified fatty acid, and cholesterol and increased the lipolytic activity of in the perirenal fat tissues.[149] Adipogenesis was inhibited in 3T3-L1 preadipocytes treated with 20 μM of evodiamine via the phosphorylation of epidermal growth factor and downstream protein kinase-Cα and extracellular signal-regulated kinase.[150] This alkaloid given orally to Sprague–Dawley rats at a daily dose of 40 mg/kg for 25 days lowered food intake by 13% and decreased body weight by about 10%.[150] This alkaloid evoked an increase of serum leptin from 1.9 to 2.1 ng/mL.[150] This treatment induced a fall of orexigenic peptides neuropeptide Y and agouti-gene related peptide in hypothalamus arcuate nucleus.[150] This alkaloid did not interfere with the expression of anorexigenic peptides pro-opiomelanocortin (POMC), cocaine- and amphetamine-regulated transcript (CART), and melanocortin receptor-4 in hypothalamus arcuate nucleus.[151] It must be recalled that neurons in the arcuate nucleus secrete neuropeptides regulating food intake and body weight.[152] Leptin lowers food intake and increases energy expenditure by inhibiting the orexigenic neuropeptide Y (NPY)/agouti gene-related protein (AgRP) neurons and stimulation of anorexigenic POMC/CART neurons in the arcuate nucleus.[153] Rutaecarpine given at a dose of 20 mg/kg for 4 week reduced the intake of food of C57BL/6 and spontaneous type 2 diabetic obese *ob/ob* mice by 20.1% and 22.1%, respectively. This regimen resulted in reduction in body weight by 32% in C57BL/6 mice on high-fat diet and a 35% bodyweight decrease in *ob/ob* mice fed a normal diet.[151] In C57BL/6 mice, rutaecarpine at a dose of 20 mg/kg reduced plasma cholesterol, triglycerides, glucose insulinemia, and leptinaemia.[151] In the same experiment, rutaecarpine reduced the expression of orexigenic AgRP and NPY in the arcuate nucleus of treated rodents.[151] Type 2 diabetic obese KK-Ay mice given intraperitoneally at a daily dose of 3 mg/kg for 7 days evodiamine, had lowered body weight gain without diminishing food intake, reduced glycemia, reduced insulinemia by 40%, and reduced mass of inguinal and retroperitoneal white adipose tissues.[154] In the same experiment, the glycemia of evodiamine-treated rodents remained constant, whereby the insulinemia was halved implying that insulin sensitivity was increased.[154] Histopathological examination of liver tissues demonstrated a reduction of fat accumulation in hepatocytes.[154] Analysis of the white inguinal adipose tissues evidenced a decrease in phosphorylated mTOR and phosphorylated S6K, decrease in phosphorylated Akt, and decrease in phosphorylated insulin receptor substrate-1.[154] In diabetic rodents, evodiamine increased the phosphorylation of adenosine monophosphate-activated protein kinase in adipose tissues and liver.[154]

4.47 *Murraya koenigii* (L.) Spreng.

Synonyms: *Bergera koenigii* L.; *Chalcas koenigii* (L.) Kurz
Common names: tiao liao jiu li xiang (Chinese); karivepu (India); daun kari (Malay); curry leaf tree
Subclass Rosidae, Superorder Rutanae, Order Rutales, Family Rutaceae
Medicinal use: indigestion (India)

In type 2 diabetes, the ability of insulin to stimulate to tyrosine phosphorylation of insulin receptor substrate-1,[155] insulin-induced association between phosphoinositide 3-kinase and insulin receptor substrate-1 and substrate-2 and activation of phosphoinositide 3-kinase[156] and insulin-induced translocation of glucose transporter-4 to the plasma membrane are impaired.[157] *Murraya koenigii* (L.) Spreng. (Figure 4.6) elaborates carbazole alkaloids that increase insulin sensitivity. One such alkaloid is mahanimbine which decrease plasma cholesterol and triglycerides in rodents.[158] Mahanine at 35 μg/mL prevented insulin resistance in L6-myotubes induced by palmitate as evidenced by sustained phosphorylation of insulin receptor, phosphorylated phosphoinositide 3-kinase, phosphorylated phosphoinositide-dependent kinase-1, phosphorylated Akt, phosphorylated glucose transporter-4, and increased uptake of glucose.[159] This alkaloid reduced palmitate-induced phosphorylation of protein kinase C_ε and subsequent nuclear translocation and inhibition of high mobility group promoted insulin receptor expression.[159] Mahanine also inhibited palmitate-induced activation of nuclear factor-κB which accounts for insulin resistance.[159] This carbazole given orally at a dose of 0.6 mg/kg/day for 15 days to high-fat diet-induced type 2 diabetic Golden hamsters lowered glycemia by 55%.[159] Mahanine, mahanimbine, and 8,8′-biskoenigine from this plant inhibited the enzymatic activity of protein-tyrosine phosphatase 1B with IC_{50} values equal to 1.7, 1.8, and 2.2 μM, respectively.[160] Ethanol extract of leaves of *Murraya koenigii* (L.) Spreng. given orally to dexamethasone-induced

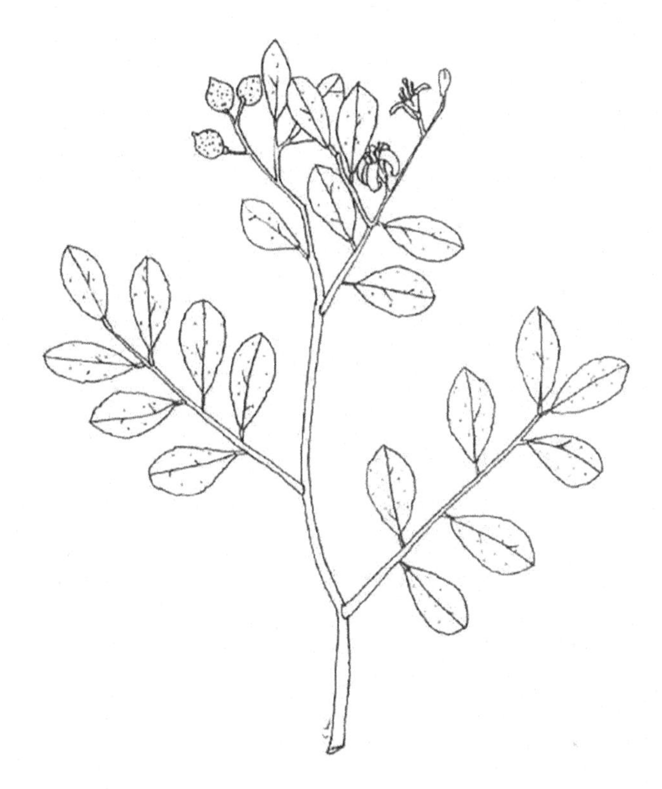

FIGURE 4.6 *Murraya koenigii* (L.) Spreng.

hyperglycemic Swiss albino mice at a daily dose of 100 mg/kg for 8 days lowered glycemia from 123.7 to 71.5 mg/dL.[161] This regimen lowered weight loss induced by dexamethasone-induced catabolism and improved glucose tolerance by 27.6% in oral glucose tolerance tests.[161] In skeletal muscles, the extract enhanced insulin-dependent Akt phosphorylation that was reduced by dexamethasone.[161] In L6-GLUT4myc myotubes, this extract at a concentration of 25 μg/mL boosted glucose uptake and glucose transporter-4 translocation, and this effect was increased in the presence of insulin.[161] In obese patients with insulin resistance, plasma fatty acids hampers glucose uptake by skeletal muscles by activating nuclear factor-κB and protein kinase $C_θ$ that phosphorylate insulin receptor substrate-1 reducing insulin receptor substrate-1 associated phosphoinositide 3-kinase activity.[162]

4.48 *Toddalia asiatica* (L.) Lam

Synonyms: *Paullinia asiatica* L.; *Toddalia floribunda* Wall.; *Toddalia tonkinensis* Guillaumin
Common names: fei long zhang xue (Chinese); kaka todali (India)
Subclass Rosidae, Superorder Rutanae, Order Rutales, Family Rutaceae
Medicinal use: pain in bowels (India)

Watanabe et al. provided evidence that aculeatin, toddaculin, and toddalolactone isolated from the stems of *Toddalia asiatica* (L.) Lam (Figure 4.7) increased at a concentration of 50 μM the accumulation of triglycerides in differentiating 3T3-L1 adipocytes.[163] Out of these coumarins, aculeatin increased the expression peroxisome proliferator-activated receptor-γ, and targets glycerophosphate dehydrogenase, fatty acid-binding protein-2, fatty acid translocase, and glucose transporter-4. The expression of monocyte chemoattractant protein-1 and interleukin-6 were

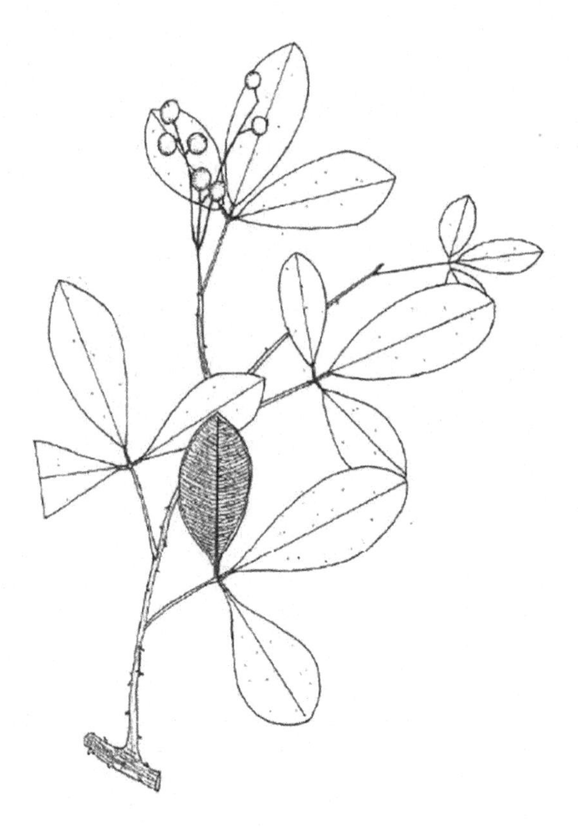

FIGURE 4.7 *Toddalia asiatica* (L.) Lam.

downregulated.[163] In mature 3T3-L1 adipocytes, aculeatin at 100 μM stimulated lipolysis as well as glucose uptake.[163]

4.49 *Zanthoxylum piperitum* (L.) DC.

Common name: Japanese pepper
Subclass Rosidae, Superorder Rutanae, Order Rutales, Family Rutaceae
Medicinal use: promotes digestion (China)

Ethanol extract of fruits of *Zanthoxylum piperitum* (L.) DC. at a concentration of 0.75 μg/mL inhibited the differentiation of 3T3-L1 preadipocytes *in vitro* by 80%.[164] This extract inhibited the expression of adipogenic transcription factors peroxisome proliferator-activated receptor-γ and CCAAT/enhancer-binding protein-α as well as sterol regulatory element-binding protein-2 and fatty acid synthetase.[164] In C57BL/6 mice on high-fat diet with 0.5% of extract for 6 weeks a lower weight gain was observed, food intake was unchanged and epididimal white adipose tissue weight was lowered from 1.4 to 1.2 g (normal diet: 0.6 g).[164] This regimen lowered serum triglycerides to normal values, reduced plasma cholesterol from 166.7 to 155.7 mg/dL (normal: 91 mg/dL), lowered low-density lipoprotein–cholesterol from 95.1 to 75.5 mg/dL (normal: 30 mg/dL).[164] This treatment lowered (but not normalized) glycemia by 11%, reduced insulinemia by 59%, and reduced serum leptin by 54%.[164] This treatment reduced triglyceride accumulation and cholesterol levels in the liver as well as liver weight.[164] In the liver of treated animals, peroxisome proliferator-activated receptor-γ and CCAAT/enhancer-binding protein-α, sterol regulatory element-binding protein-2, fatty acid synthetase, acetyl-CoA carboxylase expression were inhibited when compared with untreated group.[164]

4.50 *Swietenia humilis* Zucc.

Synonyms: *Swietenia bijuga* P. Preuss; *Swietenia cirrhata* S.F. Blake
Common name: Pacific mahogany
Subclass Rosidae, Superorder Rutanae, Order Rutales, Family Meliaceae

Aqueous extract of seeds of *Swietenia humilis* Zucc. given at a single oral dose of 100 mg/kg to Sprague–Dawley rats fed with fructose lowered 30 minutes peak glycemia from 87.4 to 49.2 mg/dL in oral glucose tolerance test.[165] This extract given at 100 mg/kg/day for 2 weeks had no major effects on glycemia but lowered fasting insulin to 0.6 ng/mL to normal (0.8 ng/mL, standard diet) value suggesting an improvement of insulin sensitivity.[165] This regimen increased hepatic weight and hepatic glycogen contents and decreased abdominal fat mass to normal values (8.9 g).[165] In streptozotocin-nicotinamide-induced type 2 diabetic mice, the extract given orally at a single dose of 100 mg/kg halved peak 30 minutes glycemia in oral glucose tolerance test.[165] From the seeds, 2-hydroxy-des-tigloyl-6-deoxyswietenine acetate and methyl-2-hydroxy-3-β-tigloyloxy-1-oxomelia-8(30)-enate lowered at a single oral dose of 31,6 mg/kg 30 minutes peak glycemia by 41% and 51%, respectively.[165] Humulin B lowered 30 minutes glycemia by 60% at a dose of 10 mg/kg.[165] The mode of action of these limonoids is apparently unknown, and it is tempting to speculate the increased uptake of glucose from skeletal muscles.[165]

4.51 *Swietenia macrophylla* King.

Synonyms: *Swietenia belizensis* Lundell; *Swietenia candollei* Pittier; *Swietenia krukovii* Gleason; *Swietenia tessmannii* Harms
Common name: big-leaf mahogany
Subclass Rosidae, Superorder Rutanae, Order Rutales, Family Meliaceae
Medicinal use: fever (Phillipines)

Swietenine isolated from the seeds of *Swietenia macrophylla* King. and given orally to neonatal streptozotocin-induced type 2 diabetic Wistar rats at a daily dose of 50 mg/kg/day for 5 days lowered fasting blood glucose from 177.1 mg/dL to 121.3 mg/Kg (normal: 73.6 mg/dL; glibenclamide at 1 mg/Kg: 115.1 mg/dL).[166] This regimen lowered diabetic plasma cholesterol from 93.4 to 69.1 mg/dL (normal: 67.2 mg/dL), triglycerides from 103.7 to 72.2 mg/dL (normal: 58.1 mg/dL), and replenished liver glycogen content.[166] The mode of action of swietenine is unknown but replenishment of hepatic glycogen suggest either increased insulin secretion or improved insulin sensitivity. With regard to the improvement of insulin sensitivity, Lau et al. provided evidence that 6-*O*-acetylswietenolide, diacetyl swietenolide, and swietenine from the seeds were able to promote the binding of peroxisome proliferator-activated receptor-γ to its co-activator *in vitro*.[167] At a concentration of 10 μM, these limonoids induced the accumulation of triglycerides in 3T3-L1 preadipocytes.[167] In differentiated 3T3-L1 adipocytes, 6-*O*-acetylswietenolide at 2 μM, diacetyl swietenolide at 2 μM, and swietenine at 50 nM increased the expression of peroxisome proliferator-activated receptor-γ and downstream targets glucose transporter-4, adiponectin, adipsin, and similarly to rosiglitazone at 1 μM.[167] Similarly, 6-*O*-acetylswietenolide, diacetyl swietenolide, and swietenine at 2 μM increased the translocation of glucose transporter-4 in C1C12 myotubes and increased glucose uptake.[167] It should be noted that agonists of peroxisome proliferator-activated receptor-γ, such as rosiglitazone, improve the sensitivity of adipocyte tissues to insulin by promoting insulin-mediated glucose uptake by inducing the expression of glucose transporter-4, phosphoinositide 3-kinase activity, insulin receptor substrate-1, and insulin receptor substrate-2. The seeds of *Swietenia macrophylla* King. elaborate swietemacrophin, humilinolide F, 3,6-*O*, *O*-diacetylswietenolide, 3-*O*-tigloylswietenolide, and swietemahonin E that inhibited *in vitro* the generation of superoxide anion radicals by human neutrophils challenged with fMet-Leu-Phe with

IC_{50} values below 50 μM and swietenine was inactive.[168] Being anti-inflammatory and promoting glucose uptake at low doses, limonoids produced by this plant could improve the glycemia of streptozotocin-induced diabetic rodents. These compounds could be too toxic for therapeutic use.

4.52 *Swietenia mahagoni* (L.) Jacq.

Synonyms: *Cedrela mahagoni* L.; *Cedrus mahagani* (L.) Mill.; *Swietenia acutifolia* Stokes; *Swietenia fabrilis* Salisb.; *Swietenia mahogoni* Lam.
Common name: tao hua xin mu (Chinese)
Subclass Rosidae, Superorder Rutanae, Order Rutales, Family Meliaceae
Medicinal use: hypertension (Malaysia)

Methanol extract of stem bark of *Swietenia mahagoni* (L.) Jacq. given orally at a daily dose of 25 mg/kg/day for 15 days to streptozotocin-induced diabetic Wistar albino rats lowered weight loss and fasting glycemia from 296.5 (diabetic) to 74.5 mg/dL (normal: 75 mg/dL).[169] This regimen corrected hepatic and renal oxidation, glutathione, as well as catalase activity.[169] Limonoids with the ability to increase insulin sensitivity are probably involved.

4.53 *Toona sinensis* (A. Juss.) M. Roem.

Synonyms: *Cedrela glabra* C. DC.; *Cedrela longiflora* Wall. ex C. DC.; *Cedrela serrata* Royle; *Cedrela serrulata* Miq.; *Surenus sinensis* (A. Juss.) Kuntze; *Toona glabra* (C. DC.) Harms; *Toona serrata* (Royle) M. Roem.; *Toona serrulata* (Miq.) Harms
Common names: xiang chun (Chinese); Chinese toon
Subclass Rosidae, Superorder Rutanae, Order Rutales, Family Meliaceae
Medicinal use: indigestion (China)
Pharmacological target: insulin resistance

Aqueous extract of leaves of *Toona sinensis* (A. Juss.) M. Roem. given orally at a daily dose of 1 g/kg to alloxan-induced diabetic rats for 5 weeks lowered fasting plasma glucose from 557.8 to 173 mg/dL (normal: 104.4 mg/dL) and increased plasma insulin.[170] In oral glucose tolerance test, the extract at 1 g/kg had no effect after 180 minutes, whereby in diabetic rats, this extract lowered glycemia by 30%.[170] Nonpolar extract of leaves of *Toona sinensis* (A. Juss.) M. Roem. given orally at a dose of 150 mg/kg/day for 56 days to streptozotocin high-fat diet-induced type 2 diabetic C57BL6J mouse decreased glycemia to about 300 mg/dL (untreated diabetic 600 mg/dL; normal: 150 mg/dL).[171] This regimen lowered insulin in diabetic rodent from 1.3 to 0.6 ng/mL (normal: 1.4 ng/mL), decreased insulin resistance, decreased plasma triglycerides to normal values, increased plasma adiponectin, and halved creatinine, whereby serum cholesterol was unchanged.[171] Histological evaluation evidenced reduced sized adipocytes and the absence of hepatosteatosis.[171] Rutin isolated from *Toona sinensis* (A. Juss.) M. Roem. given at a single oral dose of 25 mg/kg to insulin receptor antagonist S961-induced diabetic C57BL/6 mice halved glucose area under the curve and blood glucose peak glycemia.[172] *In vitro*, this flavonoid glycoside at a concentration of 100 μM was not able by itself to induce the phosphorylation of insulin receptor of L6 myotubes, neither promoted the phosphorylation of insulin receptor upon insulin stimulation but sustained insulin receptor phosphorylation in presence of S961.[172] In L6 myotubes, rutin at the same concentration normalized Akt phosphorylation despite S961.[172] In C2C12 myotubes, rutin at a concentration of 100 μM prevented S961-induced inhibition of glucose transporter-4 expression.[172] In L6 myotubes, this flavonoid at 10 μM increased glucose uptake induced by insulin and prevented S961-induced inhibition of glucose entry upon insulin stimulation.[172] Rutin at a concentration of 5 μg/mL induced glucose uptake by L6 rat muscle cells *in vitro* and in the absence of insulin as efficiently as pioglitazone at a concentration of 0.1 μg/mL.[111]

4.54 *Aphanamixis grandifolia* Bl.

Synonym: *Amoora aphanamyxis* Roem. & Schult.
Subclass Rosidae, Superorder Rutanae, Order Rutales, Family Meliaceae
Medicinal use: flue (Indonesia)

In adipocytes, absorbed fatty acids from very low density lipoproteins or chylomicrons are re-esterified to glycerol via a reaction involving glycerol-3-phosphate acyltransferase, 1-acylglycerol-3-phosphate acyl transferase, phosphatidate phosphohydrolase, and diacylglycerol acyltransferase-1.[173] Diacylglycerol acyltransferase-1 in adipocytes catalyzes the acylation of 1,2-diacylglycerol into triglyceride which is a rate limiting step in triglyceride synthesis in adipose tissues.[173] Aphadilactone A and aphadilactone C (Figure 4.8) isolated from the leaves of *Aphanamixis grandifolia* Bl. inhibited diacylglycerol acyltransferase-1 activity by 2 5.5% and 85.9% at 10 μM *in vitro*.[174]

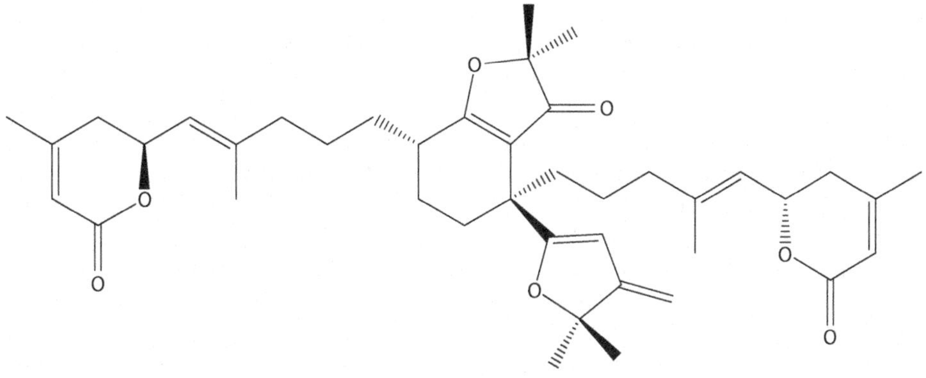

FIGURE 4.8 Aphadilactone C.

4.55 *Quassia amara* L.

Common names: kayuh pahit (Indonesia); corales (Philippines); Surinam Quassia
Subclass Rosidae, Superorder Rutanae, Order Rutales, Family Simaroubaceae

Methanol extract of stem wood of *Quassia amara* L. given orally to streptozotocin-nicotinamide type 2 diabetes Charles Foster rats at a dose of 200 mg/kg lowered peak glycemia at 30 minutes in oral glucose tolerance test from about 145 to 115 mg/dL.[175] In normoglycemics rats, this extract had no effects on glucose tolerance. This extract given 14 days at a daily dose of 200 mg/mL prevented weight loss, lowered fasting blood glucose level from 349.1 mg/dL (diabetic untreated) to 105.1 mg/dL (normal: 79.7 mg/dL).[175] This regimen increased plasma insulin from 7.5 μIU/mL (diabetic) to 8.7 IU/mL (normal: 17.9 μIU/mL) and increased hepatic glycogen contents from about 12 to 22 mg/g (normal: 30 mg/dL).[175] Furthermore, this extract lowered total cholesterol from 143.2 to 98.2 mg/dL (normal: 79.5 mg/dL), triglycerides from 115.3 to 73.8 mg/dL (normal: 46.8 mg/dL), low-density lipoprotein–cholesterol from 95.7 to 54.5 mg/dL (normal: 31.4 mg/dL), and increased high-density lipoprotein–cholesterol from 24.4 to 28.9 mg/dL (normal: 31.4 mg/dL).[175] Note that the lack of hypoglycemic activity of methanol extract of stem wood in normal rats and the absence of robust increased plasma insulin in streptozotocin-nicotinamide type 2 diabetes Charles Foster rats suggests, at least, the presence of natural products, which after oral absorption and first-pass metabolism improve glucose uptake by skeletal muscles.

4.56 *Commiphora mukul* (Hook. ex Stocks) Engl.

Synonym: *Commiphora wightii* (Arn.) Bhandari
Common names: guggul (India); Indian Bdellium
Subclass Rosidae, Superorder Rutanae, Order Rutales, Family Burseraceae
Medicinal use: ulcers (India)

In COS-7 cells, ethyl acetate fraction of gum of *Commiphora mukul* (Hook. ex Stocks) Engl. and commipheric acid induced the expression of peroxisome proliferator-activated receptor-α with EC_{50} values equal to 0.5 and 0.8 μM, respectively.[176] Commipheric acid induced the expression of peroxisome proliferator-activated receptor-γ with an EC_{50} equal to 0.6 μM and guggulipid evoked the expression of liver X receptor with an EC_{50} equal to 0.6 μg/mL.[176] E- and Z-guggulsterones had no effect. Guggulipid at dose of 5 μg/mL and commipheric acid at a dose of 5 μM induced the differentiation of 3T3-L1 preadipocytes, whereby guggulsterone had no effect.[176] Guggulipid given in food at a dose of 20 g/kg of C57B1/6 *Lep^{ob}/Lep^{ob}* mice for 7 weeks improved glucose tolerance, mildly increased plasma insulin, lowered weight gain without lowering food intake, and lowered triglycerides from 2.7 to 2.5 mM, whereby commipheric acid was inactive.[176] In a parallel study, guggulsterone E/Z isolated from the oleoresin of *Commiphora mukul* (Hook. ex Stocks) Engl. given orally at a dose of 75 mg/kg/day for 8 weeks to Wistar rats on high-fat diet kept glycemia and lipid profile close to the values obtained from rodents on normal diet, and lowered plasma insulin from 8.3 to 4.5 ng/mL (normal: 3.3 ng/mL).[177] This treatment increased, but not normalized, hepatic and skeletal muscle glycogen contents, and lowered the activity of glucose-6-phosphatase.[177] *In vitro*, guggulsterone at a concentration of 100 μg/mL induced, with or without concomitant addition of insulin, an increase in glucose uptake by medium psoas muscle isolated from rodents on high-fat diet.[177] Ethanol extract of gum resin of *Commiphora mukul* (Hook. ex Stocks) Engl. given orally at a dose of 200 mg/kg/day for 60 days to Wistar rats on high fructose diet lowered body weight from 117 to 103 g (rats on normal diet 100 g), prevented increase in fasting glycemia, lowered fasting plasma insulinemia by 67.4%, and rectified insulin resistance.[178] This treatment lowered serum cholesterol, low-density lipoprotein–cholesterol, and atherogenic index to normal group values, normalized triglycerides, very low-density lipoprotein, and increased high-density lipoprotein.[178] Besides, the extract reinstated normal hepatic lipid peroxidation, superoxide dismutase, catalase and glutathione peroxidase.[178]

4.57 *Canarium odontophyllum* Miq.

Common names: dabai (Sarawak); Sibu olive
Subclass Rosidae, Superorder Rutanae, Order Rutales, Family Burseraceae
Nutritional use: fruits eaten in Sarawak

Oil of fruits of *Canarium odontophyllum* Miq. given to healthy New Zealand white rabbits as part of diet for 4 weeks increased food intake, inhibited body weight gain, lowered low-density lipoprotein–cholesterol, triglycerides, and lipid peroxidation while boosting total cholesterol and high-density lipoprotein–cholesterol.[179] This regimen increased the enzymatic activity of superoxide dismutase and glutathione peroxidase.[179] Ethanol extract of deffated fruits (containing catechin, epicatechin, vanillic acid, ferulic acid, and apigenin) given at a dose of 600 mg/kg/day for 4 weeks to streptozotocin high-fat diet type 2 diabetic Sprague–Dawley rats evoked a decrease of plasma glucose, improved oral glucose tolerance, had no effect on plasma insulin and increased insulin sensitiviry.[180] This regimen decreased plasma cholesterol, triglycerides, and low-density lipoprotein–cholesterol. It also increased high-density lipoprotein–cholesterol toward normal values.[180]

4.58 *Mangifera indica* L.

Common names: am (India); mango
Subclass Rosidae, Superorder Rutanae, Order Rutales, Family Anacardiaceae
Medicinal use: diabetes (India)
Pharmacological target: insulin resistance

The gallotannin 1,2,3,4,6-penta-*O*-galloyl-β-D-glucose isolated from the leaves of *Mangifera indica* L. given 21 days to C57BL/6 mice on high-fat diet at a dose of 50 mg/kg/day for 21 days evoked a reduction of body weight, normalized plasma glucose, lowered total cholesterol from 189.2 to 71.4 mg/dL (normal: 69.4 mg/dL) and triglycerides from 254.3 to 126.6 mg/dL (normal: 109.6 mg/dL).[181] This regimen improved glucose tolerance and lowered plasma insulin from 126.7 to 70 ng/dL (normal: 69 ng/dL). This gallotannin could, at least, account to the fact that ethanol extract of leaves given to spontaneous type 2 diabetic obese KK-Ay mice at a daily dose of 500 mg/kg for 8 weeks evoked a mild decrease in mesentery fat compared with untreated group.[182] This regimen lowered fasted serum glycemia, triglycerides, and total cholesterol by 27%, 22%, and 36%, respectively.[182] This regimen lowered fasted liver glycemia, triglycerides, and total cholesterol by 27%, 22%, and 36%, respectively.[182] The extract lowered fasting liver triglycerides by 36% and attenuated hepatic free fatty acids total cholesterol and low-density lipoprotein.[182] Soleus muscle isolated from treated mice, the extract evoked the phosphorylation of adenosine monophosphate-activated protein kinase an increased expression phosphorylated phosphoinositide 3-kinase, Akt and glycogen synthetase-1.[182] *Mangifera indica* L. shelters mangiferin, which at 1 μM evoked an increase in glucose transporter-4 expression, evoked the phosphorylation of adenosine monophosphate-activated protein kinase and repressed peroxisome proliferator-activated receptor-γ.[183] The increased uptake of glucose by this xanthone glycoside in L6-myotubes was inhibited by peroxisome proliferator-activated receptor-γ antagonist GW9662.[183]

4.59 *Rhus coriaria* L.

Common names: sumak (India); tanning sumac
Subclass Rosidae, Superorder Rutanae, Order Rutales, Family Anacardiaceae
Medicinal use: dysentry (India)

Methanol extract of seeds of *Rhus coriaria* L. given to streptozotocin-induced diabetic Wistar rats at a daily dose of 400 mg/kg for 5 weeks decreased glycemia from 324.6 to 117 mg/dL (normoglycemic rodents 97.1 mg/dL).[184] This regimen lowered plasma insulin from 24 mU/L to 15.8 mU/mL, a value close to normal (13.1 mU/mL) and reduced insulin resistance least.[184] The natural products involved here are unknown but one could suggest, at gallotannins that abound in this plant. In fact, alloxan-induced diabetic Wistar rats treated for 28 days with daily oral administration of aqueous extract at 250 mg/kg evoked a reduction in blood glucose and a decrease in hepatic and renal malonaldehyde and increased activity of hepatorenal catalase.[185] Type 2 diabetic patients taking 3 g daily of sumac for 3 months had a decrease in plasma insulin from 7 to 5.3 μU/mL, decrease insulin resistance and a decrease in plasma malonaldehyde.[186] One could draw an inference that metabolites of tannins in sumac, such as gallic acid could improve glucose tolerance by lowering insulin resistance in skeletal muscles.

4.60 *Geranium thunbergii* Siebold ex Lindl. & Paxton

Common names: zhong ri lao guan cao (Chinese); gennoshoko (Japanese)
Subclass Rosidae, Superorder Rutanae, Order Geraniales, Family Geraniaceae
Medicinal use: dysentry (Japan)

Ethanol extract of *Geranium thunbergii* Siebold ex Lindl. & Paxton given orally at a dose of 400 mg/kg to C57BL/6J mice for 6 weeks, decreased body weight, lowered epididymal and retroperitoneal white adipose tissue mass and adipocyte size.[187] This regimen lowered triglycerides from 88.7 to 41.2 mg/dL (normal diet: 78 mg/dL), total cholesterol from 164.8 to 140 mg/dL (normal: 106.7 mg/dL), low-density lipoprotein–cholesterol from to 6.2 to 3.8 mg/dL (normal diet: 4.8 mg/dL).[187] This regimen decreased serum leptin by 55% and increased serum adiponectin above normal fed group values.[187] In adipocytes of treated animals, this extract inhibited the expression of peroxisome proliferator-activated receptor-γ, fatty acid-binding protein as well as sterol regulatory element-binding protein-2 and fatty acid synthetase.[187] The principle involved here is not known. It must be recalled that *Geranium thunbergii* Siebold ex Lindl. contains ellagitannins such as geraniin,[188] and it is reasonable to suggest that metabolites of this tannin could, at least, be involved.

4.61 *Pelargonium graveolens* L'Hér. ex Aiton

Synonym: *Pelargonium intermedium* Kunth
Common name: rose Geranium
Subclass Rosidae, Superorder Rutanae, Order Geraniales, Family Geraniaceae

Essential oil from the leaves of *Pelargonium graveolens* given orally to alloxan-induced diabetic Wistar rats (glycemia >10 mmol/dm^3) at a concentration of 150 mg/kg/day for 1 month decreased glycemia from about 14 to 8 mM (glibenclamide 600 μg/kg: 10 mM; normal: 7 mM).[189] This supplementation increased hepatic glycogen.[189] At the hepatic and renal level, the essential oil increased the enzymatic activities of catalase, superoxide dismutase and glutathione peroxidase and lowered oxidation reactive species more potently than glibenclamide.[189]

4.62 *Salacia oblonga* Wall.

Subclass Rosiidae, Superorder Celastranae, Order Celastrales, Family Celastraceae
Medicinal use: diabetes (India)

Aqueous extract of roots of *Salacia oblonga* at 75 μg/mL induced the uptake of glucose by L6-myotubes and 3T3-L1 fibroblasts cultured *in vitro*; this effect was accentuated by insulin.[190] In L6-myotubes, this effect was induced by the expression of glucose transporter-4.[190] The plant contains mangiferin.

4.63 *Salacia reticulala* Wight.

Common name: pitika (India)
Subclass Rosidae, Superorder Celastranae, Order Celastrales, Family Celastraceae
Medicinal use: diabetes (India)

(–)-Epicatechin and 3β,22β-dihydroxyolean-12-en-29-oic acid isolated from a root extract of *Salacia reticulata* Wight. inhibited rat adipose tissue lipoprotein lipase *in vitro*.[191] A tannin fraction of roots inhibited *in vitro* glycerophosphate dehydrogenase.[191] Extract of *Salacia reticulata* Wight. at a concentration of 50 μg/mL reduced triacylglycerol accumulation in differentiating 3T3-L1 cells and decreased peroxisome proliferator-activated receptor-γ, fatty acid-binding protein, glycerophosphate dehydrogenase, and adiponectin secretion.[192]

4.64 *Taxillus chinensis* (DC) Danser

Synonyms: *Loranthus chinensis* DC.; *Loranthus estipitatus* Stapf; *Scurrula chinensis* (DC.) G. Don; *Taxillus estipitatus* (Stapf) Danser
Common name: guang ji sheng (Chinese)
Subclass Rosidae, Superorder Santalanae, Order Santalales, Family Loranthaceae
Medicinal use: liver ailments (China)

Ethanol extract of stems of *Taxillus chinensis* (DC) Danser given at a daily dose of 25 mg to mice on high-fat diet for 10 days evoked a loss of body weight by 9.8% and a reduction of food intake compared with untreated animals.[193] It inhibited fatty acid synthetase.[193]

4.65 *Weigela subsessilis* L.H. Bailey

Common names: korai-yabu-utugi (Japan); byeong-kkot-na-mu (Korea); Korean weigela
Subclass Asteridae, Superorder Cornanae, Order Dipsacales, Family Caprifoliaceae

The leaves of *Weigela subsessilis* L.H. Bailey contain corosolic acid, ursolic acid, ilekudinol B, and pomolic which induced glucose uptake by L6 myotubes in the absence or presence of insulin *in vitro* with a maximal activity at 50 μM.[194] Ilekudinol A, Ilekudinol B, and ursolic inhibited protein-tyrosine phosphatase 1B activity with IC_{50} of 29.1, 5.3, and 3.6 μM (RK-682: 4.5 μM), respectively.[195] The hypoglycemic properties of corosolic acid have been discussed elsewhere.

4.66 *Acanthopanax senticosus* (Rupr. ex Maxim.) Harms

Synonyms: *Eleutherococcus senticosus* (Rupr. ex Maxim.) Maxim.; *Hedera senticosa* Rupr. ex Maxim.
Common names: ci wu jia (Chinese); Siberian ginseng
Subclass Cornanae, Superorder Cornanae, Order Apiales, Family Araliaceae
Medicinal use: fatigue (China)

Aqueous extract of roots of *Acanthopanax senticosus* (Rupr. ex Maxim.) Harms given orally at a dose of 150 mg/kg, 3 times per day for 3 days, to streptozotocin-fructose induced type 2 diabetic Wistar rats (glycemia ≥20 mmol/L) improved insulin sensitivity as evidenced by a reduction of postprandial glycemia from about 190 to 155 mg/dL after 30 minutes in oral glucose tolerance test.[196] The extract given 3 times per day at a dose of 150 mg/kg for 28 days lowered fasting plasma glucose from 10.4 to 8.3 mmol/L, lowered fasting plasma insulin, and increased the hypoglycemic effect induced by insulin injection.[196] These effects, could at least, be mediated by acanthoic acid, ent-kaur-16-en-19-oic acid (Figure 4.9), and 16αH,17-isovaleryloxy-ent-kauran-19-oic acid that inhibited the enzymatic activity

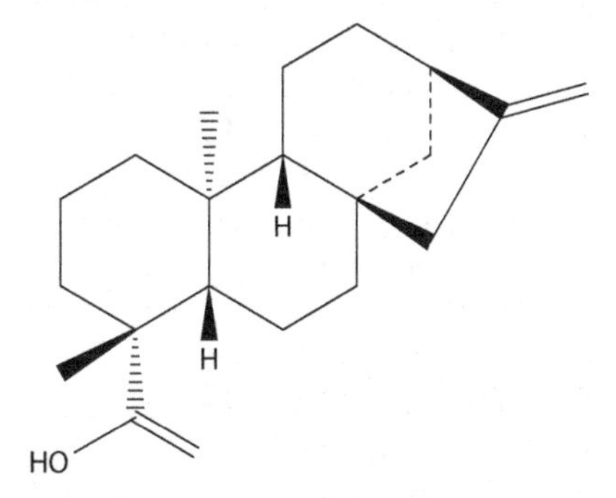

FIGURE 4.9 ent-Kaur-16-en-19-oic acid.

of protein-tyrosine phosphatase 1B with IC_{50} of 23.5, 20.2, and 7.1 μM, respectively.[197] Other principles could involve (7*S*,8*R*)-3-hydroxyl-4-methoxyl-balanophorin, (7*S*,8*R*)-5-methoxyl-balanophorin, balanophorin, curcasinlignan A, curcasinlignan B, and curcasinligan C that inhibited *in vitro* the enzymatic activity of protein-tyrosine phosphatase 1B with IC_{50} values below 30 μM, respectively.[198]

4.67 *Panax ginseng* C.A. Meyer

Synonyms: *Aralia ginseng* (C.A. Mey.) Baill.; *Panax chin-seng* Nees; *Panax quinquefolius* var. *ginseng* (C.A. Mey.) Regel & Maack
Common names: ren shen (Chinese); ginseng
Subclass Cornanae, Superorder Cornanae, Order Apiales, Family Araliaceae
Medicinal use: tonic (China)
History: The plant was listed in the Pent' sao Kangmu, sixteenth century Chinese Pharmacopoeia
Pharmacological target: insulin resistance

Roots of *Panax ginseng* C.A. Meyer given orally for 4 weeks to leptin-deficient (B6.V-Lepob, "*ob/ob*") mice at a dose of 200 mg/kg/day evoked a decrease in body weight by 50% compared with untreated animals, lowered blood glucose concentration by about 30%, and decreased the size of adipocytes.[199] This supplementation increased the expression of peroxisome proliferator-activated receptor-γ and consequently increased the expression of lipoprotein lipase in adipocytes of treated animals.[199] This supplementation also induced the expression of insulin receptor and glucose transporter-4 in the skeletal muscle.[199] Steamed and dried roots of *Panax ginseng* C.A. Meyer given to Otsuka Long-Evans Tokushima Fatty rats given at a dose of 200 mg/kg/day for 42 weeks increased insulinemia, normalized total cholesterol, and decreased low-density lipoprotein–cholesterol from 0.8 to 0.6 mmol/L (normal: 0.5 mmol/L).[200] After 40 weeks, this regimen lowered visceral mass gain by about 20%; it did not modify food intake significantly and improved insulin and glucose tolerance compared with untreated group. The supplementation induced the phosphorylation of adenosine monophosphate-activated protein kinase and acetyl-CoA carboxylase in the gastrocnemius muscle of rats.[200] This supplementation increased the expression of glucose transporter-4 and peroxisome proliferator-activated receptor coactivator-1α.[200] Roots powder of *Panax ginseng* C.A. Meyer given orally at a dose of 200 mg/kg/day to Sprague–Dawley rats on high-fat diet for 12 weeks evoked a decrease in body weight and total fat mass by 7.7% and 30%, respectively, and food intake was not changed.[201] This regimen improved insulin sensitivity. High-fat diet evoked an increase of leptin in rodents and that increase was inhibited by the root powder treatment. The regimen prevented high-fat diet induced reduction of the phosphorylation of glucose transporter-4, insulin receptor substrate-1, Akt, and glycogen synthase kinase-3.[201] Ginsenoside Rg1 at a concentration of 20 μM boosted the intake of glucose by cultured insulin-resistant C2C12 cells by increasing the phorphorylation of adenosine monophosphate-activated protein kinase and subsequent translocation of glucose transporter-4.[202] In skeletal muscle cells, activation of adenosine monophosphate-activated protein kinase inhibits glycogen synthethase blocking thus glycogen synthesis and induce glucose transporter-4 promoting the entry of glucose as discussed elsewhere. Treatment of 3T3-L1 cells by ginsenoside Rg3 at a concentration of 40 μM inhibited triglyceride accumulation during differentiation (induced by dexamethasone-insulin-3-isobutyl-1-methylxanthine) by about 40% by activating of adenosine monophosphate-activated protein kinase hence repression of peroxisome proliferator-activated receptor-γ.[203] Ginsenoside Rg3 at a concentration of 20 μM increased by about 10% the intake of glucose by differentiated 3T3-L1 adipocytes and induced the expression of insulin receptor substrate-1, phosphoinositide 3-kinase, glucose transporter-4.[204] In a subsequent study, rodents given a diet supplemented with 5% ginseng, for 13 weeks had lower body weight by 21% compared with untreated animals whereby food intake was not changed.[205] This regimen lowered total adipose tissue mass by 30%, reduced the size of adipocytes by 49% and abrogated the accumulation of fats in the liver.[205] With regard to lipid profile, the supplementation lowered triglycerides by

29% and unesterified fatty acids by 44%.[205] This supplementation lowered serum insulin by 72% and glycemia by 14% (v). In adipose tissues of treated animals the expression of peroxisome proliferator-activated receptor γ coactivator 1-α and consequently uncoupling protein-2, carnitine palmitoyl-transferase-1 and median-chain acylcoenyme A dehydrogenase was increased suggesting increase fatty acids β-oxidation.[205] One could observe that saponins from medicinal plants often activate adenosine monophosphate-activated kinase. One could speculate that saponins produced by *Panax ginseng* C.A. Meyer have the ability to activate adenosine monophosphate-activated kinase, *in vitro* improving thereby insulin sensitivity and fatty acid β-oxidation.

4.68 *Angelica gigas* (Miq.) Franch. & Sav.

Synonym: *Peucedanum decursivum* Miq.
Common names: chao xian dang gui (Chinese); oninodake (Japan); zam dan gui (Korean); Korean danggui
Medicinal use: anemia (Korea)

Decursin (Figure 4.10) isolated from the arial part of *Angelica gigas* (Miq.) Franch. & Sav. inhibited at a concentration of 20 μg/mL by approximately 40% the accumulation of triglycerides during the differentiation of 3T3-L1 preadipocytes (induced by insulin-dexamethasone-3-isobutyl-1-methylxanthine) *in vitro* and lowered the expression of fatty acid synthetase.[206] When the differentiation of preadipocytes into adipocytes is induced experimentally *in vitro* by insulin, dexamethasone, and the phosphodiesterase inhibitor 3-isobutyl-1-methylxanthine, the expression of positive regulators of adipogenesis is increased.[207] Once such regulator is sterol regulatory element-binding protein-1c with induces the expression of fatty acyl synthetase which catalyzes the synthesis of fatty acids. Decursin given at a daily dose of 200 mg/kg to C57BL/6J mice as part of high-fat diet for 7 weeks lowered body weight gain despite unchanged food intake.[206] This prenylated coumarin brought serum triglycerides from 112 mg/dL to 87.3 mg/dL (normal: 97.2 mg/dL), and lowered total cholesterol from 253.6 to 216.3 mg/dL (normal: 230.3 mg/dL).[206] This regimen improved glucose tolerance in parenteral glucose administration. This prenylated coumarin lowered the production of leptin, resistin, interleukin-6, and monocyte chemoattractant protein-1 induced by high-fat diet.[206] Leptin, resistin, interleukin-6, and monocyte chemoattractant protein-1 are negatively regulated by peroxisome proliferator-activated receptor-γ suggesting that decursin and/or first-pass metabolite(s) could act as peroxisome proliferator-activated receptor-γ agonist.

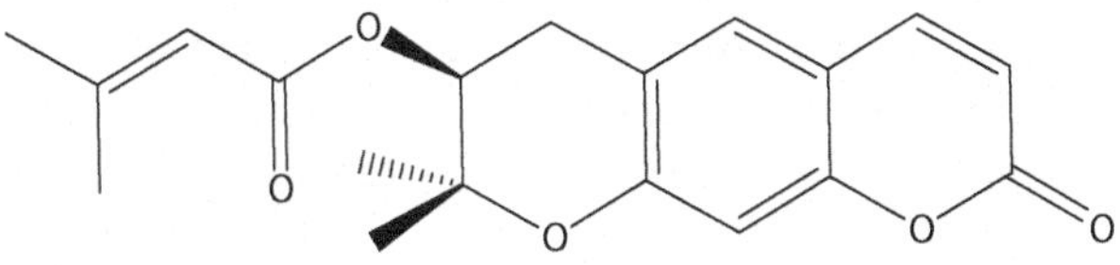

O O O O O

FIGURE 4.10 Decursin.

4.69 *Angelica keiskei* Koidz.

Synonym: *Archangelica keiskei* Miq.
Common names: ashitaba (Japanese); Japanese angelica
Subclass Cornanae, Superorder Cornanae, Order Apiales, Family Apiaceae
Nutritional use: vegetable (Japan)

4-Hydroxyderricin (Figure 4.11) and xanthoangelol from *Angelica keiskei* Koidz. at a concentration of 30 μM commanded glucose uptake by rat skeletal muscle L6 myotube cells similarly with insulin at

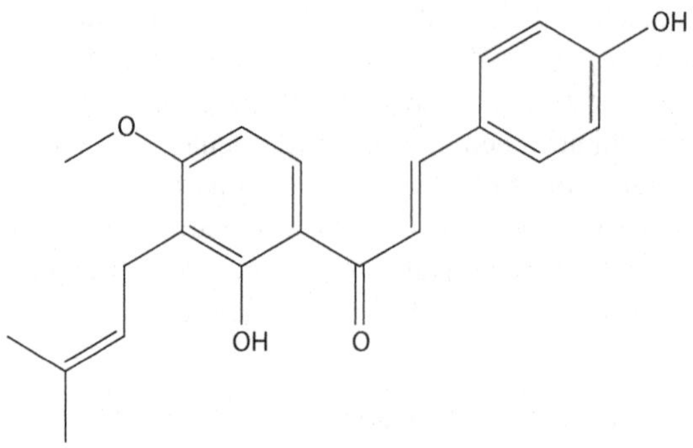

FIGURE 4.11 4-Hydroxyderricin.

0.1 μM.[208] These prenylated chalcones induced the translocation of glucose transporter-4, and this effect was independent of protein kinase C, Akt and adenosine monophosphate-activated protein kinase phosphorylation.[208] In oral glucose tolerance test using mice an extract of *Angelica keiskei* given at a single oral dose of 250 mg/kg lowered glycemia at 30 minutes from 223 to 175 mg/dL.[208] Ethanol extract of roots at 0.03% induced triglyceride accumulation in 3T3-L1 differentiating adipocytes and evoked intake of glucose by differentiated.[209] From this extract, 4-hydroxyderricin and xanthoangelol at a concentration of 10 μM induced triglyceride accumulation in 3T3-L1 differentiating adipocytes and evoked intake of glucose.[209] The triglyceride accumulation in 3T3-L1 activity induced by these prenylated xanthones was not induced via peroxisome proliferator-activated receptor-γ activation.[209] 4-Hydroxyderricin or xanthoangelol added at 0.15% of diet of KK-Ay/Ta mice for 4 weeks evoked a decrease in glycemia by about 50% and 33%, respectively.[209] 4-Hydroxyderricin or xanthoangelol added at 0.15% of diet of KK-Ay/Ta mice for 7 weeks evoked a decrease in glycemia, water intake by about 40%, 25% and had no effect on body weight.[209] It must be recalled that activation of peroxisome proliferator-activated receptor-γ in adipocytes command the expression of glucose transporter type 4 hence increase plasma glucose uptake and decreased glycemia. In a subsequent study, 4-hydroxyderricin and xanthoangelol inhibited the differentiation of 3T3-L1 pre-adipocytes by approximately 20%.[210] These chalcones enhanced the activation of adenosine monophosphate-activated protein kinase by phosphorylation of α-subunit inducing a subsequent phosphorylation of acetyl-CoA carboxylase, downregulated glycerol-3-phosphate acyltransferase-1, upregulated carnitine palmitoyltransferase-1, inhibited the expression of CCAAT/enhancer-binding protein-α, and peroxisome proliferator-activated receptor-γ and consequently glucose transporter-4.[210] 4-Hydroxyderricin and xanthoangelol inhibited the expression of CCAAT/enhancer-binding protein-β, which is upstream of CCAAT/enhancer-binding protein-α.[210] These chalcones also induced the phosphorylation of extracellular signal-regulated kinase-1/2 and JNK and inhibition of extracellular signal-regulated kinase-1/2 and JNK phosphorylation by PD98059 and SP600125, abrogated the flavones activity.[210] The phosphorylation of p38 mitogen-activated protein kinase by activated adenosine monophosphate-activated protein kinase triggers peroxisome proliferator-activated receptor-γ phosphorylation and inhibits adipogenesis.[211]

4.70 *Cnidium monnieri* (L.) Cusson

Synonym: *Selinum monnieri* L.
Common name: she chuang (Chinese)
Subclass Cornanae, Superorder Cornanae, Order Apiales, Family Apiaceae
Medicinal use: tonic (China)
Pharmacological target: insulin resistance

Osthole from the fruits of *Cnidium monnieri* (L.) Cusson at a concentration of 25 μM induced the phosphorylation of adenosine monophosphate-activated protein kinase and subsequent phosphorylation (inhibition) of acetyl-CoA carboxylase in mouse C2C12 skeletal myoblasts and rat L6 skeletal myoblasts.[212] In L6 myotubes, osthole at 50 μM evoked the translocation of glucose transporter-4 and glucose uptake more efficiently than insulin, and this effect was annihilated by adenosine monophosphate-activated protein kinase inhibitor compound C.[212] Osthole increased AMP:ATP ratio in L6 myotubes, evoked the phosphorylation of Akt and subsequent inhibition of glycogen synthase kinase-3, activation of AS160 whence glucose transporter-4 translocation and glucose uptake.[212] Given orally to streptozotocin-induced diabetic mice 5 days per week for 8 weeks at a dose of 100 mg/kg this prenylated coumarin lowered glycemia from approximately 220 to 150 mg/dL.[212] Given orally at a dose of 10 mg/kg/day to Sprague–Dawley rats on high-fat high-sucrose diet for 4 weeks osthol lowered plasma cholesterol from 2.5 to 1.6 mmol/L (normal: 1.4 mmol/L), normalized triglycerides from 0.5 to 0.1 mmol/L, and normalized plasma unesterified fatty acids from 1178.5 to 628.3 μmol/L.[213] At the hepatic level, osthol lowered triglycerides from 17.9 to 11.6 mg/g and brought unesterified fatty acids to 3.8 μmol/g (normal value of 9.6 μmol/g).[213] The supplementation lowered serum insulin to normal value, lowered fasting glycemia from 7.3 to 6.5 mmol/L (normal: 6 mmol/L), increased insulin sensitivity, and augmented plasma adiponectin.[213] Osthol at a concentration of 120 μM decreased intracellular triacylglycerol and unesterified fatty acid in 3T3L1 adipocytes and lowered free fatty concentration in cellular supernatant.[214] This treatment increased peroxisome proliferator-activated receptor-γ expression and repressed sterol regulatory element-binding protein-1c and subsequently fatty acid synthetase and diacylglycerol acyltransferase expression.[214] The chromone glycoside cnidimoside B from the fruits of *Cnidium monnieri* (L.) Cusson at a concentration of 300 μM inhibited 3-isobutyl-1-methyl-xanthine-insulin-dexamethasone-induced 3T3L1 preadipocyte differentiation by 85.3%.[215]

4.71 *Peucedanum japonicum* Thunb.

Synonym: *Anethum japonicum* (Thunb.) Koso-Pol.
Common name: bin hai qian hu (Chinese)
Subclass Asteridae, Superorder Cornanae, Order Apiales, Family Apiaceae
Medicinal use: cough (Japan)
Pharmacological target: insulin resistance

Ethanol extract of leaves and stems of *Peucedanum japonicum* Thunb. given to C57BL/6 mice as part of 0.8% of diet for 4 weeks had no effect on food intake, lowered white adipose tissue weight from 8.3 to 5 g.[216] Liver triglyceride were reduced from 34.9 to 21.4 mg/dL and fecal triglycerides were increased from 0.3 to 0.5 mg/day, whereas fecal bile acids were decreased from 1.5 to 0.7 mg/day.[216] This extract inhibited the activity of pancreatic lipase by 70% at a concentration of 3 mg/mL.[216] In the liver, the supplementation induced the expression of farnesoid X receptor and peroxisome proliferator-activated receptor-α[216] In adipose tissue, the extract induced the expression of peroxisome proliferator-activated receptor-γ. In the soleus muscle of treated rodent, the extract evoked the expression of carnitine palmitoyltransferase-1 and uncoupling protein-3.[216] Pteryxin isolated from *Peucedanum japonicum* Thunb. inhibited at a concentration of 20 μg/mL the dexamethasone-insulin-3-isobutyl-1-methylxanthine induced differentiation of 3T3-L1 pre-adipocytes as evidenced by a 57.4% inhibition of triglyceride accumulation.[217] In adipocytes, this prenylated coumarin induced the expression of peroxisome proliferator-activated receptor-γ, lipoprotein lipase, fatty acid-binding protein, as well as uncoupling protein-2.[217] In adipocytes, pteryxin reduced the expression of sterol regulatory element-binding protein-1c and acetylCoA carboxylase, pyruvate dehydrogenase kinase 4, glucose transporter-4, and insulin receptor substrate-1.[217] It must be recalled that peroxisome proliferator-activated receptor-γ promotes insulin-mediated glucose uptake by inducing the expression of glucose transporter-4, phosphoinositide 3-kinase activity, insulin receptor substrate-1, and substrate-2 as discussed elsewhere.

4.72 *Platycodon grandiflorus* (Jacq.) A. DC.

Synonym: *Platycodon glaucum* (Thunb.) Nak.
Common name: jie geng (Chinese)
Medicinal use: cough (Korea)
Pharmacological target: insulin resistance

The WNT/β-catenin pathway is a negative regulator of adipogenesis, which is inhibited during adipogenic differentiation.[218] When the WNT/β-catenin is inactive, β-catenin binds to glycogen synthase kinase-3, adenomatous polyposis coli, and AXIN to be subsequently degraded by proteasomes.[218,219] When WNT/β-catenin is activated, disheveled promotes the dissociation of the destruction complex resulting in the stabilization of β-catenin and translocation in the nucleus.[220] The transcriptional coactivator β-catenin binds and activates the transcription factors and lymphoid-enhancing factors to induce the expression of target genes such as cyclin D and peroxisome proliferator-activated receptor-δ and inhibits the transcriptional activity of peroxisome proliferator-activated receptor-γ by direct interaction.[220,221] *Platycodon grandiflorus* (Jacq.) A. DC. elaborates platycodin D which at a concentration of 10 μM inhibited the differentiation of 3T3-L1 preadipocytes (induced by dexamethasone-insulin-3-isobutyl-1-methylxanthine) with reduced expression of peroxisome proliferator-activated receptor-γ and CCAAT/enhancer-binding protein-α and downstream fatty acid-binding protein.[222] This saponin increased the expression of low-density lipoprotein-related protein-6, increased the expression of disheveled-2 which stabilizes β-catenin, and lowered the expression of AXIN (responsible for β-catenin degradation was decreased) and evoked nuclear accumulation of β-catenin.[222] This saponin upregulated the expression of β-catenin target genes peroxisome proliferator-activated receptor-δ and cyclin D1.[222] In a subsequent study, this saponin at a concentration of 5 μM inhibited the differentiation of 3T3-L1 adipocytes with a decrease in lipid content by 62.4%, induced the phosphorylation of adenosine monophosphate-activated protein kinase and acetyl-CoA carboxylase, blocked the expression of peroxisome proliferator-activated receptor-γ and CCAAT/enhancer-binding protein-α and downstream fatty acid synthetase and fatty acid-binding protein.[223] This saponin given orally to C57BL/6 at a dose of 15 mg/kg/day for 8 weeks had no effect on food intake, lowered weight gain, lowered cholesterol, triglycerides, low-density lipoprotein, and increased high-density lipoprotein.[223] In white adipose tissue of treated rodent, platycodin D lowered the expression of peroxisome proliferator-activated receptor-γ, CCAAT/enhancer-binding protein-α, increased the expression of phosphorylated adenosine monophosphate-activated protein kinase, and phosphorylated acetyl-CoA carboxylase.[223] Ethanol extract of this plant at a dose of 100 mg/kg/day to C57BL/6J mice on high-fat diet for 16 weeks had no effect on food intake, lowered body weight gain and lowered plasma total cholesterol.[224] At a concentration of 100 μg/mL, the extract inhibited 3T3-L1 differentiation.[224] In differentiated adipocytes, the extract evoked an increase in Sirtuin 1 (SIRT1), phosphorylation of adenosine monophosphate-activated protein kinase, repressed the expression of peroxisome proliferator-activated receptor-γ and C, CCAAT/enhancer-binding protein-α, and subsequently adiponectin.[224] From this extract, platycodin D at 5 μM activated adenosine monophosphate-activated protein kinase and decreased the expression of peroxisome proliferator-activated receptor-γ and CCAAT/enhancer-binding protein-α. Platycodin D given orally to Lepr$^{-/-}$ (*db/db*) orally at a dose of 5 mg/kg for 5 weeks reduced expression of peroxisome proliferator-activated receptor-γ and CCAAT/enhancer-binding protein-α in inguinal white adipose tissue and epididimal white adipose tissue and increased the expression of peroxisome proliferator-activated receptor coactivator-1α in interscapular brown adipose tissue.[224]

4.73 *Artemisia capillaris* Thunb.

Synonyms: *Artemisia sachalinensis* Tilesius ex Besser; *Oligosporus capillaris* (Thunb.) Poljakov
Common names: yin chen hao (Chinese); capillary Artemisia
Subclass Asteridae, Superorder Asteranae, Order Asterales, Family Asteraceae
Medicinal use: toothache (Japan)
Pharmacological target: insulin resistance

Quercetin, isorhamnetin, arcapillin and the coumarins esculetin and 6-methoxy artemicapin C from *Artemisia capillaris* Thunb. inhibited protein-tyrosine phosphatase 1B with IC_{50} values below 50 μM (positive standard ursolic acid: IC_{50} of 4 μM).[225] From the same plant 1,5-dicaffeoylquinic acid, 3,4-dicaffeoylquinic acid, 3,5-dicaffeoyl quinic acid, 3,5-dicaffeoylquinic acid methyl ester, 4,5-dicaffeoylquinic acid, and 3-caffeoylquinic acid inhibited protein-tyrosine phosphatase 1B with IC_{50} values below 20 μM.[208] In fact, the flavonol isorhamnetin (Figure 4.12) at a concentration of 30 μM induced glucose uptake by L6 myotubes *in vitro* as efficiently as insulin at 0.1 μM.[208] The plant contains scoparone which at 100 μg/mL inhibited triglyceride accumulation in differentiating 3T3-L1 preadipocytes as a result of downregulation of sterol regulatory element-binding protein-1c, CCAAT/enhancer-binding protein-α, peroxisome proliferator-activated receptor-γ, and downstream target fatty acid-binding protein, fatty acid translocase, and lipoprotein lipase.[226]

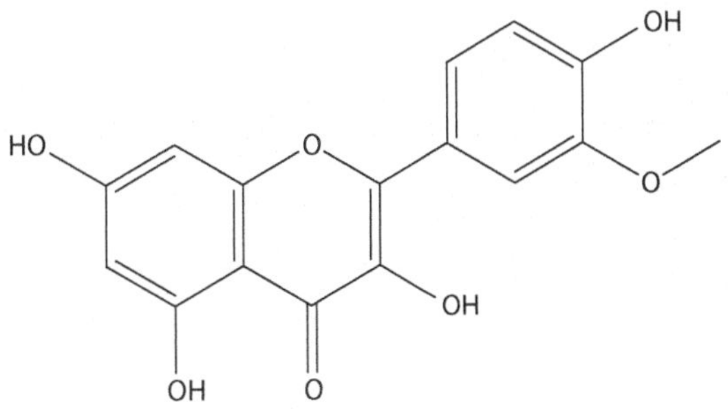

FIGURE 4.12 Isorhamnetin.

4.74 *Artemisia scoparia* Maxim.

Synonym: *Artemisia scoparia* Waldst. & Kit.
Common name: zhu mao hao (Chinese)
Subclass Asteridae, Superorder Asteranae, Order Asterales, Family Asteraceae
Medicinal use: hepatoprotective (Pakistan)
Pharmacological target: insulin resistance

The flavan blumeatin, the flavones cirsilineol and cirsiliol, the flavanonol 7 methoxytaxifolin isolated from the aerial parts of *Artemisia scoparia* Maxim. at a concentration of 30 μM inhibited triglyceride accumulation in 3-isobutyl-1-methyl-xanthone-dexamethasone-insulin-induced 3T3-L1 adipocytes differentiation by 38.9%, 37.2%, 77% and 49.8%, respectively.[227] These flavonoids inhibited glucose uptake in 3T3-L1 cells by 44.4%, 67.4%, 63.1%, and 19.8%, respectively.[227]

4.75 *Artemisia dracunculus* L.

Synonyms: *Artemisia aromatica* A. Nelson; *Artemisia glauca* Pall. ex Willd.
Common names: long hao (Chinese); tarhun (Turkey); tarragon
Subclass Asteridae, Superorder Asteranae, Order Asterales, Family Asteraceae
Nutritional use: seasoning (Turkey)
Pharmacological target: insulin resistance

Ethanol extract of *Artemisia dracunculus* L. given to streptozotocin-induced diabetic mice at a dose of 500 mg/kg/day for 7 days lowered glycemia by 20% (insulin 1 IU/kg: 42%).[228] This extract given to nondiabetic mice had no effect[228] (suggesting, at least, increased peripheral intake of glucose by skeletal muscles given to genetically type 2 diabetic mice (KK-Ay/TaJcl) a dose of 500 mg/kg/day for 7 days, the extract lowered plasma glucose by 24% (metformin 300 mg/kg/day: 41%).[228] In diabetic animals, the extract lowered the expression of phosphoenolpyruvate carboxy-kinase.[228] It must be recalled that in adipocytes, peroxisome proliferator-activated receptor-γ promotes triglycerides storage by increasing the expression of phosphoenolpyruvate carboxy-kinase.[79] Ethanol extract of tarragon at a concentration of 5 μg/mL increased insulin-induced serine 473-phosphorylation of Akt in human skeletal muscle myoblasts of lean, obese or diabetic subjects *in vitro* challenged with TNFα and interleukin-6[229] For all subjects, TNFα and interleukin-6 induced the activation of extracellular signal-regulated kinase-1/2 and this effect was attenuated by the extract.[229] Tumor necrosis factor-α evoked the activation of nuclear factor-κB in human skeletal muscle myoblasts from lean, obese or diabetic subjects, whereas interleukin-6 did not and that effect was inhibited by the extract.[229] 2′,4′-dihydroxy-4-methoxydihydrochalcone (Figure 4.13) isolated from this herb at a concentration of 10 μg/mL prevented palmitic acid (which notably abounds in palm oil)[230] induced decreased in glycogen content in L6 myotube upon insulin stimulation via increased Akt 1 and Akt 2 phosphorylation.[231] The consumption of tarragon in metabolic syndrome could be beneficial.

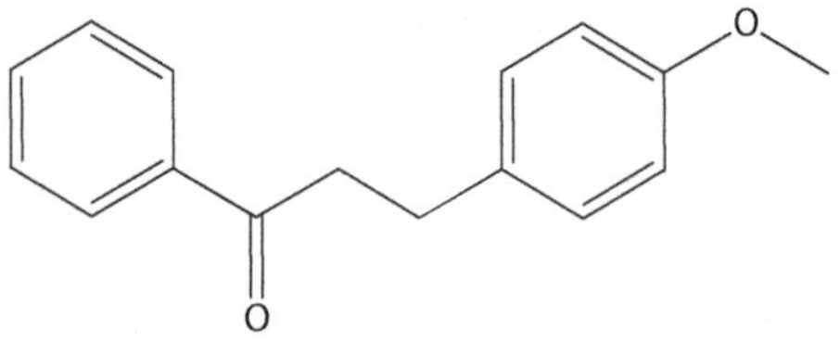

FIGURE 4.13 2′,4′-Dihydroxy-4-methoxydihydrochalcone.

4.76 *Artemisia indica* Willd.

Synonym: *Artemisia vulgaris* var. *indica* (Willd.) Maxim.
Common names: wu yue ai (China); mastam (India)
Subclass Asteridae, Superorder Asteranae, Order Asterales, Family Asteraceae
Medicinal use: spasms (India)

Chloroform fraction of aerial parts of *Artemisia indica* Willd. given to streptozotocin-induced diabetic Sprague–Dawley rats orally at a dose f 200 mg/kg/day for 15 days prevented body weight loss, lowered glycemia from 516.7 to 220.8 mg/dL (normal: 91.5 mg/dL; glibenclamide 0.5 mg/Kg; 206.2 mg/dL).[232] This treatment lowered total cholesterol, triglycerides, increased high-density lipoprotein, and lowered low-density lipoprotein.[232] The fraction lowered serum alanine aminotransferase, aspartate aminotransferase, and alkaline phosphatase in diabetic rodents.[232]

4.77 *Artemisia princeps* Pamp.

Synonym: *Artemisia parvula* Pamp.
Common names: kui hao (Chinese); yomogi (Japanese); Japanese mugwort
Subclass Asteridae, Superorder Asteranae, Order Asterales, Family Asteraceae
Nutritional use: seasoning (Japan)
Pharmacological target: insulin resistance

C57BL/KsJ-*db/db* mice are genetically engineered to develop insulin resistance.[233] Eupatilin (Figure 4.14) from *Artemisia princeps* Pamp. given to C57BL/Ksj-*db/db* mice at 0.02g/100 g of sucrose-enriched diet for 6 weeks had no effect on food intake, lowered glycemia by 19.7%, and doubled plasma insulin and plasma adiponectin.[234] At the hepatic level, this flavone increased the activity of glucokinase, and lowered gluconeogenic glucose-6-phosphatase and phosphoenol-pyruvate carboxy-kinase by 31% and 28%, respectively. In intraperitoneal glucose tolerance test, this flavonoid improved plasma glucose clearance.[234] At the pancreatic level, eupatilin treatment increased insulin level from 101.8 to 151.2 ng/mg.[234] This trimethoxyflavone could account to the fact that ethanol extract of aerial parts of *Artemisia princeps* Pamp. given orally to C57BL/6 mice on high-fat diet for 14 weeks as 1% of diet had no effect on food intake, lowered body weight, liver weight, and mesenteric white adipose tissue weight.[235] This supplementation lowered plasma insulin, glucose, cholesterol from 180 to 122 mg/dL, attenuated nonesterified fatty acids but had no effect on plasma triglycerides.[235] The extract improved oral glucose tolerance test week 12 with a decrease in postprandial glucose area under the curve by about 30% in oral glucose tolerance test.[235] This regimen lowered hepatic cholesterol and triglycerides and decreased the activity of fatty acid synthetase.[235] The supplementation lowered plasma leptin by 73.3% and had no effect on plasma adiponectin.[235] Clinical trials are warranted.

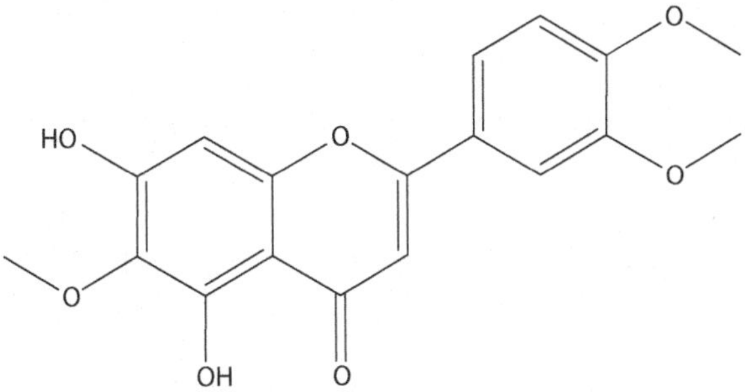

FIGURE 4.14 Eupatilin.

4.78 *Atractylodes macrocephala* Koidz.

Synonym: *Atractylis macrocephala* (Koidz.) Hand.-Mazz.
Common name: bai shu (Chinese)
Subclass Asteridae, Superorder Asteranae, Order Asterales, Family Asteraceae
Medicinal use: fatigue (China)

Aqueous extract of rhizome of *Atractylodes macrocephala* Koidz. given orally at a dose of 500 mg/k/day to Sprague–Dawley rats on high-fat diet for 6 weeks had no effect on food intake, lowered body weight, plasma triglycerides (suggesting lipase inhibition) from 162 to about 75 mg/dL (normal: 111 mg/dL), and had no effect on high-density lipoprotein–cholesterol and low-density lipoprotein–cholesterol.[236] *In vitro*, the extract at a concentration of 25 μg/mL attenuated by about 20% 3T3-L1 preadipocytes differentiation induced by isobutylmethylxanthine-dexamethasone-insulin via reduction of phosphorylated Akt and had no effect on perilipin expression.[236]

4.79 *Cichorium intybus* L.

Common names: ju ju (Chinese); kaasani (India); chicory
Subclass Asteridae, Superorder Asteranae, Order Asterales, Family Asteraceae
Medicinal use: jaundice (India)
Pharmacological target: insulin resistance

Ethanol extract of *Cichorium intybus* L. given to streptozotocin-induced diabetic Sprague–Dawley rats (glycemia >300 mg/dL) at a single oral dose of 125 mg/kg lowered 30 minutes glycemia in oral glucose tolerance test.[237] The extract given orally and daily at 125 mg/kg for 14 days lowered plasma glucose by 20% (metformin at 500 mg/kg/day: 22%) and had no effect on plasma insulin.[237] This regimen lowered serum triglycerides by about 91% and total cholesterol by 16%.[237] The extract lowered hepatic glucose-6-phosphatase activity by 28%.[237] The plant contains chlorogenic acid which at a concentration of 100 μg/mL increased insulin-induced glucose uptake by L6 myotube by 16.1%.[238] In a subsequent study, methanolic extract given orally at a dose of 125 mg/kg/day to high-fat diet-streptozotocin-induced diabetic rats lowered polyphagia; limited body weight loss; lowered plasma glucose from 278 to 116 mg/dL, cholesterol from 205.5 to 124.2 mg/dL, and triglycerides from 391 to 125.3 mg/dL; and increased insulin from 0.2 to 0.3 pg/mL.[239] This extract lowered the expression of protein tyrosine phosphatase 1B in skeletal muscles, increased glycogen contents of liver and skeletal muscles.[239] In the liver the extract decreased glucose-6-phosphatase and fructose-1,6-diphosphatase activities.[239] From this extract, chlorogenic acid at a concentration of 10 ng/mL increased insulin-dependent glucose uptake by 3T3-L1 adipocytes.[239] Chlorogenic acid inhibited protein tyrosine phosphatase 1B with an IC_{50} value of 3.1 μg/mL.[239] This phenolic acid at concentration of 10 ng/mL inhibited 3T3-L1 preadipocytes differentiation induced by dexamethasone-insulin-3-isobutyl-1-methylxanthine with decrease in expression pf CCAAT/enhancer-binding protein-α, peroxisome proliferator-activated receptor-γ and sterol-regulatory element-binding protein-1c.[239] Ethanol extract of seeds of *Cichorium intybus* L. given to Wistar rats as part of high cholesterol-high fructose diet for 2 weeks at 1.5 g/100 g for 2 weeks lowered plasma glucose from 13.2 to 10.8 mmol/L (normal: 9.9 mmol/L) and atherogenic index from 0.3 to 0.2. This supplementation had no effect on total cholesterol and triglycerides.[240] Healthy volunteers drinking 300 mL/day of ground roasted chicory roots (20 g) for a week lowered the aggregation of isolated platelets induced *in vitro* by ADP and had no effect on collagen-induced aggregation nor epinephrine induced aggragation.[241] This regimen decreased blood viscosity in volunteers and decreased serum macrophage migration inhibitory factor levels suggesting preventive effects against monocyte adhesion and migration in vascular endothelium during atherosclerosis.[241] In France, chicory roots have been used as substitute for coffee. Such beverages could be of value in metabolic syndrome.

4.80 *Chromolaena odorata* (L.) R.M. King & H. Rob.

Synonym: *Eupatorium odoratum* L.
Common names: fei ji cao (Chinese); Siam weed
Subclass Asteridae, Superorder Asteranae, Order Asterales, Family Asteraceae
Medicinal use: diabetes (India)

Ethanol extract of leaves of *Chromolaena odorata* (L.) R.M. King & H. Rob. given once orally at a dose of 400 mg/kg to streptozocin-diabetic Wistar rats evoked a decrease in glycemia after 6 hours by 58.8% whereby glibenclamide at a dose of 10 mg/kg inhibited glycemia by 38.2%.[242] Administration of the extract for 8 weeks at a dose of 400 mg/kg/day or glibenclamide reduced

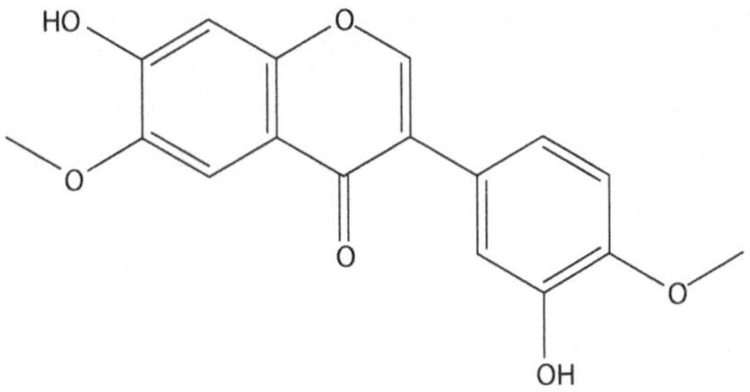

FIGURE 4.15 Odoratin.

further glycemia by 60%.[242] On the fifth week of treatment, oral glucose tolerance test revealed glycemias below 300 mg/dL in the diabetic rats 60 minutes after glucose challenge as well as a mild increase in insulinemia compared with untreated diabetic animals.[242] In addition, diabetic rodents after 8 weeks of treatment exhibited lens opacity index of 0.7 compared with 3.7 (in untreated animals).[242] With regard to lipid profile resulting from the 8 weeks of treatment, the serum concentration of high-density lipoprotein– cholesterol, very low-density lipoprotein–cholesterol, low-density lipoprotein–cholesterol were corrected to near normal values.[242] Furthermore, hepatic glutathione and catalase were elevated to near normal values.[242] Odoratin (Figure 4.15) isolated from this plant induced the activation of peroxisome proliferator-activated receptor-γ in transfected L02 human hepatocytes with an EC_{50} value of 3.1 μM (rosiglitazone EC_{50} value of 0.9 μM).[243]

4.81 *Chrysanthemum morifolium*

Synonyms: *Dendranthema grandiflorum* (Ramat.) Kitam.; *Tanacetum morifolium* Kitam.
Common names: ju hua (Chinese); chrysanthemum
Subclass Asteridae, Superorder Asteranae, Order Asterales, Family Asteraceae
Medicinal use: fever (China)

Aqueous extract of *Chrysanthemum morifolium* given orally to KK-Ay mice at 5% of AIN-93 M diet for 5 weeks had no effect on food intake, body weight, fat mass, liver weight and plasma alanine aminotransferase, and aspartate aminotransferase.[244] This supplementation lowered glycemia from about 500 to 300 mg/dL and improved insulin hypoglycemic activity in insulin tolerance test.[244] This supplementation decreased insulin resistance by almost 50%.[244] This supplementation increased plasma adiponectin and increased the expression of adiponectin in adipose.[244] In epididymal white adipose tissues, the extract evoked the expression of peroxisome proliferator-activated receptor-γ and induced peroxisome proliferator-activated receptor-γ downstream targets fatty acid-binding protein, glucose transporter-4 and lipoprotein lipase.[244] In epididymal white adipose tissues, the extract evoked a reduction of proinflammatory adipocytokines monocyte chemoattractant peptide-1 and tumor necrosis factor-α and lowered the expression of the macrophage marker F4/80.[244] Clinical trials are warranted.

4.82 *Cirsium japonicum* DC.

Common names: ji (Chinese); Japanese thistle
Subclass Asteridae, Superorder Asteranae, Order Asterales, Family Asteraceae
Medicinal use: fever (China)

Pectolinarin and 5,7-dihydroxy-6,4-dimethoxyflavone isolated from *Cirsium japonicum* DC. at a concentration of 50 μM induced the activation of peroxisome proliferator-activated receptor-γ in

3T3-L1 preadipocytes.[245] These flavones at the same concentration increased the accumulation of triglycerides in dexamethasone-insulin-3-isobutyl-1-methylxanthine -induced 3T3-L1 preadipocyte differentiation by 98% and 87%, respectively.[245] In fully differentiated adipocytes, pectolinarin and 5,7-dihydroxy-6,4-dimethoxyflavone increased insulin-induced glucose uptake by 131% and 125%, respectively.[245]

4.83 *Gynura divaricata* (L.) DC.

Synonyms: *Cacalia ovalis* Ker Gawl.; *Gynura hemsleyana* H. Lév.; *Gynura ovalis* (Ker Gawl.) DC.; *Senecio divaricatus* L.
Common name: bai zi cai (Chinese)
Subclass Asteridae, Superorder Asteranae, Order Asterales, Family Asteraceae
Medicinal use: bleeding (Malaysia)

3,4-dicaffeoylquinic acid, 4,5-dicaffeoyl quinic acid, methyl 3,4-dicaffeoylquinate and methyl 4,5-dicaffeoylquinate isolated from the aerial parts *Gynura divaricata* inhibited yeast α-glucosidase with IC_{50} values of 187.2, 130.8, 12.2, and 13 μM, respectively (acarbose: 867.4 μM).[246] From the same plant, 3,4-dicaffeoylquinic acid, 4,5-dicaffeoyl quinic acid, 5-*O*-p-coumaroylquinic acid and p-coumaric acid inhibited protein-tyrosine phosphatase 1B activity *in vitro*.[246] Lyophylized stems and leaves of *Gynura divaricata* (L.) DC. given to high-fat diet streptozotocin-induced diabetic mice as 4.8% of diet for 4 weeks attenuated body weight loss, lowered fasting blood glucose from about 17.5 mmol/L to normal (about 7 mmol/L).[247] This supplementation increased hepatic glutathione peroxidase and superoxide dismutase and lowered hepatic malondialdehyde and increased glycogen contents.[247] The regimen lowered plasma insulin from about 15 mU/L to normal (about 2 mU/L).[247] The supplementation lowered plasma cholesterol and triglycerides and increased high-density lipoprotein–cholesterol levels, increased pancreatic expression of phosphoinositide 3-kinase, phosphorylated Akt and phosphoinositide-dependent kinase-1 in diabetic mice.[247]

4.84 *Matricaria chamomilla* L.

Synonyms: *Chamomilla recutita* (L.) Rauschert; *Chamomilla vulgaris* Gray; *Matricaria recutita* L.
Common names: mu ju (Chinese); papatya (Turkey); chamomile
Subclass Asteridae, Superorder Asteranae, Order Asterales, Family Asteraceae
Medicinal use: indigestion (Turkey)

Infusions of 3 g of flowers of *Matricaria chamomilla* in 150 mL of water given to type 2 diabetes patients 3 times a day for 8 weeks post-prandially had no effects on glycemia, lowered plasma insulin from 15.9 to 10.6 μIU/mL and decreased insulin resistance.[248] This supplementation lowered total cholesterol by 9.5%, triglycerides by 18.3% and lowered low-density lipoprotein–cholesterol by 8.8%, whereby high-density lipoprotein–cholesterol was unchanged.[248] Ethanolic extract of chamomile flowers given orally to high-fat diet-fed C57/BL6 mice for 6 weeks at a dose of 200 mg/kg/day lowered plasma glucose by 13%, increased plasma insulin by 23%, and decreased insulin resistance.[249] In oral glucose tolerance test after 2 weeks, the extract lowered peak postprandial glycemia at 15 minutes as well as 15 minutes peak plasma insulin.[249] The extract evoked the expression of peroxisome proliferator-activated receptor-γ coactivator 1α and 1β, ATP citrate lyase and phosphoenolpyruvate carboxy-kinase 1 in visceral white adipose.[249] The extract given for 6 weeks at similar dose lowered plasma concentration of free fatty acids by 53% and triglycerides by 35%, lowered plasma cholesterol and low-density lipoprotein–cholesterol/very low-density lipoprotein ratio. High-density lipoprotein was unchanged.[249] The extract given concurrently with high-fat diet to C57BL/6 mice for 20 week prevented rise in glycemia and insulin resistance and increased free fatty acids and triglycerides in the plasma.[249] This regimen protected liver against hepatosteatosis as evidence by a 64% decrease in plasma alanine aminotransferase, a 72% decrease in hepatic triglycerides and 64% decrease in hepatic free fatty acids.[249] At the hepatic level, the extract induced fatty acid oxidation via the induction of carnitine palmitoyltransferase-2, acyl-CoA dehydrogenase and peroxisome proliferator-activated receptor-γ coactivator 1α.[249]

4.85 *Petasites japonicus* (Siebold & Zucc.) Maxim.

Synonyms: *Nardosmia japonica* Siebold & Zucc.; *Petasites albus* A. Gray; *Petasites liukiuensis* Kitam.; *Petasites spurius* Miq.; *Tussilago petasites* Thunb.
Common name: feng dou cai (Chinese)
Medicinal use: fractures (China)

Petasin from *Petasites japonicus* (Siebold & Zucc.) Maxim. at a concentration of 10 μM lowered oxygen consumption by mitochondria hence increased AMP/ATP ratio, activation of adenosine monophosphate-activated protein kinase and downstream acetyl-CoA carboxylase in H4IIE cells, C2C12 cells and 3T3L1 cells.[250] This sesquiterpene given at a single oral dose of 200 mg/kg evoked after 2 hours an increased expression of phosphorylated adenosine monophosphate-activated protein kinase and phosphorylated acetyl-CoA carboxylase in the liver, epididymal white adipose tissue and quadriceps skeletal muscles in C57BL/6J mice.[250] In oral glucose tolerance test, petasin at a single oral dose of 200 mg/kg lowered 15 minutes peak glycemia from about 200 to 150 mg/dL and lowered glucose area under the curve by about 50%.[250]

4.86 *Siegesbeckia pubescens* (Makino) Makino

Common name: xi xian cao (Chinese)
Subclass Asteridae, Superorder Asteranae, Order Asterales, Family Asteraceae
Medicinal use: hypertension (China)

Siegesbeckia pubescens (Makino) Makino contains *ent*-16β H,17-isobutyryloxy-kauran-19-oic acid which inhibited protein-tyrosine phosphatase 1B activity IC_{50} values of 8.7 μM.[251]

4.87 *Solidago virgaurea* L.

Common names: mao gao yi zhi huang hua (Chinese); common golden rod
Subclass Asteridae, Superorder Asteranae, Order Asterales, Family Asteraceae
Medicinal use: diuretic (China)

Kaempferol-3-*O*-rutinoside, chlorogenic acid and protocatechuic acid isolated from *Solidago virgaurea* L. at a concentration of 60 μg/mL inhibited the accumulation of triglycerides in isobutylmethylxanthine-insulin-dexamethasone-induced differentiating 3T3-L1 adipocytes by 48.2%, 63.9%, and 66.8%, respectively.[252] Kaempferol-3-*O*-rutinoside decreased the expression of peroxisome proliferator-activated receptor-γ and CCAAT/enhancer-binding protein-α during 3T3-L1 adipocytes differentiation.[252]

4.88 *Tithonia diversifolia* (Hemsl.) A. Gray

Synonym: *Mirasolia diversifolia* Hemsl.
Common names: zhong bing ju (Chinese); nitobegiku (Japanese)
Subclass Asteridae, Superorder Asteranae, Order Asterales, Family Asteraceae
Medicinal use: diabetes (Taiwan)

Ethanol extract of *Tithonia diversifolia* (Hemsl.) A. Gray given at a single dose of 500 mg/kg to diabetic ddY mice lowered glycemia from about 550 to 300 mg/dL after 7 hours. This extract given to normal mice did not lower glycemia.[253] The extract at 500 mg/kg/day to diabetic ddY mice for 3 weeks lowered glycemia from about 525 to 350 mg/dL and had no effect on body weight.[253] This supplementation lowered plasma insulin and improved insulin-induced glucose plasma clearance in insulin tolerance test.[253] Tagitinin G, tagitinin I, 1β-hydroxydiversifolin-3-*O*-methyl ether, 1β-hydroxytirotundin-3-*O*-methyl ether isolated from the aerial parts of *Tithonia diversifolia* at a concentration of 10 μg/mL induced glucose uptake by 3T3-L1 adipocytes similarly with pioglitazone at 10 μg/mL.[254] The sesquiterpene tirotundin (Figure 4.16) and taginin at a concentration of 10 μM induced liver X receptor-dependant CYP7A1, sterol regulatory element-binding protein-1c and ABCA1 gene promoter transactivation as well as the transactivation of farnesoid X receptor-dependent SHP gene in HepG2 cells.[255] Liver X receptor increases glucose uptake by induction of glucose transporter-4 expression in skeletal muscles and adipocytes.[256]

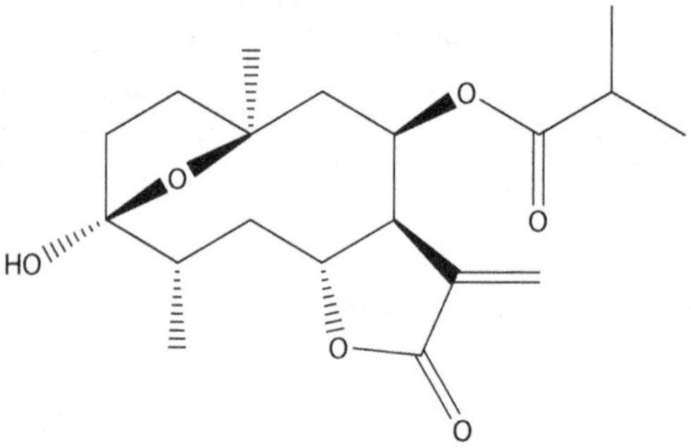

FIGURE 4.16 Tirotundin.

4.89 *Morinda citrifolia* L.

Common names: hai bin mu ba (Chinese); mengkudu (Malay); bankoro (Philippines); Indian mulberry
Superorder Lamianae, Order Rubiales, Family Rubiaceae
Medicinal use: diabetes (Indonesia)

(−)-Pinoresinol, lirioresinol B, lirioresinol B dimethyl ether, and ursolic acid from *Morinda citrifolia* inhibited protein-tyrosine phosphatase 1B with IC_{50} values of 18.6, 15, 16.8, and 4.1 μM, respectively.[257] In 3T3-L1 adipocytes, lirioresinol B at a concentration of 20 μM, lirioresinol B dimethyl ether at 40 μM, and ursolic acid at 40 μM induced glucose uptake closely to rosiglitazone at 400 μg/mL, whereas (−)-pinoresinol was inactive.[257]

4.90 *Rubia cordifolia* L.

Synonym: *Rubia pratensis* (Maximowicz) Nakai.
Common name: qian cao (Chinese)
Subclass Lamiidae, Superorder Lamianae, Order Rubiales, Family Rubiaceae
Medicinal use: rheumatism (Korea)

Mollugin (Figure 4.17) isolated from the roots of *Rubia cordifolia* L. at a concentration of 20 μM inhibited dexamethasone-insulin-3-isobutyl-1-methylxanthine-induced differentiation of preadipocytes into mature adipocytes as evidenced by a 80% reduction in triglyceride contents with repression of CCAAT/enhancer-binding protein-α and peroxisome proliferator-activated receptor-γ expression.[258]

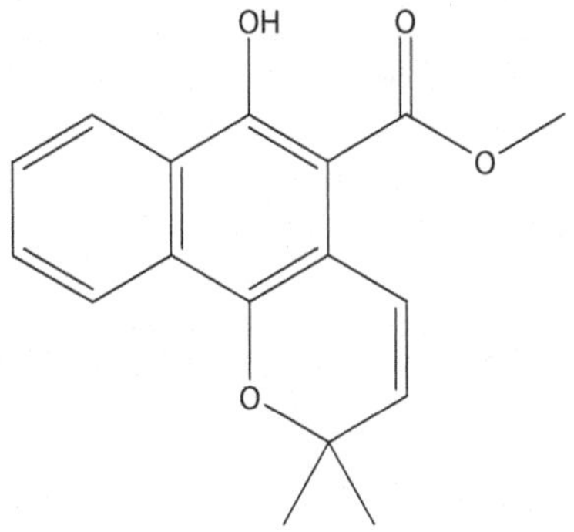

OH O O O

FIGURE 4.17 Mollugin.

4.91 *Enicostemma littorale* Blume

Synonym: *Enicostemma hyssopifolium* (Willd.) I. Verd.
Subclass Lamiidae, Superorder Lamianae, Order Rubiales, Family Gentianaceae
Medicinal use: diabetes (India)

Aqueous extract of the *Enicostemma littorale* Blume given to streptozotocin-induced diabetic Sprague–Dawley rats orally at a dose of 1 g/kg/day for 3 weeks attenuated body weight loss, food intake and water intake.[259] This extract lowered glycemia from approximately 380 to 250 mg/dL.[259] This regimen lowered serum cholesterol, serum creatinine, serum aspartate aminotransferase and serum alanine aminotransferase.[259] The extract mitigated kidney damages including glomerular hypertrophy.[259] From this plant, swertiamarin given at 50 mg/kg/day for 3 weeks had similar effects with the extract although being less potent with regard to serum glucose, serum creatinine, serum alanine aminotransferase and aspartate aminotransferase and renal histoprotection and more potent in reducing serum urea and serum cholesterol.[259] Swertiamarin at 100 μg/mL had no effect on promoting the differentiation of 3T3L1 mouse pre-adipocytes but increased adiponectin expression.[260] It must be noted that swertiamarin is metabolized into gentianine by intestinal bacteria[261] gentanine enhanced isobutyl-1-methylxanthin-dexamethasone-insulin-indiced differentiation of 3T3L1 mouse preadipocytes at a dose of 100 μg/mL by increasing triglycerides accumulation by 40% and enhancing the expression of peroxisome proliferator-activated receptor-γ, glucose transporter-4, and adiponectin.[260,261]

4.92 *Capsicum minimum* Mill.

Synonyms: *Capsicum fastigiatum* Bl.; *Capsicum minimum* Roxb.
Common name: gna yoke (Burma); oosi mulagai (India); bird's eye pepper
Subclass Lamiidae, Superorder Lamianae, Order Solanales, Family Solanaceae

Capsaicin at a dose of 0.014% of high fat diet given to rodents for 10 days evoked a decrease in peripheral fat pad weight from 2.2 to 1.6 g, a decrease the epididymal fat pad weight from 2.2 to 1.9 g and induced a decrease in triglyceridemia from 41.7 to 29.3 mg/dL but cholesterolemia was unchanged.[262] In the same experiment, capsaicin evoked a decrease in total hepatic lipids from 82.5 to 77.7 mg/dL.[262] An increase of glycemia from 103 to 118.9 mg/dL was observed.[262] The enzymatic activity of glucose-6-phosphate dehydrogenase was increase in the liver from 0.7 to 1 U/g, and in the perirenal adipose tissue, the activity of lipoprotein lipase increased from 20.5 to 25.5 μmol/μg.[262] Capsaicin at a concentration of 10 μM reduced triglyceride contents of differentiated 3T3-L1 adipocytes by 33% and increased glycerol secretion after 24 hours of exposure.[263] This alkaloid increased the expression of hormone sensitive lipase, carnitine palmitoyltransferase-Iα, and uncoupling protein-2.[263] Capsaicin at a dose of 0.015% of high-fat diet reduced the volume of adipose tissue of rodents.[264] In the same experiment, capsaicin mitigated glycemia and insulinemia, serum free fatty acids, leptin. Besides, capsaicin evoked an increase in insulin receptor substrate-1 and glucose transporter-4.[264] Capsaicin also reduced tumor necrosis-α, interleukin-6 and monocyte chemoattractant peptide-1 in adipose tissues with a decrease of circa 25% of macrophages in the adipose tissues by activation of the transient receptor potential vanilloid type 1 and evoked peroxisome proliferator-activated receptor-γ in hepatocytes.[264] This alkaloid at a concentration of 100 μM induced an increase in glucose uptake by C2C12 myocytes via stimulation of adenosine monophosphate-activated protein kinase and p38 mitogen-activated protein kinase on account of reactive oxygen species generation.[265]

4.93 *Lycopersicon esculentum* Mill.

Synonyms: *Lycopersicon lycopersicum* (L.) H. Karst.; *Solanum lycopersicum* L.
Common names: fan qie (Chinese); tomato
Subclass Lamiidae, Superorder Lamianae, Order Solanales, Family Solanaceae
Medicinal use: laxative (Cambodia, Laos, and Vietnam)

Naringenin chalcone from the peel of fruits of *Lycopersicon esculentum* Mill. boosted the secretion of adiponectin by differentiating 3T3-L1 adipocytes at a concentration of 50 μM via peroxisome proliferator-activated receptor-γ activation.[266] Aqueous extract of unripe fruits of *Lycopersicon esculentum* Mill. given to C57BL/6 mice at 2% of high-fat diet for 13 weeks did not reduced food intake, and lowered body weight by 9.9%.[134] This regimen decreased epididymal adipose tissue weight by about 27.5% and liver weight by 19.5%.[134] This regimen lowered serum total cholesterol and low-density lipoprotein–cholesterol by 6.2% and 16.7%, respectively. The extract decreased liver cholesterol by about 53.6%, to a level similar to rodents on normal chow.[134] High-fat diet decreased the expression of phosphorylated adenosine monophosphate-activated protein kinase, acetyl-CoA carboxylase in the liver and the extract restored the expression of phosphorylated adenosine monophosphate-activated protein kinase and acetyl-CoA carboxylase.[134] High-fat diet increased the expression 3-hydroxy-3- methylglutaryl-coenzyme A reductase which as lowered upon treatment with the extract.[134] In epididymal adipose tissues, high-fat diet increased the expression of peroxisome proliferator-activated receptor-γ, CCAAT/enhancer-binding protein-α and periplin and the extract decreased peroxisome proliferator-activated receptor-γ, CCAAT/enhancer-binding protein-α and periplin to normal.[134] Tomatine at a concentration of 10 μM and trigonelline at a concentration of 300 μM from the extract inhibited the accumulation of triglycerides in 3T3-L1 adipocytes exposed to dexamethasone-insulin-3-isobutyl-1-methylxanthine by 78.6% and 23.4%, respectively.[134] Steroidal alkaloids esculeogenin A and tomatidine isolated from this plant inhibited the accumulation of cholesteryl ester from acetyl low-density lipoprotein in human monocyte-derived macrophages *in vitro*.[267] In CHO cells tomatidine inhibited the enzymatic activity of acyl-CoA:cholesterol acyl-transferase-1 and acyl-CoA:cholesterol acyl-transferase-2.[267] In apoE-deficient mice fed with tomatidine orally at a dose of 50 mg/kg/day as part of diet for 70 days, weight was unchanged, total serum cholesterol was reduced by about 20%, low-density lipoprotein–cholesterol by about 25%, and lowered atherosclerotic lesions by about 65%.[267]

4.94 *Physalis peruviana* L.

Synonyms: *Boberella peruviana* (L.) E.H.L. Krause; *Boberella pubescens* (L.) E. H. L. Krause
Common names: deng long guo (Chinese); winter-cherry
Subclass Lamiidae, Superorder Lamianae, Order Solanales, Family Solanaceae
Nutritional use: fruits eaten (China); winter-cherry
Pharmacological target: insulin resistance

Visceral-adipose tissues secrete proinflammatory cytokines including tumor necrosis factor-α and interleukin-6. Adipocytes and macrophages (in a paracrine manner) activate Jun-N-terminal kinase and nuclear factor-κB, which phosphorylate insulin receptor substrate and reduce the sensibility of insulin receptor to create a state of insulin resistance in adipocytes and neighboring skeletal muscle cells.[1] *Physalis peruviana* L. elaborates 4β-hydroxywithanolide E[268] (Figure 4.18), which inhibited monocyte chemoattractant peptide-1 production with half maximal inhibitory concentrations of 1.5 μM in 3T3-L1/RAW264.7 macrophages coculture.[269] In adipocytes, tumor necrosis factor-α induced the nuclear translocation of nuclear factor-κB and these effects were inhibited by 4β-hydroxywithanolide E at a concentration of 10 μM.[269] This sterol inhibited the expression of monocyte chemoattractant peptide-1, tumor necrosis factor-α, and interleukin-6[269] 4β-Hydroxywithanolide E given orally to *db/db* mice for 2 weeks orally at a dose of 30 mg/kg/day lowered macrophages infiltration in epididymal adipose tissue and inhibited the expression of proinflammatory cytokines monocyte chemoattractant peptide-1, tumor necrosis factor-α and interleukin-6 expression.[269] In addition, this regimen improve doral glucose tolerance with a 30 minutes decrease of postprandial glycemia by 20%.[269] This sterol lowered basal glycemia by 40% and evoked a mild increase in insulinemia.[269] 4β-Hydroxywithanolide E lowered plasma monocyte chemoattractant peptide-1, free fatty acids and increased plasma adiponectin.[269] Adiponectin inhibits the

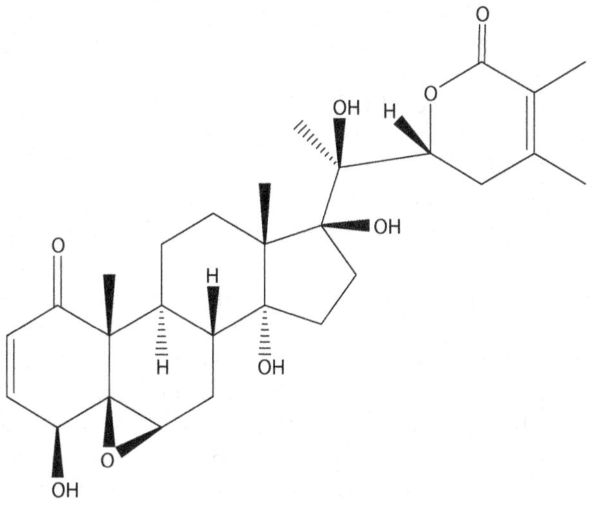

FIGURE 4.18 4β-Hydroxywithanolide E.

transformation of macrophages into foam cells and reduces intracellular cholesteryl ester by suppressing the expression of class A scavenger receptors in atheromas.[270]

4.95 *Solanum xanthocarpum* Schrad. & Wendl.

Synonyms: *Solanum surattense* Burm. f.; *Solanum virginianum* L.
Common names: kantakari (India); wild egg plant
Subclass Lamiidae, Superorder Lamianae, Order Solanales, Family Solanaceae
Medicinal use: diabetes (India)

Aqueous extract of fruits of *Solanum xanthocarpum* Schrad. & Wendl. given to Wistar rats orally at a single dose of 200 mg/kg lowered glycemia from 90.1 to 79.4 mg/dL after 8 hours (glibenclamide at 2.5 mg/kg: 69.5 mg/dL).[271] The extract given to alloxan-induced diabetic (glycemia >200 mg/dL) orally at a dose of 200 mg/kg lowered glycemia from 308.3 to 139.2 mg/dL at 8 hours (glibenclamide at 2.5 mg/kg: 122.5 mg/dL).[271] The extract given daily at a dose of 200 mg/kg for 9 days to Wistar rats lowered glycemia by about 35%. The extract given daily at a dose of 200 mg/kg for 21 days to alloxan-induced diabetic Wistar rats lowered glycemia by about 61%.[271] The extract increased hepatic glycogen as potently a glibenclamide, lowered total cholesterol, triglycerides, increased high-density lipoprotein, lowered very low-density lipoprotein and low-density lipoprotein.[271] *In vitro*, the extract evoked uptake of glucose by isolated rat hemidiaphragm.[271] In addition.[271] The plant contains steroidal saponins solasonine, solamargine, and khasianine.[272] Solacongestidine, solafloridine, and solasodine inhibited the synthesis of cholesterol from 24,25-dihydrolanosterol.[273]

4.96 *Withania coagulans* (Stocks) Dunal

Synonym: *Puneeria coagulans* Stocks
Common name: cheese maker
Subclass Lamiidae, Superorder Lamianae, Order Solanales, Family Solanaceae
Medicinal use: facilitate digestion (India)
Pharmacological targets: atherogenic hyperlipidemia; insulin resistance

Coagulin C, 17β-hydroxywithanolide K, withanolide F, (17S,20S,22R)-14α,15α,17β,20β-tetrahydroxy-1-oxowitha-2,5,24-trienolide and coagulin L isolated from the dried fruits of *Withania coagulans* (Stocks) Dunal given at a single oral dose of 100 mg/kg to Sprague–Dawley rats and streptozotocin-induced diabetic Sprague-Dawley rats (glycemia between 144 and 270 mg/dL) 30 minutes before sucrose loading lowered postprandial glycemia by more than 30%.[274] Coagulin L given to C57BL/Ksj-*db/db* mice at a dose of 50 mg/kg/days for 10 days lowered postprandial blood glucose by 22.7%. This regimen lowered triglycerides by 14.7%, total cholesterol by 25.7%, increased high-density lipoprotein–cholesterol by 24.7%, lowered low-density lipoprotein–cholesterol by 21.2% and lowered very low-density lipoprotein–cholesterol by 15.6%.[274] Aqueous extract of fruits of *Withania coagulans* (Stocks) Dunal given to albino rabbits on cholesterol-enriched diet orally at a dose of 250 mg/kg/day for 6 weeks lowered plasma cholesterol from 362.1 to 157.8 mg/dL (normal diet group: 88.2 mg/dL), low density lipoproteins and increased high-density lipoproteins.[275] This regimen lowered serum lipid peroxides, increased glutathione and superoxide dismutase activity.[275] In the liver of treated animals, the extract lowered lipid contents, increased 3-hydroxy-3-methylglutaryl-coenzyme A CoA reductase activity, lowered acetyl-CoA carboxylase activity, lowered lipid peroxidation and increased glutathione and increased the enzymatic activity of superoxide dismutase.[275] Coagulin L at a concentration of 15 μM inhibited dexamethasone-insulin-3-isobutyl-1-methylxanthine -induced differentiation of 3T3-L1 preadipocytes into adipocytes.[276] This saponin inhibited the expression of sterol regulatory element-binding protein-1c, peroxisome proliferator-activated receptor-γ, CCAAT/enhancer-binding protein-α downstream target fatty acid-binding protein, lipoprotein lipase, fatty acid synthetase, and glucose transporter-4.[276]

4.97 *Withania somnifera* (L.) Dunal

Synonyms: *Physalis somnifera* L.; *Withania kansuensis* Kuang & A. M. Lu; *Withania microphysalis* Suess.
Common names: shui qie (Chinese); ashwagandha (India); winter cherry
Subclass Lamiidae, Superorder Lamianae, Order Solanales, Family Solanaceae
Medicinal use: diuretic (India)
History: the plant was known to Sushruta

Aqueous root extract of *Withania somnifera* (L.) Dunal given to streptozotocin-neonatal-induced type 2 diabetic Wistar rats (fasting glycemia ≥200 mg/dL) orally at a dose of 400 mg/kg/day for 5 weeks lowered glycemia from 324.6 to 121.2 mg/dL (normal: 97.1 mg/dL).[277] The extract had no effect on the glycemia or plasma insulin of normal rats suggesting increase intake of glucose by skeletal muscles.[277] In streptozotocin-neonatal-induced type 2 diabetic Wistar rats, the extract lowered insulinemia by about 30% and increased insulin sensitivity to normal. It had no effect on plasma insulin of normal rats.[277] The extract improved glucose tolerance in oral glucose tolerant test with a decrease in postprandial glycemia from 342.9 to 150.3 mg/dL at 60 minutes.[277] In subsequent experiment, methanol extract of leaves *Withania somnifera* (L.) Dunal at a concentration of 100 μg/mL increased the intake of glucose by L6 myotubes and 3T3-L1 adipocytes by 50% and 30%, respectively.[278] From this extract, withaferin A at a concentration of 10 μM increased glucose uptake in L6 myotubes by 54%.[278]

4.98 *Campsis grandiflora* (Thunb.) K. Schum.

Synonyms: *Bignonia chinensis* Lam.; *Bignonia grandiflora* Thunb.; *Campsis adrepens* Lour.; *Campsis chinensis* (Lam.) Voss; *Tecoma chinensis* (Lam.) K. Koch; *Tecoma grandiflora* Loisel.
Common name: ling xiao (Chinese)
Subclass Lamiidae, Superorder Lamianae, Order Lamiales, Family Bignoniaceae
Medicinal use: haemorages (Japan)

The flowers of *Campsis grandiflora* (Thunb.) K. Schum. contain maslinic acid, corosolic acid, 23-hydroxyursolic acid, arjunolic acid (Figure 4.19) which at a concentration of 100 μg/mL inhibited acyl-CoA: cholesterol acyltransferase-1 by 46.2%, 46.7%, 41.5%, and 60.8%, respectively.[279]

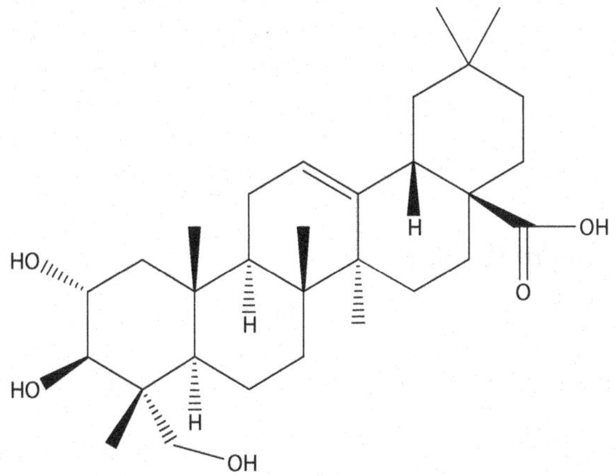

FIGURE 4.19 Arjunolic acid.

4.99 *Kigelia pinnata* (Jacq.) DC.

Synonyms: *Bignonia africana* Lam; *Crescentia pinnata* Jacq.; *Kigelia pinnata* (Jacq.) Rev. Bign.
Common name: African sausage tree
Subclass Lamiidae, Superorder Lamianae, Order Lamiales, Family Bignoniaceae

7-hydroxy eucommiol, catalpol (Figure 4.20), specioside and verminoside isolated from *Kigelia pinnata* (Jacq.) DC. induced at a concentration of 10 μM the translocation of glucose transporter-4 in L6 skeletal muscle cells by 36.3%, 50.0%, 41.8%, and 45.6%.[280] Rosiglitazone increased by 54% glucose transporter-4 membrane translocation at 10 μM respectively.[280]

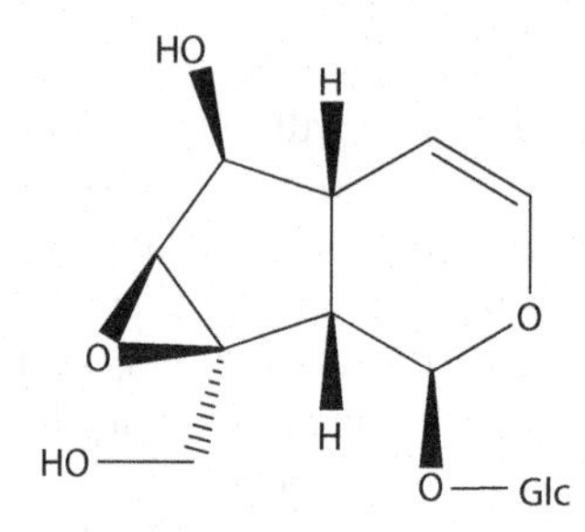

FIGURE 4.20 Catalpol.

4.100 *Catalpa bignonioides* Walter

Common name: catalpa
Subclass Lamiidae, Superorder Lamianae, Order Lamiales, Family Bignoniaceae
Medicinal use: anthelminthic (Pakistan)

Catalpa bignonioides Walter shelters catalpic acid[281] which given to C57BL/6J mice as part of high-fat diet (1 g/100 g) for 78 days had no effect on body weight, lowered white adipose tissue from 1.4 to 0.6 g (normal: 0.8 g), glycemia from 301.3 to 205.4 mg/dL (normal: 153.8 mg/dL), and lowered insulin from 2.2 to 0.7 ng/mL (normal: 0.9 ng/mL).[282] In oral glucose tolerance test, this trans-9,trans-11,cis-13-octadecetrienoic acid fatty acid improved glucose tolerance.[282] In white adipose tissue, this supplementation increased the expression of peroxisome proliferator-activated receptor-α, stearoyl-CoA desaturase and enoyl-CoA hydratatase.[282] In *db/db* mice, catalpic acid in high-fat diet (1 g/100 g) for 28 days lowered white adipose tissue weight, increased high-density lipoprotein–cholesterol and lowered plasma triglycerides.[282] In white adipose tissue, this supplementation increased the expression of peroxisome proliferator-activated receptor-α, stearoyl-CoA desaturase, whereby enoyl-CoA hydratase was not affected.[282] It must be recalled that peroxisome proliferator-activated receptors are activated by long chain fatty acids.

4.101 *Oroxylum indicum* (L.) Kurz.

Synonyms: *Bignonia indica* L.; *Bignonia pentandra* Lour.; *Calosanthes indica* (L.) Blume *Spathodea indica* (L.) Pers.
Common names: mu hu die (Chinese); arala (India); bongelai kayu (Malay); trumpet flower

Oroxylin A from *Oroxylum indicum* (L.) Kurz. at a concentration of 20 μM abrogated the differentiation of 3T3 preadipocytes (induced by dexamethasone-insulin-3-isobutyl-1-methylxanthine) as evidenced by a decreased triglyceride accumulation by 61.2% and adiponectin secretion by 31.4%.[283]

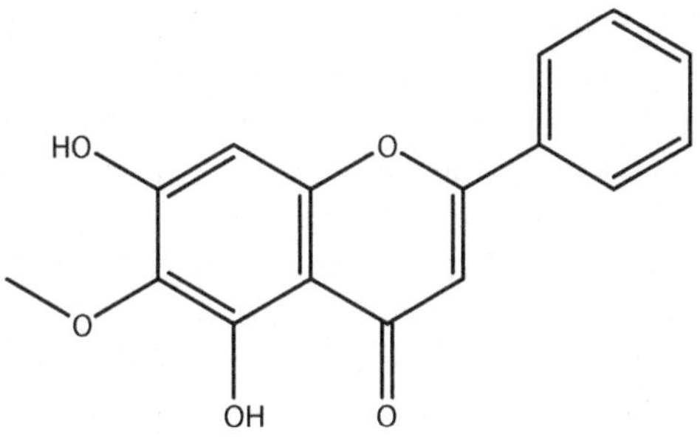

FIGURE 4.21 Oroxylin A.

This flavone (Figure 4.21) decreased the expression of peroxisome proliferator-activated receptor-γ and downstream fatty acid synthetase and lipoprotein lipase.[283] In mature 3T3-L1 adipocytes, this flavone induced lipolysis as evidenced by increased expression of glycerol, it increased tumor necrosis factor-α secretion and decreased Akt phosphorylation.[283]

4.102 *Tecoma stans* (L) Juss. ex Kunth

Synomyms: *Bignonia stans* L.; *Stenolobium stans* (L.) Seem.
Common name: Yellow bells
Subclass Lamiidae, Superorder Lamianae, Order Lamiales, Family Bignoniaceae
Medicinal use: diuretic (India)

Aqueous extract of leaves of *Tecoma stans* (L) Juss. ex Kunth at a concentration of 70 μg/mL increased glucose uptake by 3T3 adipocytes by 193%.[284] In 3T3 adipocytes rendered insulin-resistant by tumor necrosis factor-α, the extract at a concentration of 70 μg/mL evoked an increase of glucose uptake by 3T3 adipocytes by 94%.[284] Tecomine and/or boschniakine could account for these effects as discussed elsewhere.

4.103 *Borago officinalis* L.

Common name: borrage
Subclass Lamiidae, Superorder Lamianae, Order Boraginales, Family Boraginaceae
Medicinal use: Turkey
Pharmacological target: insulin resistance

Evidence suggests that dietary saturated dietary fatty acids such as palmitic acid (in palm oil), promotes the production of interleukin-6 and tumor necrosis factor-α, which accounts for the development of insulin resistance.[285] Linoleic acid, linolenic acid and γ-linolenic acid at a concentration of 100 μM protected mouse C2C12 myoblasts against palmitic acid-induced expression of interleukin-6 and tumor necrosis factor-α.[286] These unsaturated fatty acids inhibited palmitic acid-induced phosphorylation of protein kinase C_θ extracellular signal-regulated kinase-1/2, and mitogen-activated protein kinase.[286] Linoleic acid, linolenic acid and γ-linolenic acid at a concentration of 100 μM prevented inhibition of phosphorylation of Akt and AS160 as well as glucose transporter-4 translocation and glucose uptake by mouse C2C12 myoblasts. These fatty acids inhibited palmitic acid-induced nuclear factor-κB activation.[286] Oil of *Borago officinalis* L. enriched with γ-linolenic acid given orally to streptozotocin high-fat diet-induced diabetic

C57/6JNarl mice (glycemia >500 mg/dL) at a dose of 150 μg/mL every other day for 16 weeks lowered the expression of interleukin-6 and tumor necrosis factor-α. It also inhibited the phosphorylation of extracellular signal-regulated kinase-1/2, p38 mitogen-activated protein kinase and protein kinase C_θ, and inhibited the activation of nuclear factor-κB in the gastroctnemius.[287] Linoleic acid is a peroxisome proliferator-activated receptor-γ agonist.[288] Borage oil added at 6% of diet for 32 weeks to In F344/DuCrj rats reduced the occurrence of ventricular tachycardia and inhibited the duration of ventricular tachycardia induced by experimental and acute coronary artery occlusion compared with rodents receiving sheep fat at 6% of diet.[287] This regimen reduced the occurrence and duration of ventricular fibrillation in ischemic state.[287] Allantoin at a concentration of 1 μM induced glucose intake by C2C12 cells via the activation of imidazoline I_{2B} receptor as efficiently as insulin at the same concentration.[289] This effect was accompanied by an increased expression of phosphorylated adenosine monophosphate-activated protein kinase.[289] Borage is well known to be hepatotoxic on account alkaloids and has no place in therapeutic. Oil of seeds however could be of value.

4.104 *Lithospermum erythrorhizon* Siebold & Zucc.

Synonym: *Lithospermum officinale* var. *erythrorhizon* (Siebold & Zucc.) Maxim.
Common names: zi cao (Chinese); murasaki (Japanese); Chinese groomwell
Medicinal use: wounds (Japan)
Pharmacological profile: atherogenic hyperlipidemia

Kawabata et al. provided evidence that shikonin from *Lithospermum erythrorhizon* Siebold & Zucc. at a concentration of 2 μM inhibited 3-isobuty-1-methylzanthine-dexamethasone-insulin-induced differentiation of 3T3-L1 mouse fibroblast cells into adipocytes as evidenced by a 76.4% decrease of lipid droplet accumulation.[208] This prenylated naphthoquinone inhibited the expression of peroxisome proliferator-activated receptor-γ, CCAAT/enhancer-binding protein-α, and downstream target fatty acid-binding protein.[208] Exposure of 3T3-L1 mouse fibroblasts to 3-isobuty-1-methylzanthine-dexamethasone-insulin evoked an increased in extracellular signal-regulated kinase-1/2 phosphorylation that was inhibited by shikonin.[208] In fact, shikonin at a concentration of 30 μM induced glucose uptake by L6 myotubes *in vitro* as efficiently as insulin at 0.1 μM.[208] In a subsequent study, ethanol extract of roots given to orally at a to C57BL/6 mice at 0.5% of high-fat diet for 8 weeks lowered body weight by 10%, whereas food consumption was not altered.[290] This regimen lowered the mass of liver, epididymal adipose tissues, and retroperitoneal adipose tissues.[290] This regimen decreased plasma triglycerides from 84.4 to 58.9 mg/dL (normal: 84.4 mg/dL), cholesterol from 214 to 183 mg/dL (normal: 131 mg/dL), high-density lipoprotein–cholesterol was not affected, plasma unesterified fatty acids were decreased from 1419 to 1101 μEq/L (normal: 1418 μEq/L), leptin from 51.7 to 23.8 ng/mL (normal: 9.4 ng/mL), and insulin from 558 to 377 pg/mL (normal: 345 pg/mL).[290] This extract lowered hepatic lipid contents, and decreased the expression of hepatic sterol regulatory element-binding protein-1c, fatty acid synthetase and peroxisome proliferator-activated receptor-γ.[290] In adipocytes, the supplementation reduced the expression peroxisome proliferator-activated receptor-γ and CCAAT/enhancer-binding protein-α as well as lipoprotein lipase and fatty acid-binding protein.[290] From this extract, acetylshikonin at a concentration of 0.5 μM inhibited 3-isobuty-1-methylzanthine-dexamethasone-insulin-induced differentiation of 3T3-L1 mouse fibroblast cells and inhibited the expression of expression of transcription factors peroxisome proliferator-activated receptor-γ and CCAAT/enhancer-binding protein-α as well as enzymes lipoprotein lipase and fatty acid-binding protein.[290] This naphtoquinone suppressed extracellular signal-regulated kinase-1/2 phosphorylation during the early stages of adipocyte differentiation in 3T3-L1 cells.[291]

4.105 *Barleria lupulina* L.

Synonyms: *Barleria monostachya* Bojer ex Bouton; *Barleria norbertii* Benoist
Subclass Lamiidae, Superorder Lamianae, Order Lamiales, Family Acanthaceae
Medicinal use: diabetes (India)

Methanol extract of aerial part of *Barleria lupulina* L. given orally to streptozotocin-induced diabetic Wistar rats (plasma glucose >250 mg%) at a single dose of 300 mg/kg lowered glycemia from 265 to 225 mg% (untreated: 265 mg%; glibenclamide 10 mg/kg/day: 215.8 mg%) after 12 hours.[292] The principles involved here are apparently unknown. In normal Wistar rats, the extract had no activity[292] indicating, at least, a mode of action independent on insulin secretion.[292] In normal rats, hypoglycemia is compensated by hepatic neoglucogenesis. Glucacon stimulates the production and secretion of glucose in the plasma by the liver in order to maintain glycemia within normal range.[293] This hormone promotes glyconenolysis by activating glycogen phosphorylase, phosphorylase kinase, glucose-6-phosphatase, and inhibiting glycogen synthetase.[294] *Barleria lupulina* L. contains shanzhiside methyl ester, 8-*O*-acetylshanzhiside methyl ester (barlerin), 6,8-*O*, *O*-diacetylshanzhiside methyl ester (acetylberlerin), 6-*O*-acetylshanzhiside methyl ester, and ipolamiidoside.[295] The hypoglycemic properties of these iridoid glycosides need to be examined.

4.106 *Clerodendrum colebrookianum* Walp.

Synonym: *Clerodendrum glandulosum* Coleb.
Common names: East Indian Glory Bower (English), Kuthap (India) xian mo li (Chinese)
Subclass Lamiidae, Superorder Lamiidae, Order Lamiales, Family Verbenaceae
Medicinal use: hypertension (India)

Methanol extract of leaves of *Clerodendrum colebrookianum* Walp. given to Sprague–Dawley rats on high-fat diet for 7 days at a dose of 40 mg/kg/day lowered plasma cholesterol and low-density lipoprotein by 41% and 85% and increased high-density lipoprotein by 38%.[296] Aqueous extract of leaves given at 3% of high-fat diet to C57BL/6J mice lowered plasma triglycerides and free fatty acids.[297] This regimen decreased to normal hepatic triglycerides and lowered hepatic fatty acids.[297] This extract increased hepatic activity of superoxide dismutase, catalase, and glutathione peroxidase and increased glutathione contents toward normal values.[297] The supplementation prevented hepatocyte ballooning and parenchymatous fat accumulation and resaturated normal histoarchitecture in high-fat diet.[297] Aqueous extract of leaves given to C57BL/6J mice as part of 1% of high-fat diet for 20 weeks had no effect on food intake and decreased body weight gain.[298] This supplementation lowered plasma triglycerides from 191 to 52 mg/dL (normal: 46.8 mg/mL), free fatty acids from 125 to 64.7 mg/dL (normal: 46.7 mg/dL), and leptin from 43.3 to 20 ng/L (normal: 11.8 ng/L).[298] This regimen decreased abdominal, epididymal, and perirenal adipose weights.[298] It also decreased the expression of peroxisome proliferator-activated receptor-γ, sterol regulatory element-binding protein-1c, fatty acid synthetase, and leptin, while carnitine palmitoyltransferase-1 was increased in epididymal fat.[298] At a concentration of 200 μg/mL, the extract lowered inhibited the differentiation of 3T3-L1 mouse preadipocytes induced by dexamethasone-insulin-3-isobutyl-1-methylxanthine.[298] In differentiated 3T3-L1 adipocytes, the extract inhibited the secretion of leptin and triglycerides accumulation, increased the release of glycerol, and decreased glyceraldehyde 3-phosphate dehydrogenase activity.[298] It would be interesting to identify the principles involved here.

4.107 *Ruellia tuberosa* L.

Common names: lu li cao (Chinese); menow-weed
Subclass Lamiidae, Superorder Lamianae, Order Lamiales, Family Acanthaceae
Medicinal use: diuretic (India)

Ruellia tuberosa L. produces verbascoside, isoverbascoside, and forsythoside B.[299] Verbascoside at a concentration of 108 μg/mL reduced by about 30% the accumulation of triglycerides in mature 3T3-L1 adipocytes challenged with high glucose. This caffeic acid derivative evoked a decrease of generation of reactive oxygen species, and protected mitochondrial membrane potential.[300] This glycoside induced the phosphorylation of adenosine monophosphate-activated protein kinase, increased the expression of peroxisome proliferator-activated receptor-α, and inhibited the expression of fatty acid synthetase.[300] Verbascoside reduced the expression of interleukin-1β, interleukin-6, tumor necrosis factor-α, and monocyte chemoattractant peptide-1.[300]

4.108 *Tectona grandis* L.f.

Synonyms: *Tectona theka* Lour.; *Theka grandis* (L.f.) Lam.
Common names: you mu (Chinese); malapangit (Philippins); teak
Subclass Lamiidae, Superorder Lamiidae, Order Lamiales, Family Verbenaceae
Medicinal use: sore throat (Philippines)
Pharmacological target: insulin resistance

Ethanol extract of bark of *Tectona grandis* L.f. given at a dose of 200 mg/kg/day for 15 days to dexamethasone-induced hyperglycemic albino mice lowered glycemia from 76.4 to 62.3 mg/dL (normal: 54.4 mg/dL; Pioglitazone 2 mg/kg: 55.2 mg/dL), triglycerides from 144.3 to 110.2 mg/dL (normal: 83.6 mg/dL; Pioglitazone 2 mg/kg/day: 85.4 mg/dL), and prevented body weight loss.[301] The extract increased glutathione, increased the activity of superoxide dismutase and catalase and brought close to normal lipid peroxidation in the liver.[301] Isolated hemidiaphragm of dexamethasone-induced hyperglycemic albino mice treated with the extract, in absence of insulin, had the ability to uptake glucose by 18.6 mg/g/30 min, whereby hemidiaphragm from untreated rodents could uptake 5.1 mg/g/30 min of glucose (normal: 8.8 mg/g/30 min of glucose; (Pioglitazone 2 mg/kg/day: 7.4 mg/g/30 min).[301] Methanol extract of flowers, containing phenolics including ellagic acid, ferulic acid, rutin, and quercetin, and given orally to nicotinamide-streptozotocin-induced type 2 diabetic Wistar rats (plasma glucose >200 mg/dL) at a dose of 200 mg/kg/day for 4 weeks lowered glycemia by about 70% and this effect was comparable with glibenclamide 5 mg/kg/day.[111] This regimen increased body weight in diabetic animals, increased serum insulin close to normal, lowered total cholesterol from 132.4 to 67 mg/dL (normal: 64.9 mg/dL; glibenclamide 5 mg/kg/day: 62.3 mg/dL), triglycerides from 146.8 to 98.7 mg/dL (normal: 91.3 mg/dL; glibenclamide 5 mg/kg/day: 87.7 mg/dL), and increased high-density lipoprotein from 33.6 to 51.2 mg/dL (normal: 54.8 mg/dL; glibenclamide 5 mg/kg/day: 49.8 mg/dL).[111] Gallic acid, quercetin, and rutin at a concentration of 5 μg/mL induced glucose uptake by L6 rat muscle cells *in vitro* in the absence of insulin as efficiently as pioglitazone at a concentration of 0.1 μg/mL.[111]

4.109 *Calamintha officinalis* Moench

Synonyms: *Calamintha nepeta* (L.) Savi; *Melissa calamintha* L.
Common name: calamint
Subclass Lamiidae, Superorder Lamiidae, Order Lamiales, Family Lamiaceae
Nutritional use: seasoning (Turkey)

Aqueous extract of *Calamintha officinalis* Moench given orally to streptozotocin-induced diabetic Wistar rats (fasting glycemia >16.5 mmol/L) and normal Wistar rats at a single dose of 20 mg/kg lowered glycemia by approximately 20% at 4 hours.[302] Aqueous extract of *Calamintha officinalis* Moench given orally to streptozotocin-induced diabetic Wistar rats (fasting glycemia >16.5 mmol/L) at a dose of 20 mg/kg/day for 2 weeks induced harmful fall of glycemia (from about 25 to 3 mmol/L).[302] In normal rats, the extract given orally a dose of 20 mg/kg/day for 2 weeks evoked a reduction from approximately 6 to 4.5 mmol/L suggesting a hypoglycemic mechanism based on peripheral glucose consumption.[302] This regimen had no effect on plasma insulin for both normal and diabetic rats.[302]

4.110 *Marrubium vulgare* L.

Synonym: *Marrubium hamatum* Kunth
Common names: ou xia zhi cao (Chinese); horehound
Subclass Lamiidae, Superorder Lamianae, Order Lamiales, Family Lamiaceae
Medicinal use: colds (China)
History: the plant was known to Pliny the elder

Type 2 diabetic patients on glibenclamide, taking daily an infusion of 1 g of powder of *Marrubium vulgare* L. for 21 days, 3 times daily before each meal, had lowered fasting plasma glucose by 0.6%, cholesterol by 4.1% and triglycerides by 5.7%.[303] 6-Octadecynoic acid isolated from a methanol extract of this plant at a concentration of 50 μM increased the accumulation of triglycerides in isobutylmethylxanthine-dexamathasone, insulin-induced differentiating 3T3-L1 preadipocytes, and this effect was inhibited by peroxisome proliferator-activated receptor-γ antagonist GW9662.[304]

4.111 *Melissa officinalis* L.

Synonym: *Melissa bicornis* Klokov
Common names: xiang feng hua (Chinese); lemon balm
Subclass Lamiidae, Superorder Lamiidae, Order Lamiales, Family Lamiaceae
Medicinal use: arteriosclerosis (Turkey)

Ethanol extract of *Melissa officinalis* L. activated *in vitro* peroxisome proliferator-activated receptor-γ.[305] The extract at 0.6 mg/mL induced in primary human adipocytes the expression of peroxisome proliferator-activated receptor-γ, fatty acid-binding protein 4, (which is a marker of adipogenesis and triglycerides accumulation), fatty acid transport protein 4 and fatty acid translocase, (which mediate fatty acid transport into adipocytes) pyruvate dehydrogenase kinase-4 (indicating intensified entry of acetyl-CoA into the tricarboxylic acid cycle) as well as liver X receptor α and stearoyl-CoA desaturase.[305] In C57BL/6 mice on high-fat diet-induced mild insulin resistance, the extract at a dose of 200 mg/kg/for 2 weeks had no effect on food intake and lowered fasting blood glucose by 14% (as efficiently as rosiglitazone at a dose of 4 mg/kg/day).[305] This regimen had no effects of plasma insulin and decreased insulin resistance by 35% (rosiglitazone at 4 mg/kg/day: 75%).[305] In a subsequent study, essential oil of lemon balm given to human APOE2 (R158C) transgenic mice orally at a dose of 12.5 μg/day for 2 weeks lowered plasma triglycerides by 36% and plasma cholesterol by 24% without effect on low-density lipoprotein–cholesterol and hepatic cholesterol.[306] Oral administration of linalool at a dose of 120 mg/mouse/day for 6 weeks to C57BL/6J mice on high-fat diet lowered lipid plasma triglycerides and low-density lipoprotein–cholesterol by 40% and 43%, respectively.[306] Essential oil of *Melissa officinalis* L. given to streptozotocin-induced diabetic Wistar rats (plasma glucose >250 mg/mL) given orally at a dose of 0.04 mg/day for 4 weeks prevented body weight loss and lowered glycemia from 366.1 to 119.7 mg/dL (normal: 91.6 mg/dL).[307]

4.112 *Prunella vulgaris* L.

Synonym: *Prunella asiatica* Nak.
Common names: xia ku cao (Chinese); heal-all
Subclass Lamiidae, Superorder Lamiidae, Order Lamiales, Family Lamiaceae
Medicinal use: diarrhea (Korea)

Oleanolic acid, ursone, 2α,3α-dihydroxylurs-12-en-28-oic acid, 2α,3β-dihydroxyolean-12-en-28-oic acid, 2α,3α-24-trihydroxyurs-12-en-28-oic acid, 2α,3α-24-trihydroxyolean-12-en-28-oic acid, 2α,3α-24-trihydroxyurs-12,20(30)-dien-28-oic acid, 2α,3β, 19α-trihydroxyurs-12-en-28-oic acid and 2α,3α, 24-trihydroxyurs-12-en-28-oic acid-28-β-D-glucopyranoside isolated from the spikes of *Prunella vulgaris* L. inhibited the enzymatic activity of rabbit muscle glycogen phosphorylase with IC_{50} values below 70 μM (positive standard caffeine: IC_{50} value 74.5 μM).[308] In 3T3-L1 differentiated adipocytes, oleanolic acid, 2α,3β-dihydroxyolean-12-en-28-oic acid, 2α,3α-24-trihydroxyurs-12-en-28-oic acid, and 2α,3α, 24-trihydroxyurs-12-en-28-oic acid-28-β-D-glucopyranoside at a concentration of 10 μM enhanced insulin-induced glucose consumption.[308] 2α,3α,24-Trihydroxyurs-12-en-28-oic acid at 10 μM was as efficient as metformin at a concentration of 1 mM.[308]

4.113 *Perilla frutescens* (L.) Britton

Synonym: *Ocimum frutescens* L.; *Perilla ocymoides* L.
Common names: zi su (Chinese); deul gae (Korean); perilla
Subclass Lamiidae, Superorder Lamiidae, Order Lamiales, Family Lamiaceae
Nutritional use: salad (Korea)
Pharmacological target: atherogenic hyperlipidemia

Ethanol extract of *Perilla frutescens* (L.) Britton as part of 3% of high-fat diet C57BL/6 mice for 4 weeks decreased body weight gain from 8.9 to 3.6 g/day (normal diet: 2.6 g/day), lowered liver weight toward normal values and decreased epididymal fat tissue from 51.6 to 41.2 mg/g (normal: 14.7 mg/g).[309] This supplementation lowered triglycerides from 213.6 to 106.3 mg/dL (normal: 90.6 mg/dL), total cholesterol from 330.9 to 144.4 mg/dL (normal: 93.1 mg/dL), low-density lipoprotein–cholesterol from 254.4 to 72.6 mg/dL (normal: 51.5 mg/dL) and increased high-density lipoprotein–cholesterol from 33.8 to 59.3 mg/dL (normal: 40.9 mg/dL).[309] The extract lowered aspartate aminotransferase and alanine aminotransferase and normalized hepatic cytoarchitecture.[309] The extract decreased the expression of acetyl-CoA carboxylase, glycerol-3-phosphate dehydrogenase and peroxisome proliferator-activated receptor γ in epididymal fat tissues.[309] This plant contains luteolin[310] which at a concentration of 30 μM induced glucose uptake by L6 myotubes *in vitro* as efficiently as insulin at 0.1 μM.[208] The plant contains scutellarin and rosmarinic acid,[311] tormentic acid, oleanolic acid, and ursolic acid.[312] Oleanolic acid given to Swiss mice at a dose of 10 mg/kg/day for 7 days lowered 30 minutes postprandial glycemia by 37% in oral glucose tolerance test.[313] This oleanane triterpene given as 0.05% drinking water to Swiss mice on high-fat diet for 15 weeks had no effect on food intake and lowered body weight by 11.1%.[313] This supplementation lowered plasma cholesterol and triglycerides and decreased visceral fat accumulation by 50% and this effect was superior to sibutramine (0.05% of drinking water).[313] Oleanolic acid lowered to normal liver weight as well as plasma aspartate aminotransferase and alkaline phosphatase and improved hepatic cytoarchitecture.[313] This triterpene lowered glycemia from 136.3 to 96 mg/dL (normal: 86.1 mg/dL; sibutramine: 137.4 mg/dL) and had no effect on insulin.[313] This triterpene lowered ghrelin and increased leptin.[313] Ghrelin is secreted from the stomach to stimulate appetite.[314]

4.114 *Salvia miltiorrhiza* Bunge

Common name: dan shen (Chinese)
Subclass Lamiidae, Superorder Lamiidae, Order Lamiales, Family Lamiaceae
Medicinal use: heart diseases (China)

Cryptotanshinone from *Salvia miltiorrhiza* Bunge at a concentration of 8 μM inhibited the differentiation of 3T3-L1 mouse embryonic fibroblasts (induced by dexamethasone-insulin-3-isobutyl-1-methylxanthine) as evidenced by a reduction of triglyceride accumulation by 48%.[315] During differentiation, the expression of CCAAT/enhancer-binding protein-β increased on day 1 and cryptotanshinone reduced the expression of CCAAT/enhancer-binding protein-β by 35.9%.[315] During the differentiation, the expression of peroxisome proliferator-activated receptor-γ increased to a peak on day 4 and it was lowered by 49.6% by cryptotanshinone.[315] During differentiation, the expression of CCAAT/enhancer-binding protein-α was maximum on day 6 and it was lowered by 53.4% by cryptotanshinone. During the differentiation, the expression of fatty acid-binding protein was increased to a peak at day 6 and cryptotanshinone reduced the expression by 80.4%.[315] During the differentiation, the expression of adiponectin was increased to a peak at day 7 and this quinoid diterpene reduced the expression by about 95%.[315] Growing evidence suggests that the phosphorylation of STAT3 induces the expression of CCAAT[316]/enhancer-binding protein-β and peroxisome proliferator-activated receptor-γ.[316] Cryptotanshinone inhibited STAT3 phosphorylation during the first hours of differentiation.[315,316]

4.115 *Acorus calamus* L.

Synonyms: *Acorus americanus* (Raf.) Raf.; *Acorus angustatus* Raf.; *Acorus angustifolius* Schott; *Acorus asiaticus* Nakai; *Acorus cochinchinensis* (Lour.) Schott; *Acorus griffithii* Schott; *Acorus spurius* Schott; *Acorus triqueter* Turcz. ex Schott; *Acorus verus* (L.) Houtt.; *Orontium cochinchinense* Lour.
Common names: chang pu; sweet-flag
Class Liliopsida, Subclass Alismatidae, Superorder Aranae, Order Arales, Family Acoraceae
Medicinal use: diabetes (Indonesia)

Ethylacetate fraction of roots of *Acorus calamus* L. at a concentration of 25 μg/mL increased insulin-dependent glucose uptake by L6 rate skeletal muscle cells from 3.1 to 4.7 mmol/L.[322] Given orally to C57BL/Ks *db/db* mice at a dose of 100 mg/kg/day for 3 weeks, this extract lowered food and water intake, lowered serum glucose from 27.3 to 12.4 mmol/L (rosiglitazone 10 mg/kg/day: 12.4 mmol/L), lowered total cholesterol from 5.6 to 4.6 mmol/L.[322] This regimen had no effect on plasma insulin, lowered free fatty acids and increased adiponectin.[322] β-Asarone (Figure 4.22) from *Acorus calamus* L. at a concentration of 0.2 mM reduced the accumulation of triglycerides

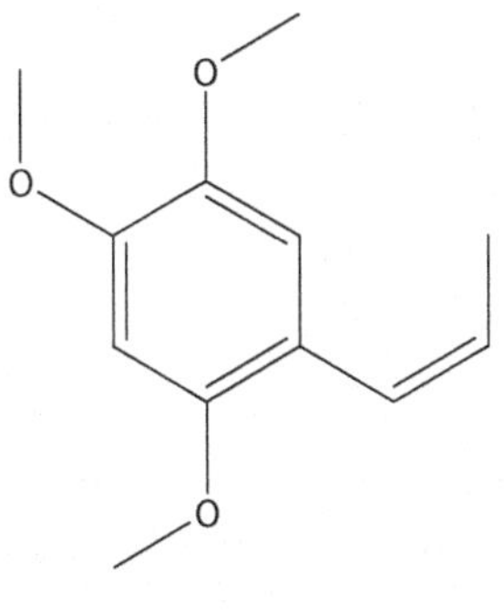

O O O

FIGURE 4.22 β-Asarone.

in dexamethasone-insulin-3-isobutyl-1-methylxanthine-insulin-induced differentiation of 3T3-L1 preadipocytes by about 60%.[323] β-asarone inhibited the phosphorylation of extracellular signal-regulated kinase-1/2 and therefore CCAAT/enhancer-binding protein-β, and downregulation of CCAAT/enhancer-binding protein-β. This phenyl propanoid attenuated the expression of peroxisome proliferator-activated receptor-γ and CCAAT/enhancer-binding protein-α to inhibits adipocyte differentiation.[323]

4.116 *Pinellia ternata* (Thunb.) Makino

Synonym: *Arum ternatum* Thunb.
Common name: ban xia (Chinese)
Class Liliopsida, Subclass Aranae, Order Arales, Family Araceae
Medicinal use: phlegm (China)

Aqueous extract of *Pinellia ternata* (Thunb.) Makino given to obese Zucker rats orally at a daily dose of 400 mg/kg for 6 weeks had no effect on body weight, food intake, plasma cholesterol, lowered plasma free fatty acids from 0.5 to 0.4 mmol/L and plasma triglycerides from 583.6 to 440.2 mg/dL. In the brown adipose tissue, the regimen induced the expression of uncoupling protein-1 (only found in brown adipose tissues) but no increase was observable with peroxisome proliferator-activated receptor-α and peroxisome proliferator-activated receptor coactivator-1α.[324] In white adipose tissue, the extract induced the expression of peroxisome proliferator-activated receptor-α and peroxisome proliferator-activated receptor coactivator-1α.[324] *Pinellia ternata* (Thunb.) Makino contains alkaloids of which ephedrine and trigonelline.[325] Ephedrine at a dose of 0.1% of a diet given to obese A^{vy}/a mice to 39 days evoked a weight loss of 6.7 g in the first week of the regimen and 3.2 g during the second week, and no further weight lost for the remaining days of the treatment. During the first 2 weeks the food consumption of mice receiving ephedrine increased from 4 to 4.6 g/mouse/day.[326] Subcutaneous injection of ephedrine at a dose of 40 mg/Kg twice a day for 11 days induced a weight loss of 6.5 g and a reduction of food intake of 7 g/mouse/day.[326] Furthermore, 60 mg/Kg given subcutaneously twice a day evoked another 3.5 g in weight loss with a reduction of food intake of 5g g/mouse/day.[326] In normal mice, ephedrine at 0.08% induced a fall in body weight by 3 g after 10 days with concomitant increase in food consumption.[326] After 39 days of treatment with 0.1% of ephedrine the A^{vy}/a mice had a reduction of body weight of 8 g, a reduction of cholesterolemia from 142 to 123.6 mg/dL, a reduction of triglyceridemia from 179.9 to 159 mg/dL and a reduction of hepatic triglycerides from 104.2 to 32.5 μmol/g.[326] Ephedrine in healthy volunteers induces increase in thermogenesis, heart rate, systolic blood pressure and plasma glucose by activating β-receptors expressed by adipocytes.[327] This alkaloid has been incorporated in some slimming products and taken at high dose and reported to induce hypertension, palpitations, tachycardia, cardiac arrest, stroke, and seizure.[328,329] This is an supplementary example to suggest that the use of infusion or decoctions of non toxic medicinal plants should be preferred as pure concentrated isolated natural products.

4.117 *Spirodela polyrhiza* (L.) Schleid.

Common name: zi ping (Chinese)
Synonym: *Lemna polyrrhiza* L.
Class Liliopsida, Subclass Aranae, Order Arales, Family Araceae
Medicinal use: swelling (China)

Spirodela polyrhiza (L.) Schleid. contains vitexin and orientin which is at a concentration of 100 μM inhibited the accumulation of triglycerides in dexamethasone-insulin-3-isobutyl-1-methylxanthine-induced differentiation of 3T3-L1 preadipocytes by 39.4% and 33%, respectively.[330] These flavonoids decreased CCAAT/enhancer-binding protein-α and γ expression in 3T3-L1 cells.[330]

4.118 *Liriope platyphylla* F.T. Wang & T. Tang

Synonyms: *Liriope muscari* (Decne.) L.H. Bailey; *Ophiopogon spicatus* Hook.
Common name: kuo ye shan mai dong (Chinese)
Class Liliopsida, Subclass Liliidae, Superorder Lilianae, Order Liliales, Family Liliaceae
Medicinal use: fatigue (China)

Binding of insulin to its receptor activates phosphoinositide 3-kinase which catalyzes the formation of phosphatidylinositol-3,4,5-trisphosphate, an allosteric activator of phosphoinositide-dependent kinase.[331] Targets of phosphoinositide-dependent kinase include Akt and atypical protein kinase C.[331] To fully activate Akt protein ser-473 activated Akt and thr-308 must be sequentially phosphorylated commands glucose transporter-4 translocation.[331] Homoisoflavones fraction of *Liriope platyphylla* F.T. Wang and T. Tang at a concentration of 1 μg/mL enhanced glucose uptake induced by insulin in 3T3-L1 adipocytes via increased tyrosine phosphorylation of insulin receptor substrate-1, increased phosphorylation of Akt on ser-473 and increased glucose transporter-4 membrane translocation.[332] Homoisoflavones present in this plant induce methylophiopogonone A, ophiopogonone A, methyophiopogonanone A, and ophiopogonanone A.[332]

4.119 *Gastrodia elata* Blume

Synonyms: *Gastrodia mairei* Schltr.; *Gastrodia viridis* Makino
Common name: tian ma (Chinese)
Class Liliopsida, Subclass Liliidae, Superorder Lilianae, Order Orchidales, Family Orchidaceae
Medicinal use: epilepsy (China)

Gastrodia elata Blume given to Sprague–Dawley rats on high-fat diet at a daily dose of 300 mg/kg/day for 8 weeks decreased food intake, lowered body weight gain from 108.3 to 79.2 g and epididymal fat pads from 5.8 to 4.1 g.[333] This regimen attenuated plasma glucose from 105 to 98 mg/dL (normal diet: 101 mg/dL), normalized plasma insulin, lowered triglycerides from 115 to 94 mg/dL (low fat diet: 73 mg/dL), increased adiponectin near to normal diet values, and lowered leptin to.[333] The extract lowered area under the curve of plasma glucose in oral glucose tolerance test by 17% and lowered insulin resistance in hyperinsulinemic-euglycemic clamp by 24%.[333] At the hepatic level the extract lowered triglyceride accumulation and increased glycogen contents[333] implying increased hepatic sensitivity. This regimen decreased hypothalamic leptin resistance induced by high-fat diet by increasing the phosphorylation of STAT3[333]. It also attenuated the phosphorylation of adenosine monophosphate-activated protein kinase in the hypothalamus lowering thereby the expression of long-form leptin receptor (Ob-Rb) and orexigenic neuropeptides NPY and AgRP. 4-Hydroxybenzaldehyde and vanillin from this plant at a concentration of 20 μM increased insulin-induced uptake of glucose by 3T3-L1 adipocytes, increased the expression of carnitine palmitoyltransferase-1, decreased glycerol-3-phosphate acyl transferase-1 expression, decreased CCAAT/enhancer-binding protein-α expression but had no effect on peroxisome proliferator-activated receptor-γ.[333] This further examplifies the importance of simple phenolic natural products.

4.120 *Crocus sativus* L.

Common names: fan hong hua (Chinese); kesar (India); Saffron
Class Liliopsida, Subclass Liliidae, Superorder Lilianae, Order Iridales, Family Iridaceae
Medicinal use: enlargement of the liver (India)
History: the plant was known as being as aphrodisiac (*venerem stimulat*) by Pedanius Dioscorides (circa 40–90 AD), a Greek physician and during the time of the Emperor Nero

Safranal from *Crocus sativus* L. inhibited protein-tyrosine phosphatase 1B with an IC_{50} of about 10 μM.[334] In mouse myoblast cell line (C2C12) treated with safranal at a concentration of 10 μM inhibited protein-tyrosine phosphatase-1B by about 30% and increased glucose transporter-4 translocation.[334] This a monoterpene aldehyde at 20 μM increased glucose uptake in C2C12 by about 260%.[334] Safranal given orally to type 2 diabetic KK-Ay mice at a daily dose of 20 mg/kg for 15 days had no effect on body weight, but lowered plasma glucose from 140 to 119 mg/dL.[334] In oral glucose tolerance test, the treatment lowered 30 minutes peak glycemia from 436 to 380 mg/dL.[334]

4.121 *Anemarrhena asphodeloides* Bunge

Common name: zhi mu (Chinese)
Subclass Liliidae, Superorder Lilianae, Order Asparagales, Family Asparagaceae
Medicinal use: fever (fever)

Ethanol fraction of rhizomes of *Anemarrhena asphodeloides* Bunge given to KK-Ay mice (mouse model of type 2 diabetes) for 8 weeks lowered glycemia from 20.7 to 8.5 mM (pioglitazone 30 mg/kg/day: 6.5 mM), lowered insulin level and reduced insulin resistance.[335] This regimen improved the cytoarchitecture of pancreas, liver and kidney in KK-Ay mice.[335] In line, in Sprague–Dawley rats with insulin resistance (induced by BCG vaccine) the extract at a dose of 180 mg/kg improved insulin sensitivity in hyperinsulinemic-euglycemic clamp experiment.[335] In streptozotocin-induced diabetic mice, with very low levels of plasma insulin, this fraction at a dose of 270 mg/kg/day for 7 days had no effect on fasting plasma glucose but in combination with exogenous insulin improved glycemia.[335] Natural products with the ability to increase insulin sensitivity do not evoke hypoglycemia in normal rodents because of liver neoglucogenesis. Likewise, in state of profound insulin depletion (because of complete destruction of pancreatic β-cells) induced by alloxan or streptozotocin, such natural products are inactive. Ethanol fraction of rhizomes of *Anemarrhena asphodeloides* Bunge, evoked the phosphorylation of adenosine monophosphate-activated protein kinase in 3T3-L1 preadipocytes and increased the phosphorylation of the adenosine monophosphate-activated protein kinase and acetyl-CoA carboxylase via calcium calmodulin-dependent protein kinase-β[335] implying, at least an increase of cytosolic calcium contents. Note that saponins *in vitro* have a tendency to stimulate adenosine monophosphate-activated protein kinase via increased cytosolic contents.

4.122 *Polygonatum falcatum* A. Gray

Common names: huang-ching (China); naruko-yuri (Japan); jin huang jeong (Korean); deer bamboo
Subclass Liliidae, Superorder Lilianae, Order Asparagales, Family Asparagaceae
Nutritional use: food (China)

Kaempferol isolated from the rhizome of *Polygonatum falcatum* A. Gray at a concentration of 40 μM inhibited the accumulation of triglycerides in differentiating 3T3-L1 adipocytes by about 80%.[336] This flavonol decreased the expression of sterol regulatory element-binding protein-1c, and peroxisome proliferator-activated receptor-γ and downstream targets: adipsin and lipoprotein lipase.[336]

4.123 *Polygonatum odoratum* (Mill.) Druce

Synonyms: *Convallaria odorata* Mill.; *Polygonatum japonicum* C. Morren & Decne.; *Polygonatum maximowiczii* F. Schmidt; *Polygonatum vulgare* Desf.
Common name: yu zhu (Chinese)

Subclass Liliidae, Superorder Lilianae, Order Asparagales, Family Asparagaceae
Medicinal use: cough (China)
Pharmacological target: insulin resistance

Ethanol extract of rhizome of *Polygonatum odoratum* (Mill.) Druce given to C57BL/6 mice as 1% part of high-fat diet for 8 weeks prevented body weight gain, had no effect on food intake and lowered plasma glucose from about 6 to 5 mmol/L.[337] This regimen improved glucose tolerance in intraperitoneal glucose tolerance test and improved glucose clearance during intraperitoneal insulin tolerance test.[337] This regimen lowered plasma cholesterol and triglycerides and had no effect on low-density lipoprotein–cholesterol and high-density lipoprotein–cholesterol.[337] This extract lowered plasma insulin to normal, doubled plasma adiponectin levels, and normalized serum leptin.[337] The extract given 2 weeks at 1% of high-fat diet to C57BL/6 mice after 12 weeks of high-fat diet lowered glycemia and triglycerides.[337] In both preventive and therapeutic treatment, the extract lowered liver triglycerides and had no effect on liver cholesterol.[337] In the preventive treatment, the extract increased hepatic peroxisome proliferator-activated receptor-γ, acetyl-CoA carboxylase and uncoupling protein-2.[337] In white adipose tissue, the extract in preventive study increased peroxisome proliferator-activated receptor coactivator-1α, acetyl-CoA carboxylase, acyl-CoA-oxidase, peroxisome proliferator-activated receptor-γ, fatty acid-binding protein, lipoprotein lipase, uncoupling protein-2, glucose transporter-4, and decreased tumor necrosis factor-α.[337] In preventive treatment the extract increased body temperature and brown adipose tissue expression of uncoupling protein-2.[337] In therapeutic treatment, the extract increased hepatic peroxisome proliferator-activated receptor-γ, -α and fatty acid translocase.[337] In white adipose tissue, the extract in therapeutic study increased peroxisome proliferator-activated receptor-γ and target fatty acid-binding protein, peroxisome proliferator-activated receptor coactivator-1α, acetyl-CoA carboxylase, lipoprotein lipase, uncoupling protein-2, glucose transporter-4, and decreased tumor necrosis factor-α.[337] Clinical trials are warranted.

4.124 *Dioscorea oppositifolia* L.

Common name: sarpakhya (India)
Subclass Liliidae, Superorder Dioscoreanae, Order Dioscoreales, Family Dioscoreaceae
Medicinal use: swelling (India)

Ethanol extract of *Dioscorea opposita* L. given orally to Sprague–Dawley rats at a dose of 300 mg/kg/day for 2 weeks prevented insulin resistance induced by subcutaneous injection of dexamethasone as evidenced by a decrease of plasma glucose from 9.5 to 2.8 mM (normal: 4.5 mM) and insulinemia from 0.7 to 0.2 nM (normal: 0.09 nM), respectivley.[338] The extract lowered 30 minutes peak glycemia in oral glucose tolerance test.[338] The extract at a concentration of 50 μg/mL increased insulin-mediated glucose uptake by 3T3-L1 adipocytes by 5 folds together with increased expression of glucose transporter-4.[338] (4*E*,6*E*)-1,7-bis(4-hydroxyphenyl)-4,6-heptadien-3-one, (3R,5R)-3,5-dihydroxy-1,7-bis(4-hydroxyphenyl)-3,5-heptanediol, batatasin I, (1*E*,4*E*,6*E*)-1,7-bis(4-hydroxyphenyl)-1,4,6-heptatrien-3-one isolated from the rhizomes of *Dioscorea oppositifolia* L. at a concentration of 20 μM inhibited the accumulation of triglycerides in dexamethasone-insulin-3-isobutyl-1-methylxanthine-induced differentiating 3T3-L1 preadipocytes concurrent decreased expression of peroxisome proliferator-activated receptor-γ.[339] Batatasin I and (1*E*,4*E*,6*E*)-1,7-bis(4-hydroxyphenyl)-1,4,6-heptatrien-3-one down-regulated the expression of CCAAT/enhancer-binding protein-α.[338] (4*E*,6*E*)-1,7-bis(4-hydroxyphenyl)-4,6-heptadien-3-one and batatasin I induced the phosphorylation of adenosine monophosphate-activated protein kinase.[338] Batasin I boosted the expression of carnitine palmitoyltransferase-1.[338]

4.125 *Dioscorea batatas* Decne.

Synonym: *Dioscorea polystachya* Turcz.
Common name: shu yu (Chinese); nagaimo (Japanese); Chinese yam
Subclass Liliidae, Superorder Dioscoreanae, Order Dioscoreales, Family Dioscoreaceae
Medicinal use: rheumatism (China)

Ethanol extract of rhizomes of *Dioscorea batatas* given orally to high-fat diet fed C57BL/6J mice at a dose of 200 mg/kg/day for 14 weeks evoked a mild decrease of body and visceral fat weight.[340] This extract lowered plasma cholesterol from 217.9 to 168.1 mg/dL (normal: 140.4 mg/dL), triglycerides from 140.8 to 102.9 mg/dL (normal: 94.3 mg/dL) and increased high-density lipoprotein from 98.7 to 123.8 mg/dL (normal: 119.9 mg/dL).[340] The extract lowered plasma leptin and prevented hepatic steatosis.[340] In epididymal adipose tissue, the extract decreased the expression of CCAAT/enhancer-binding protein-α as well as fatty acid translocase, and decreased levels of tumor necrosis factor-α, interleukin-6 and monocyte chemoattractant peptide-1.[340]

4.126 *Areca catechu* L.

Synonyms: *Areca catechu* Willd.; *Areca faufel* Gaertn.; *Areca himalayana* Griff. ex H. Wendl.; *Areca hortensis* Lour.; *Areca nigra* Giseke ex H. Wendl.
Common names: bin lang (Chinese); puga (India); buah pinang (Malay); betel nut
Subclass Arecidae, Superorder Arecanae, Order Arecales, Family Arecaeae
Medicinal use: anxiolytic (Malaysia)
History: the plant was known to Sushruta

Arecoline (Figure 4.23) at a concentration of 60 µM inhibited adipogenesis in 3T3-L1 cells by 82% with reduction of CCAAT/enhancer-binding protein-α, -γ and downstream targets fatty acid-binding protein, glucose transporter-4, lipoprotein lipase.[341] In differentiated 3T3-L1 adipocytes, this alkaloid at 500 µM evoked lipolysis as evidenced by a burst in glycerol release (on probable account of cyclic adenosine monophosphate/adenylyl cyclase) and abrogated insulin-dependent glucose uptake by 18%.[341] This alkaloid at a much lower concentration (20 µg/mL) reduced the accumulation of lipid droplet in 3T3-L1 adipocytes via downregulation of CCAAT/enhancer-binding protein-β and peroxisome proliferator-activated receptor-γ.[342] In the same experiment, arecoline at 20 µg/ml downregulated insulin receptor, insulin receptor substrate-1, glucose transporter-4, fatty acid synthetase, perilipin, and adipophilin.[342] Arecoline is a M_3 muscarinic agonist.[343] Yang et al. observed that acetylcholine at 1 µM attenuated insulin-stimulated glucose uptake and the release of glycerol from adipose tissues isolated from Wistar rats via M3 muscarinic receptor activation.[344] It must be recalled that arecoline is toxic for the liver and testicles.[345]

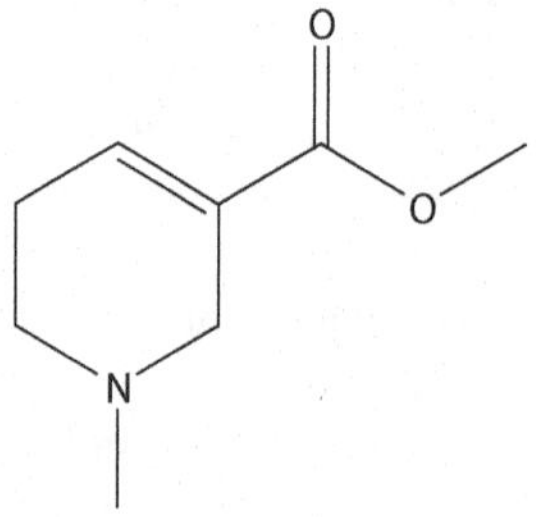

FIGURE 4.23 Arecoline.

4.127 *Alpinia galanga* (L.) Willd.

Synonym: *Maranta galanga* L.
Common names: hong dou kou (Chinese); greater galangal
Subclass Commelinidae, Superorder Zingiberanae, Order Zingiberales, Family Zingiberaceae
Medicinal use: food poisoning (Vietnam)
History: the plant was known to Rhazes (860–932 AD), surgeon in Bagdad

Methanol extract of rhizomes of *Alpinia galanga* (L.) Willd. given orally to New Zealand rabbits at a dose equivalent to 4 g/kg of powdered rhizomes lowered plasma glucose from 101.4 to 73.8 mg/dL after 4 hours and this effect was similar to gliclazide at 80 mg/kg (76.4 mg/kg).[346] In alloxan-induced diabetic rodents, the extract had no ability to lower plasma glucose similarly to gliclazide.[346] Sulfonylureas like gliclazide have been reported not to decrease the blood glucose levels of alloxan diabetic animals with complete destruction of β-cells. When all β-cells are destroyed by alloxan on streptozotocin, there is no more insulin and insulinotropic agent do no impose hypoglycemia. Galangin from the rhizome of *Alpinia galanga* (L.) Willd. given orally at a dose of 100 mg/kg/day for 16 days orally to fructose-fed Wistar rats lowered glucose from 120.6 to 73.8 mg/dL (normal: 69.8 mg/dL), triglycerides from 163.4 to 98.5 mg/dL (normal: 91.6 mg/dL), decreased plasma insulin from 62.6 to 33.5 μU/mL (normal: 34.2 μU/mL), and increased insulin sensitivity an improvement of glucose clearance by skeletal muscles and/or adipocytes.[348] This flavone at a concentration of 30 μM induced glucose uptake by L6 myotubes *in vitro* as efficiently as insulin at 0.1 μM.[208]

4.128 *Alpinia katsumadae* Hayata

Synonym: *Alpinia hainanensis* K. Schum.; *Languas katsumadae* (Hayata) Merr.
Common name: cao dou kou (Chinese)
Subclass Commelinidae, Superorder Zingiberanae, Order Zingiberales, Family Zingiberaceae
Medicinal use: indigestion (China)
Pharmacological target: insulin resistance

Cardamonin, alpinetin, and pinocembrin isolated from the seeds of *Alpinia katsumadae* Hayata induced at a concentration of 30 μM glucose uptake by L6 skeletal muscle cells.[349] The effect of cardamonin (Figure 4.24) was superior to insulin at 0.1 μM.[349] This chalcone induced the translocation of glucose transporter-4 and this effect was not inhibited by wortmannin indicating the noninvolvement of phosphoinositide 3-kinase. Cardamonin did not induce the phosphorylation of Akt or protein kinase λ/ζ nor it had effect on the phosphorylation of adenosine monophosphate-activated protein kinase.[349] Flavonoids have the ability (at high concentration) to generate reactive oxygen species hence activation of adenosine monophosphate activated protien kinase.

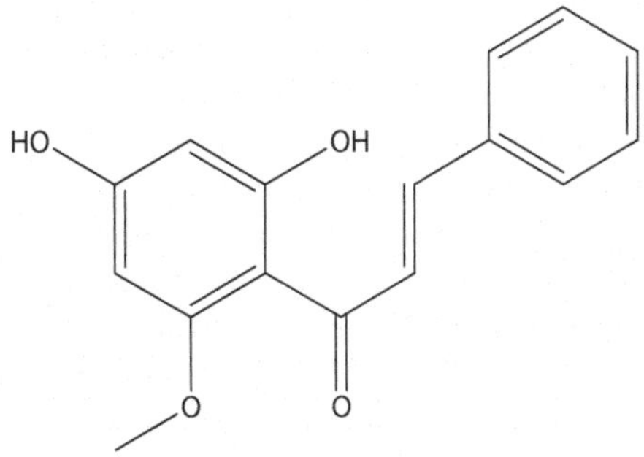

HO OH O O

FIGURE 4.24 Cardamonin.

4.129 *Alpinia officinarum* Hance

Synonym: *Languas officinarum* (Hance) Farw.
Common names: gao liang jiang (Chinese); lesser galangal
Subclass Commelinidae, Superorder Zingiberanae, Order Zingiberales, Family Zingiberaceae
Medicinal use: indigestion (China)
Pharmacological target: insulin resistance

Ethanol extract of *Alpinia officinarum* Hance given to C57BL/6J mice at 0.5% of high-fat diet at a for 8 weeks lowered body weight gain, and epididymal and perirenal fat pad weight.[350] This regimen decreased plasma insulin to normal and doubled plasma leptin.[350] This supplementation lowered hepatic triglycerides contents by 17% and attenuated hepatic cholesterol.[350] The extract decreased the expression of sterol regulatory element-binding protein-1, CCAAT/enhancer-binding protein-α, peroxisome proliferator-activated receptor-γ in the epididymal fat pads and suppressed hepatic expression of CCAAT/enhancer-binding protein-α, peroxisome proliferator-activated receptor-γ and fatty acid synthetase in the liver.[350] From this extract galangin at 50 μM inhibited the accumulation of triglycerides in differentiating 3T3-L1 adipocytes with decreased expression of sterol regulatory element-binding protein-1, CCAAT/enhancer-binding protein-α, peroxisome proliferator-activated receptor-γ, and fatty acid synthetase.[350]

4.130 *Alpinia zerumbet* (Pers.) B.L. Burtt & R.M. Sm.

Synomyms: *Alpinia fluviatilis* Hayata; *Alpinia schumanniana* Valeton; *Alpinia speciosa* (J.C. Wendl.) K. Schum.; *Costus zerumbet* Pers.; *Languas speciosa* (J.C. Wendl.) Small; *Zerumbet speciosum* J.C. Wendl.
Common names: yan shan jiang (Chinese); getto (Japan); shell-flower galangal
Subclass Commelinidae, Superorder Zingiberanae, Order Zingiberales, Family Zingiberaceae
Medicinal use: stomachache (China)

Catecholamines released during stress, glucagon during fast and serotonin, are brought by the plasma to the surface of adipocytes where they bind to membrane receptor, activating adenylate cyclase increasing levels of cyclic adenosine monophosphate.[351] This secondary messenger activates protein kinase A which in turn on both hormone sensitive lipase and perilipin-1.[351] The phosphorylation of perilipin-1 induces the on adipocyte triglyceride lipase to convert triglycerides into diacylglycerol whereas hormone sensitive lipase catalyze the hydrolysis of diacylglycerol to monoacylglycerol. Monoacyl glycerol is hydrolyzed into glycerol and unesterified fatty acid by monoacylglycerol lipase.[352] Unesterified fatty acids are transported to the cytoplasmic membrane by adipocyte fatty acid-binding protein.[352] Hormone-sensitive lipase is inhibited by insulin and, in obese patients with insulin resistance the activation of hormone-sensitive lipase evokes an increase of secretion of non-esterified free fatty acids promoting the synthesis of very low-density lipoprotein by the liver, β-cells dysfunction and worsening of insulin resistance.[353] Natural products with the ability to stimulate adipocyte hormone-sensitive lipase may conceptually be of usefulness to boost hydrolysis of stored triglycerides in adipocytes and to reduce adipose mass. One such compound is 5,6-dehydrokawain (Figure 4.25) from *Alpinia zerumbet* B.L. Burtt & R.M. Sm.[354] which at 250 μg/mL reduced triglyceride contents in 3T3-L1 adipocytes by 63.4%, increased cytoplasmic cyclic adenosine monophosphate by 56.9%, inhibited glycerol-3-phosphate dehydrogenase by 90.5%, and induced glycerol release by 225% with concurrent increased of cytoplasmic cyclic adenosine monophosphate.[355] This pyrone reduced triglyceride accumulation by 63.4%.[355] *In vitro*, 5,6-dehydrokawain inhibited porcine pancreatic lipase activity with an IC_{50} of 74.4 μg/mL (quercetin: 38.5 μg/mL).[355] Terpinen-4-ol isolated from ginger given intravenously evoked an immediate and dose-dependent reduction of mean aortic blood pressure in deoxycorticosterone acetate-salt hypertensive Wistar rats by about 30% at 5 mg/kg.[356]

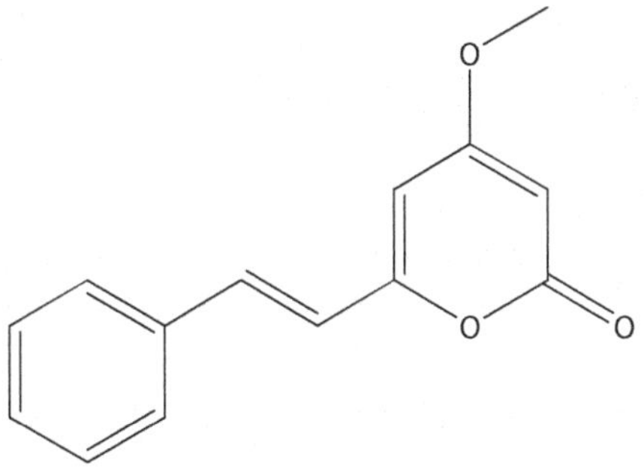

FIGURE 4.25 5,6-Dehydrokawain.

4.131 *Amomum xanthioides* Wall ex. Baker

Synonym: *Amomum villosum* var. *xanthioides* (Wall. ex Baker) T.L. Wu & S.J. Chen
Common name: sha ren (Chinese)
Subclass Commelinidae, Superorder Zingiberanae, Order Zingiberales, Family Zingiberaceae
Medicinal use: sore throat (China)

Aqueous extract of seeds of *Amomum xanthioides* Wall ex. Baker at a concentration of 0.5 mg/mL increased glucose uptake by 3T3-L1 adipocytes (in absence of insulin) and doubled glucose uptake by 3T3-L1 adipocytes in the presence of insulin.[357] The active principles involved here are apparently unknown. Phenolic compounds inducing adenosine monophosphate-activated protein kinase could be involved.

4.132 *Boesenbergia pandurata* (Roxb.) Schltr.

Synonyms: *Boesenbergia rotunda* (L.) Mansf.; *Kaempferia pandurata* Roxb.
Common name: ao chun jiang (Chinese)
Subclass Commelinidae, Superorder Zingiberanae, Order Zingiberales, Family Zingiberaceae
Medicinal use: indigestion (China)
Pharmacological target: atherogenic hyperlipidemia

Ethanol extract of *Boesenbergia pandurata* (Roxb.) Schltr. at a concentration of 25 μg/mL inhibited the accumulation of triglycerides in differentiating 3T3-L1 preadipocytes and insulin-induced hepatic accumulation of triglycerides.[358] These effect was accompanied with decreased expression of acetyl-CoA carboxylase, fatty acid synthetase, sterol regulatory element-binding protein-1c, and peroxisome proliferator-activated receptor-γ in both cell types.[358] The extract increased the phosphorylation of adenosine monophosphate-activated protein kinase and acetyl-CoA carboxylase and increased the phosphorylation of p38 mitogen-activated protein kinase in both cell types.[358] The extract given to C57BL/6J mice at a dose of 200 mg/kg/day for 8 weeks had no effect on food intake and lowered weight gain by 60%, lowered plasma cholesterol, low-density lipoprotein–cholesterol and triglycerides by 19%, 55%, and 24%, respectively, without effect on high-density lipoprotein–cholesterol.[358] This regimen decreased subcutaneous fat mass by 40%. In white adipose tissues, the extract activated adenosine monophosphate-activated protein kinase, lowered the expression of acetyl-CoA carboxylase, fatty acid synthetase, sterol regulatory element-binding protein-1c, and peroxisome proliferator-activated receptor-γ. Conversely, the extract intake increased the expression of lipolytic peroxisome proliferator-activated receptor-α, peroxisome proliferator-activated receptor coactivator-1α, and uncoupling protein-2.[358] In the liver of treated rodent, the expression of acetyl-CoA carboxylase, fatty acid synthetase, and sterol regulatory element-binding protein-1c were reduced, uncoupling protein-2 was increased, and phosphorylation of adenosine monophosphate-activated protein kinase was observed.[358]

4.133 *Curcuma longa* L.

Synonym: *Curcuma domestica* Valeton
Common names: jiang huang (Chinese); dilau (Philippines); turmeric
Subclass Commelinidae, Superorder Zingiberanae, Order Zingiberales, Family Zingiberaceae
Medicinal use: diabetes (Philippines)

Ethanol extract of rhizome of *Curcuma longa* L. given to genetically diabetic KK-Ay mice at 1g/100g of diet for 4 weeks had no effect on food intake or body weight, lowered plasma glucose from about 22.5 to 12.5 mM.[359] This extract increased glycerol secretion by dexamethasone-insulin-3-isobutyl-1-methylxanthine-induced differentiating human preadipocytes at 20 μg/mL.[359] In transfected CV-1 cells, the extract at 10 μg/mL elicited a binding activity with peroxisome proliferator-activated receptor-γ which was more potent than troglitazone at 0.4 μg/mL.[359] From this extract, Ar-turmerone, curcumin, demethoxycurcumin and bisdemethoxycurcumin at 5 mg/mL elicited peroxisome proliferator-activated receptor-γ ligand binding activity in transfected CV-1 cells.[359]

4.134 *Kaempferia parviflora* Wall. ex Baker

Synonyms: *Kaempferia rubromarginata* (S.Q. Tong) R. J. Searle; *Stahlianthus rubromarginatus* S.Q. Tong
Common names: kra-chai dam (Thailand); Thai ginseng
Subclass Commelinidae, Superorder Zingiberanae, Order Zingiberales, Family Zingiberaceae
Medicinal use: gastric ulcer (Thailand)

3,5,7-Trimethoxyflavone, 3,5,7,4′-tretramethoxyflavone and 3,5,7,3′,4′-pentamethoxyflavone isolated from the rhizomes of *Kaempferia parviflora* Wall. ex Baker at a concentration of 30 μM promoted the accumulation of triglycerides in differentiating 3T3-L1 preadipocytes.[360] 3,5,7,4′-Tretramethoxyflavone and 3,5,7,3′,4′-pentamethoxyflavone lowered the expression of GATA binding protein 2, increased the expression of CCAAT/enhancer-binding protein-β and δ, peroxisome proliferator-activated receptor-γ and target adiponectin.[360] In adipocytes, of GATA binding protein 2 and 3 inhibits peroxisome proliferator-activated receptor-γ and CCAAT/enhancer-binding protein-α.[82]

4.135 *Zingiber officinale* Roscoe

Synonyms: *Amomum zingiber* L.; *Zingiber aromaticum* Noronha; *Zingiber sichuanense* Z.Y. Zhu, S.L. Zhang & S.X. Chen
Common names: jiang (Chinese); aderuck (India); halia (Malay); ginger
Subclass Commelinidae, Superorder Zingiberanae, Order Zingiberales, Family Zingiberaceae
Medicinal use: indigestion (India)
History: the plant was known to Dioscorides
Pharmacological target: atherogenic hyperlipidemia, insulin resistance

8-Gingerol and 6-gingerol isolated from the rhizome of *Zingiber officinale* Roscoe at a concentration of 40 and 160 μM increased uptake of glucose by L6 myoblasts in absence of insulin *in vitro*.[361] Gingerol at 40 μM evoked a mild increased expression of glucose transporter-4 and increased cell surface distribution in L6 myoblasts.[361] Methanol extracts of *Zingiber officinale* Roscoe inhibited the enzymatic activity of protein-tyrosine phosphatase 1B with IC_{50} values of 14 μg/mL. From this extract, (5*R*)-2,6,9-humulatrien-5-ol-8-one, kaempferol-3,4′-di-*O*-methyl ether and 6-gingerol protein-tyrosine phosphatase-1B with IC_{50} values below 30 μM, respectively.[362] 6-Gingerol at a concentration of 150 μM increased glucose uptake by L6 rat myoblast on account of increased cytosolic contents in calcium and subsequent activation of adenosine monophosphate-activated protein kinase and phosphorylation of acetyl-CoA carboxylase.[363] This enzyme catalyzes the carboxylation of acetyl-CoA to form malonyl-CoA, the accumulation of which inhibits carnitine palmitoyltransferase-1 and related transfer long-chain fatty acyl-CoA from cytosol into mitochondria for β-oxidation hence fatty acid-induced insulin resistance.[363] Phosphorylation of Ser79 in acetyl-CoA carboxylase will inactivate the enzyme, leading to a switch of the cellular metabolism from energy storage to expenditure.[363] Ethanol extract of rhizome of *Zingiber officinale* Roscoe given at a dose of 200 mg/kg/day to Sprague–Dawley rats on high-fructose diet for 10 weeks had no effect on food intake, attenuated mildly body weight gain, and lowered plasma glucose from 4.2 to 2.9 mmol/L (normal: 3.9 mmol/L; metformin at 200 mg/kg/day: 4 mmol/L).[364] This extract lowered plasma insulin from 0.7 to 0.3 ng/mL (normal: 0.1 ng/mL; metformin at 200 mg/kg/day: 0.1 ng/mL) and decreased insulin resistance as efficiently as metformin).[364] In oral glucose tolerance performed at the end of the treatment, the extract lowered plasma glucose area under the curve in oral glucose tolerance test more efficiently than metformin.[364] In skeletal muscles of treated rodents, the extract increased the expression of phosphorylated adenosine monophosphate-activated protein kinase by 46%.[364] In L6 rat myoblast cells, 6-gingerol at concentration of 150 μM increased the phosphorylation of adenosine monophosphate-activated protein kinase and peroxisome proliferator-activated receptor coactivator-1α expression.[364] Ethanol extract of rhizomes of ginger given to C57BL/6J mice on high-fat diet at 0.3% of diet for 18 weeks had no effect on food intake, decreased body and adipose tissue weight.[365] This supplementation lowered plasma leptin and cholesterol, it had no effect on plasma glucose, decreased insulinemia from 0.4 to 0.1 pg/mL and decreased insulin resistance.[365] From this extract, 6-gingerol and 6-shoagol at 2 μM increased peroxisome proliferator-activated receptor-δ in cotransfected HEK293 cells and had no effect on peroxisome proliferator-activated receptor-α and receptor-γ.[365] In human myoblast cells, the extract at 0.002% increased the expression of carnitine palmitoyltransferase-1 and PDK4.[365] 6-Gingerol at a concentration of 15 μg/mL inhibited the differentiation of 3T3-L1 preadipocytes (induced by dexamethasone-insulin-3-isobutyl-1-methylxanthine) differentiation as evidenced by a decrease in triglycerides accumulation by 32.5%.[366] This effect was accompanied with a decreased expression of peroxisome proliferator-activated receptor-γ, CCAAT/enhancer-binding protein-α, fatty acid synthetase and fatty acid-binding protein.[366] 6-Gingerol also decreased the expression of phosphorylated Akt and phosphorylated glycogen synthase kinase-3.[366]

4.136 *Cyperus rotundus* L.

Synonym: *Chlorocyperus rotundus* (L.) Palla
Common names: xiang fu zi (Chinese); musta (India); nut-grass
Subclass Commelinidae, Superorder Juncanae, Order Juncales, Family Cyperaceae
Medicinal use: diuretic (India)

Ethanol extract of rhizomes of *Cyperus rotundus* L. given orally to alloxan-induced diabetic Sprague–Dawley rats at a dose of 500 mg/kg/day for 7 days decreased glycemia from about 190 to 100 mg/dL and this effect was comparable with metformin (given at a dose of 450 mg/kg/day).[367] Hexane extract of rhizomes added to diet at 0.3% and given to Zucker rats evoked a mild reduction of body weight gain, had no effect on food intake, lowered plasma insulin from 162.2 to 140.9 μU/mL, had no effect on plasma glucose, increased triglycerides, increased alkaline phosphatase, and had no effect on blood pressure.[368] *In vitro*, the extract inhibited the binding of cyanopindolol to β3-adrenergic receptor by 47.6% at a concentration of 50 μg/mL. At 250 μg/mL the extract induced lipolysis in 3T3-F442 A cells.[368]

APPENDIX

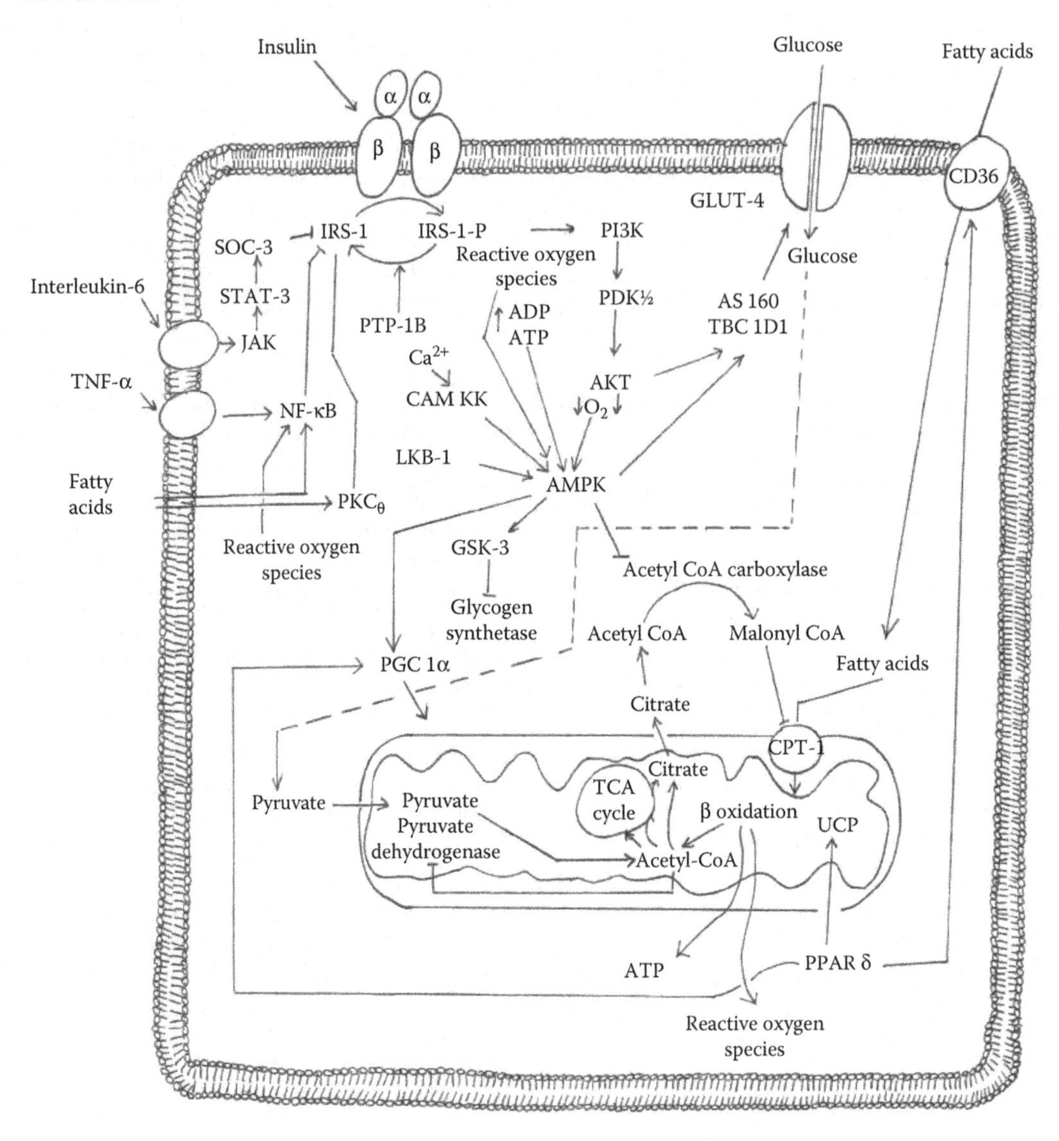

APPENDIX 4.1 Skeletal muscles metabolism.

Akt: protein kinase B; AMPK: adenosine monophosphate-activated protein kinase; AS160: Akt substrate of 160 kDa (TBC1D); CAMKK: Calmodulin-dependent protein kinase; CD36: Fatty Acid Translocase (FAT/CD36); CPT-1: carnitine palmitoyl transferase-I; GLUT-4: glucose transporter-4; GSK-3: Glycogen synthase kinase-3; IRS-1: Insulin receptor substrate-1; IRS: Insulin receptor substrate-1; JAK: c-jun NH2 terminal *kinase*; LKB1: liver kinase B1; mTOR: mechanistic target of rapamycin; NF-κB: Nuclear factor κB; PDK1/2: phosphionsitide-dependent kinase 1/2; PGC-1*a*: peroxisome proliferator-activated receptor-gamma coactivator-1 alpha; PTP-1B: Protein-tyrosine phosphatase-1B PI3K: phosphatidylinositol 3-kinase; PDK1/2: phosphoinositide-dependent kinase 1; $PKC_θ$: protein kinase θ; UCP: uncoupling protein; $PKC_θ$: protein kinase Cq; S6K1: ribosomal S6 kinase 1; SOCS-3: suppressor of cytokine signaling 3; STAT-3: Signal transducer and activator of transcription-3; TNF-α: tumor necrosis-α.

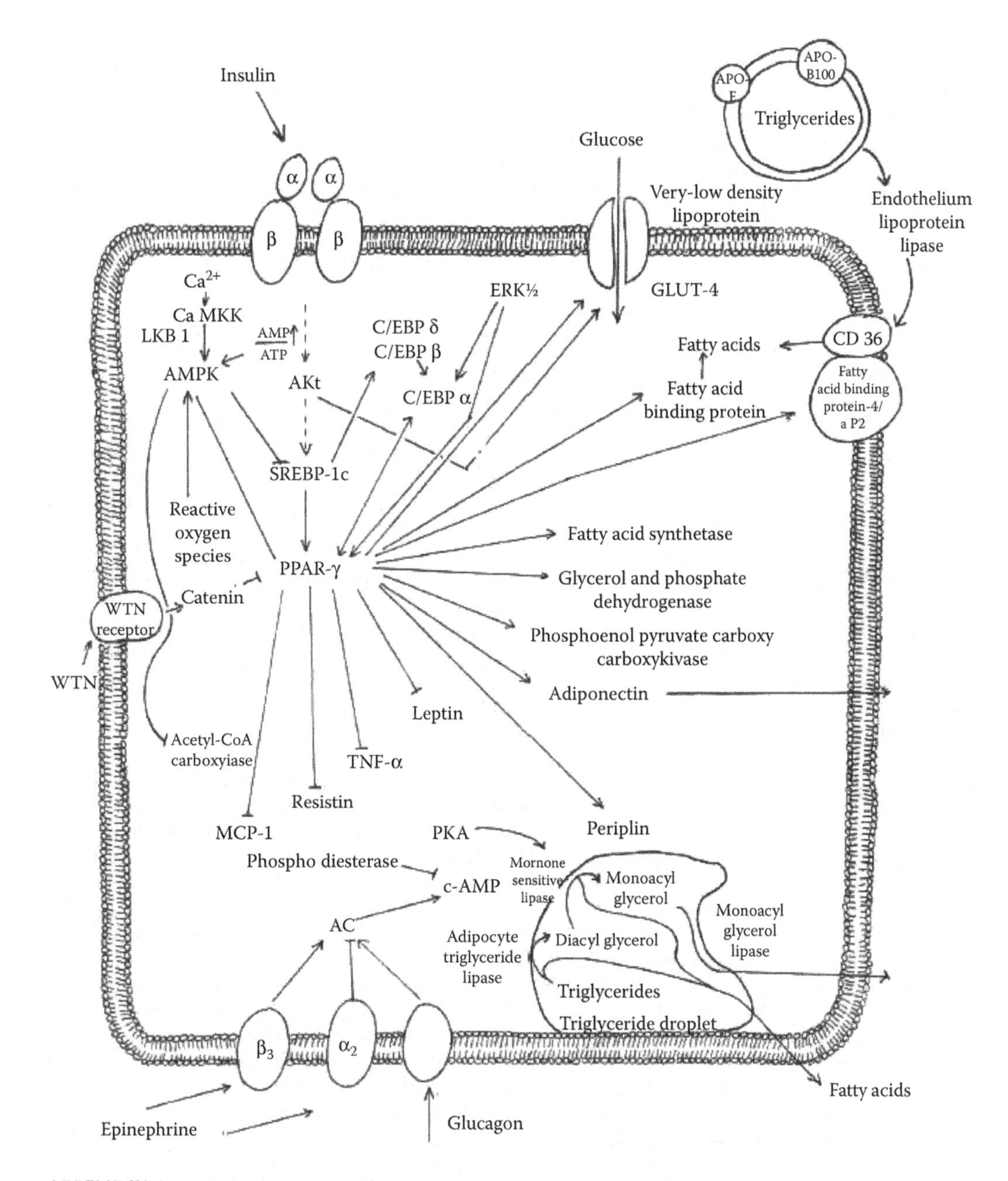

APPENDIX 4.2 Adipocyte metabolism.

α2: alpha-2 receptor; Akt: protein kinase B; AC: Adenylate cyclase; AMPK: adenosine monophosphate-activated protein kinase; ApoB100: Apolipoprotein-B100; ApoE: Apolipoprotein-E; β1-3: beta receptors 1-3; C/EBP α: CCAAT/enhancer-binding protein-alpha; C/EBP β: CCAAT/enhancer-binding protein-beta; C/EBP δ: CCAAT/enhancer-binding protein-delta; CAMKK: Calmodulin-dependent protein kinase; CD36: Fatty Acid Translocase (FAT/CD36); ERK1/2: Extracellular-signal regulated kinase ; GLUT-4: glucose transporter-4; MCP-1: Monocyte chemoattractant protein-1; PPAR-γ: Peroxisome proliferator-activated receptor-gamma coactivator-1 alpha; PKA: Protein kinase A; PKC_{θ}: protein kinase θ; SREBP-1c: sterol regulatory element-binding protein-1c; TNF-α: Tumor necrosis factor-α; UCP: uncoupling protein.

REFERENCES

1. Eckel, R.H., Grundy, S.M. and Zimmet, P.Z., 2005. The metabolic syndrome. *The Lancet*, 365(9468), 1415–1428.
2. WHO, 2006. Definition and diagnostic of diabetes mellitus and intermediate hyperglycemia: Report of a WHO/IDF consultation. Geneva, Switzerland: WHO.
3. Hotamisligil, G.S., Shargill, N.S. and Spiegelman, B.M., 1993. Adipose expression of tumor necrosis factor-alpha: Direct role in obesity-linked insulin resistance. *Science*, 259, 87.
4. Zimmermann, R., Strauss, J.G., Haemmerle, G., Schoiswohl, G., Birner-Gruenberger, R., Riederer, M. et al., 2004. Fat mobilization in adipose tissue is promoted by adipose triglyceride lipase. *Science*, 306(5700), 1383–1386.
5. Karpe, F., Dickmann, J.R. and Frayn, K.N., 2011. Fatty acids, obesity, and insulin resistance: Time for a reevaluation. *Diabetes*, 60(10), 2441–2449.
6. Shulman, G.I., 2000. Cellular mechanisms of insulin resistance. *The Journal of Clinical Investigation*, 106(2), 171–176.
7. Keay, S. and Grossberg, S.E., 1980. Interferon inhibits the conversion of 3T3-L1 mouse fibroblasts into adipocytes. *Proceedings of the National Academy of Sciences*, 77(7), 4099–4103.
8. Hsieh, Y.H. and Wang, S.Y., 2013. Lucidone from *Lindera erythrocarpa* Makino fruits suppresses adipogenesis in 3T3-L1 cells and attenuates obesity and consequent metabolic disorders in high-fat diet C57BL/6 mice. *Phytomedicine*, 20(5), 394–400.
9. Rosen, E.D., Hsu, C.H., Wang, X., Sakai, S., Freeman, M.W., Gonzalez, F.J. and Spiegelman, B.M., 2002. C/EBPalpha induces adipogenesis through PPAR-gamma: A unified pathway. *Genes & Development*, 16(1), 22–26.
10. Keller, H., Dreyer, C., Medin, J., Mahfoudi, A., Ozato, K. and Wahli, W., 1993. Fatty acids and retinoids control lipid metabolism through activation of peroxisome proliferator-activated receptor-retinoid X receptor heterodimers. *Proceedings of the National Academy of Sciences*, 90(6), 2160–2164.
11. Kadowaki, T., Yamauchi, T., Kubota, N., Hara, K., Ueki, K. and Tobe, K., 2006. Adiponectin and adiponectin receptors in insulin resistance, diabetes, and the metabolic syndrome. *The Journal of Clinical Investigation*, 116(7), 1784–1792.
12. Cao, Z., Umek, R.M. and McKnight, S.L., 1991. Regulated expression of three C/EBP isoforms during adipose conversion of 3T3-L1 cells. *Genes & Development*, 5(9), 1538–1552.
13. Freise, C., Trowitzsch-Kienast, W., Erben, U., Seehofer, D., Kim, K.,Y., Zeitz, M., Ruehl, M. and Somasundaram, R., 2013. (+)-Episesamin inhibits adipogenesis and exerts anti-inflammatory effects in 3T3-L1 (pre)adipocytes by sustained Wnt signaling, down-regulation of PPARγ and induction of iNOS. *Journal of Nutrition Biochemistry*, 24(3), 550–555.
14. Browning, J.D. and Horton, J.D., 2004. Molecular mediators of hepatic steatosis and liver injury. *The Journal of Clinical Investigation*, 114(2), 147–152.
15. Seung, W.L., Rho, M.C., Hye, R.P., Choi, J.H., Ji, Y.K., Jung, W.L., Kim, K., Hyun, S.L. and Young, K.K., 2006. Inhibition of diacylglycerol acyltransferase by alkamides isolated from the fruits of *Piper longum* and *Piper nigrum*. *Journal of Agricultural and Food Chemistry*, 54(26), 9759–9763.
16. Kim, K.J., Lee, M.S., Jo, K. and Hwang, J.K., 2011. Piperidine alkaloids from *Piper retrofractum* Vahl. protect against high-fat diet-induced obesity by regulating lipid metabolism and activating AMP-activated protein kinase. *Biochemical and Biophysical Research Communications*, 411(1), 219–225.
17. Park, U.H., Jeong, H.S., Jo, E.Y., Park, T., Yoon, S.K., Kim, E.J., Jeong, J.C. and Um, S.J., 2012. Piperine, a component of black pepper, inhibits adipogenesis by antagonizing PPARγ activity in 3T3-L1 cells. *Journal of Agricultural and Food Chemistry*, 60(15), 3853–3860.
18. Ono, Y., Hattori, E., Fukaya, Y., Imai, S. and Ohizumi, Y., 2006. Anti-obesity effect of *Nelumbo nucifera* leaves extract in mice and rats. *Journal of Ethnopharmacology*, 106(2), 238–244.
19. Yu, B., Cook, C. and Santanam, N., 2009. The aporphine alkaloid boldine induces adiponectin expression and regulation in 3T3-L1 cells. *Journal of Medicinal Food*, 12(5), 1074–1083.
20. Rosen, E.D., Hsu, C.H., Wang, X., Sakai, S., Freeman, M.W., Gonzalez, F.J. and Spiegelman, B.M., 2002. C/EBPα induces adipogenesis through PPARγ: a unified pathway. *Genes & Development*, 16(1), 22–26.
21. Ahn, J.H., Kim, E.S., Lee, C., Kim, S., Cho, S.H., Hwang, B.Y. and Lee, M.K., 2013. Chemical constituents from *Nelumbo nucifera* leaves and their anti-obesity effects. *Bioorganic & Medicinal Chemistry Letters*, 23(12), 3604–3608.
22. Punitha, I.S.R., Shirwaikar, A. and Shirwaikar, A., 2006. Antidiabetic activity of benzyl tetra isoquinoline alkaloid berberine in streptozotocin-nicotinamide induced type 2 diabetic rats. *Diabetologia Croatica*, 34(4), 117–128.

23. Turner, N., Li, J.Y., Gosby, A., To, S.W., Cheng, Z., Miyoshi, H., Taketo, M.M. et al., 2008. Berberine and its more biologically available derivative, dihydroberberine, inhibit mitochondrial respiratory complex I: A mechanism for the action of berberine to activate AMP-activated protein kinase and improve insulin action. *Diabetes*, 57(5), 1414–1418.
24. Chen, Y., Li, Y., Wang, Y., Wen, Y. and Sun, C., 2009. Berberine improves free-fatty-acid-induced insulin resistance in L6 myotubes through inhibiting peroxisome proliferator-activated receptor gamma and fatty acid transferase expressions. *Metabolism*, 58(12), 1694–1702.
25. Chen, C., Zhang, Y. and Huang, C., 2010. Berberine inhibits PTP1B activity and mimics insulin action. *Biochemical and Biophysical Research Communications*, 397(3), 543–547.
26. Yang, T.C., Chao, H.F., Shi, L.S., Chang, T.C., Lin, H.C. and Chang, W.L., 2014. Alkaloids from *Coptis chinensis* root promote glucose uptake in C2C12 myotubes. *Fitoterapia*, 93, 239–244.
27. Yin, J., Gao, Z., Liu, D., Liu, Z. and Ye, J., 2008. Berberine improves glucose metabolism through induction of glycolysis. *American Journal of Physiology - Endocrinology and Metabolism*, 294(1), E148–E156.
28. Zhou, J. and Zhou, S., 2010. Berberine regulates peroxisome proliferator-activated receptors and positive transcription elongation factor b expression in diabetic adipocytes. *European Journal of Pharmacology*, 649(1–3), 390–397.
29. Hu, Y. and Davies, G.E., 2009. Berberine increases expression of GATA-2 and GATA-3 during inhibition of adipocyte differentiation. *Phytomedicine*, 16(9), 864–873.
30. Cok, A., Plaisier, C., Salie, M.J., Oram, D.S., Chenge, J. and Louters, L.L., 2011. Berberine acutely activates the glucose transport activity of GLUT1. *Biochimie*, 93(7), 1187–1192.
31. Lu, H.L., Wang, H.W., Wen, Y., Zhang, M.X. and Lin, H.H., 2006. Roles of adipocyte derived hormone adiponectin and resistin in insulin resistance of type 2 diabetes. *World Journal of Gastroenterology*, 12(11), 1747.
32. Maeda, N., Takahashi, M., Funahashi, T., Kihara, S., Nishizawa, H., Kishida, K., Nagaretani, H. et al., 2001. PPARγ ligands increase expression and plasma concentrations of adiponectin, an adipose-derived protein. *Diabetes*, 50(9), 2094–2099.
33. Zhou, L., Wang, X., Yang, Y., Wu, L., Li, F., Zhang, R., Yuan, G., Wang, N., Chen, M. and Ning, G., 2011. Berberine attenuates cAMP-induced lipolysis via reducing the inhibition of phosphodiesterase in 3T3-L1 adipocytes. *Biochimica et Biophysica Acta - Molecular Basis of Disease*, 1812(4), 527–535.
34. Li, Y., Wang, P., Zhuang, Y., Lin, H., Li, Y., Liu, L., Meng, Q., Cui, T., Liu, J. and Li, Z., 2011. Activation of AMPK by berberine promotes adiponectin multimerization in 3T3-L1 adipocytes. *FEBS Letters*, 585(12), 1735–1740.
35. Klip, A. and Pâquet, M.R., 1990. Glucose transport and glucose transporters in muscle and their metabolic regulation. *Diabetes Care*, 13(3), 228–243.
36. Holman, G.D. and Kasuga, M., 1997. From receptor to transporter: Insulin signalling to glucose transport. *Diabetologia*, 40(9), 991–1003.
37. Sangeetha, M.K., Priya, C.D. and Vasanthi, H.R., 2013. Anti-diabetic property of Tinospora cordifolia and its active compound is mediated through the expression of Glut-4 in L6 myotubes. *Phytomedicine*, 20(3–4), 246–248.
38. Hevener, A.L., He, W., Barak, Y., Le, J., Bandyopadhyay, G., Olson, P., Wilkes, J., Evans, R.M. and Olefsky, J., 2003. Muscle-specific Pparg deletion causes insulin resistance. *Nature Medicine*, 9(12), 1491–1497.
39. Hosseini, S., Jamshidi, L., Mehrzadi, S., Mohammad, K., Najmizadeh, A.R., Alimoradi, H. and Huseini, H.F., 2014. Effects of Juglans regia L. leaf extract on hyperglycemia and lipid profiles in type two diabetic patients: A randomized double-blind, placebo-controlled clinical trial. *Journal of Ethnopharmacology*, 152(3), 451–456.
40. Pitschmann, A., Zehl, M., Atanasov, A.G., Dirsch, V.M., Heiss, E. and Glasl, S., 2014. Walnut leaf extract inhibits PTP1B and enhances glucose-uptake in vitro. *Journal of Ethnopharmacology*, 152(3), 599–602.
41. Hundal, H.S., Ramlal, T., Reyes, R., Leiter, L.A. and Klip, A., 1992. Cellular mechanism of metformin action involves glucose transported translocation from an intracellular pool to the plasma membrane in L6 muscle cells. *Endocrinology*, 131, 1162–1173.
42. Hundal, R.S. and Inzucchi, S.E., 2003. Metformin: New understandings, new uses. *Drugs*, 63(18), 1879–1894.
43. Vetrichelvan, T. and Jegadeesan, M., 2002. Anti-diabetic activity of alcoholic extract of *Aerva lanata* (L.) Juss. ex Schultes in rats. *Journal of Ethnopharmacology*, 80(2–3), 103–107.
44. Agrawal, R., Sethiya, N.K. and Mishra, S.H., 2013. Antidiabetic activity of alkaloids of Aerva lanata roots on streptozotocin-nicotinamide induced type-II diabetes in rats. *Pharmaceutical Biology*, 51(5), 635–642.

45. Riya, M.P., Antu, K.A., Pal, S., Chandrakanth, K.C., Anilkumar, K.S., Tamrakar, A.K., Srivastava, A.K. and Raghu, K.G., 2015. Antidiabetic property of Aerva lanata (L.) Juss. ex Schult. is mediated by inhibition of alpha glucosidase, protein glycation and stimulation of adipogenesis. *Journal of Diabetes*, 7(4), 548–561.
46. Sasaki, T., Li, W., Higai, K. and Koike, K., 2015. Canthinone alkaloids are novel protein tyrosine phosphatase 1B inhibitors. *Bioorganic & Medicinal Chemistry Letters*, 25(9), 1979–1981.
47. Deutschländer, M.S., Lall, N., Van de Venter, M. and Hussein, A.A., 2011. Hypoglycemic evaluation of a new triterpene and other compounds isolated from *Euclea undulata* Thunb. var. myrtina (Ebenaceae) root bark. *Journal of Ethnopharmacology*, 133(3), 1091–1095.
48. Lee, S.H., Kim, B., Oh, M.J., Yoon, J., Kim, H.Y., Lee, K.J., Lee, J.D. and Choi, K.Y., 2011. *Persicaria hydropiper* (L.) spach and its flavonoid components, isoquercitrin and isorhamnetin, activate the Wnt/β-catenin pathway and inhibit adipocyte differentiation of 3T3-L1 cells. *Phytotherapy Research*, 25(11), 1629–1635.
49. Kim, Y.S., Lee, Y.M., Kim, J.H. and Kim, J.S., 2013. *Polygonum cuspidatum* inhibits pancreatic lipase activity and adipogenesis via attenuation of lipid accumulation. *BMC Complementary and Alternative Medicine*, 13, 282.
50. Dalen, K.T., Schoonjans, K., Ulven, S.M., Weedon-Fekjaer, M.S., Bentzen, T.G., Koutnikova, H., Auwerx, J. and Nebb, H.I., 2004. Adipose tissue expression of the lipid droplet–associating proteins S3-12 and perilipin is controlled by peroxisome proliferator–Activated receptor-γ. *Diabetes*, 53(5), 1243–1252.
51. Choi, S.Z., Lee, S.O., Jang, K.U., Chung, S.H., Park, S.H., Kang, H.C., Yang, E.Y., Cho, H.J. and Lee, K.R., 2005. Antidiabetic stilbene and anthraquinone derivatives from *Rheum undulatum. Archives of Pharmacal Research*, 28(9), 1027–1030.
52. Tzeng, T.F., Lu, H.J., Liou, S.S., Chang, C.J. and Liu, I.M., 2012. Emodin protects against high-fat diet-induced obesity via regulation of AMP-activated protein kinase pathways in white adipose tissue. *Planta Medica*, 78(10), 943–950.
53. Lee, W., Yoon, G., Hwang, Y.R., Kim, Y.K. and Kim, S.N., 2012. Anti-obesity and hypolipidemic effects of *Rheum undulatum* in high-fat diet-fed C57BL/6 mice through protein tyrosine phosphatase 1B inhibition. *BMB Reports*, 45(3), 141–146.
54. Umar, A., Ahmed, Q.U., Muhammad, B.Y., Dogarai, B.B. and Soad, S.Z., 2010. Anti-hyperglycemic activity of the leaves of *Tetracera scandens* Linn. Merr. (Dilleniaceae) in alloxan induced diabetic rats. *Journal of Ethnopharmacology*, 131(1), 140–145.
55. Hayashi, S. and Sakaguchi, T., 1976. Effects of epinephrine and theophylline on lipolytic response in hereditary diabetic mice. *Tohoku Journal of Experimental Medicine*, 119(1), 91–100.
56. Alhomida, A.S., 2001. Evaluation of theophylline-stimulated changes in carnitine palmitoyltransferase activity in skeletal muscle and liver of rats. *Journal of Enzyme Inhibition*, 16(2), 177–183.
57. Uddin, M.N., Sharma, G., Yang, J.L., Choi, H.S., Lim, S.I., Kang, K.W. and Oh, W.K., 2014. *Oleanane triterpenes* as protein tyrosine phosphatase 1B (PTP1B) inhibitors from *Camellia japonica. Phytochemistry*, 103, 99–106.
58. Xu, Y., Zhang, M., Wu, T., Dai, S., Xu, J. and Zhou, Z., 2015. The anti-obesity effect of green tea polysaccharides, polyphenols and caffeine in rats fed with a high-fat diet. *Food & Function*, 6(1), 297–304.
59. Knight, S.F. and Imig, J.D., 2007. Obesity, insulin resistance, and renal function. *Microcirculation*, 14(4–5), 349–362.
60. Singer, G. and Granger, N., 2007. Inflammatory responses underlying the microvascular dysfunction associated with obesity and insulin resistance. *Microcirculation*, 14(4–5), 375–387.
61. Maffei, M., Halaas, J., Ravussin, E., Pratley, R.E., Lee, G.H., Zhang, Y., Fei, H. et al., 1995. Leptin levels in human and rodent: measurement of plasma leptin and ob RNA in obese and weight-reduced subjects. *Nature Medicine*, 1(11), 1155–1161.
62. Lee, Y.M., Kim, Y.S., Lee, Y., Kim, J., Sun, H., Kim, J.H. and Kim, J.S., 2012. Inhibitory activities of pancreatic lipase and phosphodiesterase from Korean medicinal plant extracts. *Phytotherapy Research*, 26(5), 778–782.
63. Bnouham, M., Merhfour, F.Z., Ziyyat, A., Aziz, M., Legssyer, A. and Mekhfi, H., 2010. Antidiabetic effect of some medicinal plants of Oriental Morocco in neonatal non-insulin-dependent diabetes mellitus rats. *Human & Experimental Toxicology*, 29(10), 865–871.
64. Choi, Y.H., Zhou, W., Oh, J., Choe, S., Kim, D.W., Lee, S.H. and Na, M., 2012. Rhododendric acid A, a new ursane-type PTP1B inhibitor from the endangered plant *Rhododendron brachycarpum* G. Don. *Bioorganic & Medicinal Chemistry Letters*, 22(19), 6116–6119.

65. Kwon, J.H., Chang, M.J., Seo, H.W., Lee, J.H., Min, B.S., Na, M., Kim, J.C. et al., 2008. Triterpenoids and a sterol from the stem-bark of *Styrax japonica* and their protein tyrosine phosphatase 1B inhibitory activities. *Phytotherapy Research*, 22(10), 1303–1306.
66. Li, Y.F., Li, J., Shen, Q. and Hu, L.H., 2007. Benzoquinones from *Ardisia japonica* with inhibitory activity towards human protein tyrosine phosphatase 1B (PTP1B). *Chemistry & Biodiversity*, 4(5), 961–965.
67. Singh, D., Singh, R., Singh, P. and Gupta, R.S., 2009. Effects of embelin on lipid peroxidation and free radical scavenging activity against liver damage in rats. *Basic & Clinical Pharmacology & Toxicology*, 105(4), 243–248.
68. Naik, S.R., Niture, N.T., Ansari, A.A. and Shah, P.D., 2013. Anti-diabetic activity of embelin: involvement of cellular inflammatory mediators, oxidative stress and other biomarkers. *Phytomedicine*, 20(10), 797–804.
69. Gandhi, G.R., Stalin, A., Balakrishna, K., Ignacimuthu, S., Paulraj, M.G. and Vishal, R., 2013. Insulin sensitization via partial agonism of PPARγ and glucose uptake through translocation and activation of GLUT4 in PI3K/p-Akt signaling pathway by embelin in type 2 diabetic rats. *Biochimica et Biophysica Acta*, 1830(1), 2243–2255.
70. DeFronzo, R.A., Tripathy, D., Schwenke, D.C., Banerji, M., Bray, G.A., Buchanan, T.A., Clement, S.C. et al., 2011. Pioglitazone for diabetes prevention in impaired glucose tolerance. *New England Journal of Medicine*, 364(12), 1104–1115.
71. Saltiel, A.R. and Olefsky, J.M., 1996. Thiazolidinediones in the treatment of insulin resistance and type II diabetes. *Diabetes*, 45(12), 1661–1669.
72. Singh, G., Gupta, P., Rawat, P., Puri, A., Bhatia, G. and Maurya, R., 2007. Antidyslipidemic activity of polyprenol from *Coccinia grandis* in high-fat diet-fed hamster model. *Phytomedicine*, 14(12), 792–798.
73. Bunkrongcheap, R., Hutadilok-Towatana, N., Noipha, K., Wattanapiromsakul, C., Inafuku, M. and Oku, H., 2014. Ivy gourd (*Coccinia grandis* L. Voigt) root suppresses adipocyte differentiation in 3T3-L1 cells. *Lipids in Health and Disease*, 13, 88.
74. Spiegelman, B.M., 1998. PPAR-gamma: Adipogenic regulator and thiazolidinedione receptor. *Diabetes*, 47(4), 507–514.
75. Fortis-Barrera, Á., Alarcón-Aguilar, F.J., Banderas-Dorantes, T., Díaz-Flores, M., Román-Ramos, R., Cruz, M. and García-Macedo, R., 2013. *Cucurbita ficifolia* Bouché (Cucurbitaceae) and D-chiro-inositol modulate the redox state and inflammation in 3T3-L1 adipocytes. *Journal of Pharmacy and Pharmacology*, 65(10), 1563–1576.
76. Lee, J., Kim, D., Choi, J., Choi, H., Ryu, J.H., Jeong, J., Park, E.J., Kim, S.H., and Kim, S., 2012. Dehydrodiconiferyl alcohol isolated from *Cucurbita moschata* shows anti-adipogenic and anti-lipogenic effects in 3T3-L1 cells and primary mouse embryonic fibroblasts. *Journal of Biological Chemistry*, 287(12), 8839–8851.
77. Tontonoz, P., Hu, E. and Spiegelman, B.M., 1995. Regulation of adipocyte gene expression and differentiation by peroxisome proliferator activated receptor γ. *Current Opinion in Genetics & Development*, 5(5), 571–576.
78. Schoonjans, K., Staels, B. and Auwerx, J., 1996. Role of the peroxisome proliferator-activated receptor (PPAR) in mediating the effects of fibrates and fatty acids on gene expression. *Journal of Lipid Research*, 37(5), 907–925.
79. Brown, J.M. and McIntosh, M.K., 2003. Conjugated linoleic acid in humans: Regulation of adiposity and insulin sensitivity. *The Journal of Nutrition*, 133(10), 3041–3046.
80. Seo, C.R., Yang, D.K., Song, N.J., Yun, U.J., Gwon, A.R., Jo, D.G., Cho, J.Y. et al., 2013. Cucurbitacin B and cucurbitacin I suppress adipocyte differentiation through inhibition of STAT3 signaling. *Food and Chemical Toxicology*, 64, 217–224.
81. MacDougald, O.A. and Mandrup, S., 2002. Adipogenesis: Forces that tip the scales. *Trends in Endocrinology & Metabolism*, 13(1), 5–11.
82. Rosen, E.D. and MacDougald, O.A., 2006. Adipocyte differentiation from the inside out. *Nature Reviews Molecular Cell Biology*, 7(12), 885–896.
83. Shih, C.C., Lin, C.H., Lin, W.L. and Wu, J.B., 2009. *Momordica charantia* extract on insulin resistance and the skeletal muscle GLUT4 protein in fructose-fed rats. *Journal of Ethnopharmacology*, 123(1), 82–90.
84. Hotta, K., Funahashi, T., Bodkin, N.L., Ortmeyer, H.K., Arita, Y., Hansen, B.C. and Matsuzawa, Y., 2001. Circulating concentrations of the adipocyte protein adiponectin are decreased in parallel with reduced insulin sensitivity during the progression to type 2 diabetes in rhesus monkeys. *Diabetes*, 50(5), 1126–1133.

85. Tomas, E., Kelly, M., Xiang, X., Tsao, T.S., Keller, C., Keller, P., Luo, Z. et al., 2004. Metabolic and hormonal interactions between muscle and adipose tissue. Proceedings of the Nutrition Society, *63*(2), 381–385.
86. Yamauchi, T., Kamon, J., Minokoshi, Y.A., Ito, Y., Waki, H., Uchida, S., Yamashita, S. et al., 2002. Adiponectin stimulates glucose utilization and fatty-acid oxidation by activating AMP-activated protein kinase. *Nature Medicine*, 8(11), 1288–1295.
87. Purohit, A. and Vyas, K.B., 2006. Hypolipidaemic efficacy of Capparis decidua fruit and shoot extracts in cholesterol fed rabbits. *Indian Journal of Experimental Biology*, 43(10), 863–866.
88. Sharma, B., Salunke, R., Balomajumder, C., Daniel, S. and Roy, P., 2010. Anti-diabetic potential of alkaloid rich fraction from *Capparis decidua* on diabetic mice. *Journal of Ethnopharmacology*, 127(2), 457–462.
89. Kanaujia, A., Duggar, R., Pannakal, S.T., Yadav, S.S., Katiyar, C.K., Bansal, V., Anand, S., Sujatha, S. and Lakshmi, B.S., 2010. Insulinomimetic activity of two new gallotannins from the fruits of *Capparis moonii. Bioorganic & Medicinal Chemistry*, 18(11), 3940–3945.
90. Kazemian, M., Abad, M., Haeri, M.R., Ebrahimi, M. and Heidari, R., 2015. Anti-diabetic effect of *Capparis spinosa* L. root extract in diabetic rats. *Avicenna Journal of Phytomedicine*, 5(4), 325–332.
91. Huseini, H.F., Hasani-Rnjbar, S., Nayebi, N., Heshmat, R., Sigaroodi, F.K., Ahvazi, M., Alaei, B.A. and Kianbakht, S., 2013. *Capparis spinosa* L. (Caper) fruit extract in treatment of type 2 diabetic patients: A randomized double-blind placebo-controlled clinical trial. *Complementary Therapies in Medicine*, 21(5), 447–452.
92. Chand, K., Akankshaa, Rahuja, N., Mishra, D.P., Srivastava, A.K. and Maurya, R., 2011. Major alkaloidal constituent from *Impatiens niamniamensis* seeds as antihyperglycemic agent. *Medicinal Chemistry Research*, 20(9), 1505–1508.
93. Irudayaraj, S.S., Stalin, A., Sunil, C., Duraipandiyan, V., Al-Dhabi, N.A. and Ignacimuthu, S., 2016. Antioxidant, antilipidemic and antidiabetic effects of ficusin with their effects on GLUT4 translocation and PPARγ expression in type 2 diabetic rats. *Chemico-Biological Interactions*, 256, 85–93.
94. Hünigsmann, H., 1986. Psoralen photochemotherapy-mechanisms, drugs, toxicity. In *Therapeutic Photomedicine*, Hönigsmann, H. and Stingl, G. (Eds.). Basel, Switzerland: Karger Publishers, pp. 52–66.
95. Ahn, J.H., Liu, Q., Lee, C., Ahn, M.J., Yoo, H.S., Hwang, B.Y. & Lee, M.K., 2012. A new pancreatic lipase inhibitor from *Broussonetia kanzinoki. Bioorganic & Medicinal Chemistry Letters*, 22(8), 2760–2763.
96. Jocken, J.W., Langin, D., Smit, E., Saris, W.H., Valle, C., Hul, G.B., Holm, C., Arner, P. and Blaak, E.E., 2007. Adipose triglyceride lipase and hormone-sensitive lipase protein expression is decreased in the obese insulin-resistant state. *The Journal of Clinical Endocrinology & Metabolism*, 92(6), 2292–2299.
97. Kim, Y.S., Lee, Y.M., Kim, H., Kim, J., Jang, D.S., Kim, J.H. and Kim, J.S., 2010. Anti-obesity effect of *Morus bombycis* root extract: Anti-lipase activity and lipolytic effect. *Journal of Ethnopharmacology*, 130(3), 621–624.
98. Naowaboot, J., Chung, C.H., Pannangpetch, P., Choi, R., Kim, B.H., Lee, M.Y., and Kukongviriyapan, U., 2012. Mulberry leaf extract increases adiponectin in murine 3T3-L1 adipocytes. *Nutrition Research*, 32(1), 39–44.
99. Yang, Z.G., Matsuzaki, K., Takamatsu, S. and Kitanaka, S., 2011. Inhibitory effects of constituents from *Morus alba* var. multicaulis on differentiation of 3T3-L1 cells and nitric oxide production in RAW264.7 cells. *Molecules*, 16(7), 6010–6022.
100. Hu, X., Ji, J., Wang, M., Wu, J.W., Zhao, Q.S., Wang, H.Y. and Hou, A.J., 2011. New isoprenylated flavonoids and adipogenesis-promoting constituents from *Morus notabilis. Bioorganic & Medicinal Chemistry Letters*, 21(15), 4441–4446.
101. Chongsa, W., Radenahmad, N. and Jansakul, C., 2014. Six weeks oral gavage of a Phyllanthus acidus leaf water extract decreased visceral fat, the serum lipid profile and liver lipid accumulation in middle-aged male rats. *Journal of Ethnopharmacology*, 155(1), 396–404.
102. Park, S.J., Park, J.H., Han, A., Davaatseren, M., Kim, H.J., Kim, M.S., Hur, H.J. et al., 2015. Euphorbiasteroid, a component of *Euphorbia lathyris* L., inhibits adipogenesis of 3T3-L1 cells via activation ofAMP-activated protein kinase. *Cell Biochemistry and Function*, 33(4), 220–225.
103. Wagstaff, D.J., 2008. *International Poisonous Plants Checklist: An Evidence-based Reference*. Boca Raton, FL: CRC Press.
104. Jouad, H., Maghrani, M. and Eddouks, M., 2002. Hypoglycaemic effect of *Rubus fructicosis* L. and *Globularia alypum* L. in normal and streptozotocin-induced diabetic rats. *Journal of Ethnopharmacology*, 81(3), 351–356.
105. Kameswara Rao, B., Renuka Sudarshan, P., Rajasekhar, M.D., Nagaraju, N. and Appa Rao, C., 2003. Antidiabetic activity of *Terminalia pallida* fruit in alloxan induced diabetic rats. *Journal of Ethnopharmacology*, 85(1), 169–172.

106. Karam, J.H., 1998. Pancreatic hormones and antidiabetic drugs. In *Basic and Clinical Pharmacology*, Katzung, B.G. (Ed.). New York: Appleton and Lange, p. 684.
107. Shaw, R.J., Lamia, K.A., Vasquez, D., Koo, S.H., Bardeesy, N., DePinho, R.A., Montminy, M. and Cantley, L.C., 2005. The kinase LKB1 mediates glucose homeostasis in liver and therapeutic effects of metformin. *Science*, 310(5754), 1642–1646.
108. Banihani, S.A., Makahleh, S.M., El-Akawi, Z., Al-Fashtaki, R.A., Khabour, O.F., Gharibeh, M.Y., Saadah, N.A., Al-Hashimi, F.H. and Al-Khasieb, N.J., 2014. Fresh pomegranate juice ameliorates insulin resistance, enhances β-cell function, and decreases fasting serum glucose in type 2 diabetic patients. *Nutrition Research*, 34(10), 862–867.
109. Mena, P., García-Viguera, C., Navarro-Rico, J., Moreno, D.A., Bartual, J., Saura, D. and Martí, N., 2011. Phytochemical characterisation for industrial use of pomegranate (*Punica granatum* L.) cultivars grown in Spain. *Journal of the Science of Food and Agriculture*, 91(10), 1893–1906.
110. Mason, S.A., Della Gatta, P.A., Snow, R.J., Russell, A.P., Wadley, G.D., 2016. Ascorbic acid supplementation improves skeletal muscle oxidative stress and insulin sensitivity in people with type 2 diabetes: Findings of a randomized controlled study. *Free Radical Biology and Medicine*, 93, 227–238.
111. Ramachandran, S. and Rajasekaran, A., 2014. Blood glucose-lowering effect of *Tectona grandis* flowers in type 2 diabetic rats: A study on identification of active constituents and mechanisms for antidiabetic action. *Journal of Diabetes*, 6(5), 427–437.
112. Gandhi, G.R., Jothi, G., Antony, P.J., Balakrishna, K., Paulraj, M.G., Ignacimuthu, S., Stalin, A. and Al-Dhabi, N.A., 2014. Gallic acid attenuates high-fat diet fed-streptozotocin-induced insulin resistance via partial agonism of PPARγ in experimental type 2 diabetic rats and enhances glucose uptake through translocation and activation of glucose transporter-4 in PI3K/p-Akt signaling pathway. *European Journal of Pharmacology*, 745, 201–216.
113. Xing, Y., Yan, F., Liu, Y., Liu, Y. and Zhao, Y., 2010. Matrine inhibits 3T3-L1 preadipocyte differentiation associated with suppression of ERK1/2 phosphorylation. *Biochemical and Biophysical Research Communications*, 396(3), 691–695.
114. Bost, F., Aouadi, M., Caron, L., Even, P., Belmonte, N., Prot, M., Dani, C. et al., 2005. The extracellular signal–regulated kinase isoform ERK1 is specifically required for in vitro and in vivo adipogenesis. *Diabetes*, 54(2), 402–411.
115. Prusty, D., Park, B.H., Davis, K.E. and Farmer, S.R., 2002. Activation of MEK/ERK signaling promotes adipogenesis by enhancing peroxisome proliferator-activated receptor γ (PPARγ) and C/EBPα gene expression during the differentiation of 3T3-L1 preadipocytes. *Journal of Biological Chemistry*, 277(48), 46226–46232.
116. Weidner, C., de Groot, J.C., Prasad, A., Freiwald, A., Quedenau, C., Kliem, M., Witzke, A. et al., 2012. Amorfrutins are potent antidiabetic dietary natural products. *Proceedings of the National Academy of Sciences U S A*, 109(19), 7257–7262.
117. Li, P., Fan, W., Xu, J., Lu, M., Yamamoto, H., Auwerx, J., Sears, D.D. et al., 2011. Adipocyte NCoR knockout decreases PPARγ phosphorylation and enhances PPARγ activity and insulin sensitivity. *Cell*, 147(4), 815–826.
118. Kumar, D.S., Prashanthi, G., Avasarala, H. and Banji, D., 2013. Antihypercholesterolemic effect of *Macrotyloma uniflorum* (Lam.) Verdc (Fabaceae) extract on high-fat diet-induced hypercholesterolemia in Sprague-Dawley rats. *Journal of Dietary Supplements*, 10(2), 116–128.
119. Panda, V., Laddha, A., Nandave, M. and Srinath, S., 2016. Dietary phenolic acids of Macrotyloma uniflorum (horse gram) protect the rat heart against isoproterenol-induced myocardial infarction. *Phytotherapy Research*, 30, 1146–1155.
120. Konishi, Y., Hitomi, Y. and Yoshioka, E., 2004. Intestinal absorption of p-coumaric and gallic acids in rats after oral administration. *Journal of Agricultural and Food Chemistry*, 52(9), 2527–2532.
121. Yoon, S.A., Kang, S.I., Shin, H.S., Kang, S.W., Kim, J.H., Ko, H.C. and Kim, S.J., 2013. p-Coumaric acid modulates glucose and lipid metabolism via AMP-activated protein kinase in L6 skeletal muscle cells. *Biochemical and Biophysical Research Communications*, 432(4), 553–557.
122. Birnbaum, M.J., 2005. Activating AMP-activated protein kinase without AMP. *Molecular Cell*, 19(3), 289–290.
123. Steinberg, G.R. and Kemp, B.E., 2009. AMPK in health and disease. *Physiological Reviews*, 89(3), 1025–1078.
124. Hardie, D.G. and Sakamoto, K., 2006. AMPK: A key sensor of fuel and energy status in skeletal muscle. *Physiology*, 21(1), 48–60.

125. Muoio, D.M., Seefeld, K., Witters, L.A. and Coleman, R.A., 1999. AMP-activated kinase reciprocally regulates triacylglycerol synthesis and fatty acid oxidation in liver and muscle: Evidence that sn-glycerol-3-phosphate acyltransferase is a novel target. *Biochemical Journal*, 338(3), 783–791.
126. Scalber, A., Morand, C., Manach, C. and Remesy, C., 2002. Absorption and metabolism of polyphenols in the gut and impact on health. *Biomedicine & Pharmacotherapy*, 56, 276–282.
127. Anandharajan, R., Pathmanathan, K., Shankernarayanan, N.P., Vishwakarma, R.A. and Balakrishnan, A., 2005. Upregulation of Glut-4 and PPAR gamma by an isoflavone from *Pterocarpus marsupium* on L6 myotubes: A possible mechanism of action. *Journal of Ethnopharmacology*, 97(2), 253–260.
128. Maurya, R., Singh, R., Deepak, M., Handa, S.S., Yadav, P.P. and Mishra, P.K., 2004. Constituents of *Pterocarpus marsupium*: An ayurvedic crude drug. *Phytochemistry*, 65(7), 915–920.
129. Rawat, P., Kumar, M., Rahuja, N., Lal Srivastava, D.S., Srivastava, A.K. and Maurya, R., 2011. Synthesis and antihyperglycemic activity of phenolic C-glycosides. *Bioorganic & Medicinal Chemistry Letters*, 21(1), 228–233.
130. Grover, J.K., Vats, V. and Yadav, S.S., 2005. *Pterocarpus marsupium* extract (Vijayasar) prevented the alteration in metabolic patterns induced in the normal rat by feeding an adequate diet containing fructose as sole carbohydrate. *Diabetes, Obesity and Metabolism*, 7(4), 414–420.
131. Mohankumar, S.K., O'Shea, T. and McFarlane, J.R., 2012. Insulinotrophic and insulin-like effects of a high molecular weight aqueous extract of *Pterocarpus marsupium* Roxb. hardwood. *Journal of Ethnopharmacology*, 141(1), 72–79.
132. Hamza, N., Berke, B., Cheze, C., Le Garrec, R., Umar, A., Agli, A.N., Lassalle, R., Jové, J., Gin, H. and Moore, N., 2012. Preventive and curative effect of *Trigonella foenum-graecum* L. seeds in C57BL/6J models of type 2 diabetes induced by high-fat diet. *Journal of Ethnopharmacology*, 142(2), 516–522.
133. Yoshinari, O., Sato, H. and Igarashi, K., 2009. Anti-diabetic effects of pumpkin and its components, trigonelline and nicotinic acid, on Goto-Kakizaki rats. *Bioscience, Biotechnology, and Biochemistry*, 73(5), 1033–1041.
134. Choi, K.M., Lee, Y.S., Shin, D.M., Lee, S., Yoo, K.S., Lee, M.K., Lee, J.H. et al., Green tomato extract attenuates high-fat-diet-induced obesity through activation of the AMPK pathway in C57BL/6 mice. *Journal of Nutritional Biochemistry*, 24(1), 335–342.
135. Ilavenil, S., Arasu, M.V., Lee, J.C., Kim, D.H., Roh, S.G., Park, H.S., Choi, G.J., Mayakrishnan, V. and Choi, K.C., 2013. Trigonelline attenuates the adipocyte differentiation and lipid accumulation in 3T3-L1 cells. *Phytomedicine*, 21, 758–765.
136. Son, Y., Nam, J.S., Jang, M.K., Jung, I.A., Cho, S.I. and Jung, M.H., 2013. Antiobesity activity of *Vigna nakashimae* extract in high-fat diet-induced obesity. *Bioscience, Biotechnology, and Biochemistry*, 77(2), 332–338.
137. Narender, T., Shweta, S., Tiwari, P., Papi Reddy, K., Khaliq, T., Prathipati, P., Puri, A., Srivastava, A.K., Chander, R., Agarwal, S.C. and Raj, K., 2007. Antihyperglycemic and antidyslipidemic agent from *Aegle marmelos*. *Bioorganic & Medicinal Chemistry Letters*, 17(6), 1808–1811.
138. Sharma, A.K., Bharti, S., Goyal, S., Arora, S., Nepal, S., Kishore, K., Joshi, S., Kumari, S. and Arya, D.S., 2011. Upregulation of PPARγ by *Aegle marmelos* ameliorates insulin resistance and β-cell dysfunction in high fat diet fed-streptozotocin induced type 2 diabetic rats. *Phytotherapy Research*, 25(10), 1457–1465.
139. Karmase, A., Jagtap, S. and Bhutania, K.K., 2013. Antiadipogenic activity of *Aegle marmelos* Correa. *Phytomedicine*, 20(14), 1247–1271.
140. Saravanan, M., Pandikumar, P., Saravanan, S., Toppo, E., Pazhanivel, N. and Ignacimuthu, S., 2014. Lipolytic and antiadipogenic effects of (3,3-dimethylallyl) halfordinol on 3T3-L1 adipocytes andhigh fat and fructose diet induced obese C57/BL6J mice. *European Journal of Pharmacology*, 740, 714–721.
141. Prince, P.S. and Rajadurai, M., 2005. Preventive effect of *Aegle marmelos* leaf extract on isoprenaline-induced myocardial infarction in rats: Biochemical evidence. *Journal of Pharmacy and Pharmacology*, 57(10), 1353–1357.
142. Jing, L., Zhang, Y., Fan, S., Gu, M., Guan, Y., Lu, X., Huang, C. and Zhou, Z., 2013. Preventive and ameliorating effects of citrus D-limonene on dyslipidemia and hyperglycemia in mice with high-fat diet-induced obesity. *European Journal of Pharmacology*, 715(1–3), 46–55.
143. Kang, S.I., Shin, H.S., Kim, H.M., Hong, Y.S., Yoon, S.A., Kang, S.W., Kim, J.H., Kim, M.H., Ko, H.C. and Kim, S.J., 2012. Immature Citrus sunki peel extract exhibits antiobesity effects by β-oxidation and lipolysis in high-fat diet-induced obese mice. *Biological and Pharmaceutical Bulletin*, 35(2), 223–230.
144. Yoshida, H., Watanabe, W., Oomagari, H., Tsuruta, E., Shida, M. and Kurokawa, M., 2013. *Citrus* flavonoid naringenin inhibits TLR2 expression in adipocytes. *Journal of Nutrition Biochemistry*, 24(7), 1276–1284.

145. Xu, H., Barnes, G.T., Yang, Q., Tan, G., Yang, D., Chou, C.J., Sole, J. et al., 2003. Chronic inflammation in fat plays a crucial role in the development of obesity-related insulin resisiatnce. *The Journal of Clinical Investigation*, 112(12), 1821–1830.
146. Deshmane, S.L., Kremlev, S., Amini, S. and Sawaya, B.E., 2009. Monocyte chemoattractant protein-1 (MCP-1): An overview. *Journal of Interferon & Cytokine Research*, 29(6), 313–326.
147. Takahashi, H., Hara, H., Goto, T., Kamakari, K., Wataru, N., Mohri, S., Takahashi, N., Suzuki, H., Shibata, D. and Kawada, T., 2015. 13-Oxo-9(Z),11(E),15(Z)-octadecatrienoic acid activates peroxisome proliferator-activated receptor γ in adipocytes. *Lipids*, 50(1), 3–12.
148. Lehrke, M. and Lazar, M.A., 2005. The many faces of PPARγ. *Cell*, 123(6), 993–999.
149. Kobayashi, Y., Nakano, Y., Kizaki, M., Hoshikuma, K., Yokoo, Y. and Kamiya, T., 2001. Capsaicin-like anti-obese activities of evodiamine from fruits of Evodia rutaecarpa, a vanilloid receptor agonist. *Planta Medica*, 67(7), 628–633.
150. Kim, S.-J., Lee, S.-J., Lee, S., Chae, S., Han, M.-D., Mar, W. and Nam, K.-W., 2009. Rutecarpine ameliorates bodyweight gain through the inhibition of orexigenic neuropeptides NPY and AgRP in mice. *Biochemical and Biophysical Research Communications*, 389(3), 437–442.
151. Shi, J., Yan, J., Lei, Q., Zhao, J., Chen, K., Yang, D., Zhao, X. and Zhang, Y., 2009. Intragastric administration of evodiamine suppresses NPY and AgRP gene expression in the hypothalamus anddecreases food intake in rats. *Brain Research*, 1247, 71–78.
152. Schwartz, M.W., Woods, S.C., Porte, D., Seeley, R.J. and Baskin, D.G., 2000. Central nervous system control of food intake. *Nature*, 404(6778), 661–671.
153. Davidowa, H., Li, Y. and Plagemann, A., 2003. Altered responses to orexigenic (AGRP, MCH) and anorexigenic (α-MSH, CART) neuropeptides of paraventricular hypothalamic neurons in early postnatally overfed rats. *European Journal of Neuroscience*, 18(3), 613–621.
154. Wang, T., Kusudo, T., Takeuchi, T., Yamashita, Y., Kontani, Y., Okamatsu, Y., Saito, M., Mori, N. and Yamashita, H., 2013. Evodiamine inhibits insulin-stimulated mTOR-S6K activation and IRS1 serine phosphorylation in adipocytes and improves glucose tolerance in obese/diabetic mice. *PLoS One*, 8(12), e83264.
155. Thies, R.S., Molina, J.M., Ciaraldi, T.P., Freidenberg, G.R. and Olefsky, J.M., 1990. Insulin-receptor autophosphorylation and endogenous substrate phosphorylation in human adipocytes from control, obese, and NIDDM subjects. *Diabetes*, 39(2), 250–259.
156. Cusi, K., Maezono, K., Osman, A., Pendergrass, M., Patti, M.E., Pratipanawatr, T., DeFronzo, R.A., Kahn, C.R. and Mandarino, L.J., 2000. Insulin resistance differentially affects the PI 3-kinase–and MAP kinase–mediated signaling in human muscle. *The Journal of Clinical Investigation*, 105(3), 311–320.
157. Kelley, D.E., Mintun, M.A., Watkins, S.C., Simoneau, J.A., Jadali, F., Fredrickson, A., Beattie, J. and Théríault, R., 1996. The effect of non-insulin-dependent diabetes mellitus and obesity on glucose transport and phosphorylation in skeletal muscle. *Journal of Clinical Investigation*, 97(12), 2705.
158. Subramanian, S. and Arulselvan, P., 2009. Evaluation of hypolipidemic properties of *Murraya koenigii* leaves studied in streptozotocin-induced diabetic rats. *Biomedicine*, 29(3), 220–225.
159. Biswas, A., Bhattacharya, S., Dasgupta, S., Kundu, R., Roy, S.S., Pal, B.C. and Bhattacharya, S., 2010. Insulin resistance due to lipid-induced signaling defects could be prevented by mahanine. *Molecular and Cellular Biochemistry*, 336(1–2), 97–107.
160. Ma, Q., Tian, J., Yang, J., Wang, A., Ji, T., Wang, Y. and Su, Y., 2013. Bioactive carbazole alkaloids from *Murraya koenigii* (L.) Spreng. *Fitoterapia*, 87(1), 1–6.
161. Pandey, J., Maurya, R., Raykhera, R., Srivastava, M.N., Yadav, P.P. and Tamrakar, A.K., 2014. *Murraya koenigii* (L.) Spreng. ameliorates insulin resistance in dexamethasone-treated mice by enhancing peripheral insulin sensitivity. *Journal of the Science of Food and Agriculture*, 94(11), 2282–2288.
162. Boden, G., 1997. Role of fatty acids in the pathogenesis of insulin resistance and NIDDM. *Diabetes*, 46(1), 3–10.
163. Watanabe, A., Kato, T., Ito, Y., Yoshida, I., Harada, T., Mishima, T., Fujita, K., Watai, M., Nakagawa, K. and Miyazawa, T., 2014. Aculeatin, a coumarin derived from *Toddalia asiatica* (L.) Lam., enhances differentiation and lipolysis of 3T3-L1 adipocytes. *Biochemical and Biophysical Research Communications*, 453(4), 787–792.
164. Gwon, S.Y., Ahn, J.Y., Kim, T.W. and Ha, T.Y., 2012. *Zanthoxylum piperitum* DC ethanol extract suppresses fat accumulation in adipocytes and high fat diet-induced obese mice by regulating adipogenesis. *Journal of Nutritional Science and Vitaminology* (*Tokyo*), 58(6), 393–401.
165. Ovalle-Magallanes, B., Medina-Campos, O.N., Pedraza-Chaverri, J., Mata, R., 2015. Hypoglycemic and antihyperglycemic effects of phytopreparations and limonoids from Swietenia humilis. *Phytochemistry*, 110, 111–119.

166. Dewanjee, S., Maiti, A., Das, A.K., Mandal, S.C. and Dey, S.P., 2009. Swietenine: A potential oral hypoglycemic from *Swietenia macrophylla* d. *Fitoterapia*, 80(4), 249–251.
167. Lau, W.K., Goh, B.H., Kadir, H.A., Shu-Chien, A.C., Muhammad, T.S., 2015. Potent PPARγ ligands from Swietenia macrophylla are capable of stimulating glucose uptake in muscle cells. *Molecules*, 20(12), 22301–22314.
168. Chen, L.C., Liao, H.R., Chen, P.Y., Kuo, W.L., Chang, T.H., Sung, P.J., Wen, Z.H. and Chen, J.J., 2015. Limonoids from the seeds of Swietenia macrophylla and their anti-inflammatory activities. *Molecules*, 20(10), 18551–18564.
169. Panda, S.P., Haldar, P.K., Bera, S., Adhikary, S. and Kandar, C.C., 2010. Antidiabetic and antioxidant activity of Swietenia mahagoni in streptozotocin-induced diabetic rats. *Pharmaceutical Biology*, 48(9), 974–979.
170. Wang, P.H., Tsai, M.J., Hsu, C.Y., Wang, C.Y., Hsu, H.K. and Weng, C.F., 2008. Toona sinensis Roem (Meliaceae) leaf extract alleviates hyperglycemia via altering adipose glucose transporter 4. *Food and Chemical Toxicology*, 46(7), 2554–2560.
171. Hsieh, T.J., Tsai, Y.H., Liao, M.C., Du, Y.C., Lien, P.J., Sun, C.C., Chang, F.R. and Wu, Y.C., 2012. Anti-diabetic properties of non-polar *Toona sinensis* Roem extract prepared by supercritical-CO2 fluid. *Food and Chemical Toxicology*, 50(3–4), 779–789.
172. Hsu, C.Y., Shih, H.Y., Chia, Y.C., Lee, C.H., Ashida, H., Lai, Y.K. and Weng, C.F., 2014. Rutin potentiates insulin receptor kinase to enhance insulin-dependent glucose transporter 4 translocation. *Molecular Nutrition & Food Research*, 58(6), 1168–1176.
173. Karantonis, H.C., Nomikos, T. and Demopoulos, C.A., 2009. Triacylglycerol metabolism. *Current Drug Targets*, 10(4), 302–319.
174. Liu, J., He, X.F., Wang, G.H., Merino, E.F., Yang, S.P., Zhu, R.X., Gan, L.S. et al., 2014. Aphadilactones A-D, four diterpenoid dimers with DGAT inhibitory and antimalarial activities from a Meliaceaeplant. *Journal of Organic Chemistry*, 79(2), 599–607.
175. Husain, G.M., Singh, P.N., Singh, R.K. and Kumar, V., 2011. Antidiabetic activity of standardized extract of Quassia amara in nicotinamide-streptozotocin-induced diabeticrats. *Phytotherapy Research*, 25(12), 1806–1812.
176. Cornick, C.L., Strongitharm, B.H., Sassano, G., Rawlins, C., Mayes, A.E., Joseph, A.N., O'Dowd, J. et al., 2009. Identification of a novel agonist of peroxisome proliferator-activated receptors alpha and gamma that may contribute to the anti-diabetic activity of guggulipid in Lep(ob)/Lep(ob) mice. *Journal of Nutritional Biochemistry*, 20(10), 806–815.
177. Sharma, B., Salunke, R., Srivastava, S., Majumder, C. and Roy, P., 2009. Effects of guggulsterone isolated from *Commiphora mukul* in high fat diet induced diabetic rats. *Food and Chemical Toxicology*, 47(10), 2631–2639.
178. Ramesh, B. and Saralakumari, D., 2012. Antihyperglycemic, hypolipidemic and antioxidant activities of ethanolic extract of Commiphora mukul gum resin in fructose-fed male Wistar rats. *Journal of Physiology and Biochemistry*, 68(4), 573–582.
179. Shakirin, F.H., Azlan, A., Ismail, A., Amom, Z. and Yuon, L.C., 2012. Protective effect of pulp oil extracted from *Canarium odontophyllum* Miq. Fruit on blood lipids, lipid peroxidation, and antioxidant status in healthy rabbits. *Oxidative Medicine and Cellular Longevity*, 2012, 840973.
180. Mokiran, N.N., Ismail, A., Azlan, A., Hamid, M. and Hassan, F.A., 2014. Effect of dabai (*Canarium odontophyllum*) fruit extract on biochemical parameters of induced obese–diabetic rats. *Journal of Functional Foods*, 8, 139–149.
181. Mohan, C.G., Viswanatha, G.L., Savinay, G., Rajendra, C.E. and Halemani, P.D., 2013. 1,2,3,4,6 Penta-*O*-galloyl-β-D-glucose, a bioactivity guided isolated compound from *Mangifera indica* inhibits 11β-HSD-1 and ameliorates high fat diet-induced diabetic in C57BL/6 mice. *Phytomedicine*, 20(5), 417–426.
182. Zhang, Y., Liu, X., Han, L., Gao, X., Liu, E. and Wang, T., 2014. Regulation of lipid and glucose homeostasis by mango tree leaf extract is mediated by AMPK and PI3K/AKT signaling pathways. *Food Chemistry*, 141(3), 2896–2905.
183. Girón, M.D., Sevillano, N., Salto, R., Haidour, A., Manzano, M., Jiménez, M.L., Rueda, R. and López-Pedrosa, J.M., 2009. *Salacia oblonga* extract increases glucose transporter 4-mediated glucose uptake in L6 rat myotubes: Role of mangiferin. *Clinical Nutrition*, 28(5), 565–574.
184. Anwer, T., Sharma, M., Khan, G., Iqbal, M., Ali, M.S., Alam, M.S., Safhi, M.M. and Gupta, N., 2013. *Rhus coriaria* ameliorates insulin resistance in non-insulin-dependent diabetes mellitus (NIDDM) rats. *Acta Poloniae Pharmaceutica*, 70(5), 861–867.

185. Salimi, Z., Eskandary, A., Headari, R., Nejati, V., Moradi, M. and Kalhori, Z., 2015. Antioxidant effect of aqueous extract of sumac (*Rhus coriaria* L.) in the alloxan-induced diabetic rats. *Indian Journal of Physiology and Pharmacology*, 59(1), 87–93.
186. Rahideh, S.T., Shidfar, F., Khandozi, N., Rajab, A., Hosseini, S.P. and Mirtaher, S.M., 2014. The effect of sumac (*Rhus coriaria* L.) powder on insulin resistance, malondialdehyde, high sensitive C-reactive protein and paraoxonase 1 activity in type 2 diabetic patients. *Journal of Research in Medical Science*, 19(10), 933–938.
187. Sung, Y.Y., Yoon, T., Yang, W.K., Kim, S.J. and Kim, H.K., 2011. Anti-obesity effects of *Geranium thunbergii* extract via improvement of lipid metabolism in high-fat diet-induced obese mice. *Molecular Medicine Reports*, 4(6), 1107–1113.
188. Okuda, T., Yoshida, T. and Nayeshiro, H., 1977. Constituents of *Geranium thunbergii* Sieb. et Zucc. IV. Ellagitannins. 2. structure of geraniin. *Chemical & Pharmaceutical Bulletin*, 25, 1862–1869.
189. Boukhris, M., Bouaziz, M., Feki, I., Jemai, H., El Feki, A. and Sayadi, S., 2012. Hypoglycemic and antioxidant effects of leaf essential oil of *Pelargonium graveolens* L'Hér. in alloxan induced diabetic rats. *Lipids in Health and Disease*, 11, 81.
190. Girón, M.D., Sevillano, N., Salto, R., Haidour, A., Manzano, M., Jimenez, M.L., Rueda, R. and López-Pedrosa, J.M., 2009. *Salacia oblonga* extract increases glucose transporter 4-mediated glucose uptake in L6 rat myotubes: Role of mangiferin. *Clinical Nutrition*, 28(5), 565–574.
191. Yoshikawa, M., Shimoda, H., Nishida, N., Takada, M. and Matsuda, H., 2002. *Salacia reticulata* and its polyphenolic constituents with lipase inhibitory and lipolytic activities have mild antiobesity effects in rats. *Journal of Nutrition*, 132(7), 1819–1824.
192. Shimada, T., Nagai, E., Harasawa, Y., Watanabe, M., Negishi, K., Akase, T., Sai, Y., Miyamoto, K. and Aburada, M., 2011. *Salacia reticulata* inhibits differentiation of 3T3-L1 adipocytes. *Journal of Ethnopharmacology*, 136(1), 67–74.
193. Wang, Y., Deng, M., Zhang, S.Y., Zhou, Z.K. and Tian, W.X., 2008. *Parasitic loranthus* from Loranthaceae rather than Viscaceae potently inhibits fatty acid synthase and reduces body weight in mice. *Journal of Ethnopharmacology*, 118(3), 473–478.
194. Lee, M.S. and Thuong, P.T., 2010. Stimulation of glucose uptake by triterpenoids from *Weigela subsessilis*. *Phytotherapy Research*, 24(1), 49–53.
195. Na, M., Thuong, P.T., Hwang, I.H., Bae, K., Kim, B.Y., Osada, H. and Ahn, J.S., 2010. Protein tyrosine phosphatase 1B inhibitory activity of 24-norursane triterpenes isolated from *Weigela subsessilis*. *Phytotherapy Research*, 24(11), 1716–1719.
196. Liu, T.P., Lee, C.S., Liou, S.S., Liu, I.M. and Cheng, J.T., 2005. Improvement of insulin resistance by *Acanthopanax senticosus* root in fructose-rich chow-fed rats. *Clinical and Experimental Pharmacology and Physiology*, 32(8), 649–654.
197. Na, M., Oh, W.K., Kim, Y.H., Cai, X.F., Kim, S., Kim, B.Y. and Ahn, J.S., 2006. Inhibition of protein tyrosine phosphatase 1B by diterpenoids isolated from *Acanthopanax koreanum*. *Bioorganic & Medicinal Chemistry Letters*, 16(11), 3061–3064.
198. Li, J.L., Li, N., Xing, S.S., Zhang, N., Li, B.B., Chen, J.G., Ahn, J.S. and Cui, L., 2015. New neo-lignan from *Acanthopanax senticosus* with protein tyrosine phosphatase 1B inhibitory activity. *Archives of Pharmacal Research*.
199. Mollah, M.L., Kim, G.S., Moon, H.K., Chung, S.K., Cheon, Y.P., Kim, J.K. and Kim, K.S., 2009. Antiobesity effects of wild ginseng (*Panax ginseng* C.A. Meyer) mediated by PPAR-gamma, GLUT4 and LPL in ob/ob mice. *Phytotherapy Research*, 23(2), 220–225.
200. Lee, H.J., Lee, Y.H., Park, S.K., Kang, E.S., Kim, H.J., Lee, Y.C., Choi, C.S. et al., 2009. Korean red ginseng (*Panax ginseng*) improves insulin sensitivity and attenuates the development of diabetes in Otsuka Long-Evans Tokushima fatty rats. *Metabolism*, 58(8), 1170–1177.
201. Lee, S.H., Lee, H.J., Lee, Y.H., Lee, B.W., Cha, B.S., Kang, E.S., Ahn, C.W. et al., 2012. Korean red ginseng (*Panax ginseng*) improves insulin sensitivity in high fat fed Sprague-Dawley rats. *Phytotherapy Research*, 26(1), 142–147.
202. Lee, H.M., Lee, O.H., Kim, K.J. and Lee, B.Y., 2012. Ginsenoside Rg1 promotes glucose uptake through activated AMPK pathway in insulin-resistant muscle cells. *Phytotherapy Research*, 26(7), 1017–1022.
203. Hwang, J.T., Lee, M.S., Kim, H.J., Sung, M.J., Kim, H.Y., Kim, M.S. and Kwon, D.Y., 2009. Antiobesity effect of ginsenoside Rg3 involves the AMPK and PPAR-gamma signal pathways. *Phytotherapy Research*, 23(2), 262–266.
204. Lee, O.H., Lee, H.H., Kim, J.H. and Lee, B.Y., 2011. Effect of ginsenosides Rg3 and Re on glucose transport in mature 3T3-L1 adipocytes. *Phytotherapy Research*, 25(5), 768–773.

205. Sun, C., Chen, Y., Li, X., Tai, G., Fan, Y. and Zhou, Y., 2014. Anti-hyperglycemic and anti-oxidative activities of ginseng polysaccharides in STZ-induced diabetic mice. *Food & Function*, 5(5), 845–848.
206. Hwang, J.T., Kim, S.H., Hur, H.J., Kim, H.J., Park, J.H., Sung, M.J., Yang, H.J. et al., 2012. Decursin, an active compound isolated from *Angelica gigas*, inhibits fat accumulation, reduces adipocytokine secretion and improves glucose tolerance in mice fed a high-fat diet. *Phytotherapy Research*, 26(5), 633–638.
207. Moustaid, N., Lasnier, F., Hainque, B., Quignard-Boulange, A. and Pairault, J., 1990. Analysis of gene expression during adipogenesis in 3T3-F442A preadipocytes: Insulin and dexamethasone control. *Journal of Cellular Biochemistry*, 42(4), 243–254.
208. Kawabata, K., Sawada, K., Ikeda, K., Fukuda, I., Kawasaki, K., Yamamoto, N. and Ashida, H., 2011. Prenylated chalcones 4-hydroxyderricin and xanthoangelol stimulate glucose uptake in skeletal muscle cells byinducing GLUT4 translocation. *Molecular Nutrition & Food Research*, 55(3), 467–475.
209. Enoki, T., Ohnogi, H., Nagamine, K., Kudo, Y., Sugiyama, K., Tanabe, M., Kobayashi, E., Sagawa, H. and Kato, I., 2007. Antidiabetic activities of chalcones isolated from a Japanese Herb, *Angelica keiskei*. *Journal of Agricultural and Food Chemistry*, 55(15), 6013–6017.
210. Zhang, T., Sawada, K., Yamamoto, N. and Ashida, H., 4-Hydroxyderricin and xanthoangelol from Ashitaba (Angelica keiskei) suppress differentiation of preadiopocytes to adipocytes via AMPK and MAPK pathways. *Molecular Nutrition & Food Research*, 57(10), 1729–1740.
211. Lee, Y.S., Kim, W.S. and Kim, K.H., 2006. Berberine, a natural plant product, activates AMP-activated protein kinase with beneficial metabolic effects in diabetic and insulin-resistant states. *Diabetes*, 55, 2256–2264.
212. Lee, W.H., Lin, R.J., Lin, S.Y., Chen, Y.C., Lin, H.M. and Liang, Y.C., 2011. Osthole enhances glucose uptake through activation of AMP-activated protein kinase in skeletal muscle cells. *Journal ofAgricultural and Food Chemistry*, 59(24), 12874–12881.
213. Qi, Z., Xue, J., Zhang, Y., Wang, H. and Xie, M., 2011. Osthole ameliorates insulin resistance by increment of adiponectin release in high-fat and high-sucrose-induced fatty liver rats. *Planta Medica*, 77(3), 231–235.
214. Zhong, W., Shen, H., Zhou, F., Xue, J. and Xie, M.-L., 2014. Osthol inhibits fatty acid synthesis and release via PPARα/γ- mediated pathways in 3T3-L1 adipocytes. *Phytochemistry Letters*, 8(1), 22–27.
215. Kim, S.B., Ahn, J.H., Han, S.B., Hwang, B.Y., Kim, S.Y. and Lee, M.K., 2012. Anti-adipogenic chromone glycosides from *Cnidium monnieri* fruits in 3T3-L1 cells. *Bioorganic & Medicinal Chemistry Letters*, 22(19), 6267–6271.
216. Nukitrangsan, N., Okabe, T., Toda, T., Inafuku, M., Iwasaki, H. and Oku, H., 2012. Effect of *Peucedanum japonicum* Thunb extract on high-fat diet-induced obesity and gene expression in mice. *Journal of Oleo Science*, 61(2), 89–101.
217. Nugara, R.N., Inafuku, M., Takara, K., Iwasaki, H. and Oku, H., 2014. Pteryxin: A coumarin in *Peucedanum japonicum* Thunb leaves exerts antiobesity activity through modulation of adipogenic gene network. *Nutrition*, 30(10), 1177–1184.
218. Prestwich, T.C. and MacDougald, O.A., 2007. Wnt/β-catenin signaling in adipogenesis and metabolism. *Current Opinion in Cell Biology*, 19(6), 612–617.
219. Peifer, M. and Polakis, P., 2000. Wnt signaling in oncogenesis and embryogenesis—a look outside the nucleus. *Science*, 287(5458), 1606–1609.
220. Behrens, J., von Kries, J.P., Kuhl, M. and Bruhn, L., 1996. Functional interaction of beta-catenin with the transcription factor LEF-1. *Nature*, 382(6592), 638.
221. Freytag, S.O. and Geddes, T.J., 1992. Reciprocal regulation of adipogenesis by Myc and C/EBP (Alpha). *Science*, 256(5055), 379.
222. Lee, H., Bae, S., Kim, Y.S. and Yoon, Y., 2011. WNT/β-catenin pathway mediates the anti-adipogenic effect of platycodin D, a natural compound found in *Platycodon grandiflorus*. *Life Sciences*, 89(11–12), 388–394.
223. Lee, E.J., Kang, M. and Kim, Y.S., 2012. Platycodin D inhibits lipogenesis through AMPKα-PPARγ2 in 3T3-L1 cells and modulates fat accumulation in obese mice. *Planta Medica*, 78(14), 1536–1542.
224. Kim, H.L., Park, J., Park, H., Jung, Y., Youn, D.H., Kang, J., Jeong, M.Y., Um, J.Y., 2015. *Platycodon grandiflorum* A. De Candolle ethanolic extract inhibits adipogenic regulators in 3T3-L1 cells and induces mitochondrial biogenesis in primary brown preadipocytes. *Journal of Agricultural and Food Chemistry*, 63(35), 7721–7730.
225. Nurul Islam, M., Jung, H.A., Sohn, H.S., Kim, H.M. and Choi, J.S., 2013. Potent α-glucosidase and protein tyrosine phosphatase 1B inhibitors from *Artemisia capillaris*. *Archives of Pharmacal Research*, 36(5), 542–552.

226. Noh, J.R., Kim, Y.H., Hwang, J.H., Gang, G.T., Yeo, S.H., Kim, K.S., Oh, W.K., Ly, S.Y., Lee, I.K. and Lee, C.H., 2013. Scoparone inhibits adipocyte differentiation through down-regulation of peroxisome proliferators-activated receptor γ in 3T3-L1 preadipocytes. *Food Chemistry*, 141(2), 723–730.
227. Yahagi, T., Yakura, N., Matsuzaki, K. and Kitanaka, S., 2014. Inhibitory effect of chemical constituents from *Artemisia scoparia* Waldst. et Kit. on triglyceride accumulation in 3T3-L1 cells and nitric oxide production in RAW 264.7 cells. *Journal of Natural Medicines*, 68(2), 414–420.
228. Ribnicky, D.M., Poulev, A., Watford, M., Cefalu, W.T. and Raskin, I., 2006. Antihyperglycemic activity of Tarralin, an ethanolic extract of Artemisia dracunculus L. *Phytomedicine*, 13(8), 550–557.
229. Vandanmagsar, B., Haynie, K.R., Wicks, S.E., Bermudez, E.M., Mendoza, T.M., Ribnicky, D., Cefalu, W.T. and Mynatt, R.L., 2014. *Artemisia dracunculus* L. extract ameliorates insulin sensitivity by attenuating inflammatory signalling in human skeletal muscle culture. *Diabetes, Obesity and Metabolism*, 16(8), 728–738.
230. Przybylski, R. and McDonald, B.E., 1995. *Development and Processing of Vegetable Oils for Human Nutrition*. Champaign, IL: The American Oil Chemists Society.
231. Obanda, D.N., Ribnicky, D.M., Raskin, I. and Cefalu, W.T., 2014. Bioactives of *Artemisia dracunculus* L. enhance insulin sensitivity by modulation of ceramide metabolism in rat skeletal muscle cells. *Nutrition*, 30(7–8 Suppl), S59–S66.
232. Ahmad, W., Khan, I., Khan, M.A., Ahmad, M., Subhan, F. and Karim, N., 2014. Evaluation of antidiabetic and antihyperlipidemic activity of *Artemisia indica* linn (aeriel parts) in Streptozotocin induced diabetic rats. *Journal of Ethnopharmacology*, 151(1), 618–623.
233. Kodama, H., Fujita, M. and Yamaguchi, I., 1994. Development of hyperglycaemia and insulin resistance in conscious genetically diabetic (C57BL/KsJ-db/db) mice. *Diabetologia*, 37(8), 739–744.
234. Kang, Y.J., Jung, U.J., Lee, M.K., Kim, H.J., Jeon, S.M., Park, Y.B., Chung, H.G. et al., 2008. Eupatilin, isolated from *Artemisia princeps* Pampanini, enhances hepatic glucose metabolism and pancreatic beta-cell function in type 2 diabetic mice. *Diabetes Research and Clinical Practices*, 82(1), 25–32.
235. Yamamoto, N., Kanemoto, Y., Ueda, M., Kawasaki, K., Fukuda, I., Ashida, H., 2011. Anti-obesity and anti-diabetic effects of ethanol extract of *Artemisia princeps* in C57BL/6 mice fed a high-fat diet. *Food & Function*, 2(1), 45–52.
236. Kim, C.K., Kim, M., Oh, S.D., Lee, S.M., Sun, B., Choi, G.S., Kim, S.K., Bae, H., Kang, C. and Min, B.I., 2011. Effects of *Atractylodes macrocephala* Koidzumi rhizome on 3T3-L1 adipogenesis and an animal model of obesity. *Journal of Ethnopharmacology*, 137(1), 396–402.
237. Pushparaj, P.N., Low, H.K., Manikandan, J., Tan, B.K. and Tan, C.H., 2007. Anti-diabetic effects of *Cichorium intybus* in streptozotocin-induced diabetic rats. *Journal of Ethnopharmacology*, 111(2), 430–434.
238. Tousch, D., Lajoix, A.D., Hosy, E., Azay-Milhau, J., Ferrare, K., Jahannault, C., Cros, G. and Petit, P., 2008. Chicoric acid, a new compound able to enhance insulin release and glucose uptake. *Biochemical and Biophysical Research Communications*, 377(1), 131–135.
239. Muthusamy, V.S., Saravanababu, C., Ramanathan, M., Bharathi Raja, R., Sudhagar, S., Anand, S. and Lakshmi, B.S., 2010. Inhibition of protein tyrosine phosphatase 1B and regulation of insulin signalling markers by caffeoyl derivatives of chicory (Cichorium intybus) salad leaves. *British Journal of Nutrition*, 104(6), 813–823.
240. Jurgoński, A., Juśkiewicz, J., Zduńczyk, Z. and Król, B., 2012. Caffeoylquinic acid-rich extract from chicory seeds improves glycemia, atherogenic index, and antioxidant status in rats. *Nutrition*, 28(3), 300–306.
241. Schumacher, E., Vigh, E., Molnár, V., Kenyeres, P., Fehér, G., Késmárky, G., Tóth, K. and Garai, J., 2011. Thrombosis preventive potential of chicory coffee consumption: A clinical study. *Phytotherapy Research*, 25(5), 744–748.
242. Onkaramurthy, M., Veerapur, V.P., Thippeswamy, B.S., Reddy, T.N., Rayappa, H. and Badami, S., 2013. Anti-diabetic and anti-cataract effects of *Chromolaena odorata* Linn., in streptozotocin-induced diabetic rats. *Journal of Ethnopharmacology*, 145(1), 363–372.
243. Zhang, M.L., Irwin, D., Li, X.N., Sauriol, F., Shi, X.W., Wang, Y.F., Huo, C.H., Li, L.G., Gu, Y.C. and Shi, Q.W., 2012. PPARγ agonist from *Chromolaena odorata*. *Journal of Natural Products*, 75(12), 2076–2081.
244. Yamamoto, J., Tadaishi, M., Yamane, T., Oishi, Y., Shimizu, M., Kobayashi-Hattori, K., 2015. Hot water extracts of edible *Chrysanthemum morifolium* Ramat. exert antidiabetic effects in obese diabetic KK-Ay mice. *Bioscience, Biotechnology, and Biochemistry*, 79(7), 1147–1154.
245. Liao, Z., Wu, Z. and Wu, M., 2012. *Cirsium japonicum* flavones enhance adipocyte differentiation and glucose uptake in 3T3-L1 cells. *Biological and Pharmaceutical Bulletin*, 35(6), 855–860.

246. Chen, J., Mangelinckx, S., Ma, L., Wang, Z., Li, W. and De Kimpe, N., 2014. Caffeoylquinic acid derivatives isolated from the aerial parts of *Gynura divaricata* and their yeast α-glucosidase and PTP1B inhibitory activity. *Fitoterapia*, 99, 1–6.
247. Xu, B.Q., Yang, P. and Zhang, Y.Q., 2015. Hypoglycemic activities of lyophilized powder of *Gynura divaricata* by improving antioxidant potential and insulin signaling in type 2 diabetic mice. *Food & Nutrition Research*, 59, 29652.
248. Rafraf, M., Zemestani, M., and Asghari-Jafarabadi, M., 2015. Effectiveness of chamomile tea on glycemic control and serum lipid profile in patients with type 2 diabetes. *Journal of Endocrinological Investigation*, 38(2), 163–170.
249. Weidner, C., Wowro, S.J., Rousseau, M., Freiwald, A., Kodelja, V., Abdel-Aziz, H., Kelber, O. and Sauer, S., 2013. Antidiabetic effects of chamomile flowers extract in obese mice through transcriptional stimulation of nutrient sensors of the peroxisome proliferator-activated receptor (PPAR) family, *PLoS One* 8(11), e80335.
250. Adachi, Y., Kanbayashi, Y., Harata, I., Ubagai, R., Takimoto, T., Suzuki, K., Miwa, T. and Noguchi, Y., 2014. Petasin activates AMP-activated protein kinase and modulates glucose metabolism. *Journal of Natural Products*, 77(6), 1262–1269.
251. Kim, S., Na, M., Oh, H., Jang, J., Sohn, C.B., Kim, B.Y., Oh, W.K. and Ahn, J.S., 2006. PTP1B inhibitory activity of kaurane diterpenes isolated from *Siegesbeckia glabrescens. Journal of Enzyme Inhibition and Medicinal Chemistry*, 21(4), 379–383.
252. Jang, Y.S., Wang, Z., Lee, J.M., Lee, J.Y. and Lim, S.S., 2016. Screening of Korean natural products for anti-adipogenesis properties and isolation of kaempferol-3-O-rutinoside as a potent anti-adipogenetic compound from Solidago virgaurea. *Molecules*, 21(2), 226.
253. Lin, H.R., 2013. Identification of liver X receptor and farnesoid X receptor dual agonists from *Tithonia diversifolia. Medicinal Chemistry Research*, 22(7), 3270–3281.
254. Miura, T., Nosaka, K., Ishii, H. and Ishida, T., 2005. Antidiabetic effect of Nitobegiku, the herb *Tithonia diversifolia*, in KK-Ay diabetic mice. *Biological and Pharmaceutical Bulletin*, 28(11), 2152–2154.
255. Zhao, G., Li, X., Chen, W., Xi, Z. and Sun, L., 2012. Three new sesquiterpenes from Tithonia diversifolia and their anti-hyperglycemic activity. *Fitoterapia*, 83(8), 1590–1597.
256. Bechmann, L.P., Hannivoort, R.A., Gerken, G., Hotamisligil, G.S., Trauner, M. and Canbay, A., 2012. The interaction of hepatic lipid and glucose metabolism in liver diseases. *Journal of Hepatology*, 56(4), 952–964.
257. Nguyen, P.H., Yang, J.L., Uddin, M.N., Park, S.L., Lim, S.I., Jung, D.W., Williams, D.R. and Oh, W.K., 2013. Protein tyrosine phosphatase 1B (PTP1B) inhibitors from Morinda citrifolia (Noni) and their insulin mimetic activity. *Journal of Natural Products*, 76(11), 2080–2087.
258. Jun do, Y., Han, C.R., Choi, M.S., Bae, M.A., Woo, M.H. and Kim, Y.H., 2011. Effect of mollugin on apoptosis and adipogenesis of 3T3-L1 preadipocytes. *Phytotherapy Research*, 25(5), 724–731.
259. Sonawane, R.D., Vishwakarma, S.L., Lakshmi, S., Rajani, M., Padh, H. and Goyal, R.K., 2010. Amelioration of STZ-induced type 1 diabetic nephropathy by aqueous extract of *Enicostemma littorale* Blume and swertiamarin in rats. *Molecular and Cellular Biochemistry*, 340(1–2), 1–6.
260. Vaidya, H., Goyal, R.K. and Cheema, S.K., 2013. Anti-diabetic activity of swertiamarin is due to an active metabolite, gentianine, that upregulates PPAR-γ gene expression in 3T3-L1 cells. *Phytotherapy Research*, 27(4), 624–627.
261. El-Sedawy, A.I., Shu, Y.Z., Hattori, M., Kobashi, K. and Namba, T., 1989. Metabolism of swertiamarin from *Swertia japonica* by human intestinal bacteria. *Planta Medica*, 55(2), 147–150.
262. Kawada, T., Hagihara, K. and Iwai, K., 1986. Effects of capsaicin on lipid metabolism in rats fed a high fat diet. *Journal of Nutrition*, 116(7), 1272–1278.
263. Lee, M.S., Kim, C.T., Kim, I.H. and Kim, Y., 2011. Effects of capsaicin on lipid catabolism in 3T3-L1 adipocytes. *Phytotherapy Research*, 25(6), 935–939.
264. Kang, J.-H., Tsuyoshi, G., Han, I.-S., Kawada, T., Kim, Y.M. and Yu, R., 2010. Dietary capsaicin reduces obesity-induced insulin resistance and hepatic steatosis in obese mice fed a high-fat diet. *Obesity*, 18(4), 780–787.
265. Kim, S.H., Hwang, J.T., Park, H.S., Kwon, D.Y. and Kim, M.S., 2013. Capsaicin stimulates glucose uptake in C2C12 muscle cells via the reactive oxygen species (ROS)/AMPK/p38 MAPK pathway. *Biochemical and Biophysical Research Communications*, 439(1), 66–70.
266. Horiba, T., Nishimura, I., Nakai, Y., Abe, K. and Sato, R., 2010. Naringenin chalcone improves adipocyte functions by enhancing adiponectin production. *Molecular and Cellular Endocrinology*, 323(2), 208–214.

267. Fujiwara, Y., Kiyota, N., Tsurushima, K., Yoshitomi, M., Horlad, H., Ikeda, T., Nohara, T., Takeya, M. and Nagai, R., 2012. Tomatidine, a tomato sapogenol, ameliorates hyperlipidemia and atherosclerosis in apoE-deficient mice by inhibiting acyl-CoA:cholesterol acyl-transferase (ACAT). *Journal of Agricultural and Food Chemistry*, 60(10), 2472–2479.
268. Sakurai, K., Ishii, H., Kobayashi, S. and Iwao, T., 1976. Isolation of 4 beta hydroxywithanolide E, a new withanolide from *Physalis peruviana* L. *Chemical & Pharmaceutical Bulletin*, 24, 1403–1405.
269. Takimoto, T., Kanbayashi, Y., Toyoda, T., Adachi, Y., Furuta, C., Suzuki, K., Miwa, T. and Bannai, M., 2014. 4β-Hydroxywithanolide E isolated from *Physalis pruinosa* calyx decreases inflammatory responses by inhibiting the nuclear factor-κB signaling in diabetic mouse adipose tissue. *International Journal of Obesity (London)*, 38(11), 1432–1439.
270. Ouchi, N., Kihara, S., Arita, Y., Nishida, M., Matsuyama, A., Okamoto, Y., Ishigami, M. et al., 2001. Adipocyte-derived plasma protein, adiponectin, suppresses lipid accumulation and class A scavenger receptor expression in human monocyte-derived macrophages. *Circulation*, 103(8), 1057–1063.
271. Kar, D.M., Maharana, L., Pattnaik, S. and Dash, G.K., 2006. Studies on hypoglycaemic activity of *Solanum xanthocarpum* Schrad. & Wendl. fruit extract in rats. *Journal of Ethnopharmacology*, 108(2), 251–256.
272. Shanker, K., Gupta, S., Srivastava, P., Srivastava, S.K., Singh, S.C. and Gupta, M.M., 2011. Simultaneous determination of three steroidal glycoalkaloids in *Solanum xanthocarpum* by high performance thin layer chromatography. *Journal of Pharmaceutical and Biomedical Analysis*, 54(3), 497–502.
273. Kusano, G., Takahashi, A., Nozoe, S., Sonoda, Y. and Sato, Y., 1987. Solanum alkaloids as inhibitors of enzymatic conversion of dihydrolanosterol into cholesterol. *Chemical and Pharmaceutical Bulletin*, 35(10), 4321–4323.
274. Maurya, R., Akanksha, Jayendra, Singh, A.B., Srivastava, A.K., 2008. Coagulanolide, a withanolide from *Withania coagulans* fruits and antihyperglycemic activity. *Bioorganic & Medicinal Chemistry Letters*, 18(24), 6534–6537.
275. Shukla, K., Dikshit, P., Shukla, R., Sharma, S. and Gambhir, J.K., 2014. Hypolipidemic and antioxidant activity of aqueous extract of fruit of *Withania coagulans* (Stocks) Dunal incholesterol-fed hyperlipidemic rabbit model. *Indian Journal of Experimental Biology*, 52(9), 870–875.
276. Beg, M., Chauhan, P., Varshney, S., Shankar, K., Rajan, S., Saini, D., Srivastava, M.N., Yadav, P.P. and Gaikwad, A.N., 2014. A withanolide coagulin-L inhibits adipogenesis modulating Wnt/β-catenin pathway and cell cycle in mitotic clonal expansion. *Phytomedicine*, 21(4), 406–414.
277. Anwer, T., Sharma, M., Pillai, K.K. and Iqbal, M., 2008. Effect of *Withania somnifera* on insulin sensitivity in non-insulin-dependent diabetes mellitus rats. *Basic & Clinical Pharmacology & Toxicology*, 102(6), 498–503.
278. Gorelick, J., Rosenberg, R., Smotrich, A., Hanuš, L. and Bernstein, N., 2015. Hypoglycemic activity of withanolides and elicitated *Withania somnifera*. *Phytochemistry*, 116, 283–289.
279. Kim, D.H., Han, K.M., Chung, I.S., Kim, D.K., Kim, S.H., Kwon, B.M., Jeong, T.S., Park, M.H., Ahn, E.M. and Baek, N.I., 2005. Triterpenoids from the flower of *Campsis grandiflora* K. Schum. as human acyl-CoA: cholesterol acyltransferase inhibitors. *Archives of Pharmacal Research*, 28(5), 550–556.
280. Khan, M.F., Dixit, P., Jaiswal, N., Tamrakar, A.K., Srivastava, A.K. and Maurya, R., 2012. Chemical constituents of *Kigelia pinnata* twigs and their GLUT4 translocation modulatory effect in skeletalmuscle cells. *Fitoterapia*, 83(1), 125–129.
281. Chollet, M.M., 1946. Some constituents of the catalpa tree Catalpa bignonioides. I. Catalposide and catalpic acid. *Bulletin de la Societe de Chimie Biologique*, 28, 668–671.
282. Hontecillas, R., Diguardo, M., Duran, E., Orpi, M. and Bassaganya-Riera, J., 2008. Catalpic acid decreases abdominal fat deposition, improves glucose homeostasis and upregulates PPAR alpha expression in adipose tissue. *Clinical Nutrition*, 27(5), 764–772.
283. Singh, J. and Kakkar, P., 2014. Oroxylin A, a constituent of *Oroxylum indicum* inhibits adipogenesis and induces apoptosis in 3T3-L1 cells. *Phytomedicine*, 21(12), 1733–1741.
284. Alonso-Castro, A.J., Zapata-Bustos, R., Romo-Yañez, J., Camarillo-Ledesma, P., Gómez-Sánchez, M. and Salazar-Olivo, L.A., 2010. The antidiabetic plants *Tecoma stans* (L.) Juss. ex Kunth (Bignoniaceae) and *Teucrium cubense* Jacq(Lamiaceae) induce the incorporation of glucose in insulin-sensitive and insulin-resistant murine and humanadipocytes. *Journal of Ethnopharmacology*, 127(1), 1–6.
285. Weigert, C., Brodbeck, K., Staiger, H., Kausch, C., Machicao, F., Häring, H.U. and Schleicher, E.D., 2004. Palmitate, but not unsaturated fatty acids, induces the expression of interleukin-6 in human myotubes through proteasome-dependent activation of nuclear factor-κB. *Journal of Biological Chemistry*, 279(23), 23942–23952.

286. Chen, P.Y., Wang, J., Lin, Y.C., Li, C.C., Tsai, C.W., Liu, T.C., Chen, H.W., Huang, C.S., Lii, C.K. and Liu, K.L., 2015. 18-carbon polyunsaturated fatty acids ameliorate palmitate-induced inflammation and insulin resistance inmouse C2C12 myotubes. *Journal of Nutritional Biochemistry*, 26(5), 521–531.
287. Charnock, J.S., 2000. Gamma-linolenic acid provides additional protection against ventricular fibrillation in aged rats fed linoleic acidrich diets. *Prostaglandins, Leukotrienes and Essential Fatty Acids*, 62(2), 129–134.
288. Bassaganya-Riera, J., Reynolds, K., Martino-Catt, S., Cui, Y., Hennighausen, L., Gonzalez, F., Rohrer, J., Benninghoff, A.U. and Hontecillas, R., 2004. Activation of PPAR γ and δ by conjugated linoleic acid mediates protection from experimental inflammatory bowel disease. *Gastroenterology*, 127(3), 777–791.
289. Chen, M.F., Yang, T.T., Yeh, L.R., Chung, H.H., Wen, Y.J., Lee, W.J. and Cheng, J.T., 2012. Activation of imidazoline I-2B receptors by allantoin to increase glucose uptake into C2C12 cells. *Hormone and Metabolic Research*, 44(4), 268–272.
290. Gwon, S.Y., Ahn, J.Y., Chung, C.H., Moon, B. and Ha, T.Y., 2012. *Lithospermum erythrorhizon* suppresses high-fat diet-induced obesity, and acetylshikonin, a main compound of *Lithospermum erythrorhizon*, inhibits adipocyte differentiation. *Journal of Agricultural and Food Chemistry*, 60(36), 9089–9096.
291. Gwon, S.Y., Ahn, J.Y., Jung, C.H., Moon, B.K. and Ha, T.Y., 2013. Shikonin suppresses ERK 1/2 phosphorylation during the early stages of adipocyte differentiation in 3T3-L1 cells. *BMC Complementary and Alternative Medicine*, 13, 207.
292. Suba, V., Murugesan, T., Arunachalam, G., Mandal, S.C. and Saha, B.P., 2004. Anti-diabetic potential of *Barleria lupulina* extract in rats. *Phytomedicine*, 11(2–3), 202–205.
293. Brass, B.J., Abelev, Z., Liao, E.P. and Poretsky, L., 2010. Endocrine pancreas. In *Principles of Diabetes Mellitus*, Poretsky, L. (Ed.). New York: Springer, pp. 37–55.
294. Rizza, R.A., Gerich, J.E., Haymond, M.W., Westland, R.E., Hall, L.D., Clemens, A.H. and Service, F.J., 1980. Control of blood sugar in insulin-dependent diabetes: Comparison of an artificial endocrine pancreas, continuous subcutaneous insulin infusion, and intensified conventional insulin therapy. *New England Journal of Medicine*, 303(23), 1313–1318.
295. Tuntiwachwuttikul, P., Pancharoen, O. and Taylor, W.C., 1998. Iridoid glucosides of *Barleria lupulina*. *Phytochemistry*, 49(1), 163–166.
296. Devi, R. and Sharma, D.K., 2004. Hypolipidemic effect of different extracts of *Clerodendrum colebrookianum* Walp in normal and high-fat diet fedrats. *Journal of Ethnopharmacology*, 90(1), 63–68.
297. Jadeja, R.N., Thounaojam, M.C., Dandekar, D.S., Devkar, R.V. and Ramachandran, A.V., 2010. *Clerodendrum glandulosum.* Coleb extract ameliorates high fat diet/fatty acid induced lipotoxicity in experimentalmodels of non-alcoholic steatohepatitis. *Food and Chemical Toxicology*, 48(12), 3424–3431.
298. Jadeja, R.N., Thounaojam, M.C., Ramani, U.V., Devkar, R.V. and Ramachandran, A.V., 2011. Anti-obesity potential of *Clerodendrum glandulosum.* Coleb leaf aqueous extract. *Journal of Ethnopharmacology*, 135(2), 338–343.
299. Herranz-López, M., Barrajón-Catalán, E., Segura-Carretero, A., Menéndez, J.A., Joven, J. and Micol, V., 2015. Lemon verbena (*Lippia citriodora*) polyphenols alleviate obesity-related disturbances in hypertrophic adipocytes through AMPK-dependent mechanisms. *Phytomedicine*, 22(6), 605–614.
300. Phakeovilay, C., Disadee, W., Sahakitpichan, P., Sitthimonchai, S., Kittakoop, P., Ruchirawat, S. and Kanchanapoom, T., 2013. Phenylethanoid and flavone glycosides from *Ruellia tuberosa* L. *Journal of Natural Medicines*, 67(1), 228–233.
301. Ghaisas, M., Navghare, V., Takawale, A., Zope, V., Tanwar, M. and Deshpande, A., 2009. Effect of Tectona *grandis* Linn. on dexamethasone-induced insulin resistance in mice. *Journal of Ethnopharmacology*, 122(2), 304–307.
302. Lemhadri, A., Zeggwagh, N.A., Maghrani, M., Jouad, H., Michel, J.B. and Eddouks, M., 2004. Hypoglycaemic effect of *Calamintha officinalis* Moench. in normal and streptozotocin-induced diabetic rats. *Journal of Pharmacy and Pharmacology*, 56(6), 795–799.
303. Herrera-Arellano, A., Aguilar-Santamaría, L., García-Hernández, B., Nicasio-Torres, P. and Tortoriello, J., 2004. Clinical trial of *Cecropia obtusifolia* and *Marrubium vulgare* leaf extracts on blood glucose and serum lipids in type 2 diabetics. *Phytomedicine*, 11(7–8), 561–566.
304. Ohtera, A., Miyamae, Y., Nakai, N., Kawachi, A., Kawada, K,, Han, J., Isoda, H. et al., 2013. Identification of 6-octadecynoic acid from a methanol extract of *Marrubium vulgare* L. as a peroxisome proliferator-activated receptor γ agonist. *Biochemical and Biophysical Research Communications*, 440(2), 204–209.

305. Weidner, C., Wowro, S.J., Freiwald, A., Kodelja, V., Abdel-Aziz, H., Kelber, O. and Sauer, S., 2014. Lemon balm extract causes potent antihyperglycemic and antihyperlipidemic effects in insulin-resistant obese mice. *Molecular Nutrition & Food Research*, 58(4), 903–907.
306. Jun, H.J., Lee, J.H., Jia, Y., Hoang, M.H., Byun, H., Kim, K.H. and Lee, S.J., 2012. *Melissa officinalis* essential oil reduces plasma triglycerides in human apolipoprotein E2 transgenic mice by inhibiting sterol regulatory element-binding protein-1c-dependent fatty acid synthesis. *Journal of Nutrition*, 142(3), 432–440.
307. Hasanein, P. and Riahi, H., 2015. Antinociceptive and antihyperglycemic effects of *Melissa officinalis* essential oil in an experimental model of diabetes. *Medical Principles and Practice*, 24(1), 47–52.
308. Yu, Q., Qi, J., Wang, L., Liu, S.J. and Yu, B.Y., 2015. Pentacyclic triterpenoids from spikes of *Prunella vulgaris* L. inhibit glycogen phosphorylase and improve insulinsensitivity in 3T3-L1 adipocytes. *Phytotherapy Research*, 29(1), 73–79.
309. Kim, M.J. and Kim, H.K., 2009. Perilla leaf extract ameliorates obesity and dyslipidemia induced by high-fat diet. *Phytotherapy Research*, 23(12), 1685–1690.
310. Ueda, H., Yamazaki, C. and Yamazaki, M., 2002. Luteolin as an anti-inflammatory and anti-allergic constituent of *Perilla frutescens*. *Biological and Pharmaceutical Bulletin*, 25(9), 1197–1202.
311. Makino, T., Furuta, Y., Fujii, H., Nakagawa, T., Wakushima, H., Saito, K.I. and Kano, Y., 2001. Effect of oral treatment of *Perilla frutescens* and its constituents on type-I allergy in mice. *Biological and Pharmaceutical Bulletin*, 24(10), 1206–1209.
312. Chen, J.H., Xia, Z.H. and Tan, R.X., 2003. High-performance liquid chromatographic analysis of bioactive triterpenes in *Perilla frutescens*. *Journal of Pharmaceutical and Biomedical Analysis*, 32(6), 1175–1179.
313. de Melo, C.L., Queiroz, M.G., Fonseca, S.G., Bizerra, A.M., Lemos, T.L., Melo, T.S., Santos, F.A. and Rao, V.S., 2010. Oleanolic acid, a natural triterpenoid improves blood glucose tolerance in normal mice and ameliorates visceral obesity in mice fed a high-fat diet. *Chemico-Biological Interactions*, 185(1), 59–65.
314. Klok, M.D., Jakobsdottir, S. and Drent, M.L., 2007. The role of leptin and ghrelin in the regulation of food intake and body weight in humans: A review. *Obesity Reviews*, 8, 21–34.
315. Rahman, N., Jeon, M., Song, H.Y., Kim, Y.S., 2016. Cryptotanshinone, a compound of Salvia miltiorrhiza inhibits pre-adipocytes differentiation by regulation ofadipogenesis-related genes expression via STAT3 signaling. *Phytomedicine*, 23(1), 58–67.
316. Hu, E., Tontonoz, P. and Spiegelman, B.M., 1995. Transdifferentiation of myoblasts by the adipogenic transcription factors PPAR gamma and C/EBP alpha. *Proceedings of the National Academy of Sciences*, 92(21), 9856–9860.
317. Liu, S., Chen, Z., Wu, J., Wang, L., Wang, H. and Zhao, W., 2013. Appetite suppressing pregnane glycosides from the roots of *Cynanchum auriculatum*. *Phytochemistry*, 93, 144–153.
318. Jang, E.J., Kim, H.K., Jeong, H., Lee, Y.S., Jeong, M.G., Bae, S.J., Kim, S., Lee, S.K. and Hwang, E.S., 2014. Anti-adipogenic activity of the naturally occurring phenanthroindolizidine alkaloid antofine via direct suppression of PPARγ expression. *Chemistry & Biodiversity*, 11(6), 962–969.
319. Morikawa, T., Imura, K., Miyake, S., Ninomiya, K., Matsuda, H., Yamashita, C., Muraoka, O., Hayakawa, T. and Yoshikawa, M., 2012. Promoting the effect of chemical constituents from the flowers of *Poacynum hendersonii* on adipogenesis in 3T3-L1 cells. *Journal of Natural Medicines*, 66(1), 39–48.
320. Jiang, L., Zhang, N.X., Mo, W., Wan, R., Ma, C.G., Li, X., Gu, Y.L., Yang, X.Y., Tang, Q.Q. and Song, H.Y., 2008. *Rehmannia* inhibits adipocyte differentiation and adipogenesis. *Biochemical and Biophysical Research Communications*, 371(2), 185–190.
321. Nishimura, H., Sasaki, H., Morota, T., Chin, M. and Mitsuhashi, H., 1989. Six iridoid glycosides from *Rehmannia glutinosa*. *Phytochemistry*, 28(10), 2705–2709.
322. Wu, H.S., Zhu, D.F., Zhou, C.X., Feng, C.R., Lou, Y.J., Yang, B. and He, Q.J., 2009. Insulin sensitizing activity of ethyl acetate fraction of *Acorus calamus* L. in vitro and in vivo. *Journal of Ethnopharmacology*, 123(2), 288–292.
323. Lee, M.-H., Chen, Y.-Y.,Tsai, J.-W., Wang S.C, Watanabe, T. and Tsai, Y.-C., 2011. Inhibitory effect of β-asarone, a component of *Acorus calamus* essential oil, on inhibition of adipogenesis in 3T3-L1 cells. *Food Chemistry*, 126(1), 1–7.
324. Kim, Y.J., Shin, Y.O., Ha, Y.W., Lee, S., Oh, J.K. and Kim, Y.S., 2006. Anti-obesity effect of *Pinellia ternata* extract in Zucker rats. *Biological and Pharmaceutical Bulletin*, 29(6), 1278–1281.
325. Liu, Y.L., Toubro, S., Astrup, A. and Stock, M.J., 1995. Contribution of beta 3-adrenoceptor activation to ephedrine-induced thermogenesis in humans. *International Journal of Obesity and Related Metabolic Disorders: Journal of the International Association for the Study of Obesity*, 19(9), 678–685.
326. Yen, T.T., McKee, M.M. and Bemis, K.G., 1980. Ephedrine reduces weight of viable yellow obese mice (Avy/a). *Life Sciences*, 28, 119–128.

327. Liu, Y., Liang, Z. and Zhang, Y., 2010. Induction and in vitro alkaloid yield of calluses and protocorm-like bodies (PLBs) from *Pinellia ternata*. *In Vitro Cellular & Developmental Biology-Plant*, 46(3), 239–245.
328. White, L.M., Gardner, S.F., Gurley, B.J., Marx, M.A., Wang, P.L. and Estes, M., 1997. Pharmacokinetics and cardiovascular effects of ma-huang (*Ephedra sinica*) in normotensive adults. *Journal of Clinical Pharmacology*, 37(2), 116–122.
329. Haller, C.A. and Benowitz, N.L., 2000. Adverse cardiovascular and central nervous system events associated with dietary supplements containing *Ephedra* alkaloids. *New England Journal of Medicine*, 343, 1833–1838.
330. Kim, J., Lee, I., Seo, J., Jung, M., Kim, Y., Yim, N. and Bae, K., 2010. Vitexin, orientin and other flavonoids from *Spirodela polyrhiza* inhibit adipogenesis in 3T3-L1 cells. *Phytotherapy Research*, 24(10), 1543–1548.
331. Alessi, D.R. and Cohen, P., 1998. Mechanism of activation and function of protein kinase B. *Current Opinion in Genetics & Development*, 8(1), 55–62.
332. Choi, S.B., Wha, J.D. and Park, S., 2004. The insulin sensitizing effect of homoisoflavone-enriched fraction in *Liriope platyphylla* Wang et Tang via PI3-kinase pathway. *Life Sciences*, 75(22), 2653–2664.
333. Park, S., Kim, D.S. and Kang, S., 2011. *Gastrodia elata* Blume water extracts improve insulin resistance by decreasing body fat in diet-induced obese rats: Vanillin and 4-hydroxybenzaldehyde are the bioactive candidates. *European Journal of Nutrition*, 50(2), 107–118.
334. Maeda, A., Kai, K., Ishii, M., Ishii, T. and Akagawa, M., 2014. Safranal, a novel protein tyrosine phosphatase 1B inhibitor, activates insulin signaling in C2C12 myotubes and improves glucose tolerance in diabetic KK-Ay mice. *Molecular Nutrition & Food Research*, 58(6), 1177–1189.
335. Han, J., Yang, N., Zhang, F., Zhang, C., Liang, F., Xie, W. and Chen, W., 2015. Rhizoma Anemarrhenae extract ameliorates hyperglycemia and insulin resistance via activation of AMP-activated protein kinase in diabetic rodents. *Journal of Ethnopharmacology*, 172, 368–376.
336. Park, U.H., Jeong, J.C., Jang, J.S., Sung, M.R., Youn, H., Lee, S.J., Kim, E.J. and Um, S.J., 2012. Negative regulation of adipogenesis by kaempferol, a component of *Rhizoma Polygonati falcatum* in 3T3-L1 cells. *Biological and Pharmaceutical Bulletin*, 35(9), 1525–1533.
337. Gu, M., Zhang, Y., Fan, S., Ding, X., Ji, G. and Huang, C., 2013. Extracts of *Rhizoma polygonati odorati* prevent high-fat diet-induced metabolic disorders in C57BL/6 mice. *PLoS One*, 8(11), e81724.
338. Gao, X., Li, B., Jiang, H., Liu, F., Xu, D. and Liu, Z., 2007. Dioscorea opposita reverses dexamethasone induced insulin resistance. *Fitoterapia*, 78(1), 12–15.
339. Yang, M.H., Chin, Y.W., Chae, H.S., Yoon, K.D. and Kim, J., 2014. Anti-adipogenic constituents from *Dioscorea opposita* in 3T3-L1 cells. *Biological and Pharmaceutical Bulletin*, 37(10), 1683–1688.
340. Gil, H.W., Lee, E.Y., Lee, J.H., Kim, Y.S., Lee, B.E., Suk, J.W. and Song, H.Y., 2015. *Dioscorea batatas* extract attenuates high-fat diet-induced obesity in mice by decreasing expression of inflammatory cytokines. *Medical Science Monitor*, 21, 489–495.
341. Hsu, H.-F., Tsou, T.-C., Chao, H.-R., Shy, C.-G., Kuo, Y.-T., Tsai, F.-Y., Yeh, S.-C. and Ko, Y.-C., 2010 Effects of arecoline on adipogenesis, lipolysis, and glucose uptake of adipocytes-A possible role of betel-quid chewing in metabolic syndrome. *Toxicology and Applied Pharmacology*, 245(3), 370–377.
342. Hsieh, T.-J., Hsieh, P.-C., Wu, M.-T., Chang, W.-C., Hsiao, P.-J., Lin, K.-D., Chou, P.-C. and Shin, S.-J., 2011. Betel nut extract and arecoline block insulin signaling and lipid storage in 3T3-L1 adipocytes. *Cell Biology and Toxicology*, 27(6), 397–411.
343. Xie, D., Chen, L., Liu, C., Zhang, C., Liu, K. and Wang, P.S., 2004. Arecoline excites the colonic smooth muscle motility via M~ 3 receptor in rabbits. *Chinese Journal of Physiology*, 47(2), 89.
344. Yang, T.T., Chang, C.K., Tsao, C.W., Hsu, Y.M., Hsu, C.T. and Cheng, J.T., 2009. Activation of muscarinic M-3 receptor may decrease glucose uptake and lipolysis in adipose tissue of rats. *Neuroscience Letters*, 451(1), 57–59.
345. Zhou, J., Sun, Q., Yang, Z. and Zhang, J., 2014. The hepatotoxicity and testicular toxicity induced by arecoline in mice and protective effects of vitamins C and E. *Korean Journal of Physiology & Pharmacology*, 18(2), 143–148.
346. Akhtar, M.S., Khan, M.A. and Malik, M.T., 2002. Hypoglycaemic activity of *Alpinia galanga* rhizome and its extracts in rabbits. *Fitoterapia*, 73(7–8), 623–628.
347. Kahn, C.R. and Schechter, Y., 1993. Insulin and oral hypoglycaemic agents. In *Pharmacological Basis of Therapeutics*, 8th ed., Goodman, L.S. and Gilman, A.G. (Eds.). New York: Maxwell-Macmillan Publishing Co, p. 1463.
348. Sivakumar, A.S. and Anuradha, C.V., 2011. Effect of galangin supplementation on oxidative damage and inflammatory changes in fructose-fed rat liver. *Chemico-Biological Interactions*, 193(2), 141–148.

349. Yamamoto, N., Kawabata, K., Sawada, K., Ueda, M., Fukuda, I., Kawasaki, K., Murakami, A. and Ashida, H., 2011. Cardamonin stimulates glucose uptake through translocation of glucose transporter-4 in L6 myotubes. *Phytotherapy Research*, 25(8), 1218–1224.
350. Jung, C.H., Jang, S.J., Ahn, J., Gwon, S.Y., Jeon, T., Kim, T.W. and Ha, T.Y., 2012. *Alpinia officinarum* inhibits adipocyte differentiation and high-fat diet-induced obesity in mice through regulation of adipogenesis and lipogenesis. *Journal of Medicinal Food*, 15(11), 959–967.
351. Sztalryd, C., Xu, G., Dorward, H., Tansey, J.T., Contreras, J.A., Kimmel, A.R. and Londos, C., 2003. Perilipin A is essential for the translocation of hormone-sensitive lipase during lipolytic activation. *The Journal of Cell Biology*, 161(6), 1093–1103.
352. Melmed, S., Polonsky, K.S., Larsen, P.R. and Kronenberg, H.M., 2015. *Williams Textbook of Endocrinology*. Philadelphia, PA: Elsevier Health Sciences.
353. Banerji, M.A. and Chaiken, R.L., 2010. Insulin resistance and the metabolic syndrome. In *Principles of Diabetes Mellitus*, Poretsky, L. (Ed.). New York: Springer, pp. 531–555.
354. Mpalantinos, M.A., Soares de Moura, R., Parente, J.P. and Kuster, R.M., 1998. Biologically active flavonoids and kava pyrones from the aqueous extract of *Alpinia zerumbet*. *Phytotherapy Research*, 12(6), 442–444.
355. Tu, P.T. and Tawata, S., 2014. Anti-obesity effects of hispidin and Alpinia zerumbet bioactives in 3T3-L1 adipocytes. *Molecules*, 19(10), 16656–16671.
356. Lahlou, S., Interaminense, L.F., Leal-Cardoso, J.H. and Duarte, G.P., 2003. Antihypertensive effects of the essential oil of *Alpinia zerumbet* and its main constituent, terpinen-4-ol, in DOCA-salt hypertensive conscious rats. *Fundamental & Clinical Pharmacology*, 17(3), 323–330.
357. Kang, Y. and Kim, H.Y., 2004. Glucose uptake-stimulatory activity of Amomi Semen in 3T3-L1 adipocytes. *Journal of Ethnopharmacology*, 92(1), 103–105.
358. Kim, D.Y., Kim, M.S., Sa, B.K., Kim, M.B. and Hwang, J.K., 2012. *Boesenbergia pandurata* attenuates diet-induced obesity by activating AMP-activated protein kinase and regulating lipid metabolism. *International Journal of Molecular Sciences*, 13(1), 994–1005.
359. Kuroda, M., Mimaki, Y., Nishiyama, T., Mae, T., Kishida, H., Tsukagawa, M., Takahashi, K., Kawada, T., Nakagawa, K. and Kitahara, M., 2005. Hypoglycemic effects of turmeric (*Curcuma longa* L. rhizomes) on genetically diabetic KK-Ay mice. *Biological and Pharmaceutical Bulletin*, 28(5), 937–939.
360. Horikawa, T., Shimada, T., Okabe, Y., Kinoshita, K., Koyama, K., Miyamoto, K., Ichinose, K., Takahashi, K. and Aburada, M., 2012. Polymethoxyflavonoids from *Kaempferia parviflora* induce adipogenesis on 3T3-L1 preadipocytes by regulating transcription factors at an early stage of differentiation. *Biological and Pharmaceutical Bulletin*, 35(5), 686–692.
361. Li, Y., Tran, V.H., Duke, C.C. and Roufogalis, B.D., 2012. Gingerols of *Zingiber officinale* enhance glucose uptake by increasing cell surface GLUT4 in cultured L6 myotubes. *Planta Medica*, 78(14), 1549–1555.
362. Saifudin, A., Kadota, S. and Tezuka, Y., 2013. Protein tyrosine phosphatase 1B inhibitory activity of Indonesian herbal medicines and constituents of *Cinnamomum burmannii* and *Zingiber aromaticum*. *Journal of Natural Medicine*, 67(2), 264–270.
363. Li, Y., Tran, V.H., Koolaji, N., Duke, C. and Roufogalis, B.D., 2013. (S)-[6]-Gingerol enhances glucose uptake in L6 myotubes by activation of AMPK in response to [Ca2+] i. *Journal of Pharmacy & Pharmaceutical Sciences*, 16(2), 304–312.
364. Li, Y., Tran, V.H., Kota, B.P., Nammi, S., Duke, C.C. and Roufogalis, B.D., 2014. Preventative effect of *Zingiber officinale* on insulin resistance in a high-fat high-carbohydrate diet-fed rat model and its mechanism of action. *Basic & Clinical Pharmacology & Toxicology*, 115(2), 209–215.
365. Misawa, K., Hashizume, K., Yamamoto, M., Minegishi, Y., Hase, T. and Shimotoyodome, A., 2015. Ginger extract prevents high-fat diet-induced obesity in mice via activation of the peroxisome proliferator-activated receptor δ pathway. *Journal of Nutritional Biochemistry*, 26(10), 1058–1067.
366. Tzeng, T.F. and Liu, I.M., 2013. 6-gingerol prevents adipogenesis and the accumulation of cytoplasmic lipid droplets in 3T3-L1 cells. *Phytomedicine*, 20(6), 481–487.
367. Raut, N.A. and Gaikwad, N.J., 2006. Antidiabetic activity of hydro-ethanolic extract of *Cyperus rotundus* in alloxan induced diabetes in rats. *Fitoterapia*, 77(7–8), 585–588.
368. Lemaure, B., Touché, A., Zbinden, I., Moulin, J., Courtois, D., Macé, K. and Darimont, C., 2007. Administration of *Cyperus rotundus* tubers extract prevents weight gain in obese Zucker rats. *Phytotherapy Research*, 21(8), 724–730.

5 Inhibiting Low-Density Lipoproteins Intimal Deposition and Preserving Nitric Oxide Function in the Vascular System

Vascular endothelial cells release nitric oxide which relaxes and suppresses abnormal proliferation vascular smooth muscle cells and inhibit low-density lipoprotein oxidation involved in atherogenesis.[1] In Metabolic Syndrome, hyperglycemia, advanced glycation end products, low-density lipoproteins, proinflammatory cytokines, and reactive oxygen species synergistically decrease the bioavailability of nitric oxide in the vascular system resulting in what is termed "vascular endothelial dysfunction".[2] Insulin resistance itself decreases the ability of endothelial cells to produce nitric oxide.[3] Besides, hyperlipidemia in Metabolic Syndrome is associated with the chronic deposition and oxidation of low-density lipoproteins in the arterial intima.[4] Oxidized intimal low-density lipoprotein induce vascular endothelial cells to express adhesion molecules that accounts for the entry of monocytes in the subendothelial space.[4] Monocytes in contact with oxidized low-density lipoprotein differentiate in macrophages that accumulate cholesterol to form foam cells initiating thus atherosclerosis and subsequent risks of stroke and coronary heart diseases.[4–7] Thus, inhibiting low-density lipoproteins intimal deposition and oxidation and preserving nitric oxide function in arteries constitutes one therapeutic strategy to prevent cardiovascular pathologies associated with diabetes and obesity.

5.1 *Myristica fragrans* Hout.

Synonyms: *Myristica aromatica* Lam.; *Myristica moschata* Thunb.; *Myristica officinalis* L.f.
Common names: buah pala (Malay); ru du ku (Chinese); nutmeg
Subclass Magnoliidae, Superorder Magnolianae, Order Myristicales, Family Myristicaceae
Medicinal use: facilitate digestion (Malaysia)
History: the plant was known to Avicenna (980–1037 AD), physician in Bagdad

In endothelial cells, adiponectin from adipose tissues promotes vascular sensitivity by activating endothelial nitric oxide synthetase via adenosine monophosphate-activated protein kinase phosphorylation.[8] Another stimulus is an increase of cytosolic calcium translate into calcium calmodulin-dependent protein kinase II activation, stimulation of adenosine monophosphate-activated protein kinase, activation of endothelial nitric oxide synthetase resulting in increased nitric oxide production.[9] Nitric oxide is the principal endothelium-derived smooth muscle relaxing factor in larger arteries, whereas prostacyclin and endothelium-derived hyperpolarizing factor are predominant in smaller

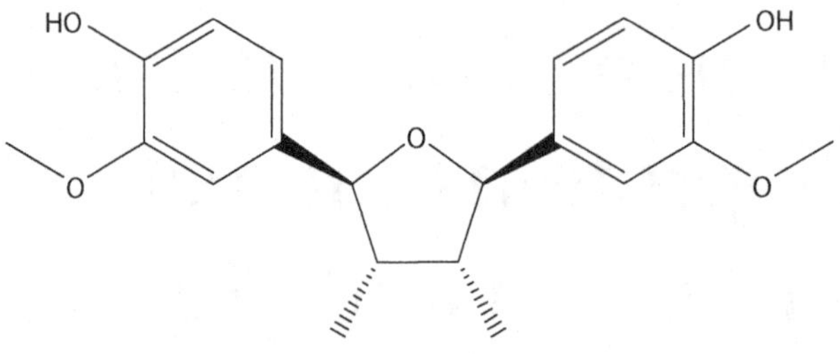

FIGURE 5.1 Nectandrin B.

vessels such as mesenteric vessels, coronary arteries, and peripheral resistance vessels.[10] Nectandrin B (Figure 5.1) at a concentration of 10 μg/mL induced endothelial nitric oxide synthetase in human umbilical endothelial cells and boosted the secretion of nitric oxide by ECV 304 cells.[11] In ECV 304 cells, this lignan induced at a concentration of 10 μg/mL of the phosphorylation of calcium calmodulin-dependent protein kinase II, adenosine monophosphate-activated protein kinase, and acetyl-CoA carboxylase and Akt.[11]

5.2 *Illigera luzonensis* (C. Presl) Merr.

Synonyms: *Gronovia ternata* Blanco; *Henschelia luzonensis* C. Presl; *Illigera meyeniana* Kunth ex Walp.; *Illigera pubescens* Merr.; *Illigera ternata* (Blanco) Dunn
Common name: tai wan qing teng (China)
Subclass Magnoliodae, Superorder Lauranae, Order Laurales, Family Hernandiaceae Blume
Medicinal use: headache (Philippines)

The aggregation of platelets is essential for hemostatic plug formation and vessel wall repair, but plays an important role in the pathophysiology of myocardial infarction and cerebrovascular stroke in metabolic syndrome.[12] Actinodaphnine, *N*-methylactinodaphnine, dicentrine, *O*-methylbulbocapnine, and liriodenine from *Illigera luzonensis* (C. Presl) Merr. at a concentration of 100 μg/mL inhibited the aggregation of platelets evoked by ADP, arachidonic acid, and collagen by more than 50%.[13] Actinodaphnine inhibited at a concentration of 100 μg/mL the tonic contractions of isolated aorta induced by norepinephrine by 84.5%.[13] Transient elevation of serum cholesterol in Metabolic Syndrome causes vasoconstriction and/or increased reactivity to vasoactive agents in several vascular beds.[14] Dicentrine is an α1-adrenoceptor blocker. At a concentration of 10 μM this alkaloid abrogated the contractions of aortic rings evoked by noradrenaline and phenylephrine.[15] It might be for that reason that at a concentration of 150 μM, this aporphine alkaloid completely blocked the aggregation of platelets and release of ATP induced by adrenaline.[15] It also reduced the aggregation of platelets exposed to arachidonic acid and collagen by circa 80% and 45%, respectively.[16] Furthermore, dicentrine inhibited the release of ATP, cytoplasmic calcium increase, and thromboxane B2 formation by platelets exposed to arachidonic acid.[16] Dicentrine given orally at a dose of 10 mg/kg twice a day for 4 weeks lowered blood pressure, cholesterolemia, and triglyceridemia, low-density lipoprotein–cholesterol, very low-density lipoprotein and apolipoprotein B of spontaneously hypertensive rats given high-fat and high-cholesterol diet. High density lipoprotein–cholesterol was increased.[17] In spontaneously hypertensive rats, this aporphine alkaloid evoked a decrease in mean arterial blood pressure, total cholesterol, and triglycerides.[17] Furthermore, dicentrine attenuated the contractility of aortic arch and abdominal aorta of fat-challenged rodents but did not prevent early atherosclerosis development.[17] Dicentrine treatment had no effects on aortic relaxation induced by acetylcholine nor nitroprusside.[17] Liriodenine converted polymorphic ventricular tachyarrhythmia induced by the ischemia-reperfusion of isolated hearts of rats with an EC_{50} of 0.3 μM. It also increased the force of contractions of right ventricular strips with an IC_{50} value of 5.5 μM through the inhibition of myocytes sodium channel.[18] This aporphine alkaloid given prophylactically at a dose of 1 μM to hearts isolated from male Sprague–Dawley rats subjected to 30 minutes of ischemia and 120 minutes of reperfusion of the left ventricle in all experimental groups reduced the size of infarcted tissues from 48.6% to 21.2%. This effect was nullified with nitric oxide synthetase inhibitor N^{ω}-nitro-L-arginine methyl ester.[18] Besides, this alkaloid boosted coronary flow by 22.7%, sustained the production of nitric oxide, and increased the expression of endothelial nitric oxide synthetase in myocytes.[18] In human umbilical vein endothelial cells deprived of serum, liriodenine at 1 μM increased cell viability from 40.4% to 63.2%, decreased apoptosis by more than 30%, and the latter effect was annihilated by N^{ω}-nitro-L-arginine methyl ester.[19] In hypercholesterolemic patients, increased platelet aggregation favors cardiovascular diseases and one can speak of a "prothrombic state."[20] Liriodenine inhibited the aggregation of platelets evoked by arachidonic acid, adenosine diphosphate, and collagen *in vitro*.[21] These aporphines are too toxic to be used in therapeutic.

5.3 *Cinnamomum cassia* (L.) J. Presl

Synonyms: *Cinnamomum aromaticum* Nees; *Laurus cassia Nees* & T. Nees
Common names: kayuh manis (Malay); Chinese cinnamon
Subclass Magnoliidae, Superorder Lauranae, Order Laurales, Family Lauraceae
Medicinal use: tonic (Malaysia)

A feature of type 2 diabetes is a chronic inflammatory state during which proinflammatory cytokines such as tumor necrosis factor-α, interleukin-6, and interleukin-1β generate vascular endothelial cells dysfunction.[22–24] 2-Methoxycinnamaldehyde (Figure 5.2) isolated from *Cinnamomum cassia* (L.) J. Presl at 50 μM induced the translocation of Nrf-2 and the expression of heme oxygenase-1 in human umbilical vein endothelial cells.[25] 2-Methoxycinnamaldehyde at 50 μM inhibited the production of vascular adhesion molecule-1 via nuclear factor-κB by human endothelial cells challenged with tumor necrosis factor-α.[25] Sprague–Dawley rats given prophylactically and intraveinously 200 μg/mL of 2-methoxycinnamaldehyde were protected from experimental cardiac infarction as evidenced by decreased plasma levels of malondialdehyde.[25] Malondialdehyde in the serum of rodents subjected to ischemia/reperfusion treated with 2-methoxycinnamaldehyde were decreased, and superoxide dismutase was increased evidencing reduction in oxidative stress.[25] Furthermore, the number of infiltrating neutrophils in myocardial tissues were reduced by the treatment.[25]

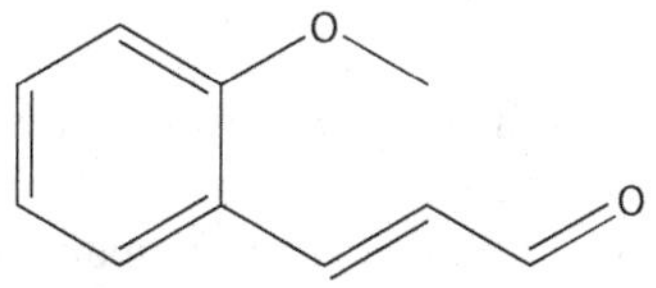

O
O

FIGURE 5.2 2-Methoxycinnamaldehyde.

5.4 *Lindera obtusiloba* Bl.

Synonym: *Benzoin obtusilobum* (Blume) Kuntze
Common name: san ya wu yao (Chinese)
Subclass Magnoliidae, Superorder Lauranae, Order Laurales, Family Lauraceae
Medicinal use: promotes liver function (Korea)

Located between vascular lumen and vascular smooth muscle cells of the vessel wall, vascular endothelium cells behave like barometers which secrete nitric oxide (also known as endothelium-derived relaxing factor) that relaxes vascular smooth muscles when blood pressure increases.[26] Ethanol extract of *Lindera obtusiloba* Bl. induced the relaxation of intact aortic rings by 80% at a concentration of 10 μg/mL and mitigated aortic contractions induced by phenylephrine or angiotensin II.[27] Competitive inhibitor of nitric oxide synthase, N^{ω}-nitro-L-arginine, blunted the aortic relaxation induced by the extract suggesting the involvement of nitric oxide.[27] In bovine aortic endothelial cells, this extract commanded the phosphorylation of Akt and endothelial nitric oxide synthetase.[27] Physiologically, insulin signaling increases endothelial nitric oxide synthetase phosphorylation through the phosphoinositide 3-kinase-Akt pathway. In a state of insulin resistance, Akt kinase activity is decreased, resulting in decreased endothelial nitric oxide synthetase activity.[3] Sprague–Dawley rats receiving infusion of angiotensin II were protected against high systolic blood pressure by oral administration of this extract for 14 days at a dose of 100 mg/kg/day.[27] Vascular smooth muscle cells from treated rats evidenced an inhibition of NADPH oxidase activity.[27]

5.5 *Persea americana* Mill.

Synonyms: *Laurus persea* L.; *Persea edulis* Raf.; *Persea gratissima* C.F. Gaertn.
Common names: e li (Chinese); avocado
Subclass Magnoliidae, Superorder Lauranae, Order Laurales, Family Lauraceae
Medicinal use: diarrhea (Philippines)

Consumption of fried fatty meals generates a transient postprandial "plasmatic oxidative stress" translating into activation of nuclear factor-κB in endothelial cells and subsequent expression of adhesion molecules such as vascular cell adhesion molecule-1 and intercellular adhesion molecule-1.[28,29] Peripheral blood mononuclear cells isolated from volunteers after fatty meal mixed with avocado pulp had a slight elevation of IkB-α suggesting a decrease in nuclear factor-κB activity and decrease in serum interleukin-6 compared with untreated group.[30] Consumption of avocado could be of value in Metabolic Syndrome.

5.6 *Litsea cubeba* (Lour.) Pers.

Synonyms: *Litsea citrata* Bl.; *Laurus cubeba* Lour.; *Daphnidium cubeba* Nees
Common name: shan ji jiao (Chinese)
Subclass Magnoliidae, Superorder Lauranae, Order Laurales, Family Lauraceae
Medicinal use: facilitates digestion (China)

In response to platelet-derived growth factor, transforming growth factor-β, cGMP and oxidative stress, vascular smooth muscle cells proliferate and migrate from the tunica media to the intima to deposit extracellular matrix protein and fibrous elements inducing intimal thickening and luminal stenosis contributing to atherogenesis.[31] Litebamine from *Litsea cubeba* (Lour.) Pers. inhibited the adhesion of smooth muscle cells to collagen at a concentration of 5 μM.[32] Besides, this phenanthrene alkaloid at a concentration of 20 μM inhibited platelet-derived growth factor induced smooth muscle cells migration.[32]

5.7 *Piper taiwanense* Lin & Lu

Common name: tai wan hu jiao (Chinese)
Subclass Magnoliidae, Superorder Piperanae, Order Piperales, Family Piperaceae
Medicinal use: tonic (Taiwan)

The detachment of thrombus from atheroma and subsequent blockage of coronary and cerebral vascular system account for myocardial infarction and cerebrovascular stroke during metabolic syndrome.[33] Maintaining the fluidity of plasma and inhibiting platelet aggregation are means to prevent the aforementioned complications.[33] Piperolactam E isolated from *Piper taiwanense* Lin & Lu at a concentration of 100 μg/mL reduced the aggregation of platelets exposed to arachidonic acid and collagen by 50% and 100%, respectively.[34] From the same plant, piperolactam B at 100 μM inhibited the aggregation of platelets evoked by thrombin, arachidonic acid collagen, and platelet-activating factor by more than 60%.[34]

5.8 *Nelumbo nucifera* Gaertn.

Synonyms: *Nelumbium nuciferum* Gaertn.; *Nelumbo speciosa* Willd.; *Nymphaea nelumbo* L.
Common names: lian (Chinese); sacred lotus
Subclass Ranunculidae, Superorder Proteanae, Order Nelumbonales, Family Nelumbonaceae
Medicinal use: anxiety (China)

The progressive deposition of cholesteryl ester and fibrous elements in arterial intima leads to atherosclerosis.[35] Low-density lipoproteins penetrate the intima, accumulate within extracellular connective proteoglycans and collagen, and are oxidized to oxidized low-density lipoprotein, which are phagocyted by macrophages.[35] Boldine at 2.5 μM prevented *in vitro* the formation of conjugated diene by low-density lipoprotein oxidized with copper or hydrogen peroxide.[36] Given at a dose of 1 or 5 mg orally, once a day, 5 days a week, for 12 weeks to low-density lipoprotein receptor knockout rodents fed with a high-fat diet, this aporphine alkaloid did not decrease serum cholesterol, low-density lipoprotein, and high-density lipoprotein or triglycerides.[36] However, 1 and 5 mg regimen of boldine reduced the formation of atherosclerotic lesions by 22% and 40%, respectively Antioxidant activity of blood drawn for the treated rodent exhibited reduced oxidizability to copper.[36] In type 2 diabetes; increased plasmatic reactive oxygen species lead to endothelial dysfunction, increased contractility and hypertension, and increased the risk of coronary heart disease, stroke, and nephropathy.[37] Hyperglycemia increases blood pressure via the generation of reactive oxygen species that are associated with an increase in NADPH oxydase activity.[38] Besides, increased glycemia is associated with inhibition of nitric oxide synthetase activity resulting in decreased levels of nitric and increased blood pressure.[39] Boldine at a concentration of 1 μM protected rat aortic endothelial cells against high glucose induced upregulation of NADPH subunits, NOX2, and $p47^{phox}$ and therefore reactive oxygen species.[40] In the same experiment, boldine at a dose of 20 mg/kg/day given intraperitoneally mitigated the glycemia of streptotozocin-induced diabetic rodents, whereby plasma triglycerides, high-density lipoprotein, and cholesterol were unaffected.[40] Boldine at a concentration of 1 μM facilitated the endothelium-dependent relaxation of aortic rings oxidatively challenged by β-NADPH.[40] In addition, boldine at a dose of 20 mg/kg/day given intraperitoneally normalized the acetylcholine-induced endothelium-dependent relaxation of aortic rings of diabetic rodents.[40] In a concurrent study, boldine given once a day orally at a dose of 50 mg/kg for 10 weeks evoked a reduction of glycemia in streptozocin-diabetic rodents from 371 to 253 mg/dL and a correction of mean blood pressure from 147 to 119 mmHg.[41] Boldine regimen protected the renal function of streptozocin-diabetic rodents as evidenced by a normalization of proteinuria/creatininuria from 14.8 to 3.4, (1.7 observed in normal rodents).[41] Malonyldialdehyde in renal tissue was reduced in diabetic rodents from 4.1 to 2.9 μmol/g.[41] From this plant, neferine *in vitro* abrogated the aggregation of platelets challenged with collagen or thrombin, and inhibited dense granule secretion.[42] At 3 μM, neferine disaggregated collaged or thrombine aggregates by 20%.[42] Isoproterenol is a β-receptor agonist used to cause severe myocardial necrosis in rodents.[43] This neferine given orally at a dose of 10 mg/kg/daily to Wistar rats for 30 days before isoproterenol challenge prevented myofibrillary degeneration and normalized the enzymatic activity of serum lactate dehydrogenase, creatinine kinase, and aspartate transaminase.[44] The reduced enzymatic activities of superoxide dismutase, gluthatione peroxidase, catalase, and depletion of glutathione upon isoproterenol intoxication were replenished by neferine treatment.[44] Furthermore, this phenolic alkaloid brought close to nonpathological values the serum concentration of cholesterol and triglycerides, low-density lipoprotein high-density lipoprotein and very low-density lipoprotein.[44]

5.9 *Piper kadsura* (Choisy) Ohwi

Synonyms: *Piper arboricola* C. DC.; *Piper futokadsura* Siebold; *Piper subglaucescens* C. DC.
Common name: feng teng (Chinese)
Subclass Magnoliidae, Superorder Piperanae, Order Piperales, Family Piperaceae
Medicinal use: promotes digestion (China)

Piperlactam S from *Piper kadsura* (Choisy) Ohwi (Figure 5.3) at 20 μM blocked the production of conjugated diene from low-density lipoproteins exposed to copper.[45] At a 10 μM, piperlactam S inhibited the production of malondialdehyde by bovine arterial endothelial cells challenged with ferrum.[45] In aortic rings, a pretreatment by this alkaloid at a concentration of 1 μM mitigated the deleterious effects of hydrogen peroxide and ferrum against acetylcholine-induced vasorelaxation.[45]

FIGURE 5.3 *Piper kadsura* (Choisy) Ohwi.

5.10 *Piper longum* L.

Synonym: *Chavica roxburghii* Miq.
Common names: bi ba (Chinese); long pepper
Subclass Magnoliidae, Superorder Piperanae, Order Piperales, Family Piperaceae
Medicinal use: facilitates digestion (China)

Following injury of endothelium platelets bind to the von Willebrand factor and secrete ADP and thromboxane A2.[46] Thromboxane A2 is a potent inducer of platelet aggregation and vasoconstrictor, and antagonists of thromboxane A2 could be of value for the management of Metabolic Syndrome.[47] Crude ethanolic extract of seeds of *Piper longum* L. at a concentration of 200 μg/mL reduced by 100% aggregation of platelets commanded by U46619 *in vitro*.[48] The extract inhibited U46619-induced platelet aggregation with an IC_{50} equal to 70 μg/mL on account of receptor antagonism and inhibition of phosphoinositide hydrolysis.[48] Thromboxane A2 binds to thromboxane A2 receptor to induce G protein-coupled phospholipase Cβ, leading to platelet activation by increasing cytoplasmic contents in calcium and activating protein kinase C.[47] The same extract at a concentration of 200 μg/mL did not inhibit the aggregation of platelets evoked by thrombin, nor ADP nor serotonin.[48] Piperine (110 μM), piperlongumine (14 μM), or β-(3,4,5-trimethoxyphenyl) propionic acid (7.5 μM) did not inhibit U46619-induced platelet aggregation.[48] Monocytes penetrate in the intima where they differentiate into macrophages which take up oxidized low-density lipoproteins via scavenger receptors including the class A scavenger receptor, class B scavenger receptor, class B receptor type-I, and lectin-like oxidized low-density lipoprotein receptor-1.[49] In macrophages, cholesterol derived from oxidized low-density lipoprotein is esterified by acyl CoA: cholesterol acyl transferase into cholesteryl ester and accumulation of cholesteryl ester in macrophages lead to the formation of foam cells.[50] Piperine at a concentration of 19 μM reduced the accumulation of lipid droplets in macrophages exposed to liposomes.[51] In the same experiment, piperine inhibited the synthesis of cholesteryl ester with an IC_{50} value of 18 μM.[51] Furthermore, piperine inhibited the enzymatic activity of acyl CoA: cholesterol acyl transferase.[51] This amide alkaloid administered intraperitoneally at a dose of 30 mg/kg to rodents evoked a decrease of mean arterial blood pressure of 34.6% and superior to the effect of verapamil at a dose of 0.3 mg/kg.[52] At 1 μM, piperine reduced the force and rate of ventricular contraction and coronary flow in isolated heart preparations of rodents with EC_{50} value below 10 μM.[52] Piperine commanded the relaxation of isolated aorta of rodents with or without endothelium with EC_{50} values equal to about 20 μM.[52] Piperlongumine at a concentration of 50 μM applied to exposed adventitia of the left partially ligated carotid artery of apolipoprotein E knockout mice fed with high-fat diet reduced after 2 weeks lesion area and plaque size by more than 70%.[53] Furthermore, piperlongumine inhibited at 5 μM vascular smooth muscle cells sprouting by 89.7%. This piperidinic amide inhibited the activation of platelet-derived growth factor receptor-β, phospholipase C (PLC), extracellular signal-regulated kinase-1/2, Akt, and nuclear factor-κB in vascular smooth muscle cells stimulated by platelet-derived growth.[53]

5.11 *Stephania cephalantha* Hayata

Synonyms: *Stephania disciflora* Hand.-Mazz.; *Stephania tetrandra* var. *glabra* Maxim.
Common name: fen fang ji (Chinese)
Subclass Ranunculidae, Superorder Ranunculanae, Order Menispermales, Family Menispermaceae
Medicinal use: tuberculosis (Taiwan)

Collagen and ADP raise cytoplasmic calcium concentration in platelet.[54] Cepharantine from *Stephania cephalantha* Hayata at a concentration of 40 μM inhibited the aggregation of platelets challenged by collagen and at 100 μM, blocked arachidonic acid-induced platelet aggregation.[55] This bisisoquinoline alkaloid at 50 μM inhibited calcium influx in platelets exposed to collagen.[55]

Furthermore, cepharantine reduced the release of arachidonic acid from platelets exposed to collagen from 26.8% to 7%.[55] This alkaloid is toxic.

5.12 *Stephania tetrandra* S. Moore

Synonyms: *Stephania disciflora* Hand.-Mazz.; *Stephania tetrandra* var. *glabra* Maxim.
Common name: fen fang ji (Chinese)
Subclass Ranunculidae, Superorder Ranunculanae, Order Menispermales, Family Menispermaceae
Medicinal use: tuberculosis (Taiwan)

Stephania tetrandra S. Moor contains tetrandrine and fangchinoline which at a concentration of 64 μM inhibited the aggregation of platelets exposed to platelet-activating factor by about 60% and 80%, respectively.[56] Tetrandrine and fangchinoline inhibited platelet activating factor-induced platelet aggregation with IC_{50} values equal to 28.6 and 21.7 μM, respectively.[56] Tetrandrine and fangchinoline inhibited platelet activating facts (PAF) receptor by 34.6% and 20.0% at 160 μM.[56] Extract of roots or tetrandrine given at concentrations of 18.6 and 18.7 μg/mL, respectively, to isolated hearts of Sprague–Dawley rats subjected to myocardial ischemia and reperfusion reduced infarct size to 17.5% and 19.6%, respectively.[57] The ventricular arrhythmia that occurred during the first 10 minutes after reperfusion exhibited a reduction of premature ventricular contractions under the extract or alkaloid influence, and these effects were superior to verapamil at 0.1 μM.[57] Tetrandrine given orally at a dose of 50 mg/kg/day for 10 weeks protected rodents against body weight loss, cardiac hypertrophy, hypertension, and reduction of coronary flow induced by left nephrectomization and sodium deoxycorticosterone poisoning.[58] Furthermore, tetrandrine at a dose of 2 mg/kg reduced the size of experimental cardiac infarct from 34% to 19% and lowered reperfusion-induced arrhythmia by 30%.[58] Fangchinoline at 3 μM prophylactically inhibited the proliferation and DNA synthesis of primary cultured rat aortic smooth muscle cells *in vitro* stimulated by platelet-derived growth factor by more than 50%.[59] In addition, this alkaloid blocked cells in G_0/G_1 phase and inhibited the activation of extracellular signal-regulated kinase-1/2 induced by platelet-derived growth factor by 74.2%.[59] Extract of roots at a concentration of 91 μg/mL inhibited neovascularization of retinal and choroidal capillaries in streptozotocin-induced diabetic rats.[60] From the extract, tetrandrine at 10 μM inhibited neovascularization of the retinal and choroidal capillaries in streptozotocin-induced diabetic rats.[60] This bisisoquinoline alkaloid given orally once a day, for 3 days, at a dose of 100 mg/kg decreased blood pressure of spontaneously hypertensive rodents from 177.4 to 128.7 mmHg.[61]

5.13 *Sinomenium acutum* (Thunb.) Rehder & E.H. Wilson

Synonyms: *Cocculus diversifolius* Miq.; *Cocculus heterophyllus* Hemsl. & E.H. Wilson; *Menispermum acutum* Thunb.; *Menispermum diversifolium* (Miq.) Gagnep.; *Sinomenium diversifolium* Diels
Common name: feng long (Chinese)
Subclass Ranunculidae, Superorder Ranunculanae, Order Menispermales, Family Menispermaceae
Medicinal use: arthritis (China)

Hypertension is one of the greatest public health problems, and it is estimated that by 2025 about 60% of the world population will suffer from it.[62] Single intraperitoneal injection of sinomenine at a dose of 10 mg/kg evoked a fall of systemic blood pressure in spontaneously hypertensive rodents from about 200 to 160 mmHg after 30 minutes. It was inactive against normotensive rodents.[63] At 10 μmol/L, sinomenine inhibited the contraction forces of isolated aortic rings precontracted with phenylephrine or potassium to 23.6% and 22.4%, respectively.[63] Phenylephrine-induced contraction is mediated by an increase of calcium influx via receptors-operated channels[63] and voltage-sensitive channels.[64] ATP-sensitive potassium channel blocker inhibited the vasorelaxing effect of sinomenine on aortic rings contractions induced by phenylephrine or potassium.[63] A7r5 rat aortic smooth muscle cells treated with phenylephrine or potassium and sinomenine at a concentration of 10 μmol/L halved intracellular calcium concentration.[63] This morphinan at a concentration of 200 μM prevented the activation of vascular smooth muscle cells isolated from the thoracic of Sprague–Dawley rats upon platelet-derived growth factor.[65] This alkaloid also moderated the proliferation of vascular smooth muscle cells.[65] Sinomenine inhibited the phosphorylation of platelet-derived growth factor receptor-β, extracellular signal-regulated kinase-1/2, p38 adenosine monophosphate-activated protein kinase, Akt, glycogen synthase kinase-3, and STAT3 induced by platelet-derived growth factor.[65] Sinomenine at a dose of 150 mg/kg/day given orally for 2 weeks protected male C57/BL6 mice against the formation of neointima after carotid artery injury.[65] Sinomenine has a chemical structure close to morphine and is known to activate μ-opioid receptors and to induce histamine release.[66,67]

5.14 *Cocculus hirsutus* (L.) Diels

Synonyms: *Cocculus villosus* DC.; *Menispermum hirsutum* L.
Common names: kattukodi (India); ink-berry
Subclass Ranunculidae, Superorder Ranunculanae, Order Menispermales, Family Menispermaceae
Medicinal use: diuretic (India)

Current drugs used to treat hypertension include diuretics such as furosemide.[68] Ethanol extract of leaves of *Cocculus hirsutus* (L.) Diels given once to male Wistar rats orally at a dose of 400 mg/kg commanded urine output after 6 hours and with potencies superior to as furosemide (at a dose of 10 mg/kg).[69] The extract given at the same dose once a day for 28 days elicited diuretic effects superior to daily administration of furosemide (at a dose of 10 mg/kg).[69] Chronic administration of extract evoked an increase in natriuria and kaliuria suggesting a natriuretic effect. It evoked an increase in creatine in urine.[69] Loop diuretics like furosemide are the most powerful of all diuretics, and these inhibit the sodium–potassium–chlorine cotransporter system in the thick ascending loop of the nephron, thereby increasing natriuresis and kaliuresis.[70] These diuretics also cause acidification of urine.[71]

5.15 *Berberis wallichiana* DC

Common names: hoang mu (Vietnamese); Wallich's barberry.
Subclass Ranunculidae; Superorder Ranunculanae; Order Berberidales; Family Berberidaceae
Medicinal use: diarrhoea (Vietnam)

Berberine given parenterally at a single dose of 10 mg/kg produced a reduction of mean arterial blood pressure of 28% mmHg and a reduction of heart rate by 18% in anesthetized rodents.[72] Natural products able to inhibit platelet aggregation induced by several types of procoagulant factors target common cellular step of signal transduction. Berberine at a dose of 30 mg/kg given intravenously to rodents reduced after 1 hour platelet aggregation evoked by ADP, collagen, arachidonic acid, and calcium ionophore A23187 by 50%, 52.5%, 34.4%, and 24.6%, respectively.[73] *In vitro*, berberine at 1 mmol/L inhibited the synthesis of Thromboxane B2 (TXB2) by platelets challenged with ADP, collagen, and arachidonic acid by 75%, 80.6%, and 28.8%, respectively.[73] The administration of berberine intravenously at a dose of 50 mg/kg induced a decrease of prostaglandin I2 (PGI2) synthesis by 44.7% after 1 hour.[73] Berberine at a dose of 100 mg/kg reduced the activity of renal aldose reductase in diabetic rodents.[74] In mesangial cells, berberine at 90 μM

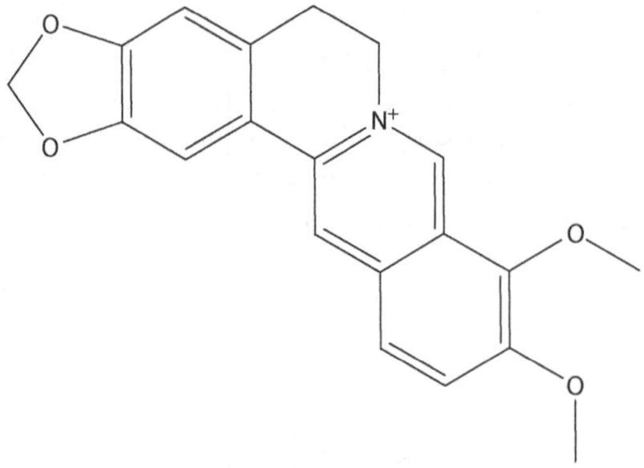

FIGURE 5.4 Berberine.

halved the enzymatic activity of aldose reductase.[74] Berberine given intraperitoneally at a dose of 5 mg/kg/day to ApoE$^{-/-}$ mice fed with a high-fat diet reduced food intake and induced weight loss. After 15 weeks induced atherosclerotic lesions 50% larger than untreated rodents on account of foam cells formation.[75] The formation of foam cells from macrophages by berberine was confirmed *in vitro* (Figure 5.4). In RAW264.7 cells, it induced SR-A induction as a result of peroxisome proliferator-activated receptor-γ phosphorylation and activation of Akt.[75] However, in other study Berberine at a concentration of 20 μM attenuated lipid accumulation in macrophages exposed to oxidized low-density lipoprotein for 24 hours with simultaneous efflux of cholesterol, induction of liver X receptor-α, and expression of (ABCA1).[76] It must be recalled that high-density lipoproteins are secreted by the liver and account for the collection of unesterified cholesterol from peripheral tissues and macrophages via ATP binding cassette A1 transporter.[77] Berberine at 1 mmol/L in drinking water of ApoE$^{-/-}$ mice on high-fat diet prevented atherosclerosis via activation of adenosine monophosphate-activated protein kinase and acetyl-CoA carboxylase, expression of nuclear respiratory factor-1 reduction in reactive oxygen species, vascular cell adhesion molecule-1 and intercellular adhesion molecule-1 in aortic tissues.[78] Berberine at 50 μM inhibited the proliferation of A7r5 vascular smooth muscle cells exposed to platelet-derived growth factor (PDGF-BB) via blockade in G0/G1 phase; decreased in Skp2, CDK2, cyclinD1, and CDK4; and increased in p27 and p21 expression.[79] Myocardial infarction in Metabolic Syndrome refers to a condition in which a portion of the myocardium undergoes damage due to deprivation of oxygen for a certain period of time followed by reperfusion of plasma leading to irreversible cellular damages.[80] Ischemia of the myocardium causes the development of arrhythmias and may lead to cardiac necrosis.[80] *Berberis vulgaris* L. elaborates berbamine which given prophylactically at 100 nmol/L to isolated hearts from Sprague–Dawley rats subjected to ischemia and reperfusion increased of the recovery of left ventricular developed pressure and reduced myocardial infarction and lactate dehydrogenase release.[81] Furthermore, in cardiomyocytes, this isoquinolone alkaloid preserved sarcoplasmic reticulum calcium-ATPase 2a during ischemia reperfusion.[81] Furthermore, the observed cardioprotection resulted from the activation of phosphoinositide 3-kinase, Akt, inhibition of glycogen synthase kinase-3, opening of the mitoKATP channel and suppressing calcium overloading, and reduction of calpain.[81]

5.16 *Caulophyllum robustum* Maxim.

Synonyms: *Caulophyllum thalictroides* (L.) Michx.; *Leontice robusta* (Maxim.) Diels
Common name: hong mao qi (Chinese)
Subclass Ranunculidae; Superorder Ranunculanae; Order Berberidales; Family Berberidaceae
Medicinal use: inflammation (China)

Caulophine from *Caulophyllum robustum* Maxim. at a concentration of 0.4 mM protected cardiomyocytes *in vitro* against hydrogen peroxide with a decrease in creatine kinase, lactate dehydrogenase, malondialdehyde. It increased superoxide dismutase activity and inhibited apoptosis.[82] Furthermore, this fluorenone alkaloid given intragastrically to Sprague–Dawley rats at a dose of 25 mg/kg/day for 5 days reduced the occurrence of cardiac infarct by 24.6% induced by ligature of the coronary. It lowered serum creatine kinase, lactate dehydrogenase, and malondialdehyde, whereas superoxide dismutase activity was increased.[82] Caulophine at 0.4 mM prophylactically sustained the viability by 95.5% of neonatal rat cardiomyocytes exposed *in vitro* to caffeine.[83] This alkaloid reduced intracellular malondialdehyde and the release of lactate dehydrogenase by myocytes.

5.17 *Coptis chinensis* Franch.

Common name: huang lian (Chinese)
Subclass Ranunculidae, Superorder Ranunculanae, Order Ranunculales, Family Ranunculaceae
Medicinal use: fever (China)

Atherogenesis encompasses the entry and chronic deposition of low-density lipoprotein in arterial intima. High-density lipoprotein inhibits intimal oxidation of low-density lipoproteins.[4] Magnoflorine (Figure 5.5) from *Coptis chinensis* Franch. prevented copper-induced oxidative insults of low-density lipoprotein *in vitro* with an IC_{50} value equal to 2.3 μM.[84] Furthermore, this alkaloid at 5 μM maintained the ability of high-density lipoprotein to protect low-density lipoprotein from copper.[84] Epiberberine, coptisine, and groenlandicine inhibited the enzymatic activity of aldose reductase with IC_{50} values equal to 33.6, 37.8, and 45.1 μg/mL, respectively.[85] A single intraperitoneal bolus of palmatine at a dose of 50 mg/kg, 1 h before ischemia followed by reperfusion increased ±dp/dt from 2928 to 3388 mmHg/s, reduced LVRDP from 12 to 8 mmHg and halved the area of myocardial infarction. A single intraperitoneal bolus of palmatine at a dose of 50 mg/kg, 1 h before ischemia followed by reperfusion for lowered serum low-density lipoprotein. Malondialdehyde was reduced by 30%.[86] In cardiac tissues, a palmatine increased the enzymatic activity of superoxide dismutase and catalase by about 30% and 20%, respectively. The expression of inducible nitric oxide and synthetase cyclooxygenase-2 were reduced.[86] In human aortic endothelial cells cultured *in vitro*, palmatine at 10 μM boosted the expression of heme oxygenase-1.[86] Coptisine given orally at a dose of 100 mg/kg for 21 days reduced the mortality of Sprague–Dawley rats with myocardium infarction induced by isoprotenolol from 76% to 40%.[87] Electrocardiograms of treated rodents evidenced a decrease in ST-segment, elevation and normalization of Q waves.[87] Furthermore, myocardial enzymes creatine kinase, aspartate transaminase, and lactate dehydrogenase were reduced.[87] Malondialdehyde was reduced, and the enzymatic activities of superoxide dismutase, catalase,

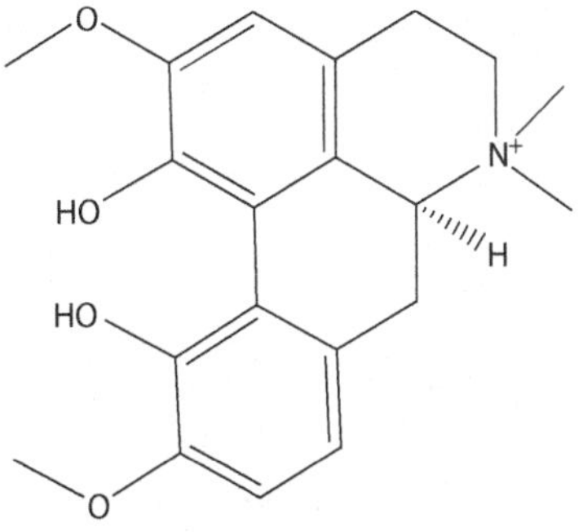

FIGURE 5.5 Magnoflorine.

and glutathione peroxidase were increased in the heart tissue.[87] Histopathological examination of heart tissues evidenced cardioprotection with reduction of necrosis, infiltration of white cells, and decrease in edema.[87] In addition, coptisine reduced the number of apoptotic cells in the heart tissue by 80% compared with the untreated group.[87]

5.18 *Thalictrum minus* L.

Synonyms: *Thalictrum caffrum* Eckl. & Zeyh.; *Thalictrum kochii* Fr.; *Thalictrum transsilvanicum* Schur
Common name: ya ou tang song cao (Chinese)
Superorder Ranunculanae, Order Ranunculales, Family Ranunculaceae
Medicinal use: fever (China)

The antithrombotic and antiatherogenic properties of nitric oxide produced by endothelial cells are due to the ability of this signaling molecule to inhibit low-density lipoprotein oxidation, intimal vascular smooth muscle cells migration, and proliferation. It also inhibit the expression of cell surface adhesion molecules.[88] Thaliporphine (Figure 5.6) from *Thalictrum minus* L. infused intravenously to Sprague–Dawley rats 15 minutes before experimental ischemia-reperfusion at a dose of 10^{-8} mol/kg evoked an increase of nitric oxide in the plasma, a reduction of lactate dehydrogenase and reduction of lethality and cardiac infarcts sizes.[89] At a dose of 10^{-7} mol/kg, this alkaloid reduced the incidence of ischemia-induced ventricular tachycardia and ventricular fibrillation.[89] *In vitro*, thaliporphine protected low-density lipoprotein against copper-induced oxidative damages with an IC_{50} equal to 15.7 μM.[89]

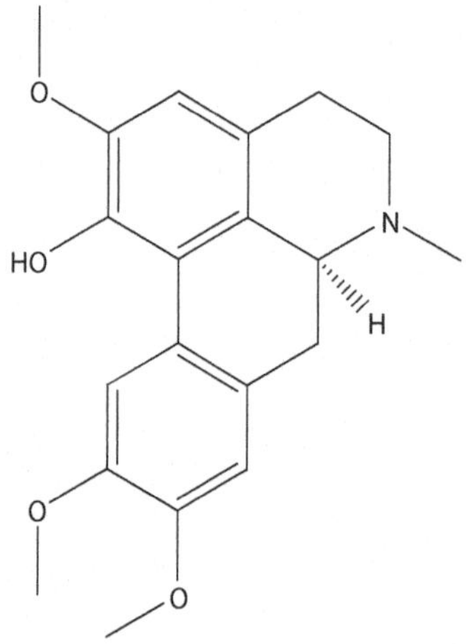

O
HO
N
H
O
O

FIGURE 5.6 Thaliporphine.

5.19 *Corydalis turtschaninovii* Besser

Common name: chi ban yan hu suo (China)
Subclass Ranunculidae; Superorder Ranunculanae; Order Papaverales; Family Papaveraceae
Medicinal use: abdominal pain (Korea)

Methanolic extract of tubers of *Corydalis turtschaninovii* Besser inhibited the enzymatic activity of aldose reductase.[90] From the same tubers, glaucine, protopine, and dehydrocorynaline inhibited the enzymatic activity of aldose reductase by 16.5%, 10.9%, and 44.5% at a dose of 50 μM.[90] Tetrahydropalmatine given at a single dose of 10 mg/kg intravenously to anesthetized rodents evoked an immediate and transient fall of mean arterial blood pressure (by 38 mmHg), a reduction of heart rate (by 43 beats/min), and a decrease in hypothalamic serotonine release (by 52%).[91] The hypotension and bradycardia evoked by tetrahydropalmatine were inhibited by the 5-HT2 receptor agonist DOI.[91] Furthermore, spinal section and bilateral vagotomy attenuated tetrahydropalmatine-induced hypotension and bradycardia, respectively.[91] Arachidonic from cytoplasmic membrane phospholipids of platelets is converted to thromboxane A2 under the combined action of cyclo-oxygenase and thromboxane A2 synthetase.[92] The plant contains tetrahydroberberine, which inhibited the aggregation of platelets challenged with arachidonic acid, adenosine diphosphate, and collagen.[93] Besides, tetrahydroberberine inhibited the generation of thromboxane B2 in platelets exposed to arachidonic acid.[93] Tetrahydroberberine at a dose of 30 mg/kg/day given intraperitoneally for 5 days negated the aggregation of platelets induced by ADP in rodents.[93] Sanguinarine (Figure 5.7) inhibited the aggregation of platelets evoked by arachidonic acid and collagen with IC_{50} values equal to 8.3 μM and 7.7 mM.[94] Against thrombin, sanguinarine inhibited platelet aggregation by 10% at a concentration of 10 μM.[94] Sanguinarine inhibited the aggregation of platelets induced by the thromboxane A2 analog U46619 with an IC_{50} of 8.6 μM.[94] The adenylate cyclase inhibitor SQ22536 attenuated the inhibitory effect of sanguinarine against the platelet aggregation evoked by arachidonic acid.[94] Sanguinarine inhibited the production of thromboxane B2 by platelets challenged with arachidonic acid with an IC_{50} of 4.5 μM and inhibited the enzymatic activity of cyclooxygenase-1 with an IC_{50} value of 28 μM.[94] This alkaloid at 300 nM inhibited the multiplication of vascular smooth muscle cells by arresting cell cycle in G1 with induction of p27, decrease in the enzymatic activities of CDK2 and CDK4, and induction of Ras and extracellular signal-regulated kinase-1/2.[95] Chelerythrine at 8 μmol/L prevented hypertrophy of neonatal ventricular myocytes of rodents exposed to high glucose levels via reduction of protein kinase Cα (PKC-α) and protein kinase β2 (PKC- β2).[96] The control of growth of vascular smooth muscle cells has critical therapeutic implication for atherosclerosis.[97]

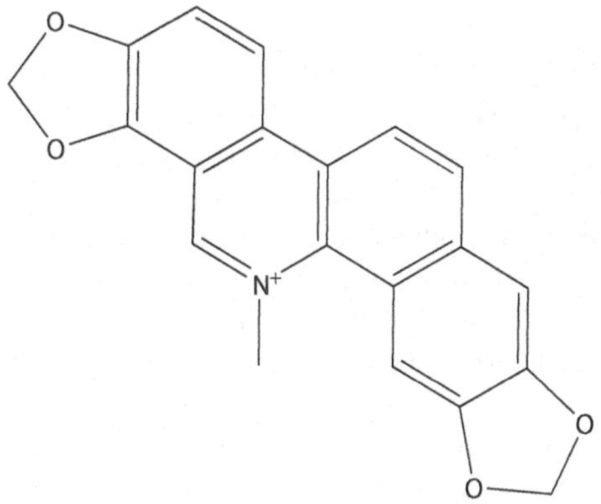

FIGURE 5.7 Sanguinarine.

5.20 *Fumaria parviflora* Lam.

Synonym: *Fumaria indica* Pugsley
Subclass Ranunculidae, Superorder Ranunculanae, Order Papaverales, Family Fumariaceae
Medicinal use: jaundice (India)

Protopine (Figure 5.8) from *Fumaria parviflora* Lam. inhibited the aggregation of human platelets induced by ADP, arachidonic acid, collagen, and platelet-activated factor with IC_{50} values equal to 9, 12, 16, and 11 μM, respectively.[98] In the same experiment, aspirin inhibited the aggregation of platelets less efficiently (with IC_{50} values equal to 430, 110, 520, and 240 μM, respectively).[98]

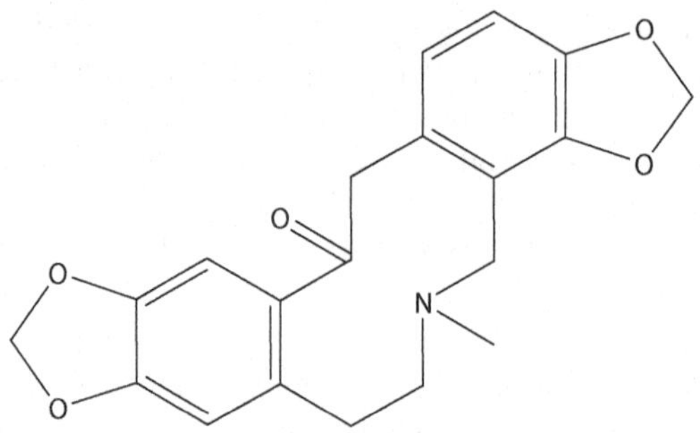

FIGURE 5.8 Protopine.

Furthermore, protopine or aspirin given intraperitoneally and prophylactically at a single dose of 100 and 50 mg/kg, respectively, protected New Zealand white rabbits against death by pulmonary embolism following intravenous administration of arachidonic acid or platelet-activated factor.[98] *In vitro*, protopine or aspirin inhibited the production of thromboxane B2 by human platelets exposed to arachidonic acid with IC_{50} values equal to 17 and 240 μM, respectively, suggesting that this protoberberine alkaloid to be a cyclo-oxygenase inhibitor.[98]

5.21 *Paeonia suffruticosa* Andrews

Synonym: *Paeonia moutan* Sims
Common name: mu dan (Chinese); tree peony
Superorder Ranunculanae, Order Paeoniales, Family Paeoniaceae
Medicinal use: fever (China)

Endothelial cells secrete nitric oxide that activates in smooth muscle guanylate cyclase hence increased cytosolic levels of cGMP, activation of protein kinase G opening of membrane potassium channels, membrane hyperpolarization, closing of voltage-dependent calcium channels and smooth muscle cells relaxation.[99] In obese patients with elevated plasma low-density lipoprotein–cholesterol, endothelium-dependent vascular relaxation is impaired because nitric oxide is less able to relax vascular smooth muscles contributing to "*endothelial dysfunction*." Paeonol (Figure 5.9) from *Paeonia suffruticosa* Andrews given orally at a daily dose of 150 mg/kg to New Zealand white rabbits receiving a high-fat diet for 12 weeks reduced the development of aortic atherosclerosis and commanded a decrease in cholesterolemia from 17.2 to 7.6 mmol/L, triglyceridemia from 11.5 to 4.6 mmol/L, and low-density lipoprotein from 0.8 to 0.5 mmol/L.[100] In addition, this phenolic compound decreased serum and aortic levels of tumor necrosis factor-α, interleukin-β, and CRP by more than 50%.[100] Immunohistochemical observation of aorta revealed an inhibition of translocation to the nucleus of the nuclear factor-κB.[100] Small phenolic molecules can be seen as "pro-drugs" of an array of first pass metabolites collectively termed "phenolic reactive substances" which are often treating "endothelial dysfunction."

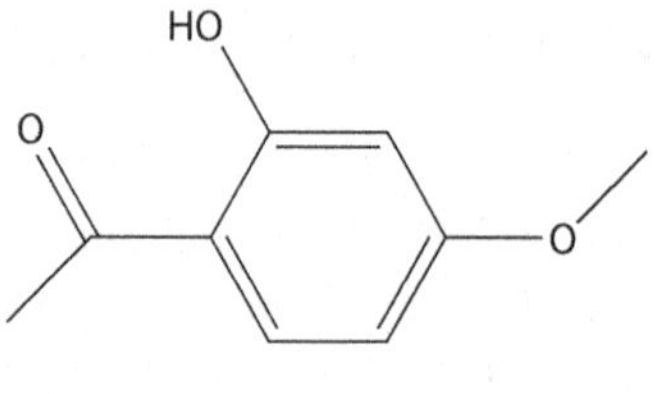

FIGURE 5.9 Paeonol.

5.22 *Juglans regia* L.

Synonyms: *Juglans duclouxiana* Dode; *Juglans fallax* Dode; *Juglans kamaonia* (C. DC.) Dode; *Juglans orientis* Dode; *Juglans sinensis* (C. DC.) Dode
Common names: hu tao (Chinese); walnut
Subclass Hamamelidae, Superorder Juglandanae, Order Juglandales, Family Juglandaceae
Medicinal use: abscesses (China)

The peel of seeds of *Juglans regia* L. contains ellagitannins, which release ellagic acid in the intestines.[101] This phenolic as well as catechin at a concentration of 1 μmol/L inhibited low-density lipoprotein oxidation induced by copper *in vitro*.[101] Individuals with persistent blood pressure above 140/90 mmHg and who are unable to lower blood pressure through lifestyle changes are treated with thiazide-like diuretics, angiotensin-converting enzyme inhibitors, calcium channel blockers, β-blockers to avoid cardiovascular diseases.[102] In obese volunteers with metabolic syndrome, the consumption of 56 g of walnut per day for 8 weeks corrected "*endothelial dysfunction*."[103] This regimen did not alter total cholesterol, low-density lipoprotein–cholesterol, high-density lipoprotein–cholesterol, and triglycerides, or fasting plasma glucose and insulin but evoked a beneficial trend in systolic blood pressure reduction.[103] ApoE$^{-/-}$ mice fed with a high-fat diet containing walnuts (1.2 g/5 g diet) for 8 weeks had lowered plasma triglycerides and cholesterol by 36% and 23%, respectively, and decreased liver triglycerides by 30%.[104] This supplementation evoked a 55% reduction in overall atherosclerotic plaques and 50% lowering of plaques fatty acid translocase staining in aortic arches.[104] Walnut intake reduced plasma prothrombin by 21%, whereby fibrinogen and factor V levels remained unaltered.[104] Oil of walnut had no effect.[104] It must be noted that results in rats and mice do not always correlate with human studies.

5.23 *Amaranthus viridis* L.

Synonyms: *Amaranthus gracilis* Desf. ex Poir.; *Amaranthus gracilis* Desf.
Common names: hin-nu-new (Burma); green amaranth
Subclass Caryophyllidae, Superorder Caryophyllanae, Order Caryophyllales, Family Amaranthaceae
Medicinal use: dysentery (China)

Methanol extract of *Amaranthus viridis* L. given orally and prophylactically to Wistar rats for 45 days decreased cardiac oxidative stress and hydrogen peroxide levels induced by isoproterenol injections.[105] This extract increased the enzymatic activity of cardiac serum superoxide dismutase and catalase; decreased serum aspartate transaminase, alanine transaminase, creatine kinase, and lactate dehydrogenase. It also reduced serum troponin by more than 50%.[105] It further elevated the levels of cardiac gluthatione peroxidase and gluthation.[105]

5.24 *Rheum rhabarbarum* L.

Synonyms: *Rheum franzenbachii* Münter; *Rheum undulatum* L.
Common names: bo ye da huang (Chinese); rhubarb
Subclass Caryophyllidae, Superorder Polygonanae, Order Polygonales, Family Polygonaceae
Medicinal use: stop bleeding (China)

In diabetic patients, abnormal platelet adhesion and aggregation, increased fibrinogen, factor VII, and increased plasminogen activator inhibitor-1 levels contribute to a procoagulant state.[106] Increased concentration of low-density lipoprotein increases the sensitivity of platelet to aggregation.[107] *Rheum rhabarbarum* L. produces desoxyrhapontigenin and rhapontigenin which a

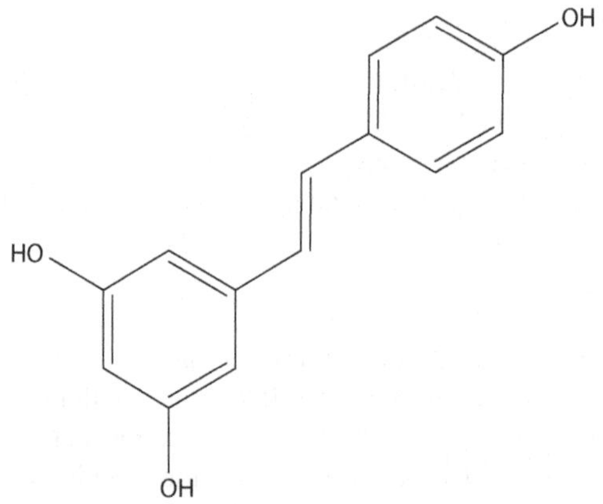

FIGURE 5.10 Resveratrol.

concentration of 0.1 mM inhibited the aggregation of platelets induced by arachidonic acid by 87% and 100%, respectively.[108] From this plant, resveratrol (Figure 5.10) inhibited the aggregation of platelets induced by collagen and ADP with IC_{50} values equal to 11.6 and 17.7 μM.[108] Methanol extract of rhizome of *Rheum rhabarbarum* L. relaxed rings of thoracic aortas precontracted with phenylephrine with EC_{50} value equal to 5.8 μg/mL.[109] From this extract, the stilbenes piceatannol, resveratrol, desoxyrhapontigenin, and ε-viniferin inhibited phenylephrine-induced contraction of rat aorta with EC_{50} values equal to 2.4, 28.6, 18.5, and 8.4 μg/mL, respectively.[109,110] Pinneceatol effect was nullified upon endothelium removal or pretreatment with N^{ω}-nitro-L-arginine methyl ester (L-NAME) suggesting endothelium and nitric oxide-dependent mechanism of vasorelaxation.[109,110] Rhein given to mice once daily to type 2 diabetic obese *db/db* mice at a dose of 150 mg/kg/day for 12 weeks reduced albuminuria, plasma glucose, cholesterol, triglycerides, low-density lipoprotein–cholesterol, and apolipoprotein E.[111] Emodin given orally at a dose of 80 mg/kg/day for 8 weeks to Wistar rats following a high-fat diet decreased body weight gain from 49.5 to 24.5 g/rats. Pioglitazone at a dose of 20 mg/kg/day lowered body weight gain to 22.3 g/rats.[112] This anthraquinone brought cholesterolemia from 228.4 to 123.7 mg/dL, whereby atherogenic index and coronary artery index were decreased from 6.2 and 8.3 to 1.3 and 2.8, respectively.[112] Chrysophanol, emodin, physcion, aloe-emodin, at a concentration of 10 μM reduced by more than 50% the secretion of interleukin-6, fibronectin, and collagen IV by rat mesangial cells exposed to high glucose concentration.[113] Anthraquinones like emodin are carcinogenic and useless in therapeutic.

5.25 *Polygonum aviculare* L.

Common names: bian xu (Chinese); prostrate knotweed
Subclass Caryophyllidae, Superorder Polygonanae, Order Polygonales, Family Polygonaceae
Medicinal use: jaundice (China)

Oxidized low-density lipoprotein in intima activates endothelial cells to express monocyte chemoattractant peptide-1 that accounts for the entry of plasma monocytes in the sub-endothelial space.[4] Monocytes in contact with oxidized low-density lipoprotein differentiate in macrophages that release interleukin-1 and tumor necrosis factor-α, which promote further expression of monocyte chemoattractant peptide-1 by endothelial cells.[4] ApoE knock-out mice on high-fat diet treated with ethanol extract of *Polygonum aviculare* L. orally at a dose of 100 mg/kg/days for 12 weeks had a mild weight gain.[114] Serum total cholesterol decreased from 1087 to 705.5 mg/dL, low-density lipoprotein from 236.8 to 107.5 mg/dL, high-density lipoprotein increased from 10.1 to 16.7 mg/dL, whereas glycemia remained elevated.[114] This treatment reduced diastolic blood pressure from 107 to 72 mg/dL and systolic blood pressure from 129 to 97 mg/dL, respectively, and reduced aortic plaque formation.[114] In aortic tissues, intercellular adhesion molecule-1, vascular cell adhesion molecule-1 and nuclear factor-κB were reduced by the treatment as well as the expression of phosphorylated-extracellular signal-regulated kinase.[114]

5.26 *Camellia sinensis* (L.) Kuntze

Synonym: *Thea chinesis* L.
Common names: cha (Chinese); tea
Subclass Dillenidae, Superorder Ericanae, Order Theales, Family Theaeae
Medicinal use: tonic (China)
History: listed in the Penst'sao Kang Mu

ADP inhibits platelet adenylyl cyclase via the stimulation of Gαi-coupled P2Y12 receptor, hence decrease in cytosolic levels of cyclic adenosine monophosphate.[115] Adenylate cyclase and guanylate cyclase activators and phosphodiesterase inhibitors increase intracellular cyclic adenosine monophosphate to inhibit platelet activation.[116] Caffeine and theophylline from *Camellia sinensis* L. inhibit cyclic nucleotide phosphodiesterase increasing thus the cytosolic contents in cyclic adenosine monophosphate.[117] Tea contains epigallocatechin, epicatechin-3-*O*-gallate, and gallocatechin-3-*O*-gallate, which at a concentration of 10^{-4} M inhibited *in vitro* the aggregation of platelets induced by ADP by 58%, 19%, and 5%, respectively.[118] Epigallocatechin at 10^{-4} M reduced *in vitro* the formation of thromboxane B_2 from exogenous and endogenous arachidonic acid by platelets by 60% and 44%, respectively.[118] Epicatechin-3-*O*-gallate and gallocatechin-3-*O*-gallate at of 10^{-4} M protected endothelial cells isolated from bovine carotid arteries against hydrogen peroxide induced injuries by 86% and 53%.[118] (+)-catechin from *Camellia sinensis* (L.) Kuntze given orally at a dose of 35 mg/day orally for 12 weeks to streptozotocin-induced diabetic Wistar rats augmented glycemia from 413 to 420 mg/dL, lowered urine volume from 166 to 137 mL, increased urine creatinine from 1 to 1.9 mg/dL, halved albuminuria and plasma creatinine. This flavonoid decreased marker of vascular endothelial dysfunction endothelin-1 in urine.[119] This provide further ground for the notion that infusions or decoctions of medicinal plants have to be preferred than pure isolated constituents. Aqueous extract of *Camellia sinensis* (L.) Kuntze given orally once a day and orally at a dose of 400 mg/kg for 29 days mitigated the cardiotoxicity of doxorubicin as evidenced by a decrease in serum aspartate aminotransferae, creatine kinase, and lactate dehydrogenase.[120] Furthermore, this extract prevented serum and cardiac lowering of glutathione and increased the activities of glutathione peroxidase, glutathione reductase, glutathione S-transferase, superoxide dismutase, and catalase in heart.[120] Consumption of green tea could be beneficial in Metabolic Syndrome.

5.27 *Calluna vulgaris* (L.) Hull

Synonym: *Erica vulgaris* L.
Common name: heather
Subclass Dillenidae, Superorder Ericanae, Order Ericales, Family Ericaceae
Medicinal use: urinary tract infection (Turkey)

Elevated glycemia during insulin resistance favors the formation of advanced glycation end products and reactive oxygen species in vessel walls which account for macrovascular and microvascular complications.[1] Ursolic acid isolated from *Calluna vulgaris* (L) Hull given orally at a dose of 50 mg/kg/day to streptozotocin-diabetic Sprague–Dawley rats for 2 months decreased glycemia from 22.1 to 17.7 mmol/L as well as glycated hemoglobin, tumor necrosis factor-α, and malondialdehyde.[121] Increased expression of nicotinamide adenine dinucleotide phosphate

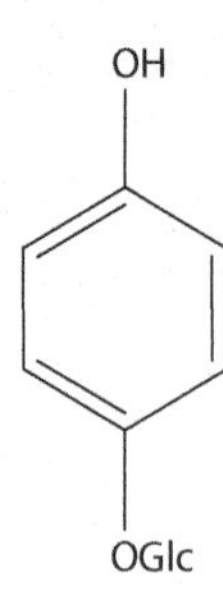

FIGURE 5.11 Arbutin.

oxidase (NADPH oxidase) and production of superoxides contribute to endothelial and renal injuries.[121] The triterpene decreased aortic contents of advanced glycation end products by 86% as well as receptors of advanced glycation end products expression, endothelial NADPH oxidase activity, and endothelial activation of nuclear factor-κB.[121] Furthermore, histological observation of aorta revealed that this triterpene was able at a dose of 50 mg/kg/day to prevent endothelial necrosis induced by diabetes.[121] The plant contains arbutin (Figure 5.11) and one could have some interest to assess the effects of this phenolic glycoside on vascular endothelium dysfunction. Clinical trials are warranted.

5.28 *Arbutus unedo* L.

Common names: kocayemiş yaprağı (Turkish); strawberry tree
Subclass Dillenidae, Superorder Ericanae, Order Ericales, Family Ericaceae
Medicinal use: kidney stones (Turkey)

Blood pressure is the result of a balance between cardiac output and peripheral vascular resistance.[122] In Metabolic Syndrome, vascular dysfunction leads to hypertension and life threatening cardiovascular and cerebrovascular diseases events.[123] Aqueous extract of roots of *Arbutus unedo* L. given daily at a dose of 150 mg/kg to Wistar rats challenged with N^{ω}-nitro-L-arginine methyl ester (L-NAME) lowered systolic blood pressure from 188 to 155 mmHg, reduced ventricular hypertrophy, and protected the cardiac baroreflex responsiveness to phenylephrine. The extract prevented relaxation induced by carbachol.[124] Thrombin binds to protease-activated receptor-1 and receptor-6 and induces platelet activation via the activation of G protein-coupled signaling.[125] Aqueous extract of leaves at a concentration of 0.05 mg/mL inhibited the aggregation of platelets induced by thrombin by 45% *in vitro* via inhibition of reactive oxygen species generation, influx of calcium and tyrosine phosphorylation.[126]

5.29 *Vaccinium myrtillus* L.

Common names: hei guo yue ju (Chinese); bilberry
Subclass Dillenidae, Superorder Ericanae, Order Ericales, Family Ericaceae
Nutritional use: food (China)

Diet containing 2% lyophilized fruits of a member of the genus *Vaccinium* L. given to stroke-prone spontaneously hypertensive rats for 12 weeks lowered mean arterial blood pressure by 27% compared to untreated animals.[127] This treatment normalized glomerular filtration rate, decreased cortical and medullary renal reactive oxygen species, increased renal gluthatione and renal catalase activity doubled urine creatinine.[127]

5.30 *Diospyros kaki* Thunb.

Common names: shi (Chinese); gam (Korean); kaki (Japanese); Japanese persimmon
Subclass Dilleniidae, Superorder Primulanae, Order Styracales, Family Ebenaceae
Medicinal use: astringent (China)

Proanthocyanidin enriched fraction from the leaves of *Persimmon kaki* Thunb. given at a single oral dose of 300 mg/kg lowered systolic blood pressure of spontaneously hypertensive rats by 48% after 4 hours and was inactive in normotensive rats.[128] *In vitro*, 10 μg/mL of this extract induced a 84.2% relaxation in intact aortic rings precontracted with norepinephrine, and this effect was abolished by the removal of aortic endothelium and inhibited by N^{ω}-nitro-L-arginine methyl ester (L-NAME).[128] Human umbilical vein endothelial cells exposed to this fraction at a concentration of 10 μg/mL evoked the phosphorylation of both Akt and endothelial nitric oxide synthetase.[128]

5.31 *Embelia ribes* Burm.f.

Common names: bai hua suan teng guo (Chinese); vidanga (India)
Subclass Dilleniidae, Superorder Primulanae, Order Primulales, Family Myrsinaceae
Medicinal use: jaundice (India)
History: *Embelia ribes* Burm.f. was known to Sushruta

Embelin from *Embelia ribes* Burm.f. given prophylactically for 3 days at a dose of 50 mg/kg/day orally protected Wistar rats against isoproterenol-induced myocardial insults as evidenced by a decreased in serum level of creatine kinase, lactate dehydrogenase, and aspartate transaminase.[129] This alkylresorcinol prevented isoproterenol-induced decrease in myocardial gluthatione, gluthatione peroxidase, gluthatione-S-transferase and gluthatione reductase, superoxide dismutase activity, catalase activity, and NADPH:quinone oxydoreductase 1.[129] This treatment prevented reduction of myocardial mitochondrial respiratory enzyme activities including NADH dehydrogenase, succinate dehydrogenase, and cytochrome c oxidase activities.[129] The expression of cytochrome c in ventricular heart tissue was reduced upon embelin treatment.[129] This treatment prevented myocardial degeneration, cytoplasmic vacuolization, and inflammatory cells infiltration.[129] This treatment reduced isoproterenol-induced elevation of total cholesterol, triglycerides, and very low-density lipoprotein–cholesterol.[129]

5.32 *Benincasa hispida* (Thunb.) Cogn.

Synonyms: *Benincasa cerifera* Savi; *Benincasa pruriens* (Sol. ex Seem.) W.J. de Wilde & Duyfjes; *Cucurbita hispida* Thunb.
Common names: dong gua (Chinese); Chinese winter melon
Subclass Dilleniidae, Superorder Violanae, Order Cucurbitales, Family Cucurbitaceae
Medicinal use: diuretic (China)

Vascular endothelium reacts to vascular lesions induced by high-plasma glucose and low-density lipoprotein by expressing macrophage chemoattractant protein-1 and interleukin-8, which participate in the development of atherosclerosis.[35] Macrophage chemoattractant protein-1 is responsible for monocyte infiltration, and interleukin-8 is a chemotactic for neutrophils.[130] Aqueous extract of *Benincasa hispida* (Thunb.) Cogn. given prophylactically at a concentration of 20 μg/mL inhibited the surface expression of intercellular-adhesion molecule-1 (ICAM-1), vascular adhesion molecule-1, and E-selectin by human umbilical vein endothelial cells challenged with glucose.[131] Hyperglycemia triggers the generation of reactive oxygen species in various cell types, which leads to nuclear factor-κB activation and atherogenesis.[130] Aqueous extract of *Benincasa hispida*

(Thunb.) Cogn. at 20 μg/mL inhibited in human umbilical vein endothelial cells translocation of nuclear factor-κB, prevented the phosphorylation of IκBα, and inhibited the expression of macrophage chemoattractant protein-8 and interleukin-8.[131]

5.33 *Melothria maderaspatana* (L.) Cogn.

Synonyms: *Cucumis maderaspatanus* L.; *Mukia maderaspatana* (L.) M. Roem.
Common names: mao er gua (Chinese); agumaki (India); bobontengan (Indonesia)
Subclass Dilleniidae, Superorder Violanae, Order Cucurbitales, Family Cucurbitaceae
Medicinal use: facilitate digestion (Indonesia)

In obese patients, arterial hypertension characterized by systolic blood pressure above 140 mmHg and diastolic blood pressure above 90 mmHg contribute to the pathophysiology of heart attack, cardiac and renal insufficiency, and vascular endothelium insults.[132] Current drugs used to treat hypertension include calcium channel blockers, angiotensin II blockers, angiotensin-converting enzyme inhibitors, sympatholytic drugs, and diuretics.[102] Ethyl acetate fraction of leaves *Melothria maderaspatana* (L.) Cogn. given orally at a dose of 60 mg/kg/day for 6 weeks to Wistar rats with hypertension (due to: kidney removal, twice weekly subcutaneous injection of deoxycorticosterone acetate and drinking water containing sodium chloride) reduced systolic blood pressure from 215.6 to 129.7 mmHg.[133] The same regimen lowered diatolic blood pressure from 172.5 to 92.8 mmHg.[133] Natural products with hypotensive activity are unknown yet from this plant.

5.34 *Crateva nurvala* Buch.-Ham

Synonyms: *Crateva lophosperma* Kurz; Crateva magna DC.
Subclass Dillenidae, Superorder Capparanae, Order Capparales, Family Capparaceae
Medicinal use: indigestion (Malaysia)

Lupeol from *Crateva nurvala* Buch.-Ham given orally for 15 days to Wistar rats on high-fat diet reduced myocardial cholesterol and triglycerides.[134] This triterpene lowered lactate dehydrogenase; increased the activities of sodium-potassium-ATPase, calcium-ATPase, and magnesium-ATPase enzymes; and mildly reverted thickening of cardiac cells and cardiac muscle hypertrophy.[134]

5.35 *Lepidium sativum* L.

Common names: jia du xing cai (Chinese); candsur (India); garden cress
Subclass Dillenidae, Superorder Capparanae, Order Capparales, Family Brassicaeae
Medicinal use: diuretic (India)

Aqueous extract of seeds of *Lepidium sativum* L. given orally at a dose of 20 mg/kg/day to spontaneously hypertensive rats lowered systolic blood pressure by 15% after 21 days of treatment without.[135] In normotensive rats, the same treatment had no effect on arterial blood pressure but boosted glomerular filtration rate as well as urine secretion.[135] Transforming growth factor-β1 induces extracellular matrix accumulation within the glomerular mesangium of kidneys and accounts for the development of renal failure in type 2 diabetes. Aqueous extract of seeds of *Lepidium sativum* L. given orally at a dose of 20 mg/kg/day to streptozocin-diabetic Wistar rats for 15 days reduced fasting blood glucose by 80%, urine glucose by 54%, and transforming growth factor-β1 by 38%.[136] When given at a single dose of 10 mg/kg intraperitoneally, this extract evoked a fall of glycemia in both normal and diabetic animals after 1 hour without any effect on insulinaemia.[136] Clinical trials are warranted.

5.36 *Abelmoschus manihot* (L.) Medik

Synonyms: *Hibiscus manihot* L.; *Hibiscus papyferus* Salisb.
Common names: huang shu kui (Chinese); lagikway (Philippines); sunset hibiscus
Subclass Dillenidae, Superorder Malvanae, Order Malvales, Family Malvaceae
Medicinal use: intestinal inflammation (China)

Flavone glycoside fraction of flowers of *Abelmoschus manihot* (L.) Medik given orally to streptozotocin-induced diabetic Sprague–Dawley rats at a dose of 200 mg/kg/day for 24 weeks halved urinary microalbumin/creatinine, glomerular cells apoptosis and reduced urine total protein by 40%.[137] From this extract, hyperoside at a concentration of 200 μg/mL inhibited the apoptosis of podocytes induced by advanced glycation end products by inhibition of caspase 3 and 8.[137] High glucose-induced reactive oxygen species. This promote podocyte insults and ultimately kidney failure.

5.37 *Morus alba* L.

Synonyms: *Morus atropurpurea* Roxb.; *Morus australis* Poir.; *Morus indica* L.
Common names: sang (Chinese); white mulberry
Subclass Dillenidae, Superorder Malvanae, Order Urticales, Family Moraceae
Medicinal use: diuretic (China)

Aqueous extract of *Morus alba* L. at a concentration of 0.1% inhibited the expression of lectin-like oxidized low-density lipoprotein receptor-1 by bovine aortic endothelial cells challenged with tumor necrosis factor-α via inactivation of nuclear factor-κB.[139] Morusinol (Figure 5.12) isolated from the root bark of *Morus alba* L. inhibited the aggregation of platelets induced by collagen and arachinonic acid with IC_{50} values equal to 13.4 and 19.8 μg/mL, respectively, with concomitant decrease in thromboxane B_2 formation.[140] Of note, this prenylated flavonoid given orally to Wistar rats at a dose of 20 mg/kg for 3 days prevented ferrum-induced arterial thrombus formation as evidenced by an increase in occlusion time from 22.5 to 42.8 minutes. Aspirin given orally at a dose of 20 mg/kg increased occlusion time from 22.5 to 29.3 minutes.[140] From this plant, morusin and kuwanon C at a concentration of 100 μM nullified platelet aggregation induced by arachidonic acid and platelet activating factor and inhibited by 95% platelet aggregation induced by collagen.[141]

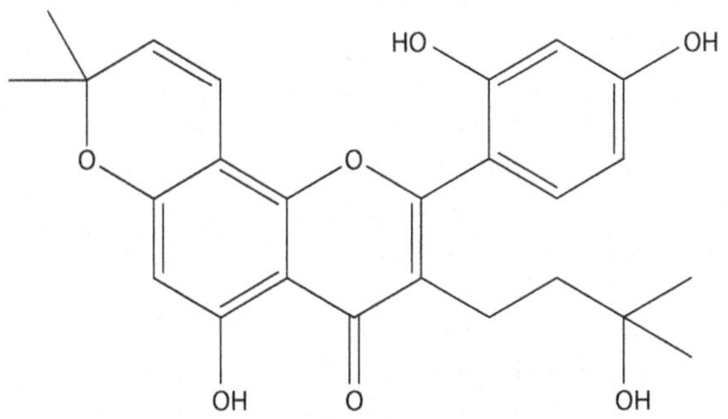

HO OH O O OH O OH

FIGURE 5.12 Morusinol.

5.38 *Cudrania tricuspidata* (Carrière) Bureau ex Lavalle

Synonym: *Maclura tricuspidata* Carrière
Subclass Dillenidae, Superorder Malvanae, Order Urticales, Family Moraceae
Medicinal use: tonic (China)

Prenylated xanthones from the root bark of *Cudrania tricuspidata* (Carr.) Bureau protected low-density lipoprotein against copper-induced peroxidation with IC_{50} values equal to 15 μM. They inhibited the enzymatic activity of acyl CoA: Cholesteryl acyl transferase 1 and 2 values inferior to 100 μM. At 50 μM, these flavonoids inhibited the expression of inducible nitric oxide synthetase by RAW cells 264.7 challenged with lipopolysacharides.[142] The root bark shelters cudraflavanone A (Figure 5.13), which at 1 μM inhibited the growth of rat aortic vascular smooth muscle cells stimulated by platelet-derived growth factor by 81%. This prenylated flavone increased cells in G1 phase, and inhibited the activation of Akt by 82%.[143]

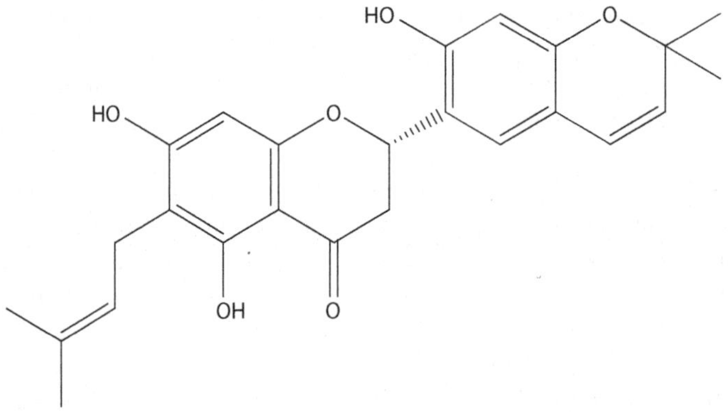

HO HO O O OH O

FIGURE 5.13 Cudraflavanone A.

5.39 *Phyllanthus emblica* L.

Synonym: *Emblica officinalis* Gaertn.
Common names: yu gan zi (Chinese); amla (India); Indian gooseberry
Subclass Dillenidae, Superorder Euphorbianae, Order Euphorbiales, Family Phyllanthaceae
Medicinal use: diabetes (Bangladesh)

Increased blood pressure is a risk factor for coronary heart disease in Metabolic Syndrome. Medications are recommended for hypertensives with blood pressure above 140/90 mmHg. These synthetic molecules can cause headaches, dizziness, tachycardia, fatigue, and sexual dysfunction.[144] Methanol extract of leaves of *Phyllanthus emblica* L. given at a dose of 10 mg/kg orally to spontaneously hypertensive rodents decreased systolic blood pressure by 6%, 3 hours after administration.[145] Aqueous extract of fruits given at a dose of 500 mg twice a day for 12 weeks type 2 diabetic patients improved endothelial function, reduced malondialdehyde levels, increased glutathione and nitric oxide levels.[146] Aqueous extract of fruits of *Phyllanthus emblica* L. given orally at a dose of 500 mg to overweight volunteers for twice a day for 12 weeks evoked a decrease of low-density lipoprotein–cholesterol and reduced the aggregability of platelets challenged *in vitro* with adenosine diphosphate and collagen.[147] Clinical trials are warranted.

5.40 *Euphorbia hirta* L.

Synonym: *Euphorbia pilulifera* L.
Common names: fei yang cao (Chinese); bobi (Philippines); hairy spurge
Subclass Dillenidae, Superorder Euphorbianae, Order Euphorbiales, Family Euphorbiaceae
Medicinal use: asthma (Philippines)

Ethanol extract of leaves of *Euphorbia hirta* L. given orally for 15 days to streptozotocin-induced diabetic albino mice at a daily dose of 500 mg/kg reduced glycemia from 192.5 to 77.6 mg/dL (normal is 75.5 mg/dL), normalized cholesterol and creatinine. This regimen lowered triglycerides from 256.1 to 142.1 mg/dL, increased high-density lipoprotein from 28.1 to 51.9 mg/dL (normal 34.1 mg/dL), and prevented weight loss.[148]

5.41 *Rhodiola rosea* L.

Synonym: *Sedum rosea* (L.) Scop.
Common names: hong jing tian (Chinese); rhodiola
Subclass Rosidae, Superorder Rosanae, Order Saxifragales, Family Crassulaceae
Medicinal use: tonic (China)

Roots of *Rhodiola rosea* given to Sprague–Dawley rats at a dose of 50 mg/kg prolonged exhaustive swimming time by 10 minutes and increased remaining ATP contents in skeletal muscles mitochondria.[149] Aqueous extract of roots given orally at a single dose of 75 mg/kg lowered systolic blood pressure in spontaneously hypertensive rats and normal rats by 18.6% and 12.5%, respectively, whereas heart rate was unchanged.[150] In spontaneous hypertensive rats, the extract increased the secretion of β-endorphins and intravenous injection of selective opioid μ-receptor antagonist abolished the hypotensive effect of this extract.[150] In diabetic patients, early pathophysiological abnormalities include increased thickness of glomerular basement membrane, augmentation of glomerular extracellular and mesangial matrix, microalbuminuria (urinary albumin excretion rate ≥20 μg/min or albumin:creatinine ratio ≥30 mg/g) due to basement membrane injuries, and increased glomerular filtration rate (GFR >90 mL/min/1.73 m^2).[151] Extract of *Rhodiola rosea* L. given orally to streptozotocin-induced diabetic Wistar rats at a dose of 20 g/kg/day for 8 weeks lowered fasting blood glucose, cholesterol, and triglycerides by 43%, 53%, and 49%, respectively.[152] This regimen lowered urine microalbumin and creatinine clearance rate by 52% and 38%, respectively.[152] Renal index, glomerular area, glomerular enlargement, and mesangial expansion as well as transforming growth factor-β1 were decreased.[152] The treatment evoked an improvement in renal structure and reduced the expression of transforming growth factor-β1.[152] Salidroside from *Rhodiola rosea* at a concentration of 10 μM protected human umbilical vein endothelial cells against hydrogen peroxide.[153] This glycoside 10 μM mitigated hydrogen peroxide-induced impaired endothelium-dependent relaxation of aortic rings of Wistar rats.[153] Furthermore, salidroside attenuated the activation of endothelial nitric oxide synthetase, adenosine monophosphate-activated protein kinase, and Akt. It also reduced nuclear factor-κB transcriptional activity.[153] In addition, salidroside increased the expression of Peroxisome proliferator-activated receptor γ-coactivator 1-α as well as mitochondrial mass and protected mitochondria against hydrogen peroxide induced insults.[153]

5.42 *Terminalia arjuna* (Roxb. ex DC.) Wight & Arn.

Synonym: *Pentaptera arjuna* Roxb.
Common names: arjun (India); arjuna myrobalan
Subclass Rosidae, Superorder Myrtanae, Order Myrtales, Family Combretaceae
Medicinal use: heart diseases (India)
Pharmacological target: vascular endothelium dysfunction

Postprandial hyperglycemia contributes to the development of atherosclerosis in diabetic and nondiabetic subjects. In the presence of high glucose concentration, endothelial cells produce proinflammatory cytokines of which interleukin-6 and tumor necrosis factor-α.[154] Ethanol extract of bark of *Terminalia arjuna* (Roxb. ex DC.) Wight & Arn. given orally to streptozotocin-induced diabetic Wistar rats at a dose of 500 mg/kg/day for 30 days lowered serum cholesterol, triglycerides, normalized high-density lipoprotein, and low-density lipoprotein.[155] The extract improved serum catalase, malondialdehyde, glutathione, and superoxide dismutase but did not improve glycemia.[155] This treatment lowered serum tumor necrosis factor-α by 43.7%, interleukin-6 by 31.3%, and lowered marker of vascular endothelial dysfunction endothelin-1.[155] This treatment mitigated myocardial subendothelial necrosis and vacuolation.[155] Methylene chloride extract of stem bark of a member of the genus *Terminalia* L. given at a dose of 150 mg/kg/day for 5 weeks to spontaneously hypertensive rats lowered systolic blood pressure from about 200 to 150 mmHg but had no effect in on normotensive rats.[156] This extract abated urinary 8-iso-$PGF_{2\alpha}$ in spontaneously hypertensive rats.[156] *In vitro* N^{ω}-nitro-L-arginine methyl ester inhibited relaxation of aortic rings of spontaneously hypertensive rats evoked by acetylcholine, and this effect was enhanced with the extract a suggesting an improvement of endothelial nitric oxide synthetase activation.[156]

5.43 *Punica granatum* L.

Common names: shi liu (Chinese); dhalim (India); pomegranate
Subclass Rosidae, Superorder Myrtanae, Order Myrtales, Family Lythraceae
Medicinal use: diabetes (India)
History: the plant was known to Serapion
Pharmacological target: vascular endothelium dysfunction

Juice of pomegranate mixed to drinking water (12.5 mL/100 mL) of Wistar rats for 30 days lowered isoproterenol induced plasma raise of creatine kinase and lactate dehydrogenase. It increased cardiac superoxide dismutase activity, cardiac catalase activity, cardiac glutathione contents.[157] Further this supplementation increased cardiac sodium-potassium-ATPase activity and reduced heart weight and infarct size.[157] Juice of *Punica granatum* L. given to streptozotocin-induced diabetic Wistar rats at a dose of 300 mg/kg orally for 4 weeks lowered weight loss from 52.2 to 34.8 g, urine volume from 16.2 to 8.8 mL, water intake and kidney weight.[158] This supplementation reduced serum glucose from 346.9 to 113.6 mg/dL, reduced urinary proteins and brought glucosuria down to nondiabetic level.[158] This extract reduced hypertension induced by subcutaneous injection of angiotensin II by 32%.[158] This treatment normalized superoxide dismutase, catalase, increased glutathion and lowered reactive oxygen species in kidneys.[158] In the kidneys, reactive oxygen species lead to endothelial dysfunction, increased contractility, vascular smooth muscle cell growth, lipid peroxidation, 8-iso-prostaglanin F2α production and deposition of extracellular matrix proteins that contribute to renal damage.[158] Pomegranate juice reduced diabetic expansion of mesangial matrix in glomeruli, microangiopathy and tubular degenerative changes.[158] This treatment normalized pancreatic superoxide dismutase, catalase, gluthathione contents and reactive oxygen species.[158] Ellagic acid from *Punica granatum* L. at a concentration of 10^{-4} M relaxed isolated thoracic aortic rings of Wistar rats precontracted with phenylephrine, and this effect was attenuated by removal of endothelium.[159] Endothelium-dependent relaxation of aortic rings by ellagic acid was inhibited by pretreatment with nitric oxide synthetase inhibitor N^{ω}-nitro-L-arginine methyl ester whereby pretreatment with prostaglandin had no effects. This phenolic at 10^{-4} M prevented calcium-induced contractions of artic rings suggesting inhibition of L-type calcium channels.[159] Intake of pomogranate juice could be beneficial in Metabolic Syndrome.

5.44 *Lagerstroemia speciosa* (L.) Pers.

Synonyms: *Lagerstroemia flos-reginae* Retz.; *Lagerstroemia reginae* Roxb.; *Munchausia speciosa* L.
Common names: banaba (Philippines); Queen crape-myrtle
Subclass Rosidae, Superorder Myrtanae, Order Myrtales, Family Lythraceae
Medicinal use: diabetes (Philippines).

In diabetics, accumulation of sorbitol in cells leads to oxidative stress culminating in tissue injury including cataracts and retinopathy.[160] Aldose reductase, the key enzyme in the polyol pathways, reduces excess of glucose into sorbitol using NADPH as a cofactor. This enzyme is involved in the pathophysiology of diabetic complications and cataract formation.[161] Number of aldose reductase inhibitors have been tested to combat diabetic complications but as for yet no therapeutic aldose reductase inhibitor has been marketed because of undesirable side effects.[162] Thus, there is a need for developing natural aldose reductase inhibitors with better activity and less side effects. Corosolic acid from *Lagerstroemia speciosa* (L.) Pers. inhibited rat leans aldose reductase, rat kidney aldose reductase and human recombinant aldose reductase activities with IC_{50} values equal to 4.6, 5.7, and 3.5 μg/mL, respectively.[162] In Wistar albino rats fed with galactose for 14 days, daily administration of this triterpene at a dose of 10 mg/kg reduced the lens aldose reductase catalyzed production of galactitol by 83%.[162]

5.45 *Syzygium cumini* (L.) Skeels

Synonyms: *Eugenia cumini* (L.) Druce; *Eugenia jambolana* Lam.; *Myrtus cumini* L.; *Syzygium jambolanum* (Lam.) DC.
Common names: wu mo (Chinese); jamu (India); duhat (Philippines); java plum
Subclass Rosidae, Superorder Myrtanae, Order Myrtales, Family Myrtaceae
Medicinal use: diabetes (India)

Ellagic acid from *Syzygium cumini* (L.) Skeels inhibited rat lens aldose reductase, rat kidney aldose reductase. and human recombinant aldose reductase activities with IC_{50} values equal to 4.9, 5.7, and 3.6 μg/mL, respectively.[162] In Wistar albino rats fed with galactose for 14 days, daily administration of ellagic acid at a dose of 10 mg/kg reduced the lens aldose reductase catalyzed production of galactitol by 81%.[162]

5.46 *Cassia fistula* L.

Common names: la chang shu (Chinese); praagraha (India); golden shower
Subclass Rosidae, Superorder Fabanae, Subclass Rosiidae, Family Fabaceae
Medicinal use: laxative (India)
History: the plant was known to Dioscorides (De Materia Medica) "*urinam cit, oculorum medicinae, bibitur contra interiores inflammationes*"

Aqueous decoction of flowers of *Cassia fistula* L. (Figure 5.14) in given orally to alloxan-diabetic Wistar rats at a dose of 10 mL/kg/day for 15 days evoked a cardiac decrease of lipid peroxidation and increased glutathione contents.[163] In addition, this extract normalized the enzymatic activity of superoxide dismutase, catalase, gluthatione reductase increased the enzymatic activity of glutathione peroxidase.[163] Clinical trials are warranted.

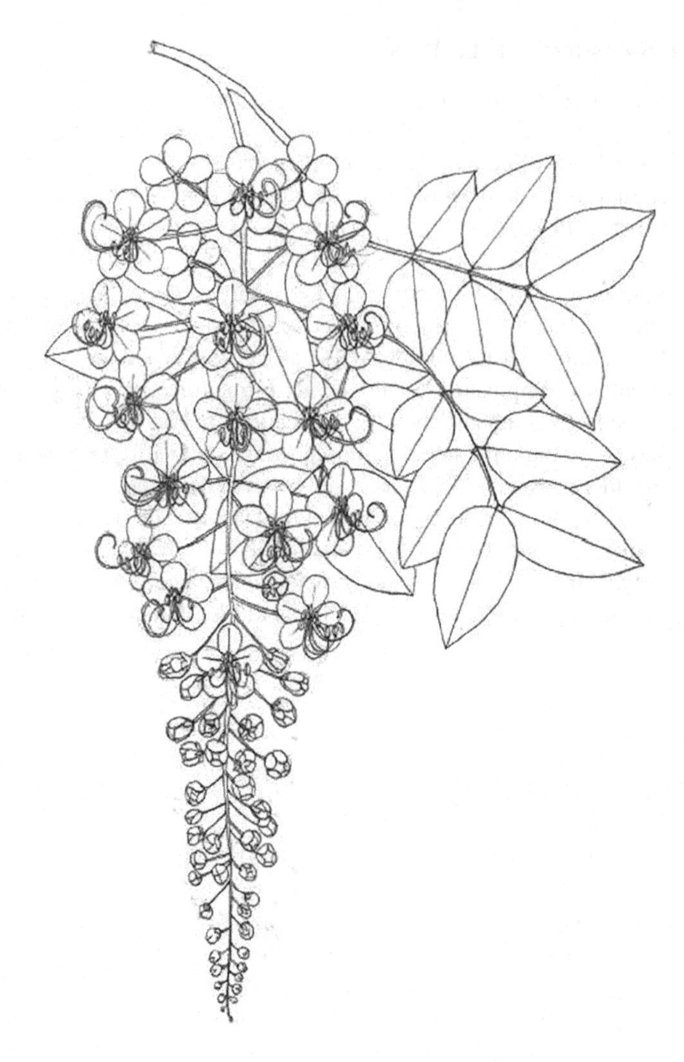

FIGURE 5.14 *Cassia fistula* L.

5.47 *Desmodium gangeticum* (L.) DC.

Synonyms: *Desmodium cavaleriei* H. Lév.; *Hedysarum gangeticum* L.
Common names: da ye shan ma huang (Chinese); saalpernie
Subclass Rosidae, Superorder Fabanae, Subclass Rosiidae, Family Fabaceae
Medicinal use: difficulties of breathing (India)
History: The plant was known to Sushruta

Aqueous extract of roots of *Desmodium gangeticum* (L.) DC. (Figure 5.15) given prophylactically at a dose of 3 mL/100 g to Sprague–Dawley rats for 1 month reduced isoproterenol-induced myocardial infarction with a decrease in cardiac creatinine phosphokinase activity, serum lactate dehydrogenase, aspartate transaminase, and alanine transaminase activities.[164] The cardiac contents of total cholesterol, high-density lipoprotein–cholesterol and low-density lipoprotein–cholesterol increased by isoproterenol were reduced by this treatment.[164] Triglycerides in the liver were reduced from 700 mg/100 g to 442.2 mg/100 g. Cardiac gluthatione reductase catalase activities were increased, whereby reactive oxygen species were decreased.[164] Clinical trials are warranted.

FIGURE 5.15 *Desmodium gangeticum* (L.) DC.

5.48 *Sophora flavescens* Aiton

Synonyms: *Sophora angustifolia* Siebold & Zucc.; *Sophora macrosperma* DC.; *Sophora tetragonocarpa* Hayata
Common name: ku shen (Chinese)
Subclass Rosidae, Superorder Fabanae, Subclass Rosiidae, Family Fabaceae
Medicinal use: jaundice (China)

Matrine given orally at a dose of 200 mg/kg daily for 10 days protected rodents against myocardial infarction induced by isoproterenol as evidenced by reduced serum lactate dehydrogenase and creatine kinase from.[165] Besides, this alkaloid at 200 mg/kg corrected LVSP, mean blood pressure, and heart rate.[165] This quinolizidinie alkaloid increased the enzymatic activity of superoxide dismutase, catalase and glutathione peroxidase in cardiac tissues.[165] Furthermore, histological observation of cardiac tissues revealed that matrine protected cardiac tissues against necrosis, inflammation and interstitial edema.[165] Oxymatrine given parenterally to Wistar rats 10 minutes before experimental myocardial infarction decreased the infarct size of the left ventricle by about 10%.[166] Myocytes from the ischemic zone from treated rats exhibited milder apoptosis, had lower fatty acid synthetase and higher Bcl-2 expression and reduced cytosolic calcium contents compared with untreated animals.[166] In a subsequent study, oxymatrine given intravenously at a dose of 20 mg/kg to Wistar rats

10 minutes before left anterior coronary descending artery occlusion reduced the duration of ventricular arrhythmia and the scores of arrhythmia from 6.7 to 4.8 minutes and 4.7 to 3.3, respectively. This alkaloid delayed the onset of arrhythmia by 1 minute.[167] Oxymatrine given orally at a dose of 60 mg/kg daily for 21 weeks to hypertensive rodents reduced the mean arterial pressure from about 150 to 120 mmHg.[168] Histological observation revealed a decrease of collagen accumulation in the interstitial and perivascular space of left ventricle.[168] The myocardium of hypertensive rodents receiving this alkaloid orally at a dose of 60 mg/kg daily for 21 weeks presented reduced amounts type I and type III collagen fibres.[168] Besides, oxymatrine given orally at a dose of 60 mg/kg daily for 21 weeks reduced serum norepinephrine from about 0.3 to 0.1 ng/mL in hypertensive rodents.[168] In addition, myocardium angiotensin II concentration was decreased from about 160 pg/mg.prot to 80 pg/mg.prot in hypertensive rodents receiving oxymatrine. Furthermore, the phosphorylation of extracellular signal-regulated kinase-1/2, in hypertensive rodents was decreased as well as the phosphorylation of JNK and p38 mitogen-activated protein kinase.[168]

5.49 *Tamarindus indica* L.

Common names: jiu ceng pi guo (Chinese); asam jawa (Malay); tamarin
Subclass Rosidae, Superorder Fabanae, Subclass Rosiidae, Family Fabaceae
Medicinal use: laxative (Malaysia)
History: the plant was known to Serapion

Ethanol extract of *Tamarindus indica* L. (Figure 5.16) given as 5% in tap water to Golden Syrian hamster on hypercholesterolemic diet for 10 weeks lowered total cholesterol from 259 to 129.7 mg/dL,

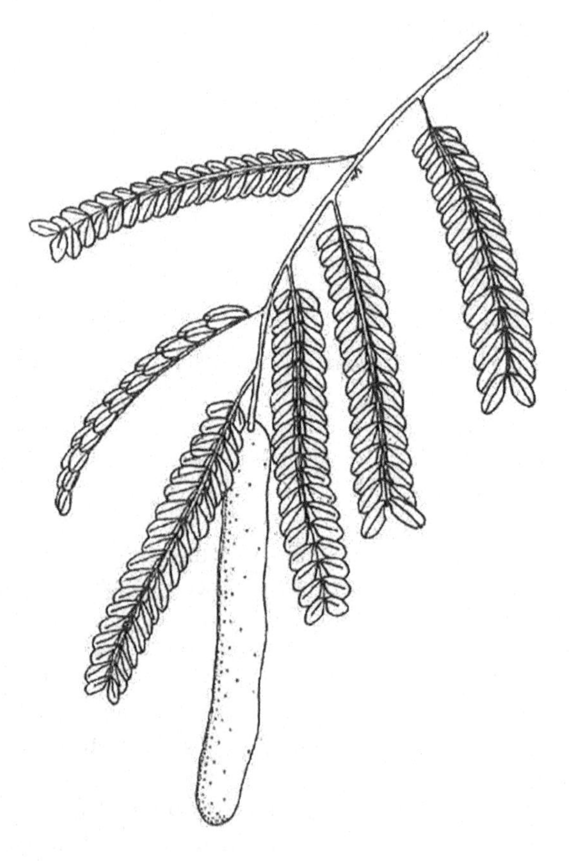

FIGURE 5.16 *Tamarindus indica* L.

triglycerides from 520 to 204 mg/dL, and increased high-density lipoprotein–cholesterol from 45.2 to 73.2 mg/dL.[169] Fasting glucose was reduced from 114.5 to 97.3 mg/dL.[169] This supplementation lowered serum reactive oxygen species, increased serum and liver glutathione peroxidase, superoxide dismutase, and catalase activities.[169] Aortae of rodents on hypercholesterolemic diet and treated with tamarind pulp extract showed a reduction in lesions with lipid accumulation as well as reduced invasion of intima.[169] Intake of tamarin could be beneficial in Metabolic Syndrome.

5.50 *Trigonella foenum-graecum* L.

Synonym: *Trigonella tibetana* (Alef.) Vassilcz.
Common names: hu lu ba (Chinese); halba (Malay); fenugreek
Subclass Rosidae, Superorder Fabanae, Order Fabales, Family Fabaceae
Medicinal use: promote urination (Malaysia)

Wistar albino rats given orally trigonelline daily for 20 days at a dose of 40 mg/kg/day before isopretenol induced cardiac infarction reduced the heart rate from 460 beats/min to normal (330 beats/min), decreased ST elevation, lowered the area of myocardial infarction from 70.2% to 23.2% and reduced serum creatine kinase and lactate dehydrogenase.[170] Furthermore, trigonelline repleted the enzymatic activity of superoxide dismutase, catalase, glutathione peroxidase, increased glutathione S-transferase and reduced malondialdehyde.[170] This alkaloid evoked a downregulation of glucose-regulated protein (Grp58), αB-crystallin and Hsp27 as well as dB and dc isoforms of calcium-calmodulin-dependent protein kinase II.[170] In diabetes, there is an increased glycosylation of hemoglobin and HbA1c is an index of glycemic control. High HbA1c levels are indicative of risks of development and/or progression of nephropathy and retinopathy.[171] In diabetic patients, hyperglycemia decreases retinal glutathione, catalase, and superoxide dismutase resulting into intense oxidative insults.[172] Seeds of fenugreek given orally to streptozotocin-induced diabetic Wistar albino rats at a dose of 200 mg/kg/day for 24 weeks lowered glycemia from 544.7 to 360.2 mg/dL as well as HbA1c from 9.1% to 5.9%.[173] This treatment increased retinal contents of glutathione and activities of superoxide dismutase and catalase. It reduced tumor necrosis factor-α and interleukin-1β levels, and mitigated microvascular retinal leakage and dilation.[173] In addition, retinae contents in angiogenic vascular endothelium growth factor and protein kinase Cβ were reduced as well as basal retinal membrane thickness.[173] In kidney cells exposed to elevated concentration of glucose there is an increase in diacylglycerol and protein kinase C activity leading to the activation of aldose reductase and sorbitol dehydrogenase and production of reactive oxygen species.[174]

5.51 *Glycosmis parviflora* (Sims) Little

Synonyms: *Citrus erythrocarpa* Hayata; *Glycosmis citrifolia* (Willd.) Lindl.; *Limonia citrifolia* Willd.
Common name: xiao hua shan xiao ju (Chinese)
Subclass Rosidae, Superorder Rutanae, Order Rutales, Family Rutaceae
Medicinal use: skin itchiness (Taiwan)

Rutaceae produce quinoleic alkaloids that have the tendency to inhibit platelet aggregation *in vitro*. *Glycosmis parviflora* (Sims) Little elaborates citracridone, atalaphyllidine, des-*N*-methylnoracronycine, 5-nydroxynoracronycine, des-*N*-methylacronycine, pyranofolidine and 4,8-dimethoxy-1-methyl-3(3-methyl-2-enyl)2-quinolone which at a concentration of 100 μg/mL inhibited the aggregation of platelets induced by arachidonic acid by more than 90%.[175] Citracridone, atalaphyllidine, des-*N*-methylnoracronycine, des-*N*-methylacronycine, pyranofolidine and 4,8-dimethoxy-1-methyl-3(3-methyl-2-enyl)2-quinolone at 100 μg/mL inhibited collagen-induced aggregation by more than

80%.[175] Citracridone and pyranofolidine inhibited further the aggregation of platelets exposed to platelet factor by about 95%.[175]

5.52 *Murraya paniculata* (L.) Jack.

Synonym: *Murraya exotica* L.
Common names: qian li xiang (Chinese); kamuning (Philippines); jasmine orange
Subclass Rosidae, Superorder Rutanae, Order Rutales, Family Rutaceae
Medicinal use: diarrhea (Philippines)

Angiotensin II increases glomerular capillary permeability to proteins, stimulates mesangial cells proliferation and accumulation of mesangial matrix.[176] Angiotensin-converting enzyme inhibitors like captopril are used for the treatment of diabetic nephropathy and for the treatment of hypertension.[177] Flavonoid fraction of leaves of *Murraya paniculata* (L.) Jack. given orally to high-fat diet streptozotocin-induced type-2 diabetic Wistar rats (fasting blood glucose > 11.1 mmol/L) at a dose of 70 mg/kg/day for 13 weeks lowered urinary albumin from 339.8 to 250.3 mg/L (normal: 212 mg/L; captopril at 10 mg/kg: 251 mg/L), and lowered urinary albumin excretion rate from 15.8 to 10.7 μg/min (normal: 5.2 μg/min; captopril at 10 mg/kg: 10.2 μg/min).[178] This regimen lowered plasmatic creatininine from 112.8 to 78.1 μmol/L (normal: 81.6 μmol/L; captopril at 85.3 μmol/L) and increased creatinine clearance from 0.6 to 0.9 mmol/L (normal: 1 mmol/L; captopril at 10 mg/kg: 0.9 mmol/L).[178] This treatment lowered fasting glucose from about 18 to 12 mmol/L, normalized total cholesterol from 2.1 to 1.6 mmol/L (normal: 1.6 mmol/L; captopril at 10 mg/kg: 1.6 mmol/L), normalized triglycerides from 0.4 to 0.2 mmol/L (normal: 0.2 mmol/L; captopril at 10 mg/kg: 0.3 mmol/L), increased high-density lipoprotein–cholesterol and lowered low-density lipoprotein–cholesterol.[178] The fraction lowered plasma interleukin-6 increased the enzymatic activities superoxide dismutase and glutathione peroxidase in the renal cortex and lowered malondialdehyde.[178] This treatment improved renal cortex glomerula histology and lowered the expression of transforming growth factor-1β by glomeruli and renal tubules.[178]

5.53 *Citrus grandis* (L.) Osbeck

Synonyms: *Citrus decumana* L.; *Citrus maxima* Merr.
Common names: you (Chinese); grapefruit
Subclass Rosidae, Superorder Rutanae, Order Rutales, Family Rutaceae
Medicinal use: facilitate digestion (China)

Epidemiological evidence suggests that consumption of *Citrus* fruits decrease the risk of stroke on account of their flavanone contents.[179,180] In volunteers with Metabolic Syndrome, the consumption a fresh grapefruit before each meal (3 times daily) for 6 weeks evoked a decrease in urinary F2-isoprostanes which is a marken of oxidative stress.[181] Naringenin given orally at a dose of 25 mg/kg/day to Sprague–Dawley rats with balloon-induced arterial injury for 14 days reduced neointimal hyperplasia as evidenced by a decrease of intima area/media area ration from 0.9 to 0.3. This flavanone also decreased expression on intimal nuclear factor-κB.[182] This flavanone lowered plasma 8-iso-PGF2α from 479.7 to 169.3 pg/mL.[182] *In vitro*, naringenin at a concentration of 100 μM prevented angiotensin II-induced proliferation and migration of vascular smooth muscle cells. It decreased reactive oxygen species production induced by angiotensin II by increasing superoxide dismutase activity and decreasing NADPH oxidase activity.[182] This flavonoid inhibited angiotensin-induced reactive oxygen species increase and downstream induced expression of extracellular signal-regulated kinase-1/2, p38 mitogen-activated protein kinase and nuclear factor-κB.[182]

5.54 *Citrus iyo* Tanaka

Common name: iyomikan (Japan)
Subclass Rosidae, Superorder Rutanae, Order Rutales, Family Rutaceae
Nutritional use: fruit (Japan)

Visceral adiposity and type 2 diabetes are associated with an imbalance between endothelium-derived relaxing and constricting factors. Enhanced oxidating stress decreases the availability of nitric oxide. It is a major risk factor the development of cardiovascular diseases.[183] Vascular ageing is associated with increase of superoxide in aortic and mesenteric which lowers the availability of nitric oxide arteries.[184] The juice of fruits of *Citrus iyo* Tanaka given for 2 weeks at 10% of drinking water to C57BL6 mice prevented by 47% neointima formation induced by polyethylene cuff placement around the femoral artery. This regimen evoked a reduction of cell proliferation, superoxide anion production in the femoral artery and inhibition of neointimal expression of phosphorylated extracellular signal-regulated kinase.[185] This regimen attenuated the accumulation of macrophages and neutrophils in injured arterial tissues.[185]

5.55 *Citrus reticulata* Blanco

Common names: gan ju (Chinese); naranga (Sanskrit); ponkan fruit
Synonyms: *Citrus chrysocarpa* Lush.; *Citrus daoxianensis* S.W. He & G.F. Liu; *Citrus erythrosa* Yu. Tanaka; *Citrus nobilis* Lour.; *Citrus sunki* hort. ex Tanaka; *Citrus unshiu* Marcov.
Subclass Rosidae, Superorder Rutanae, Order Rutales, Family Rutaceae
Medicinal use: heart diseases (India)

In Streptozotocin-induced diabetic Wistar rats receiving a normal diet with 1% freeze-dried juice of fruits of *Citrus reticulata* Blanco for 10 weeks no significant impact on body weight or glycemia was observed, Serum triglycerides was decreased from 638.9 to 505.9 mg/dL (nondiabetic control: 155.7 mg/dL), low-density lipoprotein–cholesterol from 90.7 to below 67.2 mg/dL (nondiabetic group value), cholesterol from 266.4 to 208 mg/dL (nondiabetic group: 145 mg/dL). High-density lipoprotein–cholesterol was not affected.[186] Aortic rings isolated from treated rodents were more respondent to endothelium-dependent acetycholine induced vasorelaxation that untreated diabetic animals with IC_{50} values of 63.7 and 83.3 nM, respectively. Endothelium-independent relaxation induced by sodium nitroprusside was not improved by juice supplementation suggesting a protective effect on endothelium function in diabetic rats.[186] In healthy volunteers consuming fatty meals, the simultaneous intake of 1 L of red orange juice lowered 2 hours postprandial triglycerides from 173.4 to 147 mg/dL, cholesterol from 219.1 to 213.3 mg/dL, glucose from 93 to 85.7 mg/dL. This Regimen lowered arterial stiffness, neutrophil myeloperoxidase degranulation, and lowered diatolic blood pressure from 72.1 to 70.6 mmHg.[187] In hypertensive patients, there is an increase of calcium concentration is vascular smooth muscle cells[188] and calcium channel blockers such as nifedipine are among the most widely used drugs for treatment of hypertension. These drugs evokes reduction of vasoconstrictor tonus of the smooth muscle fibers in arterioles.[189] Aurapten from *Citrus* spp. given at a single intravenous dose of 500 μg/kg to anesthetized normotensive rats evoked a 20 mmHg fall in mean arterial blood pressure (nifedipine 63 μg/Kg, about 30 mmHg) and an increase of heart rate from about 30 about 42 beats/min (same value for nifedipine 63 μg/kg).[190]

5.56 *Euodia rutaecarpa* (Juss.) Hook. f. & Thoms.

Synonym: *Boymia rutaecarpa* Juss.
Common name: wu zhu yu (China)
Subclass Rosidae, Superorder Rutanae, Order Rutales, Family Rutaceae
Medicinal use: vomiting (China)

A single intravenous administration of dehydroevodiamine from *Euodia rutaecarpa* (Juss.) Hook. f. & Thoms. at a dose of 10 mg/kg to of anesthetized rodents commanded an immediate fall in mean arterial blood pressure and heart rate by about 30%.[191] At a dose of 20 mg/kg, the hypotension evoked by that alkaloid was transient whereby the bradycardia was sustained for circa 1 hour.[191] Rutaecarpine at a dose of 0.1 mM relaxed mesenteric arterial rings with or without endothelium precontracted with phenylephrine by 90.7% and 33%, respectively.[192] The nitric oxide synthase inhibitor N^{ω}-nitro-L-arginine methyl ester inhibited the vasorelaxing activity of rutaecarpine from 87.8% to 30.6%. The guanylyl cyclase inhibitor methylene blue reduced the effects of that alkaloid from 90.2% to 37.9%. This suggests that the release of nitric oxide from vascular endothelium and subsequent activation of vascular smooth muscle guanylyl-cyclase and myorelaxation.[192] Rutaecarpine at a dose of 50 μg/g reduced the mortality of rodents by acute pulmonary embolism poisoned with ADP from 81% to 35%.[193] Furthermore, a bolus of rutaecarpine at a dose of 50 μg/g, prolonged bleeding by 2 minutes compared with saline group.[193] Rodents receiving rutaecarpine at a dose of 5 μg/g/min for 10 minutes had, compared with saline group, a fall of mean blood pressure from 94.5 to 63.2 mmHg, a reduction of blood platelets from 1070×10^9/L to 970×10^9/L.[193] Evodiamine given intravenously at a dose of 60 μg/kg to anesthetized Sprague–Dawley rats 10 minutes before coronary artery clipping induced ischemia-reperfusion reduced infarct size by about 43%. This alkaloid reduced serum creatine kinase activity by about 17%, serum tumor necrosis factor-α by 23% and these effects were nullified by capsazepine, a competitive vanilloid receptor antagonist.[194] Dehydroevodiamine at 10^{-4} M reduced the beating rate of isolated rodents hearts.[191] Rutaecarpine at 120 μM reduced the aggregation of platelets evoked by adrenaline and arachidonic acid by 80% whereas the aggregation evoked by ADP or collagen was inhibited by 40%.[191] Rutaecarpine at 200 μM inhibited the release of thromboxane B2 from platelets challenged with collagen from 124.5 to 70.1 ng/mL.[193] Angiotensin receptor blockers antagonize vasoconstriction, cardiomyocytes and vascular smooth muscle hypertrophy, aldosterone release, sympathetic activity, sodium reabsorption resulting in decreased blood pressure.[195] Evocarpin, 1-methyl-2-[(4Z,7Z)-4,7-tridecadienyl]-4(1H)-quinolone, and 1-methyl-2-[(6Z,9Z)-6,9-pentadecadienyl]-4(1H)-quinolone isolated from the fruits of *Euodia rutaecarpa* (Juss.) Hook. f. & Thoms. inhibited the binding of ^{3}H labeled angiotensin II to angiotensin II receptor *in vitro* with IC_{50} values equal to 43.4, 34.1, and 48.2 μM, respectively.[196] The enzyme 5-lipoxygenase catalyzes the synthesis of leukotriene B4 from arachidonic acid[197]. Leukotriene B4 is one of the first leukocyte chemoattractants generated to promotes leukocytes migration in response to tissue injuries and contributes to the development of type 2 diabetes in obese.[198] 1-Methyl-2-nonyl-4(1H)-quinolinone, 1-methyl-2-(6Z)-6-undecenyl-4)(1H)-quinolinone, 1-methyl-2-(4Z,7Z)-4,7-tridecadienyl-4(1H)-quinolinone, evocarpine and 1-methyl-2-(6Z,9Z)-6,9-pentadecadienyl-4(1H)-quinolinone isolated from the fruits of *Evodia rutaecarpa* inhibited 5-lipoxygenase catalyzed leukotriene B4 synthesis by activated granulocytes with IC_{50} values below 15 μM (positive control zileuton: 10.4 μM).[199] Evodiamine at 10^{-3} M abrogated both basal and angiotensin II-induced secretion of aldosterone zona glomerulosa cells isolated from Sprague–Dawley rats via inhibition of the enzymatic activity of 11-β-hydroxylase.[200] Rutaecarpine and evodiamine at 10 μM inhibited the migration of monocytes.[201] In diabetic patients, hyperglycemia induces the glycosylation of β-crystalline of lens, the generation of reactive oxygen species in retina due to auto-oxidation of glucose, and the formation of polyols. Aldose reductase is the first and rate limiting enzyme in the polyol pathway which catalyzes the conversion of glucose to sorbitol which is converted into fructose by sorbitol dehydrogenase.[202] Aqueous extract of this plant inhibited aldose reductase activity.[203] From this extract, rhetsinine inhibited aldose reductase with an IC_{50} equal to 24.1 μM and reduced sorbitol accumulation in cultured human erythrocytes challenged with high glucose by 79.3% at a concentration of 100 μM.[204]

5.57 *Melicope triphylla* (Lam.) Merr.

Synonyms: *Euodia triphylla* (Lam.) DC.; *Zanthoxylum triphyllum* (Lam.) G. Don
Common name: san ye mi zhu yu (Chinese)
Subclass Rosidae, Superorder Rutanae, Order Rutales, Family Rutaceae

Melicope triphylla (Lam.) Merr. produces the furoquinoline alkaloids, dictamine, evolitrine, and pteleine, which at 100 μg/mL, inhibited the aggregation of platelets exposed to arachidonic acid.[204] The same alkaloids at the same concentration inhibited the aggregation of platelets challenged with collagen by 55%, 89%, and 87%, respectively.[204] Dictamine, evolitrine, and pteleine inhibited at 100 μg/mL platelet-activating factor induced aggregation by 31%, 29%, and 28% and inhibited the aggregation evoked by ADP by 54%, 81%, and 44%, respectively.[204] From the plant, skimmianine and kokusaginine inhibited the aggregation of platelets evoked by arachidonic acid at 50 μg/mL by 68.9% and 100%, respectively.[204] Quinoline alkaloids of Rutaceae often inhibit platelet aggregation *in vitro*.

5.58 *Ruta angustifolia* Pers.

Synonym: *Ruta hortensis* Mill.
Common names: garuda (Malay); godong minggu (Indonesia)
Subclass Rosidae, Superorder Rutanae, Order Rutales, Family Rutaceae
Medicinal use: convulsion (Indonesia)

Dictamine, skimmianine, psoralen, chalepensin, clausidine, and graveolinine isolated from a member of the genus *Ruta* L. at a concentration of 100 μM inhibited the aggregation of rabbit platelets challenged with arachidonic acid by more than 80%.[205] Dictamine, chalepensin, and graveolinine a concentration of 100 μM inhibited the aggregation of rabbit platelets challenged with collagen by more than 80%.[205] Methanol extract of a member of the genus *Ruta* given orally at a dose of 20 mg/kg/day to Sprague–Dawley rats on cholesterol-enriched diet for 90 days lowered the activity of alanine transaminase and aspartate transaminase. This Regimen decreased total cholesterol from 107.4 to 87.6 mg/dL (normal diet: 73.9 mg/dL), lowered low-density lipoprotein–cholesterol from 72.3 to 51.6 mg/dL (normal diet: 34.8 mg/dL), increased high-density lipoprotein–cholesterol from 17.5 to 22.6 mg/dL (normal diet: 26.1 mg/dL), and lowered atherogenic index from 4.2 to 2.2 (normal diet: 1.3).[206] This treatment increased the activities of superoxide dismutase, catalase and glutathione peroxidase, and increased glutathione in liver and heart of treated animals.[206] In monocytes, this extract lowered the enzymatic activity of cyclo-oxygenase and 15-lipoxygenase by about 35% and 50%, respectively. It also decreased serum myeloperoxidase activity, decreased serum C-reactive protein as well as white blood cell count.[206] Histological observation of aorta of treated animals revealed a preventive effect on early atherogenesis.[206] It would be of interest to explore the effect of *Ruta angustifolia* Pers. (Figure 5.17) on high-fat diet in rodent.

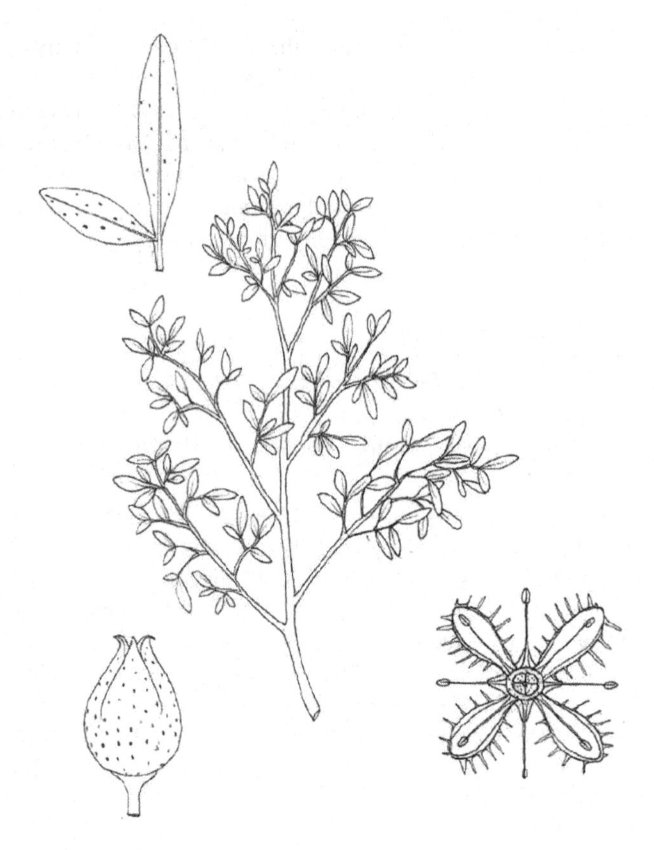

FIGURE 5.17 *Ruta angustifolia* Pers.

5.59 *Murraya euchrestifolia* Hayata

Synonym: *Clausena euchrestifolia* (Hayata) Kaneh.
Common name: dou ye jiu li xiang (Chinese)
Subclass Rosidae, Superorder Rutanae, Order Rutales, Family Rutaceae
Medicinal use: headache (China)

Murraya euchrestifolia Hayata elaborates murrayafoline A which at a concentration of 50 μg/mL abrogated the aggregation of platelets exposed to collagen.[207] Mukoeic acid, murrayazolidine and murrayamine-M at 100 μg/mL, inhibited the aggregation of platelets induced by collagen.[207] Murrayafoline A (Figure 5.18) at 20 μg/mL abrogated the aggregation of platelets exposed to

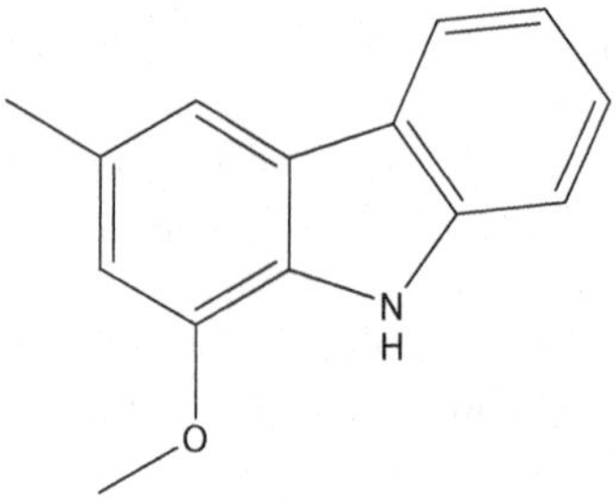

FIGURE 5.18 Murrayafoline A.

arachidonic acid. From the same plant, 3-methycarbazole, murrayanine and mukoeic acid, at 100 μg/mL inhibited the aggregation of platelets exposed to arachidonic.[207] Murrayamine-M at 50 μg/mL abrogated the aggregation of platelets evoked by arachidonic acid and platelet activating factor.[207] Murrayafoline A inhibited at a concentration of 5 μM the platelet-derived growth factor-induced proliferation of rat aortic vascular smooth muscle cells by about 40% *in vitro*.[208] This carbazole alkaloid inhibited DNA synthesis, increased cellular population in G_0/G_1 phase as a result of expression inhibition of cyclin D1, cyclin E, CDK2, CDK4, and subsequent phosphorylation of pRb in vascular smooth muscle cells.[208]

5.60 *Toddalia asiatica* (L.) Lam

Synonyms: *Paullinia asiatica* L.; *Toddalia aculeata* Pers.; *Toddalia angustifolia* Lam.; *Toddalia floribunda* Wall.; *Toddalia nitida* Lam.; *Toddalia rubricaulis* Roem. & Schult.; *Toddalia tonkinensis* Guillaumin
Common names: fei long zhang xue (Chinese); dauag (Philippines); wild orange tree
Subclass Rosidae, Superorder Rutanae, Order Rutales, Family Rutaceae
Medicinal use: tonic (Burma)

Chelerythrine isolated from the wood of *Toddalia asiatica* (L.) Lam at a concentration of 100 μg/mL inhibited by 100% platelet aggregation induced by ADP, arachidonic acid, collagen and platelet activating factor.[209] From the same plant, 2,6-dimethoxy-p-benzoquinone (Figure 5.19) at a concentration of 100 μg/mL inhibited by 100% platelet aggregation induced by arachidonic acid, collagen and platelet activating factor.[209] Isopimpinellin at a concentration of 50 μg/mL inhibited the aggregation of platelets challenged with arachidonic acid.[209] Chelerythrine a concentration of 25 μg/mL abrogated the aggregation of platelets induced by thrombin.[209]

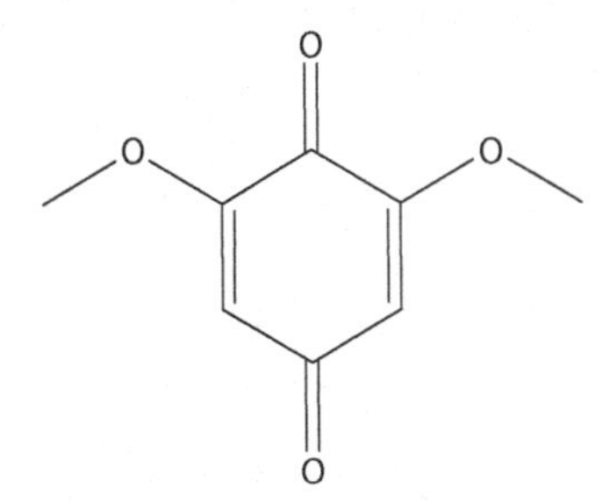

FIGURE 5.19 2,6-Dimethoxy-p-benzoquinone.

5.61 *Zanthoxylum bungeanum* Maxim.

Common name: hua jiao (Chinese)
Subclass Rosidae, Superorder Rutanae, Order Rutales, Family Rutaceae
Medicinal use: facilitate digestion (China)

α-Linolenic acid isolated from the seeds of *Zanthoxylum bungeanum* Maxim given orally at a dose of 250 mg/kg/day for 10 days to mice increased bleeding time as efficiently as aspirin at a dose of 100 mg/kg.[210] This regimen afforded a complete protection for mice challenged with thrombus induced by intravenous injection of collagen and adrenaline, and this protective effect was superior to aspirin at 100 mg/kg.[210] This fatty acid inhibited platelet expression of GP IIb/IIIa.[210] Platelets of treated animal had lowered expression of Akt and phosphoinositide 3-kinase.[210]

5.62 *Zanthoxylum schinifolium* Siebold & Zucc.

Synonyms: *Fagara pteropoda* (Hayata) Y.C. Liu; *Fagara schinifolia* (Siebold & Zucc.) Engl.; *Zanthoxylum mantschuricum* Benn.; *Zanthoxylum pteropodum* Hayata
Common name: qing hua jiao (Chinese)
Subclass Rosidae, Superorder Rutanae, Order Rutales, Family Rutaceae
Medicinal use: asthma (China)

Acetoxycollinin, anisocoumarin H, and platydesmine isolated from the root bark of *Zanthoxylum schinifolium* Siebold & Zucc. blocked at 100 μM rabbit platelet aggregation induced by arachidonic acid.[211] From the same plant, 8-methoxyanisocoumarin H at 100 μM inhibited platelet aggregation evoked by collagen by 90.7%.[211] Acetoxycollinin, anisocoumarin H, and platydesmine at 100 μM inhibited platelet aggregation evoked by collagen by 94.6%, 94.6%, and 96%, respectively.[211] Acetoxycollinin at 100 μM inhibited platelet aggregation evoked by platelet activating factor.[211]

5.63 *Zanthoxylum simulans* Hance

Synonym: *Zanthoxylum bungei* Planch.
Common name: ye hua jiao (Chinese)
Subclass Rosidae, Superorder Rutanae, Order Rutales, Family Rutaceae
Medicinal use: indigestion (China)

Zanthoxylum simulans Hance (Figure 5.20) produces benzosimuline, zanthobungeanine, (–)-acetylnornuciferine, *N*-acetydehydroanonaine, decarine, and skimmianine that nullified aggregation of

FIGURE 5.20 *Zanthoxylum simulans* Hance.

age platelets induced by arachidonic acid.[212] Benzosimuline, zanthobungeanine, (–)-acetylnornuciferine, *N*-acetydehydroanonaine, decarine, and skimmianine inhibited the aggregation induced by collagen.[212] Benzosimuline, zanthobungeanine, (–)-acetylnornuciferine, *N*-acetydehydroanonaine, decarine, and skimmianine inhibited the aggregation induced by platelet-activating factor.[212] Zanthobungeanine, (–)-acetylnornuciferine, and decarine inhibited the aggregation induced by thrombin.[212]

5.64 *Picrasma quassiodes* (D. Don) Benn.

Synonym: *Simaba quassioides* D. Don
Common names: ku shu (Chinese); bitter wood
Subclass Rosidae, Superorder Rutanae, Order Rutales, Family Simaroubaceae
Medicinal use: indigestion (Korea)

Extract of stems of *Picrasma quassiodes* (D. Don) Benn. given intragastrically to spontaneously hypertensive rats at a dose of 200 mg/kg/day for 6 weeks lowered systolic blood pressure from about 205 to 180 mmHg and this effect was comparable with captopril (12.5 mg/kg).[213] This regimen boosted plasma superoxide dismutase and nitric oxide and reduced plasma malondialdehyde by 42%.[213] Furthermore, immunohistochemical analysis of treated hypertensive rats evidenced the induction of expression of endothelial nitric oxide synthetase in vascular endothelial cells and smooth muscle cells.[213]

5.65 *Commiphora mukul* (Hook. ex Stocks) Engl.

Synonym: *Commiphora wightii* (Arn.) Bhandari
Common names: guggul (India); Indian Bdellium
Subclass Rosidae, Superorder Rutanae, Order Rutales, Family Burseraceae
Medicinal use: ulcers (India)

Hydro-alcoholic extract of *Commiphora mukul* (Hook. ex Stocks) Engl. given orally at a daily dose of 400 mg/kg/day for 31 days to Wistar rats intoxicated with isoproterenol reduced systolic arterial blood pressure from 16 to 7 mmHg, reduced heart beats from 129 to 15 beats/min.[214] This extract maintained left ventricular function and prevented myocardial necrosis.

5.66 *Boswellia serrata* Roxb. ex Colebr.

Common name: salai (India); Indian olibanum
Subclass Rosidae, Superorder Rutanae, Order Rutales, Family Burseraceae
Medicinal use: diuretic (India)

Boswellic acid from *Boswellia serrata* Roxb. ex Colebr. inhibited rat lens aldose reductase, rat kidney aldose reductase and human recombinant aldose reductase activities with IC_{50} values equal to 5, 6, and 3.9 μg/mL, respectively.[162] In Wistar albino rats fed with galactose for 14 days, daily administration of boswellic acid at a dose of 10 mg/kg reduced lens aldose reductase catalyzed production of galactitol by 74%.[162] Methanol extract of oleogum resin from another member of the genus *Boswellia* Roxb. ex Colebr. given orally to Wistar rats at a dose of 400 mg/kg/day for 30 days before subcutaneous administration of isoprenaline-lowered serum aspartate transaminase, lactate dehydrogenase, and creatinine by 30%, 45%, and 60%.[215] The extract reduced lipid peroxidation in myocardium by about 60%.[215] Besides, this treatment prevented myonecrosis and myocardial edema and lowered.[215]

5.67 *Mangifera indica* L.

Common names: am (India); mango
Subclass Rosidae, Superorder Rutanae, Order Rutales, Family Anacardiaceae
Medicinal use: diabetes (India)

Hyperglycemia, advanced glycosylation end products and reactive oxygen species, are responsible for renal damages in diabetics.[174] Mangiferin given orally to streptozotocin-induced diabetic Wistar rats for 12 weeks at a daily dose of 45 mg/kg evoked a mild reduction of glycemia, reduced serum advanced glycation end products by 25%, red blood cells sorbitol by 44%, serum malondialdehyde by 38%, and decreased renal albumin excretion by 36%.[216] This xanthone increased creatinine clearance rate, serum gluthatione peroxidase and superoxide dismutase.[216] This treatment lowered transforming growth factor-β1 expression in kidney of diabetic rodents by 27% and prevented glomerular mesangial matrix expansion.[216] *In vitro*, mangiferin at a concentration of 10^{-4} M reduced the expression of collagen type IV by rat mesangial cells by 21%.[216] Clinical trials are warranted.

5.68 *Rhus verniciflua* Stokes

Synonym: *Toxicodendron vernicifluum* (Stokes) F.A. Barkley
Common name: qi shu (Chinese)
Subclass Rosidae, Superorder Rutanae, Order Rutales, Family Anacardiaceae
Medicinal use: bleeding (China)

Fisetin at 50 μM reduced the permeability of human umbilical vein endothelial cells challenged with high glucose concentration.[217] In C57BL/6 mice given high glucose intravenously, fisetin given parenterally at a dose of 28.5 μg/mouse lowered vascular permeability.[217] In addition, this flavone at 50 μM inhibited the expression of vascular adhesion molecule-1, and E-selectin by human umbilical vein endothelial cells challenged by high glucose. This flavonoid lowered adhesion of monocytes to endothelial cells.[217] Fisetin inhibited the expression of interleukin-8 and monocyte chemoattractant peptide-1 by human umbilical vein endothelial cells. It decreased cytoplasmic reactive species and downstream activation of nuclear factor-κB.[217]

5.69 *Peganum harmala* L.

Common names: luo tuo peng (Chinese); hurmul (India)
Subclass Rosidae, Superorder Rutanae, Order Zygophyllales, Family Peganaceae
Medicinal use: tonic (India)

The seeds of *Peganum harmala* L. elaborate harmine and harmaline which, at a concentration of 10 μM, inhibited the formation of conjugated diene from low-density lipoprotein oxidized by copper by 30% and 50%, respectively.[218] These indole alkaloids inhibited the aggregation of platelets exposed to collagen.[219] Harmine and harman inhibited the phosphorylation of phospholipase Cγ2 induced by collagen by 80.5% and 98.3% at 200 μM with concomitant inhibition of intracellular increase calcium.[219]

5.70 *Euonymus alatus* (Thunb.) Siebold

Synonyms: *Celastrus alatus* Thunb.; *Celastrus striatus* Thunb.
Common name: wei mao (Chinese)
Subclass Rosidae, Superorder Celastranae, Order Celastrales, Family Celastraceae
Medicinal use: promotes blood circulation (China)

Celastrol from *Euonymus alatus* (Thunb.) Siebold at a concentration of 100 nM inhibited the expression and secretion of resistin by monocytes *in vitro*.[220] This triterpene at 100 nM inhibited the migration of human vascular smooth muscle cells induced by resistin *in vitro* by inhibiting via TLR-4 the expression of collagen I/IV and attenuating MMP-2 activity by vascular smooth muscle cells.[220] Celastrol inhibited resistin-induced adhesion of smooth muscle cells to collagen I together with concomitant inhibition of integrin β2/β3 expression.[220]

5.71 *Salacia oblonga* Wall.

Subclass Rosidae, Superorder Celastranae, Order Celastrales, Family Celastraceae
Medicinal uses: diabetes (India)

Kotalagenin 16 acetate, maytenfolic acid, 3β,22α-dihydroxyoleanane-12-en-29-oic acid, 19-hydroxyferruginol, and lambertic acid from *Salacia oblonga* Wall. inhibited at a concentration of 100 μM the enzymatic activity of aldose reductase.[221] Aqueous extract of roots given orally at a dose of 100 mg/kg/day for 6 weeks prevented the development of interstitial fibrosis and perivascular fibrosis in the left ventricle of obese Zucker rats.[222] This regimen inhibited the expression of cardiac transforming growth factor-βs1 and βs3.[222] This extract reduced nonfasting glycemia after 4 weeks whereas fasting glycemia was not changed and glucose tolerance was not improved.[222]

5.72 *Viscum articulatum* Burm. f.

Synonyms: *Viscum dichotomum* D. Don; *Viscum nepalense* Spreng.
Common name: bian zhi hu ji sheng (Chinese)
Subclass Rosidae, Superorder Santalanae, Order Santalales, Family Viscaceae
Medicinal use: tuberculosis (China)

Oleanolic acid from *Viscum articulatum* Burm.f. given at a dose of 60 mg/kg/day intraperitoneally for 10 days to Wistar rats poisoned with dexamethasone prevented rise of blood pressure, attenuated weight loss and normalized serum nitrate/nitrite and cardiac lipid peroxides.[223] Methanol extract of *Viscum articulatum* Burm. f. given at a dose of 400 mg/kg/day for 4 weeks to Wistar rats prevented nitric oxide synthetase inhibitor N^{ω}-nitro-L-arginine methyl ester-induced increase of systolic blood pressure.[224] This treatment normalized N^{ω}-nitro-L-arginine methyl ester-induced cholesterol elevation and serum creatinine and improved myocardial and glomerular histoarchitecture.[224]

5.73 *Hippophae rhamnoides* L.

Synonyms: *Elaeagnus rhamnoides* (L.) A. Nelson; *Rhamnoides hippophae* Moench
Common names: sha ji (Chinese); sea-buckthorn
Subclass Rosidae, Superorder Rhamnanae, Order Rhamnales, Family Eleagnaceae
Medicinal use: stomach pain (Tibet)

Flavonoid fraction of seeds of *Hippophae rhamnoides* L. given to Sprague–Dawley rats on high-sucrose diet at a 150 mg/kg/day for 8 weeks reduced systolic blood pressure from 136.1 to 113.3 mmHg, lowered insulinemia by 17.7%, triglycerides by 28%, cholesterol by 24.4%, free fatty acids by 26%, and increased high-density lipoprotein.[225] This extract increased plasmatic angiotensin II.[225] Macrophages phagocyting oxidized low-density lipoprotein secrete cytokines which stimulate further endothelial cells to express vascular adhesion molecule-1. Leukocytes in turn secrete tumor necrosis factor-α and interleukin-6 and migrate to nascent atheroma.[226] Isorhamnetin (Figure 5.21) from *Hippophae rhamnoides* L. at 20 μM increased the viability of macrophages challenged with oxidized low-density lipoprotein and halved lipid deposition.[227] Oral administration of this flavone

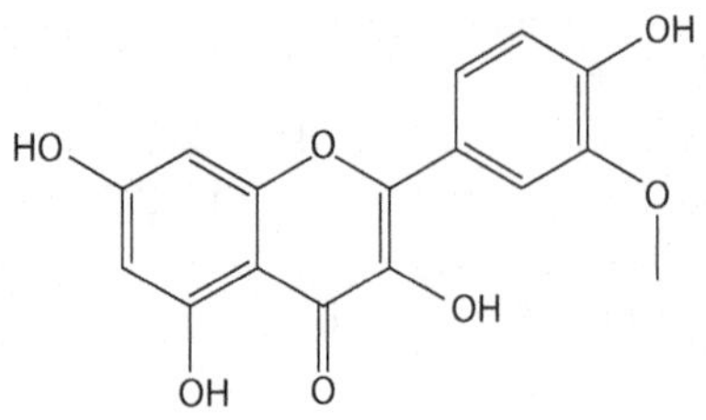

FIGURE 5.21 Isorhamnetin.

at a dose of 20 mg/kg/day for 8 weeks to $ApoE^{-/-}$ mice on high-fat diet reduced the formation of atherosclerotic aortic lesions and macrophages accumulation in lesions.[227]

5.74 *Cornus officinalis* Siebold & Zucc.

Synonyms: *Macrocarpium officinale* (Siebold & Zucc.) Nakai
Common name: shan zhu yu (Chinese)
Subclass Asteridae, Superorder Cornanae, Order Cornales, Family Cornaceae
Medicinal use: tonic (China)

Vascular endothelial cells produce endothelin-1 which binds to vascular smooth muscle cells G protein coupled receptor ET_A to induce contraction.[228] Ether extract of fruits of *Cornus officinalis* Siebold & Zucc. given orally to streptozotocin-induced diabetic Sprague–Dawley rats at a dose of 50 mg/kg/day for 4 weeks evoked a mild increase in body weight, insulinemia, slightly decreased glycemia and lowered serum endothelin-1.[229] This regimen lowered left ventricle weigh normalized left ventricle end diastolic pressure. It increased myocardial sarcoplasmic reticulum calcium-ATPase 2a expression and normalized myocardial malondialdehyde, superoxide dismutase and gluthatione peroxidase.[229] Loganin (Figure 5.22) from this plant given orally at a dose of 10 mg/kg/day to streptozotocin-induced diabetic Sprague–Dawley rats for 16 weeks decreased albuminuria from 3.5 mg/24 h to 2.4 mg/24h, serum cystatin C from 62.6 to 42.5 ng/mL and connective tissue growth factor from 923 to 611 pM.[230] This treatment evoked a mild increase of body weight, had no effect on glycemia but improved renal histoarchitecture.[230] *In vitro*, loganin at a concentration of 1 μM inhibited connective tissue growth factor expression in HK-2 cells challenged to high glucose.[230]

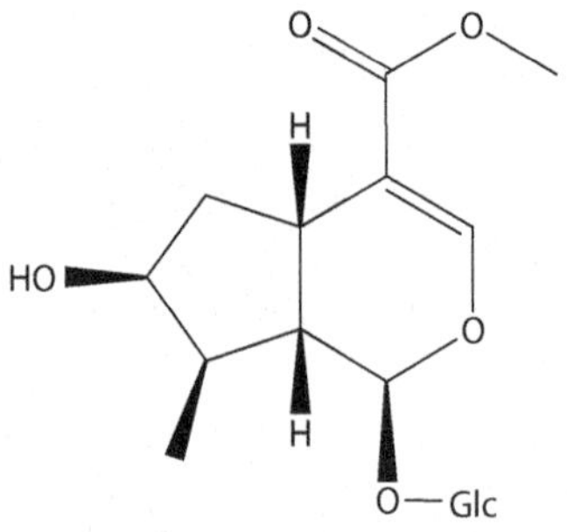

FIGURE 5.22 Loganin.

5.75 *Lonicera japonica* Thunb.

Common name: ren dong (Chinese)
Subclass Asteridae, Superorder Cornanae, Order Cornales, Family Caprifoliaceae
Medicinal use: dysentery (China)

A feature of type 2 diabetes and visceral adiposity is a chronic inflammatory state during which inflammatory cytokines such as tumor necrosis factor-α, interleukin-6 and interleukin-1β generate nephropathy and retinopathy.[231,232] Ethanol extract of flowering parts of *Lonicera japonica* Thunb. (containing 65.3 μg/g of cholorogenic acid) given orally to streptozotocin-induced diabetic rats (plasma glucose ≥350 mg/dL) orally at a dose of 200 mg/kg/day for 8 weeks attenuated body weight loss, lowered glycemia from 421.5 to 303.8 mg/dL (normal: 93.6 mg/dL; rosiglitazone: 256.3 mg/dL) and HbA_{1c} from 14.1% to 10.8% (normal: 4.8%; rosiglitazone: 4.8%).[233] This regimen lowered urine volume, proteinuria, serum creatinine from 95.3 to 68.6 μmol/L (normal: 38.1 μmol/L; rosiglitazone: 59.2 μmol/L).[233] This extract lowered renal contents of interleukin-6, tumor necrosis factor-α and transforming growth factor-β1 and increased interleukin-10 towards normal.[233] Interleukin-10 induces insulin receptor substrate-1 tyrosine phosphosphorylation acting as an insulin-sensitizing adipokine in adipose tissues.[232] This extract prevented glomerular hypertrophy and expansion of the mesangial in diabetic rodents, decreased by about 30% CD4+ and CD8+ T cell expressions in kidneys and reduced the phosphorylation of p38 mitogen-activated protein kinase in renal tissues.[233] Clinical trials are warranted.

5.76 *Panax ginseng* C.A. Meyer

Synonyms: *Aralia ginseng* (C.A. Mey.) Baill.; *Panax chin-seng* Nees; *Panax quinquefolius* var. *ginseng* (C.A. Mey.) Regel & Maack
Common names: ren shen (Chinese); ginseng
Subclass Cornanae, Superorder Cornanae, Order Apiales, Family Araliaceae
Medicinal use: tonic (China)

Ginsenoside Rg3 from *Panax ginseng* C. A. Meyer inhibited rat platelets aggregation induced by thrombin and collagen, with IC_{50} values of 40.2 and 35.2 μM, respectively.[234] Cyclic adenosine monophosphate stimulates protein kinase A that mediates the phosphorylation of an ATP-dependent calcium pump that decrease cytosolic calcium by accumulation of calcium into the dense tubular system.[235] Decreased calcium in platelets blocks not only platelets aggregation but also the secretion of pro-aggregating substances, changes of shape and adhesion.[235] The synthetic derivative dihydroginsenoside Rg3 from *Panax ginseng* C. A. Meyer blocked rat platelets aggregation induced by thrombin and collagen, with IC_{50} values of 18.8 and 20 μM, respectively.[234] This saponin evoked an increase of cytoplasmic cyclic adenosine monophosphate in platelets and this effect was not inhibited by phosphodiesterase inhibitor.[234] Dihydroginsenoside Rg3 inhibited thrombin-induced phosphorylation of extracellular signal-regulated kinase-1/2.[234] Aqueous extract of ginseng given to New Zealand White rabbits on high-cholesterol diet at a daily dose of 200 mg/kg/day for 8 weeks lowered cholesterol from by 20% (lovastatin at 2 mg/kg: 43%).[236] Plasma triglycerides were decreased from 60.2 to 42.7 mg/dL, and low-density lipoprotein–cholesterol was decreased from 557.5 to 462.5 mg/dL. High-density lipoprotein–cholesterol was unchanged.[236] This regimen reduced the formation of atheroma from 60% to 46% (lovastatin 2 mg/kg: 37%) and decreased the aggregation of platelets induced by collagen and thrombin by 98% and 54%, respectively.[236] Platelets of treated animals challenged with collagen had lower secretion of diacylglycerol than untreated animals suggesting inhibition of diacylglycerol production by the extract.[236] In platelets, phospholipase C cleaves phosphatidylinositol 4,5-biphosphate into diacylglycerol, which activates protein kinase C and inositol 1,4,5-triphosphate which increase cytoplasmic concentration of calcium. Activation of calmodulin-dependent kinases.[237] Ethanol extract of *Panax ginseng* C. A. Meyer roots given orally to spontaneously hypertensive rats at a single dose of 500 mg/kg lowered systolic blood pressure from about 185 to 155 mmHg after 6 hours.[238] This extract given daily dose of 500 mg/kg for 4 weeks lowered systolic blood pressure from about 200 to 165 mmHg.[238] This treatment induced the phosphorylation of Akt and endothelial nitric oxide synthetase in aortic endothelium of treated rodents and reduced aortic thinckness.[238] Further, this extract lowered the expression of intercellular adhesion molecule-I and increased the expression of plasminogen expression in arteries of treated animals.[238]

5.77 *Panax notoginseng* (Burkill) F.H. Chen ex C.H. Chow

Synonyms: *Aralia quinquefolia* var. *notoginseng* Burkill; *Panax pseudoginseng* var. *notoginseng* (Burkill) C. Ho & C.J. Tseng
Common names: san qi (Chinese); san-chi ginseng
Subclass Asteridae, Superorder Cornanae, Order Apiales, Family Araliaceae
Medicinal use: bleeding (China)

Aqueous solution of *Panax notoginseng* (Burkill) F.H. Chen ex C.H. Chow given orally at a daily dose of 12 mg/kg to ApoE deficient C57BL/6L mice on high-fat diet for 8 weeks reduced the development of atherosclerotic lesions from 8.7% to 2.9%.[239] This supplementation evoked a reduction of total cholesterol from 1051.5 to 861.8 mg/dL, low-density lipoprotein from 726.8 to 577 mg/dL and lowered high-density lipoprotein from 308.8 to 263.1 mg/dL.[239] *In vitro*, this extract at a concentration of 300 μg/mL lowered the tumor necrosis factor-α-induced adherence of monocytes to coronary artery endothelial cells by about 25%.[239] The extract at a concentration of 300 μg/mL inhibited the tumor necrosis factor-α-induced expression of vascular cell adhesion molecule-1 and intercellular adhesion molecule-1 by coronary artery endothelial cells.[239]

5.78 *Angelica acutiloba* (Siebold & Zucc.) Kitag.

Synonym: *Ligusticum acutilobum* Siebold & Zucc.
Common names: dong dang gui (Chinese); Japanese angelica
Subclass Asteridae, Superorder Cornanae, Order Apiales, Family Apiaceae
Medicinal use: stimulate blood circulation (China)

Ethanol extract of roots of *Angelica acutiloba* (Siebold & Zucc.) Kitag. (Figure 5.23) given orally to streptozotocin-induced diabetic Wistar rats (glycemia ≥350 mg/dL) at a daily dose of 200 mg/kg for 8 weeks did not prevent body weight loss, lowered food and water intake, lowered glycemia from 418.6

FIGURE 5.23 *Angelica acutiloba* (Siebold & Zucc.) Kitag.

to 324.2 mg/dL (normal: 93.6 mg/dL) and had no effect on lipid profile or systolic blood pressure.[240] This regimen decreased serum creatinine from 0.7 to 0.4 mg/dL (normal: 0.2 mg/dL).[240] The extract decreased creatinine clearance from 1.8 to 1.3 mL/min/kg (normal: 0.6 mL/min/kg), urinary albumin excretion rate from 19.4 μg/24 h to 13.3 μg/24 h (normal: 2.8 μg/24 h) and reduced kidney weight from 2.9 to 2 g (normal: 1.5 g).[240] Advance glycation end products and reactive oxygen species levels were attenuated in the kidneys of treated animals by approximately 27% and 35%, respectively.[240] Histological observation of kidneys of treated rat evidenced a reduction in mesangial matrix formation compared with untreated diabetic group.[240] Furthermore, the extract decreased the expression of fibronectin, nuclear factor-κB, transforming growth factor-β1, receptor for advanced glycation end products by renal cortex.[240]

5.79 *Angelica dahurica* (Fisch.) Benth. & Hook. f.

Common names: bai zhi (Chinese); yoroi-gusa (Japan)
Subclass Asteridae, Superorder Cornanae, Order Apiales, Family Apiaceae
Medicinal use: headache (China)

Imperatonin from *Angelica dahurica* (Fisch.) Benth. & Hook. f. given to spontaneously hypertensive rats orally at a dose of 25 mg/kg for 13 weeks lowered systolic blood pressure from 198.4 to 162.6 mmHg and lowered diastolic blood pressure from 127.5 to 93.8 mmHg on account of vascular L-type calcium channel antagonism.[241] This furanocoumarin given to spontaneously hypertensive rats orally for 12 weeks at a dose of 25 mg/kg/day lowered systolic blood pressure and diastolic blood pressure by 18% and 26%, respectively.[242] This coumarin reduced cortical collagen content in the renal cortical tissues. It also reduced urinary 8-iso-PGF2α by approximately 75% lowered renal and urinary reactive oxygen species, and increased renal superoxide dismutase.[242] Histologically, imperatonin reduced mesangial matrix expansion of glomeruli in spontaneous hypertensive rats.[242] In renal cortical tissue, this coumarin inhibited the enzymatic activity of NADPH oxidase by approximately 25%.[242] This treatment lowered extracellular signal-regulated kinase-1/2, p38 mitogen-activated protein kinase, and Akt activities in renal cortical tissues.[242] Reactive oxygen species activate extracellular signal-regulated kinase, p38mitogen-activated protein kinase, and c-Jun N terminal kinase activities in renal cortical tissues.[243]

5.80 *Bupleurum falcatum* L.

Synonym: *Bupleurum rossicum* Woronow
Common names: chai hu (Chinese); Sickle-leafed hare's ear
Medicinal use: fever (China)

Saikosaponin a isolated from *Bupleurum falcatum* L. inhibited ADP-induced aggregation of platelets by 73% at a concentration of 10^{-4} M *in vitro*. It inhibited thromboxane B_2 formation from arachidonic acid (10^{-4} M: 38%).[244] Bupleurumin isolated from the aerial parts inhibited the aggregation of rat platelets induced by collagen and arachidonic acid with IC_{50} values of 47.5 and 41.3 μM, respectively. Acetylsalicylic acid inhibited the aggregation of platelets induced by collagen and arachidonic acid with IC_{50} values of 420 and 66 μM, respectively.[245] This lignan was inactive against adenosine diphosphate-induced platelet aggregation.[245]

5.81 *Carum carvi* L.

Common names: ge lu zi (Chinese); caraway
Subclass Asteridae, Superorder Cornanae, Order Apiales, Family Apiaceae
Medicinal use: promote digestion (Taiwan)
History: the plant was known to Galen

Aqueous extract of fruits of *Carum carvi* L. given orally to Wistar rats at a daily dose of 100 mg/kg increased urine secretion from 5.8 to 20.2 mL.[246] Aqueous extract fruits given orally to streptozotocin-induced diabetic Wistar rats (glycemia >280 mg/dL) at a dose of 60 mg/kg for 60 days decreased urine volume, reduced body weight loss, lowered glucose from 284.2 to 148.9 mg/dL (normal: 130.6 mg/mL), decreased serum urea, serum creatinine, microalbuminuria urinary protein and improved glomerular cytoarchitecture.[247] Clinical trials are warranted.

5.82 *Cuminum cyminum* L.

Common names: zi ran qin (Chinese); jirakam (India); cumin
Subclass Asteridae, Superorder Cornanae, Order Apiales, Family Apiaceae
Medicinal use: promote digestion (India)
History: the plant was known to Dioscorides

In Wistar rats, a single intraperitoneal injection of streptozotocin at a dose of 65 mg/kg into the femoral vein provokes an increase of levels in blood glucose concentration ≥15 mM.[248] Streptozotocin has no effect on atherosclerosis and blood pressure even after 24 weeks.[248] Bilateral cataracts develop at 4 weeks in these rats with blindness in all eyes 16 weeks posttreatment.[248] Seeds of *Cuminum cyminum* given to streptozotocin-induced diabetic Wistar rats (glycemia ≥145 mg/dL) at 0.5% of diet for 8 weeks evoked a 50% protection on lenses and 50% stage 1 cataract. [249] Fasting glycemia was lowered by cumin supplementation from 290 to 240 mg/dL (normal: 80 mg/dL).[249] The supplementation prevented glycation of total soluble protein and α-crystallin in diabetic lens.[249] Clinical trials are warranted.

5.83 *Ligusticum wallichii* Franch.

Synonym: *Ligusticum striatum* DC.
Common name: tiao wen gao ben (Chinese)
Subclass Asteridae, Superorder Cornanae, Order Apiales, Family Apiaceae
Medicinal use: promote blood circulation (China)

Ligustrazine from *Ligusticum wallichii* Franch. given intravenously at a dose of 10 mg/kg to Sprague–Dawley rats 5 minutes before myocardial infarction (induced by ligation of the left anterior descending artery followed by reperfusion) reduced infarct size by 25% and this effect was abrogated by co-administration with either phosphoinositide 3-kinase inhibitor wortmannin or nitric oxide synthetase inhibitor N^{ω}-nitro-L-arginine methyl ester.[250] The treatment reduced the content of malondialdehyde, increased the enzymatic activity of superoxide dismutase, elevated contents of nitric oxide and lowered the apoptotic index from 25.2% to 15% in myocardial tissues.[250] Further caspase 3 activity was decreased by ligustrazine pretreatment by more than half in cardiac tissues.[250] In addition, this tetramethyl pyrazine induced the activation of Akt and endothelial nitric oxide synthetase in cardiac tissue.[250]

5.84 *Peucedanum japonicum* Thunb.

Synonym: *Anethum japonicum* (Thunb.) Koso-Pol.
Common name: bin hai qian hu (Chinese)
Subclass Asteridae, Superorder Cornanae, Order Apiales, Family Apiaceae
Medicinal use: cough (Japan)

3′,4′-Diisovalerylkhellactone diester isolated from *Peucedanum japonicum* Thunb. inhibited at a concentration of 200 μM the aggregation of platelets induced by platelet aggregating factor and collagen with IC_{50} values of 56.3 and 89.4 μM, respectively.[251] This coumarin inhibited thromboxane B2 formation caused by collagen and platelet aggregating factor.[251] 3′,4′-Diisovalerylkhellactone diester abrogated the increase of cytoplasmic calcium induced by platelet aggregating factor and phospholipase A2.[251] In platelets, phospholipase A2 releases arachidonic acid which is the precursor of thromboxane A2, a potent pro-aggregating agent.[252]

5.85 *Trachyspermum ammi* (L.) Sprague

Synonym: *Ammi copticum* L.; *Bunium copticum* (L.) Spreng.; *Carum copticum* (L.) Benth. & Hook. f.; *Carum copticum* C.B. Clarke; *Daucus copticus* (L.) Pers.; *Ptychotis coptica* (L.) DC.; *Sison ammi* L.; *Trachyspermum copticum* (L.) Link
Common names: xi ye cao guo qin (Chinese); yamani (India); ajwain
Subclass Asteridae, Superorder Cornanae, Order Apiales, Family Apiaceae
Medicinal use: facilitates digestion (India)

Thymol (Figure 5.24) isolated from the seeds of *Trachyspermum ammi* (L.) Sprague injected intravenously to anesthetized Wistar rats at doses ranging from 1 to 10 mg/kg caused a dose dependent and transient fall in mean arterial blood pressure and a slight increase in heart rate.[253] This effect was not abolished by atropine or β-blocking agent propranolol.[253] In isolated preparations of rabbit aorta, thymol at concentrations ranging from 10 to 300 μg/mL dose dependently relaxed contractions induced by norepinephrine or potassium.[253] The relaxation of rabbit aorta preparation evoked by thymol upon norepinephrine or potassium precontraction was not affected by endolthelium removal suggesting at least a mechanism involving the blockage of Ca^{2+} channels.[253] Thymol inhibited the aggregation of platelets induced by collagen, arachidonic acis and thrombin. It was less active against ADP induced aggregation suggesting thymol to by a cyclo-oxygenase inhibitor.[254] Clinical trials are warranted.

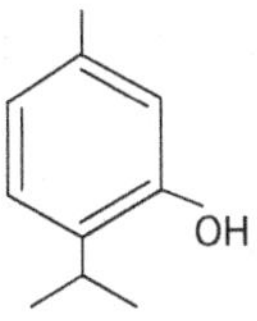

OH

FIGURE 5.24 Thymol.

5.86 *Platycodon grandiflorus* (Jacq.) A. DC.

Synonym: *Platycodon glaucum* (Thunb.) Nak.
Common name: jie geng (Chinese)
Medicinal use: cough (Korea)

Saponin fraction of roots of *Platycodon grandiflorus* (Jacq.) A. DC. (Figure 5.25) at a concentration of 0.2 mg/mL protected human umbilical vein endothelial cells from decrease of nitric oxide production induced by oxidized low-density lipoprotein. It also decreased the production of malondialdehyde induced by oxidized low-density lipoprotein and lowered the expression of vascular adhesion molecule-1.[255] These effects were comparable with simvastatin (at a concentration of 0.2 mg/mL).[255] The fraction inhibited monocyte adhesion to human umbilical vein endothelial cells as efficiently as simvastatin.[255]

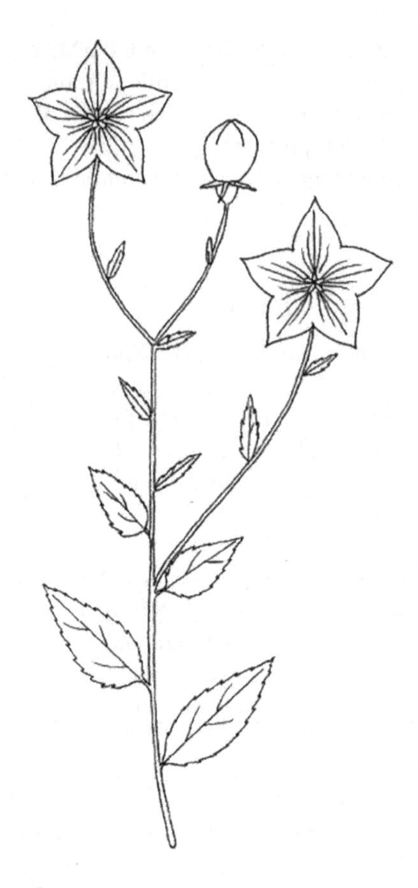

FIGURE 5.25 *Platycodon grandiflorus* (Jacq.) A. DC.

5.87 *Achillea millefolium* L.

Synonym: *Chamaemelum millefolium* (L.) E.H.L. Krause
Common names: shi (Chinese); daun seribu (Indonesian); common yarrow
Medicinal use: colds (Indonesia)

The angiotensin-converting enzyme converts angiotensin I to angiotensin II. Angiotensin II is vasoconstrictor and stimulates the secretion of aldosterone from adrenal cortex increasing thus blood pressure.[256] Artemetin (Figure 5.26) isolated from the aerial parts of *Achillea millefolium* L. given orally at a single dose of 1.5 mg/kg to Wistar evoked a reduction of 8 mmHg in mean arterial blood pressure after 3 hours.[257] This methoxylated flavone given at a single intravenous dose of 1.5 mg/kg evoked a 11.4 mmHg reduction in mean arterial blood pressure.[257] Injection

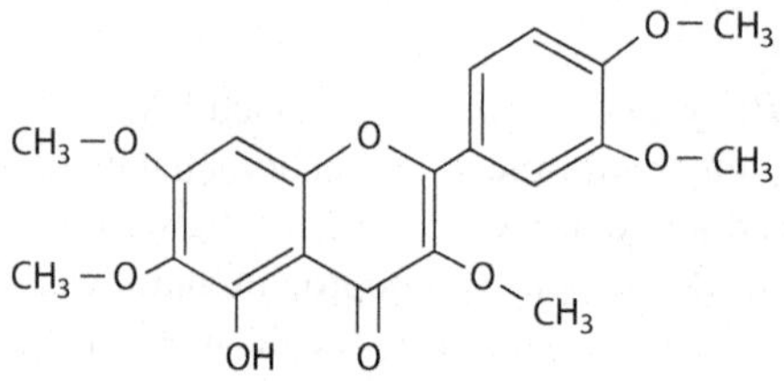

FIGURE 5.26 Artemetin.

of artemetin at a dose of 0.75 mg/kg lowered the rise of mean arterial blood pressure induced by angiotensin I by about 50% and had no effect on angiotensin II-induced hypertension.[257] This flavonoid decreased the activity of angiotensin-converting enzyme.[257] Aortic rings isolated from Wistar rats given this pentamethoxyflavone at 1.5 mg/kg orally were less sensitive to angiotensin I-induced contractions.[257]

5.88 *Actium lappa* L.

Synonym: *Arctium chaorum* Klokov; *Arctium leiospermum* Juz. & Ye. V. Serg.; *Arctium majus* (Gaertn.) Bernh.; *Lappa major* Gaertn.; *Lappa vulgaris* Hill
Common names: niu bang (Chinese); great burdock
Subclass Asteridae, Superorder Asteranae, Order Asterales, Family Asteraceae
Medicinal use: fever (China)
History: the plant was known of Galen

Lignan fraction isolated from the fruits of *Arctium lappa* L. inhibited rat lens aldose *in vitro* activity by 90% and this effect was as potent as positive standard (epalrestat at 20 μg/mL).[258] Ethanol extract of seeds of *Arctium lappa* L. given to high-fat diet Sprague–Dawley rats at a dose of 200 mg/kg/day for 6 weeks had no effect on food intake, lowered triglycerides from 67.2 to 27.2 mg/dL, increased high-density lipoprotein from 20.8 to 42.6 mg/dL. This Regimen had no effect on low-density lipoprotein or plasma glucose.[259] This regimen lowered systolic blood pressure to normal, and improved vascular endothelium sensitivity to acetylcholine induced relaxation of isolated aortic rings from treated animals.[259] Histological examination of arterial specimen of treated rodents evidenced a reduction in the thickening of the tunica intima, decreased size of atherosclerotic lesions and maintained the smoothness of the intimal endothelial layer.[259] This extract reduced the levels of aortic expression of vascular adhesion molecule-1.[259] Clinical trials are warranted.

5.89 *Aster koraiensis* Nakai

Subclass Asteridae, Superorder Asteranae, Order Asterales, Family Asteraceae
Common name: Korean aster
Nutritional use: food (Korea)

Advanced glycosylation end-products bind to vascular endothelial cells and macrophages leading to oxidative stress, release of cytokines, and production of vascular endothelial growth factor leading to vascular permeability and angiogenesis.[260] Ethanol extract of aerial parts of *Aster koraiensis* Nakai (containing 1.2% of chlorogenic acid and 2.2% of 3,5-di-*O*-caffeoylquinic acid) given to streptozotocin-induced diabetic rats (plasma glucose >300 mg/dL) for 13 weeks at a dose of 200 mg/kg/day had no effect on body weight loss, attenuated glycemia from 587.3 to 474.2 mg/dL and had no effect on HbA1c.[261] The extract reduced plasma and kidney advanced glycation end products formation and prevented renal mesangial matrix expansion, proteinuria and albuminuria in diabetic rodents.[261] The extract inhibited renal podocytes apoptosis.[261] 3,5-di-*O*-caffeoylquinic acid, 4,5-di-*O*-caffeoylquinic acid, isoquercetrin, quercetin-3-*O*-α-L-arabinopyranoside and larycitrin-3-O-α-L-rhamnoside isolated from the aerial parts of *Aster koraiensis* Nakai inhibited *in vitro* advanced glycation end products formation with IC_{50} values ranging between 6 and 10 μM (aminoguanidine: IC_{50}: 965.9 μM).[262] 3,5-di-*O*-Caffeoylquinic acid, 4,5-di-*O*-caffeoylquinic acid, isoquercetrin, quercetin-3-*O*-α-L-arabinopyranoside and larycitrin-3-O-α-L-rhamnoside inhibited rat lens aldose reductase with IC_{50} values between 0.3 to 8 μM (epalrestat: IC_{50}: 0.06 μM).[262]

5.90 *Bidens pilosa* L.

Synonyms: *Bidens alba* (L.) DC.; *Bidens hirsuta* Nutt.; *Bidens odorata* Cav.
Common names: gui zhen cao (Chinese); pisau-pisau (Philippines); hairy beggarticks
Subclass Asteridae, Superorder Asteranae, Order Asterales, Family Asteraceae
Medicinal use: boils (Philippines)

Methanol extract of leaves of *Bidens pilosa* L. given to spontaneously hypertensive rat orally at a single oral dose of 150 mg/kg decreased systolic pressure from 176 to 156 mmHg at 2 hours and this effect was equivalent to a single oral dose of nifedipine (10 mg/kg).[263] This extract given at a dose of 150 mg/kg/day for 31 days lowered systolic blood pressure by 26%, an effect close to nifedipine at 10 mg/kg/day.[263] Compared with nifedipine, the extract had no effect on cardiac frequency and it did not evoke any diuretic effects.[263] Methanol fraction of leaves given orally at a single dose of 150 mg/kg to fructose-induced hypertensive Wistar rats lowered systolic blood pressure after 6 hours by 21% and this effect was superior to as nifedipine at 10 mg/kg.[264] Heart rate of treated animal was unchanged.[264] The extract given 3 weeks at a daily dose of 150 mg/kg to Wistar rats given fructose-enriched drinking water lowered systolic blood pressure by 11%. This Regimen had no effect on plasma glucose, decreased insulin from 346.6 to 87.5 pM (normal: 105.8 pM), lowered cholesterol to normal level and increased (like nifedipine) triglycerides from 1 to 2.5 mM (normal: 0.9 mM).[264] The extract given 3 weeks at a daily dose of 150 mg/kg to fructose-induced hypertensive Wistar rats lowered systolic blood pressure from 146 to 121 mmHg, lowered plasma insulin by 50.6%.[264] Active constituents need to be identified.

5.91 *Cosmos caudatus* Kunth

Synonyms: *Bidens berteriana* Spreng.; *Bidens caudata* (Kunth) Sch. Bip.;
Common name: ulam Rajah (Malay)
Subclass Asteridae, Superorder Asteranae, Order Asterales, Family Asteraceae
Medicinal use: hypertension (Indonesia)

Aqueous extract of leaves of *Cosmos caudatus* Kunth given to Wistar rats orally at a dose of 1000 mg/kg prevented tachycardia induced by intraperitoneal injection of adrenaline with a decrease in heart beat frequency from about 2.8 to 1.7 Hz.[265] This extract at a dose of 1000 mg/kg evoked after 24 hours a pooled urine volume of 13.3 mL and this effect was equivalent to furosemide (1.8 mg/kg) with a pooled volume of 13.5 mL.[265] The plant is used as "ulam" in Malayasia. Clinical trials are warranted.

5.92 *Erigeron annuus* (L.) Pers.

Synonyms: *Aster annuus* L.; *Aster stenactis* E.H.L. Krause; *Erigeron heterophyllus* Muhl. ex Willd.; *Stenactis annua* (L.) Cass. ex Less.
Common names: yi nian peng (Chinese); sweet-scabious
Subclass Asteridae, Superorder Asteranae, Order Asterales, Family Asteraceae

Caffeic acid, and 3,5-di-*O*-caffeoylquinic acid methyl ester isolated from the flowers of *Erigeron annuus* (L.) Pers. inhibited the formation of advanced glycation end products with IC_{50} values below 15 µM (aminoguanidine: IC_{50}: 961 µM).[266] Caffeic acid, and 3,5-di-*O*-caffeoylquinic acid methyl ester inhibited rat lens aldose reductase with IC_{50} values of 25.8, and 0.8 µM (epalrestat: IC_{50}: 0.07 µM).[266] Apigenin, luteolin, apigenin-7-*O*-β-D-glucuronide, apigenin-7-*O*-β-D-glucuronide methyl ester, kaempferol, quercetin, astragalin, and quercitrin from the flowers of this plant inhibited rat lens aldose reductase with IC_{50} values of ranging from 1 to 10 µM.[267] Scutellarin which is common in the genus *Erigeron* L., inhibited the adherence of monocytes to ECV304 cells induced by high-glucose from 69% to 38%. It also reduced the expression of monocyte chemoattractant peptide-1 as well as nuclear translocation of nuclear factor-κB.[268] In alloxan-induced diabetic Kunming mice this flavonoid at a dose of 50 mg/kg/day for 3 weeks had no effect on plasma glucose but lowered monocyte chemoattractant peptide-1.[268]

5.93 *Gynura japonica* (Thunb.) Juel

Synonyms: *Cacalia pinnatifida* Lour.; *Gynura flava* Hayata; *Gynura segetum* (Lour.) Merr.; *Gynura pinnatifida* (Lour.) DC.; *Senecio japonicus* Thunb.
Common name: ju san qi (Chinese)
Subclass Asteridae, Superorder Asteranae, Order Asterales, Family Asteraceae
Medicinal use: promote blood circulation (China)

6-Acetyl-2,2-dimethylchroman-4-one, vanillin and 2,6-dimethoxy-1,4-benzoquinone isolated from the rhizome of *Gynura japonica* (Thunb.) Juel at concentrations of 100, 100, and 50 μg/mL inhibited rabbit platelet aggregation induced by arachidonic acid by 100% (aspirin at 50 μg/mL: 100%).[269] 2,6-Dimethoxy-1,4-benzoquinone at a concentration of 50 μg/mL inhibited rabbit platelet aggregation induced by collagen, thrombin or platelet activating factor by 100% (aspirin at 100 μg/mL: less than 5%).[269] Phenolic molecules have the tendency to inhibit cyclo-oxygenase 2.

5.94 *Gynura procumbens* (Lour.) Merr.

Synonyms: *Cacalia procumbens* Lour.; *Gynura buntigii* S. Moore; *Gynura cavalerei* H. Lév.; *Gynura emeiensis* Z.Y. Zhu; *Gynura sarmentosa* (Blume) DC.
Common names: ping wo ju san qi (Chinese); daun dewa (Indonesia); prakham dee khwaai (Thai)
Subclass Asteridae, Superorder Asteranae, Order Asterales, Family Asteraceae
Medicinal use: hypertension (Indonesia)

Aqueous extract of leaves of *Gynura procumbens* (Lour.) Merr. given to spontaneously hypertensive rats orally at a dose of 500 mg/kg/day for 4 weeks lowered systolic blood pressure from 191.7 to 172.7 mmHg.[270] This regimen lowered serum lactate dehydrogenase and creatine phosphokinase by 34% and 48%, respectively.[270] The extract increased serum nitric oxide by 60.7%.[270] The leaves are eaten as salad in Malayasia. Could it be a functional food?

5.95 *Inula racemosa* Hook. f.

Synonyms: *Inula repanda* Turcz.; *Limbarda japonica* (Thunb.) Raf.
Common name: zong zhuang tu mu xiang (Chinese)
Medicinal use: diarrhea (Korea)

Foam cells in atheromas become apoptotic releasing cholesterol and cellular debris resulting into the formation of a necrotic core. Proteases released by foam cells destabilize the atheroma leading to plaque rupture during which extracellular matrix molecules such as collagen, elastin, tissue factor, and von Willebrand factor promote thrombotic event.[271] Ethanol extract of roots of *Inula racemosa* Hook. f. given to high-cholesterol diet Hartley Guinea-pigs at a daily dose of 100 mg/kg/day for 90 days lowered plasma cholesterol from 91.1 to 56 mg/dL (normal: 27.6 mg/dL; atorvastatin at 10 mg/kg/day: 65.5 mg/dL), plasma triglycerides from 75.5 to 42.2 mg/dL (normal: 66 mg/dL; atorvastatin at 10 mg/kg/day: 31.2 mg/dL), increased high-density lipoprotein, lowered low-density lipoprotein, and reduced atherogenic index from 14.7 to 3.6 (normal: 3.2; atorvastatin at 10 mg/kg/day: 1.5).[272] The extract protected coronary artery endothelium from foam cell accumulation.[272]

In aorta isolated from rodents treated with the extract, lipid deposit was reduced from about 15% to 6.5% (atorvastatin at 10 mg/kg/day: about 11%).[272]

5.96 *Puchea indica* (L.) Less

Synonyms: *Baccharis indica* L.; *Erigeron denticulatum* Burm. f.
Common names: beluntas (Malay); luntas (Indonesia); tulo-lalaki (Philippines); Indian fleabane
Subclass Asteridae, Superorder Asteranae, Order Asterales, Family Asteraceae
Medicinal use: dysentery (Indonesia)

Vascular endothelial cells secrete the collagenase MMP-1 and the gelatinase MMP-2 while macrophages secrete the gelatinase MMP-9.[273] MMP-2 and MMP-9 are actively synthesized in atheromatous plaques and prevalent in rupture-prone regions.[273] MMP-2 and MMP-9 are activated by reactive oxygen species.[273] 1,3,4,5-tetra-*O*-caffeoylquinic acid, 3,4,5-tri-*O*-caffeoyl quinic acid and quercetin isolated from the leaves of *Puchea indica* (L.) Less inhibited the activity of collagenase with IC_{50} value of 1.5, 6.3, and 15.9 μM, respectively, *in vitro* (positive standard phosphramidon IC_{50} value: 7.4 μM).[274] 1,3,4,5-tetra-*O*-caffeoylquinic acid and 3,4,5-tri-*O*-caffeoyl quinic acid inhibited MMP-2 with IC_{50} values of 2.5 and 18.4 μM, respectively, and inhibited MMP9 with IC_{50} values of 6.4 and 16.8 μM, respectively.[274]

5.97 *Siegesbeckia pubescens* (Makino) Makino

Common name: xi xian cao (Chinese)
Subclass Asteridae, Superorder Asteranae, Order Asterales, Family Asteraceae
Medicinal use: hypertension (China)

Ent-16β,17-dihydroxy-kauran-19-oic acid isolated from the aerial parts of *Siegesbeckia pubescens* (Makino) inhibited platelet aggregation induced by ADP with an IC_{50} value of 513 μg/mL and inhibited thrombin induced aggregation with an IC_{50} value of 745 μg/mL.[275] This diterpene had mild effect on arachidonic-induced aggregation. ADP evoked a decrease of cyclic adenosine monophosphate in platelets and Ent-16β,17-dihydroxy-kauran-19-oic acid at a concentration of 800 μg/mL increased cyclic adenosine monophosphate.[275] This diterpene given to Wistar rats at a single intravenous dose of 20 mg/kg inhibited arteriovenous-induced thrombus growth by 42.6% (aspirin 18 mg/kg: 46.9%). It had no effect on plasma thromboxane A2 and plasma 6-keto-$PGF_{1\alpha}$, suggesting the noninvolvement of arachidonic cascade.[275]

5.98 *Smallanthus sonchifolius* (Poeppig) H. Robinson

Synonym: *Polymnia sonchifolia* Poeppig
Common names: ju shu (Chinese); yacon
Subclass Asteridae, Superorder Asteranae, Order Asterales, Family Asteraceae
Nutritional use: Food (Japan)

Aqueous extract of leaves of *Smallanthus sonchifolius* (Poeppig) H. Robinson given to streptozotocin-induced diabetic Wistar rats (plasma glucose: ≥350 mg/dL) at a dose of 70 mg/kg/day for 4 weeks reduced body weight loss, lowered plasma glucose from 546.3 to 243.2 mg/dL (normal: 101.1 mg/dL), normalized plasma cholesterol, and lowered triglycerides.[276] The extract increased plasma insulin.[276] As a consequence of insulin replenishment, this regimen decreased kidney weight, and urinary albumin. Excession of albumin in urine is a hallmark of glomerular dysfunction in diabetes.[276] Histological examination of kidneys of treated rodents evidenced improved glomerular cytoarchitecture; decreased expression of laminin-1, fibronectin, collagen IV, and collagen type-III.[276] The extract lowered the expression of transforming growth factor-β1 in kidneys and phosphorylated Smad2/3.[276]

5.99 *Sphaeranthus indicus* L.

Synonyms: *Sphaeranthus hirtus* Willd.; *Sphaeranthus mollis* Roxb.
Common names: rong mao dai xing cao (Chinese); mundi (India); East Indian globe-thistle
Subclass Asteridae, Superorder Asteranae, Order Asterales, Family Asteraceae
Medicinal use: fatigue (Burma)

Methanol extract of fruits of *Sphaeranthus indicus* L. given orally to golden Syrian hamsters on high fat-diet at a daily dose of 200 mg/kg for 60 days had no effect on body weight gain total cholesterol, triglycerides, low-density lipoprotein–cholesterol, and high-density lipoprotein–cholesterol.[277] However, this extract decreased plasma glucose from about 155 to 115 mg/dL and reduced aortic lesions by more than 40% (fenofibrate 100 mg/kg/day: by more than 80%).[277] In C57Bl/6J LDLr$^{-/-}$ mice on high-fat cholesterol diet, the extract at a daily dose of 100 mg/kg for 8 weeks reduced aortic lipid deposition by 22% (fenofibrate 100 mg/kg: 26%) and decreased plasma interleukin-6 and monocyte chemoattractant peptide-1.[277] The aortic expression of monocyte chemoattractant peptide-1 was decreased by 40% by the extract.[277] *In vitro*, the extract at 25 μg/mL inhibited lipopolysaccharide-induced production of tumor necrosis factor-α, interleukin-6, interleukin-8, and interleukin-1β by human peripheral monocytic cells.[277] From this extract, 7-hydroxy frullanolide (Figure 5.27) inhibited tumor necrosis factor-α induced expression of vascular adhesion molecule-1 and E-selectin, in human umbilical vein endothelial cells *in vitro*.[277] This sesquiterpene inhibited tumor necrosis factor-α-induced nuclear translocation of nuclear factor-κB in human umbilical vein endothelial cells.[277]

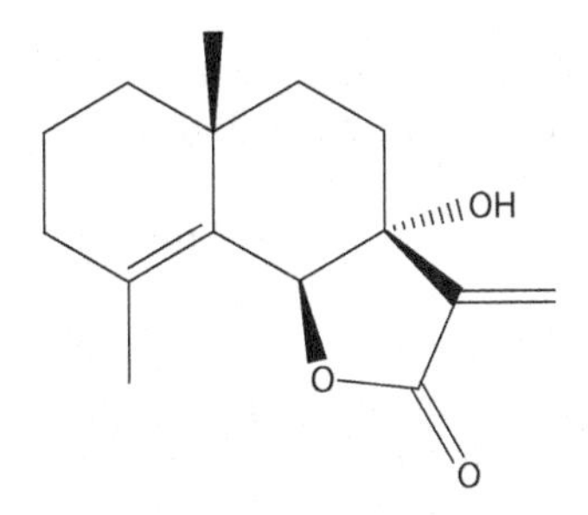

FIGURE 5.27 7-Hydroxy frullanolide.

5.100 *Spilanthes acmella* (L.) L.

Synonyms: *Bidens acmella* (L.) Lam.; *Bidens ocymifolia* Lam.; *Pyrethrum acmella* (L.) Medik.; *Spilanthes ocymifolia* (Lam.) A.H. Moore; *Verbesina acmella* L.
Common names: hin ka la (Burmese); krishnarjaka (Sri Lanka); pokok getang kerbau (Malay); biri (Philippines); tooth ache plant
Subclass Asteridae, Superorder Asteranae, Order Asterales, Family Asteraceae
Medicinal use: diuretic (Sri Lanka)

Ratnasooriya et al. provided evidence that aqueous extract of flowers of *Spilanthes acmella* (L.) L. given to albino rats orally at a single oral dose of 1500 mg/kg increased total cumulative urine output from 10.2 to 53.4 mL after 5 hours (furosemide total cumulative urine output 13 mg/kg:

8.1 mL).[278] The urine of treated animals had increases in both urinary sodium and potassium contents and a reduction in osmolarity suggesting loop diuretic activity.[279] Clinical trials are warranted.

5.101 *Swertia punicea* Hemsl.

Common names: zi hong zhang ya cai (Chinese); senburi (Japanese)
Subclass Lamiidae, Superorder Lamianeae, Order Rubiales, Family Gentianaceae
Medicinal use: promotes digestion (Japan)

Methylswetianin and bellidifolin isolated from *Swertia punicea* Hemsl. given orally to high-fat and high-fructose diet streptozotocin-induced type 2 diabetic BALB/c mice (fasting glycemia >7.8 mmol/L) at a dose 200 mg/kg/day for 28 days lowered glycemia by 44%, plasma insulin and insulin resistance.[280] The treatments increased the hepatic contents of glycogen, lowered the enzymatic activity of glucose-6-phosphatase, and increased glucokinase activity lowered serum cholesterol, triglycerides, and low-density lipoprotein–cholesterol.[280] These xanthones increased the levels of, insulin receptor substrate-1, and phosphoinositide 3-kinase expression in kidneys.[280]

5.102 *Gardenia jasminoides* J. Ellis

Synonyms: *Gardenia augusta* Merr.; *Gardenia florida* L.; *Varneria augusta* L.
Common names: zhi zi (Chinese); karinga (India); cape jasmine
Subclass Lamiidae, Superorder Lamianae, Order Rubiales, Family Rubiaceae
Medicinal use: jaundice (China)

Crocetin isolated from *Gardenia jasminoides* J. Ellis given to stroke-prone spontaneously hypertensive rats orally at a dose of 25 mg/kg/day for 3 weeks had no effect on body weight but lowered systolic blood pressure from about 240 to 210 mmHg.[281] The regimen reduced cerebral thrombogenesis, reduced urinary secretion of oxidative stress marker 8-hydroxy-2′-deoxyguanosine, and increased nitric oxide metabolites urinary excretion.[281] This regimen improved the sensitivity of isolated endothelial aortic preparation to acetylcholine-induced vasorelaxation.[281] Geniposide from this plant at a concentration of 20 μg/mL inhibited the adhesion of monocytes to human umbilical cord cells challenged with high-glucose concentration.[282] This iridoid glycoside inhibited glucose-induced increase of reactive oxygen species concentration in human umbilical cord cells, prevented cytoplasmic IkB degradation and nuclear factor-κB translocation and the expression of vascular adhesion molecule-1 and E-selectin.[282] Clinical trials are warranted.

5.103 *Knoxia valerianoides* Thorel ex Pit.

Synonym: *Knoxia roxburghii* (Spreng.) M.A. Rau
Common names: hung ya ta chi (Chinese); knoxia
Subclass Lamiidae, Superorder Lamianae, Order Rubiales, Family Rubiaceae
Medicinal use: ulcers (China)

Lucidin, 1,3,6-trihydroxy-2-methoxymethylanthraquinone, 3,6-dihydroxy-2-hydroxymethyl-9,10-anthraquinone, 1,3,6-trihydroxy-2-hydroxymethyl-9,10-anthraquinone, and 3-*O*-β-primeveroside isolated from the roots of *Knoxia valerianoides* Thorel ex Pit. inhibited rat lenses aldose reductase with IC_{50} values below 5 μM (epalrestate IC_{50}: 0.06 μM).[283]

5.104 *Uncaria rhynchophylla* (Miq.) Miq. ex Havil.

Synonyms: *Nauclea rhynchophylla* Miq.; *Ourouparia rhynchophylla* (Miq.) Matsum.
Common name: gou teng (Chinese)
Subclass Lamiidae, Superorder Lamianae, Order Rubiales, Family Rubiaceae

Rhynchophylline inhibited the aggregation of platelets challenged with arachidonic acid, collagen, and adenosine diphosphate. Given intravenously at a dose of 20 mg/kg this alkaloid protected rodents against thrombosis.[284] This oxindole alkaloid at 100 μM abrogated the contraction of endothelium-denuded human mesenteric arteries exposed to potassium via inhibition of influx of extracellular calcium via L-type calcium channel.[285] Isorhynchophylline given prophylaxycally at a dose of 30 mg/kg to Guinea pigs inhibited ouabain-induced ventricular contraction.[286] Furthermore, this alkaloid at a dose of 30 mg/kg to Wistar rats poisoned with intravenous injection of calcium chloride delayed the onset time and decreased the duration of ventricular arrhythmia and reduced mortality rate by 59%.[286] In line, this alkaloid at 30 μM reduced cardiac calcium currents and action potential duration in isolated Guine pig ventricular cells and isolated rat cardiomyocytes.[286] Isorhynchophylline from *Uncaria rhynchophylla* (Miq.) Miq. ex Havil. protected rats against monocrotaline, induced increase in right ventricle systolic pressure. It decreased right ventricular hypertrophy, and reduced the number of fully muscularized small arterioles via inhibition of pulmonary arterial smooth muscle cells proliferation.[287] *In vitro* this oxindole alkaloid inhibited the proliferation of human pulmonary arterial smooth muscle cells induced by platelet derived growth factor by preventing cyclin D1 and CDK6 expression and inducing the expression of p27Kip1.[287] Further, this oxindole alkaloid inhibited the phosphorylation of platelet-derived growth factor receptor, and subsequent activation of Akt and glycogen synthase kinase-3 and inhibited the activation of extracellular signal-regulated kinase-1/2 and STAT3.[287] Hirsutine from this plant at a concentration of 10 μM increased the viability of neonatal rat cardiomyocytes cultured under oxygen deprivation from approximately 40% to 85.6%.[288] This indole alkaloid evoked a decrease in lactate dehydrogenase.[288] Furthermore, hirsutine reversed the decrease in superoxide dismutase and increase in malondialdehyde observed during hypoxia to values close to normoxic cells.[288]

5.105 *Cynanchum wilfordii* Franch. & A. Sav.

Common names: ge shan xiao (Chinese); koikema (Japanese)
Subclass Lamiidae, Superorder Lamianeae, Order Rubiales, Family Apocynaceae
Medicinal use: tonic (China)

Ethanol extract of tubers of *Cynachum wilfordii* Franch. & A. Sav. given to apoE$^{-/-}$ C57BL6J mice on high-fat diet at a dose of 200 mg/kg/day for 10 weeks decreased plasma low-density lipoprotein–cholesterol, increased triglycerides and high-density lipoprotein.[289] This regimen normalized systolic blood pressure from about 120 to 110 mmHg and improved the sensitivity of isolated thoracic aorta to acetylcholine.[289] The extract increased the expression of endothelial nitric oxide synthetase in aortic section by 223% and decreased the expression cellular adhesion molecules E-selectin, vascular adhesion molecule-1, and intracellular adhesion molecule-1.[289] The extract lowered endothelin-1 (ET-1) expression in aortic sections by 58%.[289]

5.106 *Vinca minor* L.

Common names: hua ye man chang chun hua (Chinese); lesser periwinkle
Subclass Lamiidae, Superorder Lamianeae, Order Rubiales, Family Apocynaceae

Vinpocetine is a synthetic derivative of vincamine isolated from *Vinca minor* L.[290] FVB/NJ mice with ligated left carotid artery intraperitoneally injected with 5 mg/kg vinpocetine daily for 14 days had reduced vascular smooth muscle cell proliferation. This indole alkaloid increased luminal area and prevented thrombosis and intraplaque haemorrhage.[291] This alkaloid at a concentration of 100 μM reduced the wall thickness developed by human saphenous vein cultured *ex vivo* for 7 days.[291] *In vitro*, vinpocetine at 70 μM inhibited the proliferation of smooth muscle cells via cell cycle inhibition via downregulation of cyclin D1 and upregulation of p27Kip1.[291] This alkaloid inhibited the migration of vascular smooth muscle cells induced by platelet-derived growth factor at 50 μM, and inhibited collagen synthesis. It limited platelet-derived growth factor-induced reactive oxygen species production, and consequently activated extracellular signal-regulated kinase-1/2.[291] This molecule could be of value in Metabolic Syndrome.

5.107 *Gymnema montanum* (Roxb.) Hook. f.

Synonym: *Asclepias montana* Roxb.
Subclass Lamiidae, Superorder Lamianeae, Order Rubiales, Family Asclepiadaceae
Medicinal use: inflammation (India)

Ethanol extract of leaves of *Gymnema montanum* (Roxb.) Hook. f. given to alloxan-induced diabetic Wistar rats at a dose of 200 mg/kg/day for 3 weeks prevented weight loss and reduced water intake.[292] This regimen brought glycemia from 298 to 86 mg/dL (normal: 81 mg/dL; glibenclamide at 600 mg/kg/day: 118 mg/dL) and increased insulinemia from 5.5 to 12 (normal:13.6; glibenclamide at 600 μg/kg/day: 10 μU/mL).[292] As a consequence of increased plasma insulin, at the renal level, the extract lowered serum creatinine from 2 to 1.2 mg/dL (normal: 0.8 mg/dL; glibenclamide at 600 μg/kg/day: 1.1 mg/dL) and lowered urinary proteins from 13.8 to 11.4 mg/dL (normal: 11.2; glibenclamide: 11.4 mg/dL).[292] Furthermore, this regimen lowered renal reactive oxygen species.[292] In diabetic rodents, the extract increased renal catalase, superoxide dismutase, gluthathione peroxidase, glutathione S transferase, and levels of glutathione.[292]

5.108 *Solanum muricatum* Aiton

Common names: xian gua qie (Chinese); pepino
Subclass Lamiidae, Superorder Lamianeae, Order Rubiales, Family Apocynaceae
Nutritional use: vegetable (China)

Aqueous extract of fruits of *Solanum muricatum* Aiton given to streptozotocin-induced diabetic Balb/cA mice (fasting blood glucose >14 mmol/L) at 4% of diet for 5 weeks reduced water intake, food intake and increased body weight.[293] This extract reduced glycemia and increased plasma insulin.[293] On account of increased insulin in plasma, the regimen reduced lipid peroxidation in the kidneys of diabetic animals and evoked a decrease in renal reactive oxygen species, decreased glutathione disulfide formation, increased gluthatione levels and maintained gluthatione peroxidase and catalase activities.[293] This extract reduced interleukin-6 and tumor necrosis factor-α, interleukin-1β and monocyte chemoattractant peptide-1.[293] The regimen reduced the activity of aldose reductase and sorbitol dehydrogenase in kidneys.[293] In kidneys, reactive oxygen species invert glutathione (GSH) to glutathione disulfide, which, in turn, is reduced in a NADPH-dependent reaction catalyzed by glutathione reductase.[294] Glutathione depletion weakens the ability of cells to neutralize reactive oxygen species.[294]

5.109 *Withania coagulans* (Stocks) Dunal

Synonym: *Puneeria coagulans* Stocks
Common names: Ashvaganda (India); cheese maker

Subclass Lamiidae, Superorder Lamianae, Order Solanales, Family Solanaceae
Medicinal use: facilitate digestion (India)

Aqueous extract of fruits of *Withania coagulans* (Stocks) Dunal given orally to streptozotocin-induced diabetic Wistar rats (glycemia >350 mg/dL) at a dose of 10 mg/kg/day for 3 weeks prevented body weight loss, and reduced kidney/body weight ratio evidencing a reduction of kidney hypertrophy.[295] This extract lowered random glycemia by about 30%.[295] At the renal level, this treatment increased glutathione, lowered lipid peroxidation, and evoked a mild reduction in myeloperoxidase activity.[295] This extract decreased interleukin-1β, interleukin-6, interleukin-4, tumor necrosis factor-α, and interferon-γ.[295] Interferon-γ produced by activated T cells and NK cells in conjunction with proinflammatory cytokines activate macrophages and stimulates chemokines production that result in pathological lesions of diabetic.[296] Interleukin-1β in kidney increase the expression of chemotactic factors and adhesion.[296] Interleukin-6 alters endothelial permeability, induces proliferation, and increases fibronectin expression.[296]

5.110 *Solanum lyratum* Thunb. ex Murray

Synonyms: *Solanum kayamae* T. Yamaz.; *Solanum dichotomum* Lour.; *Solanum cathayanum* C.Y. Wu & S.C. Huang
Common name: bai ying (Chinese)
Medicinal use: fever (China)

Ethanol extract of *Solanum lyratum* Thunb. ex Murray at a concentration of 10 μg/mL inhibited the oxidation of low-density lipoprotein, induced by copper by about 75%.[297] The extract at 20 μg/mL protected endothelial cells against oxidized low-density lipoprotein-induced inhibition of nitric oxide synthetase in human umbilical vein endothelial cells.[297] This extract inhibited oxidized low-density lipoprotein-induced adhesion of monocytes to human umbilical vein endothelial cells. It also inhibited oxidized low-density lipoprotein -induced human umbilical vein endothelial cells expression of vascular adhesion molecule-1, E-selectin and monocyte chemoattractant peptide-1.[297]

5.111 *Ipomoea batatas* (L.) Lam.

Synonyms: *Batatas edulis* (Thunb.) Choisy; *Convolvulus batatas* L.; *Convolvulus edulis* Thunb.
Common names: mitha alu (India); ubi keledek (Malay), kamote (Philippines); man thet (Thai); sweet potato
Subclass Lamiidae, Superorder Lamianae, Order Solanales, Family Convolvulaceae
Medicinal use: diabetes (India)

Anthocyanins from *Ipomoea batatas* (L.) Lam. at a concentration of 4 μM enhanced the resistance of low-density lipoprotein to copper-induced oxidation.[298] Anthocyanin fraction containing peonidin-acylglucoside or pelargonidin-acylglucoside from this plant given to spontaneous hypertensive rats at 1% of diet for 11 weeks had no effects on body weight, evoked a mild decrease of heart rate and lowered systolic blood pressure from about 230 to 180 mmHg.[299] In a subsequent study, anthocyanins from this plant given to apolipoprotein E-deficient mice as 1% part of a high-fat diet lowered atherosclerotic plaque area whole aorta by 46%.[300] This regimen lowered lipid peroxidation in plasma by about 25% and reduced oxidative stress in liver and kidneys by about 30% and 20%, respectively.[300] The fraction lowered plasma vascular adhesion molecule-1 by about 25% and had no effects on plasma intracellular adhesion molecule-1 and monocyte chemoattractant peptide-1.[300] The treatment did not modify lipid parameters.[300] Anthocyanins from *Ipomoea batatas* (L.) Lam. given orally at a dose of 700 mg/kg to mice on high-fat diet for 20 weeks lowered fasting blood glucose by about 35%, and improved glucose tolerance.[301] This supplementation lowered the ratio of urine creatinine to albumin in 24 hours by about 45% and reduced collagen-IV expression in the kidneys by about 30%.[301] The regimen lowered reactive oxygen species and advanced glycation products in kidneys of mice.[301] High-fat diet evokes the phosphorylation of IKKβ, degradation ikBα, and translocation of nuclear factor-κB in the nucleus to express cyclo-oxygenase-2 and inducible nitric oxide synthetase and the treatment reversed these changes.[301] Anthocyanins decreased interleukin-1β in kidneys as well as renal thioredoxin interacting protein and oxidative stress-associated advanced glycation end products receptor (RAGE) induced by high-fat diet.[301]

5.112 *Cuscuta japonica* Choisy

Synonyms: *Cuscuta systyla* Maxim.; *Monogynella japonica* (Choisy) Hadač & Chrtek
Common name: tu si zi (Chinese)
Subclass Lamiidae, Superorder Lamianae, Order Solanales, Family Cuscutaceae
Medicinal use: vertigo (China)

3,5-di-*O*-caffeoylquinic acid, methyl-3,5-di-*O*-caffeoylquinate, 3,4-di-*O*-caffeoylquinic acid and methyl-3,4-di-*O*-caffeoylquinate isolated from the seeds of *Cuscuta japonica* Choisy inhibited *in vitro* rat angiotensin I converting enzyme with IC_{50} values of 596, 483, 534, and 460 μM, respectively.[302]

5.113 *Campsis grandiflora* (Thunb.) K. Schum.

Synonyms: *Bignonia chinensis* Lam.; *Bignonia grandiflora* Thunb.; *Campsis adrepens* Lour.; *Campsis chinensis* (Lam.) Voss; *Tecoma chinensis* (Lam.) K. Koch; *Tecoma grandiflora* Loisel.
Common name: ling xiao (Chinese)
Subclass Lamiidae, Superorder Lamianae, Order Lamiales, Family Bignoniaceae
Medicinal use: haemorages (Japan)

Oleanolic acid, hederagenin, ursolic acid, tormentic acid, and muriatic acid isolated from the leaves of *Campsis grandiflora* (Thunb.) K. Schum. inhibited the aggregation of platelet induced by epinephrine with IC_{50} values of 45.3, 32.8, 82.6, 42.9, and 46.2 μM, respectively (acetyl salicylic acid IC_{50}: 57 μM). This triterpens did not inhibit aggregation induced by adenosine diphosphate, collagen, arachidonic acid or the thromboxane A2 analogue U46619.[303] Ursolic acid inhibited rat lens aldose reductase, rat kidney aldose reductase and human recombinant aldose reductase activities with IC_{50} values equal to 4.4, 5.3, and 2 μg/mL, respectively.[162] In Wistar albino rats fed with galactose for 14 days, treated with ursolic acid at a dose of 10 mg/kg/day reduced the lens aldose reductase catalyzed production of galactitol by 85%.[162]

5.114 *Bacopa monnieri* (L.) Wettst.

Synonym: *Bramia indica* Lam. *Bramia monnieri* (L.) Pennell; *Lysimachia monnieri* L.; *Herpestis monnieri* (L.) Kunth
Common names: pa chi t'ien (China); berimi (Malaysia); brahmi (India); water-hyssop
Subclass Lamiidae, Superorder Lamianae, Order Lamiales, Family Scrophulariaceae
Medicinal use: sedative (India)
History: the plant was known to Sushruta

Intake of cholesterol-enriched meals causes a decrease of renal blood flow, glomerular filtration rate and elevation of glomerular capillary pressure.[304] Ethanol extract of *Bacopa monnieri* (L.) Wettst. given orally at a dose of 40 mg/kg/day for 45 days to Wistar rats on a diet enriched with cholesterol lowered renal cholesterol and triglycerides.[305] High-cholesterol diet reduced the activity of renal antioxidant enzymes superoxide dismutase, catalase, glutathione peroxidase and glutathione

S-transferase as well as the nonenzymatic antioxidant gluthatione and the extract normalized.[305] The extract lowered serum creatinine from 1.3 to 1.1 mg/dL (normal: 1 mg/dL).[305] High-cholesterol diet lowered the expression of endothelial nitric oxide synthetase in renal artery and the extract corrected the expression of this enzyme to normal.[305] The extract reduced swelling of renal tubules and tubular epithelial denudation and glomerular hypertrophy.[305]

5.115 *Rehmannia glutinosa* (Gaertn.) Libosch. ex Fisch. & C.A. Mey.

Synonym: *Rehmannia chinensis* Libosch. ex Fisch. & C.A. Mey.
Subclass Lamiidae, Superorder Lamianae, Order Lamiales, Family Scrophulariaceae
Common name: di huang (Chinese)
Medicinal use: tonic (China)

Catalpol isolated from *Rehmannia glutinosa* (Gaertn.) Libosch. ex Fisch. & C.A. Mey. given orally to high-fat diet streptozotocin-induced type 2 diabetic Sprague–Dawley rats (glycemia ≥16.7 mmol/L) at a dose of 60 mg/kg/day for 10 weeks lowered kidney weight.[306] This iridoid decreased glycemia from 27.4 to 17.7 mmol/L (normal: 5.3 mmol/L; metformin at 200 mg/kg: 19.1 mmol) as well as glycated serum protein.[306] Catalpol also lowered plasma urea from 23.3 to 13.2 mmol/L (normal: 7.2 mmol/L; metformin at 200 mg/kg/day: 12.5 mmol/L). It had no effect on plasma creatinine and improved glomerular cytoarchitecture.[306] Catalpol inhibited renal cortex expression of transforming growth factor-β1 and connective tissue growth factor and reduced the concentration of fibronectin and collagen type-IV.[306]

5.116 *Andrographis paniculata* (Burm. f.) Wall. ex Nees

Synonym: *Justicia paniculata* Burm. f.
Common names: chuan xin lian (Chinese); hempedu bumi (Malay/Indonesian); kalmegh (India); creat
Subclass Lamiidae, Superorder Lamianae, Order Lamiales, Family Acanthaceae
Medicinal use: diabetes (Malaysia)

Mitogen-activated protein kinases contribute to platelet aggregation induced by various stimuli. For instance, binding of ADP to P2Y receptor result in extracellular signal-regulated kinase-2 phosphorylation. In platelets, extracellular signal-regulated kinase is required for the activation of store-mediated calcium entry and phosphorylation of phospholipase A_2.[307] Extracellular signal-regulated kinases are also activated after stimulation by collagen and thrombin.[308] Extracellular signal-regulated kinase-2 is also activated during platelet aggregation induced by collagen.[309] Andrographolide and 14-deoxy-11,12-didehydroandrographolide at a concentration of 100 μM inhibited the aggregation of platelets induced by thrombin *in vitro* by about 70% and 95%, respectively. Both diterpenes inhibited extracellular signal-regulated kinase-1/2.[310] This ent-labdane diterpenes given orally at a dose of 20 mg/kg/day for 14 days to Wistar rats fed with galactose lowered galactinol lens concentration from 20.6 to 3.5 μg/mL, whereby glibenclamide at a dose of 20 mg/kg was inactive.[311] Clinical trials are warranted.

5.117 *Clerodendrum bungei* Steud.

Synonym: *Clerodendrum foetidum* Bunge
Common name: xiu mu dan (Chinese)
Subclass Lamiidae, Superorder Lamiidae, Order Lamiales, Family Lamiaceae
Medicinal use: hypertension (China)

The abietane diterpene 15-dehydrocyrtophyllone A isolated from the roots of *Clerodendrum bungei* Steud. inhibited angiotensin-converting enzyme with an IC_{50} of 42.7 μM (captopril: IC_{50} of 13.3 μM).[312]

5.118 *Rosmarinus officinalis* L.

Common names: mi die xiang (Chinese); romero (Philippines); rosemary
Subclass Lamiidae, Superorder Lamiidae, Order Lamiales, Family Lamiaceae
Medicinal use: tonic (Philippines)
Pharmacological target: vascular endothelium dysfunction

Rosmarinic acid relaxed endothelium thoracic aorta ring of rats precontracted with phenylephrine, and this effect was inhibited by N^{ω}-nitro-L-arginine methyl ester (inhibition of nitric oxide synthetase).[313] This phenolic compound given orally to streptozotocin-induced diabetic Wistar rats (glycemia >15 mmol/L) at a dose of 50 mg/kg/day for 10 weeks had no effect on body weight gain, glycemia, blood pressure, heart rate, nor urination.[314] This phenolic compound however attenuated water intake and plasma reactive oxygen species in diabetic rodents.[314] Isolated thoracic aorta of diabetic animals precontracted with phenylephrine were less responsive to endothelium-dependent relaxation induced by increasing concentrations of acetylcholine. Rosmarinic acid treatment brought aortic sensitivity to acetylcholine close to normal.[314] Endothelium-independent relaxation induced by sodium nitroprusside was not corrected by rosmarinic treatment.[314] Rosmarinic reduced aortic contents in interleukin-1β, endothelin receptor type A and B and endothelin converting enzyme-1.[314] Carnosic acid from *Rosmarinus officinalis* L. inhibited concentration-dependently rabbit platelets aggregation induced by collagen, arachidonic acid, U46610 and thrombin with IC_{50} values of 39, 34, 29, and 48 μM, respectively with concomitant inhibition of cytoplasmic calcium mobilization.[315] This phenolic abietane diterpene inhibited collagen-induced liberation of arachidonic acid by platelets with an IC_{50} value of 27.5 μM.[315] Clinical trials are warranted.

5.119 *Salvia miltiorrhiza* Bunge

Common name: dan shen (Chinese)
Subclass Lamiidae, Superorder Lamiidae, Order Lamiales, Family Lamiaceae
Medicinal use: heart diseases (China)

Aqueous extract of roots of *Salvia miltiorrhiza* Bunge at a 1:1000 dilution (5 g raw material/mL as stock solution) induced the expression of endothelial nitric oxide synthetase and nitric oxide production in human umbilical vein endothelial cells.[316] From this extract, ursolic acid at a concentration of 10 μM induced endothelial nitric oxide synthetase in human umbilical vein endothelial cells with downregulation of NADPH oxidase subunit Nox4 and decrease of reactive oxygen species.[316] Dihydrotanshinone from *Salvia miltiorrhiza* Bunge relaxed with an IC_{50} value of 10.3 μM rings of left anterior descending coronary artery of rats precontracted with 5-hydroxytryptamine.[317] This effect was unaltered by cyclooxygenase inhibitor flurbiprofen, nitric oxide synthetase inhibitor N^{ω}-nitro-L-arginine methyl ester (inhibition of nitric oxide synthetase), or nonselective muscarinic receptor antagonist atropine or endothelium removal indicating an endothelium-independent mechanism.[317] On denuded rings, the relaxing effect of dihydrotanshinone was halved by the guanylyl cyclase inhibitor ODQ.[317] Aqueous extract of roots given to diabetic patients with chronic heart disease at a dose of 5 g twice a day for 60 days decreased plasma vascular adhesion molecule-1 compared with untreated group without altering renal or hepatic function.[318] Tanshinone IIA isolated from *Salvia miltiorrhiza* Bunge at a concentration of 20 μM inhibited the adhesion of monocytes to human umbilical vein endothelial cells challenged with tumor necrosis factor-α by about 70%.[319] This diterpene quinone inhibited the activation of nuclear factor-κB and downstream expression of vascular adhesion molecule-1 and intracellular adhesion molecule-1 and in human umbilical vein endothelial cells challenged with tumor necrosis factor-α.[319] Tanshinone IIA at a concentration of 50 μM inhibited adenosine diphosphate and collagen-induced platelets aggregation by approximately 15% and 10%, respectively (acetylsalicylic acid: about 50%).[320] ADP induced phosphorylation of extracellular signal-regulated kinase-2 and tanshinone IIA inhibited that effect.[320] This diterpene given prophylactically to mice intraperitoneally increased bleeding time.[320] Salvianolic acid B (Figure 5.28) from *Salvia miltiorrhiza* Bunge inhibited copper-induced oxidation of low-density lipoprotein with IC_{50} value of 0.1 μg/mL (probucol IC_{50} of 2.7 μg/mL).[321] This phenolic compound given orally to cholesterol-enriched diet to New Zealand white rabbits at a dose of 1.5 g/kg/day for 6 weeks had no effect on plasma cholesterol.[321] Thoracic aorta isolated from rabbits on cholesterol-enriched diet treated with salvianolic acid B had lower atheroma area and lowered intimal thickness compared with untreated rodents.[321] It should be recalled that phenolics from medicinal plants are metabolized in the gut(bacteria) and the liver to yield what can be termed "reactive phenolics species" which often improve vascular function. Salvianolic acid B inhibited the aggregation of platelets induced by ADP and thrombin with IC_{50} values of 312.6 and 379.7 μg/mL, respectively.[322] This caffeic acid derivative at a concentration of 600 μg/mL inhibited soluble P-selectin secretion from platelets challenged with ADP or thrombin.[322] At a concentration of 10 μg/mL, salvianolic acid B inhibited the adhesion of ADP-activated platelets to human endothelial cells E.Ehy926 and subsequent activation of nuclear factor-κB and expression of intracellular adhesion molecule-1, interleukin-1β, interleukin-6, interleukin-8 and monocyte chemoattractant peptide-1.[322] Lithospermic acid B isolated from *Salvia miltiorrhizae* Bunge given to type 2 diabetic Otsuka Long-Evans Tokushima Fatty rats at a daily dose of 20 mg/kg for 28 weeks had no effect on body weight and had no effect on glucose tolerance.[323] This stereoisomer of salvianolic acid B lowered systolic blood pressure and reduced albuminuria.[323] Histological examination of kidneys of Long-Evans Tokushima Fatty rats treated with lithospermic acid B evidenced decreased glomerular hypertrophy and mesangial expansion.[323] Lithospermic acid B lowered collagen deposition on

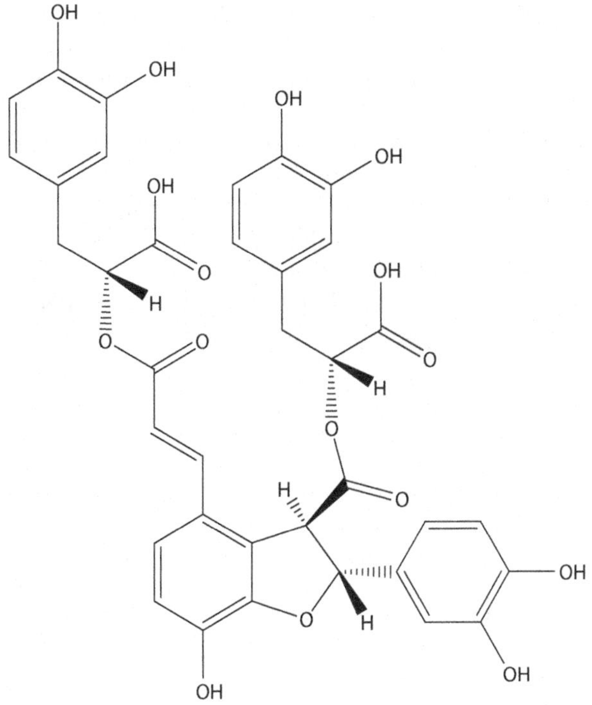

FIGURE 5.28 Salvianolic acid B.

renal cortex of treated animals it decreased transforming growth factor-β1 and malondialdehyde contents in renal cortex and lowered monocyte chemoattractant peptide-1 expression kidneys.[323] Salvianolic acid A inhibited matrix metalloproteinase 2, 8, 9, and 13 with IC_{50} values of 48.4, 73.6, 47.7, and 88 μM, respectively.[324] In angiotensin II-induced aortic aneurysm in apolipoprotein E-deficient ($ApoE^{-/-}$) mice, salvianolic acid A given intraperitoneally once a day for 3 weeks at a dose of 20 mg/kg reduced whole aorta from heart to kidney diameter with concomitant inhibition of plasma metalloproteinase-9.[324] This phenolic compound had no effect on blood pressure, total cholesterol, and triglycerides.[324] Salvianolic acid A inhibited the infiltration of mononuclear cells and macrophages in aorta and prevented elastin fragmentation.[324] Ethanol extract of roots of *Salvia miltiorrhiza* Bunge given prophylactically at a dose of 59.5 mg/kg/day for 5 days to Sprague–Dawley rats prevented isoproterenol-induced ST-segment elevation.[325] This treatment lowered markers of acute myocardial injury such as lactate dehydrogenase and creatine kinase, decreased malondialdehyde, and increased superoxide dismutase and glutathione peroxidase activities in plasma.[325] The extract reduced left ventricular dysfunction induced by isoproterenol and diminished subendocardial necrosis, capillary dilatation and leukocyte infiltration induced by isoproterenol.[325] Clinical trials are warranted.

5.120 *Ocimum basilicum* L.

Synonym: *Ocimum thyrsiflorum* L.
Common names: luo le (China); arjaka (India); daun selaseh (Malay); sweet basil
Subclass Lamiidae, Superorder Lamiidae, Order Lamiales, Family Lamiaceae
Medicinal use: diuretic (India)
History: the plant was known to Aulus Cornelius Celsus (25 BC–50 AD), Roman physician

Aqueous extract of the plant given orally at a dose of 0.5 g/kg/day to Wistar rats on high-cholesterol diet for 10 weeks prevented body weight gain. Thoracic aorta of rodent fed with high-cholesterol diet was less sensitive to carbachol-induced endothelium-dependent relaxation. Lack of responsiveness was corrected to normal in rodents treated the extract.[326] *In vitro*, the extract inhibited rat platelet aggregation induced by ADP and thrombin.[326] Sprague–Dawley rats with experimental left renal partial occlusion-induced hypertension (systolic blood pressure >150 mmHg) receiving *Ocimum basilicum* L. aqueous extract at a dose of 400 mg/kg/day for 4 weeks had a fall of systolic blood pressure from 201.4 to 180 mmHg (normal: 138 mmHg; captopril 30 mg/kg/day: 165 mmHg).[327] The extract lowered heart weight, plasma endothelin and plasma angiotensin II close to normal values.[327] Circulating and intraglomerular vascular endothelial growth factor is responsible for interstitial fibrosis.[328] *Ocimum basilicum* L. contains xanthomicrol[329] which at a concentration of 10 μM inhibited the proliferation of human umbilical vein endothelial cells by 51%.[330] Xanthomicrol (Figure 5.29) at concentrations of 10 μg/mL

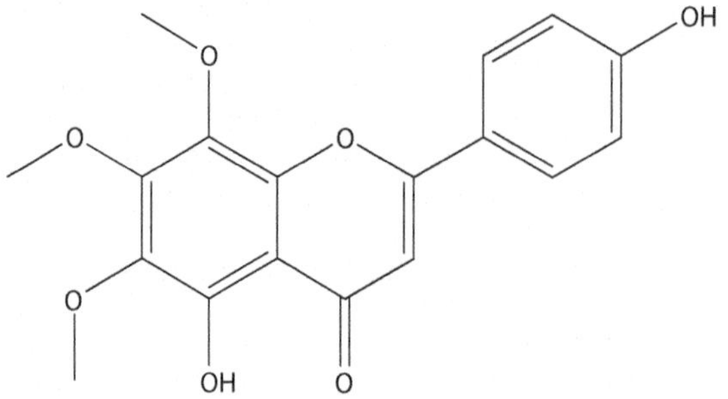

FIGURE 5.29 Xanthomicrol.

inhibited microvessel outgrowth from aortic preparations *in vitro* by 100%.[330] This flavone at 10 μg/mL reduced vascular endothelium growth factor expression by 38% in human umbilical vein endothelial cells.[330]

5.121 *Ocimum sanctum* L.

Synonym: *Ocimum tenuiflorum* L.
Common names: sheng luo le (Chinese); tulasi (India); holy basil
Subclass Lamiidae, Superorder Lamiidae, Order Lamiales, Family Lamiaceae
Medicinal use: heart diseases (India)

Aqueous extract of *Ocimum sanctum* L. given orally at a dose of 590 mg/kg/day to Wistar rats on high-cholesterol diet for 3 weeks lowered total cholesterol from 133 to 96 mg/dL (normal: 38 mg/dL) and triglycerides from 50 to 26 mg/dL (normal: 25 mg/dL). This regimen increased high-density lipoprotein–cholesterol from 17 to 20 mg/dL (normal: 20 mg/dL), low-density lipoprotein–cholesterol from 105 to 71 mg/dL (normal: 13 mg/dL), and atherogenic index from 6.9 to 4 (normal: 0.9).[331] This extract had no effect on food intake and body weight.[331] This regimen increased fecal cholic acid and deoxycholic acid excretion from 0.5 to 0.8 mg/g feces (normal: 0.2 and 0.1 mg/g feces, respectively).[331] The extract lowered to normal plasma lactate dehydrogenase and creatine kinase. It reduced reactive oxygen species in the heart.[331] Histological observation of liver, heart and thoracic aorta evidenced an improvement of cytoarchitecture compared.[331] Methanol extract of leaves of *Ocimum sanctum* L. given to Sprague–Dawley rats at a dose of 150 mg/kg/day for 30 days before myocardial insults (induced by subcutaneous injection of isoproterenol) lowered plasma lactate dehydrogenase and creatine kinase. These enzymes are markers of myofibril degeneration and myocyte necrosis.[332] Isoproterenol lowered the activity of superoxide dismutase in the heart of rodents but pretreatment with the extract brought the activity of superoxide dismutase and lipoperoxidation to normal values.[332] This extract decreased the expression of nuclear factor-κB (which is induced by reactive oxygen species) lowered the activities of phospholipase A, C and D in heart, lowered, leukotriene B4 and thromboxane in plasma.[332] The extract improved the cytoarchitecture of myocytes close to normal rodents.[332] Clinical trials are warranted.

5.122 *Orthosiphon stamineus* Benth.

Synonyms: *Clerodendranthus spicatus* (Thunb.) C.Y. Wu ex H.W. Li; *Orthosiphon aristatus* (Blume) Miq.
Common names: misai kunching (Malay); Java tea
Subclass Lamiidae, Superorder Lamiidae, Order Lamiales, Family Lamiaceae
Medicinal use: diuretic (Malaysia)

Methylripariochromene from *Orthosiphon stamineus* Benth. administered subcutaneously to spontaneously hypertensive rats at a single dose of 100 mg/kg evoked a decrease in mean blood pressure via a mechanism involving L-type calcium channel inhibition.[333] Methanol extract of *Orthosiphon stamineus* Benth., a plant used to lower blood pressure and to promote diuresis in Malaysia, given at a single oral dose of 2 g/kg to Sprague–Dawley rats had no diuretic effect.[334] Oral administration at a dose of 0.5 g/kg/day for 7 days increased cumulative urine volume from about 25 mL/100 g to 32.5 mL/100 g with increase in sodium and potassium excression.[334] In another study, aqueous extract of the plant administered orally at a dose of 10 mg/kg to Spraque–Dawley rats increased urine output without increase of urinary sodium and chlorine but increased urinary secretion of potassium.[335] Clinical trials are warrented.

5.123 *Perilla frutescens* (L.) Britton

Synonym: *Ocimum frutescens* L.; *Perilla ocymoides* L.
Common names: zi su (Chinese); deul gae (Korean); perilla
Subclass Lamiidae, Superorder Lamiidae, Order Lamiales, Family Lamiaceae
Medicinal use: indigestion (Korea)

Aqueous extract of leaves of *Perilla frutescens* (L.) Britton at a concentration of 50 μg/mL induced the secretion of nitrite by vascular smooth muscle cells.[336] The extract at a concentration of 25 μg/mL abrogated platelet derived growth factor or tumor necrosis factor-α induced vascular smooth muscle cells proliferation. The nitric synthetase inhibitor N^{ω}-nitro-L-arginine methyl ester abrogated tumor necrosis factor-α induced proliferation suggesting that the extract inhibit this proliferative effect by inhibit nitric oxide production.[336]

5.124 *Thymus linearis* Benth.

Common name: ban ajwain (India); wild thyme
Subclass Lamiidae, Superorder Lamiidae, Order Lamiales, Family Lamiaceae
Medicinal use: indigestion (India)

Methanol extract of aerial parts of *Thymus linearis* Benth. given at a single oral dose of 500 mg/kg to Sprague–Dawley rats lowered mean arterial blood pressure from 105.2 to 86.6 mmHg and heart rate from 388 to 320 beats/min after 6 hours.[337] In cholesterol-enriched diet-induced hypertensive Sprague–Dawley rats, the extract at a dose of 500 mg/kg/day for 9 days lowered the mean arterial blood pressure from 108.6 to 83.6 mmHg (untreated 124.4 mmHg) and lowered heart rate from 381 to 300.2 beats/min (untreated 455 beats/min).[337] In Sprague–Dawley rats given 10% glucose solution for 21 days, the extract at a dose of 500 mg/kg/day lowered the mean arterial blood from 108.2 to 77 mmHg (untreated 127.1 mmHg) and lowered heart rate from 391.6 to 296.6 beats/min (untreated 410.4 beats/min).[337] The extract increased alanine aminotransferase, lowered aspartate aminotransferase and alkanine phosphatase, and decreased serum triglycerides cholesterol and low-density lipoprotein.[337]

5.125 *Leonorus sibiricus* L.

Synonym: *Leonurus manshuricus* Yabe
Common names: chung wei (Chinese); Siberian motherworth
Medicinal use: tonic (China)

Ethanol extract of *Leonorus sibiricus* L. (Figure 5.30) given to atherogenic diet fed C57BL/6 mice orally at a dose of 0.01 g/20 g for 14 weeks lowered cholesterol from 213.9 to 155.5 mg/dL (normal: 96.6 mg/dL), increased high-density lipoprotein–cholesterol from 38.8 to 50.1 mg/dL (normal: 43.2 mg/dL), and decreased atherogenic index from 4.5 to 2.1 (normal: 1.2).[338] *In vitro*, the extract at a concentration of 400 μg/mL inhibited the production of reactive oxygen species by human umbilical vein endothelial cells induced by tumor necrosis factor-α by 73%. This extract lowered lectin-like oxidized low-density lipoprotein receptor-1 expression by 27% as well as the expression of vascular adhesion molecule-1 by 94%.[338] Leonurine (Figure 5.31) given orally at a dose of 15 mg/kg per day for 8 weeks protected Sprague–Dawley rats against chronic myocardial ischemia resulting from left ventricular infarction. It was evidenced by a reduction of heart and lung edema, lowering of the left ventricular end-diastolic pressure from 15.5 to 11.5 mmHg, and increase the mean aortic pressure from 69 to 107.8 mmHg.[339] Furthermore, this guanidine alkaloid induced the phosphorylation of Akt and the expression of hypoxia-inducible factor 1-alpha. It also induced the expression of Bcl-2,

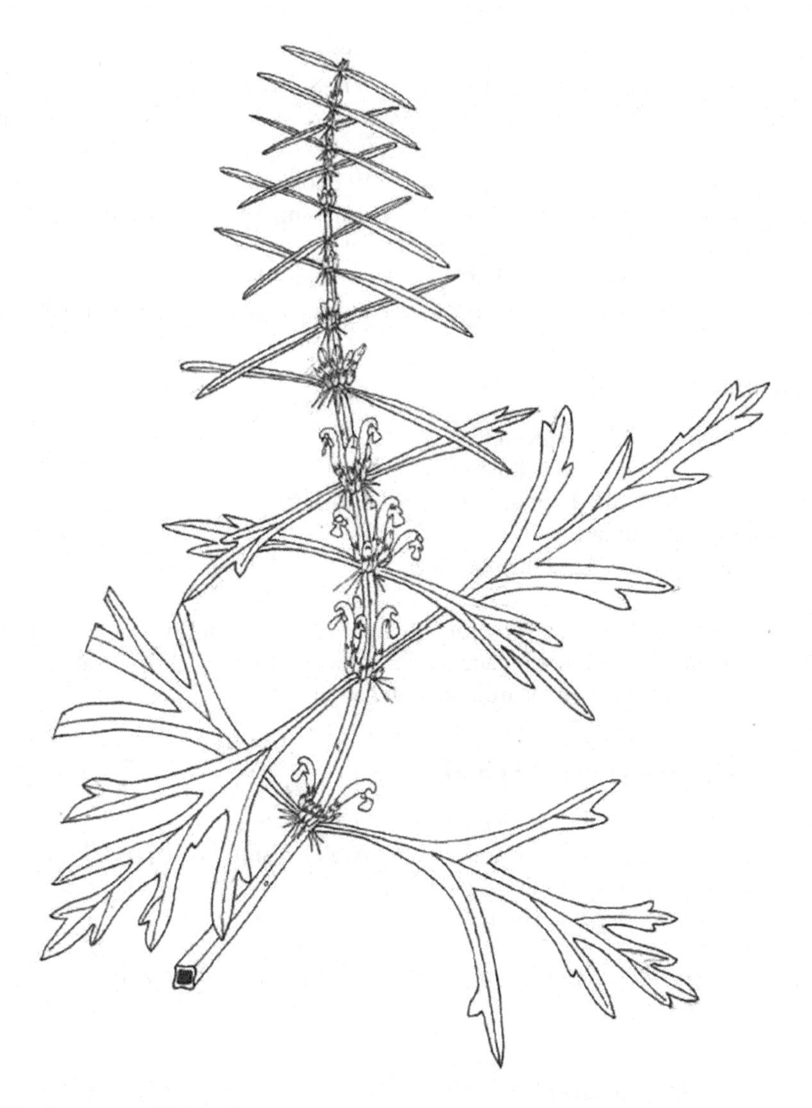

FIGURE 5.30 *Leonorus sibiricus* L.

FIGURE 5.31 Leonurine.

vascular endothelium growth factor, and downregulated Bax in postischemic myocytes.[339] Leonurine prevented alpha smooth muscle actin expression as well as type I and type III collagen synthesis by neonatal rat cardiac fibroblasts stimulated with angiotensin II via inhibition of Nox4 and reactive oxygen species generation.[339] Further, leonurine abrogated the expression of MMP-2/9 by neonatal rat cardiac fibroblasts stimulated with angiotensin II.[340] Leonurine given orally for 6 weeks at a dose of 30 mg/kg/day to Sprague–Dawley rats with myocardial infarction (induced by ligation of the left anterior descending artery) prevented the generation of reactive oxygen species and the expression of Nox4. It also inhibited the translocation of nuclear factor-κB, the expression of inducible nitric oxide synthetase and tumor necrosis factor-α and reduced fibrosis in the myocardium.[340]

5.126 *Marrubium vulgare* L.

Synonym: *Marrubium hamatum* Kunth
Common names: ou xia zhi cao (Chinese); horehound
Superorder Lamianae, Order Lamiales, Family Lamiaceae
Medicinal use: colds (China)
History: the plant was known to Pliny the elder

Methanol extract of aerial parts of *Marrubium vulgare* L. given orally at a dose of 40 mg/kg to Wistar rats attenuated isoproterenol induced myocardial infection.[341] The extract lowered cardiac necrosis by 60% and serum and myocardial malondialdehyde.[341]

5.127 *Colocacia esculenta* (L.) Schott

Synonyms: *Arum esculentum* L.; *Colocasia antiquorum* Schott
Subclass Alismatidae, Superorder Aranae, Order Arales, Family Araceae
Common names: yu (Chinese); alukam (India); taro
Medicinal use: internal bleeding (India)

The flavonoids orientin, isoorientin, isovitexin, luteolin-7-*O*-glucoside, and luteolin-7-*O*-rutinoside from the leaves of *Colocacia esculenta* (L.) Schott inhibited rat lens aldose reductase *in vitro*.[342]

5.128 *Veratrum dahuricum* (Turcz.) Loes.

Synonym: *Veratrum album* var. *dahuricum* Turcz.
Subclass Liliidae, Superorder Lilianae, Order Melianthales, Family Melianthaceae
Medicinal use: epilepsy (China)

Veratrum dahuricum (Turcz.) Loes. produces jervine which at 1000 μmol/L inhibited the aggregation of platelets induced by arachidonic acid by 92.7%.[343] Besides, the plant shelters 3, 15-diangeloylgermine which at a dose of 500 μmol/L inhibited platelet aggregation factor and arachidonic acid induced platelet aggregation by 73.6% and 60.4%, respectively.[343] From the same plant, 15-angeloylgermine abrogated platelet aggregation in the presence of arachidonic acid at 200 μmol/L.[343] The plant is poisonous and has no place in therapeutic.

5.129 *Anemarrhena asphodeloides* Bunge

Common name: zhi mu (Chinese)
Subclass Liliidae, Superorder Lilianae, Order Asparagales, Family Asparagaceae
Medicinal use: fever (fever)
Pharmacological target: vascular endothelium dysfunction

Ethanol extract of rhizomes of *Anemarrhena asphodeloides* Bunge given to streptozotocin-induced diabetic Wistar rats at a dose of 540 mg/kg/day for 12 weeks had no effect on body weight, glycemia or plasma insulin. However, it increased serum activity of gluthatione peroxidase and superoxide dismutase and lowered malondialdehyde.[344] This regimen decreased lens sorbitol contents and serum advanced glycation end products.[344] This extract lowered microvascular retinal proliferation and the occurrence rate of cataract.[344] Mangiferin and neomangiferin from this extract at a concentration of 4×10^{-6} M prevented high glucose-induced subnormal growth of pericytes.[344] Total polyphenol fraction of rhizomes of *Anemarrhena asphodeloides* Bunge at a concentration of 100 μg/mL evoked the phosphorylation of adenosine monophosphate-activated protein kinase in E.A hy926 human umbilical vein endothelial cells.[345] This fraction induced nitric oxide secretion and inhibited endothelin-1 secretion and these effect were inhibited by adenosine monophosphate-activated protein kinase inhibitor compound C.[345] Advanced glycation end products-induced reduction of endothelial nitric oxide synthetase expression was prevented by the fraction.[345] This fraction inhibited reactive oxygen species generation induced by advanced glycation end products and mitochondrial insults.[345] Adiponectin, which is a known activator of adenosine monophosphate-activated protein kinase inhibits tumor necrosis factor-α and subsequent expression of interleukin-8, vascular cell adhesion molecule 1 (via inhibition of nuclear factor-κB) and monocytes recruitment.[346] Adiponectin inhibits the formation of atherosclerotic lesions and decreases the expression of class A scavenger receptor, tumor necrosis factor and vascular adhesion molecule in aorta.[347] Adiponectin induces endothelial cells migration and differentiation to form capillary-like structures and prevents endothelial cell apoptosis via adenosine monophosphate-activated protein kinase activation.[8] Clinical trials are warranted.

5.130 *Liriope platyphylla* F.T. Wang & T. Tang

Synonyms: *Liriope muscari* (Decne.) L.H. Bailey; *Ophiopogon spicatus* Hook.
Common name: kuo ye shan mai dong (Chinese)
Subclass Liliidae, Superorder Lilianae, Order Liliales, Family Liliaceae
Medicinal use: fatigue (China)

(+)-Platyphyllarin A, (−)-liriopein B, (3R)-3-(4′-hydroxybenzyl)-5,7-dihydroxy-6-methyl-chroman-4-one, (3R)-3-(2′,4′-dihydroxybenzyl)-5,7-dihydroxychroman-4-one, and (3R)-3-(2′,4′-dihydroxybenzyl)-5,7-dihydroxy-methyl-chroman-4-one from the roots of *Liriope platyphylla* F.T. Wang & T. Tang inhibited the aggregation of human platelets induced by collagen with IC_{50} values below 30 μM.[348]

5.131 *Smilax glabra* Roxb.

Synonyms: *Smilax blinii* H. Lév.; *Smilax dunniana* H. Lév.; *Smilax hookeri* Kunth; *Smilax mengmaensis* R.H. Miao; *Smilax trigona* Warb.
Common name: tu fu ling (Chinese)
Subclass Liliidae, Superorder Lilianae, Order Smilacales, Family Smilacaceae
Medicinal use: venereal diseases (China)

In type-2 diabetes, advanced glycation end-products interact with receptors of advanced glycation end products expressed by kidney cells. This leads to the activation of protein kinase C, mitogen activated protein kinases and nuclear factor-κB favoring the production of transforming growth factor-β1 and subsequent mesangial hypertrophy and glomerular sclerosis associated with fribronectin synthesis.[349] Ethanol extract of rhizomes of *Smilax glabra* Roxb. prevented advanced glycation end product-induced apoptosis of human umbilical vein endothelial cells *in vitro*.[350] This extract lowered advanced glycation end product-induced reactive oxygen species generation, increased superoxide dismutase activity and decreased malondialdehyde contents.[350] This extract decreased advanced glycation end product receptor expression in human umbilical vein endothelial cells, decreased transforming growth factor-β1, and inhibited advanced glycation end product-induced extracellular signal-regulated kinase-1/2 and nuclear factor-κB.[350]

5.132 *Anoectochilus roxburghii* (Wall.) Lindl.

Synonyms: *Anoectochilus yungianus* S.Y. Hu; *Chrysobaphus roxburghii* Wall.; *Zeuxine roxburghii* (Wall.) M. Hiroe
Common name: jin xian lan (China)
Subclass Liliidae, Superorder Lilianae, Order Orchidales, Family Orchidaceae
Medicinal use: tuberculosis (China)

Aqueous extract of *Anoectochilus roxburghii* (Wall.) Lindl. given to alloxan-induced diabetic Kunming mice (glycemia >16.8 mmol/L) at a dose of 2 g/kg/day for 14 days lowered glycemia from 27.8 to 17.2 mmol/L.[351] The extract increased hepatic and kidney glutathione peroxidase.[351] Kinsenoside isolated from this orchid given to high-fat diet streptozotocin-induced diabetic mice at a daily dose of 100 mg/kg for 21 days lowered plasma glucose from 31.7 to 21.8 mmol/L (normal: 5.6 mmol/L).[352] This glycoside normalized plasma triglycerides from 1.5 to 1 mmol/L, lowered cholesterol from 2.8 to 1.4 mmol/L (normal 1.1 mmol/L), serum malondialdehyde and increased superoxide dismutase activity.[352] The treatment inhibited thickening of thoracic aorta due to deposition of collagen in the tunica media.[352] *In vitro*, kinsenoside at a concentration of 50 μg/mL increased activities of antioxidant enzymes in human umbilical vein endothelial cells challenged with high glucose concentration.[352] This glycoside inhibited the expression of nuclear factor-κB and lowered the expression of MMP2 and MMP9.[352]

5.133 *Orchis mascula* L.

Synonyms: *Orchis pinetorum* Lacaita; *Orchis signifera* Vest; *Orchis speciosa* Heuff. ex Rochel; *Orchis wanjkowii* E. Wulff
Common names: salib misri (India); early purple orchis
Subclass Liliidae, Superorder Lilianae, Order orchidales, Family Orchidaceae
Medicinal use: sexual dysfunction (Pakistan)
History: the plant was known to Theophrastus (371–287 BC), Greek philosopher and botanist, as male aphrodisiac

Roots of *Orchis mascula* L. given to spontaneously hypertensive rats at 3% of diet for 8 weeks lowered systolic blood pressure from 203.4 to 174.2 mmHg, low-density lipoprotein–cholesterol from 21.9 to 5.9 mg/dL, triglycerides from 93.8 to 29.2 mg/dL, atherogenic index from 0.3 to 0.09, and had no effect on glycemia.[353] This regimen prevented the reduction of sensitivity of spontaneously hypertensive rats aortae to acetylcholine.[353] Intravenous injection of methanol extract of roots to normal rats evoked a 32% fall in blood pressure.[353] In isolated rabbit aorta, the extract inhibited concentrations induced by phenylephrine and high potassium via a mechanism involving calcium channel blockade.[353] High potassium-induced contraction in smooth muscle is mediated by cell membrane depolarization and an increase in calcium influx through voltage-gated calcium channels.[354] Natural products that selectively relax smooth muscle contractions induced by low potassium are considered as potassium ATP channel opener while natural products inhibiting at both high and low potassium induced contractions are calcium antagonists.[355]

5.134 *Dendrobium chrysotoxum* Lindl.

Synonyms: *Callista chrysotoxa* (Lindl.) Kuntze; *Dendrobium suavissimum* Rchb. f.
Common name: gu chui shi hu (Chinese)
Subclass Liliidae, Superorder Lilianae, Order orchidales, Family Orchidaceae
Nutritional use: tea (China)

Ethanol extract of *Dendrobium chrysotoxum* Lindl. given orally at a dose of 300 mg/kg/day for 2 months to streptozotocin-induced diabetic Sprague–Dawley rats (glycemia >16.5 mmol/L) had no effect on body weight loss and plasma glucose.[356] This regimen lowered the retinal expression of endothelial cell adhesion molecule, and lowered the number of vessels in retina.[356] The extract lowered plasma vascular endothelial growth factor, MMP2 and MMP9 as well as retinal expression of vascular endothelial growth factor.[356] The treatment lowered serum platelet-derived growth factor levels as well as insulin like growth factor-1 levels, interleukin-6, and interleukin-1β.[356] In retina, the extract decreased the expression of intracellular adhesion molecule-1 and the translocation of nuclear factor-κB.[356] At the renal level this regimen increased the content of glutathione and lowered lipid peroxidation together with an increase in enzymatic activity of superoxide dismutase and catalase.[356] Clinical trials are warranted.

5.135 *Nervilia plicata* (Andrews) Schltr.

Synonyms: *Epipactis plicata* Roxb.; *Nervilia discolor* (Blume) Schltr; *Nervilia purpurea* (Hayata) Schltr.; *Nervilia velutina* (Parish & Rchb. f.) Schltr.
Common names: mao ye yu lan (Chinese); i tian hong (Taiwan)
Subclass Liliidae, Superorder Lilianae, Order Orchidales, Family Orchidaceae
Medicinal use: hypertension (Taiwan)

Ethanol extract of stems of *Nervilia plicata* (Andrews) Schltr. given to nicotinamide-streptozotocin-induced type 2 diabetic Wistar rats daily for 10 days at a dose of 5 mg/kg lowered glycemia from 312.16 to 88.1 mg/dL (normal: 86.3 mg/dL; glibenclamide 5 mg/kg/day: 100.3 mg/dL).[357] The extract given for further 20 days at intervals of 5 days improved kidney histoarchitecture, lowered plasma urea by 61.4% and serum creatinine by 69.5%.[357] This regimen lowered kidney and pancreas lipid peroxidation by 70.5% and 77.4%, respectively.[357]

5.136 *Belamcanda chinensis* (L.) Redouté

Synonyms: *Belamcanda punctata* Moench; *Ixia chinensis* L.; *ardanthus chinensis* (L.) Ker Gawl.
Common names: she gan (Chinese); leopard flower
Subclass Liliidae, Superorder Lilianae, Order Orchidales, Order Iridales, Family Iridaceae
Medicinal use: fever (China)

Tectoridin and tectorigenin isolated from the rhizome of *Belamcanda chinensis* (L.) Redouté given to streptozotocin-induced diabetic Sprague–Dawley rats orally at a dose of 100 mg/kg/day lowered sorbitol contents of lenses by 27.7% and 50.7%, respectively (epalrestat at 100 mg/kg/day: 57.7%).[358] *In vitro*, tectorigenin and tectoridin (Figure 5.32) inhibited rat aldose reductase with IC_{50} values of 1.1 and 1 μM, respectively.[358] Iristectorigenin B isolated from rhizome at a concentration of 20 μM stimulated the transactivation of liver X receptor in cotransfected HEK293 cells *in vitro*.[358] At a concentration of 10 μM, this isoflavone increased cholesterol efflux, lowered intracellular cholesterol concentration, and increased liver X receptor targets ATP-binding-cassette A1 and G1.[359]

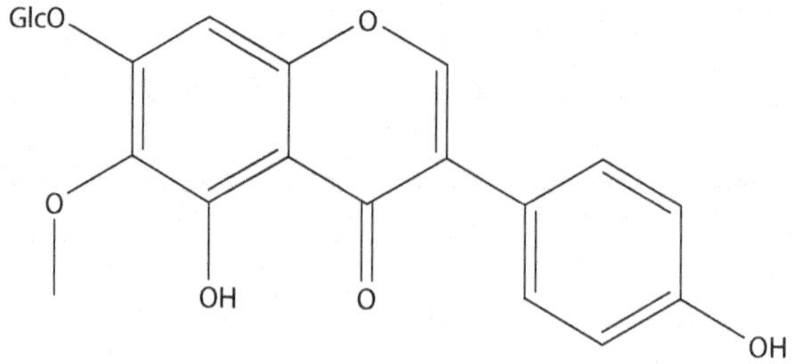

GlcO O OH O OH

FIGURE 5.32 Tectoridin.

5.137 *Asparagus racemosus* Willd.

Common names: chang ci tian men dong (Chinese); indivari (India); India asparagus
Subclass Lillidae, Superorder Lilianae, Order Asparagales, Family Asparagaceae
Medicinal use: male infertility (India)

Ethanol extract of roots of *Asparagus racemosus* Willd. given to streptozotocin-induced diabetic Wistar rats (plasma glucose >300 mg/dL) at dose of 250 mg/kg/day for 4 weeks attenuated body weight loss, decreased plasma glucose from 362.8 to 165.8 mg/dL (normal: 79.2 mg/dL) plasma creatinine and cholesterol and plasma triglycerides.[360] This regimen decreased kidney weight, urinary secretion, and kidney malondialdehyde; increased catalase activity; and improved kidney cytoarchitecture.[360]

5.138 *Alpinia officinarum* Hance

Synonym: *Languas officinarum* (Hance) Farw.
Common names: gao liang jiang (Chinese); galangal
Subclass Commelinidae, Superorder Zingiberanae, Order Zingiberales, Family Zingiberaceae
Medicinal use: indigestion (China)

6-Hydroxy-1,7-diphenyl-4-en-3-heptanone,1,7-diphenyl-4-en-3-heptanone, and 1,7-diphenyl-5-methoxy-3-heptanone isolated from *Alpinia officinarum* Hance elicited platelet activating factor receptor binding antagonistic activities with IC_{50} values of 1.3, 5, and 1.6 μM, respectively.[361]

5.139 *Curcuma wenyujin* Y.H Chen & C. Ling

Common name: wen yu jin (China)
Subclass Commelinidae, Superorder Zingiberanae, Order Zingiberales, Family Zingiberaceae
Medicinal use: blood stasis (China)

Curdione isolated from the essential oil of *Curcuma wenyujin* Y.H Chen & C. Ling (Figure 5.33) inhibited platelet-activated factor-induced platelet aggregation with an IC_{50} of 60 μM with decrease of cytoplasmic calcium concentration and cyclic adenosine monophosphate (c-AMP).[362] This germacrane sesquiterpene (Figure 5.34) given orally to Kunming mice at a single dose of 200 mg/kg prevented carrageenan-induced tail thrombus formation by 60.6% (aspirin at 100 mg/kg: 70.1%).[362]

FIGURE 5.33 *Curcuma wenyujin* Y.H. Chen & C. Ling.

FIGURE 5.34 Curdione.

5.140 *Kaempferia parviflora* Wall. ex Baker

Synonyms: *Kaempferia rubromarginata* (S.Q. Tong) R. J. Searle; *Stahlianthus rubromarginatus* S.Q. Tong
Common names: krai chai dam (Thai); black galingale
Subclass Commelinidae, Superorder Zingiberanae, Order Zingiberales, Family Zingiberaceae
Medicinal use: sexual impotence (Thailand)
Pharmacological target: vascular endothelium dysfunction

Methoxyflavone fraction of rhizomes of *Kaempferia parviflora* Wall. ex Baker given at a dose of 100 mg/kg to Wistar rats orally twice a day for 6 weeks had no effect of food intake and lowered body weight. This regimen decreased epididymis, retroperitoneal, and subcutaneous adipose tissue weight. It lowered glycemia from 128 to 117.4 mg/dL and plasma triglycerides from 85 to 64.2 mg/dL.

Plasma cholesterol, high-density lipoprotein–cholesterol and low-density lipoprotein–cholesterol were unaltered.[363] This regimen increased the EC_{50} of contractile response of thoracic aortic rings to phenylephrine from 85.6 to 142.4 nM. It decreased EC_{50} values of the vasodilatory responses to acetylcholine from 287.2 to 145.2 nM.[363] Extract intake increased the aortic expression of endothelial nitric oxide synthetase.[363] The extract had no effect of systolic and diastolic blood pressure.[363]

5.141 *Zingiber mioga* (Thunb.) Roscoe

Synonyms: *Amomum mioga* Thunb.; *Zingiber echuanense* Y.K. Yang; *Zingiber oligophyllum* K. Schum.
Common names: jang ho (Chinese); yang ha (Korean); myoga ginger
Subclass Commelinidae, Superorder Zingiberanae, Order Zingiberales, Family Zingiberaceae
Medicinal use: malaria (China)

Miogatrial and miogadial, isolated from the flower buds of *Zingiber mioga* (Thunb.) Roscoe inhibited platelet aggregation induced by arachidonic acid with IC_{50} values of 11.9 and 10.7 μM, respectively.[364] From the plant, miogatrial, miogadial, and E-labda-8(17),12-diene-15,16-dial inhibited platelet aggregation induced by ADP with IC_{50} values of 3.5, 3, and 3.2 μM, respectively.[364] Miogatrial, miogadial, and E-labda-8(17),12-diene-15,16-dial inhibited 5-lipoxygenase with IC_{50} value of 7.5, 4, and 18.9 μM, respectively.[364]

5.142 *Zingiber officinale* Roscoe

Synonyms: *Amomum zingiber* L.; *Zingiber aromaticum* Noronha; *Zingiber sichuanense* Z.Y. Zhu, S.L. Zhang & S.X. Chen
Common names: jiang (Chinese); aderuck (India); halia (Malay); ginger
Subclass Commelinidae, Superorder Zingiberanae, Order Zingiberales, Family Zingiberaceae
Medicinal use: indigestion (India)
History: the plant was known to Dioscorides

6-Dehydrogingerdione, 11-isodehydrogingerdione, 6-gingerol, 8-gingerol, 10-gingerol, 6-shoagol from the rhizomes of *Zingiber officinale* Roscoe inhibited *in vitro* the aggregation of platelets induced by arachidonic acid or collagen.[365] From these, shogaol inhibited platelet activated factor-induced platelet aggregation.[365] 10-Gingerol at a concentration of 285.7 μM inhibited potassium-induced contraction of aortic rings in isolated aortic rings.[365] High potassium-induced contraction in smooth muscle cells is owed to cytoplasmic membrane depolarization and an increase in calcium influx through voltage-operated calcium channels.[366] Potassium-induced contractions are inhibited by natural products blocking the calcium channels acting in way similar to verapamil and nifedipine.[367]

5.143 *Zingiber zerumbet* (L.) Roscoe ex Sm.

Synonym: *Amomum zerumbet* L.; *Zingiber sylvestre* (L.) Garsault
Common names: hong qiu jiang (Chinese); lempoyang (Malay); shampoo ginger
Medicinal use: fatigue (China)

In kidneys, hyperglycemia activates p38 phosphorylation and downstream expression of transforming growth factor-β1 hence fibronectin deposition and nephropathy.[368] High-plasma glucose also enhance inflammatory cytokines, monocyte chemoattractant peptide-1 to activate macrophages. Hyperglycemia also stimulates the expression of intracellular adhesion molecule-1 conveying monocytes and macrophages infiltration in kidneys.[328] Infiltrating monocytes and macrophages and renal endothelial, mesangial, dendritic, and epithelial cells produce proinflammatory cytokines such as Interleukin-1β, Interleukin-6 and tumor necrosis factor-α leading to renal injuries.[328] Zerumbone from

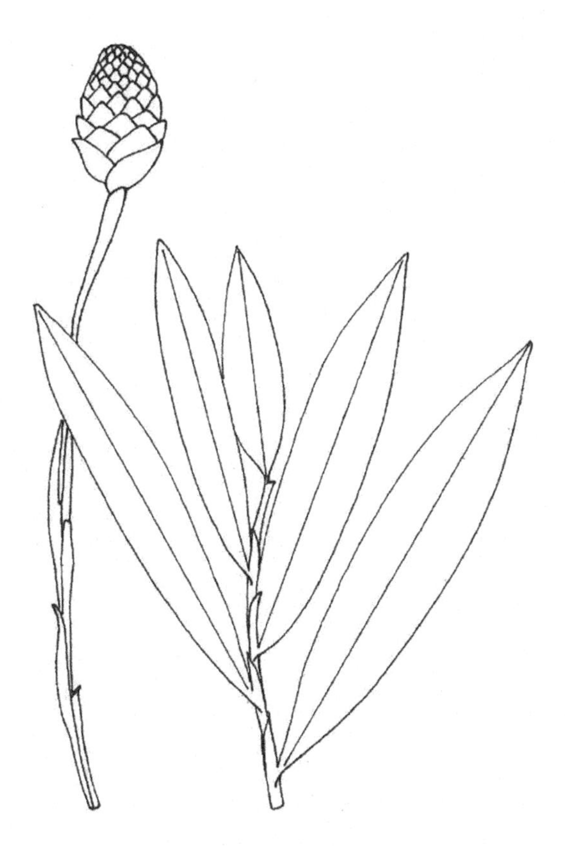

FIGURE 5.35 *Zingiber zerumbet* (L.) Roscoe ex Sm.

Zingiber zerumbet (L.) Roscoe ex Sm. (Figure 5.35) given to streptozotocin-induced diabetic Wistar rats (glycemia ≥350 mg/dL) once daily for 8 weeks at a dose of 40 mg/kg attenuated body weight loss, lowered plasma glucose from 445 to 317 mg/dL (normal: 95.7 mg/dL), HbA1c from 14.5% to 11.2% (normal: 4.8%), serum creatinine from 87.8 μmol/L to 64.1 μU/mL (normal: 35.1 μU/mL whereas insulinema was not affected.[369] This regimen improved the cytoarchitecture of kidneys with decreased focal mesangial matrix expansion and decreased macrophage infiltration.[369] This monocyclic sesquiterpene ketone decreased tumor necrosis factor-α, interleukin-1β, and interleukin-6 renal expression as well as intracellular adhesion molecule-1 and monocyte chemoattractant peptide-1 expression.[369] The treatment decreased renal profibrotic factor transforming growth factor-β1 by 43.7% and renal expression of phosphorylated p38 mitogen-activated protein kinase.[369]

5.144 *Dioscorea batatas* Decne.

Synonym: *Dioscorea polystachya* Turcz.
Common names: shu yu (Chinese); nagaimo (Japanese); Chinese yam
Superorder Dioscoreanae, Order Dioscoreales, Family Dioscoreaceae
Medicinal use: rheumatism (China)

Hexane extract of edible part of *Dioscorea batatas* Decne. given at a dose 200 mg/kg, 3 times per week for 21 weeks to high-fat diet $ApoE^{-/-}$ mice decreased plasma cholesterol, and oxidized low-density lipoprotein levels.[370] The extract decreased intimal lesion in aorta by about 30% and inhibited interleukin 6 expression.[370] From this extract β-sitosterol at a concentration of 100 μM and ethyl linoleate at a concentration of 10 μM inhibited the expression of vascular adhesion molecule-1 induced by tumor necrosis factor-α by mouse vascular smooth muscle cells by about 30% and 20%, respectively.[370]

APPENDIX

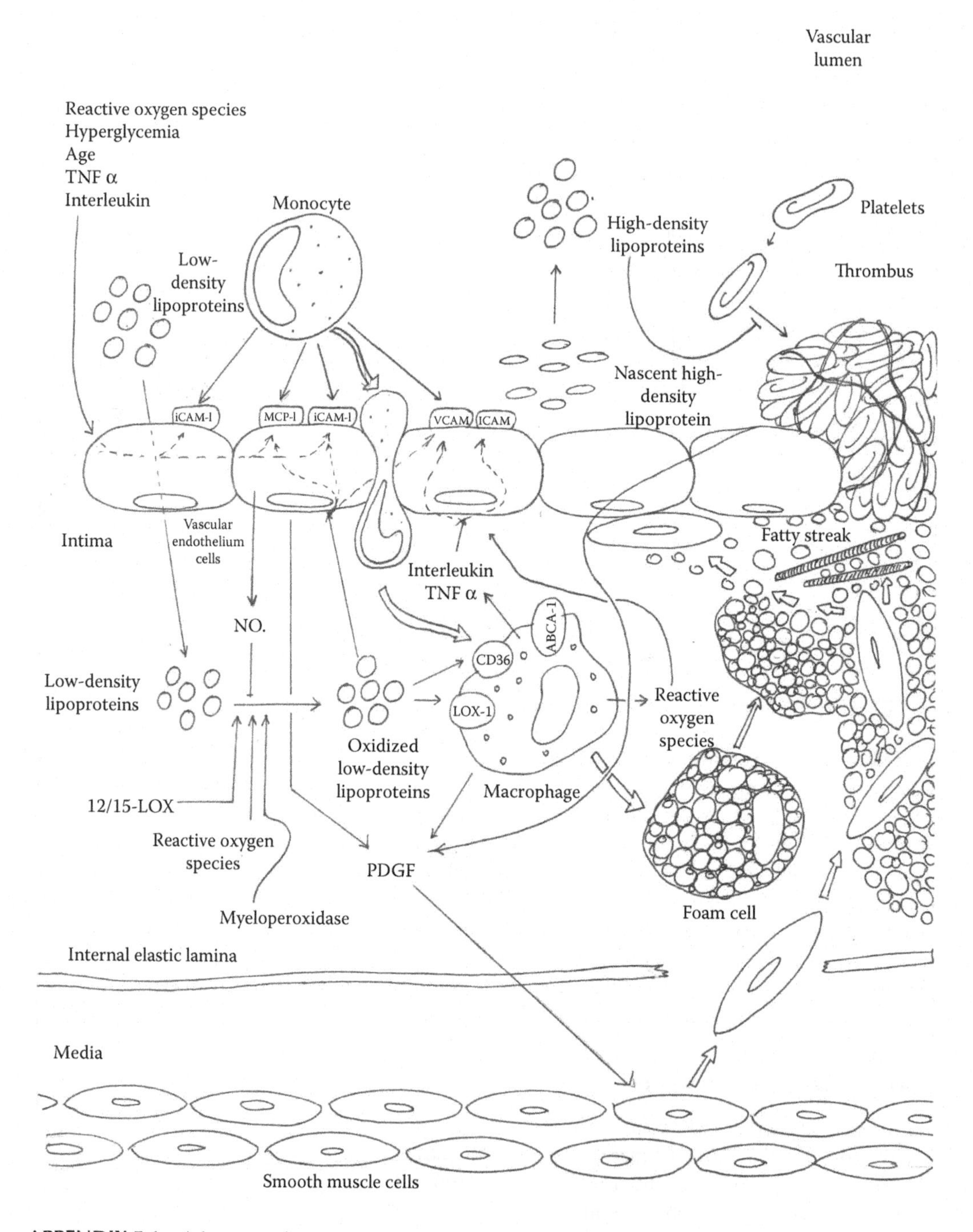

APPENDIX 5.1 Atherogenesis.

ABCA1: ATP binding cassette transporter-1; CD36: Fatty Acid Translocase (FAT/CD36); ICAM-1: intercellular adhesion molecule-1;LOX-1: Lectin-Like Oxidized LDL. Receptor. 1; 12/15-LOX: 12/15-Lipoxygenase; NO: Nitric oxide; PDGF: Platelet-derived growth factor; VCAM: Vascular cell adhesion molecule-1; TNF-α: Tumor necrosis factor-α.

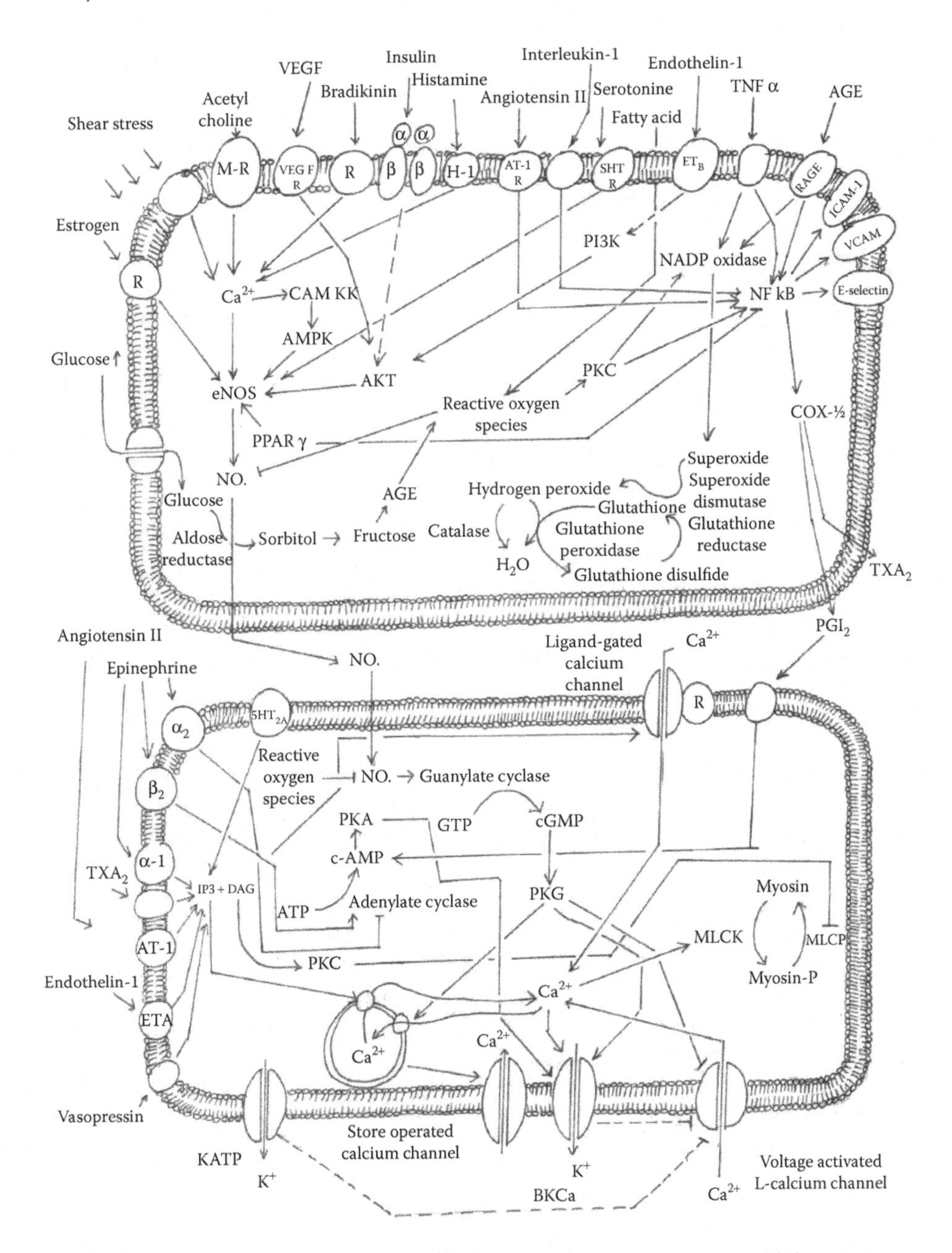

APPENDIX 5.2 Vascular endothelium and smooth muscle cell.

α1: alpha receptor type 1; α2: alpha receptor type 2; AGE: Advanced glycation end-products; AMPK: adenosine monophosphate-activated protein kinase; AT-1 R: Angiotensine-II receptor type-1; β2: beta receptor type 2; CAMKK: Calmodulin-dependent protein kinase kinase; COX-1/2: Cyclooxygenase ½; eNOS: endothelial nitric oxide synthetase; ET-A: Endothelin receptor-A; ET-B: Endothelin receptor B; H1: Histamine receptor type-1; 5HT R: Serotonine receptor; ICAM-1: intercellular adhesion molecule-1; MLCK: myosin light chain kinase; MLCP: myosin light chain phosphatase; MR: Muscarinic receptor; NO: Nitric oxide; PGI2: Prostaglandin I2; PI3K: phosphoinositide-3-kinase; PKA: Protein kinase A; PKC: Protein kinase C; R: Receptor; RAGE: Recepor for advanced glycation end-products; VCAM: Vascular cell adhesion molecule-1; VEGF R: VEGF receptor; VEGF: Vascular endothelium growth factor; TNF-α: Tumor necrosis factor-α; TXA2: Thromboxane A2.

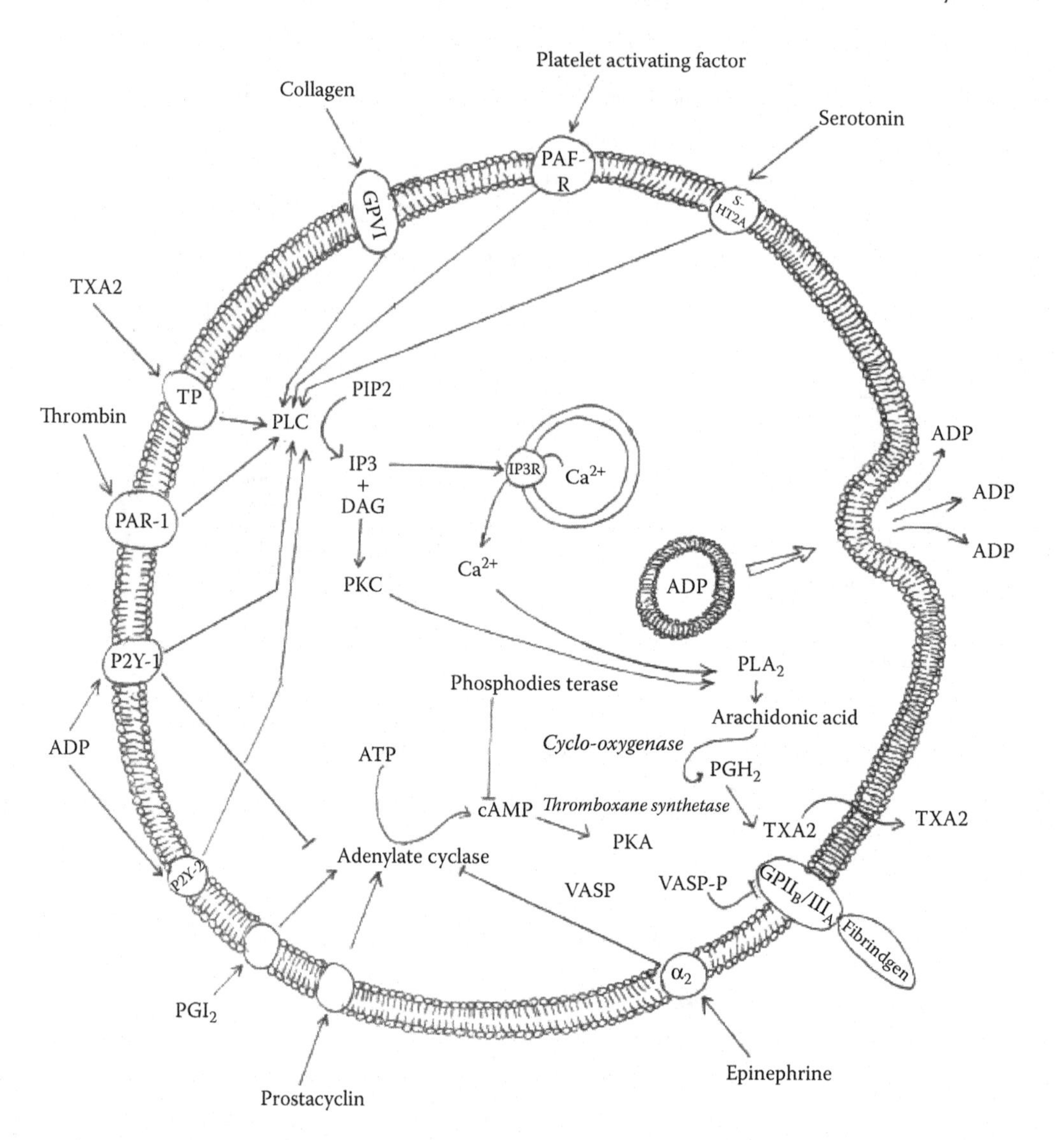

APPENDIX 5.3 Platelet aggregation.

α2: Alpha-2 receptor; P: Prostacyclin; PAF-R: Platelet activating factor receptor; PKA: protein kinase A; TXA2: Thromboxane A2; VASP: vasodilator-stimulated phosphoprotein.

REFERENCES

1. Davignon, J. and Ganz, P., 2004. Role of endothelial dysfunction in atherosclerosis. *Circulation*, *109*(23 suppl 1), III27–32.
2. Gimbrone, M.A., Topper, J.N., Nagel, T., Anderson, K.R. and Garcia-Cardeña, G., 2000. Endothelial dysfunction, hemodynamic forces, and atherogenesisa. *Annals of the New York Academy of Sciences*, *902*(1), 230–240.
3. Kim, J.A., Montagnani, M., Koh, K.K. and Quon M.J., 2006. Reciprocal relationships between insulin resistance and endothelial dysfunction: molecular and pathophysiological mechanisms. *Circulation*, *113*, 1888–1904.
4. Wallace, D. and Hahn, B.H., 2012. *Dubois' Lupus Erythematosus and Related Syndromes: Expert Consult-Online*. London: Elsevier Health Sciences.
5. Huang, P.L., 2005. Unraveling the links between diabetes, obesity, and cardiovascular disease. *Circulation Research*, *96*, 1129–1131.
6. Liu, S. and Manson, J.E., 2001. Dietary carbohydrates, physical inactivity, obesity, and the "metabolic syndrome" as predictors of coronary heart disease. *Current Opinion in Lipidology*, *12*(4), 395–404.
7. Kenchaiah, S., Evans, J.C., Levy, D., Wilson, P.W., Benjamin, E.J., Larson, M.G., Kannel, W.B. and Vasan, R.S., 2002. Obesity and the risk of heart failure. *New England Journal of Medicine*, *347*(5), 305–313.
8. Li, R., Wang, W.Q., Zhang, H., Yang, X., Fan, Q., Christopher, T.A., Lopez, B.L. et al., 2007. Adiponectin improves endothelial function in hyperlipidemic rats by reducing oxidative/nitrative stress and differential regulation of eNOS/iNOS activity. *American Journal of Physiology-Endocrinology and Metabolism*, *293*(6), E1703–E1708.
9. Huang, P.L., 2009. eNOS, metabolic syndrome and cardiovascular disease. *Trends in Endocrinology & Metabolism*, *20*(6), 295–302.
10. Shimokawa, H., Yasutake, H., Fujii, K., Owada, M.K., Nakaike, R., Fukumoto, Y., Takayanagi, T. et al., 1996. The importance of the hyperpolarizing mechanism increases as the vessel size decreases in endothelium-dependent relaxations in rat mesenteric circulation. *Journal of Cardiovascular Pharmacology*, *28*(5), 703–711.
11. Hien, T.T., Oh, W.K., Nguyen, P.H., Oh, S.J., Lee, M.Y. and Kang, K.W., 2011. Nectandrin B activates endothelial nitric-oxide synthase phosphorylation in endothelial cells: Role of the AMP-activated protein kinase/estrogen receptor α/phosphatidylinositol 3-kinase/Akt pathway. *Molecular Pharmacology*, *80*(6), 1166–1178.
12. Van Zanten, G.H., De Graaf, S., Slootweg, P.J., Heijnen, H.F., Connolly, T.M., De Groot, P.G. and Sixma, J.J., 1994. Increased platelet deposition on atherosclerotic coronary arteries. *Journal of Clinical Investigation*, *93*(2), 615.
13. Chen, K.S, Wu, Y.C, Teng, C.M., Ko, F.N. and Wu, T.S., 1997. Bioactive alkaloids from *Illigera luzonensis*. *Journal of Natural Products*, *60*(6), 645–647.
14. Kaplan, R., Aynedjian, H.S., Schlondorff, D. and Bank, N., 1990. Renal vasoconstriction caused by short-term cholesterol feeding is corrected by thromboxane antagonist or probucol. *Journal of Clinical Investigation*, *86*(5), 1707.
15. Teng, C.M., Yu, S.M., Ko, F.N., Chen, C.C., Huang, Y.L. and Huang, T.F., 1991. Dicentrine, a natural vascular alpha 1-adrenoceptor antagonist, isolated from *Lindera megaphylla*. *British Journal of Pharmacology*, *104*(3), 651–656.
16. Yu, S.M., Chen, C.C., Ko, F.N., Huang, Y.L., Huang, T.F. and Teng, C.M., 1992. Dicentrine, a novel antiplatelet agent inhibiting thromboxane formation and increasing the cyclic AMP level of rabbit platelets. *Biochemical Pharmacology*, *43*(2), 323–329.
17. Yu, S.M., Kang, Y.F., Chen, C.C. and Teng, C.M., 1993. Effects of dicentrine on haemodynamic, plasma lipid, lipoprotein level and vascular reactivity in hyperlipidaemic rats. *British Journal of Pharmacology*, *108*(4), 1055–1061.
18. Chang, G.J., Wu, M.H., Wu, Y.C. and Su, M.J., 1996. Electrophysiological mechanisms for antiarrhythmic efficacy and positive inotropy of liriodenine, a natural aporphine alkaloid from Fissistigma glaucescens. *British Journal of Pharmacology*, *118*(7), 1571–1583.
19. Chang, W.L., Chung, C.H., Wu, Y.C. and Su, M.J., 2004. The vascular and cardioprotective effects of liriodenine in ischemia-reperfusion injury via NO-dependent pathway. *Nitric Oxide*, *11*(4), 307–315.
20. Lacoste, L., Lam, J.Y., Hung, J., Letchacovski, G., Solymoss, C.B. and Waters, D., 1995. Hyperlipidemia and coronary disease. *Circulation*, *92*(11), 3172–3177.

21. Moharam, B.A., Jantan, I., Jalil, J. and Shaari, K., 2010. Inhibitory effects of hylligenin and quebrachitol isolated from mitrephora vulpina on platelet activating factor receptor binding and platelet aggregation. *Molecules*, *15*(11), 7840–7848.
22. Schenk, S., Saberi, M. and Olefsky, J.M., 2008. Insulin sensitivity: Modulation by nutrients and inflammation. *The Journal of Clinical Investigation*, *118*(9), 2992–3002.
23. Chang, Y.C. and Chuang, L.M., 2010. The role of oxidative stress in the pathogenesis of type 2 diabetes: from molecular mechanism to clinical implication. *American Journal of Translational Research*, *2*(3), 316–331.
24. Gao, Y.J., 2007. Dual modulation of vascular function by perivascular adipose tissue and its potential correlation with adiposity/lipoatrophy-related vascular dysfunction. *Current Pharmaceutical Design*, *13*(21), 2185–2192.
25. Hwa, J.S., Jin, Y.C., Lee, Y.S., Ko, Y.S., Kim, Y.M., Shi, L.Y., Kim, H.J. et al., 2012. 2-Methoxycinnamaldehyde from Cinnamomum cassia reduces rat myocardial ischemia and reperfusion injury in vivo due to HO-1 induction. *Journal of Ethnopharmacology*, *139*(2), 605–615.
26. Beckman, J.A., Ganz, J., Creager, M.A., Ganz, P. and Kinlay, S., 2001. Relationship of clinical presentation and calcification of culprit coronary artery stenoses. *Arteriosclerosis, Thrombosis, and Vascular Biology*, *21*(10), 1618–1622.
27. Lee, J.O., Oak, M.H., Jung, S.H., Park, D.H., Auger, C., Kim, K.R., Lee, S.W. and Schini-Kerth, V.B., 2011. An ethanolic extract of *Lindera obtusiloba* stems causes NO-mediated endothelium-dependent relaxations in rat aortic rings and prevents angiotensin II-induced hypertension and endothelial dysfunction in rats. *Naunyn-Schmiedebergs Archives of Pharmacology*, *383*(6), 635–645.
28. Maruni, N., Offermann, M.K., Swerlick, R., Kunsch, C., Rosen, C.A., Ahmad, M., Alexander, R.W. and Medford, R.M., 1993. Vascular cell adhesion molecule-1 (VCAM-1) gene transcription and expression are regulated through an antioxidant-sensitive mechanism in human vascular endothelial cells. *Journal of Clinical Investigation*, *92*, 1866–1874.
29. Collins, T., Read, M.A., Neish, A.S., Whitley, M.Z., Thanos, D. and Maniatis, T., 1995. Transcriptional regulation of endothelial cell adhesion molecules: Nuclear factor-κB and cytokine-inducible enhamcers. *FASEB Journal*, *9*, 899–909.
30. Li, Z., Wong, A., Henning, S.M., Zhang, Y., Jones, A., Zerlin, A., Thames, G., Bowerman, S., Tseng, C.H. and Heber, D., 2013. Hass avocado modulates postprandial vascular reactivity and postprandial inflammatory responses to a hamburger meal in healthy volunteers. *Food & Function*, *4*(3), 384–391.
31. Bauters, C. and Isner, J.M., 1997. The biology of restenosis. *Progress in Cardiovascular Diseases*, *40*(2), 107–116.
32. Huang, C.-H., Huang, W.-J., Wang, S.-J., Wu, P.-H. and Wu, W.-B., 2008. Litebamine, a phenanthrene alkaloid from the wood of *Litsea cubeba*, inhibits rat smooth muscle cell adhesion and migration on collagen. *European Journal of Pharmacology*, *596*(1–3), 25–31.
33. Fuster, V., Badimon, L., Badimon, J.J. and Chesebro, J.H., 1992. The pathogenesis of coronary artery disease and the acute coronary syndromes. *New England Journal of Medicine*, *326*(4), 242–250.
34. Chen, Y.C., Chen, J.J., Chang, Y.L., Teng, C.M., Lin, W.Y., Wu, C.C. and Chen, I.S., 2004. A new aristolactam alkaloid and anti-platelet aggregation constituents from *Piper taiwanense*. *Planta Medica*, *70*(2), 174–177.
35. Ross, R., 1999. Atherosclerosis—An inflammatory disease. *New England Journal of Medicine*, *340*(2), 115–126.
36. Santanam, N., Penumetcha, M., Speisky, H. and Parthasarathy, S., 2004. A novel alkaloid antioxidant, Boldine and synthetic antioxidant, reduced form of RU486, inhibit the oxidation of LDL *in-vitro* and atherosclerosis *in vivo* in LDLR-/- mice. *Atherosclerosis*, *173*(2), 203–210.
37. Ceriello, A. and Motz, E., 2004. Is oxidative stress the pathogenic mechanism underlying insulin resistance, diabetes, and cardiovascular disease? The common soil hypothesis revisited. *Arteriosclerosis, Thrombosis, and Vascular Biology*, *24*(5), 816–823.
38. Toyoshi, I., Ping, L., Fumio, U., Hai, Y.Y., Maiko, K., Minako, I., Tsuyoshi, A. et al., 2000. High glucose level and free fatty acid stimulate reactive oxygen species production through protein kinase C–dependent activation of NAD(P)H oxidase in cultured vascular cells. *Diabetes*, *49*, 1939–1945.
39. Muniyappa, R., Srinivas, P.R., Ram, J.L., Walsh, M.F. and Sower, J.R., 1998. Calcium and protein kinase C mediate high-glucose-induced inhibition of inducible nitric oxide synthase in vascular smooth muscle cells. *Hypertension*, *31*, 289–295.
40. Lau, Y.S.,Tian, X.Y., Huang, Y., Murugan, D., Achike, F.I., and Mustafa, M.R., 2013. Boldine protects endothelial function in hyperglycemia-induced oxidative stress through an antioxidant mechanism. *Biochemical Pharmacology*, *85*(3), 367–375.

41. Hernández-Salinas, R., Vielma, A.Z., Arismendi, M.N., Boric, M.P., Sáez, J.C. and Velarde, V., 2013. Boldine prevents renal alterations in diabetic rats. *Journal of Diabetes Research, 2013*, 593–672.
42. Zhou, Y.J., Xiang, J.Z., Yuan, H., Liu, H., Tang, Q., Hao, H.Z., Yin, Z., Wang, J. and Ming, Z.Y., 2013. Neferine exerts its antithrombotic effect by inhibiting platelet aggregation and promoting dissociation of platelet aggregates. *Thrombosis Research, 132*(2), 202–210.
43. Acosta, D. (Ed.), 2003. *Cardiovascular Toxicology*. Boca Raton, FL: CRC Press.
44. Lalitha, G., Poornima, P., Archanah, A. and Padma, V.V., 2013. Protective effect of neferine against isoproterenol-induced cardiac toxicity. *Cardiovascular Toxicology, 13*(2), 168–179.
45. Tsai, J.-Y., Chou, C.-J., Chen, C.-F. and Chiou, W.-F., 2003. Antioxidant activity of piperlactam S: Prevention of copper-induced LDL peroxidation and amelioration of free radical-induced oxidative stress of endothelial cells. *Planta Medica, 69*(1), 3–8.
46. Jenkins, P.V., Pasi, K.J. and Perkins, S.J., 1998. Molecular modeling of ligand and mutation sites of the type A domains of human von Willebrand factor and their relevance to von Willebrand's disease. *Blood, 91*(6), 2032–2044.
47. Moncada, S. and Vane, J.R., 1978. Pharmacology and endogenous roles of prostaglandin endoperoxides, thromboxane A2, and prostacyclin. *Pharmacological Reviews, 30*(3), 293–331.
48. Iwashita, M., Saito, M., Yamaguchi, Y., Takagaki, R. and Nakahata, N., 2007. Inhibitory effect of ethanol extract of *Piper longum* L. on rabbit platelet aggregation through antagonizing thromboxane A2 receptor. *Biological and Pharmaceutical Bulletin, 30*(7), 1221–1225.
49. Mehta, J.L., Chen, J., Hermonat, P.L., Romeo, F. and Novelli, G., 2006. Lectin-like, oxidized low-density lipoprotein receptor-1 (LOX-1): A critical player in the development of atherosclerosis and related disorders. *Cardiovascular Research, 69*(1), 36–45.
50. Leon, C., Hill, J.S. and Wasan, K.M., 2005. Potential role of acyl-coenzyme A: Cholesterol transferase (ACAT) inhibitors as hypolipidemic and antiatherosclerosis drugs. *Pharmaceutical Research, 22*(10), 1578–1588.
51. Matsuda, D., Ohte, S., Ohshiro, T., Jiang, W., Rudel, L., Hong, B., Si, S. and Tomoda, H., 2008. Molecular target of piperine in the inhibition of lipid droplet accumulation in macrophages. *Biological and Pharmaceutical Bulletin, 31*(6), 1063–1066.
52. Taqvi, S.I., Shah, A.J. and Gilani, A.H., 2008. Blood pressure lowering and vasomodulator effects of piperine. *Journal of Cardiovascular Pharmacology, 52*(5), 452–458.
53. Son, D.J., Kim, S.Y., Han, S.S., Kim, C.W., Kumar, S., Park, B.S., Lee, S.E., Yun, Y.P., Jo, H. and Park, Y.H., 2012. Piperlongumine inhibits atherosclerotic plaque formation and vascular smooth muscle cell proliferation by suppressing PDGF receptor signaling. *Biochemical and Biophysical Research Communications, 427*(2), 349–354.
54. Rink, T.J. and Sage, S.O., 1990. Calcium signaling in human platelets. *Annual Review of Physiology, 52*(1), 431–449.
55. Kometani, M., Kanaho, Y., Sato, T. and Fujii, T., 1985. Inhibitory effect of cepharanthine on collagen-induced activation in rabbit platelets. *European Journal of Pharmacology, 111*(1), 97–105.
56. Kim, H.S., Zhang, Y.H., Fang, L.H., Yun, Y.P. and Lee, H.K., 1999. Effects of tetrandrine and fangchinoline on human platelet aggregation and thromboxane B2 formation. *Journal of Ethnopharmacology, 66*(2), 241–246.
57. Yu, X.C., Wu, S., Wang, G.Y., Shan, J., Wong, T.M., Chen, C.F. and Pang, K.T., 2001. Cardiac effects of the extract and active components of radix stephaniae tetrandrae. II. Myocardial infarct, arrhythmias, coronary arterial flow and heart rate in the isolated perfused rat heart. *Life Sciences, 68*(25), 2863–2872.
58. Yu, X.C., Wu, S., Chen, C.F., Pang, K.T. and Wong, T.M., 2004. Antihypertensive and anti-arrhythmic effects of an extract of Radix Stephaniae Tetrandrae in the rat. *Journal of Pharmacy and Pharmacology, 56*(1), 115–122.
59. Zhang, Y.H., Fang, L.H. and Ku, B.S., 2003. Fangchinoline inhibits rat aortic vascular smooth muscle cell proliferation and cell cycle progression through inhibition of ERK1/2 activation and c-fos expression. *Biochemical Pharmacology, 66*(9), 1853–1860.
60. Liang, X.C., Hagino, N., Guo, S.S., Tsutsumi, T. and Kobayashi, S., 2002. Therapeutic efficacy of *Stephania tetrandra* S. Moore for treatment of neovascularization of retinal capillary (retinopathy) in diabetes--in vitro study. *Phytomedicine, 9*(5), 377–384.
61. Zhang, J., Yu, B., Zhang, X.Q., Sheng, Z.F., Li, S.J., Wang, Z.J., Cui, X.Y., Cui, S.Y. and Zhang, Y.H., 2014. Tetrandrine, an antihypertensive alkaloid, improves the sleep state of spontaneously hypertensive rats (SHRs). *Journal of Ethnopharmacology, 151*(1), 729–732.
62. Kearney, P.M., Whelton, M., Reynolds, K., Muntner, P., Whelton, P.K. and He, J., 2005. Global burden of hypertension: Analysis ofworlwide data. *Lancet, 365*, 217–223.

63. Lee, P.Y., Chen, W., Liu, I.M. and Cheng, J.T., 2007. Vasodilatation induced by sinomenine lowers blood pressure in spontaneously hypertensive rats. *Clinical and Experimental Pharmacology and Physiology*, *34*(10), 979–984.
64. Wesselman, J.P., VanBavel, E., Pfaffendorf, M. and Spaan, J.A., 1996. Voltage-operated Ca2+ channels are essential for the myogenic responsiveness of cannulated rat mesenteric small arteries. *Journal of Vascular Research*, *33*, 32–41.
65. Zhu, L., Hao, Y., Guan, H., Cui, C., Tian, S., Yang, D., Wang, X., Zhang, S., Wang, L. and Jiang, H., 2013. Effect of sinomenine on vascular smooth muscle cell dedifferentiation and neointima formation after vascular injury in mice. *Molecular and Cellular Biochemistry*, *373*(1–2), 53–62.
66. Mayeda, H., 1953. The release of histamine by sinomenine. *The Japanese Journal of Pharmacology*, *3*(1), 62–72.
67. Wang, M.H., Chang, C.K., Cheng, J.H., Wu, H.T., Li, Y.X. and Cheng, J.T., 2008. Activation of opioid μ-receptor by sinomenine in cell and mice. *Neuroscience Letters*, *443*(3), 209–212.
68. Calhoun, D.A., Jones, D., Textor, S., Goff, D.C., Murphy, T.P., Toto, R.D., White, A. et al., 2008. Resistant hypertension: Diagnosis, evaluation, and treatment. *Circulation*, *117*(25), e510–e526.
69. Badole, S.L., Bodhankar, S.L., Patel, N.M. and Bhardwaj, S., 2009. Acute and chronic diuretic effect of ethanolic extract of leaves of *Cocculus hirsutus* (L.) Diles in normal rats. *Journal of Pharmacy and Pharmacology*, *61*(3), 387–393.
70. Na, K.Y., Oh, Y.K., Han, J.S., Joo, K.W., Lee, J.S., Earm, J.H., Knepper, M.A. and Kim, G.H., 2003. Upregulation of Na+ transporter abundances in response to chronic thiazide or loop diuretic treatment in rats. *American Journal of Physiology-Renal Physiology*, *284*(1), F133–F143.
71. Osorio, F.V. and Teitelbaum, I., 1996. Mechanisms of defective hydroosmotic response in chronic renal failure. *Journal of Nephrology*, *10*(5), 232–237.
72. Chun, Y.T., Yip, T.T., Lau, K.L., Kong, Y.C. and Sankawa, U., 1979. A biochemical study on the hypotensive effect of berberine in rats. *General Pharmacology*, *10*(3), 177–182.
73. Huang, C.G., Chu, Z.L., Wei, S.J., Jiang, H. and Jiao, B.H., 2002. Effect of berberine on arachidonic acid metabolism in rabbit platelets and endothelial cells. *Thrombosis Research*, *106*(4–5), 223–227.
74. Liu, W., Liu, P., Tao, S., Deng, Y., Li, X., Lan, T., Zhang, X. et al., 2008. Berberine inhibits aldose reductase and oxidative stress in rat mesangial cells cultured under high glucose. *Archives of Biochemistry and Biophysics*, *475*(2), 128–134.
75. Li, K., Yao, W., Zheng, X. and Liao, K., 2009. Berberine promotes the development of atherosclerosis and foam cell formation by inducing scavenger receptor A expression in macrophage. *Cell Research*, *19*(8), 1006–1017.
76. Lee, T.S., Pan, C.C., Peng, C.C., Kou, Y.R., Chen, C.Y., Ching, L.C., Tsai, T.H., Chen, S.F., Lyu, P.C. and Shyue, S.K., 2010. Anti-atherogenic effect of berberine on LXRα-ABCA1-dependent cholesterol efflux in macrophages. *Journal of Cellular Biochemistry*, *111*(1), 104–110.
77. Oram, J.F. and Lawn, R.M., 2001. ABCA1: The gatekeeper for eliminating excess tissue cholesterol. *Journal of Lipid Research*, *42*(8), 1173–1179.
78. Wang, Q., Zhang, M., Liang, B., Shirwany, N., Zhu, Y. and Zou, M.H., 2011. Activation of AMP-activated protein kinase is required for berberine-induced reduction of atherosclerosis in mice: The role of uncoupling protein 2. *PLoS ONE*, *6*(9), e25436.
79. Liu, J., Xiu, J., Cao, J., Gao, Q., Ma, D. and Fu, L., 2011. Berberine cooperates with adrenal androgen dehydroepiandrosterone sulfate to attenuate PDGF-induced proliferation of vascular smooth muscle cell A7r5 through Skp2 signaling pathway. *Molecular and Cellular Biochemistry*, *355*(1–2), 127–134.
80. Thygesen, K., Alpert, J.S., Jaffe, A.S., Simoons, M.L., Chaitman, B.R. and White, H.D., 2012. Third universal definition of myocardial infarction. *Circulation*, CIR-0b013e31826e1058.
81. Zhang, C.M., Gao, L., Zheng, Y.J. and Yang, H.T., 2012. Berbamine protects the heart from ischemia/reperfusion injury by maintaining cytosolic Ca2+ homeostasis and preventing calpain activation. *Circulation Journal*, *76*(8), 1993–2002.
82. Si, K., Liu, J., He, L., Li, X., Gou, W., Liu, C. and Li, X., 2010. Caulophine protects cardiomyocytes from oxidative and ischemic injury. *Journal of Pharmacological Sciences*, *113*(4), 368–377.
83. Si, K.W., Liu, J.T., He, L.C., Li, X.K., Gou, W., Liu, C.H. and Li, X.Q., 2010. Effects of caulophine on caffeine-induced cellular injury and calcium homeostasis in rat cardiomyocytes. *Basic & Clinical Pharmacology & Toxicology*, *107*(6), 976–981.
84. Hung, T.M., Lee, J.P., Min, B.S., Choi, J.S., Na, M., Zhang, X., Ngoc, T.M., Lee, I. and Bae, K., 2007. Magnoflorine from Coptidis Rhizoma protects high density lipoprotein during oxidant stress. *Biological and Pharmaceutical Bulletin*, *30*(6), 1157–1160.

85. Jung, H.A., Yoon, N.Y., Bae, H.J., Min, B.-S. and Choi, J.S., 2008. Inhibitory activities of the alkaloids from Coptidis Rhizoma against aldose reductase. *Archives of Pharmacal Research, 31*(11), 1405–1412.
86. Kim, Y.M., Ha, Y.M., Jin, Y.C., Shi, L.Y., Lee, Y.S., Kim, H.J., Seo, H.G. et al., 2009. Palmatine from Coptidis rhizoma reduces ischemia-reperfusion-mediated acute myocardial injury in the rat. *Food and Chemical Toxicology, 47*(8), 2097–2102.
87. Gong, L.L., Fang, L.H., Wang, S.B., Sun, J.L., Qin, H.L., Li, X.X., Wang, S.B. and Du, G.H., 2012. Coptisine exert cardioprotective effect through anti-oxidative and inhibition of RhoA/Rho kinase pathway on isoproterenol-induced myocardial infarction in rats. *Atherosclerosis, 222*(1), 50–58.
88. Garg, U.C. and Hassid, A., 1989. Nitric oxide-generating vasodilators and 8-bromo-cyclic guanosine monophosphate inhibit mitogenesis and proliferation of cultured rat vascular smooth muscle cells. *Journal of Clinical Investigation, 83*(5), 1774.
89. Hung, L.-M., Lee, S.-S., Chen, J.-K., Huang, S.-S. and Su, M.-J., 2001. Thaliporphine protects ischaemic and ischaemic-reperfused rat hearts via an NO-dependent mechanism. *Drug Development Research, 52*, 446–453.
90. Kubo, M., Matsuda, H., Tokuoka, K., Kobayashi, Y., Ma, S. and Tanaka, T., 1994. Studies of anti-cataract drugs from natural sources. I. Effects of a methanolic extract and the alkaloidal components from *Corydalis* tuber on *in vitro* aldose reductase activity. *Biological and Pharmaceutical Bulletin, 17*(3), 458–459.
91. Chueh, F.Y., Hsieh, M.T, Chen, C.F. and Lin, M.T., 1995. Hypotensive and bradycardic effects of dl-tetrahydropalmatine mediated by decrease in hypothalamic serotonin release in the rat. *Japanese Journal of Pharmacology, 69*(2), 177–180.
92. Carey, F., Menashi, S. and Crawford, N., 1982. Localization of cyclo-oxygenase and thromboxane synthetase in human platelet intracellular membranes. *Biochemical Journal, 204*(3), 847–851.
93. Xuan, B., Wang, W. and Li, D.X., 1994. Inhibitory effect of tetrahydroberberine on platelet aggregation and thrombosis. *Zhongguo Yao Li Xue Bao, 15*(2), 133–135.
94. Jeng, J.H., Wu, H.L., Lin, B.R., Lan, W.H., Chang, H.H., Ho, Y.S., Lee, P.H. et al., 2007. Antiplatelet effect of sanguinarine is correlated to calcium mobilization, thromboxane and cAMP production. *Atherosclerosis, 191*(2), 250–258.
95. Lee, B., Lee, S.-J., Park, S.-S., Kim, S.-K., Kim, S.-R., Jung, J.-H., Kim, W.-J. and Moon, S.-K., 2008. Sanguinarine-induced G1-phase arrest of the cell cycle results from increased p27KIP1 expression mediated via activation of the Ras/ERK signaling pathway in vascular smooth muscle cells. *Archives of Biochemistry and Biophysics, 471*(2), 224–231.
96. Zhang, W.B., Wang, M., Zhou, B.Q., Zhu, J.H. and Fu, G.S., 2009. The effect of chelerythrine on the hypertrophy of cardiac myocytes of neonatal rats induced by different glucose levels and its mechanism. *Yao Xue Xue Bao, 44*(2), 115–120.
97. Ip, J.H., Fuster, V., Badimon, L., Badimon, J., Taubman, M.B. and Chesebro, J.H., 1990. Syndromes of accelerated atherosclerosis: Role of vascular injury and smooth muscle cell proliferation. *Journal of the American College of Cardiology, 15*(7), 1667–1687.
98. Saeed, S.A., Gilani, A.H., Majoo, R.U. and Shah, B.H., 1997. Anti-thrombotic and anti-inflammatory activities of protopine. *Pharmacological Research, 36*(1), 1–7.
99. Friebe, A. and Koesling, D., 2003. Regulation of nitric oxide-sensitive guanylyl cyclase. *Circulation Research, 93*(2), 96–105.
100. Li, H., Dai, M. and Jia, W., 2009. Paeonol attenuates high-fat-diet-induced atherosclerosis in rabbits by anti-inflammatory activity. *Planta Medica, 75*(1), 7–11.
101. Anderson, K.J., Teuber, S.S., Gobeille, A., Cremin, P., Waterhouse, A.L. and Steinberg, F.M., 2001. Walnut polyphenolics inhibit in vitro human plasma and LDL oxidation. *Journal of Nutrition, 131*(11), 2837–2842.
102. Mancia, G., De Backer, G., Dominiczak, A., Cifkova, R., Fagard, R., Germano, G., Grassi, G. et al., 2007. 2007 Guidelines for the management of arterial hypertension. *European Heart Journal, 28*(12), 1462–1536.
103. Katz, D.L., Davidhi, A., Ma, Y., Kavak, Y., Bifulco, L. and Njike, V.Y., 2012. Effects of walnuts on endothelial function in overweight adults with visceral obesity: a randomized, controlled, crossover trial. *Journal of the American College of Nutrition, 31*(6), 415–423.
104. Nergiz-Ünal, R., Kuijpers, M.J., de Witt, S.M., Heeneman, S., Feijge, M.A., Garcia Caraballo, S.C., Biessen, E.A., Haenen, G.R., Cosemans, J.M. and Heemskerk, J.W., 2013. Atheroprotective effect of dietary walnut intake in ApoE-deficient mice: involvement of lipids and coagulation factors. *Thrombosis Research, 131*(5), 411–417.

105. Saravanan, G., Ponmurugan, P., Sathiyavathi, M., Vadivukkarasi, S. and Sengottuvelu, S., 2013. Cardioprotective activity of *Amaranthus viridis* Linn: Effect on serum marker enzymes, cardiac troponin and antioxidant system in experimental myocardial infarcted rats. *International Journal of Cardiology, 165*(3), 494–498.
106. Colwell, J.A., 1997. Pharmacological strategies to prevent macrovascular disease in NIDDM. *Diabetes, 46*(Supplement 2), S131–S134.
107. Relou, I.A., Hackeng, C.M., Akkerman, J.W. and Malle, E., 2003. Low-density lipoprotein and its effect on human blood platelets. *Cellular and Molecular Life Sciences, 60*(5), 961–971.
108. Aburjai, T.A., 2000. Anti-platelet stilbenes from aerial parts of *Rheum palaestinum. Phytochemistry, 55*(5), 407–410.
109. Ko, S.K., Lee, S.M. and Whang, W.K., 1999. Anti-platelet aggregation activity of stilbene derivatives from *Rheum undulatum. Archives of Pharmacal Research, 22*(4), 401–403.
110. Yoo, M.Y., Oh, K.S., Lee, J.W., Seo, H.W., Yon, G.H., Kwon, D.Y., Kim, Y.S., Ryu, S.Y. and Lee, B.H., 2007. Vasorelaxant effect of stilbenes from rhizome extract of rhubarb (*Rheum undulatum*) on the contractility of rat aorta. *Phytotherapy Research, 21*(2), 186–189.
111. Gao, Q., Qin, W.S., Jia, Z.H., Zheng, J.M., Zeng, C.H., Li, L.S. and Liu, Z.H., 2010, Rhein improves renal lesion and ameliorates dyslipidemia in db/db mice with diabetic nephropathy. *Planta Medica, 76*(1), 27–33.
112. Tzeng, T.F., Lu, H.J., Liou, S.S., Chang, C.J. and Liu, I.M., 2012. Emodin protects against high-fat diet-induced obesity via regulation of AMP-activated protein kinase pathways in white adipose tissue. *Planta Medica, 78*(10), 943–950.
113. Yang, Y., Yan, Y.M., Wei, W., Luo, J., Zhang, L.S., Zhou, X.J., Wang, P.C., Yang, Y.X. and Cheng, Y.X., 2013. Anthraquinone derivatives from Rumex plants and endophytic Aspergillus fumigatus and their effects on diabetic nephropathy. *Bioorganic & Medicinal Chemistry Letters, 23*(13), 3905–3909.
114. Haeng Park, S., Sung, Y.Y., Jin Nho, K. and Kyoung Kim, H., 2014. Anti-atherosclerotic effects of *Polygonum aviculare* L. ethanol extract in ApoE knock-out mice fed a Western diet mediated via the MAPK pathway. *Journal of Ethnopharmacology, 151*(3), 1109–1115.
115. André, P., Delaney, S.M., LaRocca, T., Vincent, D., DeGuzman, F., Jurek, M., Koller, B., Phillips, D.R. and Conley, P.B., 2003. P2Y 12 regulates platelet adhesion/activation, thrombus growth, and thrombus stability in injured arteries. *The Journal of Clinical Investigation, 112*(3), 398–406.
116. Radomski, M.W., Palmer, R.M.J. and Moncada, S., 1987. Comparative pharmacology of endothelium-derived relaxing factor, nitric oxide and prostacyclin in platelets. *British Journal of Pharmacology, 92*(1), 181–187.
117. Lam, S.C., Guccione, M.A., Packham, M.A. and Mustard, J.F., 1982. Effect of cAMP phosphodiesterase inhibitors on ADP-induced shape change, cAMP and nucleoside diphosphokinase activity of rabbit platelets. *Thrombosis and Haemostasis, 47*(2), 90–95.
118. Chang, W.C. and Hsu, F.L., 1991. Inhibition of platelet activation and endothelial cell injury by flavan-3-ol and saikosaponin compounds. *Prostaglandins Leukot Essent Fatty Acids, 44*(1), 51–56.
119. Chennasamudram, S.P., Kudugunti, S., Boreddy, P.R., Moridani, M.Y. and Vasylyeva, T.L., 2012. Renoprotective effects of (+)-catechin in streptozotocin-induced diabetic rat model. *Nutrition Research, 32*(5), 347–356.
120. Khan, G., Haque, S.E., Anwer, T., Ahsan, M.N., Safhi, M.M. and Alam, M.F., 2014. Cardioprotective effect of green tea extract on doxorubicin-induced cardiotoxicity in rats. *Acta Poloniae Pharmaceutica, 71*(5), 861–868.
121. Xiang, M., Wang, J., Zhang, Y., Ling, J. and Xu, X., 2012. Attenuation of aortic injury by ursolic acid through RAGE-Nox-Nuclear factor-κB pathway in streptozotocin-induced diabetic rats. *Archives of Pharmacal Research, 35*(5), 877–886.
122. Vikrant, S. and Tiwari, S.C., 2001. Essential hypertension–pathogenesis and pathophysiology. *Journal, Indian Academy of Clinical Medicine, 2*(3), 140–161.
123. Ogunniyi, A. and Talabi, O., 2000. Cerebrovascular complications of hypertension. *Nigerian Journal of Medicine: Journal of the National Association of Resident Doctors of Nigeria, 10*(4), 158–161.
124. Afkir, S., Nguelefack, T., B., Aziz, M., Zoheir, J., Cuisinaud, G., Bnouham, M., Mekhfi, H., Legssyer, A., Lahlou, S. and Ziyyat, A., 2008. *Arbutus unedo* prevents cardiovascular and morphological alterations in L-NAME-induced hypertensive rats Part I: Cardiovascular and renal hemodynamic effects of *Arbutus unedo* in L-NAME-induced hypertensive rats. *Journal of Ethnopharmacology, 116*(2), 288–295.
125. Coughlin, S.R., 2000. Thrombin signalling and protease-activated receptors. *Nature, 407*(6801), 258–264.
126. El Haouari, M., López, J.J., Mekhfi, H., Rosado, J.A. and Salido, G.M., 2007. Antiaggregant effects of *Arbutus unedo* extracts in human platelets. *Journal of Ethnopharmacology, 113*(2), 325–331.

127. Elks, C.M., Reed, S.D., Mariappan, N., Shukitt-Hale, B., Joseph, J.A., Ingram, D.K. and Francis, J., 2011. A blueberry-enriched diet attenuates nephropathy in a rat model of hypertension via reduction in oxidative stress. *PLoS One*, *6*(9), e24028.
128. Kawakami, K., Aketa, S., Sakai, H., Watanabe, Y., Nishida, H. and Hirayama, M., 2011. Antihypertensive and vasorelaxant effects of water-soluble proanthocyanidins from persimmon leaf tea in spontaneously hypertensive rats. *Bioscience, Biotechnology, and Biochemistry*, *75*(8), 1435–1439.
129. Sahu, B.D., Anubolu, H., Koneru, M., Kumar, J.M., Kuncha, M., Rachamalla, S.S. and Sistla, R., 2014. Cardioprotective effect of embelin on isoproterenol-induced myocardial injury in rats: possible involvement of mitochondrial dysfunction and apoptosis. *Life Sciences*, *107*(1–2), 59–67.
130. Koch, A.E., Kunkel, S.L., Pearce, W.H., Shah, M.R., Parikh, D., Evanoff, H.L., Haines, G.K., Burdick, M.D. and Strieter, R.M., 1993. Enhanced production of the chemotactic cytokines interleukin-8 and monocyte chemoattractant protein-1 in human abdominal aortic aneurysms. *The American Journal of Pathology*, *142*(5), 1423.
131. Moon, M.K., Kang, D.G., Lee, Y.J., Kim, J.S. and Lee, H.S., 2009. Effect of *Benincasa hispida* Cogniaux on high glucose-induced vascular inflammation of human umbilical vein endothelial cells. *Vascular Pharmacology*, *50*(3–4), 116–122.
132. Zieman, S.J., Melenovsky, V. and Kass, D.A., 2005. Mechanisms, pathophysiology, and therapy of arterial stiffness. *Arteriosclerosis, Thrombosis, and Vascular Biology*, *25*(5), 932–943.
133. Veeramani, C., Al-Numair, K.S., Chandramohan, G., Alsaif, M.A., Alhamdan, A.A. and Pugalendi, K.V., 2012. Antihypertensive effect of *Melothria maderaspatana* leaf fractions on DOCA-salt-induced hypertensive rats and identification of compounds by GC-MS analysis. *Journal of Natural Medicines*, *66*(2), 302–310.
134. Sudhahar, V., Kumar, S.A., Sudharsan, P.T. and Varalakshmi, P., 2007. Protective effect of lupeol and its ester on cardiac abnormalities in experimental hypercholesterolemia. *Vascular Pharmacology*, *46*(6), 412–418.
135. Maghrani, M., Zeggwagh, N.A., Michel, J.B. and Eddouks, M., 2005. Antihypertensive effect of *Lepidium sativum* L. in spontaneously hypertensive rats. *Journal of Ethnopharmacology*, *100*(1–2), 193–197.
136. Eddouks, M. and Maghrani, M., 2008. Effect of *Lepidium sativum* L. on renal glucose reabsorption and urinary TGF-beta 1 levels in diabetic rats. *Phytotherapy Research*, *22*(1), 1–5.
137. Zhou, L., An, X.F., Teng, S.C., Liu, J.S., Shang, W.B., Zhang, A.H., Yuan, Y.G. and Yu, J.Y., 2012. Pretreatment with the total flavone glycosides of Flos Abelmoschus manihot and hyperoside prevents glomerular podocyte apoptosis in streptozotocin-induced diabetic nephropathy. *Journal of Medicinal Food*, *15*(5), 461–468.
138. Hoshi, S., Shu, Y., Yoshida, F., Inagaki, T., Sonoda, J., Watanabe, T., Konodo, K. and Nagata, M., 2002. Podocyte injury promotes progressive nephropathy in zucker diabetic fatty rats. *Laboratory Investigation*, *82*, 25–35.
139. Shibata, Y., Kume, N., Arai, H, Hayashida, K., Inui-Hayashida, A., Minami, M., Mukai, E. et al., 2007. Mulberry leaf aqueous fractions inhibit TNF-alpha-induced nuclear factor kappaB (NF-kappaB) activation and lectin-like oxidized LDL receptor-1 (LOX-1) expression in vascular endothelial cells. *Atherosclerosis*, *193*(1), 20–27.
140. Lee, J.J., Yang, H., Yoo, Y.M., Hong, S.S., Lee, D., Lee, H.J., Lee, H.J., Myung, C.S., Choi, K.C. and Jeung, E.B., 2012. Morusinol extracted from *Morus alba* inhibits arterial thrombosis and modulates platelet activation for the treatment of cardiovascular disease. *Journal of Atherosclerosis and Thrombosis*, *19*(6), 516–522.
141. Ko, H.H., Yu, S.M., Ko, F.N., Teng, C.M. and Lin, C.N., 1997. Bioactive constituents of *Morus australis* and *Broussonetia papyrifera*. *Journal of Natural Products*, *60*(10), 1008–1011.
142. Park, K.H., Park, Y.D., Han, J.M., Im, K.R., Lee, B.W., Jeong, I.Y., Jeong, T.S. and Lee, W.S., 2006. Anti-atherosclerotic and anti-inflammatory activities of catecholic xanthones and flavonoids isolated from *Cudrania tricuspidata*. *Bioorganic & Medicinal Chemistry Letters*, *16*(21), 5580–5583.
143. Han, H.J., Kim, T.J., Jin, Y.R., Hong, S.S., Hwang, J.H., Hwang, B.Y., Lee, K.H., Park, T.K. and Yun, Y.P., 2007. Cudraflavanone A, a flavonoid isolated from the root bark of *Cudrania tricuspidata*, inhibits vascular smooth muscle cell growth via an Akt-dependent pathway. *Planta Medica*, *73*(11), 1163–1168.
144. Houston, M.C., 1989. New insights and new approaches for the treatment of essential hypertension: selection of therapy based on coronary heart disease risk factor analysis, hemodynamic profiles, quality of life, and subsets of hypertension. *American Heart Journal*, *117*(4), 911–951.
145. Higashino, H. and Suzuki, A., 2005. Hypotensive effects of the Siamese plants on SHRSP: *Imperata cylindrica* and *Phyllanthus emblica* extracts. *Japanese Heart Journal*, *34*(4), 519.

146. Usharani, P., Fatima, N. and Muralidhar, N., 2013. Effects of *Phyllanthus emblica* extract on endothelial dysfunction and biomarkers of oxidative stress in patients with type 2 diabetes mellitus: A randomized, double-blind, controlled study. *Diabetes, Metabolic Syndrome and Obesity*, *6*, 275–284.
147. Khanna, S., Das, A., Spieldenner, J., Rink, C. and Roy, S., 2015. Supplementation of a standardized extract from *Phyllanthus emblica* improves cardiovascular risk factors andplatelet aggregation in overweight/class-1 obese adults. *Journal of Medicinal Food*, *18*(4), 415–420.
148. Kumar, S., Rashmi and Kumar, D., 2010. Evaluation of antidiabetic activity of *Euphorbia hirta* L. in streptozocin induced diabetic mice. *Indian Journal of Natural Products and Resources*, *1*(2), 200–203.
149. Abidov, M., Crendal, F., Grachev, S., Seifulla, R. and Ziegenfuss, T., 2003. Effect of extracts from *Rhodiola rosea* and *Rhodiola crenulata* (Crassulaceae) roots on ATP content in mitochondria of skeletal muscles. *Bulletin of Experimental Biology and Medicine*, *136*(6), 585–587.
150. Lee, W.J., Chung, H.H., Cheng, Y.Z., Lin, H.J. and Cheng, J.T., 2012. Rhodiola-water extract induces β-endorphin secretion to lower blood pressure in spontaneously hypertensive rats. *Phytotherapy Research*, *27*(10), 1543–1547.
151. Nosadini, R., Velussi, M., Brocco, E., Bruseghin, M., Abaterusso, C., Saller, A., Dalla Vestra, M. et al., 2000. Course of renal function in type 2 diabetic patients with abnormalities of albumin excretion rate. *Diabetes*, *49*(3), 476–484.
152. Wang, Z.S., Gao, F., and Lu, F.E., 2013. Effect of ethanol extract of *Rhodiola rosea* on the early nephropathy in type 2 diabetic rats. *Journal of Huazhong University of Science and Technology* [*Medical Sciences*], *33*(3), 375–378.
153. Xing, S., Yang, X., Li, W., Bian, F., Wu, D., Chi, J., Xu, G., Zhang, Y, and Jin, S., 2014. Salidroside stimulates mitochondrial biogenesis and protects against H_2O_2-induced endothelial dysfunction. *Oxidative Medicine and Cellular Longevity*, 1–13.
154. Mah, E. and Bruno, R.S., 2012. Postprandial hyperglycemia on vascular endothelial function: Mechanisms and consequences. *Nutrition Research*, *32*(10), 727–740.
155. Khaliq, F., Parveen, A., Singh, S., Gondal, R., Hussain, M.E. and Fahim, M., 2013. Improvement in myocardial function by *Terminalia arjuna* in streptozotocin-induced diabetic rats: Possible mechanisms. *Journal of Cardiovascular Pharmacology and Therpeutics*, *18*(5), 481–489.
156. Tom, E.N., Girard-Thernier, C., Martin, H., Dimo, T., Alvergnas, M., Nappey, M., Berthelot, A. and Demougeot, C., 2014. Treatment with an extract of *Terminalia superba* Engler & Diels decreases blood pressure and improves endothelial function in spontaneously hypertensive rats. *Journal of Ethnopharmacology*, *151*(1), 372–379.
157. Jadeja, R.N., Thounaojam, M.C., Patel, D.K., Devkar, R.V. and Ramachandran, A.V., 2010. Pomegranate (*Punica granatum* L.) juice supplementation attenuates isoproterenol-induced cardiac necrosis in rats. *Cardiovascular Toxicology*, *10*(3), 174–180.
158. Mohan, M., Waghulde, H. and Kasture, S., 2010. Effect of pomegranate juice on Angiotensin II-induced hypertension in diabetic Wistar rats. *Phytotherapy Research*, 24 Suppl 2:S196–S203.
159. Yılmaz, B. and Usta, C., 2013. Ellagic acid-induced endothelium-dependent and endothelium-independent vasorelaxation in rat thoracic aortic rings and the underlying mechanism. *Phytotherapy Research*, *27*(2), 285–289.
160. Safi, S.Z., Qvist, R., Kumar, S., Batumalaie, K. and Ismail, I.S.B., 2014. Molecular mechanisms of diabetic retinopathy, general preventive strategies, and novel therapeutic targets. *BioMed Research International*, 1–18.
161. Tomlison, D.R., Stevens, E.J. and Diemel, L.T., 1994. Aldose reductase inhibitors and their potential for the treatment of diabetes complications. *Trends in Pharmacological Sciences*, *15*, 292–298.
162. Rao, A.R., Veeresham, C. and Asres, K., 2013. In vitro and in vivo inhibitory activities of four Indian medicinal plant extracts and their major components on rat aldose reductase and generation of advanced glycation endproducts. *Phytotherapy Research*, *27*(5), 753–760.
163. Manonmani, G., Bhavapriya, V., Kalpana, S., Govindasamy, S. and Apparanantham, T., 2005. Antioxidant activity of *Cassia fistula* (Linn.) flowers in alloxan induced diabetic rats. *Journal of Ethnopharmacology*, *97*(1), 39–42.
164. Kurian, G.A., Philip, S. and Varghese, T., 2005. Effect of aqueous extract of the *Desmodium gangeticum* DC root in the severity of myocardial infarction. *Journal of Ethnopharmacology*, *97*(3), 457–461.
165. Li, X., Zhou, R., Zheng, P., Yan, L., Wu, Y., Xiao, X. and Dai, G., 2010. Cardioprotective effect of matrine on isoproterenol-induced cardiotoxicity in rats. *Journal of Pharmacy and Pharmacology*, *62*(4), 514–520.
166. Hong-Li, S., Lei, L., Lei, S., Dan, Z., De-Li, D., Guo-Fen, Q., Yan, L., Wen-Feng, C. and Bao-Feng, Y., 2008. Cardioprotective effects and underlying mechanisms of oxymatrine against Ischemic myocardial injuries of rats. *Phytotherapy Research*, *22*(7), 985–989.

167. Cao, Y.G., Jing, S., Li, L., Gao, J.Q., Shen, Z.Y., Liu, Y., Xing, Y. et al., 2010. Antiarrhythmic effects and ionic mechanisms of oxymatrine from *Sophora flavescens*. *Phytotherapy Research*, *24*(12), 1844–1849.
168. Huang, X.Y. and Chen, C.X., 2013. Effect of oxymatrine, the active component from Radix *Sophorae flavescentis* (Kushen), on ventricular remodeling in spontaneously hypertensive rats. *Phytomedicine*, *20*(3–4), 202–212.
169. Martinello, F., Soares, S.M., Franco, J.J., Santos, A.C., Sugohara, A., Garcia, S.B., Curti, C. and Uyemura, S.A., 2006. Hypolipemic and antioxidant activities from *Tamarindus indica* L. pulp fruit extract in hypercholesterolemic hamsters. *Food and Chemical Toxicology*, *44*(6), 810–818.
170. Panda, S., Biswas, S. and Kar, A., 2013. Trigonelline isolated from fenugreek seed protects against isoproterenol-induced myocardial injury through down-regulation of Hsp27 and aB-crystallin. *Nutrition*, *29*(11–12), 1395–1403.
171. Calisti, L. and Tognetti, S., 2004. Measure of glycosylated hemoglobin. *Acta Bio-medica: Atenei Parmensis*, *76*, 59–62.
172. Winegrad, A.I., 1987. Banting lecture 1986: Does a common mechanism induce the diverse complications of diabetes? *Diabetes*, *36*(3), 396–406.
173. Gupta, S.K., Kumar, B., Nag, T.C., Srinivasan, B.P., Srivastava, S., Gaur, S. and Saxena, R., 2014. Effects of *Trigonella foenum-graecum* (L.) on retinal oxidative stress, and proinflammatory and angiogenic molecular biomarkers in streptozotocin-induced diabetic rats. *Molecular and Cellular Biochemistry*, *388*(1–2), 1–9.
174. Brownlee, M., 2001. Biochemistry and molecular cell biology of diabetic complications. *Nature*, *414*(6865), 813–820.
175. Leu, Y.-L., Chan, Y.-Y., Wu, T.-S., Teng, C.-M. and Chen, K.-T., 1998. Antiplatelet aggregation principles from *Glycosmis citrifolia*. *Phytotherapy Research*, *12*(Suppl. 1), S77–S79.
176. Leehey, D.J., Singh, A.K., Alavi, N. and Singh, R., 2000. Role of angiotensin II in diabetic nephropathy. *Kidney International*, *58*, S93–S98.
177. Lewis, E.J., Hunsicker, L.G., Bain, R.P. and Rohde, R.D., 1993. The effect of angiotensin-converting-enzyme inhibition on diabetic nephropathy. *New England Journal of Medicine*, *329*(20), 1456–1462.
178. Zou, J., Yu, X., Qu, S., Li, X., Jin, Y. and Sui, D., 2014. Protective effect of total flavonoids extracted from the leaves of *Murraya paniculata* (L.) Jack on diabetic nephropathy in rats. *Food and Chemical Toxicology*, *64*, 231–237.
179. Cassidy, A., Rimm, E.B., O'Reilly, É.J., Logroscino, G., Kay, C., Chiuve, S.E. and Rexrode, K.M., 2012. Dietary flavonoids and risk of stroke in women. *Stroke*, *43*(4), 946–951.
180. Hu, D., Huang, J., Wang, Y., Zhang, D. and Qu, Y., 2014. Fruits and vegetables consumption and risk of stroke a meta-analysis of prospective cohort studies. *Stroke*, *45*(6), 1613–1619.
181. Dow, C.A., Wertheim, B.C., Patil, B.S. and Thomson, C.A., 2013. Daily consumption of grapefruit for 6 weeks reduces urine F2-isoprostanes in overweight adults with high baseline values but has no effect on plasma high-sensitivity C-reactive protein or soluble vascular cellular adhesion molecule 1. *Journal of Nutrition*, *143*(10), 1586–1592.
182. Xu, C., Chen, J., Zhang, J., Hu, X., Zhou, X., Lu, Z. and Jiang, H., 2013. Naringenin inhibits angiotensin II-induced vascular smooth muscle cells proliferation and migration and decreases neointimal hyperplasia in balloon injured rat carotid arteries through suppressing oxidative stress. *Biological and Pharmaceutical Bulletin*, *36*(10), 1549–1555.
183. Matz, R.L., Sotomayor, D., Alvarez, M., Schott, C., Stoclet, J.C. and Andriantsitohaina, R., 2000. Vascular bed heterogeneity in age-related endothelial dysfunction with respect to NO and eicosanoids. *British Journal of Pharmacology*, *131*(2), 303–311.
184. Brandes, R.P., Fleming, I. and Busse, R., 2005. Endothelial aging. *Cardiovascular Research*, *66*(2), 286–294.
185. Ohnishi, A., Asayama, R., Mogi, M., Nakaoka, H., Kan-No, H., Tsukuda, K., Chisaka, T. et al., 2015. Drinking citrus fruit juice inhibits vascular remodeling in cuff-induced vascular injury mouse model. *PLoS One*, *10*(2), e0117616.
186. Kamata, K., Kobayashi, T., Matsumoto, T., Kanie, N., Oda, S., Kaneda, A. and Sugiura, M., 2005. Effects of chronic administration of fruit extract (Citrus unshiu Marc) on endothelial dysfunction in streptozotocin-induced diabetic rats. *Biological and Pharmaceutical Bulletin*, *28*(2), 267–270.
187. Cerletti, C., Gianfagna, F., Tamburrelli, C., De Curtis, A., D'Imperio, M., Coletta, W., Giordano, L. et al., 2015. Orange juice intake during a fatty meal consumption reduces the postprandial low-grade inflammatory response in healthy subjects. *Thrombosis Research*, *135*(2), 255–259.
188. Touyz, R.M. and Schiffrin, E.L., 2000. Signal transduction mechanisms mediating the physiological and pathophysiological actions of angiotensin II in vascular smooth muscle cells. *Pharmacological Reviews*, *52*(4), 639–672.

189. Abernethy, D.R. and Schwartz, J.B., 1999. Calcium-antagonist drugs. *New England Journal of Medicine*, *341*(19), 1447–1457.
190. Imenshahidi, M., Eghbal, M., Sahebkar, A. and Iranshahi, M., 2013. Hypotensive activity of auraptene, a monoterpene coumarin from Citrus spp. *Pharmaceutical Biology*, *51*(5), 545–549.
191. Yang, M.C., Wu, S.L., Kuo, J.S. and Chen, C.F., 1990. The hypotensive and negative chronotropic effects of dehydroevodiamine. *European Journal of Pharmacology*, *182*(3), 537–542.
192. Chiou, W.F., Chou, C.J., Liao, J.F., Sham, A.Y. and Chen, C.F., 1994. The mechanism of the vasodilator effect of rutaecarpine, an alkaloid isolated from *Evodia rutaecarpa*. *European Journal of Pharmacology*, *257*(1–2), 59–66.
193. Sheu, J.R., Hung, W.C., Wu, C.H., Lee, Y.M. and Yen, M.H., 2000. Antithrombotic effect of rutaecarpine, an alkaloid isolated from *Evodia rutaecarpa*, on platelet plug formation in *in vivo* experiments. *British Journal of Haematology*, *110*(1), 110–115.
194. Rang, W.Q., Du, Y.H., Hu, C.P., Ye, F., Xu, K.P., Peng, J., Deng, H.W. and Li, Y.J., 2004. Protective effects of evodiamine on myocardial ischemia-reperfusion injury in rats. *Planta Medica*, *70*(12), 1140–1143.
195. Frauman, A.G., Johnston, C.I. and Fabiani, M.E., 2001. Angiotensin receptors: Distribution, signalling and function. *Clinical Science*, *100*(5), 481–492.
196. Lee, H.S., Oh, W.K., Choi, H.C., Lee, J.W., Kang, D.O., Park, C.S., Mheen, T.I. and Ahn, J.S., 1998. Inhibition of angiotensin II receptor binding by quinolone alkaloids from *Evodia rutaecarpa*. *Phytotherapy Research*, *12*(3), 212–214.
197. Harizi, H., Corcuff, J.B. and Gualde, N., 2008. Arachidonic-acid-derived eicosanoids: roles in biology and immunopathology. *Trends in Molecular Medicine*, *14*(10), 461–469.
198. Spite, M., Hellmann, J., Tang, Y., Mathis, S.P., Kosuri, M., Bhatnagar, A., Jala, V.R. and Haribabu, B., 2011. Deficiency of the leukotriene B4 receptor, BLT-1, protects against systemic insulin resistance in diet-induced obesity. *The Journal of Immunology*, *187*(4), 1942–1949.
199. Adams, M., Kunert, O., Haslinger, E. and Bauer, R., 2004. Inhibition of leukotriene biosynthesis by quinolone alkaloids from the fruits of Evodia rutaecarpa. *Planta Medica*, *70*(10), 904–908.
200. Hung, P.H., Lin, L.C., Wang, G.J, Chen, C.F. and Wang, P.S., 2001. Inhibitory effect of evodiamine on aldosterone release by *Zona glomerulosa* cells in male rats. *Chinese Journal of Physiology*, *44*(2), 53–57.
201. Heo, S.-K., Yun, H.-J., Yi, H.-S., Noh, E.-K. and Park, S.-D., 2009. Evodiamine and rutaecarpine inhibit migration by LIGHT via suppression of NADPH oxidase activation. *Journal of Cellular Biochemistry*, *107*(1), 123–133.
202. Yabe-Nishimura, C., 1998. Aldose reductase in glucose toxicity: A potential target for the prevention of diabetic complications. *Pharmacological Reviews*, *50*(1), 21–34.
203. Kato, A., Yasuko, H., Goto, H., Hollinshead, J., Nash, R.J. and Adachi, I., 2009. Inhibitory effect of rhetsinine isolated from *Evodia rutaecarpa* on aldose reductase activity. *Phytomedicine*, *16*(2–3), 258–261.
204. Su, T.-L., Lin, F.-W., Teng, C.-M., Chen, K.-T. and Wu, T.-S., 1998. Antiplatelet aggregation principles from the stem and root bark of *Melicope triphylla*. *Phytotherapy Research*, *12*(Suppl. 1), S74–S76.
205. Wu, T.S., Shi, L.S., Wang, J.J., Iou, S.C., Chang, H.C., Chen, Y.P., Kuo, Y.-H., Chang, Y.L. and Teng, C.M., 2003. Cytotoxic and antiplatelet aggregation principles of *Ruta graveolens*. *Journal of the Chinese Chemical Society*, *50*(1), 171–178.
206. Ratheesh, M., Shyni, G.L., Sindhu, G. and Helen, A., 2011. Inhibitory effect of *Ruta graveolens* L. on oxidative damage, inflammation and aortic pathology in hypercholesteromic rats. *Experimental and Toxicologic Pathology*, *63*(3), 285–290.
207. Wu, T.S., Chan, Y.Y., Liou, M.J., Lin, F.W., Shi, L.S. and Chen, K.T., 1998. Platelet aggregation inhibitor from *Murraya euchrestifolia*. *Phytotherapy Research*, *12*(Suppl. 1), S80–S82.
208. Han, J.H., Kim, Y., Jung, S.H., Lee, J.J., Park, H.S., Song, G.Y., Cuong, N.M., Kim, Y.H. and Myung, C.S., 2015. Murrayafoline A induces a G0/G1-phase arrest in platelet-derived growth factor-stimulated vascular smooth muscle cells. *Korean Journal of Physiology and Pharmacology*, *19*(5), 421–426.
209. Tsai, I.L., Wun, M.F., Teng, C.M., Ishikawa, T. and Chen, I.S., 1998. Anti-platelet aggregation constituents from Formosan *Toddalia asiatica*. *Phytochemistry*, *48*(8), 1377–1382.
210. Yang, Q., Cao, W., Zhou, X., Cao, W., Xie, Y. and Wang, S., 2014. Anti-thrombotic effects of α-linolenic acid isolated from *Zanthoxylum bungeanum* Maxim seeds. *BMC Complementary and Alternative Medicine*, *14*, 348.
211. Tsai, I.L., Lin, W.Y., Teng, C.M., Ishikawa, T., Doong, S.L., Huang, M.W., Chen, Y.C. and Chen, I.S., 2000. Coumarins and antiplatelet constituents from the root bark of *Zanthoxylum schinifolium*. *Planta Medica*, *66*(7), 618–623.
212. Chen, I.S., Tsai, I.W., Teng, C.M., Chen, J.J., Chang, Y.L., Ko, F.N., Lu, M.C. and Pezzuto, J.M., 1998. Pyranoquinoline alkaloids from *Zanthoxylum simulans*. *Phytochemistry*, *46*(3), 525–529.

213. Zhao, W., Yu, J., Su, Q., Liang, J., Zhao, L., Zhang, Y. and Sun, W., 2013. Antihypertensive effects of extract from *Picrasma quassiodes* (D. Don) Benn. in spontaneously hypertensive rats. *Journal of Ethnopharmacology, 145*(1), 187–192.
214. Ojha, S.K., Nandave, M., Arora, S., Mehra, R.D., Joshi, S., Narang, R. and Arya, D.S., 2008. Effect of *Commiphora mukul* extract on cardiac dysfunction and ventricular function in isoproterenol-induced myocardial infarction. *Indian Journal of Experimental Biology, 46*(9), 646–652.
215. Zaki, A.A., Hashish, N.E., Amer, M.A. and Lahloub, M.F., 2014. Cardioprotective and antioxidant effects of oleogum resin "Olibanum" from *Boswellia carteri* Birdw. (Bursearceae). *Chinese Journal of Natural Medicines, 12*(5), 345–350.
216. Li, X., Cui, X., Sun, X., Li, X., Zhu, Q. and Li, W., 2010. Mangiferin prevents diabetic nephropathy progression in streptozotocin-induced diabetic rats. *Phytotherapy Research, 24*(6), 893–899.
217. Kwak, S., Ku, S.K. and Bae, J.S., 2014. Fisetin inhibits high-glucose-induced vascular inflammation *in vitro* and *in vivo*. *Inflammation Research, 63*(9), 779–787.
218. Berrougui, H., Isabelle, M., Cloutier, M., Hmamouchi, M. and Khalil, A., 2006. Protective effects of *Peganum harmala* L. extract, harmine and harmaline against human low-density lipoprotein oxidation. *Journal of Pharmacy and Pharmacology, 58*(7), 967–974.
219. Im, J.H., Jin, Y.R., Lee, J.J., Yu, J.Y., Han, X.H., Im, S.H., Hong, J.T., Yoo, H.S., Pyo, M.Y. and Yun, Y.P., 2009. Antiplatelet activity of beta-carboline alkaloids from *Perganum harmala*: A possible mechanism through inhibiting PLCgamma2 phosphorylation. *Vascular Pharmacology, 50*(5–6), 147–152.
220. Kang, S.W., Kim, M.S., Kim, H.S., Kim, Y., Shin, D., Park, J.H. and Kang, Y.H., 2013. Celastrol attenuates adipokine resistin-associated matrix interaction and migration of vascular smooth muscle cells. *Journal of Cellular Biochemistry, 114*(2), 398–408.
221. Matsuda, H., Murakami, T., Yashiro, K., Yamahara, J. and Yoshikawa, M., 1999. Antidiabetic principles of natural medicines. IV. Aldose reductase and qlpha-glucosidase inhibitors from the roots of *Salacia oblonga* Wall. (Celastraceae): Structure of a new friedelane-type triterpene, kotalagenin 16-acetate. *Chemical and Pharmaceutical Bulletin (Tokyo), 47*(12), 1725–1729.
222. Li, Y., Peng, G., Li, Q., Wen, S., Huang, T.H., Roufogalis, B.D. and Yamahara, J., 2004. *Salacia oblonga* improves cardiac fibrosis and inhibits postprandial hyperglycemia in obese Zucker rats. *Life Sciences, 75*(14), 1735–1746.
223. Bachhav, S.S., Bhutada, M.S., Patil, S.D., Baser, B. and Chaudhari, K.B., 2012. Effect of *Viscum articulatum* Burm. (Loranthaceae) in N^{ω}-nitro-L-arginine methyl ester induced hypertension and renal dysfunction. *Journal of Ethnopharmacology, 142*(2), 467–473.
224. Bachhav, S.S., Patil, S.D., Bhutada, M.S. and Surana, S.J., 2011. Oleanolic acid prevents glucocorticoid-induced hypertension in rats. *Phytotherapy Research, 25*(10), 1435–1439.
225. Pang, X., Zhao, J., Zhang, W., Zhuang, X., Wang, J., Xu, R., Xu, Z. and Qu, W., 2008. Antihypertensive effect of total flavones extracted from seed residues of *Hippophae rhamnoides* L. in sucrose-fed rats. *Journal of Ethnopharmacology*, **117**(2), 325–331.
226. Hulsmans, M. and Holvoet, P., 2010. The vicious circle between oxidative stress and inflammation in atherosclerosis. *Journal of Cellular and Molecular Medicine, 14*(1–2), 70–78.
227. Luo, Y., Sun, G., Dong, X., Wang, M., Qin, M., Yu, Y. and Sun, X., 2015. Isorhamnetin attenuates atherosclerosis by inhibiting macrophage apoptosis via PI3K/AKT activation and HO-1 induction. *PLoS One, 10*(3), e0120259.
228. Bouallegue, A., Bou Daou, G. and Srivastava, A.K., 2007. Endothelin-1-induced signaling pathways in vascular smooth muscle cells. *Current Vascular Pharmacology, 5*(1), 45–52.
229. Qi, M.Y., Liu, H.R., Dai, D.Z., Li, N. and Dai, Y., 2008. Total triterpene acids, active ingredients from *Fructus Corni*, attenuate diabetic cardiomyopathy by normalizing ET pathway and expression of FKBP12.6 and SERCA2a in streptozotocin-rats. *Journal of Pharmacy and Pharmacology, 60*(12), 1687–1694.
230. Jiang, W.L., Zhang, S.P., Hou, J. and Zhu, H.B., 2012. Effect of loganin on experimental diabetic nephropathy. *Phytomedicine, 19*(3–4), 217–222.
231. Kampoli, A.M., Tousoulis, D., Briasoulis, A., Latsios, G., Papageorgiou, N. and Stefanadis, C., 2011. Potential pathogenic inflammatory mechanisms of endothelial dysfunction induced by type 2 diabetes mellitus. *Current Pharmaceutical Design, 17*(37), 4147–4158.
232. González-Périz, A. and Clària, J., 2010. Resolution of adipose tissue inflammation. *The Scientific World Journal, 10*, 832–856.
233. Tzeng, T.F., Liou, S.S., Chang, C.J. and Liu, I.M., 2014. The ethanol extract of *Lonicera japonica* (Japanese honeysuckle) attenuates diabetic nephropathy byinhibiting p-38 MAPK activity in streptozotocin-induced diabetic rats. *Planta Medica, 80*(2–3), 121–129.

234. Lee, W.M., Kim, S.D., Park, M.H., Cho, J.Y., Park, H.J., Seo, G.S. and Rhee, M.H., 2008. Inhibitory mechanisms of dihydroginsenoside Rg3 in platelet aggregation: Critical roles of ERK2 and cAMP. *Journal of Pharmacy and Pharmacology*, *60*(11), 1531–1536.
235. Zhang, Y.H., Chung, K.H., Ryu, C.K., Ko, M.H., Lee, M.K. and Yun, Y.P., 2001. Antiplatelt effect of 2-chloro-3-(4-acetophenuyl)-amino-1,4-naphthoquinone (NQ301): A possible mechanism through inhibition of intracellular Ca2+ mobilization. *Biological and Pharmaceutical Bulletin*, *24*, 618–622.
236. Hwang, S.Y., Son, D.J., Kim, I.W., Kim, D.M., Sohn, S.H., Lee, J.J. and Kim, S.K., 2008. Korean red ginseng attenuates hypercholesterolemia-enhanced platelet aggregation through suppression of diacylglycerol liberation in high-cholesterol-diet-fed rabbits. *Phytotherapy Research*, *22*(6), 778–783.
237. Rittenhouse-Simmons, S., 1979. Production of diglyceride from phosphatidylinositol in activated human platelets. *Journal of Clinical Investigation*, *63*(4), 580.
238. Hong, S.Y., Kim, J.Y., Ahn, H.Y., Shin, J.H. and Kwon, O., 2012. Panax ginseng extract rich in ginsenoside protopanaxatriol attenuates blood pressure elevation in spontaneously hypertensive rats by affecting the Akt-dependent phosphorylation of endothelial nitric oxide synthase. *Journal of Agricultural and Food Chemistry*, *60*(12), 3086–3091.
239. Wan, J.B., Lee, S.M., Wang, J.D., Wang, N., He, C.W., Wang, Y.T. and Kang, J.X., 2009. *Panax notoginseng* reduces atherosclerotic lesions in ApoE-deficient mice and inhibits TNF-alpha-induced endothelial adhesion molecule expression and monocyte adhesion. *Journal of Agricultural and Food Chemistry*, *57*(15), 6692–6697.
240. Liu, I.M., Tzeng, T.F., Liou, S.S. and Chang, C.J., 2011. *Angelica acutiloba* root alleviates advanced glycation end-product-mediated renal injury in streptozotocin-diabetic rats. *Journal of Food Science*, *76*(7), H165–H174.
241. Zhang, Y., Cao, Y., Wang, Q., Zheng, L., Zhang, J. and He, L., 2011. A potential calcium antagonist and its antihypertensive effects. *Fitoterapia*, *82*(7), 988–996.
242. Cao, Y., Zhang, Y., Wang, N. and He, L., 2014. Antioxidant effect of imperatorin from *Angelica dahurica* in hypertension via inhibiting NADPH oxidase activation and MAPK pathway. *Journal of the American Society of Hypertension*, *8*(8), 527–536.
243. McCubrey, J.A., LaHair, M.M. and Franklin, R.A., 2006. Reactive oxygen species-induced activation of the MAP kinase signaling pathways. *Antioxidants & Redox Signaling*, *8*(9–10), 1775–1789.
244. Chang, W.C. and Hsu, F.L., 1991. Inhibition of platelet activation and endothelial cell injury by flavan-3-ol and saikosaponin compounds. *Prostaglandins, Leukotrienes and Essential Fatty Acids*, *44*(1), 51–56.
245. Kim, S.Y. and Yun-Choi, H.S., 2007. Platelet anti-aggregating activities of bupleurumin from the aerial parts of Bupleurum falcatum. *Archives of Pharmacal Research*, *30*(5), 561–564.
246. Lahlou, S., Tahraoui, A., Israili, Z. and Lyoussi, B., 2007. Diuretic activity of the aqueous extracts of *Carum carvi* and *Tanacetum vulgare* in normal rats. *Journal of Ethnopharmacology*, *110*(3), 458–463.
247. Sadiq, S., Nagi, A.H., Shahzad, M. and Zia, A., 2010. The reno-protective effect of aqueous extract of Carum carvi (black zeera) seeds in streptozotocin induced diabetic nephropathy in rodents. *Saudi Journal of Kidney Diseases and Transplantation*, *21*(6), 1058–1065.
248. Wei, M., Ong, L., Smith, M.T., Ross, F.B., Schmid, K., Hoey, A.J., Burstow, D. and Brown, L., 2003. The streptozotocin-diabetic rat as a model of the chronic complications of human diabetes. *Heart, Lung and Circulation*, *12*(1), 44–50.
249. Kumar, P.A., Reddy, P.Y., Srinivas, P.N. and Reddy, G.B., 2009. Delay of diabetic cataract in rats by the antiglycating potential of cumin through modulation of alpha-crystallin chaperone activity. *Journal of Nutritional Biochemistry*, *20*(7), 553–562.
250. Lu, L., Jiang, S.S., Xu, J., Gong, J.B. and Cheng, Y., 2012. Protective effect of ligustrazine against myocardial ischaemia reperfusion in rats: the role of endothelial nitric oxide synthase. *Clinical and Experimental Pharmacology and Physiology*, *39*(1), 20–27.
251. Hsiao, G., Ko, F.N., Jong, T.T. and Teng, C.M., 1999. Antiplatelet action of 3',4'-diisovalerylkhellactone diester purified from Peucedanum japonicum Thunb. *Biological and Pharmaceutical Bulletin*, *21*(7), 688–692.
252. Moncada, S. and Vane, J.R., 1978. Unstable metabolites of arachidonic acid and their role in haemostasis and thrombosis. Br Med Bull, *34*(2), 129–135.
253. Aftab, K., Atta-Ur-Rahman, K. and Usmanghani K., 1995. Blood pressure lowering action of active principle from *Trachyspermum ammi* (L.) sprague. *Phytomedicine*, *2*(1), 35–40.
254. Okazaki, K., Kawazoe, K. and Takaishi, Y., 2002. Human platelet aggregation inhibitors from thyme (Thymus vulgaris L.). *Phytotherapy Research*, *16*(4), 398–399.

255. Wu, J., Yang, G., Zhu, W., Wen, W., Zhang, F., Yuan, J. and An, L., 2012. Anti-atherosclerotic activity of platycodin D derived from roots of *Platycodon grandiflorus* in human endothelial cells. *Biological and Pharmaceutical Bulletin*, *35*(8), 1216–1221.
256. Skeggs, L.T., Kahn, J.E. and Shumway, N.P., 1956. The preparation and function of the angiotensine I converting enzyme. *Journal of Experimental Medicine*, *103*, 295–299.
257. de Souza, P., Gasparotto, A., Crestani, S., Stefanello, M.É.A., Marques, M.C.A., da Silva-Santos, J.E. and Kassuya, C.A.L., 2011. Hypotensive mechanism of the extracts and artemetin isolated from *Achillea millefolium* L. (Asteraceae) in rats. *Phytomedicine*, *18*(10), 819–825.
258. Xu, Z., Yang, H., Zhou, M., Feng, Y. and Jia, W., 2010. Inhibitory effect of total lignan from Fructus Arctii on aldose reductase. *Phytotherapy Research*, *24*(3), 472–473.
259. Lee, Y.J., Choi, D.H., Cho, G.H., Kim, J.S., Kang, D.G. and Lee, H.S., 2012. *Arctium lappa* ameliorates endothelial dysfunction in rats fed with high fat/cholesterol diets. *BMC Complementary and Alternative Medicine*, *12*, 116.
260. Chappey, O., Dosquet, C., Wautier, M.P. and Wautier, J.L., 1997. Advanced glycation end products, oxidant stress and vascular lesions. *European Journal of Clinical Investigation*, *27*(2), 97–108.
261. Sohn, E., Kim, J., Kim, C.S., Kim, Y.S., Jang, D.S. and Kim, J.S., 2010. Extract of the aerial parts of *Aster koraiensis* reduced development of diabetic nephropathy via anti-apoptosis of podocytes in streptozotocin-induced diabetic rats. *Biochemical and Biophysical Research Communications*, *391*(1), 733–738.
262. Lee, J., Lee, Y.M., Lee, B.W., Kim, J.H. and Kim, J.S., 2012. Chemical constituents from the aerial parts of Aster koraiensis with protein glycation and aldose reductase inhibitory activities. *Journal of Natural Products*, *75*(2), 267–270.
263. Dimo, T., Nguelefack, T.B., Kamtchouing, P., Dongo, E., Rakotonirina, A. and Rakotonirina, S.V., 1999. [Hypotensive effects of a methanol extract of Bidens pilosa Linn on hypertensive rats]. *Comptes Rendus de I Academie des Sciences – Series III*, *322*(4), 323–329.
264. Dimo, T., Rakotonirina, S.V., Tan, P.V., Azay, J., Dongo, E. and Cros, G., 2002. Leaf methanol extract of *Bidens pilosa* prevents and attenuates the hypertension induced by high-fructose diet in Wistar rats. *Journal of Ethnopharmacology*, *83*(3), 183–191.
265. Amalia, L., Anggadiredja, K., Sukrasno, F.I. and Inggriani, R., 2012. Antihypertensive potency of Widl Cosmos (*Cosmos caudatus* Kunth, Asteraceae) leaf extract. *Journal of Pharmacology and Toxicoligy*, *7*(8), 359–368.
266. Jang, D.S., Yoo, N.H., Lee, Y.M., Yoo, J.L., Kim, Y.S. and Kim, J.S., 2008. Constituents of the flowers of *Erigeron annuus* with inhibitory activity on the formation of advanced glycation end products (AGEs) and aldose reductase. *Archives of Pharmacal Research*, *31*(7), 900–904.
267. Yoo, N.H., Jang, D.S., Yoo, J.L., Lee, Y.M., Kim, Y.S., Cho, J.H. and Kim, J.S., 2008. Erigeroflavanone, a flavanone derivative from the flowers of Erigeron annuus with protein glycation and aldose reductase inhibitory activity. *Journal of Natural Products*, *71*(4), 713–715.
268. Luo, P., Tan, Z.H., Zhang, Z.F., Zhang, H., Liu, X.F. and Mo, Z.J., 2008. Scutellarin isolated from *Erigeron multiradiatus* inhibits high glucose-mediated vascular inflammation. *Yakugaku Zasshi*, *128*(9), 1293–1299.
269. Lin, W.Y., Kuo, Y.H., Chang, Y.L., Teng, C.M., Wang, E.C., Ishikawa, T. and Chen, I.S., 2004. Anti-platelet aggregation and chemical constituents from the rhizome of Gynura japonica. *Planta Medica*, *69*(8), 757–764.
270. Kim, M.J., Lee, H.J., Wiryowidagdo, S. and Kim, H.K., 2006. Antihypertensive effects of *Gynura procumbens* extract in spontaneously hypertensive rats. *Journal of Medicinal Food*, *9*(4), 587–590.
271. Tabas, I., 1997. Free cholesterol-induced cytotoxicity: a possible contributing factor to macrophage foam cell necrosis in advanced atherosclerotic lesions. *Trends in Cardiovascular Medicine*, *7*(7), 256–263.
272. Mangathayaru, K., Kuruvilla, S., Balakrishna, K. and Venkhatesh, J., 2009. Modulatory effect of *Inula racemosa* Hook. f. (Asteraceae) on experimental atherosclerosis in guinea-pigs. *Journal of Pharmacy and Pharmacology*, *61*(8), 1111–1118.
273. Galis, Z.S., Sukhova, G.K., Lark, M.W. and Libby, P., 1994. Increased expression of matrix metalloproteinases and matrix degrading activity in vulnerable regions of human atherosclerotic plaques. *Journal of Clinical Investigation*, *94*(6), 2493.
274. Ohtsuki, T., Yokosawa, E., Koyano, T., Preeprame, S., Kowithayakorn, T., Sakai, S., Toida, T. and Ishibashi, M., 2008. Quinic acid esters from *Pluchea indica* with collagenase, MMP-2 and MMP-9 inhibitory activities. *Phytotherapy Research*, *22*(2), 264–266.
275. Wang, J.P., Xu, H.X., Wu, Y.X., Ye, Y.J., Ruan, J.L., Xiong, C.M. and Cai, Y.L., 2011. Ent-16β,17-dihydroxy-kauran-19-oic acid, a kaurane diterpene acid from *Siegesbeckia pubescens*, presentsanti-platelet and antithrombotic effects in rats. *Phytomedicine*, *18*(10), 873–878.

276. Honoré, S.M., Cabrera, W.M., Genta, S.B. and Sánchez, S.S., 2012. Protective effect of yacon leaves decoction against early nephropathy in experimental diabetic rats. *Food and Chemical Toxicology*, *50*(5), 1704–1715.
277. Prabhu, K.S., Lobo, R. and Shirwaikar, A., 2008. Antidiabetic properties of the alcoholic extract of *Sphaeranthus indicus* in streptozotocin-nicotinamidediabetic rats. *Journal of Pharmacy and Pharmacology*, *60*(7), 909–916.
278. Ratnasooriya, W.D., Pieris, K.P., Samaratunga, U. and Jayakody, J.R., 2004. Diuretic activity of *Spilanthes acmella* flowers in rats. *Journal of Ethnopharmacology*, *91*(2–3), 317–320.
279. Rang, H.P., Dale, M.M. and Ritter, J.M., 1995. *Pharmacology*. London: Churchill Livingstone, pp. 367–384.
280. Tian, L.Y., Bai, X., Chen, X.H., Fang, J.B., Liu, S.H. and Chen, J.C., 2010. Anti-diabetic effect of methylswertianin and bellidifolin from *Swertia punicea* Hemsl. and its potential mechanism. *Phytomedicine*, *17*(7), 533–539.
281. Higashino, S., Sasaki, Y., Giddings, J.C., Hyodo, K., Sakata, S.F., Matsuda, K., Horikawa, Y. and Yamamoto, J., 2014. Crocetin, a carotenoid from *Gardenia jasminoides* Ellis, protects against hypertension and cerebral thrombogenesis in stroke-prone spontaneously hypertensive rats. *Phytotherapy Research*, *28*(9), 1315–1319.
282. Wang, G.F., Wu, S.Y., Xu, W., Jin, H., Zhu, Z.G., Li, Z.H., Tian, Y.X., Zhang, J.J., Rao, J.J. and Wu, S.G., 2010. Geniposide inhibits high glucose-induced cell adhesion through the NF-kappaB signaling pathway in human umbilical vein endothelial cells. *Acta Pharmacologica Sinica*, *31*(8), 953–962.
283. Yoo, N.H., Jang, D.S., Lee, Y.M., Jeong, I.H., Cho, J.H., Kim, J.H. and Kim, J.S., 2010. Anthraquinones from the roots of *Knoxia valerianoides* inhibit the formation of advanced glycation end products and rat lens aldose reductase in vitro. *Archives of Pharmacal Research*, *33*(2), 209–214.
284. Chen, C.X., Jin, R.M., Li, Y.K., Zhong, J., Yue, L., Chen, S.C. and Zhou, J.Y., 1992. Inhibitory effect of rhynchophylline on platelet aggregation and thrombosis. *Zhongguo Yao Li Xue Bao*, *13*(2), 126–130.
285. Li, P.Y., Zeng, X.R., Cheng, J., Wen, J., Inoue, I. and Yang, Y., 2013. Rhynchophylline-induced vasodilation in human mesenteric artery is mainly due to blockage of L-type calcium channels in vascular smooth muscle cells. *Naunyn-Schmiedebergs Archives of Pharmacology*, *386*(11), 973–982.
286. Gan, R., Dong, G., Yu, J., Wang, X., Fu, S. and Yang, S., 2011. Protective effects of isorhynchophylline on cardiac arrhythmias in rats and guinea pigs. *Planta Medica*, *77*(13), 1477–1481.
287. Guo, H., Zhang, X., Cui, Y., Deng, W., Xu, D., Han, H., Wang, H., Chen, Y., Li, Y. and Wu, D., 2014. Isorhynchophylline protects against pulmonary arterial hypertension and suppresses PASMCs proliferation. *Biochemical and Biophysical Research Communications*, *450*, 729–734.
288. Wu, L.X., Gu, X.F., Zhu, Y.C. and Zhu, Y.Z., 2011. Protective effects of novel single compound, Hirsutine on hypoxic neonatal rat cardiomyocytes. *European Journal of Pharmacology*, *650*(1), 290–297.
289. Choi, D.H., Lee, Y.J., Oh, H.C., Cui, Y.L., Kim, J.S., Kang, D.G. and Lee, H.S., 2012. Improved endothelial dysfunction by *Cynanchum wilfordii* in apolipoprotein E(-/-) mice fed a high fat/cholesterol diet. *Journal of Medicinal Food*, *15*(2), 169–179.
290. Tárnok, K., Kiss, E., Luiten, P.G.M., Nyakas, C., Tihanyi, K., Schlett, K. and Eisel, U.L.M., 2008. Effects of Vinpocetine on mitochondrial function and neuroprotection in primary cortical neurons. *Neurochemistry International*, *53*(6), 289–295.
291. Cai, Y., Knight, W.E., Guo, S., Li, J.D., Knight, P.A. and Yan, C., 2012. Vinpocetine suppresses pathological vascular remodeling by inhibiting vascular smooth muscle cell proliferation and migration. *Journal of Pharmacology and Experimental Therapeutics*, *343*(2), 479–488.
292. Ramkumar, K.M., Ponmanickam, P., Velayuthaprabhu, S., Archunan, G. and Rajaguru, P., 2009. Protective effect of *Gymnema montanum* against renal damage in experimental diabetic rats. *Food and Chemical Toxicology*, 47(10), 2516–2521.
293. Hsu, C.C., Guo, Y.R., Wang, Z.H. and Yin, M.C., 2011. Protective effects of an aqueous extract from pepino (*Solanum muricatum* Ait.) in diabetic mice. *Journal of the Science of Food and Agriculture*, *91*(8), 1517–1522.
294. Fuchs, J. (Ed.), 1997. *Lipoic Acid in Health and Disease*. New York: CRC Press.
295. Ojha, S., Alkaabi, J., Amir, N., Sheikh, A., Agil, A., Fahim, M.A. and Adem, A., 2014. Withania coagulans fruit extract reduces oxidative stress and inflammation in kidneys of streptozotocin-induced diabetic rats. *Oxidative Medicine and Cellular Longevity*, 2014, 201436.
296. Kimmel, P.L. and Rosenberg, M.E. (Eds.), 2014. *Chronic Renal Disease*. UK: Elsevier.

297. Kuo, W.W., Huang, C.Y., Chung, J.G., Yang, S.F., Tsai, K.L., Chiu, T.H., Lee, S.D. and Ou, H.C., 2009. Crude extracts of *Solanum lyratum* protect endothelial cells against oxidized low-density lipoprotein-induced injury by direct antioxidant action. *Journal of Vascular Surgery*, *50*(4), 849–860.
298. Kano, M., Takayanagi, T., Harada, K., Makino, K. and Ishikawa, F., 2005. Antioxidative activity of anthocyanins from purple sweet potato, *Ipomoera batatas* cultivar Ayamurasaki. *Bioscience, Biotechnology, and Biochemistry*, *69*, 979–988.
299. Shindo, M., Kasai, T., Abe, A. and Kondo, Y., 2007. Effects of dietary administration of plant-derived anthocyanin-rich colors to spontaneously hypertensive rats. *Journal of Nutritional Science and Vitaminology (Tokyo)*, *53*(1), 90–93.
300. Miyazaki, K., Makino, K., Iwadate, E., Deguchi, Y. and Ishikawa, F., 2008. Anthocyanins from purple sweet potato Ipomoea batatas cultivar Ayamurasaki suppress the development of atherosclerotic lesions and both enhancements of oxidative stress and soluble vascular cell adhesion molecule-1 in apolipoprotein E-deficient mice. *Journal of Agricultural and Food Chemistry*, *56*(23), 11485–11492.
301. Shan, Q., Zheng, Y., Lu, J., Zhang, Z., Wu, D., Fan, S., Hu, B., Cai, X., Cai, H., Liu, P. and Liu, F., 2014. Purple sweet potato color ameliorates kidney damage via inhibiting oxidative stress mediated NLRP3 inflammasome activation in high fat diet mice. *Food and Chemical Toxicology*, *69*, 339–346.
302. Oh, H., Kang, D.G., Lee, S. and Lee, H.S., 2002. Angiotensin converting enzyme inhibitors from *Cuscuta japonica* Choisy. *Journal of Ethnopharmacology*, *83*(1–2), 105–108.
303. Jin, J.L., Lee, Y.Y., Heo, J.E., Lee, S., Kim, J.M. and Yun-Choi, H.S., 2004. Anti-platelet pentacyclic triterpenoids from leaves of Campsis grandiflora. *Archives of Pharmacal Research*, *27*(4), 376–380.
304. Keane, W.F., Kasiske, B.L., O'Donnell, M.P. and Kim, Y., 1993. Hypertension, hyperlipidemia, and renal damage. *American Journal of Kidney Diseases*, *21*(5), 43–50.
305. Kamesh, V. and Sumathi, T., 2014. Nephroprotective potential of *Bacopa monniera* on hypercholesterolemia induced nephropathy via the NO signaling pathway. *Pharmaceutical Biology*, *52*(10), 1327–1334.
306. Dong, Z. and Chen, C.X., 2013. Effect of catalpol on diabetic nephropathy in rats. *Phytomedicine*, *20*(11), 1023–1029.
307. Garcia, A., Shankar, H., Murugappan, S., Kim, S. and Kunapuli, S.P., 2007. Regulation and functional consequences of ADP receptor-mediated ERK2 activation in platelets. *Biochemical Journal*, *404*(2), 299–308.
308. Börsch-Haubold, A.G., Kramer, R.M. and Watson, S.P., 1995. Cytosolic phospholipase A2 is phosphorylated in collagen-and thrombin-stimulated human platelets independent of protein kinase C and mitogen-activated protein kinase. *Journal of Biological Chemistry*, *270*(43), 25885–25892.
309. Roger, S., Pawlowski, M., Habib, A., Jandrot-Perrus, M., Rosa, J.P. and Bryckaert, M., 2004. Costimulation of the Gi-coupled ADP receptor and the Gq-coupled TXA2 receptor is required for ERK2 activation in collagen-induced platelet aggregation. *FEBS Letters*, *556*(1–3), 227–235.
310. Thisoda, P., Rangkadilok, N., Pholphana, N., Worasuttayangkurn, L., Ruchirawat, S. and Satayavivad, J., 2006. Inhibitory effect of *Andrographis paniculata* extract and its active diterpenoids on platelet aggregation. *European Journal of Pharmacology*, *553*(1–3), 39–45.
311. Veeresham, C., Swetha, E., Rao, A.R. and Asres, K., 2013. In vitro and in vivo aldose reductase inhibitory activity of standardized extracts and the major constituent of *Andrographis paniculata*. *Phytotherapy Research*, *27*(3), 412–416.
312. Liu, Q., Hu, H.J., Li, P.F., Yang, Y.B., Wu, L.H., Chou, G.X. and Wang, Z.T., 2014. Diterpenoids and phenylethanoid glycosides from the roots of *Clerodendrum bungei* and their inhibitory effects against angiotensin converting enzyme and α-glucosidase. *Phytochemistry*, *103*, 196–202.
313. Ersoy, S., Orhan, I., Turan, N.N., Sahan, G., Ark, M. and Tosun, F., 2008. Endothelium-dependent induction of vasorelaxation by *Melissa officinalis* L. ssp. officinalis in rat isolatedthoracic aorta. *Phytomedicine*, *15*(12), 1087–1092.
314. Sotnikova, R., Okruhlicova, L., Vlkovicova, J., Navarova, J., Gajdacova, B., Pivackova, L., Fialova, S. and Krenek, P., 2013. Rosmarinic acid administration attenuates diabetes-induced vascular dysfunction of the rat aorta. *Journal of Pharmacy and Pharmacology*, *65*(5), 713–723.
315. Lee, J.J., Jin, Y.R., Lee, J.H., Yu, J.Y., Han, X.H., Oh, K.W., Hong, J.T., Kim, T.J., Yun, Y.P., 2007. Antiplatelet activity of carnosic acid, a phenolic diterpene from *Rosmarinus officinalis*. *Planta Medica*, *73*(2), 121–127.
316. Steinkamp-Fenske, K., Bollinger, L., Völler, N., Xu, H., Yao, Y., Bauer, R., Förstermann, U. and Li, H., 2007. Ursolic acid from the Chinese herb danshen (*Salvia miltiorrhiza* L.) upregulates eNOS and downregulates Nox4expression in human endothelial cells. *Atherosclerosis*, *195*(1), e104–e111.

317. Lam, F.F., Yeung, J.H., Chan, K.M. and Or, P.M., 2008. Dihydrotanshinone, a lipophilic component of *Salvia miltiorrhiza* (danshen), relaxes rat coronary artery byinhibition of calcium channels. *Journal of Ethnopharmacology*, *119*(2), 318–321.
318. Qian, S., Wang, S., Fan, P., Huo, D., Dai, L. and Qian, Q., 2012. Effect of *Salvia miltiorrhiza* hydrophilic extract on the endothelial biomarkers in diabetic patients with chronic artery disease. *Phytotherapy Research*, *26*(10), 1575–1578.
319. Chang, C.C., Chu, C.F., Wang, C.N., Wu, H.T., Bi, K.W., Pang, J.H. and Huang, S.T., 2014. The anti-atherosclerotic effect of tanshinone IIA is associated with the inhibition of TNF-α-induced VCAM-1, ICAM-1 and CX3CL1 expression. *Phytomedicine*, *21*(3), 207–216.
320. Maione, F., De Feo, V., Caiazzo, E., De Martino, L., Cicala, C. and Mascolo, N., 2014. Tanshinone IIA, a major component of *Salvia milthorriza* Bunge, inhibits platelet activation via Erk-2 signaling pathway. *Journal of Ethnopharmacology*, *155*(2), 1236–1242.
321. Yang, T.L., Lin, F.Y., Chen, Y.H., Chiu, J.J., Shiao, M.S., Tsai, C.S., Lin, S.J. and Chen, Y.L., 2011. Salvianolic acid B inhibits low-density lipoprotein oxidation and neointimal hyperplasia in endothelium-denuded hypercholesterolaemic rabbits. *Journal of the Science of Food and Agriculture*, *91*(1), 134–141.
322. Xu, S., Zhong, A., Bu, X., Ma, H., Li, W., Xu, X. and Zhang, J., 2015. Salvianolic acid B inhibits platelets-mediated inflammatory response in vascular endothelial cells. *Thrombosis Research*, *135*(1), 137–145.
323. Kang, E.S., Lee, G.T., Kim, B.S., Kim, C.H., Seo, G.H., Han, S.J., Hur, K.Y. et al., 2008. Lithospermic acid B ameliorates the development of diabetic nephropathy in OLETF rats. *European Journal of Pharmacology*, *579*(1–3), 418–425.
324. Zhang, T., Xu, J., Li, D., Chen, J., Shen, X., Xu, F., Teng, F. et al., 2014. Salvianolic acid A, a matrix metalloproteinase-9 inhibitor of *Salvia miltiorrhiza*, attenuates aortic aneurysm formation in apolipoprotein E-deficient mice. *Phytomedicine*, *21*(10), 1137–1145.
325. Zhou, R., He, L.F., Li, Y.J., Shen, Y., Chao, R.B. and Du, J.R., 2012. Cardioprotective effect of water and ethanol extract of *Salvia miltiorrhiza* in an experimental model of myocardial infarction. *Journal of Ethnopharmacology*, *139*(2), 440–446.
326. Amrani, S., Harnafi, H., Gadi, D., Mekhfi, H., Legssyer, A., Aziz, M., Martin-Nizard, F. and Bosca, L., 2009. Vasorelaxant and anti-platelet aggregation effects of aqueous *Ocimum basilicum* extract. *Journal of Ethnopharmacology*, *125*(1), 157–162.
327. Umar, A., Imam, G., Yimin, W., Kerim, P., Tohti, I., Berké, B. and Moore, N., 2010. Antihypertensive effects of *Ocimum basilicum* L. (OBL) on blood pressure in renovascular hypertensive rats. *Hypertension Research*, *33*(7), 727–730.
328. Turner, N., Goldsmith, D., Winearls, C., Lamiere, N., Himmelfarb, J. and Remuzzi, G., 2015. *Oxford Textbook of Clinical Nephrology*. UK: Oxford University Press.
329. Fatope, M.O. and Takeda, Y., 1988. The constituents of the leaves of *Ocimum basilicum*. *Planta Medica*, *54*(2), 190.
330. Abbaszadeh, H., Ebrahimi, S.A. and Akhavan, M.M., 2014. Antiangiogenic activity of xanthomicrol and calycopterin, two polymethoxylated hydroxyflavones in both in vitro and ex vivo models. *Phytotherapy Research*, *28*(11), 1661–1670.
331. Suanarunsawat, T., Ayutthaya, W.D., Songsak, T., Thirawarapan, S. and Poungshompoo, S., 2011. Lipid-lowering and antioxidative activities of aqueous extracts of *Ocimum sanctum* L. leaves in rats fed with a high-cholesterol diet. Oxidative Medicine and Cellular Longevity, 2011, 962025.
332. Kavitha, S., John, F. and Indira, M., 2015. Amelioration of inflammation by phenolic rich methanolic extract of *Ocimum sanctum* Linn. leaves in isoproterenol induced myocardial infarction. *Indian Journal of Experimental Biology*, *53*(10), 632–640.
333. Matsubara, T., Bohgaki, T., Watarai, M., Suzuki, H., Ohashi, K. and Shibuya, H., 1999. Antihypertensive actions of methylripariochromene A from *Orthosiphon aristatus*, an Indonesian traditional medicinal plant. *Biological and Pharmaceutical Bulletin*, *22*(10), 1083–1088.
334. Adam, Y., Somchit, M.N., Sulaiman, M.R., Nasaruddin, A.A., Zuraini, A., Bustamam, A.A. and Zakaria, Z.A., 2009. Diuretic properties of *Orthosiphon stamineus* Benth. *Journal of Ethnopharmacology*, *124*(1), 154–158.
335. Arafat, O.M., Tham, S.Y., Sadikun, A., Zhari, I., Haughton, P.J. and Asmawi, M.Z., 2008. Studies on diuretic and hypouricemic effects of *Orthosiphon stamineus* methanol extracts in rats. *Journal of Ethnopharmacology*, *118*(3), 354–360.
336. Makino, T., Ono, T., Muso, E. and Honda, G., 2002. Effect of *Perilla frutescens* on nitric oxide production and DNA synthesis in cultured murine vascular smoothmuscle cells. *Phytotherapy Research*, *16* (Suppl 1), S19–S23.

337. Alamgeer, Akhtar, M.S., Jabeen, Q., Khan, H.U., Maheen, S., Haroon-Ur-Rash, Karim, S., Rasool, S. et al., 2014. Pharmacological evaluation of antihypertensive effect of aerial parts of *Thymus linearis* benth. *Acta Poloniae Pharmaceutica*, *71*(4), 677–682.
338. Lee, M.J., Lee, H.S., Park, S.D., Moon, H.I. and Park, W.H., 2010. Leonurus sibiricus herb extract suppresses oxidative stress and ameliorates hypercholesterolemia in C57BL/6 mice and TNF-alpha induced expression of adhesion molecules and lectin-like oxidized LDL receptor-1 in human umbilical vein endothelial cells. *Bioscience, Biotechnology, and Biochemistry*, *74*(2), 279–284.
339. Liu, X., Pan, L., Gong, Q. and Zhu, Y., 2010. Leonurine (SCM-198) improves cardiac recovery in rat during chronic infarction. *European Journal of Pharmacology*, *649*(1–3), 236–241.
340. Liu, X.H., Pan, L.L., Deng, H.Y., Xiong, Q.H., Wu, D., Huang, G.Y., Gong, Q.H. and Zhu, Y.Z., 2013. Leonurine (SCM-198) attenuates myocardial fibrotic response via inhibition of NADPH oxidase 4. *Free Radical Biology and Medicine*, *54*, 93–104.
341. Yousefi, K., Soraya, H., Fathiazad, F., Khorrami, A., Hamedeyazdan, S., Maleki-Dizaji, N. and Garjani, A., 2013. Cardioprotective effect of methanolic extract of *Marrubium vulgare* L. on isoproterenol-induced acute myocardial infarction in rats. *Indian Journal of Experimental Biology*, *51*(8), 653–660.
342. Li, H.M., Hwang, S.H., Kang, B.G., Hong, J.S. and Lim, S.S., 2014. Inhibitory effects of Colocasia esculenta (L.) Schott constituents on aldose reductase. *Molecules*, *19*(9), 13212–13224.
343. Tang, J., Li, H.L., Shen, Y.H., Jin, H.Z., Yan, S.K., Liu, X.H., Zeng, H.W., Liu, R.H., Tan, Y.X. and Zhang, W.D., 2010. Antitumor and antiplatelet activity of alkaloids from *Veratrum dahuricum*. *Phytotherapy Research*, *24*(6), 821–826. doi:10.1002/ptr.3022.
344. Li, X., Cui, X., Wang, J., Yang, J., Sun, X., Li, X., Zhu, Q. and Li, W., 2013. Rhizome of *Anemarrhena asphodeloides* counteracts diabetic ophthalmopathy progression in streptozotocin-induced diabetic rats. *Phytotherapy Research*, *27*(8), 1243–1250.
345. Zhao, Q., Sun, Y., Ji, Y., Xu, L., Liu, K., Liu, B. and Huang, F., 2014. Total polyphenol of *Anemarrhena asphodeloides* ameliorates advanced glycation end products-induced endothelial dysfunction by regulation of AMP-Kinase. *Journal of Diabetes*, *6*(4), 304–315.
346. Kobashi, C., Urakaze, M., Kishida, M., Kibayashi, E., Kobayashi, H., Kihara, S., Funahashi, T. et al., 2005. Adiponectin inhibits endothelial synthesis of interleukin-8. *Circulation Research*, *97*(12), 1245–1252.
347. Okamoto, Y., Kihara, S., Ouchi, N., Nishida, M., Arita, Y., Kumada, M., Ohashi, K., Sakai, N., Shimomura, I., Kobayashi, H. and Terasaka, N., 2002. Adiponectin reduces atherosclerosis in apolipoprotein E-deficient mice. *Circulation*, *106*(22), 2767–2770.
348. Tsai, Y.C., Chiang, S.Y., El-Shazly, M., Wu, C.C., Beerhues, L., Lai, W.C., Wu, S.F., Yen, M.H., Wu, Y.C. and Chang, F.R., 2013. The oestrogenic and anti-platelet activities of dihydrobenzofuroisocoumarins and homoisoflavonoids fromLiriope platyphylla roots. *Food Chemistry*, *140*(1–2), 305–314.
349. Kanwar, Y.S., Wada, J., Sun, L., Xie, P., Wallner, E.I., Chen, S., Chugh, S. and Danesh, F.R., 2008. Diabetic nephropathy: Mechanisms of renal disease progression. *Experimental Biology and Medicine*, *233*(1), 4–11.
350. Sang, H.Q., Gu, J.F., Yuan, J.R., Zhang, M.H., Jia, X.B. and Feng, L., 2014. The protective effect of *Smilax glabra* extract on advanced glycation end products-induced endothelial dysfunction in human umbilical vein endothelial cells via RAGE-ERK1/2-NF-κB pathway. *Journal of Ethnopharmacology*, *155*(1), 785–795.
351. Cui, S.C., Yu, J., Zhang, X.H., Cheng, M.Z., Yang, L.W. and Xu, J.Y., 2013. Antihyperglycemic and antioxidant activity of water extract from *Anoectochilus roxburghii* in experimental diabetes. *Experimental and Toxicologic Pathology*, *65*(5), 485–488.
352. Liu, Z.L., Liu, Q., Xiao, B., Zhou, J., Zhang, J.G. and Li, Y., 2013. The vascular protective properties of kinsenoside isolated from *Anoectochilus roxburghii* under high glucose condition. *Fitoterapia*, *86*, 163–170.
353. Aziz, N., Mehmood, M.H., Siddiqi, H.S., Mandukhail, S.U., Sadiq, F., Maan, W. and Gilani, A.H., 2009. Antihypertensive, antidyslipidemic and endothelial modulating activities of *Orchis mascula*. *Hypertension Research*, *32*(11), 997–1003.
354. Godfraind, T. and Kaba, A., 1969. Blockade or reversal of the contraction induced by calcium and adrenaline in depolarized arterial smooth muscle. *British Journal of Pharmacology*, *36*(3), 549–560.
355. Gilani, A.H., Mehmood, M.H., Janbaz, K.H., Khan, A.U. and Saeed, S.A., 2008. Ethnopharmacological studies on antispasmodic and antiplatelet activities of *Ficus carica*. *Journal of Ethnopharmacology*, *119*(1), 1–5.
356. Gong, C.Y., Yu, Z.Y., Lu, B., Yang, L., Sheng, Y.C., Fan, Y.M., Ji, L.L. and Wang, Z.T., 2014. Ethanol extract of *Dendrobium chrysotoxum* Lindl ameliorates diabetic retinopathy and its mechanism. *Vascular Pharmacology*, *62*(3), 134–142.

357. Kumar, E.K. and Janardhana, G.R., 2011. Antidiabetic activity of alcoholic stem extract of *Nervilia plicata* in streptozotocin-nicotinamide induced type 2 diabetic rats. *Journal of Ethnopharmacology*, *133*(2), 480–483.
358. Jung, S.H., Lee, Y.S., Lee, S., Lim, S.S., Kim, Y.S. and Shin, K.H., 2002. Isoflavonoids from the rhizomes of *Belamcanda chinensis* and their effects on aldose reductase and sorbitol accumulation in streptozotocin induced diabetic rat tissues. *Archives of Pharmacal Research*, *25*(3), 306–312.
359. Jun, H.J., Hoang, M.H., Lee, J.W., Yaoyao, J., Lee, J.H., Lee, D.H., Lee, H.J., Seo, W.D., Hwang, B.Y. and Lee, S.J., 2012. Iristectorigenin B isolated from *Belamcanda chinensis* is a liver X receptor modulator that increases ABCA1 and ABCG1 expression in macrophage RAW 264.7 cells. *Biotechnology Letters*, *34*(12), 2213–2221.
360. Somania, R., Singhai, A.K., Shivgunde, P. and Jain, D., 2012. *Asparagus racemosus* Willd (Liliaceae) ameliorates early diabetic nephropathy in STZ induced diabetic rats. *Indian Journal of Experimental Biology*, *50*(7), 469–475.
361. Fan, G.J., Kang, Y.H., Han, Y.N. and Han, B.H., 2007. Platelet-activating factor (PAF) receptor binding antagonists from *Alpinia officinarum*. *Bioorganic & Medicinal Chemistry Letters*, *17*(24), 6720–6722.
362. Xia, Q., Wang, X., Xu, D.J., Chen, X.H. and Chen, F.H., 2012. Inhibition of platelet aggregation by curdione from *Curcuma wenyujin* essential Oil. *Thrombosis Research*, *130*(3), 409–414.
363. Yorsin, S., Kanokwiroon, K., Radenahmad, N. and Jansakul, C., 2014. Effects of *Kaempferia parviflora* rhizomes dichloromethane extract on vascular functions in middle-aged male rat. *Journal of Ethnopharmacology*, *156*, 162–174.
364. Abe, M., Ozawa, Y., Uda, Y., Morimitsu, Y., Nakamura, Y. and Osawa, T., 2006. A novel labdane-type trialdehyde from myoga (*Zingiber mioga* Roscoe) that potently inhibits human platelet aggregation and human 5-lipoxygenase. *Bioscience, Biotechnology, and Biochemistry*, *70*(10), 2494–2500.
365. Liao, Y.R., Leu, Y.L., Chan, Y.Y., Kuo, P.C. and Wu, T.S., 2012. Anti-platelet aggregation and vasorelaxing effects of the constituents of the rhizomes of *Zingiber officinale*. *Molecules*, *17*(8), 8928–8937.
366. Hudgins, A. and Weiss, G., 1968. Effect of Ca2+ removal upon vascular smooth muscle contraction induced by norepinephrine, hitamine and potassium. *Journal of Pharmacology and Experimental Therapeutics*, *159*, 91–97.
367. Karaki, H., Ozaki, H., Horri, M., Mitsui-Sato, M., Amano, K., Harada, K., Miyamamoto, S., Nakazawa, H., Won, K.J. and Sato, K., 1997. Calcium movements, distribution, and functions in smooth muscle. *Pharmacological Reviews*, *49*, 157–230.
368. Ni, W.J., Tang, L.Q. and Wei, W., 2015. Research progress in signalling pathway in diabetic nephropathy. *Diabetes/Metabolism Research and Reviews*, *31*(3), 221–233.
369. Tzeng, T.F., Liou, S.S., Chang, C.J. and Liu, I.M., 2013. Zerumbone, a tropical ginger sesquiterpene, ameliorates streptozotocin-induced diabetic nephropathy in rats by reducing the hyperglycemia-induced inflammatory response. *Nutrition & Metabolism* (*London*), *10*(1), 64.
370. Koo, H.J., Park, H.J., Byeon, H.E., Kwak, J.H., Um, S.H., Kwon, S.T., Rhee, D.K. and Pyo, S., 2014. Chinese yam extracts containing β-sitosterol and ethyl linoleate protect against atherosclerosis in apolipoprotein E-deficient mice and inhibit muscular expression of VCAM-1 in vitro. *Journal of Food Science*, *79*(4), H719–H729.

Index

Note: Page numbers followed by f refer to figures, respectively.

B

C

D

E

F

G

Q

R

S

T

U

V

W

X

Z